WATER AND SANITATION-RELATED
DISEASES AND THE ENVIRONMENT

VIDEOS

Chapter 10   Dracunculiasis (Guinea Worm Disease): Case Study Of The Effort To Reduce Guinea Worm
- **Dracunculiasis (Guinea Worm Disease): Case Study Of The Effort To Reduce Guinea Worm**
    The Carter Center
- **Guinea Worm's Last Stand: Southern Sudan**
    Cielo Productions, Courtesy of The Carter Center

Chapter 11   Onchocerciasis
- **Onchocerciasis**
- **The Crab and the Fly: River Blindness (Onchocerciasis)**
    Courtesy of The Carter Center

Chapter 13   Schistosomiasis
- **Schistosomiasis**
    The Carter Center
- **Kill or Cure: Bilharzia**
    Rockhopper TV, Courtesy of the Schistosomiasis Control Initiative (SCI)

Chapter 14   Trachoma
- **Trachoma Control Program**
    The Carter Center

Chapter 16   The Zimbabwe Cholera Epidemic of 2008–2009
- **Finding the Link between Cholera and coastal fresh water**
    Courtesy of CSIR Council of Scientific and Industrial Research
- **An interview with Dr. Rita Colwell: Water Quality is Crucial to Human Health and Security**
    Stockholm International Water Institute (SIWI)

Chapter 18   Household Water Treatment and Safe Storage in Low-Income Countries
- **Saving Lives with Safe Water: Household Treatment and Safe Storage**
    Courtesy of UNICEF and Population Services International (PSI)

Chapter 20   The Sanitation Challenge in India
- **Sulabh Freedom:** A film dedicated to the UN for Declaring the Year 2008 as International Year of Sanitation Movement: Gandhi lives
    Sulabh International Social Service Organization

Chapter 22   Household-Centered Environmental Sanitation Systems
- **Perfect Technology**
    Sulabh International Social Service Organization

Chapter 28   Global Substitution of Mercury-Based Medical Devices in the Health Sector
- **Uncommon Heroes: Health Care Without Harm**
    Produced by the Skroll Foundation Courtesy of Health Care Without Harm
- **Vapores de Mecurio**
    Courtesy of Dave Heinien, Safety and Health Coordinator Environmental Health and Safety, Bowling Green State University
- **Mercury is Hazardous to Health**
    Produced by Toxic Links
    Courtesy of Health Care Without Harm

Chapter 30   Additional Measures to Prevent, Ameliorate, and Reduce Water Pollution and Related Water Diseases: Global Water Governance
- **UNICEF: World Water Day Video**
    Producer: Rachel Bonham-Carter

SHORT-CLIP VIDEOS

Chapter 34   Ocean Pollution: Health and Environmental Impacts of Brominated Flame Retardants
- **Newborn Humpback Calf**
- **Gray Seal Underwater**
- **Harbor Release**
- **Harbor Seal Mother and Pup Swimming and Coming Ashore**
    All Courtesy of Scott Tucker of Expedition New England

WATER AND SANITATION-RELATED DISEASES AND THE ENVIRONMENT

Challenges, Interventions, and Preventive Measures

Edited by

JANINE M. H. SELENDY

WILEY-BLACKWELL

A John Wiley & Sons, Inc., Publication

Published by John Wiley & Sons, Inc., Hoboken, New Jersey
Published simultaneously in Canada

Library of Congress Cataloging-in-Publication Data:

Water and sanitation related diseases and the environment : challenges, interventions, and
preventive measures / edited by Janine M. Selendy.
        p. ; cm.
   Includes bibliographical references and index.
   ISBN 978-0-470-52785-6 (cloth)
1. Waterborne infection. 2. Water–Purification. I. Selendy, Janine M.
   [DNLM: 1. Water Supply. 2. Developing Countries. 3. Disease Outbreaks–prevention & control. 4. Sanitation. 5. Water Microbiology. 6. Water Pollution–prevention & control. WA 675]
   RA642.W3.W367 2011
   363.739'4–dc22

                                                                    2010039502

Printed in Singapore

oBook ISBN: 978-1-118-14859-4
ePDF ISBN: 978-1-118-14861-7
ePub ISBN: 978-1-118-14860-0

10 9 8 7 6 5 4 3 2 1

CONTENTS

The DVD accompanying this book presents:

- 16 videos and 4 short-clip videos, totaling nearly 4 hours of supplemental content
- 525 illustrations, tables, and maps from the text.
- Additional text content not found in the book

Updates and additional resources will be posted on a companion website: www.wiley.com/go/selendy/water

PREFACE

With each invited piece that has come in for the book, I have felt the remarkable dedication of the authors to the improvement of human health. The facts they present of the extent of the problems are often startlingly grave, but I feel their discussions of prevention and intervention measures generate hope. With this book it is my hope that what we have brought together will provide the background, guidance, and inspiration for those who are now engaged in—and those who are concerned and might join in—addressing this tremendous task.

Water and Sanitation-Related Diseases and the Environment: Challenges, Interventions, and Preventive Measures began as a syllabus on water and sanitation prepared at the invitation of Yale School of Public Health. I am grateful to Yale School of Public Health for inviting me to prepare the syllabus, to its Health Education Committee for its approval and for the opportunity to lecture to its students and faculty. Course preparation and subsequent lectures and a seminar I prepared at the invitation of then Dean Allan Rosenfield at Columbia University Mailman School of Public Health revealed that inadequate sources link the many interconnected aspects of public health, water and sanitation-related diseases, and the environment and that the topic is far more critical and extensive than one can convey through a brief introduction.

I am grateful to all of the authors who have generously contributed their expertise to make this a comprehensive interdisciplinary discussion of these crucial issues. I am grateful for the encouragement and advice of Drs. Robert Lawrence, Rita Colwell, Barry Levy, and the late Dean Allan Rosenfield, to Professor Robert Wyman, our immediate host at Yale and to Yale University for their hosting of Horizon International.

During a meeting with John Wiley & Sons, Inc. representatives to discuss the possibility of publishing a series of publications drawing on Horizon International's extensive material presented on the Horizon Solution Site (www.solutions-site.org), I mentioned my water and sanitation initiatives and concerns. Wiley responded with an invitation to produce this book with an accompanying DVD, making it a Wiley–Horizon publication; provide electronic versions with postings of the book and its accompanying DVD on a special website and on the Horizon Solutions site; and to create a place for supplemental material after initial publication.

My John Wiley & Sons editor, Karen Chambers and her colleagues who have made this book possible, deserve special credit not only for engaging me to produce this book, but for their steadfast enthusiasm about the importance of this book and its accompanying DVD and expressed desire to make sure that it reaches a significant international audience.

The financial assistance of Peter and Helen Haje Foundation toward the production of the DVD has been of inestimable help in making it a reality with the quality and substance I hoped to realize.

I dedicate this book to those whose love and encouragement have been a source of inspiration and devotion to make this book a reality: my sons, Philippe and Béla; their children, Max and Liam, and Nicolas and Linnea; daughter-in-law, Jennifer, and former daughter-in-law, Ulrika. I am thankful for the love and loving care, and ever-ready words of wisdom and humor from my partner, Charles R. Dickey. Finally, I am grateful to our little good spirit who makes everyone smile—our dog, Heather.

JANINE M. H. SELENDY

New Haven, Connecticut
March 2011

CONTRIBUTORS

Jens Aagaard-Hansen, M.D., M.P.H., Dr. TM, Specialist in General Medicine, Steno Health Promotion Center, Steno Diabetes Center, Copenhagen, Denmark

M. John Albert, Ph.D., FRCPath, FAAM, Professor, Department of Microbiology, Faculty of Medicine, Kuwait University, Kuwait City, Kuwait

Lorraine C. Backer, Ph.D., M.P.H., National Center for Environmental Health, Centers for Disease Control, Atlanta, Georgia

Michele Barry, M.D., FACP, Senior Associate Dean for Global Health, Director of Center for Global Health, Stanford University, Palo Alto, California

Michael L. Bennish, M.D., Executive Director, Mpilonhle, Mtubatuba, South Africa; Senior Associate, Department of Population, Family and Reproductive Health, Johns Hopkins University, Baltimore, Maryland

Brian G. Blackburn, M.D., Clinical Assistant Professor, Department of Internal Medicine and Division of Infectious Diseases and Geographic Medicine, Stanford University School of Medicine, Stanford, California

Boakye A. Boatin, UNICEF/UNDP/World Bank/WHO Special Programme for Research and Training in Tropical Diseases (TDR), World Health Organization, Geneva, Switzerland

Wayne W. Carmichael, Ph.D., Professor Emeritus, Department of Biological Sciences, Wright State University, Dayton, Ohio

David O. Carpenter, M.D., Director, Institute for Health and the Environment, State University at Albany, Rensselaer, New York

Marcia C. Castro, Ph.D., Assistant Professor of Demography, Department of Global Health and Population, Harvard School of Public Health, Cambridge, Massachusetts

Nikhil Chandavarkar, Ph.D., United Nations Water Secretary, Chief, Communications and Outreach Branch Division for Sustainable Development, UN Department for Economic and Social Affairs, New York, New York

Aruna Chandran, M.D., M.P.H., Assistant Scientist, Center for American Indian Health; Hib Initiative, Johns Hopkins University Bloomberg School of Public Health, Baltimore, Maryland

Thomas F. Clasen, Department of Infectious and Tropical Diseases, London School of Hygiene & Tropical Medicine, London, United Kingdom

Joseph A. Cook, M.D., M.P.H., FACP, Adjunct Epidemiology Professor, University of North Carolina at Chapel Hill School of Public Health, Chapel Hill, North Carolina, and former Executive Director of the International Trachoma Initiative

Edward Dodge, M.D., M.P.H., San Antonio, Texas, Visiting Lecturer, Faculty of Health Sciences, Africa University, Mutare, Zimbabwe and Courtesy Clinical Associate Professor, Department of Community Health and Family Medicine, University of Florida College of Medicine

Menachem Elimelech, Ph.D., Roberto C. Goizueta Professor of Environmental and Chemical Engineering, Yale University, New Haven, Connecticut

Sean Fitzwater, M.H.S., Research Associate, Johns Hopkins University Bloomberg School of Public Health, Baltimore, Maryland

Lora E. Fleming, M.D., Ph.D., Professor, Departments of Epidemiology and Public Health, and Marine Biology and Fisheries, Miller School of Medicine and Rosenstiel School of Marine and Atmospheric Sciences, University of Miami, Miami, Florida

Julius N. Fobil, Dr. PH., Department of Biological, Environmental & Occupational Health Sciences, School of Public Health, College of Health Sciences, University of Ghana, Accra, Ghana

Jeffery A. Foran, M.D., President, EHSI, LLC, Adjunct Professor, School of Public Health, University of Illinois, Chicago, Illinois

Julio Frenk, M.D., Dean of the Faculty, Harvard School of Public Health, T & G Angelopoulos Professor of Public Health and International Development, Harvard School of Public Health and Harvard Kennedy School, Cambridge, Massachusetts

Octavio Gómez-Dantés, M.D., M.P.H., Director of Analysis and Evaluation, CARSO Health Institute, Mexico City, Mexico

Jay Graham, M.B.A., Ph.D., American Association for the Advancement of Science, Science and Technology Policy Fellow, U.S. AID, Washington, DC

Jeffrey K. Griffiths, M.D., M.P.H and T.M., Director, Global Health, Public Health and Professional Degree Programs and Adjunct Associate Professor, Friedman School of Nutrition Science and Policy, Tufts University School of Medicine, Boston, Massachusetts; Chair of the Drinking Water Panel, U.S. EPA Science Advisory Board, Washington DC.

Scott B. Halstead, M.D., Director, Supportive Research and Development, Pediatric Dengue Vaccine Initiative, Seoul, South Korea

Adrian Hopkins, M.D., Director of Mectizan Donation Program, Task Force for Global Health, Atlanta, Georgia

Donald R. Hopkins, M.D., M.P.H., Associate Executive Director of Health Programs, The Carter Center, Atlanta, Georgia

Kurunthachalam Kannan, Ph.D., Professor, Department of Environmental Health Sciences, State University of New York at Albany; Chief, Laboratory of Organic Analytical Chemistry, Wadsworth Center, New York State Department of Health, Albany, New York

Joshua Karliner, International Coordinator, Health Care Without Harm, San Francisco, California

Unni Krishnan Karunakara, Deputy Director of Health, Millennium Villages Project, The Earth Institute, Assistant Clinical Professor of Population & Family Health, Mailman School of Public Health, Columbia University, New York, New York

Moses N. Katabarwa, Ph.D., M.P.H., Senior Epidemiologist, Emory University/The Carter Center, Atlanta, Georgia

Jonathan K. Kish, M.P.H., Department of Epidemiology and Public Health, University of Miami Miller School of Medicine, Miami, Florida

Margaret Kosek, M.D., Johns Hopkins Bloomberg School of Public Health Laboratory Satellite IQTLAB—Research Site, Peru

Alexander Kraemer, Department of Public Health Medicine, School of Public Health, University of Bielefeld, Bielefeld, Germany

Robert S. Lawrence, M.D., Professor, Director, Center for a Livable Future, joint appointments in Health Policy and Management and in Medicine at Johns Hopkins Bloomberg School of Public Health, Baltimore, Maryland

Barry S. Levy, M.D., M.P.H., Consultant and Adjunct Professor of Public Health, Tufts University School of Medicine, Boston, Massachusetts

Pascal Magnussen, M.D., Dr. T.M&H., Specialist in Tropical Medicine and Infectious Diseases, Center for Health Research and Development (DBL), Faculty of Life Sciences, University of Copenhagen, Copenhagen, Denmark

Silvio P. Mariotti, M.D., Ophthalmologist, Senior Medical Officer Prevention of Blindness and Deafness, World Health Organization, Geneva, Switzerland

Juergen May, Infectious Disease Epidemiology Unit, Bernhard-Notch Institute for Tropical Medicine, Hamburg, Germany

M. Danielle McDonald, Ph.D., Assistant Professor, Animal Physiology, Molecular Biology, Pharmacology and Toxicology, Marine Biology and Fisheries, Miller School of Medicine and Rosenstiel School of Marine and Atmospheric Sciences, University of Miami, Miami, Florida

Peter Orris, M.D., M.P.H., Professor and Chief of Service, Environmental and Occupational Medicine, University of Illinois Medical Center, Chicago, Illinois

Bindeshwar Pathak, Ph.D., Action Sociologist and Social Reformer, International Expert on Cost-Effective Sanitation, Biogas and Rural Development, Founder of the Sulabh International Social Service Organisation, New Delhi, India

Gretchen Loeffler Peltier, M.P.H., Ph.D., Postdoctoral Fellow, The Earth Institute, Columbia University, New York, New York

Ernesto Ruiz-Tiben, Ph.D., Director of Dracunculiasis Eradication Program, The Carter Center, Atlanta, Georgia

Uriel N. Safriel, UN Convention to Combat Desertification and Department of Evolution, Systematics and Ecology, Institute of Life Sciences, Hebrew University of Jerusalem, Jerusalem, Israel

Mathuram Santosham, M.D., M.P.H., Director, Center for American Indian Health; Professor, International Health and Pediatrics, Johns Hopkins Bloomberg School of Public Health, Baltimore, Maryland

Amy R. Sapkota, Ph.D., M.P.H., Assistant Professor, School of Public Health, Maryland Institute for Applied Environmental Health, University of Maryland, College Park, Maryland

Janine M. H. Selendy, Chairman, President and Publisher, Horizon International, New Haven, Connecticut

Kerry Shannon, M.P.H., M.D./Ph.D. candidate, Johns Hopkins School of Medicine, Baltimore, Maryland

Susan D. Shaw, Dr. PH., Director, Marine Environmental Research Institute (MERI), Center for Marine Studies, Blue Hill, Maine

Victor W. Sidel, M.D., Distinguished University Professor of Social Medicine, Montefiore Medical Center, Albert Einstein College of Medicine, and adjunct Professor of Public Health, Weill Medical College of Cornell University, New York, New York

Laura Sima, Ph.D. candidate, Yale University, New Haven, Connecticut

Burton H. Singer, Ph.D., Courtesy Professor, Emerging Pathogens institute University of Florida, Gainesville, Florida

Helena M. Solo-Gabriele, Ph.D., P.E, Professor, Civil, Architectural, and Environmental Engineering, University of Miami, Miami, Florida

Ian Stewart, Scientist, Queensland Health Forensic and Scientific Services and Adjunct Research Fellow, School of Public Health, Griffith University, Queensland, Australia

Birgitte Jyding Vennervald, M.D., Center for Health Research and Development (DBL), Faculty of Life Sciences, University of Copenhagen, Copenhagen, Denmark

Gretchen Welfinger-Smith, M.S, Institute for Health and the Environment, School of Public Health, University at Albany, Rensselaer, New York

Mary E. Wilson, M.D., Associate Professor in the Department of Global Health and Population, Harvard School of Public Health, Associate Clinical Professor of Medicine, Harvard Medical School, Boston, Massachusetts

Anson Elisabeth Wright, MS, M.S.PH, Hygiene and Sanitation Advisor, Millennium Villages Project, The Earth Institute, Columbia University, New York, New York

Robert Wyman, Ph.D., Professor of Molecular, Cellular and Developmental Biology, Yale University, New Haven, Connecticut

INTRODUCTION

Janine M. H. Selendy and Jens Aagaard-Hansen

> ...once we can secure access to clean water and to adequate sanitation facilities for all people, irrespective of the difference in their living conditions, a huge battle against all kinds of diseases will be won.
>
> Dr. Lee Jong-wook, Director-General, World Health Organization

Written by authorities from the fields of public health, medicine, epidemiology, environmental health, climate change, environmental engineering, and demography this book presents an interdisciplinary picture of the conditions responsible for water and sanitation-related diseases, the pathogens and their biology, morbidity, and mortality resulting from lack of safe water and sanitation, distribution of these diseases, and the conditions that must be met to reduce or eradicate them.

The book covers access to and maintenance of clean water, and guidelines for the safe use of wastewater, excreta and greywater, and examples of solutions, but with an emphasis on what is achievable considering that 2.6 billion individuals have no toilets.

Meeting water and sanitation needs coupled with protection of the environment and prevention of pollutants is essential to every effort to improve the health and living conditions of billions of people. Meeting these needs is fundamental not only to effectively diminish incidence of diseases that afflict a third or more of the people of the world but also to improve education and economic well-being and elevate billions of individuals out of vicious cycles of poverty.

The health statistics are startling, each number representing a baby, a girl or boy, a father or mother. They constitute, according to the World Health Organization (WHO), 1.2 billion individuals who are exposed to water-related illness from their drinking water and 2.6 billion individuals without access to any type of improved sanitation facility (1). Of those who lack adequate sanitation facilities or practice unsafe hygiene behavior, about 2 million individuals, most of whom are less than 5 years old, die every year due to diarrheal diseases. This equates to the death of one child every 15 s and exceeds the death rates from such killer diseases as malaria and tuberculosis (2).

Water and sanitation-related infections are often combined with malnutrition. For example, consider iron deficiency which is the main cause of anemia. Of the 2 billion individuals who suffer from anemia, 9 out of 10 live in developing countries according to WHO and UNICEF (3). Furthermore, anemia may contribute to up to 20% of maternal deaths (4). In addition to nutritional deficiencies, significant contributors to anemia are infections related to hygiene, sanitation, unsafe water, and inadequate water management. Those infections include malaria, schistosomiasis, and hookworm.

The difference that successful measures can make is strikingly significant. Considering diarrhea, for example, according to WHO, "...improved water supply reduces diarrhea morbidity by between 6% to 25%, if severe outcomes are included. Improved sanitation reduces diarrhea morbidity by 32%. Hygiene interventions including hygiene education and promotion of hand washing can lead to a reduction of diarrheal cases by up to 45%. Improvements in drinking-water quality through household water treatment, such as chlorination at point of use, can lead to a reduction of diarrhea episodes by between 35% and 39% (5)."

The chapters on individual diseases cover not only basic information about the disease in question, that is, causative agent and pathogenesis epidemiology, clinical manifestations, treatment, prevention and control but also distribution, prevalence and incidence, and interconnected factors such as environmental factors. These chapters, as appropriate, discuss and emphasize the significance of the close relationships among water access and quality, lack of adequate sanitation, lack of adequate hygiene, and the relationship of human activity to the diseases. The chapters address how the diseases are further aggravated because the patients often suffer from more than one infection as well as from nutritional inadequacies. Other factors they address are the importance of adequate maternal and child health care, the significance of climate change, and other environmental factors. Prevention and control are important parts of the discussions.

The preventive measures and solutions the book presents provide guidance for possible action on the local, national, and international levels. That guidance includes the importance of providing for adequate maternal and child health care and complications caused by reduced immunity to disease, for example, as caused by malnutrition. As Allan Rosenfield, former Dean of the Mailman School of Public Health stressed, "Programs of Maternal and Child Health need to focus their attention on these water related conditions." Dr. Rosenfield emphasized that "For the various diarrheal conditions, oral rehydration programs are essential; particularly for children . . . this became a major initiative of UNICEF in the 1980s based initially on research efforts in Bangladesh. In addition, of course, it is essential that mothers make every effort to protect their children from unclean water."

Environmental factors play a significant role in the prevalence of infectious diseases. The WHO states: "Large-scale and global environmental hazards to human health include climate change, stratospheric ozone depletion, loss of biodiversity, changes in hydrological systems and the supplies of freshwater, land degradation and stresses on food-producing systems . . .. Ecosystem disruption can impact on health in a variety of ways and through complex pathways. These are moreover modified by a local population's current vulnerability and their future capacity to implement adaptation measures. The links between ecosystem change and human health are seen most clearly among impoverished communities, who lack the 'buffers' that more affluent communities can afford."

The broad extent, variety, and ramifications of environmental factors are covered in many chapters throughout the book and particularly in Section V. "Worldwide, environmental factors play a role in more than 80% of adverse outcomes reported by the World Health Organization, including infectious diseases, injuries, mental retardation, and cancer," quoted by the Centers for Disease Control (CDC) (6). According to UNESCO, "Some 300–500 million tons of heavy metals, solvents, toxic sludge, and other wastes accumulate each year from industry . . . (7). In developing countries, 70% of industrial wastes are dumped untreated into waters where they pollute the usable water supply" (8).

It is essential for the understanding of the diseases covered in this book to consider the sociocultural and economic context. Human perceptions and practices strongly influence transmission patterns and access to preventive and curative measures in a multitude of ways. Thus, local notions of illness (including perceived etiology, pathogenesis, diagnosis, and preferred treatment) are often not compatible with biomedical lore and they are usually influenced by factors such as gender, age, religion, and ethnicity. Social and demographic macrotrends in society such as migration, urbanization, socioeconomic development (or financial crisis), health sector reforms, and political priorities such as privatization and level of control with, for example, environmental development projects play crucial roles in either ameliorating or (usually) aggravating the situation. In some cases populations may be directly opposed to control interventions, which from a biomedical perspective are rational, as shown by examples of cholera (9) and deworming programs addressing soil-transmitted helminths and schistosomiasis (10). In the case of cholera, environmental factors (presence of *Vibrio* bacteria in lakes and river deltas together with specific fluctuations in rainfall and water level) combined with sociocultural factors (migration and inadequate hygiene) are strong determinants for the outbreaks of epidemics.

Ameliorating conditions and taking preventive measures to reduce pollutants found in ground and surface waters are also addressed. These include toxic chemical pollutants resulting from natural sources such as arsenic and lead and anthropogenic sources, for example, mercury and persistent organic pollutants such as polychlorinated biphenyls. The toll these pollutants take is just beginning to be recognized.

Subjects not traditionally considered in discussion of disease and infectious disease that need to be addressed in developing solutions and predicting potential outbreaks are covered in many chapters. These include climate change and human alteration of ecosystems. For example, how global warming is serving to expand the ranges of diseases such as malaria and dengue and how human activities such as deforestation and mining can result in still water pools where mosquitoes breed, introducing water-related diseases to areas where they were previously unknown.

While many water and sanitation-related diseases are covered in depth, there are many others that receive more limited coverage in one or more chapters such as arsenicosis, cryptosporidiosis, and campylobacteriosis. The DVD that accompanies the book contains tables and images from each chapter of the book, supplementary images and videos. The Carter Center has provided several videos for the DVD, some of which can also be found on their Web page at www. CarterCenter.org. In addition, the DVD has multimedia material from several other sources including videos from the Sulabh International Social Service Organisation, http://www.sulabhinternational.org, on Household Centred Environmental Sanitation Systems and from the Schistosomiasis Control Initiative, http://www3.imperial.ac.uk/schisto.

Preventive measures and solutions are presented throughout the book and in more detail in a special section at the end of the book. They are also included on the book's DVD and as case studies and articles on Horizon International's Solutions site at www.solutions-site.org. Additional material and DVD including short clips from Horizon International's television programs, www.horizoninternationalty. org and updates will be found at www.wiley.com/go/selendy/water.

REFERENCES

1. Water, Sanitation, and Hygiene Links to Health: Facts and Figures—Updated November 2004. Available at http://www.who.int/water_sanitation_health/publications/facts2004/en/print.html. Accessed on April 4, 2008.

2. World Bank. The World Bank's Increased Focus on Basic Sanitation and Hygiene, 2006. Available at http://siteresources.worldbank.org/INTWSS/Resources/focusonsanitationandhygiene.pdf.

3. WHO # 4. Available at http://www.who.int/water_sanitation_health/diseases/anemia/en/. Accessed on March 16, 2011.

4. WHO #7. Available at http://www.who.int/water_sanitation_health/diseases/anemia/en/. Accessed on April 7, 2008.

5. WHO #1. Water, sanitation and hygiene links to health. Available at http://www.who.int/water_sanitation_health/publications/facts2004/en/index.html. Accessed on April 4 2008.

6. CDC#1. Available at http://www.cdc.gov/nceh/globalhealth/global.htm, Accessed on September 29, 2008.

7. UNESCO. Available at http://siwi.client.constructit.se/sa/node.asp?node=159. Accessed on September 29, 2008.

8. UNESCO. Available at http://webworld.unesco.org/water/news/newsletter/161.shtml#know. Accessed September 29, 2008.

9. Nations MK, Monte CMG. "I'm not dog, no!": cries of resistance against cholera control campaigns. *Soc Sci Med* 1996;43(6):1007–1024.

10. Parker M, Allen T, Hastings J. Resisting control of neglected tropical diseases: dilemmas in the mass treatment of schistosomiasis and soil-transmitted helminths in north-west Uganda. *J Biosoc Sci* 2008;40:161–181.

SECTION I

DEFINING THE PROBLEM

This section defines and gives details of the relationships among water access and quality, diarrheal diseases, malnutrition, undernutrition and anemia, lack of adequate sanitation, lack of adequate hygiene, environmental factors, and water and sanitation-related diseases.

1

TACKLING THE WATER CRISIS: A CONTINUING NEED TO ADDRESS SPATIAL AND SOCIAL EQUITY

JAY GRAHAM

1.1 INTRODUCTION

After decades of investment, an estimated 884 million of the world's poorest people remain with unreliable and unsafe water. Access to safe water is essential for the health, security, livelihood, and quality of life and is especially critical to women and girls, as they are more likely than men and boys to be burdened with collecting water for domestic use. Some of the trends in access to safe water globally look positive. With 87% of the world's population—nearly 5.9 billion people—using safe drinking water sources, the world is on schedule to meet the drinking water target of the Millennium Development Goals (MDGs) set for 2015 (Fig. 1.1). In China, 89% of the population of 1.3 billion has access to drinking water from improved sources, up 22% since 1990. In India, 88% of the population of 1.2 billion has access, an increase of 16% since 1990. Further, 3.8 billion people (57%) of the world's population currently get their drinking water from a piped connection that provides running water in their homes or compound. A number of spatial and social inequities, however, persist and need to be addressed. More than 8 out of 10 people without access to improved drinking water sources live in rural areas. Regionally, sub-Saharan Africa and Oceania are most behind in coverage. Just 60% of the population in sub-Saharan Africa and 50% of the population in Oceania is estimated to be using improved sources of drinking water. The poor also suffer disproportionately. A comparison of the richest and poorest population strata in sub-Saharan Africa shows that the richest 20% are two times more likely to use an improved drinking water source than the poorest 20%. Compounding the situation, many of those counted as having access are left with water systems that will be short lived. For these systems to reach sustainability, more focused efforts must be made regarding who will maintain water systems and where the money and skills to do so will come from.

1.2 ACCESS TO IMPROVED WATER SUPPLIES

1.2.1 Background

Improvements in water supply, sanitation, and hygiene have greatly advanced the health of industrialized countries (1) in places where diarrhea, cholera, and typhoid were once the leading causes of childhood illness and death. Improved water supply and sanitation interventions provide a wide range of benefits—explicit and implicit. These include higher lifespan, reduced morbidity and mortality from various diseases, augmented agriculture and commerce, higher school attendance, lower health care costs, and less physical burden. The time-savings can allow women to engage in non-illness-related tasks, and provide more time for childcare, socialization, and education activities (2). Further, when water supplies are brought closer to homes, the savings in women's energy expenditure can result in a reduction of energy (food) intake. This savings may then be transferred to children's intake of food at no extra cost (3). The implicit benefits of an improved water supply include higher quality of life due to available supply of drinking water and increased potential for communities to engage in other improvements, once they have achieved improved access to a safe water supply.

Water and Sanitation-Related Diseases and the Environment: Challenges, Interventions, and Preventive Measures, First Edition.
Edited by Janine M. H. Selendy.

FIGURE 1.1 Young girls collecting water (East Hararghe, Ethiopia). According to the 2005 Ethiopia DHS, only 8% of households report having water on their premises and more than half of the rural population reports taking more than 30 min for collecting drinking water. It was also noted that women and children shoulder the greatest burden for collecting water and spend a disproportionate amount of time hauling water over long distances.

Worldwide it is estimated that 884 million people lack access to an *improved water supply* (4), defined as one that, by nature of its construction or through active intervention, is protected from outside contamination, in particular from contamination with fecal matter.[1] Under existing trends of coverage improvement, however, the target to halve the proportion of the world's population without access to an improved water supply, as set out in the Millennium Development Goals (MDG Target 10, Goal 7), is on track (see Fig. 1.2).

This lack of basic access to improved water supply results in significant impacts to health, because of water-related diseases, as well as lost productivity. Globally, annual deaths from diarrhea—linked to lack of access to water and sanitation infrastructure and poor hygiene—were estimated at 1.87 million (95% confidence interval, 1.56 million–2.19 million), reflecting an estimated 19% of total child deaths in 2004 (5). Nearly three-quarters of those deaths occurred in just 15 countries (Table 1.1), and deaths are highly regionalized (Fig. 1.3).

Improvement to water supply, in terms of quantity, reliability, and quality, is an essential part of a country's development; however, there are a number of obstacles that limit

[1] According to WHO/UNICEF, an "improved drinking water source" includes piped water into dwelling, plot, or yard; public tap/standpipe; tubewell/borehole; protected dug well; protected spring; and rainwater.

TABLE 1.1   Countries Accounting for Three-Quarters of Diarrheal Deaths, 2004 (5)

Country	Deaths Due to Diarrhea (Thousands)
India	535
Nigeria	175
Democratic Republic of the Congo	95
Ethiopia	86
Pakistan	77
China	74
Bangladesh	69
Afghanistan	65
Indonesia	39
Angola	34
Niger	33
Uganda	28
Myanmar	26
United Republic of Tanzania	25
Mali	24
Total of 15 countries	1384

successful improvement. Rapid population growth, degradation of the environment, the increase of poverty, inequality in the distribution of resources and the misappropriation of funds are some of the factors that have prevented water supply interventions from producing better results (6). Further, numerous studies have shown that resources and time are being spent in water supply interventions that do not take into account beneficiaries' needs, preferences, customs, beliefs, ways of thinking, and socioeconomic and political structures (i.e., the enabling environment).

1.2.2   Past Efforts to Improve Access to Safe Water

Development interventions began to flourish in the 1970s as disparities, in terms of quality of life and access to basic services between wealthy and poor countries, became evident. The original motivation for providing water and sanitation to the inhabitants of less developed countries was based upon the consideration that water and sanitation is a cornerstone to public health and a basic human right (7). As a human right, those services should, therefore, be financed by the government of an individual nation, but because governments of economically developing countries did not have the resources needed to provide basic services to their entire population, it was assumed that industrialized countries and international organizations should assist in the provision of these services (8). In fact, the approach taken for the design and implementation of most of these early projects did not typically consider the preferences of beneficiaries, as it was perceived that they did not have knowledge and ability to contribute. Facilities constructed soon fell into disrepair due to lack of operation and maintenance, resulting from deficiencies in organization, training, and sense of ownership by

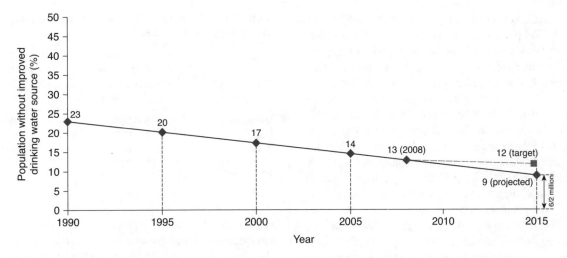

FIGURE 1.2   Based on current trends, the world is on track to meet the water target of the Millennium Development Goals.

beneficiaries. Soon after many water supply and sanitation interventions, communities often found themselves in the same conditions as they had previously known. The results were not promising, and it became evident that there was something missing in the planning.

During the International Drinking Water Supply and Sanitation Decade (1981–1990), the international community established as a common goal the provision of safe water supplies and adequate sanitation services to all the communities around the world. This meant that by 1990 every person worldwide should have his or her basic needs met. In 1981, it was estimated that 2.4 billion people would need to gain access to improved water supplies—a figure equivalent to connecting 660,000 people to a safe supply of water each day

for 10 years (9). Even though this goal was far from accomplished, an estimated 370,000 people, on average, received improved water supplies each day. Following the decade, and after two world conferences (New Delhi in 1990 and Dublin in 1992), the international community determined that water and sanitation could no longer be regarded simply as a right. After the Dublin conference there was a shift to the view of safe water as an economic good because it had an environmental and a productive value. It became clear that need was no longer a sufficient reason for the international community to provide water and sanitation to any community (7).

After the World Conference on Water and Sanitation held at The Hague, Netherlands, in March 2000, the international

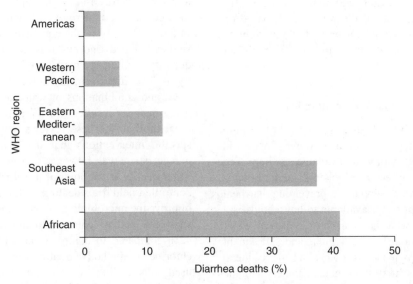

FIGURE 1.3   Distribution of deaths due to diarrhea in low- and middle-income countries in five WHO regions, 2004 (5).

community set a new common goal and published "Vision 21: Water for People." Vision 21 proposed a world in which, by 2025, everybody would know the importance of hygiene and education and enjoy safe water and appropriate sanitation services. At the United Nations Summit in September 2000, 189 UN Member States adopted the Millennium Declaration, from which emerged the aforementioned Millennium Development Goals. Target 10 of MDG 7 is to "halve by 2015 the proportion of people without sustainable access to safe drinking water and basic sanitation (over 1990 estimates)." The MDGs have been a significant force of garnering donor support and government commitment to increasing water supply and sanitation. A very important aspect of Vision 21 and the MDGs, one that reflects concerns of the international community, is the recognition of the need for a new approach to water security. This new approach emphasizes "buy-in" before the implementation of a water project in any community and a stronger focus on ensuring that improvements made be sustained. Another particular aspect of Vision 21 is the ratification of water and sanitation as a basic human right. After the Water Decade, the international community indicated that water and sanitation could not be viewed as a basic right any longer, because the beneficiaries of the projects did not value the improvements made and facilities constructed when they were not required to contribute monetarily. In other words, people will not appreciate, continue to utilize, and preserve something to which they have not contributed. The World Conference, however, concluded that the lack of a sense of ownership and commitment to project improvements on the part of the beneficiaries was due to the inadequate and often neglected inclusion of beneficiaries' preferences into project design and implementation. Further, it was noted that beneficiaries of water projects should be responsible for the costs of the operation and maintenance of the system but not for the costs of the water itself, based on the idea that every individual on earth has the right to obtain and consume enough water to guarantee his/her survival.

1.2.3 Impacts of Improved Water Supplies

There is a significant—and still growing—body of literature on the impacts associated with improved water supplies, in terms of increased quantity of water available and improved water quality. Most analyses have looked at health effects, especially the role of water supplies in preventing diarrheal disease. The quantity of water available to households is a critical component of what is meant by "improved water supply," and it is essential for the hygiene and subsequent health of a population. When assessing health benefits due to water supply programs, it is important to understand the different interactions between water quality and water quantity. For many infectious diarrheal diseases, exposure–risk

TABLE 1.2 Numbers of People Who Received Improved Water Supplies (1981–2000)

Water Supply Category	Average Number of People Receiving Services per Day (1981–1990)	Average Number of People Receiving Services per Day (1991–2000)
Urban	100,000	130,000
Rural	270,000	90,000
Total	370,000	220,000

Source: See Ref. 9.

relationship is unclear. There remains debate regarding attributable risk and interactions of specific exposures within the fecal–oral route of disease (10). Exposure risks in children with persistent diarrhea, rather than in children with acute diarrhea, accounts for an important gap in our knowledge, because persistent diarrhea affects immune competency and increases subsequent susceptibility. Thus, it may be more important for future research to characterize exposure routes in children suffering from persistent diarrhea versus acute diarrhea (11).

Between 1980 and 2000, most studies of water quality assessed only the source of water and not the point at which users actually consumed the water (point of use). In a review of 67 studies to determine the health impact of water supplies, Esrey et al. (12) found that the median reduction in diarrheal morbidity from improvements in water availability to be 25% and the median reduction based on improvements to water quality at the source, not at the point of use, to be 16%, with a range of 0–90%. Combinations of water quality at the source and water quantity resulted in a 37% median reduction in diarrheal morbidity (see Table 1.2). In 1991, the analysis was updated, covering 144 studies and looking more carefully at their content and the rigor with which they were conducted. In the 1991 analysis, the conclusion drawn by viewing only studies deemed rigorous was that improvements in water quantity resulted in a median reduction of diarrheal morbidity of 30%, improvements to water quality at the source of 15%, and combinations of water quality at the source and water quantity resulted in a 17% median reduction in diarrheal morbidity. These reviews helped set the agenda for specific interventions that the global community would pursue. There was, however, a growing interest in assessing water quality at the point of use. In 2003, an analysis of 21 controlled field trials dealing with interventions designed to improve the microbiological water quality at the point of use showed a median reduction in endemic diarrheal disease of 42% compared to control groups (13). Nine studies used chlorine as a method of treating water, five used filtering, four used solar disinfection, and three used a combination of flocculation and disinfection. This study and subsequent studies resulted in donor investments for improving drinking water quality at the point of use; a large number of economically developing countries now have point-of-use products that are being socially marketed.

In a more recent review of studies using experimental (randomized assignment) and quasiexperimental methods the impact of water, sanitation, and/or hygiene interventions on diarrhea morbidity among children in low- and middle-income countries was conducted (14). Sixty-five rigorous impact evaluations were identified for quantitative synthesis, covering 71 distinct interventions assessed on 130,000 children across 35 developing countries during the past three decades. These studies were evaluated for a range of factors, such as type of intervention, effect size and precision, internal validity, and external validity. The interventions were grouped into five categories: water supply improvements, water quality, sanitation, hygiene, and multiple interventions involving a combination of water and sanitation and/or hygiene. The results challenged the notion that interventions to improve water quality treatment at the point of use are necessarily the most efficacious and sustainable interventions for promoting reduction of diarrhea. The analysis suggests that while point-of-use water quality interventions appear to be highly effective, and generally more effective than water supply or improving water quality at the source, much of the evidence is from small trials conducted over short periods of time. The review indicated that point-of-use interventions conducted over longer periods of time demonstrated smaller effects as compliance rates fell. Interestingly, the study found that hygiene interventions, particularly the promotion of handwashing with soap, were effective in reducing diarrhea morbidity, even over longer periods of time.

Calculations of the cost-effectiveness of the interventions described above have shown point-of-use and hygiene interventions to be highly efficient for bringing about health improvements (15, 16). Estimates of cost-effectiveness from improved water supplies, in terms of the costs per disability-adjusted life year (DALY) averted, show that a community connection to a water source results in a cost aversion of 94 USD/DALY. This is less than half the figure for household water connection, but substantially higher than estimates for point-of-use water quality interventions, which are estimated at 53 USD/DALY averted, using chlorination (16). Estimates from improved hygiene and sanitation suggest that hygiene promotion is most cost effective, at 3 USD/DALY averted, followed by sanitation promotion, at 11 USD/DALY (15).

Water supply interventions have many benefits. For example, better water supplies enable improved hygiene practices, such as handwashing and better home hygiene, and there are likely considerable spillover effects in terms of environmental health benefits. In Lesotho, use of smaller quantities of water was related with higher rates of *Giardia lamblia* infection (17). In Taiwan, a reduction of 45% in rates of trachoma was noted when the water supply was attached to

the home, compared to a water supply that was 500 or more meters away (18). Time-savings associated with water supply interventions are also significant. In rural Nigeria, Blum et al. (19) estimated that the installation of water systems reduced collection time from 6 h to 45 min per household per day during the dry season, mainly benefiting adolescent girls and young women. In addition, Wang et al. (20) estimated a time-savings of 20 min per household per day from a village water supply improvement in China. In the Philippines, water quantity was strongly associated with nutritional status. Children in households that averaged less than 6 L per capita per day were significantly more malnourished than children in households that averaged 6–20 L or more than 20 L per capita per day (21). A study of Pakistan households showed that increased water quantity available at the household level was associated with reduced stunted growth in children (22).

It has also been observed that reducing water collection time can positively affect time spent on children's hygiene, food preparation, and feeding children (23). For households without a source of drinking water in their compound, it is usually women who go to the source to collect drinking water. In a recent analysis of more than 40 developing countries, women collected water for almost two-thirds of homes, versus a quarter of households where men collected water. In 12% of homes, children were responsible for collecting water, and girls under 15 years of age were twice as likely to collect water as boys of the same age category (24, 25).

The public health gains stemming from access to increased quantities of water typically occur in steps. The first step relates to overcoming a lack of basic access, where distance, time, and costs involved in water collection combine to result in volume use inadequate to support basic personal hygiene and that may be only marginally adequate for human consumption (Table 1.3). Significant health gains occur largely when water is available at the household level. Other benefits derived from the second step in improving access include increased time available for other purposes. Yet, availability of new or improved water supplies does not always translate directly into a significant increase in use. In East Africa, after new water supplies were placed in proximity to households, no increases in the amount of water used resulted if the original water source was less than 1 km from the home (26).

Incremental improvements can occur as one moves up the continuum of water supply service. However, providing a basic level of access is the priority for most water and health agencies. In fact, progress toward universal achievement of this level of service remains a focus of international policy initiatives as highlighted by the MDGs and the WHO/UNICEF Joint Monitoring Program. The most important health benefits are likely to be obtained when focus is placed on resources to ensure that all households have access to

TABLE 1.3 Level of Water Supply Service and Related Potential Hygiene and Health Effects

Service Level	Distance/Time for Water Collection	Quantity Collected	Hygiene-Related Issues	Health Risk
No access	>1 km or >30 min total collection time	Very low—typically less than 5 L/person/day	Hygiene not assured and consumption needs may be at risk	Very high
Basic access	Between 100 m and 1 km (5–30 min collection time)	Low—average is typically less than 20 L/person/day	Not all hygiene requirements may be met	Medium
Intermediate access	On-site (e.g., single tap in compound or in house)	Medium—average is typically 50 L/person/day	Most basic hygiene and consumption needs met	Low
Optimal access	Water is piped into the home through multiple taps	Varies significantly, but typically above 100 L/person/day	All uses can be met	Very low

Source: See Ref. 26.

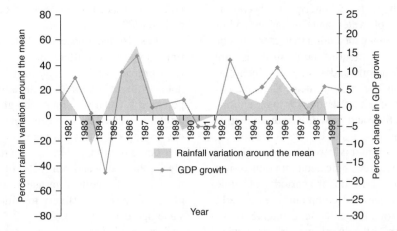

FIGURE 1.4 Rainfall variation around the mean and GDP growth in Ethiopia. Ethiopia, suffers from highly variable rainfall, both temporally and spatially, and experiences regular droughts that devastate portions of the country. It is estimated that hydrological variability currently costs the Ethiopian economy, which relies on rainfed subsistence agriculture, over one-third of its growth potential (29).

improved water sources and, in some circumstances, in directly upgrading to access at household level (27).

Water use among the poor can be an essential part of livelihood coping strategies. In practice, the use of water for domestic purposes cannot easily be distinguished from productive use, particularly among very poor communities. When communities design their own water systems, they invariably plan for multiple use water systems, and this is especially the case if the livelihoods of households depend on livestock (28). In multiple use approach interventions, it is critical that planners (i) work with the community to assess the range of water needs in collaboration with end users; (ii) examine water sources available; and (iii) match water supplies to needs based on the quantity, quality, and reliability required for various purposes. There may also be important health and social gains from ensuring adequate quality of service to support small-scale productive use, especially when this involves food production (Fig. 1.4).

Access to water used for small-scale productive activity in such areas is therefore important as part of economic growth and may deliver significant indirect health benefits as a result (27). Although water scarcity is a significant and growing problem, it should be highlighted that as a continent, Africa's water supplies are more than adequate to provide fresh drinking water for the entire population and are sufficient for their economic needs. Only 5.5% of renewable water resources are currently withdrawn, while 340 million people on the continent still lack access to safe drinking water (30). Although water resources are available, most lack the economic resources to capture and use them. In industrialized countries, 70–90% of annual renewable water resources are withdrawn, while only 3.8% of Africa's surface and groundwater is harnessed (30).

The water-related indicator used for target 10 of MDG 7 is "sustainable access to an improved water source." The technologies considered "improved," however, often do not consistently result in high-quality water. There are certain sources of water that the public health community condemns as risky (e.g., unprotected wells) and others they deem safe (e.g., protected wells). Comparing water quality from protected and unprotected supplies across countries, however, has demonstrated that in many cases protected supplies often provide lower water quality than do protected wells in other countries. This suggests that certain practices—not certain types of water sources—may be more important in improving water quality (31). As mentioned above, it is now generally accepted that providing safe water at the source does not imply that water is safe at the point of use. A study by Gundry et al. found that about 40% of water samples from microbiologically safe sources of water were contaminated at the point of consumption. Household water treatment at the point of use for most communities is an important intervention, regardless of whether the water comes from an improved source.

The situation for urban water utilities is not much better. This is also the case for many urban communities that have access to a piped supply. It is estimated that nearly one in five water utility systems in Africa, Asia, and Latin America fail to use water disinfectants (32). Reasons for this failure include cost, operations, and maintenance of equipment and concern about disinfection by-products. Water systems in many of these regions are characterized by intermittent water flows, which can affect the quality of the water due to the negative pressure in the pipes. Thus, there is no guarantee that water is clean, even when derived from a piped system.

Of the people who report treating their water, roughly 1.1 billion people say they typically boil their water at home before drinking it—four times more than the number of people who report chlorinating or filtering their drinking water. Boiling is currently one of the most accessible means for water treatment to most populations, and has been shown effective (33). However, in the absence of safe storage, water that is boiled is immediately vulnerable to recontamination; especially when poor, the environment is unhygienic. Further, this mode of treatment can have serious side effects, such as indoor air pollution and depletion of environmental resources if biofuels (e.g., wood) are used for boiling.

1.2.4   Naturally Occurring and Anthropogenic Water Pollution

In addition to microbiological contamination of water—the emphasis of this chapter—naturally occurring and anthropogenic sources of chemical pollution can pose serious human health risks. Although no published estimates are available on the global burden of disease resulting from chemically polluted water (34), a number of countries are increasingly facing water pollution challenges due to chemicals, especially where the industrial sector is developing. In addition to anthropogenic pollutants, groundwater commonly contains naturally occurring toxic chemicals, including arsenic and fluoride, which dissolve into the water from soil or rock layers. The most extensive problem of this category is arsenic contamination of groundwater, which has been observed in Argentina, Bangladesh, Chile, China, India, Mexico, Nepal, Taiwan, and parts of eastern Europe and the United States (35). Arsenic in Bangladesh's groundwater was first highlighted in 1993 and was a result of international agencies promoting protected wells in an effort to eliminate diarrheal diseases caused by fecally contaminated surface waters. Millions of shallow wells were drilled into the Ganges delta alluvium in Bangladesh, and estimates indicate that approximately 40 million people were put at risk of arsenic poisoning-related diseases because of high arsenic levels in the groundwater (36). Fluoride is another naturally occurring pollutant that causes health effects, and exposure to high levels in drinking water can detrimentally affect bone development and in some cases can cause crippling skeletal fluorosis. The burden of disease from chemical pollution in specific areas can be large. There are a number of events that have underscored the high levels of disease burden from chemical pollution, including methylmercury poisoning, chronic cadmium poisoning, and diseases of nitrate exposure, as well as lead exposure (34).

1.2.5   Resources Needed

The water supply component of the MDGs, while formally on track, is not a guaranteed success, especially if efforts are not sustained. Moreover, uneven progress exists between rural and urban populations, and the lower baseline water supply coverage in rural compared to urban areas is significant. There is a wide range of estimates for meeting the water supply target of the MDGs. Hutton and Bartram (37) estimated total spending, excluding project costs, required in developing countries to meet the water component of the MDG target to be 42 billion USD (Fig. 1.5). This translates to 8 USD per capita spending for water supply.

1.2.6   Spatial and Social Inequities in Access

"Equity" relates closely to the idea of fairness and that all members of a society have equal rights. Water supply interventions, for example, are considered equitable if they affect all parts of society equally. For example, perfect equity in intracountry budgets would be reflected in a situation where every citizen is allocated an equal amount of the investment, regardless of the part of the country where the citizen lives. Equal levels of access to clean and safe water would be an equitable outcome (38). Equity is concerned with comparing different parts of society, which is complicated by the many ways that society can be grouped. For example, geography, social or health status, gender, and ethnicity can be used for comparisons. Two categories of

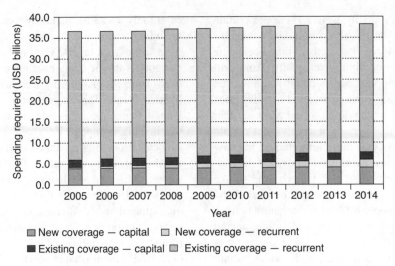

FIGURE 1.5   Spending required from 2005 until 2015 to meet water supply component of MDG target 10, excluding project costs (38).

disparities are useful for thinking about equity in water supply and sanitation (38). The first is spatial equity and includes geography, where groups are defined by where they live, such as rural versus urban, or the partitioning of a country into administrative boundaries. Social equity is concerned with groups defined by attributes linked to their identity, which traverse spatial boundaries. Particularly vulnerable groups may include women, people living with HIV/AIDS, the elderly, the disabled, orphans, and widows. The poor are also an important group that is large and critically important, but often difficult to define (38). There is obviously overlap between social and spatial inequities. For example, a large percentage of the urban population without access is also poor, and a larger proportion of the rural population who spend time collecting water are women. Additionally, equitable investments do not necessarily equate to equitable outcomes, and costs may vary according to a number of factors. For water supplies, population density, distance from places where parts are available, or the geology can affect costs (38). A number of spatial and social inequities persist and need to be addressed, and there are many challenges facing efforts to improve equitable access. Population growth is a major barrier to current efforts in the water sector to reduce the number of people living without access to safe water. In the last 40 years the population of the world has gone from 3,659 million in 1970 to roughly 6,800 million, people in 2010. In 1980, the United Nations estimated that 1,800 million people lacked access to safe water supplies; today, there are still 884 million people without access to safe water.

Spatially, more than 8 out of 10 people without access to improved drinking water sources live in rural areas (Fig. 1.6). Regionally, sub-Saharan Africa and Oceania are regions most behind in coverage (Fig. 1.7). Just 60% of the

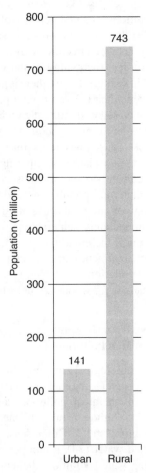

FIGURE 1.6   Number of people living in urban or rural areas without access to an improved water supply, 2008 (24).

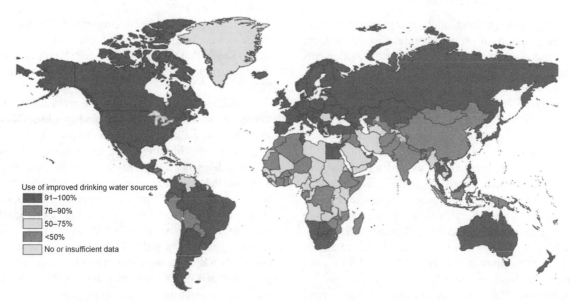

FIGURE 1.7   Countries represented by the percentage of population using improved drinking water supplies. (*See insert for color representation of this figure.*)

population in sub-Saharan Africa and 50% of the population in Oceania is estimated to use improved sources of drinking water. Coverage in 19 countries in sub-Saharan Africa increased by nearly 10% between 1990 and 2006; however, absolute numbers of unserved went up by 37 million. Compounding the situation, many of those counted as having access have nonfunctioning water systems. Improved access to rural water supply remains almost totally donor driven since most "improved options" are out of reach for users to construct at their own expense. Subsequently, improved access to rural water supply in sub-Saharan Africa has been progressing at less than 0.5% each year; the required rate to achieve the MDGs is 2.8% (31). In rural parts of Africa, for more than a quarter of the population in a variety of sub-Saharan African countries, a single trip to collect water takes longer than 30 min (Fig. 1.8) (24).

Urban areas are growing at such a pace that many municipal water facilities are unable to keep up with the increasing population (Fig. 1.9). Indeed, the provision of water to rapidly growing cities and towns continues to be an

FIGURE 1.8   Percentage of population that takes more than 30 min to collect water during one trip (29).

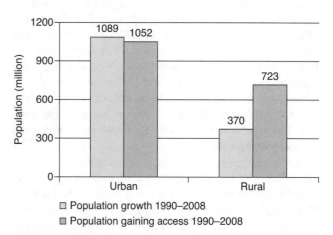

FIGURE 1.9   Increase in population growth in urban and rural sectors compared to the increase in the population achieving improved water supply coverage worldwide, between 1990 and 2008 (24).

overwhelming challenge facing municipal governments, and although urbanization can offer economies of scale for water supply systems, the growth in slum and squatter settlements makes the situation particularly difficult.

Even when a piped supply exists, typically in urban areas, it is not always reliable. Less than 10% of people in many South Asian cities receive a 24 h piped water supply. Problems arise because many municipal pipelines reach wealthiest clients first, even though they are built with aid from governments and international institutions with the goal of making water more accessible to the poor. Thus, a significant number of urban populations without utility connections must rely on alternatives, such as service from small-scale water providers (SSWPs). Currently SSWPs are most prevalent in Southeast Asia, where a quarter of households in Cebu (Philippines), Ho Chi Minh City, Jakarta, and Manila may use these services (39).

In urban areas of the developing world, governments have favored large water utilities. Unfortunately, existing tariffs and management structures have caused these systems to fail to provide piped water coverage to entire populations. Connection fees are frequently too high or total available water is insufficient to support an urban area. Many utilities choose not to equip poor neighborhoods because of the high percentage of unpaid bills, fraudulent consumption, low levels of individual consumption, and because network maintenance costs are high. Additionally, people that occupy land illegally may also be excluded from public services. In cases where water companies are allowed or mandated to serve poor households, water is not always affordable or payment schedules may not be feasible. Thus, many people are forced to illegally draw their water from "spaghetti networks" that connect to the border of a municipal grid system or to purchase expensive, and commonly contaminated, water from SSWPs.

Of further importance are the inequalities surrounding the cost of water for the urban poor. While an SSWP generally offers a more flexible payment schedule, its water is usually pricier and consumes a large portion of household expenses. It has been cited that in some cities the poor pay huge premiums to water vendors over the standard water price of those hooked up to municipal systems: 60 times more in Jakarta, Indonesia; 83 times more in Karachi, Pakistan; and 100 times more in both Port-au-Prince, Haiti and Nouakchott, Mauritania.

Additionally, because water is of unknown quality, the urban poor may pay even more in order to purify it. The United Nations Development Program estimated in 1992 that households in Jakarta, Indonesia, spent a combined total of up to 50 million USD/year to boil drinking water, an amount equivalent to 1% of the city's gross domestic product. In Bangladesh, for example, boiling water uses nearly 11% of the family income among the lowest earning 25% of all households.

Socially, the poor suffer disproportionately. A comparison of the richest and poorest population strata in sub-Saharan Africa shows that the richest 20% are two times more likely to use an improved drinking water source than the poorest 20%.

In most developing countries, the provision of water and sanitation are women's responsibility (24). Often, rural women must walk long distances to provide their families with water for drinking, cooking, domestic hygiene, and personal hygiene. Interventions to increase access often diminish the time that women spend gathering water and have provided participants with opportunities to learn new skills and spend more time cultivating crops in the time previously used for water collection. These classes of changes can have positive impacts on the local economy, especially when income-earning involves tasks such as laundry work and other types of activities that use water. By allowing for less time for water collection, new opportunities enable women to effectively contribute to the communities' economic growth (40).

1.2.7 Sustainability

Sustainability of water supplies is especially difficult in rural areas because of the lack of support through monitoring systems, training, human resource back-up support, and availability of spare parts and services. Throughout rural sub-Saharan Africa, thousands of water systems are developed every year, such as boreholes equipped with motorized or hand/foot pumps. These systems often fall into disrepair shortly after installation. It is estimated that 50,000 water supply systems are not functioning across Africa—a number representing an investment of nearly 300 million USD. This problem occurs for one reason: lack of operations and maintenance; operations and maintenance, however, is a multifaceted feature of any water system.

Many of the negative results in past interventions were linked to (i) lack of community participation; (ii) utilization of inappropriate technologies; (iii) lack of a sense of ownership on the part of the beneficiaries; (iv) failure to provide the institutional support required for the project; and (v) dissatisfaction of the community with project outcomes (41). In order to design a more effective and responsive approach for the provision of water and sanitation, development organizations and donor agencies are utilizing a series of participatory methodologies and techniques that focus on getting intended users actively involved in all stages of the project cycle. Fundamentally, community participation increases the probability of success and the sustainability of the projects implemented.

Participatory approaches evolved from disciplines such as anthropology, sociology, research on farming systems, and others, and have tried to fill in the existing gap between technology (hardware) and operations and maintenance

TABLE 1.4   Seven Characteristics of the Service Delivery Approach

1. Invests on the basis of need for the entire district, as well as investing in support services and frameworks
2. Addresses financing needs for full life cycle costs from the outset to ensure asset replacement
3. Operates on a continuous time frame, not project timeline, for service delivery
4. Allows flexibility for water systems so that different management and technical approaches can be used
5. Works to achieve full coverage within established geographic/administrative boundaries
6. Seeks to coordinate all actors to work collectively under an overarching strategy, including commonly agreed-upon model(s), depending on the service provided
7. Works with most appropriate management model for service delivery

Source: See Ref. 29.

(software). These approaches were developed based upon the flaws identified and the lessons learned while implementing the supply-driven approach for the provision of safe water and sanitation services. The underlying principle was and continues to be the involvement of all stakeholders, especially the main users of the system, in all the phases of water and sanitation programs or projects, with the intention of improving their sustainability and probability of success. The primary objective was to be more responsive to the needs and preferences of users and more appropriate to given local conditions and the environment. Another important characteristic of these participatory methodologies was the significant change in the role that users of the system played during the design, implementation, construction, operation, and maintenance of the systems. Participatory methodologies were developed to facilitate the process of empowerment and capacity-building of the communities benefiting from development interventions (42).

Community participation can bring about numerous benefits to development interventions, but such benefits must be weighed against the time and costs related to their implementation. For participation and commitment on the part of the community to be effective, financial and human resources must exist at the beginning of the process; in this way, planners may ensure success (43). It is important to note that there is no one approach toward community participation that works in all situations. The approaches utilized in the water sector have to be flexible enough to incorporate site-specific information about environmental, social, and cultural factors; in addition, stakeholders' needs and priorities into the design and implementation of water and sanitation projects must be accounted for (43).

One of the most commonly used models for developing rural water interventions involves village-level coordination and the development of a system for cost recovery for operations and maintenance. Typically, a community bank account is opened and a community member is appointed to collect the fees. The selection of the technology and personnel who have the skills to operate and maintain it are also part of the operations and maintenance system in place. Other models have been developed and experimented with and include public and private sector arrangements that aim to provide support to community systems following construction.

The community management model has brought many benefits; however, it has not always resulted in sustainable water supply at scale. It is becoming clearer that communities often cannot manage the variety of tasks that arise after the construction of water systems, such as repairs, accounting, conflict resolution, legal issues, and system replacement. A new model, the service delivery approach, was developed for improving rural water services and aims to better incorporate enabling environment factors with the aim of increasing sustainability and scale. The approach considers the whole life cycle of service, from design, day-to-day operations, and maintenance to eventual replacement (Table 1.4).

For millions of rural people, the top half of Fig. 1.10 represents a standard water supply intervention. Following construction of a new system users have access to an

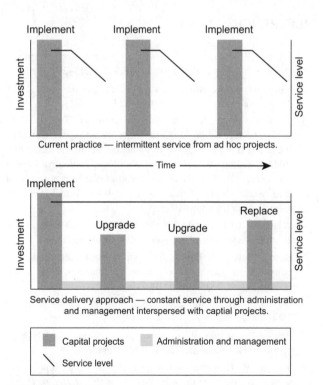

FIGURE 1.10   Current model of rural water supply interventions (top half) versus the service delivery approach (bottom half) that provides for constant service through ongoing support interspersed with capital projects (29).

improved source, but due to lack of follow-up support, the system quickly deteriorates until it is nonfunctional. In the service delivery approach, once water supply access is improved, it is maintained through a proper understanding of the full life cycle costs and institutional support needs (29).

1.2.8   Final Remarks

There is evidence that the global community is making progress toward providing all people worldwide access to a safe and reliable water supply. In 2000, for example, the number of people without a safe water supply was nearly 1.1 billion—the estimate in 2010 is 884 million. This gain is considerable given that the population of less economically developed countries went up by nearly 700 million during the decade. A number of spatial and social inequities, however, persist and certain challenges to improve equitable access are growing. Population growth—among urban areas in particular—is a major obstacle to current efforts to reduce the number of people living without access to safe water. If we are to continue moving forward, all sectors of society will need to more fully engage; these include researchers, national and local governments, NGOs, the private sector, international and bilateral agencies, and communities.

REFERENCES

1. Preston SH, Van de Walle E. Urban French mortality in the nineteenth century. *Popul Stud* 1978;32(2):275–297.

2. Esrey SA. Water, waste, and well-being: a multicountry study. *Am J Epidemiol* 1996;143(6):608–623.

3. (a) Chakravarty I, Nath KJ. *Time/Energy Savings with Improved Accessibility to Community Water Supply*. Calcutta, India: All India Institute of Hygiene and Public Health, 1994; (b) Diaz E, Esrey SA, Hurtado E. *Social and Biological Impact Following the Introduction of Household Water in Rural Guatemala*. Ottawa, Ontario, Canada: International Development Research Center, 1995.

4. WHO/UNICEF Joint Monitoring Programme, 2008. Available at http://www.wssinfo.org/en/r0_MDG2008.html.

5. Boschi-Pinto C, Velebit L, Shibuya K. Estimating child mortality due to diarrhoea in developing countries. *Bull World Health Organ* 2008;86(9):710–717.

6. Briscoe J. When the cup is half full. *Environment* 1993; 35(4):7–37.

7. Black M. *Learning What Works*. Washington, DC: UNDP—World Bank: Water and Sanitation Program, 1998.

8. Yacoob M. Community self-financing of water supply and sanitation: what are the promises and pitfalls? *Health Policy Plan* 1990;5(4):358–366.

9. Mara D, Feacham R. Taps and toilets for all—two decades already, and now a quarter century more. *Water* 2001;21:13–14.

10. Pruss A, Havelaar A. The global burden of disease study and applications in water, sanitation and hygiene. In: Fewtrell L, Bartram J, editors. *Water Quality: Standards, and Health*. London, UK: IWA Publishing, 2001, pp. 43–59.

11. Sazawal S, Bhan MK, Bhandari N, Clemens J, Bhatnagar S. Evidence for recent diarrhoeal morbidity as a risk factor for persistent diarrhoea: a case-control study. *Int J Epidemiol* 1991;20(2):540–545.

12. Esrey, S.A., Feachem, R.G., Hughes, J.M. Interventions for the control of diarrhoeal diseases among young children: improving water supplies and excreta disposal facilities. *Bulletin of the World Health Organization* 1985;63(4):757–772.

13. Clasen T, Roberts I, Rabie T, Cairncross S. Interventions to improve water quality for preventing infectious diarrhea. Cochrane Review. In: *The Cochrane Library*. Oxford: The Cochrane Collaboration, 2003.

14. Waddington H, Snilstveit B, Fewtrell L, White, H. Water, Sanitation and Hygiene Interventions to Combat Childhood Diarrhoea in Developing Countries. International Initiative for Impact Evaluation, 2009. Available at http://www.3ieimpact.org/admin/pdfs2/17.pdf.

15. Cairncross S, Valdmanis, V., Water supply, sanitation and hygiene promotion. In: Jamison DT,et al., editors. *Disease Control Priorities in Developing Countries*, 2nd edition. New York: Oxford University Press, 2006.

16. Clasen T, Cairncross S, Haller L, Bartram J, Walker, D., Cost-effectiveness of water quality interventions for preventing diarrhoeal disease in developing countries. *J Water Health* 2007;5(4):599–608.

17. Esrey SA, Collett J, Miliotis MD, Koornhof HJ, Makhale, P. The risk of infection from *Giardia lamblia* due to drinking water supply, use of water, and latrines among preschool children in rural Lesotho. *Int J Epidemiol* 1989;18(1):248–253.

18. Assaad FA, Maxwell-Lyons F, Sundaresan T. Use of local variations in trachoma endemicity in depicting interplay between socio-economic conditions and disease. *Bull World Health Organ* 1969;41(2):181–194.

19. Blum D, Emeh R, Huttly S, Dosunmu-Ogunbi O, Okeke N, Ajala M, Okoro J, Akujobi C, Kirkwood B, Feachem R. The Imo State (Nigeria) Drinking Water Supply and Sanitation Project, 1. Description of the project, evaluation methods, and impact on intervening variables. *Trans R Soc Trop Med Hyg* 1990;84(2):309–315.

20. Wang ZS, Shepard DS, Zhu YC, Cash RA, Zhao RJ, Zhu ZX, Shen FM. Reduction of enteric infectious disease in rural China by providing deep-well tap water. *Bull World Health Organ* 1989;67(2):171–180.

21. Magnani R, Tourkin S, Hartz M. *Evaluation of the Provincial Water Project in the Philippines*. Washington, DC: US Department of Commerce, 1984.

22. van der Hoek W, Feenstra SG, Konradsen F. Availability of irrigation water for domestic use in Pakistan: its impact on prevalence of diarrhoea and nutritional status of children. *J Health Popul Nutr* 2002;20(1):77–84.

23. Prost A, Négrel AD. Water, trachoma and conjunctivitis. *Bull World Health Organ* 1989;67(1):9–18.

24. WHO/UNICEF Joint Monitoring, Programme. Progress on Sanitation and Drinking-Water 2010 Update. Available at http://whqlibdoc.who.int/publications/2010/9789241563956_eng_full_text.pdf.

25. MICS and DHS surveys from 45 developing countries, 2005–2008.

26. White GF, Bradley DJ, White AU. Drawers of water: domestic water use in East Africa, 1972. *Bull World Health Organ* 2002;80(1):63–73; discussion 61-2.

27. Howard G, Bartram J. Domestic Water Quantity, Service, Level and Health. World Health Organization Report, 2003.

28. Adank M, et al. The costs and benefits of multiple uses of water: the case of *Gorogutu woreda* of East Hararghe zone, Oromiya Regional States, Eastern Ethiopia. RiPPLE Working Paper 7, 2008.

29. Lockwood H, Schouton T. 2009. Available at http://www.irc.nl/page/45530.

30. Flatt C. Business: the solution to Africa's water scarcity? *Water* 2009;21:32–33.

31. Sutton S.The risks of a technology-based MDG indicator for rural water supply. 33rd WEDC International Conference, Accra, Ghana, 2008.

32. WHO/UNICEF Joint Monitoring Programme for Water Supply and, Sanitation. Global Water Supply and Sanitation Assessment 2000 Report. Available at http://www.who.int/water_-sanitation_health/monitoring/jmp2000.pdf.

33. Rosa G, Miller L, Clasen T. Microbiological effectiveness of disinfecting water by boiling in rural Guatemala. *Am J Trop Med Hyg* 2010;82(3):473–477.

34. Kjellstrom T, Lodh M, McMichael T, Ranmuthugala G, Shrestha R, Kingsland S. Air and water pollution: burden and strategies for control. In: Jamison DT, et al., editors. *Disease Control Priorities in Developing Countries*, 2nd edition. New York: Oxford University Press, 2006.

35. World Health Organization. *Arsenic and Arsenic Compounds.* Environmental Health Criteria 224. Geneva: World Health Organization, 2001.

36. Kinniburgh DG, Smedley PA, editors. Arsenic Contamination of Groundwater in Bangladesh. BGS Technical Report WC/00/19. Keyworth, UK: British Geological Survey; Dhaka: Department of Public Health Engineering, 2001.

37. Hutton G, Bartram J. Global costs of attaining the millennium development goal for water supply and sanitation. *Bull World Health Organ* 2008;86(1):13–19.

38. Taylor B. Water: more for some . . . or some for more? WaterAid Tanzania, 2008. Available at http://www.wateraid.org/documents/plugin_documents/water__more_for_some.pdf.

39. McIntosh AC. *Asian Water Supplies: Reaching the Urban Poor.* Asian Development Bank, London: IWA Publishing.

40. Bell M, Franceys R. Improving human welfare through appropriate technology: government responsibility, citizen duty or customer choice. *Soc Sci Med* 1995;40(9): 1169–1179.

41. Elmendorf ML, Isely RB. Public and private roles of women in water supply and sanitation programs. *Hum Organ* 1983;42(3):195–204.

42. Sawyer R, Clarke L. *The PHAST Initiative, Participatory Hygiene and Sanitation Transformation.* Geneva: World Health Organization and UNDP—World Bank: Water and Sanitation Program, 1997.

43. White A. Community participation in water and sanitation: concepts, strategies and methods. Technical Paper Series No. 17. The Hague, The Netherlands: IRC International Water and Sanitation Centre, 1981.

2

SANITATION AND HYGIENE: TAKING STOCK AFTER THREE DECADES

Jay Graham

2.1 INTRODUCTION

In 1977, the international community dedicated the decade of the 1980s as the International Drinking Water Supply and Sanitation Decade. The goal was to bring clean water and sanitation to all people worldwide by 1990. Now, over 30 years later, almost 39% of the world's population—2.6 billion people—remain without improved sanitation. The lack of this most fundamental service contributes to an estimated 1.87 million annual deaths due to diarrhea—more than 90% of which are in children under 5 years of age. If current trends in sanitation coverage continue unchanged, the international community will again miss its current, and less ambitious, target that aims to halve the proportion of people without access to sanitation by 2015. Based on trends, it is anticipated that the sanitation target of the Millennium Development Goals (MDGs) Target 10, Goal 7 will be missed by 1 billion people. There are significant differences both spatially and socially between the groups who are making progress and those who are falling behind. The greatest number of people without access to improved sanitation is in southern Asia, but there are also large numbers in eastern Asia and sub-Saharan Africa. Seven out of 10 people without basic sanitation live in rural areas. There are some positive developments, however. Major gains in sanitation coverage have been made in northern Africa, southeastern Asia, and eastern Asia. Furthermore, open air defecation—the riskiest sanitation practice—is on the decline worldwide, with a global decrease from 25% in 1990 to 17% in 2008, equating to a decrease of 168 million people. This practice remains widespread,

however, in southern Asia, where an estimated 44% of the population defecates in the open. More concerted efforts, investments, and creativity—as well as persistence—are needed for sanitation if we are to lay this most critical foundation of health and development.

2.2 BACKGROUND

Sustainable and equitable access to sanitation and improved hygiene are widely acknowledged as critical, but neglected, development goals (1, 2). It has now been nearly 30 years since the international community first established as a common goal the provision of sanitation for all. In 1980, the United Nations estimated that 3.9 billion people lacked access to adequate sanitation. Currently, the number is 2.6 billion. The gains are significant (Table 2.1), especially given that the world's population increased 50% between 1981 and 2000—an increase of nearly 2.4 billion people. To achieve the global commitment to halve the fraction of the world's population without access to improved sanitation, as set out in MDG Target 10, Goal 7, it is estimated that 1 billion urban dwellers and 900 million people in rural areas must be provided with sanitation in order to reach the MDG for sanitation in 2015 (3). This seems unattainable considering that it took industrial countries more than 100 years to achieve sanitation coverage (4). Furthermore, the numbers of people with access to sanitation in urban areas is thought to be overestimated, exacerbating the challenges (5). Under existing trends, the sanitation target of the MDGs is off track by 1 billion people (Fig. 2.1).

Water and Sanitation-Related Diseases and the Environment: Challenges, Interventions, and Preventive Measures, First Edition.
Edited by Janine M. H. Selendy.
© 2011 Wiley-Blackwell. Published 2011 by John Wiley & Sons, Inc.

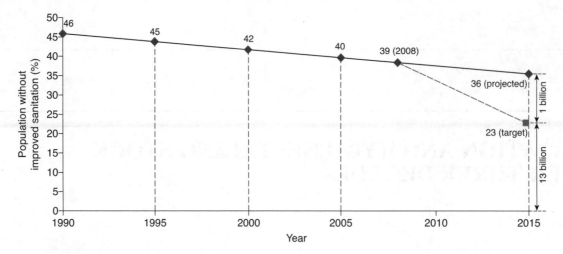

FIGURE 2.1   Based on current trends, the world is off track to meet the sanitation target of the MDGs by 1 billion people.

Globally, sanitation is recognized as a cornerstone of public health, as evidenced in a recent poll of readers of the highly regarded *British Medical Journal*, who voted sanitation the most important medical milestone since 1840 (6). Further, there is growing recognition that environmental health interventions such as sanitation and hygiene, must be more fully integrated into mainstream development policies because of the strong linkages to malnutrition and poverty reduction. Sanitation and hygiene are among the most cost-effective interventions for reducing the economic impacts of disease. Further, poor sanitation and hygiene are inextricably linked to water quality, and in most economically developing countries the main source of contaminated water is fecal contamination.

The sanitation ladder is a common framework for evaluating sanitation progress. As the name implies, the ladder provides steps that go from the lower rungs of sanitation matters, such as open defecation, to creation of more sophisticated and effective sanitation (Fig. 2.2). It is increasingly understood that there is value in motivating people to climb up the sanitation ladder, even if the resulting sanitation facility is not ideal (i.e., it does not meet the standards of the WHO/UNICEF Joint Monitoring Programme (JMP) for water supply and sanitation). As an example, many countries

are making great strides to end the practice of open defecation as a first step of their sanitation programs (Fig. 2.3).

The JMP is responsible for determining what types of sanitation facilities along the sanitation ladder constitute

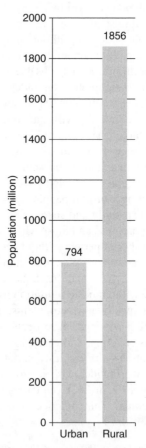

FIGURE 2.2   Number of people living in urban or rural areas that do not have access to improved sanitation.

TABLE 2.1   Numbers of People Who Received Improved Sanitation per Day (1981–2000) (4)

Sanitation Category	Average Number of People Receiving Services per Day (1981–1990)	Average Number of People Receiving Services per Day (1991–2000)
Urban	80,000	160,000
Rural	120,000	50,000
Total	200,000	210,000

(a)

(b)

FIGURE 2.3   Ethiopia is making significant advances in ending open defecation and moving people up the sanitation ladder. This latrine (interior, right side) in the Amhara Region does not meet the JMP definition of "improved" sanitation, but represents progress over the still common practice of open defecation. The gourd placed over the pit is used as fly control measure.

improved and unimproved facilities. Its coverage estimates are based on household surveys and censuses, which are often conducted under very different situations and with differing quality from household to household, and which have sometimes been controversial (7). Given the difficulties of attempting to compare data across more than 200 countries and territories, JMP has made considerable progress in monitoring approaches and has increasingly provided a

clearer picture of the nuances of sanitation coverage (Figs. 2.3–2.6) (8).

Open defecation, the lowest rung on the sanitation ladder, is practiced by an estimated 1.1 billion people (Fig. 2.7), and almost two-thirds of those who practice open defecation live in southern Asia. In sub-Saharan Africa, 221 million people are defecating in the open, the second largest total for any region. The proportion of people engaging in open

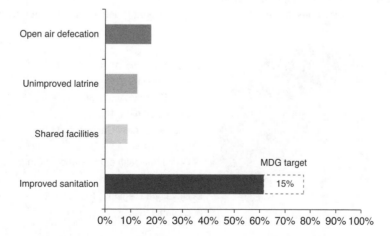

FIGURE 2.4   Proportion of the world's population using an improved, shared, unimproved sanitation facility or practicing open air defecation (2006).

FIGURE 2.5 Use of improved sanitation facilities by country (2008). (*See insert for color representation of this figure.*)

defecation, however, decreased from 25% in 1990 to 17% in 2008, and there have been declines in this practice in all regions of the world. Open defecation rates fell by 25% in sub-Saharan Africa and 20% in southern Asia during this same time period. Because of population growth, however, the population in absolute numbers who practice open defecation increased from 188 million in 1990 to 224 million in 2008. There also remains a relatively high proportion of the rural population who practice open defecation in Latin America and the Caribbean (23%) and western Asia (14%) (8). Figure 2.8 shows countries with the highest numbers of people engaging in open air defecation.

Shared sanitation facilities are typically facilities that are public or shared between two or more households. Because

of widespread lack of maintenance among this category of facilities, they are not considered "improved" sanitation. There are an estimated 751 million people worldwide who use shared facilities, a practice that is especially common in densely populated urban areas lacking space to construct household-level sanitation. Shared sanitation facilities are most widespread in urban sub-Saharan Africa, serving nearly one-third the population. In 1990, nearly 250 million people in urban areas used shared facilities versus 145 million people in rural areas. In 2008, those numbers nearly doubled to 500 million for urban populations and rose to 254 million for rural populations.

"Unimproved" sanitation facilities (e.g., open pit latrines or bucket latrines) are generally facilities considered

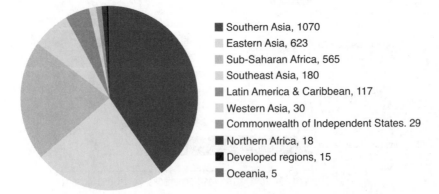

FIGURE 2.6 Regional distribution of the people (in millions) without access to improved sanitation (2008).

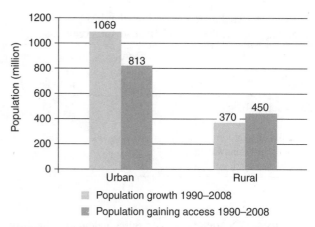

FIGURE 2.9 Increase in population growth in urban and rural sectors compared to the increase in the population achieving improved sanitation coverage worldwide between 1990 and 2008.

Nearly two-thirds (64%) of the 1.3 billion people who gained access to improved sanitation from 1990 to 2008 are urban residents. Although this statistic is ostensibly positive, urban areas are growing at such a pace that many are unable to keep up with their increasing population (Fig. 2.9) (8). Indeed, the provision of sanitation to rapidly growing cities and towns is one of the most daunting challenges facing the sanitation sector, and although urbanization potentially offers economies of scale for sanitation systems, the growth in slum and squatter settlements makes the situation particularly challenging. This growth appears to be leading to increased numbers of urban poor engaging in open air defecation.

2.2.1 Impacts Associated with Sanitation and Hygiene Interventions

Diarrhea remains the second leading cause of death in children under 5 years of age globally, and nearly one in every five child deaths, around 1.5 million a year, is due to diarrhea. It kills more children than AIDS, malaria, and measles combined, meaning more than 4,000 child deaths everyday. Additionally, nearly 1 billion people worldwide are infested with intestinal worms and as a result suffer nutritional deficiencies and poor growth. Exposure to feces—primarily human, but in some cases animal—is at the core of these diseases.

Health impacts are typically considered the most important effects associated with poor sanitation, but there are a number of other considerations, which include the following:

- Time and energy savings that derive from increased access to sanitation.
- Reduced risk from sexual violence because women have alternatives to defecating in the open.

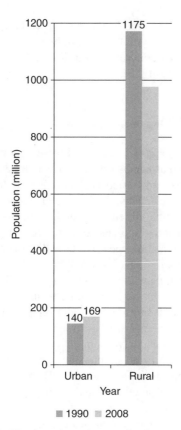

FIGURE 2.7 Number of people engaging in open defecation in urban and rural sectors worldwide (1990–2008). (*See insert for color representation of this figure.*)

ineffective at hygienically separating human excreta from human contact, and a spectrum of facilities fall into this category. One in five people in rural areas of developing regions use unimproved sanitation facilities, down just 3% since 1990. The proportion of the rural population using unimproved sanitation facilities is greater than four times that in urban areas.

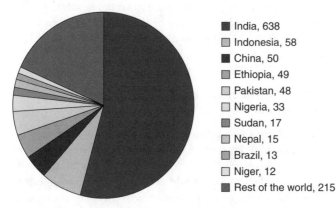

FIGURE 2.8 Distribution of people (in millions) by country engaging in open defecation.

TABLE 2.2   Median Reductions in Diarrhea Morbidity Resulting from Sanitation and Hygiene Interventions

Intervention Type	Rigorous Studies Identified	Median Reduction in Diarrhea Morbidity (%)
Sanitation (10)	5	36
Promotion of multiple hygiene messages (10, 11)	6	33–35
Promotion of handwashing with soap (12)	7	47

- Income generation, which results if sanitation facilities are operated as businesses.
- Retention of students and teachers, when schools with proper sanitation facilities are able to attract and retain more teachers and students, especially young girls who are reaching puberty (9).

The health impacts associated with sanitation and hygiene interventions have been fairly well documented, although the number of high-quality studies using experimental or quasiexperimental design are limited (Table 2.2). The heterogeneity of the effects observed in sanitation and hygiene interventions most likely stem from the complexity of the fecal–oral route of disease and the fact that interventions are embedded in specific country and social contexts, which have a strong bearing on intervention acceptance and subsequent impact.

Soil-transmitted helminths are a fundamental cause of disease associated with health and nutrition problems beyond gastrointestinal tract disturbances. Preschool age children account for 10–20% of the 2 billion people worldwide who are infected with soil-transmitted helminths: *Ascaris lumbricoides* (roundworm), *Trichuris trichiura* (whipworm), and *Ancylostoma duodenale/Necator americanus* (hookworms) (14). Of the 200–400 million preschool age children infected with roundworm, whipworm, and hookworm worldwide, a large proportion are found in the eastern Asia region. Helminths consume nutrients from children they infect, thus retarding their cognitive and physical development. They detrimentally affect tissues and organs, cause abdominal pain, diarrhea, intestinal obstruction, anemia, ulcers, and other health problems. All of these sequelae can subsequently slow cognitive development and impair learning. Deworming school children through drug treatment is a curative approach for expelling the heavy wormload, but it is often only a temporary intervention for reducing worm infection, and reinfection is frequent. Improved sanitation and hygiene are also needed control measures to prevent infection and reinfection. More recently, research suggests a subclinical disorder of the small intestine, tropical enteropathy, which is linked to exposures to large quantities of fecal bacteria, may be a significant cause of child undernutrition globally and that sanitation and hygiene may be more important to undernutrition, and thus child mortality, than previously thought (15).

2.3   HYGIENE

Strong relationships have been observed between hygiene behaviors such as handwashing at critical times (i.e., before preparing food and after defecation) and diarrheal disease (16). In a review by Huttly et al. (11) it was shown that hygiene programs targeting a single behavior were more effective than programs addressing several hygiene behaviors. In a systematic review, Curtis and Cairncross (12) also documented major reductions in diarrheal disease morbidity from handwashing with soap, with a range of 42–44% in the pooled studies. Further, household-focused hygiene interventions have been found to deliver significant health improvements even when environmental conditions and services are not conducive to improved health, such as access to piped water. As with all interventions, however, the range of health impacts is considerable. There is some debate regarding which practice—handwashing postdefecation or prior to eating—is most important. Curtis et al. (13) suggest that the critical time is after defecation rather than before eating, while other studies suggest that the reverse can be true under certain circumstances (17). Clemens and Stanton (18) observed reductions in diarrhea incidence among young children associated with mothers washing hands prior to food preparation. Soap is an essential component of handwashing behavior, and handwashing with water only has been found to provide little or no benefit (19). Others have found, however, that some reductions in contamination were found when washing with water alone, and that use of alternative rubbing agents, such as mud or ash, provided benefits similar to soap's (20). Hoque et al. (21) also found that use of mud, ash, and soap achieved the same level of reductions of hand contamination and suggest that the rubbing of hands was more important than the agent used. The authors also noted that rinsing hands with 2 L of clean water was also effective, although this practice would be unlikely if access to water were limited (21).

2.3.1   Cost-Effectiveness of Sanitation and Hygiene Interventions

Cost effectiveness ratios (US$ per disability-adjusted life year (DALY) averted) have been developed for various interventions aimed at reducing the burden of diarrhea. Considerable caution should be taken when we interpret these estimates, and significant variability can occur as a result of contextual realities, which may raise or lower the cost of the intervention or limit the effectiveness of the

TABLE 2.3 Estimates of Cost-Effectiveness Ratios for Various Interventions Targeting Diarrheal Disease (23)

Interventions Targeting Diarrheal Disease	Estimates of Cost-Effectiveness Ratios (USD/DALY averted)
Sanitation construction and promotion	≤270
Household water connection	223
Hand pump or stand post	94
Water sector regulation and advocacy	47
Household water treatment (24)	
Chlorination	46–744
Ceramic filtration	125–2637
Solar disinfection	54–861
Flocculation and disinfection	94–8754
Sanitation promotion	11
Hygiene promotion	3

Note: The range of estimates for household water treatment is for various WHO epidemiological subregions.

intervention. Given these major caveats, however, the two most cost-effective interventions for reducing the burden of diarrheal diseases are hygiene promotion and sanitation promotion (Table 2.3). Another intervention that appears to be cost effective is availability of community health clubs, which are estimated to cost 1 USD per capita (22).

2.4 NEW APPROACHES FOR IMPROVING SANITATION AND HYGIENE

In the past, the provision of sanitation was chiefly a supply-driven approach, and participation was conceived as a small contribution of the community, in cash or kind, to the implementation of a previously designed solution to their sanitation problems. These contributions did not give community members the opportunity to participate in the decision-making process, nor did they create a sense of ownership on the part of project beneficiaries (25). Subsequently, a large number of latrines went unused for the purposes they were intended, and instead were often used as storage sheds or, in some cases were locked and used only when guests visited. Undoubtedly, there are myriad reasons why people have opted not to use latrines, such as fear that latrines cause infertility; latrine faces the wrong direction for religious reasons, for example, toward Mecca; fear that latrine could give sorcerers easy access to excreta for hostile purposes; belief that defecating into the subsurface is unacceptable because ancestors are buried there; latrine limits the use of nutrients in excreta for agricultural purposes; latrine limits users from seeing their feces to find out whether they have worms; latrines disrupt time used for talking with friends; lack of view from latrine; latrine smells or has flies (26).

New approaches to sanitation are generally no- or low-subsidy programs, and many reasons against subsidies have been put forth, including the following:

- Improvements in sanitation coverage typically stop once subsidy budgets run out
- Subsidies lead to inappropriate facility designs that are often too expensive
- Subsidies are often not captured by the poor, who need sanitation most
- Subsidies can potentially destroy a developing sanitation market by creating perverse incentives
- Households often do not use and maintain latrines that are heavily subsidized.

  When subsidies are used effectively, it is usually to offer low interest loans or to remove certain administrative fees.

Participatory approaches have evolved from the mistakes observed implementing supply-driven approaches for the provision of sanitation. The underlying principle of new approaches is to involve all stakeholders in all the phases of sanitation programs or projects; this involvement's intent is greater responsiveness to the needs and preferences of the users and to provide a solution that is more appropriate to given local conditions and environment. Participatory approaches used in the sanitation sector include five participatory methodologies and methods: demand responsive approach (DRA), self-esteem, associative strength, resource-fulness, action planning, and responsibility (SARAR), participatory hygiene and sanitation transformation (PHAST), participatory learning and action (PLA), and participatory rural appraisal (PRA). Of all participatory methodologies and methods mentioned, PHAST, which originated in Africa, is the most common.

2.4.1 Community-Led Total Sanitation

Community-led total sanitation (CLTS) is a community-wide participatory approach to improve sanitation based on stimulating a collective sense of disgust and shame. The approach applies simple, visual methods to make obvious the spread of fecal–oral contamination, highlighting the fact that community members are eating each others' feces. The approach has many of the following characteristics:

- Allows community members to analyze their own sanitation situation, using transect walks, mapping of open defecation sites, and community-wide fecal load calculations (more commonly known as "shit" calculations).
- Raises the perception of risks associated with open defecation.

- Generates a collective sense of disgust and shame, which is used to mobilize communities to end open defecation.
- Hardware subsidies are usually not provided, and it generally promotes use of local sanitation options based on affordability and durability.

The approach, developed in 1999 in Bangladesh, has been transferred and adapted to nearly two dozen countries in sub-Saharan Africa, beginning around 2005 (27). Community-wide approaches such as CLTS and its variants are often referred to as *total sanitation* approaches because they aim to achieve universal use of toilets and end open defecation in the communities targeted. The total sanitation approach is rapidly gaining acceptance in rural parts of sub-Saharan Africa, and a number of organizations and governments are using the approach to jumpstart sanitation activities (28). While CLTS shows promise, it is unclear if the sanitation improvements made by households during the CLTS process will be sustained. It is also in doubt that the approach can be applied to urban and periurban communities where low-cost sanitation (i.e., pit latrines) are not options and there is potentially less social cohesion.

2.4.2 Sanitation Marketing

Sanitation marketing aims to develop an enabling environment that will facilitate the purchase of sanitation products by the poor (Fig. 2.10) (29). There is increasing evidence that development of the market is the only sustainable approach to meeting the need for sanitation in the developing world. In fact, progress in access to sanitation has mostly been achieved by the market. Indeed, most new sanitation solutions in Africa and other regions have been and continue to be privately acquired by individual households from the local marketplace, including retailers, wholesalers, importers, masons, pit diggers, house builders, pit emptiers, and others. The most effective and efficient method to meeting the need for sanitation in the developing world is to support this market. As a result, the rate of sanitation adoption has been slowly increasing, oftentimes with limited government or international donor intervention. In order to accelerate the rate of sanitation coverage, sanitation marketing aims to know who has or has not been building toilets, why, what barriers exist, and what motivates purchases (30). It is now widely understood that interventions must respond to what people want, rather than what "officials" think they should have. Market research can reveal what people are used to and comfortable with, as well as what they're willing to pay. Sanitation marketing research often results in new solutions—ones that public health officials and engineers had not considered (31). Based on marketing research, it is now recognized that sanitation and hygiene promotion are unlikely to be successful unless messages are based upon the hopes and desires of the target population. Motivating factors for people to improve hygiene- and sanitation-related conditions are rarely health driven. When asked to prioritize reasons for improving sanitation, many households indicated various reasons ahead of health: (i) ridding house of smell and flies, (ii) avoidance of discomfort of the bush, (iii) gaining prestige from visitors, (iv) privacy, and (v) avoiding dangers at night (31).

2.5 SANITATION AND HYGIENE OPTIONS

The cost of sanitation typically goes up as one moves up the sanitation ladder. One of the most inexpensive forms of sanitation is the pit latrine, which can be simple and relatively low cost. The drawbacks to pit latrines is that they cannot generally be used in crowded urban areas, on rocky ground, or where the groundwater level is high or in areas that periodically flood. They require access to open ground and typically the digging of new pits every few years. There is very little information on the levels of groundwater contamination from pit latrines; however, studies have shown microbiological impacts on groundwater up to 25 m deep (32).

Ecological sanitation (ecosan) applies the fundamentals of ecology to sanitation and is gaining popularity in areas where there are limited water resources, sewage

FIGURE 2.10   The sanitation marketing framework.

TABLE 2.4   Cost Estimates of Various Sanitation Alternatives per Household (34)

Sanitation Option	Construction Cost (US$)
Simple pit latrine[a]	26–60
VIP latrine[a]	50–57
Pour-flush latrine[a]	50–91
Ecosan toilet (35, 36)	
Rural India	96
Periurban South Africa	261–348
Urban regions	350–1200
Small bore sewerage[a]	52–112
Conventional sewerage[a]	120–160

[a] Range of median costs for Africa, Asia, Latin America, and the Caribbean.

infrastructure, or both. Ecosan often does not use water and promotes ecological balance by recycling human excreta as fertilizer (33). Human excreta have been recycled for centuries in Asia and this process continues to play a major role in agricultural production. The goals of ecosan are to provide a system that is affordable, acceptable to the community, simple to use, disease preventing, and protective of the environment. Ecosan has made slow progress, however, despite decades of promotion by donors, and the lack of demand for ecosan makes many development agencies question its viability. Further, households must be convinced that the advantages of ecosan amply outweigh any disadvantages or added costs (Table 2.4). A less expensive form of ecosan is the arborloo, which is a simple pit latrine. After its usefulness as a latrine is exhausted, a tree is planted in its fertile soil.

Simplified sewerage or small bore sewers, developed in the 1980s in Brazil, hold great promise for densely populated urban and periurban areas. They are much less expensive than conventional sewerage and are much more efficient in terms

of space and materials required (37). In many cases, people are too poor, however, to cover the costs associated with household-level sanitation, or there may be communities of homeless or street children where individual-focused sanitation is not appropriate. These communities are best served by community-level sanitation facilities (i.e., sanitation blocks), which may be owned and operated by a community cooperative, a public utility, or a private owner(s) (38).

2.6   FINANCING SANITATION

In contrast to water supply, which generally receives consistent financial resources, there is often less interest in investing in hygiene and sanitation. JMP estimates that in the 1990s sanitation received 3.1 billion USD annually from governments and donor agencies (2).

There is a lot of variability among global estimates of the financial needs to meet the MDG water supply and sanitation targets, which range from 9 billion USD to 70 billion USD annually. Hutton and Bartram (2) estimated total spending required to meet the sanitation target of the MDGs to be 142 billion USD. This translates to per capita spending of 28 USD for sanitation, translating to roughly 14 billion USD total annually. Cost studies also indicate that new financing should be targeted at rural areas of Asia and Africa.

While international aid has been increasing since the mid 1990s, the percentage going to water supply and sanitation has gone down. In 2006, total aid flows to the sector reached 6.3 billion USD (Fig. 2.11). Most of those resources, however, were not allocated to the countries most lacking in access (39). This may change, however, as increasing political pressure moves the international community to refocus efforts on these basic services.

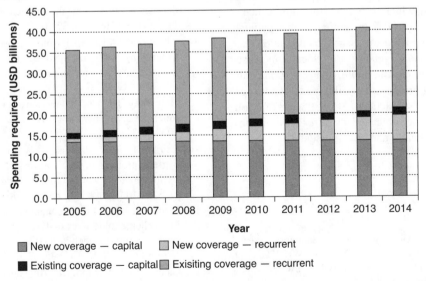

FIGURE 2.11   Spending required on new and existing sanitation facilities from 2005 until 2015 to meet Millennium Development Goal Target 10, excluding program costs.

2.7   FINAL REMARKS

Sanitation and hygiene are critical components of development, and there is mounting evidence that they are worthy investments. The most important argument for sanitation and hygiene among many leaders in economically developing countries—or at least finance ministers—is undoubtedly the fact that sanitation and hygiene investments generate excellent returns. Global and regional cost-benefit analyses have demonstrated that a 1 USD investment in sanitation, for example, can yield economic benefits of 9–34 USD (40). The economic benefits of sanitation stem mostly from the value generated as a result of time saved by not being ill. These figures received ample attention during 2008, the International Year of Sanitation, which aimed to put the global community back on track to achieve the sanitation target of the MDGs.

The best use of public resources appears to be in hygiene promotion, specifically handwashing, and sanitation promotion, which is focused on building demand for sanitation through total sanitation approaches, as well as establishing clear national policies and building local government and private sector capacity. Continued creativity and innovation is vital in the field of health promotion, as highlighted by the "No toilet, no bride" campaign, which the Haryana (India) government embarked on to create awareness about sanitation. In this campaign, the government used various communication messages that conveyed the idea that young women should not marry into a household that has no toilet. Based on media reports, it is estimated that 1.41 million toilets have been built across the mostly rural state since 2005.

There is some hope that the sanitation situation will start to improve. A greater level of political commitment was observed in 2008, when 32 African ministers signed the eThekwini declaration, which recognizes the significance of spending at least 0.5% of GDP on sanitation and hygiene. There are also efforts to increase pressure on international aid organizations and donor agencies to allocate increased spending to the sector. Furthermore, many regional and country-level working groups are developing to better organize efforts and share best practices. After 30 years of effort, more focus and persistence is needed if we are to lay this most critical foundation of health and development.

REFERENCES

1. JMP, 2008 Progress on Drinking Water and Sanitation: Special Focus on Sanitation. UNICEF, New York and WHO, Geneva, 2008. Report available at: http://www.who.int/water_sanitation_health/monitoring/jmp2008.pdf).

2. Hutton G, Bartram J. Global costs of attaining the millennium development goal for water supply and sanitation. *Bull World Health Organ* 2008;86(1):13–19.

3. WHO/UNICEF. Meeting the MDG drinking water and sanitation target: a mid-term assessment, 2004.

4. Mara D, Feacham R. Taps and toilets for all—two decades already, and now a quarter century more. *Water* 2001;21: 13–14.

5. Jonsson, Åsa and David Satterthwaite (2000) 'Overstating the provision of safe water and sanitation to urban populations: A critical review of the quality and reliability of official statistics and of the criteria used in defining what is 'adequate' or 'safe', Background paper prepared for the National Academy of Sciences, International Institute for Environment and Development, London.

6. Ferriman A. BMJ readers choose the "sanitary revolution" as greatest medical advance since 1840. *BMJ* 2007;334:111.

7. Cotton A, Bartram J. Sanitation: on- or off-track? Issues of monitoring sanitation and the role of the Joint Monitoring Programme. *Waterlines* 2008;27(1):12–29.

8. WHO/UNICEF Joint Monitoring Programme. Progress on Sanitation and Drinking-Water Progress 2010 Update. Available at http://whqlibdoc.who.int/publications/2010/9789241563956_eng_full_text.pdf.

9. Pearson J, Mcphedran K. A literature review of the non-health impacts of sanitation. *Waterlines* 2008;27(1):48–61.

10. Esrey SA, Potash JB, Roberts L, Shiff C. Effects of improved water supply and sanitation on ascariasis, diarrhea, dracunculiasis, hookworm infection, schistosomiasis, and trachoma. *Bull World Health Organ* 1991;69(5):609–621.

11. Huttly SRA, Morriss SS, Pisani V. Prevention of diarrhea in young children in developing countries. *Bull World Health Organ* 1997;75(2):165–174.

12. Curtis V, Cairncross S. Effect of washing hands with soap on diarrhea risk in the community: a systematic review. *Lancet Infect Dis* 2003;3(5):275–281.

13. Curtis V, Cairncross S, Yonli R (2000) Domestic hygiene and diarrhea – pinpointing the problem. Tropical Medicine and International Health 5, 22–32.

14. Albonico M, Allen H, Chitsulo L, Engels D, Gabrielli A-F, et al. Controlling soil-transmitted helminthiasis in pre-school-age children through preventive chemotherapy. *PLoS Negl Trop Dis* 2008;2(3):e126.

15. Humphrey JH. Child undernutrition, tropical enteropathy, toilets, and handwashing. *Lancet* 2009;374(9694): 1032–1035.

16. Cairncross S, Valdmanis V. Water supply, sanitation and hygiene promotion. In: Jamison DT, et al. editors. *Disease Control Priorities in Developing Countries*, 2nd edition. New York: Oxford University Press, 2006.

17. Birmingham ME, Lee LA, Ntakibirora M, Bizimana F, Deming MS. A household survey of dysentery in Burundi: implications for the current pandemic in sub-Saharan Africa. *Bull World Health Organ* 1997;75(1):45–53.

18. Clemens JD, Stanton BF. An educational intervention for altering water-sanitation behaviors to reduce childhood

diarrhea in urban Bangladesh. I. Application of the case-control method for development of an intervention. *Am J Epidemiol* 1987;125(2):284–291.

19. Oo KN, Aung WW, Thida M, Toe MM, Lwin HH, Khin EE. Relationship of breast-feeding and hand-washing with dehydration in infants with diarrhoea due to *Escherichia coli. J Health Popul Nutr* 2000;18(2):93–96.

20. Hoque BA, Briend A. A comparison of local handwashing agents in Bangladesh. *J Trop Med Hyg* 1991;94(1): 61–64.

21. Hoque BA, Mahalanabis D, Alam MJ, Islam MS. Post-defecation handwashing in Bangladesh: practice and efficiency perspectives. *Public Health* 1995;109(1):15–24.

22. Waterkeyn J, Cairncross S. Creating demand for sanitation and hygiene through community health clubs: a cost-effective intervention in two districts of Zimbabwe. *Soc Sci Med* 2005;61:1958–1970.

23. Laxminarayan R, Chow J, Shahid-Salles SA. Chapter 2. In: Jamison DT, et al., editors. *Disease Control Priorities in Developing Countries*, 2nd edition. New York: Oxford University Press, 2006, p. 41.

24. Clasen TF, Haller L. *Water Quality Interventions to Prevent Diarrhoea: Cost and Cost-Effectiveness*. Geneva: Public Health and the Environment World Health Organization, 2008.

25. White A. Community participation in water and sanitation; concepts, strategies and methods. Technical Paper Series No. 17. The Hague, The Netherlands: IRC International Water Sanitation Centre, 1981.

26. Pickford J. *Low-Cost Sanitation: A Survey of Practical Experience*. Intermediate Technology Publications, 1995.

27. 'AfricaSan—one year on.' *Workshop on CLTS in Africa*. Institute of Development Studies and Plan, 2009.

28. Peal A, Evans B, Van Der Voorden C. Available at: http://www.eawag.ch/forschung/siam/lehre/alltagsverhaltenII/pdf/Introduction_to_Hygiene_and_Sanitation_Software.pdf.

29. Fuertes P, Baskovich MR, Zevallos M, Brikke F. The private sector and sanitation for the poor: a promising approach for inclusive markets in Peru. *Waterlines* 2008;27(4):307–322.

30. Jenkins MW. Achieving the good life: why some people want latrines in rural Benin. *Soc Sci Med* 2005;61(6): 2446–2459.

31. Cairncross S. The case for marketing sanitation. *Water & Sanitation Program*. World Bank, 2004. Available at http://www.wsp.org/UserFiles/file/af_marketing.pdf.

32. Dzwairo B, Hoko Z, Love D, Guzha E. Assessment of the impacts of pit latrines on groundwater quality in rural areas: a case study from Marondera district, Zimbabwe. *Phys Chem Earth* 2006;31:779–788.

33. Esrey S, Gough J, Rapaport D, Sawyer R, Simpson-Hébert M, Vargas J, Winblad U. *Ecological Sanitation*, 1st edition. Stockholm: Department for Natural Resources and the Environment, Swedish International Development Cooperation Agency (Sida), 1998.

34. WHO/UNICEF. Global Water Supply and Sanitation Assessment 2000 Report, 2000.

35. Mara DD. Sanitation now: what is good practice and what is poor practice? Proceedings of the IWA International Conference on Sanitation Challenge: New Sanitation and Models of Governance, Wageningen, The Netherlands, May 19–21, 2008.

36. McCann B. The sanitation of ecosan. *Water* 2005;21:29.

37. Mara DD. Water, sanitation and hygiene for health of developing nations. *Public Health* 2003;117:452–456.

38. Mara DD, Alabaster G. A new paradigm for low-cost urban water supplies and sanitation in developing countries. *Water Policy* 2008;10:119–129.

39. Sanitation and water: why we need a global framework for, action. *WaterAid*, 2008. Report available at: http://www.wateraid.org/documents/plugin_documents/sanitation_and_water__why_we_need_a_global_framework_for_action.pdf.

40. Hutton G, Haller L, Bartram J. Global cost-benefit analysis of water supply and sanitation interventions. *J Water Health* 2007;5:481–502.

3

WATER AND HEALTH: THE DEMOGRAPHIC BACKGROUND

Robert Wyman

3.1 INTRODUCTION

Water scarcity, in many parts of the globe, is a major determinant of poor agricultural productivity, which in turn leads to malnutrition and poverty, key risk factors for disease. A rapidly increasing population exacerbates water scarcity and makes remediation of other water problems more difficult (see chapter 1). UN projections show rather steady population growth in the century from 1950 to 2050 and no significant decrease in the next century. The ratio of people living in less developed countries is ~4.5 and essentially all growth is in the less developed countries while developed countries face a decline. For the foreseeable future, the world is going to be much more crowded than it is now.

3.2 WATER SCARCITY, HEALTH, AND DEMOGRAPHY

Many chapters in this book focus on the effect of contaminated water on the epidemiology of a variety of particular diseases. This chapter focuses on what is an even greater threat to human health and well being: the scarcity of freshwater—clean or contaminated.

The World Health Organization (1) lists some basic statements about water scarcity. Among these are

(i) One in three people around the world lacks water sufficient to meet daily needs.

(ii) Almost one-fifth of the world's population (about 1.2 billion people) live in areas where water is physically scarce. One-quarter of the global population also live in developing countries that face water shortages due to a lack of infrastructure designed to fetch water from rivers and aquifers.

(iii) Globally, the problem is getting worse as cities and populations grow, and with it, the need for more water for agriculture, industry, and households.

This last aspect, population growth, is the focus of this chapter.

The primacy of concern over population is illustrated by the choice for the very first sentence of the Executive Summary of the World Bank's Water and Development: "The amount of available water has been constant for millennia, but, over time, the planet has added 6 billion people" (26). A similar concern over rapid population growth is expressed in Chapter 1 of this book.

The world's population depends on water for food, drink, and sanitation. Life is impossible without the first two and the lack of clean water is the direct cause of myriad diseases, many described in this volume.

Water, thus, has a universal effect on human health. Among the many ways water scarcity affects human health, probably the dominant one is the lack of sufficient water for high productivity agriculture in many areas of the globe. The result of poor agricultural yields is malnutrition and poverty; both are major causes of poor health.

The Global Burden of Disease (GBD) study quantified the health effects of more than 100 diseases and injuries for the world as a whole in 1990 and then again in 2001 (2, 3). Undernutrition (which has many root causes besides water

Water and Sanitation-Related Diseases and the Environment: Challenges, Interventions, and Preventive Measures, First Edition.
Edited by Janine M. H. Selendy.
© 2011 Wiley-Blackwell. Published 2011 by John Wiley & Sons, Inc.

scarcity) was the leading global cause of health loss in both 1990 and 2001 (4). The statistics are striking: The Food and Agriculture Organization (FAO) estimates that over 1 billion people were undernourished worldwide in 2009, more than at any time since 1970 (5). Malnutrition, in each of its major forms dwarfs most other diseases globally (6). Twenty-five thousand people die every day—including one child every 5 s—from hunger-related causes (UN World Food Program, 7). Maternal and child undernutrition is the underlying cause of 3.5 million deaths each year. It accounts for 35% of the disease burden in children younger than 5 years and 11% of the total global disease burden (8, 9).

The links in the causal chain of factors—from population pressure to water scarcity to low agricultural productivity to poverty and malnutrition to disease—have never been properly quantified. So many other factors interact with this causal chain that untangling it is almost impossible.

Agriculture uses ~70% of the world's available freshwater. As the world's population rises, the demand for food will increase, as will the demand for water. Many areas of the world where water is in short supply are also ones with the most rapid rate of population growth. They are also often the poorest regions with most of the labor force engaged in agriculture. They have little money to import food or to import the technology that would allow them to grow more food. Continued population growth in these regions will lead to increasing poverty and its attendant malnutrition, morbidity, and mortality. Water scarcity is not limited to the poor or the rural; places as rich as California are now having battles over the allocation of water between agriculture and increasing urban populations.

In short, as population grows, demand for water for agricultural, industrial, personal, recreational, and aesthetic uses will grow. Population pressure not only exacerbates water scarcity but also makes remediation of other water problems, such as the supply of clean water, more difficult. It behooves anyone concerned with any aspect of water supply to be aware of the increasing number of users demanding a share of this fixed-quantity asset. This short piece on the demographic background is intended to supply the basic information needed to help problem solvers start thinking about the demographic aspects of water scarcity.

3.3 POPULATION IS GROWING STEADILY AND WILL CONTINUE TO DO SO

At current rates of population growth, a billion people will be added to the globe's population in the next 13 or so years. The footprint of a billion more people would seem to overwhelm the sum effect that all conservation and restoration efforts can achieve in a similar span of time.

Contrary to the impression of an impending decline in world population provided by the popular media (20,25), the world's population has been growing rather steadily for the

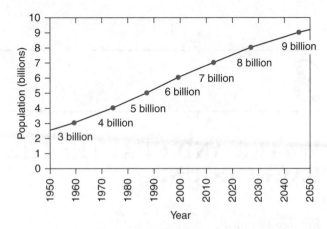

FIGURE 3.1   World population: 1950–2050. *Source*: U.S. Census Bureau, International Data Base, December 2009 update.

past 50 years and is projected to continue on that trajectory until at least 2050 (Fig. 3.1). Looked at more closely (Fig. 3.2), the number of people added each year grew from 1950 to 1990 and has now eased off that peak somewhat. (The dip just before 1960 is the result of the famine that occurred during China's Great Leap Forward.) However, global population is still increasing by the same amounts as it did in the 1960s and 1970s, during the peak of the "population explosion" scare. Nevertheless, public, scholarly, and activist concern with human population numbers has almost disappeared.

In a sense, constant growth represents progress. If one calculates population growth as percentage change from year to year, the large increase in the population base makes each numerically constant increment a smaller fraction of the whole. (Fig. 3.3). The rate as a percent is down to almost half of what it was in the late 1960s.

The story is the same for the population of the United States. Population growth in the United States has been even more linear and is projected to have the same slope as far into the future as projections go (Fig. 3.4).

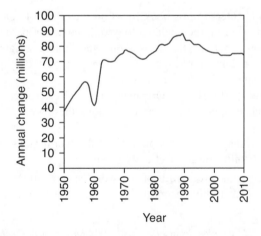

FIGURE 3.2   Annual world population change: 1950–2010. *Source*: U.S. Census Bureau, International Data Base, December 2009 update, http://www.census.gov/ipc/www/idb/worldpopchggraph.php.

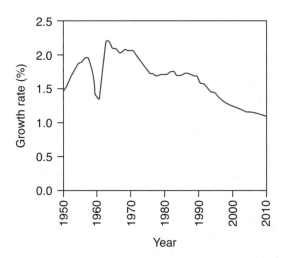

FIGURE 3.3 Percent growth rate of world population: 1950–2010. *Source*: U.S. Census Bureau, International Data Base, December 2009 update, http://www.census.gov/ipc/www/idb/ worldgrgraph.php.

There are now 2.5 births in the world for every death (20/1000 births versus 8/1000 deaths (10). The earth is a long way from population stabilization.

3.4 DIFFERENT MEASURES OF GROWTH

In considering the impact of population growth, planners have to decide whether absolute growth (numbers of people) or percentage growth (percent increase over current population) is the more relevant measure.

For most manufactured items, the supply is expandable. In most modern situations, the total economy is itself expandable. In those cases, percent growth is probably more appropriate. The reason is that, as an economy grows a percentage of production is plowed back into creating further productive capital and technological research. The larger economy allows more to be invested in growth. This positive feedback is what allows economies to grow in a somewhat exponential fashion. Constant growth percent translates into an ever-increasing absolute amount of growth. It is the same mechanism as exponential population growth. Economies of scale also allow rapid growth of production. Once fixed costs have been distributed over a large number of produced objects, only marginal costs are incurred in further production.

However, freshwater is an essentially fixed supply resource. When demand increases, there is not much ability to increase the supply. A fixed pie must be divided by an increasing number of consumers. Absolute growth in the number of end users is probably the relevant variable.

Technology acts primarily by increasing the efficiency of water use. However, we must ask whether increased conservation and technological improvement can keep up with a continually increasing population? Unfortunately, the answer is no: The trajectory of conservation and efficiency is the inverse of positively fed-back growth. At the start of an efficiency or conservation program, it is the low-hanging fruit that is picked first. Progress in the early stages of such efforts is likely to be rapid. But as time proceeds, further steps become more and more difficult, and progress slows down. In the declining exponential of the amount of water needed for a certain amount of industrial or food production, one sees that a constant percentage of improvement brings an ever-decreasing absolute amount of water saved. In short, while industrial production is often capable of accelerated growth, conservation and efficiency improvements are more likely to proceed at a decelerating pace (Fig. 3.5).

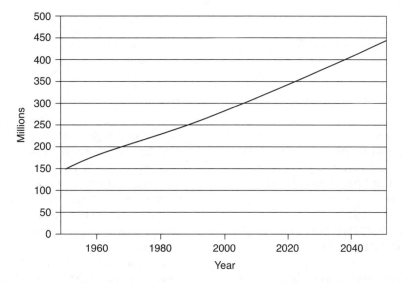

FIGURE 3.4 U.S. population: 1950–2050. *Source*: U.S. Census Bureau, http://www.census.gov/ cgi-bin/ipc/idbsprd.

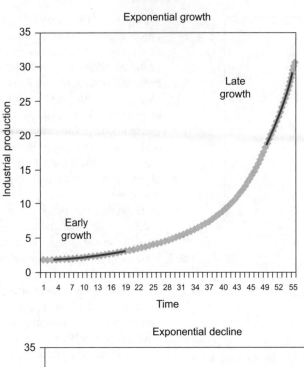

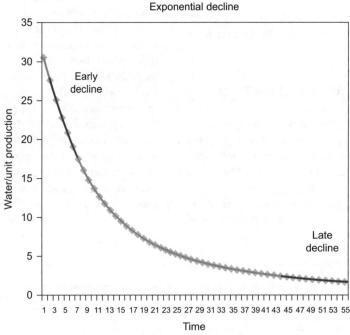

FIGURE 3.5   Exponential growth and decline.

An obvious counterexample is the possibility of desalination. If desalination could be made attractive from an economic, energy usage, and transportation cost point of view, water could be produced from a limitless resource.

3.5 AGE STRUCTURE OF THE POPULATION AND POPULATION MOMENTUM

The age structure of a population is most easily visualized in a type of graph called a *population pyramid*. Figure 3.6

characterizes Nigeria at two points in time. The number of individuals in each age cohort is depicted as a horizontal block, males on the left and females on the right. Successively older cohorts are stacked on top of the younger ones. It can be seen that each of the younger age cohorts is made up of more individuals than any of the older ones. This shape characterizes countries with an expanding and young population.

Notice that the number of women leaving their reproductive years at about age 45 is a much smaller number than those entering their reproductive years at age 15–19. The ratio can

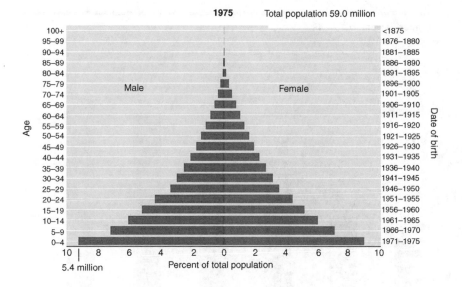

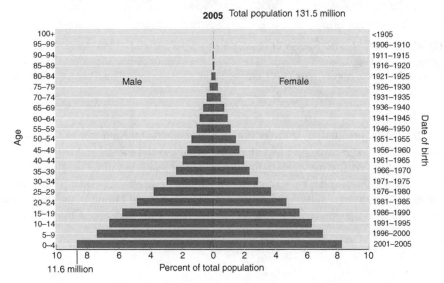

These two population profiles compare the size of different age groups in Nigeria's population in 1975 and in 2005. The bars along the left side represent males, the right side, females; each bar shows the relative size of a five-year age cohort in ascending age. In both of the years shown, Nigeria's population maintains the classic pyramid shape of a very youthful population, with progressively larger proportions among each successively younger age group.

FIGURE 3.6 Nigeria's total population and age structures, 1975 and 2005. *Source*: PAI Research Commentary 1:Nov 10, 2006.

be as high as 3:1, as can be seen in the pyramids for Nigeria in 1975 or 2005. The ratio was 3:1 just before China introduced its One-Child Policy (data from 1982 Census of China; 11). For the world as a whole, the number of women of child-bearing age has been growing rapidly and is expected to continue growing, but at a slower pace Fig. 3.7).

If the average number of children born to each woman stays constant, the number of children born each year will rise. This phenomenon is called "population momentum."

Average fertility must fall in order to keep the number of children born constant. In some cases, like China in 1980 or Nigeria now, the fertility drop must be dramatic.

What has been happening in the world is that, even though average fertility has been falling, it has not fallen fast enough to keep the number of women in childbearing age from rising. Figure 3.8 superimposes the fall in fertility (black line) on the same data for number of childbearing women shown in Fig. 3.7. Similarly, fertility has not fallen

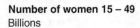

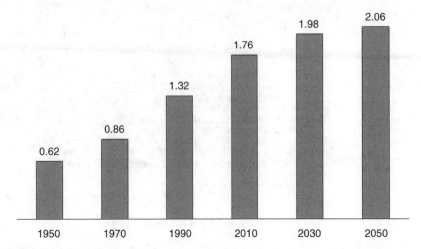

FIGURE 3.7 Number of women of childbearing age, worldwide, 1950–2050. *Source*: United Nations, *World Population Prospects: The 2004 Revision* (medium scenario), 2005. © 2006 Population Reference Bureau, http://www.prb.org/Publications/GraphicBank/PopulationTrends.aspx.

fast enough to keep the world population from rising (Fig. 3.1). The public is often confused by the fact that population has been rising while fertility has been falling.

In the recent past, the number of births has risen dramatically in a period of equally dramatic falls in fertility (Fig. 3.9). Some projections for the future (see below) envision that the number of births will level off for countries of all developmental status (Fig. 3.10).

The resolving of momentum is a very slow, multigenerational process. Consider China as an example. China's fertility started falling in the 1960s, fell dramatically in the 1970s, and the One-Child Policy was introduced in

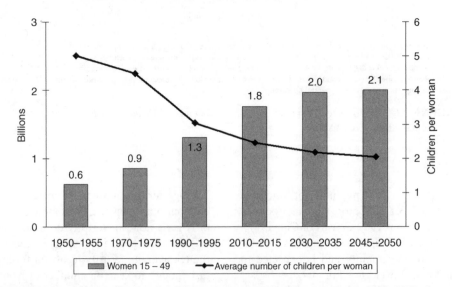

FIGURE 3.8 Number of women of childbearing age and fertility, worldwide 1950–2050. *Source*: United Nations, *World Population Prospects: The 2004 Revision* (medium scenario), 2005. © 2006 Population Reference Bureau, http://www.prb.org/Publications/GraphicBank/PopulationTrends.aspx.

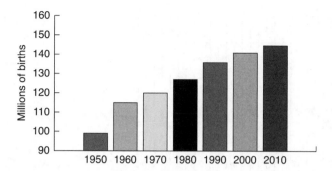

FIGURE 3.9 Numbers of births, worldwide. *Source*: United Nations, *World Population Prospects*, 1995.

1979–1980. Figure 3.11 follows China's population for 43 years after fertility started to fall. The population grows all through this period, and it is only at the very end that one can detect some slowing of the growth rate. Even though China's fertility has been below replacement since ~1990 (12), China's population is still growing by over 8 million people a year (13).

In roughly the same time period Algeria had a less precipitous fall in its fertility rate. The years 1965–1985 were politically very difficult for Algeria. Nevertheless, its fertility rate fell by 25%. However, because of momentum, the population growth rate went up by 19%; the absolute population grew by 83%, and the annual population growth increased by 217% (Fig. 3.12).

Twenty-five years further on Algeria's crude birth rate has dropped further, halving to 16.9 live births/1000 population, according to the CIA World Factbook, but its population has grown another 57%. Algeria's fertility rate is now near replacement level (UN: 2.38 births per woman for 2005–2010; 2009 U.S. Census Bureau and CIA World Factbook: 1.79 births/woman), but its popu-

lation growth rate is still high (UN: 1.5%; CIA World Factbook 1.2%). Its population is still growing by is over 1/2 million a year (13).

Almost every developing country in the world wants to slow its population growth rate (14), but momentum makes demographic stability very difficult to achieve. The World Bank projected in 1990 that Nigeria, Bangladesh, Iran, and Brazil would take 110–145 years to progress from replacement fertility to population stability (15).

Momentum currently accounts for the lion's share of world population growth and will continue to be the dominant factor for the foreseeable future. Figure 3.13 compares a scenario where fertility declines immediately to replacement to the scenario of the UN medium variant projection.

The long time scale of momentum is often used as a reason to not focus policy on fertility reduction. But changing CO_2 levels in the atmosphere is an even slower process and yet it has received a vast amount of policy attention.

The age structure of the population in developed countries can be almost the inverse of that of a developing country. This is clearly seen by comparison of the population pyramids for Uganda and Italy (Fig. 3.14). When a country drops below replacement fertility, there are generally fewer individuals in each younger cohort.

One may note that population momentum works in both directions with respect to age cohorts. In the population pyramid for Italy (Fig. 3.14), the below-reproductive-age cohorts are smaller than the reproductive cohorts. If fertility stays constant, the number of childbearing women falls and as a result, the number of children born each year also falls.

Countries with high birthrates, like Nigeria and Uganda, have "bottom-heavy" population pyramids; countries with below replacement fertility, like Italy, have "top-heavy"

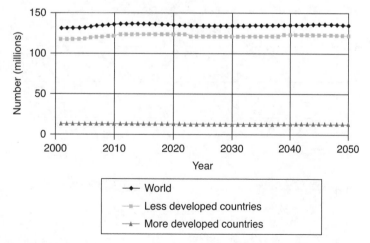

FIGURE 3.10 Projected live births by development status, 2000–2050. *Source*: U.S. Census Bureau, International Data Base, December 2008 update, http://www.census.gov/ipc/www/img/worldbirths.gif.

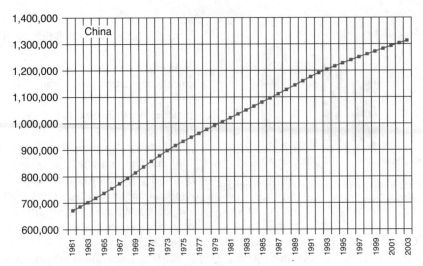

FIGURE 3.11   China: population 1961–2003. *Source*: http://www.slideshare.net/isc/china-population-polices.

Algeria: Data from: Algerian Maternal and Child Health Survey.
Reported in: Studies in Family Planning: 25: p. 191, 1985.

	1965	1985	% change
Crude birth rate/1000	50.1	38.0	−24.2
Population growth rate (%/year)	2.42%	2.87%	+18.6%
Population (millions)	11.9	21.8	+83%
Annual pop. growth	288,000	626,000	+217%

FIGURE 3.12   Algeria: population statistics, 1965 and 1985.

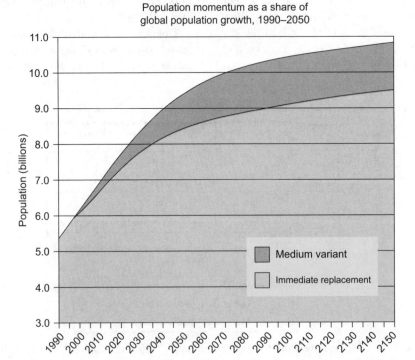

FIGURE 3.13   Momentum as a share of population growth. *Source*: United Nations, *World Population Projections to 2150*, 1996. New York, United Nations.

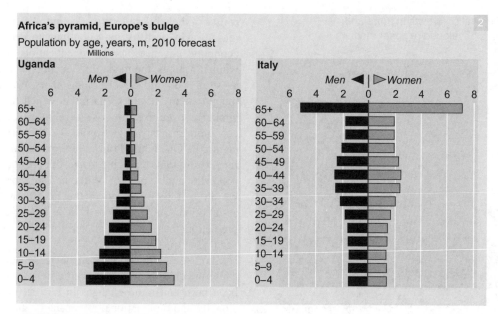

FIGURE 3.14 Contrasting population pyramids, Uganda and Italy. *Source*: United Nations, Population Division, *World Population Prospects: The 2008 Revision*, 2009. New York: United Nations.

pyramids. In a country with a stable population and a long life expectancy, the pyramid will look fairly "rectangular," except at its top, where the depredations of age take effect.

Some countries have made little progress toward rectangularizing their population pyramid. Figure 3.6 shows that for Nigeria, in the 30 years between 1975 and 2005, population more than doubled from 59.0 to 131.5 million. But the shape of its pyramid did not change. It is no closer to population stabilization now than it was in 1975.

Nevertheless, as a group, less developed countries have moved part way toward the rectangular-shaped population

pyramid of more developed countries, as Fig. 3.15 shows. This figure also indicates that the number of people living in less developed countries is vastly more than those living in developed countries. One compilation puts the ratio at 4.5:1, from the statistics of 1.232 billion in more developed countries and 5.578 billion in less developed countries (15). Of course, the factors that make up the boundary that separates "more" from "less" developed is arbitrary, and the designation for individual countries must be fluid as the economic situation of countries improves or deteriorates. Because of current high birth rates, essentially all of world population growth is taking place and will continue to take place in the

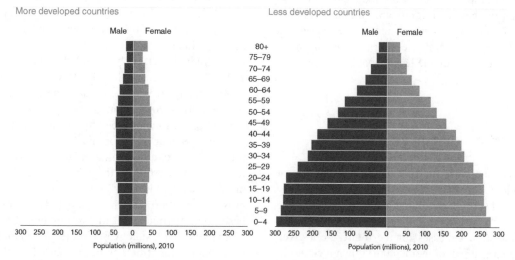

FIGURE 3.15 Age structure, more and less developed countries. *Source*: United Nations, Population Division, *World Population Prospects: The 2008 Revision*, 2009. New York: United Nations.

Population growth in more and
less developed regions: medium
projections

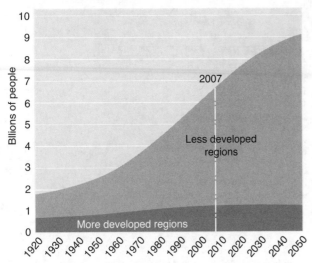

FIGURE 3.16   Share of total population, more and less developed regions. *Source*: United Nations Population Division, *World Population Prospects: The 2008 Revision*, 2007. New York: United Nations.

poor countries (Fig. 3.16). As momentum continues in opposite directions in developed and less developed countries, the demographic ratio in question becomes more extreme; an increasing fraction of the global population will live in countries with high fertility cultures. Migration from poorer to richer countries tends to equalize population change rates. Even after including migrants in host countries' populations, essentially all population growth is taking place in the poorer countries. The effects of migration on future fertility is complex for pre-existing residents and for migrants in both sending and receiving countries.

The decreasing ratio of population in developed to less developed countries has great significance for the amount of capital available for investment in water conservation and water use efficiency. Generally, countries with the most rapid population growth are the least capable of providing this capital.

3.6   PROJECTIONS OF THE FUTURE

Public health planners need to take account of the size of future populations. Population has been rising dramatically and fertility has been falling dramatically. How will these countervailing trends work out in the future? It is not known. Continued population increases for the future are a given because of the large effect of momentum. When it comes to fertility, however, scholars still argue vehemently about the

causes of the modern fertility decline. We can think of many reasons why fertility may continue to decline: continued economic advance, spread of education, improvement in women's status, the media influencing countries toward the Western two-child model, among others. We can equally well imagine many future scenarios in which fertility stops declining or even rises: the rise in religiosity, ethnic conflicts, economic disarray, rejection of the Western model, China ending its One-Child Policy, environmental problems, high fertility migrants replacing low fertility residents of receiving countries, and so on (16). Without, a solid predictive theory, we are left pretty much to the land between reasonable assumptions and crystal balls. The record of demographic predictions has had its high and low points with respect to accuracy (16, 17).

Short-term projections can be reasonably accurate because the child bearers considered in such projections have been born and rapid changes in fertility occur only occasionally and on a national rather than a global scale. If one considers childbearing to begin at age 15, then a 15-year projection at least has the correct maximum number of potential child bearers to work with; mortality can only reduce this number.

Longer term projections become increasingly uncertain (18). Very significant (and unexpected) changes occur on scales ranging from one to several decades. China's total fertility rate (TFR) dropped in half (from 5.81 to 2.75 children/women) in one decade (1970–1979) (12). This drop was not due to the One-Child Policy, as it occurred in the decade before that policy was enunciated. Other fertility drops in East Asia have happened on a similar time scale. Most of Europe's fertility decline happened in a 60-year period (1870–1930) (19).

Small changes in assumptions lead to large changes in population projections 40 or 50 years out. In 1996, the UN reduced its population projection for the year 2050 by 650 million people. This change created a media frenzy. The *New York Times Magazine* ran the headline "The Population Explosion Is Over" with subhead "More lonely people, without siblings, uncles, aunts, cousins, children, or grandchildren" (20). The wheel then turned; between the 1998 UN Revision and the 2000 Revision, the projection for 2050 rose by over 400 million. The people lost in 1996 reappeared in 2000. During this time the actual rise in population was unperturbed by the brouhaha.

Much popular attention has been focused on the UN's projections. Acknowledging the uncertainty about its ability to know the future, the UN does not label its future numbers as "predictions" but rather as "projections" and provides eight projection variants (14). Each projection is based on a particular set of assumptions about the future. The UN does not state which projection is most likely, but most public attention has been focused on the projection labeled "medium."

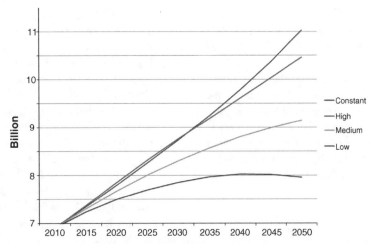

FIGURE 3.17   United Nations population projections. *Source*: United Nations, Population Division, *World Population Prospects. The 2008 Revision*, 2009. New York: United Nations. http://esa.un.org/unpp/p2k0data.asp.

The "highest" of the UN projections is perhaps the most conservative. It assumes that fertility will remain as it is now. That presumption is similar to that made by efficient market theory. All currently knowable or predictable information is already folded into people's decisions to buy or sell a stock — or to have, or not have, another child. Future increases or decreases in stock price or fertility can only be based on unpredictable future events. This scenario produces the highest line in Fig. 3.17. It projects a near doubling of population from 6 billion in 2000 to 11 billion in 2050. At constant fertility, population thereafter would keep rising with ever larger increments.

The other reasonable option is to presume the continuation of a trend. If one chooses the addition of a constant number of individuals each year as the trend then the projection is a line slanting upward at its current slope. The addition of a constant number of individuals over an increasing base gives, of course, a continual decrease in the percent rate of population growth.

Three further projections assume different fertility assumptions. The "low," "medium," and "high" United Nations projections all assume a continuation of the trend to decreasing fertility. These projections are "optimistic" in that they envision further sizeable reductions in fertility beyond the currently achieved level. The UN projections start with an estimate of the world TFR as 2.56 births per woman for 2005–2010 (14). The high scenario imagines that world fertility will continue to drop, with all countries eventually converging (at different rates and not necessarily by 2050) to a TFR of 2.35. The medium scenario assumes that fertility will drop further by 1/2 child per woman in all countries, eventually converging to level of 1.85 (i.e., an assumption that the TFR of all countries will drop significantly below the replacement level of ~2.1). The low scenario assumes that global fertility drops by another 1/2 child per woman to 1.35; that is, a level for all countries

that is close to the lowest levels currently reached by the world's most developed countries (21).

The small changes between the 2006 and the 2008 projections include slight increases in the baseline global TFR, an increase in life expectancy of 1 year, and a reduction in the number of AIDS deaths (30 million less during 2005–2020). All of these population-increasing changes are projected to be more than counteracted by an assumed faster rate of fertility decrease in the 49 least developed countries (14, 22). The net result is that the 2008 projection for 2050 is 40 million less than the 2006 projection.

None of the ideas presented here should be taken as this author's argument for or against any one of the scenarios. The considerations are merely presented as cautionary against too firm a belief in any one scenario. And, although the UN is careful not to label these as predictions nor state a preference nor a likelihood for each projection, policymakers often adopt the medium variant in their calculations.

The important point is that all the fertility projections, except the lowest, indicate a continuously rising population through the projected period. In the medium variant, population is expected to be over 9 billion in 2050 and to be increasing by over 30 million persons annually at that time (13).

Environmental change occurs over a much longer time scale than the above projections, and so there is a demand for long-term population projections. Given any particular set of assumptions about the future, computers can play out the consequences for as long as anyone likes. In 2007, the International Institute for Applied Systems Analysis (23) projected that population might peak at ~9 billion around 2070. By 2100, it projects about a 10% chance that the world population will be smaller than it is today and an equal chance it could be more than 11 billion. Starting in 1968 the UN has issued seven sets of long-range projections. In the most recent of these reports (24), the time horizon for the projections was extended from 2150 to 2300. In this projection, world population peaks at 9.22 billion in 2075 and then declines only slightly to reach a level of 8.97 billion by 2300.

The important point to be gained from these long-term projections is that, in the view of these expert teams, the world's population is going to hover around 9 billion for a long time. In sum, the experts do not envision a crash in world population like that once proffered in much of the popular media (20, 26). The world is going to be much more crowded than it is now for the foreseeable future.

REFERENCES

1. World Health Organization. 10 Facts about water scarcity, 2009 http://www.who.int/features/factfiles/water/en/index.html. Accessed Mar 26, 2010.

2. Murray CJL, Lopez AD. *The Global Burden of Disease in 1990: A Comprehensive Assessment of Mortality and Disability from Diseases, Injuries and Risk Factors in 1990 and Projected to 2020.* Cambridge: Harvard University Press, 1996.

3. Ezzati M, Lopez AD, Rodgers A, Murray CLJ, editors. *Comparative Quantification of Health Risks: Global and Regional Burden of Disease Attributable to Selected Major Risk Factors.* Geneva: World Health Organization, 2004.

4. Lopez AD, Mathers CD, Ezzati M, Jamison DT, Murray CJL. Global and regional burden of disease and risk factors, 2001: systematic analysis of population health data. *Lancet* 2006;367:1747–1757.

5. Food and Agriculture Organization, UN. *The State of Food Insecurity in the World 2009.* Rome: FAO, 2009.

6. World Health Organization. *Turning the Tide of Malnutrition: Responding to the Challenge of the 21st Century.* Geneva: WHO, 2000 (WHO/NHD/00.7).

7. Sheeran J. The challenge of hunger. *Lancet* 2008;371:180–181.

8. Food and Agriculture Organization, UN. *The State of Food Insecurity in the World 2006.* Rome: FAO, 2006.

9. Black RE, Allen HA, Bhutta ZA, Caulfield LE, de Onis M, Ezzati M, Mathers C, Rivera J. Maternal and child undernutrition. 1. Global and regional exposures and health consequences. *Lancet* 2008;371:243–260.

10. PRB. *2009 World Population Data Sheet.* Washington, DC: Population Reference Bureau, 2009.

11. Spence J. *The Search for Modern China,* 2nd edition. New York, London: W. W. Norton, 1999.

12. China Population, Development and Research Center (CPDRC), 2010, http://www.cpirc.org.cn/en/year.htm.

13. United Nations. World Population Prospects, 2008 Revision, Population Database, 2008.

14. United Nations. World Population Policies, 2009.

15. PRB. *World Population. Fundamentals of Growth.* Washington, DC: Population Reference Bureau, 1990.

16. Wyman RJ. The projection problem. *Popul Environ* 2003; 24:329–337.

17. Cohen JE. *How Many People can the Earth Support?* New York: W. W. Norton, 1995.

18. Teitelbaum MS. The media marketplace for garbled demography. *Popul Dev Rev* 2004;30:317–327.

19. Coale AJ, Watkins SC. *The Decline of Fertility in Europe.* Princeton, NJ: Princeton University Press, 1986.

20. Wattenberg BJ. *The Population Explosion Is Over.* The New York Times Magazine, Nov 23, 1997.

21. Goldstein JR, Sobotka T, Jasilioniene A. The end of the "lowest–low" fertility? *Popul Dev Rev* 2009;35:663–699.

22. Jiang L. *Smaller Population Size in the New UN Population Projection Depends on Expanded Access to Family Planning.* Washington, DC: Population Action International, 2009, http://www.populationaction.org/blog/2009/03/the-smaller-population-size-in.html.

23. Lutz W, Sanderson W, Scherbov S. IIASA's 2007 Probabilistic World Population Projections, IIASA World Population Program Online Data Base of Results 2008, http://www.iiasa.ac.at/Research/POP/proj07/index.html?sb=5

24. United Nations. *World Population to 2300.* UN, New York: Department of Economic and Social Affairs, Population Division, 2004.

25. Pearce, Fred (2010) The coming Population Crash. Beacon Press, Boston.

26. World Bank (2010): Independent Evaluation Group. Water and Development. Overview p iv http//www.worldbank.org/ieg/water.

4

WATER AND WAR: AVERTING ARMED CONFLICT AND PROTECTING HUMAN RIGHTS

BARRY S. LEVY AND VICTOR W. SIDEL

4.1 INTRODUCTION

Access to water is a basic human right. Per capita availability of and accessibility to freshwater is decreasing in many parts of the world due to population growth, industrialization, urbanization, climate change, and other factors. Conflicts over water have sharply increased in recent decades; most of these conflicts have occurred in the context of violence, but few have become violent. Violent conflict over water can be prevented by measures to increase the availability of and accessibility to freshwater, and by laws and regulations, mediation and arbitration, and proactive cooperation among nations or states or provinces within nations. The dangers that conflicts over water present can be transformed into opportunities for sustainable peace.

Ban Ki-moon, the Secretary General of the United Nations, speaking at the World Economic Forum in Davos, Switzerland, in 2008 stated: "The challenge of securing safe and plentiful water for all is one of the most daunting challenges faced by the world today .... Too often, where we need water, we find guns instead. Population growth will make the problem worse. So will climate change. As the global economy grows, so will its thirst. Many more conflicts lie just over the horizon."

As we enter the second decade of this century, the global challenges to ensure availability and accessibility of water are indeed daunting. This chapter explores these challenges and the opportunities that they provide, in the context of preventing armed conflict over water and protecting access to water as a basic human right.

4.2 AVAILABILITY AND ACCESSIBILITY OF WATER

Less than 0.01% of water is accessible to people, although water covers two-thirds of Earth's surface. Less than 3% of water worldwide is freshwater, and two-thirds of this is contained in glaciers and the polar ice caps. Of the remaining one-third of the 3% that is freshwater, about 70% is used for agricultural purposes.

About three-fifths of water flowing in all rivers is shared by two or more countries—in 263 river basins in 145 countries (1). Two-fifths of the world population lives in these river basins (1).

The World Bank estimates that, on average, people require between 100 and 200 L of water daily to meet their basic needs (for drinking, cooking, personal hygiene, and sanitation), amounting to 36–72 m^3 of water per person annually. If one includes other needs for water, such as agriculture, industry, and energy production, the total annual average requirement of water per person is 1000 m^3 (2). During 1990, six countries in Africa and five countries in the Middle East—all of which are in arid or semiarid regions—had less than 1000 m^3 of freshwater available per person (2). And, each of these 11 countries are projected to have substantially less water per person during 2025 than it had during 1990, due to major population increases (Table 4.1).

Approximately 1 billion people do not have access to safe water. In addition, an inadequate supply of water may lead to shortages in food and energy and may limit industrial development (3). These problems will become more acute as

Water and Sanitation-Related Diseases and the Environment: Challenges, Interventions, and Preventive Measures, First Edition.
Edited by Janine M. H. Selendy.
© 2011 Wiley-Blackwell. Published 2011 by John Wiley & Sons, Inc.

TABLE 4.1 Annual Per Capita Water Availability in Selected Countries of Africa and the Middle East

Country	Per Capita Water Availability (m³ per Person per Year) 1990	2025
Rwanda	880	350
Algeria	750	380
Burundi	660	280
Kenya	590	190
Tunisia	530	330
Israel	470	310
Jordan	260	80
Yemen	240	80
United Arab Emirates	190	110
Libya	160	60
Saudi Arabia	160	50

Source: See Ref. 4, p. 146.

TABLE 4.2 Population Growth Rate, 1998–2050, Selected Countries

Major Water Source	Country	Growth Rate (%)
Nile	Egypt	76
	Ethiopia	243
Jordan	Jordan	178
	Israel	54
Tigris–Euphrates	Iraq	157
	Syria	126
Indus	Pakistan	142
	India	57

Source: See Ref. 4, pp. 157 and 163.

the world population grows from approximately 6.9 billion people now to approximately 9.0 billion by 2050. The population growth rate is even higher in countries that share a major source of freshwater with one or more other countries (Table 4.2) (4). For example, the Nile River Basin is shared by 11 countries: Sudan, Ethiopia, Egypt, Uganda, Tanzania, Kenya, Congo, Rwanda, Burundi, Eritrea, and the Central African Republic.

4.3 ACCESS TO WATER AS A BASIC HUMAN RIGHT

For many years, the human right to water did not receive adequate attention, although it was specifically recognized in the Convention on the Elimination of Discrimination Against Women (1979), the Convention on the Rights of the Child (1989), and the African Charter on the Rights and Welfare of the Child (1990). In 1992, McCaffrey analyzed the human rights framework of the United Nations and related interna-

tional law and concluded that there is a right to an amount of water to sustain life. In addition, he concluded that each nation has the "due diligence obligation to safeguard these rights" (5). In 1999, Gleick built on this analysis and concluded that international law, international agreements, and evidence from the practice of nations broadly and strongly support the human right to a basic water requirement (6). Gleick has also stated the benefits of explicitly acknowledging a human right to water (7):

1. To encourage the international community and individual governments to renew their efforts to meet the basic water needs of their populations.
2. By acknowledging such a right, pressures to translate that right into specific national and international legal obligations and responsibilities are more likely.
3. To spotlight the deplorable state of water management in many parts of the world.
4. To help focus attention on the need to more widely address international watershed disputes and to help resolve conflicts over the use of shared water by identifying minimum water requirements and allocations for all basin parties.
5. Explicitly acknowledging a human right to water can help set specific priorities for water policy: meeting a basic water requirement for all humans to satisfy this right should take precedence over other water allocations and investment decisions.

4.4 CONFLICTS OVER WATER

Both within countries and between countries, conflicts over water are steadily increasing. Although most of these conflicts have arisen in areas in which violence is widespread, few have become violent (Table 4.3). Recent conflicts over water have included fights between animal herders and farmers with competing needs in Burkina Faso, Ghana, and Ivory Coast; thousands of farmers breeching security and storming the area of Hirakud Dam in India to protest allocation of water to industry; Tamil Tiger rebels cutting the water supply to government-held villages in Sri Lanka and government forces

TABLE 4.3 Water Conflicts, 1900–2007

Time Period	Number of Conflicts	Average Number per Year	Number of Violent Conflicts and Conflicts in the Context of Violence
1900–1959	22	0.4	19 +
1960–1989	38	1.3	23 +
1990–2007	83	4.6	61 +

Source: Adapted from Ref. 8.

then launching attacks on the Mavil Aru Reservoir; and clashes in Kenya and Ethiopia over water as well as livestock and grazing land, in which at least 40 people died (8).

Throughout much of the Middle East, access to oil has been the major resource concern for many years. However, there has also been a conflict over water rights in the Jordan River Basin for many years, and now access to water is a major resource concern throughout the region. Climate change is warming the Fertile Crescent, where the Tigris and Euphrates rivers are located, adversely affecting Iraq, Syria, Lebanon, and southeast Turkey. An inadequate amount of grazing land has caused thousands of herds of sheep to leave, creating many refugees. More than 160 villages in Syria have been abandoned, and 800,000 people have lost their livelihood. The network of dams in Turkey has decreased the water supply by half. Thus far, however, armed conflict has not occurred (9).

In Asia, water problems are also becoming serious. If current trends continue, by 2030 several developing countries in Asia will probably face an unprecedented crisis in water quality. Between now and 2025, some cities with more than 10 million residents, such as Karachi, Mumbai, Manila, Shanghai, and Jakarta, will face severe freshwater shortages (10).

The major underlying reasons for conflicts over water include (i) low rainfall, inadequate water supply, and dependency on one major water source; (ii) high population growth and rapid urbanization; (iii) modernization and industrialization of societies; and (iv) history of armed combat and poor relations between countries and among groups within countries (4). Water scarcity alone is infrequently the cause of armed conflict over water.

The immediately precipitating causes of conflicts over water are often sociopolitical tensions, including border disputes; disputes concerning dams, reservoirs, and other large-scale projects; and disputes concerning environmental and resource issues. According to Gleick, the relationship between water and conflict can be placed in the following four categories (11):

1. *Water as a Military Tool (State Actors)*: Where water resources, or water systems themselves, are used by a nation or state as a weapon during a military action.
2. *Water as a Military Target (State Actors)*: Where water resources or water systems are targets of military actions by nations or states.
3. *Terrorism (Nonstate Actors)*: Where water resources or water systems are the targets or tools of violence or coercion by nonstate actors.
4. *Development Disputes (State and Nonstate Actors)*: Where water resources or water systems are a major source of contention and dispute in the context of economic and social development.

4.5 HEALTH CONSEQUENCES OF VIOLENT CONFLICT

Violent conflict in general has numerous serious health consequences to individuals and to the health of the public (12):

1. Deaths, injuries, and illnesses and long-term physical and mental impairment.
2. Destruction of the health-supporting infrastructure of society, including medical care and public health services, safe food and water supply, sanitation and waste disposal systems, transportation, communication, and energy supply.
3. Forced migration, creating refugees to other countries, and internally displaced persons within countries.
4. Diversion of human and financial resources away from health and other human services.
5. Promotion of further violence as a way of resolving conflicts.

4.6 PREVENTING ARMED CONFLICTS OVER WATER

One set of approaches to preventing conflicts over water are measures to increase the availability of water. These include

1. reducing use of water, such as reducing wasteful uses and increasing efficiency of water use;
2. increasing availability of clean water, such as by reducing industrial pollution and sewage contamination of water;
3. establishing and maintaining new groundwater wells; and
4. designing and implementing improved methods of desalinization.

Another set of approaches to addressing these problems is resolving water conflicts before they "boil over"—before they become violent or develop more serious consequences. These measures include (a) laws and regulations, at the local, state or province, national, or international level; (b) proactive cooperation among nations or among states or provinces within nations; and (c) mediation and arbitration.

By resolving conflicts over water, countries and groups within countries can help to build sustainable peace. Water plays an important role in improving and maintaining health, food security, and economic prosperity. Therefore, sound water management practices help to support social, environmental, and economic stability and help prevent violent conflict over water. Sustainable water management is a prerequisite for establishing and maintaining peace.

There have been more than 3800 unilateral, bilateral, or multilateral declarations or conventions concerning water, including 286 treaties, 61 of which refer to more than 200 international river basins (13). In addition, there are many examples of cooperation and peace-building that have helped prevent or resolve conflicts over water.

There were 1831 water-related events between nations in the Middle East between 1948 and 1999 (8). Two-thirds resulted in cooperation among these nations and the vast majority of the remaining one-third did not escalate beyond verbal arguments (14). Only 37 incidents led to acute conflict (14).

One example of such cooperation has been the Good Water Neighbors Project, established in 2001, which has used transboundary water resources to build peace. It has brought together Israeli, Jordanian, and Palestinian communities to protect shared water resources and has significantly improved the local water sector and helped to build peace at the local level (15).

There have been numerous examples of cooperation elsewhere. For example, the Nile Basin Initiative, a partnership begun in 1999 among nine countries that share the Nile River, has helped to develop the Nile in a cooperative manner, share substantial socioeconomic benefits, and promote regional peace and security (16).

Another example is the Autonomous Water Authority created by Bolivia and Peru, which share Lake Titicaca. These countries have recognized how crucial it is to work together on management of water resources (17).

Yet another example is in the freshwater basin of the Aral Sea, which is shared by Kazakhstan, Uzbekistan, and four other countries. The surface of the Aral Sea had shrunk to less than one-half of its original size by diversion of water, which drained two rivers feeding it and devastated the environment. With the completion of the Kok-Aral Dam, a World Bank project, the Aral Sea has now begun to fill again (18).

The role of women is especially important in promoting and maintaining cooperation among and within countries concerning peaceful access to water. Women produce about 70% of food in most developing countries. They are major stakeholders in all development issues related to water. Yet, up until now, they have often remained on the periphery of planning and management decisions concerning water resources (19).

4.7   CONCLUSION

Despite the great challenges to peace that are posed by current and imminent conflicts over water, there is reason for hope that these dangers can be transformed into opportunities. As the United Nations has stated: "Despite widespread perceptions that water basins shared by countries tend to engender hostility rather than collaborative solutions, water is an often untapped resource of fruitful cooperation" (13).

REFERENCES

1. Wolf AT, et al. International river basins of the world. *Int J Water Resour Dev* 1999;15:387–427.

2. Gleick PH. Water and conflict. *Int Security* 1993;18(1):90, 100–101.

3. Solomon S. *Water: The Epic Struggle for Wealth, Power and Civilization*. New York: HarperCollins, 2010, p. 371.

4. Klare MT. *Resource Wars: The New Landscape of Global Conflict*. New York: Henry Holt and Company, 2001.

5. McCaffrey SC. A human right to water: domestic and international implications. *Georgetown Int Environ Law Rev* 1992;5:1–24.

6. Gleick PH. A human right to water. *Water Policy* 1999;1:497–503.

7. Gleick PH. *The World's Water, 2000–2001: The Biennial Report on Freshwater Resources*. Washington, DC: Island Press, 2000.

8. Gleick PH. Water conflict chronology. In: *The World's Water, 2008–2009: The Biennial Report on Freshwater Resources*. Washington, DC: Island Press, 2009, pp. 151–196.

9. Amos D, Mideast Water Crisis Brings Misery, Uncertainly. NPR. Available at http://www.npr.org/templates/story/story.php?storyId=122294630&ps=cprs. Accessed on Jan 8, 2010.

10. Tacio HD. Asia's Looming Water Crisis. Gaia Discovery. Available at http://www.gaiadiscovery.com/latest-planet/asias-looming-water-crisis.html. Accessed on Feb 15, 2010.

11. Gleick PH. *The World's Water: The Biennial Report on Freshwater Resources, 1998–1999*. Washington, DC: Island Press, 1998, p. 107.

12. Levy BS, Sidel VW, editors. *War and Public Health*, 2nd edition. New York: Oxford University Press, 2008.

13. United Nations. From water wars to bridges of cooperation: exploring the peace-building potential of a shared resource. 10 stories the world should hear more about. Available at http://www.un.org/events/tenstories/06/story.asp?storyID=2900. Accessed on Feb 2, 2010.

14. Postel S, Wolf AT. Dehydrating conflict. *Foreign Policy*, September/October 2001. Available at http://www.globalwaterpolicy.org/pubs/FP_Conflict.pdf. Accessed on Feb 2, 2010.

15. Kramer A. *Regional Water Cooperation and Peacebuilding in the Middle East*. Berlin: Adelphi Research, 2008.

16. Nile Basin Initiative. Available at http://www.nilebasin.org. Accessed on Feb 2, 2010.

17. Delli Priscoli J, Wolf AT. *Managing and Transforming Water Conflicts*. Cambridge, UK: Cambridge University Press, 2009, pp. 211–214.

18. The World Bank. Saving a Corner of the Aral Sea. Available at http://web.worldbank.org/WBSITE/EXTERNAL/COUNTRIES/ECAEXT/0,contentMDK:20633813~menu-PK:258604~pagePK:146736~piPK:226340~theSite-PK:258599,00.html. Accessed on Feb 2, 2010.

19. Gleick PH. *The World's Water: The Biennial Report on Freshwater Resources, 2006–2007*. Washington, DC: Island Press, 2006, p. 126.

SECTION II

WATER AND SANITATION-RELATED DISEASES

The chapters in this section on individual diseases not only cover basic information about the disease in question, that is, pathogen, epidemiology, clinical manifestations, treatment, prevention, and control but also distribution, prevalence, and incidence and interconnected factors such as environmental factors. These chapters, as appropriate, discuss and emphasize the significance of the close relationships among water access and quality, lack of adequate sanitation, lack of adequate hygiene, and the relationship of human activity to the diseases. They address how the diseases are further aggravated by nutritional inadequacies and anemia, infections, and other problems that often exist in the same person, such as diarrheal diseases. Other factors they address are the importance of adequate maternal and child health care, the significance of climate change, and other environmental factors. Prevention and control are important parts of the discussions in this section.

5

INFECTIOUS DIARRHEA

SEAN FITZWATER, ARUNA CHANDRAN, MARGARET KOSEK, AND MATHURAM SANTOSHAM

5.1 INTRODUCTION

Diarrhea is one of the leading causes of morbidity and death globally, disproportionally affecting young children and developing countries. The majority of diarrheas are gastrointestinal infections caused by viral, bacterial, and parasitic pathogens, which disrupt the gut's normal secretion and absorption functions. Most infectious diarrhea episodes are acute in nature, and can lead to rapid loss of bodily fluids, dehydration, and death if proper supportive care is not provided. Repeated bouts of acute diarrhea and chronic diarrhea can exasperate malnutrition and lead to poor growth, impaired immune function, and increased susceptibility to diarrhea as well as other infections. Fortunately, most infectious diarrhea can be successfully managed. Uncomplicated diarrhea can be easily managed with a simple regimen of oral rehydration and nutritional therapies, while severe dehydration may require intravenous (IV) fluids; for cholera or dysenteric diarrhea, such as shigellosis, antibiotics are recommended in addition to rehydration. Diarrhea can be prevented to a large extent by disrupting transmission of pathogens through vaccination against leading pathogens, nutritional interventions, and sanitary interventions including improving personal hygiene, water quality, disposal of waste, and appropriate preparation of food. Together, improved case management and prevention of diarrhea has lead to a dramatic reduction in diarrheal deaths over the last several decades, although progress has stagnated in recent years. Diarrhea morbidity has not seen similar improvements on the global scale, although targeted interventions are clearly efficacious.

5.2 EPIDEMIOLOGY

Infectious diarrhea is a leading cause of disease and death worldwide. Each year there are approximately 4.6 billion episodes of diarrhea, resulting in 2.2 million deaths (1). Globally, infectious diarrhea is responsible for 3.7% of all deaths, ranking 5th in terms of leading cause of death (1). Most of this morbidity and mortality occurs among young children (Table 5.1). In 2004, the average child under the age of 5 had a mean of 3.8 episodes of diarrhea, leading to an estimated 1.7 million deaths (1), making diarrhea the second most common cause of childhood mortality, responsible for 18% of all deaths in children under age 5 (2).

Multiple factors are associated with increased risk of diarrheal diseases, including lack of access to clean water, inadequate disposal of human waste, overcrowding, limited quality control in food handling and processing, and malnutrition (3). As a result, the burden of diarrheal illnesses varies widely globally (Table 5.2 and Fig. 5.1). The majority of diarrheal cases and highest incidence of disease occur in low- and middle-income countries, which lack basic sanitary preventive measures. Additionally deaths due to diarrhea occur almost exclusively in low- and middle-income countries due to lack of access to appropriate medical care (Fig. 5.2), countries where WHO estimates that in 2004 over 99% of deaths due to diarrhea occurred (1). In total, the 10 countries with the highest number of deaths due to diarrhea (India, Nigeria, Ethiopia, the Democratic Republic of the Congo, Afghanistan, Bangladesh, Pakistan, China, Angola, and Niger) accounted for 57% of diarrheal cases and 62% of diarrheal deaths. By comparison, less than

Water and Sanitation-Related Diseases and the Environment: Challenges, Interventions, and Preventive Measures, First Edition.
Edited by Janine M. H. Selendy.
© 2011 Wiley-Blackwell. Published 2011 by John Wiley & Sons, Inc.

TABLE 5.1   Morbidity and Mortality Associated with Diarrheal Disease Stratified by Age Group

Age Group (Years)	Deaths	Deaths (per 100,000)	Cases (Millions)	Cases (per Person)
0–5	1,742,879	280.7	2,390	3.85
6–15	59,200	4.8	750	0.61
16–60	161,769	4.1	1,282	0.33
>60	195,168	29.7	190	0.29

Adapted from Ref. 1.

14,000 deaths due to diarrhea occurred in 2004 in all high-income countries combined (1). Although diarrhea rarely causes death in high-income countries, it still accounts for a significant health problem and utilization of health care services. In the United States alone, 150,000 children are hospitalized each year with diarrhea, accounting for 13% of all hospital admissions for children under 5 years old (4), but only 30 deaths (1).

The burden of diarrheal disease varies within countries, with the highest rates of disease and deaths occurring in impoverished and neglected populations (5, 6). In high-income countries, minority populations tend to have higher rates of diarrhea than the general population due largely to poorer socioeconomic conditions, although most severe disease in all social strata are concentrated in children and elderly. In the United States, Native American and Alaskan Native infants have nearly twice the rate of hospitalizations due to diarrhea compared with the general U.S. population (7). In all countries, natural disasters and other humanitarian crises put populations at increased risk of diarrhea disease by breaking down the regular sanitation infrastructure, displacing populations, and facilitating the spread of pathogens while concurrently interrupting access to optimal case management (8). In Bangladesh, the floods of 1988, 1998, and 2004 resulted in a doubling of diarrheal cases seen at the International Center for Diarrheal Disease Research, from an average of 249 cases per day during nonflood times to 542 cases per day during flooding (9). Well-planned responses to crises have been shown to limit the mortality caused by diarrhea. A study of refugee camps in Tanzania from 2005 to 2007 found increased rates of diarrheal illness compared to the general population, but a decreased risk of death. The rate of death in refugee camps varied yearly from 0.05 to 0.07 death per 1000 diarrheal cases, while the national average was over 12 times higher (0.88 per 1000) (10).

TABLE 5.2   Morbidity and Mortality Associated with Diarrheal Disease Stratified by Income Group

Income Group	Deaths (Total)	Deaths (per 100,000)	Cases (Millions)	Cases (per Person)
High	14,713	1.5	246	0.24
Upper middle	83,378	9.0	577	0.62
Lower middle	1,112,005	31.0	2,848	0.79
Lower	948,920	104.9	941	1.04
Global	2,159,016	33.6	4,612	0.72

Adapted from Ref. 1. Income stratification is based on Work Bank classification (2008).

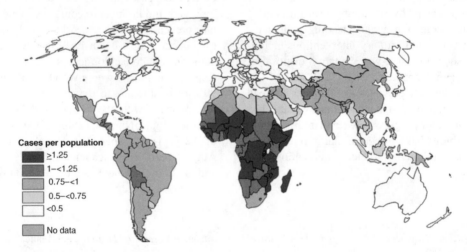

Cases per population
- ≥1.25
- 1–<1.25
- 0.75–<1
- 0.5–<0.75
- <0.5
- No data

FIGURE 5.1   Incidence on diarrhea per population. (Adapted from Ref. 1.)

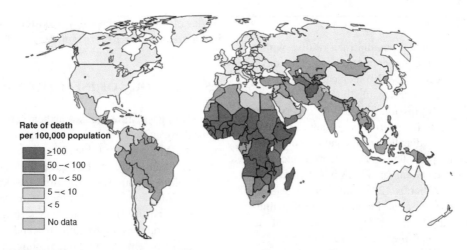

FIGURE 5.2 Rate of death due to diarrhea age per 100,000 population. (Adapted from Ref. 1.)

5.3 DEFINITION AND PHYSIOLOGY

Diarrhea is defined by the WHO as a condition in which the stricken person has at least three loose or watery stools per day (11), often accompanied by an increased sense of urgency, abdominal discomfort, and alterations in stool characteristics (12). These symptoms are caused by an alteration in the gastrointestinal tract's ability to absorb nutrients, augmented secretion of fluids, or both. Infectious agents create this state through a combination of inflammatory and noninflammatory processes (13).

Noninflammatory processes leading to diarrhea are noninvasive and are caused by enterotoxin-producing organisms or by organisms that disrupt the normal absorptive or secretory functions of the intestinal lining without causing significant inflammation or mucosal damage (13). A variety of pathogens produce enterotoxins, substances that change the permeability of the cells lining the intestinal tract by altering the function of specific cellular pathways (14–17). The specific target of an enterotoxin is pathogen dependent, but the result is an alteration to the function of cell membrane-bound transport proteins and ion channels. This leads to a dysregulated outflow of water into the lumen of the intestine, resulting in diarrhea (13). Other organisms act through less specific processes that lead to damage to the intestinal mucosa. Pathogen adherence and replication can lead to increased rate of apoptosis in enterocytes with minimal inflammation (13, 18, 19). This damage leads to a disruption in the epithelial barrier and changes the pathology of the intestinal mucosa, resulting in a reduction in the latter's ability to selectively absorb and secrete water and nutrients.

Inflammatory diarrhea is a result of pathogen-produced cytotoxins or active invasion of the intestinal tissue by pathogens (13). Cytotoxin-producing pathogens release toxins, which damage epithelial cells or disrupt the junctions between cells (17, 20–22). Some organisms directly invade the intestinal mucosa and, often with the aid of toxins, induce an acute inflammatory reaction (22–25). In both cases this leads to ulcerations in the intestinal mucosa, resulting in damage to the mucosal surface, disruption of absorption and secretion, and the presence of blood, leucocytes, and/or mucus in the stool. Furthermore, ulcerations can act as a portal for other intestinal organisms to enter the body, increasing the likelihood of sepsis.

5.4 MANIFESTATIONS OF INFECTIOUS DIARRHEA

5.4.1 Acute Watery Diarrhea

The most common manifestation of infectious diarrhea is acute watery diarrhea, which accounts for approximately 85% of all diarrheal cases globally (26). Acute watery diarrhea is categorized by WHO as a condition in which a person has a diarrheal episode with a duration of less than 14 days; it is characterized by discharges of watery feces with or without the presence of mucous (27). The main clinical complication associated with acute diarrhea is dehydration, which is often more severe in children than adults (28). Most uncomplicated cases of acute watery diarrhea will resolve with appropriate rehydration therapy and do not require specific medicinal treatments or laboratory diagnostics (12, 29). Since there is considerable loss of nutrients during diarrheal episodes, it is critical to include continued age-appropriate feeding as part of the treatment of childhood diarrhea (28). Although the case fatality rate is relatively low, <1% even in low income settings, the large number of cases results in the majority of deaths due to diarrhea being caused by acute watery diarrhea (30).

5.4.2 Persistent Diarrhea

Although most cases of infectious diarrhea resolve within 2 weeks, 3–19% of acute cases continue beyond 14 days, becoming persistent diarrhea (27, 29). Persistent diarrhea is characterized by chronic or relapsing infections with an infectious agent (31). Progression from acute to persistent diarrhea is associated with poverty, poor sanitation, and malnutrition; severe acute diarrheal cases and dysentery are more likely to become persistent (29). As with acute diarrhea, persistent diarrheal patients are at risk of dehydration if not provided rehydration therapy (28). Additionally, persons with persistent diarrhea are at an increased risk of malnutrition because of chronic malabsorption and are at an increased risk of infections spreading beyond the gut (28). The outcome and duration of persistent diarrhea can be improved if the causative pathogen is identified and treated (31). In developing countries, specific diagnostics and treatment are generally not available, resulting in relatively high risk of death; up to 12% case fatality rates have been reported for nondysenteric persistent diarrhea in rural India (30).

5.4.3 Bloody Diarrhea

Bloody diarrhea, also known as dysentery, is typically accompanied by abdominal pain, mucous in the stool, and often systemic toxicity (26, 32). Blood in the stool can accompany chronic or acute infectious diarrhea; approximately 5–10% of all diarrheal cases are dysenteric or become dysenteric (26, 30). In total, dysentery is responsible for more than 300,000 deaths in children per year (33). Bloody diarrhea is typically caused by bacterial or parasitic pathogens, although parasitic dysentery is relatively rare in developed countries (32, 34). Dysenteric patients are at risk of dehydration, but much less so than cases with acute or persistent watery diarrhea. However, since bloody stool is indicative of a break in the gastrointestinal tract, dysenteric patients have an increased risk of developing sepsis, especially severely malnourished children (28). Additionally, damage to the gastrointestinal mucosa disrupts nutrient absorption more than noncomplicated diarrhea, which puts dysenteric patients at a higher risk of developing malnutrition (28). Treatment with appropriate antibiotics in addition to rehydration is indicated (28). With proper treatment, the likelihood of death from dysentery is low, although increasing antibiotic resistance to the most common cause of dysentery, Shigella is a progressive problem (35). However, in developing countries, appropriate diagnostics and treatment are not always available, which increases the rates of additional complications, and thus persistent diarrhea may result in significant mortality. In one community-based cohort study in India, the overall case fatality of dysentery in children was 4%. However, the

subset of these patients with persistent dysenteric diarrhea had a greater than 20% risk of death (30).

5.5 ETIOLOGY OF INFECTIOUS DIARRHEA

The etiology of infectious diarrhea varies globally and is caused by a heterogeneous mix of viral, bacterial, and parasitic pathogens (Table 5.3). Few high-quality comprehensive etiology studies have been performed outside of developed countries, limiting the knowledge of the exact etiologies in developing countries. Complicating this is the ability of most microbes to inhabit the gastrointestinal tract asymptomatically and coinhabit the gut with other potentially pathogenic microbes (36, 37). In a review of case control studies of persistent diarrhea in developing countries, a known diarrheal pathogen was isolated in 75% of cases, but it was also found in 43% of healthy controls (27). Similarly, infections with multiple potential pathogens have been found in 15–20% of acute cases (5, 34, 38).

5.5.1 Viral Pathogens

The majority of viral gastrointestinal infections are caused by rotavirus, caliciviruses, astroviruses, and adenoviruses (56). Numerous other viral pathogens are known to cause diarrhea, including toroviruses, coronaviruses, picornaviruses, and pestiviruses, but are of limited (or unknown) epidemiological importance (42, 43, 56). The etiology of viral diarrhea has been best characterized in developed countries, where they represent the majority acute diarrheal cases and are responsible for most foodborne outbreaks of gastroenteritis (57). The role of specific viral pathogens is less well understood in developing countries due largely to the limited availability diagnostic testing (58). However, despite incomplete data, the high burden of viral pathogens in endemic diarrheal disease developing counties is apparent (27, 38, 43, 59). The symptoms of viral gastroenteritis are similar regardless of the specific pathogen, manifesting with acute noninflammatory watery diarrhea, fever, and vomiting, although the severity of symptoms varies by virus (60, 61). Symptoms typically resolve within a week of disease onset; specific diagnostic tests and treatment are not necessary, although supportive care using rehydration therapy is critical (62). Endemic viral diarrhea occurs mainly in young children, peaking during cooler months, although seasonality is less pronounced in tropical settings (38, 63, 64). The fecal–oral route is thought to be the main transmission route in developing countries, but there is an increasing consensus that viral pathogens are also spread by person-to-person transmission or aerosolization of viral particles in developed nations (3, 65).

TABLE 5.3   Etiology of Infectious Diarrhea

	Syndrome	Global Burden
Viruses		
Rotavirus	Acute diarrhea	29–47% of diarrhea hospitalizations (39), 114 million episodes of gastroenteritis, 24 million outpatient visits, 2.4 million hospitalizations each year (40), 475,000–580,000 deaths annually (41)
Calicivirus	Mild acute diarrhea	7–22% of childhood acute gastroenteritis (42), 68–80% of outbreaks of gastroenteritis in developed countries (43)
Astrovirus	Mild acute diarrhea	2–9% of childhood acute diarrhea (42)
Adenovirus	Mild acute and persistent diarrhea	2–6% of childhood acute gastroenteritis (42)
Measles	Disposes for severe acute and persistent diarrhea	Complication in 12–63% of all measles cases (44–48); implicated in 2–5% of diarrheal cases in unvaccinated children (48)
Bacteria		
Escherichia coli	Acute watery, persistent, and dysenteric diarrhea	30–40% of cases diarrhea (34), 23–30% of persistent diarrhea (27)
Enterotoxigenic	Acute diarrhea	15–30% of diarrhea illness (49), 380,000 deaths per year (50)
Enteropathogenic	Acute and persistent diarrhea	4–8% of diarrhea illness (49), <30% of persistent diarrhea (27, 51)
Enteroaggregative	Acute, persistent, and bloody diarrhea	<30% of persistent diarrhea (27, 51)
Enterohemorrhagic	Bloody diarrhea, hemorrhagic colitis	Approximately 100,000 cases per year in the US (52); rare outside developed countries (50)
Enteroinvasive	Bloody diarrhea	<4% of diarrheal cases (53)
Shigella	Severe bloody diarrhea	5–15% of diarrheal cases, 10% diarrheal deaths, 30–50% of dysentery (54)
Salmonella, nontyphi	Acute watery diarrhea, rarely dysentery	0–15% of diarrheal cases (49), 1.4 million cases in the US (35)
Campylobacter	Acute diarrhea, rarely bloody diarrhea	2–14% of all diarrhea disease (49)
Vibrio cholerae	Severe acute watery	120,000 deaths (55)
Parasites		
Cryptosporidium	Mild acute diarrhea, persistent diarrhea, dysentery	16% persistent diarrhea (29)
Giardia lamblia	Mild acute diarrhea, persistent diarrhea	<10% persistent diarrhea (27)
Entamoeba histolytica	Severe dysentery	50 million hospitalizations (12), 40,000 deaths (12)

5.5.1.1   *Rotavirus*

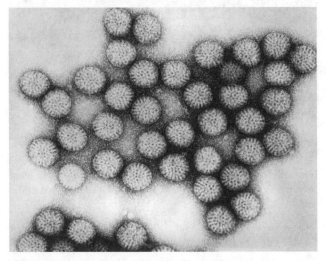

Transmission electron micrograph of rotavirus virions. (Content provider: CDC/Dr. Erskine Palmer. *Source*: CDC, Public Health Image Library, Image ID #178.) (*See insert for color representation of this figure.*)

Rotaviruses are a group of spherical nonenveloped viruses characterized by a double-stranded, segmented, RNA genome (14, 66). Rotavirus infection is the most common cause of severe diarrhea globally, resulting in an estimated 114 million episodes of gastroenteritis, 24 million outpatient visits, and 2.4 million hospitalizations each year (40). Rotavirus disease occurs almost exclusively in children and infects nearly every child by the age of 5 years (67). The rate of rotavirus illness is similar in both developed and developing countries; rotavirus is responsible for 29 to 45% of all hospitalizations due to diarrhea, with higher proportions seen in countries with a higher average income (39). In total, there were over 500,000 deaths attributed to rotavirus in 2004, resulting in 5% of all deaths in children <5 years of age (41, 68). The main transmission route in developing countries is thought to be fecal–oral, as viral particles are shed profusely in feces of patients (14). However, the prevalence of rotavirus in developed countries with excellent sanitation suggests that nonfecal routes play a large role in transmission (3, 69).

The viral capsid, composed of a glycoprotein (G-type antigen) coat with protease-sensitive protein (P-type antigen) surface spikes, is the major antigenic target in humans. Over 40 G and P antigen combinations have been observed in humans, although globally, 6 G/P combinations account for 57 to 100% or rotavirus disease, depending on region (70). This diversity results in decreased, but does not eliminate, cross protection from other rotavirus serotypes after infection with one serotype (14). Rotavirus causes disease by disrupting gastrointestinal function through a variety of strategies including reducing expression of digestive enzymes, altering tight junctions between cells, and stimulating the secretion of chlorine ions into the intestinal lumen using a potent enterotoxin (14). After a brief incubation period of 1–3 days, the disruption to the gastrointestinal tract leads to the nonspecific symptoms of viral diarrheal infection (14, 67). No pathogen-specific treatment exists, but most cases resolve in 2–7 days after onset of symptoms. Severity varies widely from case to case, but compared to other causes of viral diarrhea, rotavirus is associated with an increased severity of gastroenteritis symptoms (60, 61). However, repeat infection with rotavirus provides increasing protection against disease regardless of serotype, with two natural infections providing protection against severe disease and three infections protecting against any rotavirus diarrhea (71).

5.5.1.2 Calicivirus

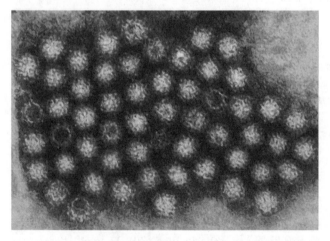

Transmission electron micrograph of norovirus virions. (Content provider: CDC/Charles D. Humphrey. *Source*: CDC, Public Health Image Library, Image ID # 10708.) (*See insert for color representation of this figure.*)

The viral family Caliciviridae is a diverse group of single-stranded RNA viruses (43). Norovirus and sapovirus are the most well-known members of the group,

although they themselves are composites of many antigenically distinct subtypes (58). The clinical presentations of calicivirus infections are indistinguishable from those of other viral causes of acute gastroenteritis. However, on average, symptoms are less severe than what are found in rotavirus disease, so there is a lesser risk of hospitalization (63, 72). The burden of caliciviruses is most well described in the context of outbreaks in developed countries, where it is reported to be responsible for 68–80% of gastroenteritis outbreaks, predominately due to contaminated food (43). However, there is a growing acceptance that caliciviruses play an important role in endemic diarrhea in children. A Finnish study following infants enrolled in a rotavirus vaccine trial found 21% of gastroenteritis cases to be positive for a calicivirus (63). Similarly, a multiyear study in China of winter diarrhea demonstrated norovirus to be present in 24% of children who tested negative for rotavirus, while a longitudinal community cohort study in Peru found norovirus in 21% of children with diarrhea (72, 73). The commonality of calicivirus infections in children is supported by high antibody seroprevalence in children, with approximately 50% of children in developed nations having antibodies against norovirus by the age of 5 (63).

5.5.1.3 Astrovirus

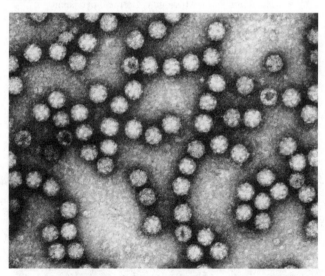

Transmission electron micrograph of Astrovirus virions. (Content provider: Graham Colm. *Source*: http://en.wikipedia.org/wiki/File: Astrovirus_4.jpg.)

Astroviruses are a ubiquitous group of small non-enveloped, single-stranded RNA viruses closely related to caliciviruses (59). Astroviruses are the third most common cause of viral diarrhea in children, accounting for 2–16%

of children hospitalized with diarrhea (61). Additionally, astroviruses are an important cause of nosocomial outbreaks among children and immunocompromised patients (59). Clinical manifestations are generally mild symptoms compared to rotavirus or norovirus and of shorter duration. As a result, astrovirus infections are more common in diarrhea identified in the community compared to hospitalized patients with diarrhea (43, 61). Most people are exposed to astroviruses in childhood. Greater than 90% of children develop antibodies against astrovirus serotype 1 in the United States, Japan, and the Netherlands by the age of 9 years (61). However, different serotypes of astrovirus result in different patterns of infection, with specific serotypes predominating in infants while others are more common in older children (74). Astroviruses have a more variable seasonality than other viral causes of gastroenteritis: cases can peak in the winter or early summer depending on geography (59).

5.5.1.4 Adenovirus

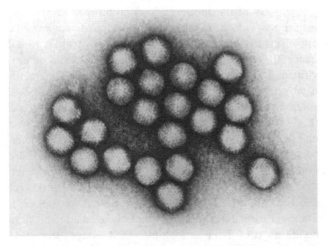

Transmission electron micrograph of adenovirus virions. (Content provider: CDC/Dr. G. William Gary, Jr. *Source*: CDC, Public Health Image Library, Image ID # 10010.) (*See insert for color representation of this figure.*)

Adenovirus is a group of lytic nonenveloped DNA viruses that cause a wide array of diseases (56, 75). Although many serotypes of adenovirus can cause diarrhea, most acute gastroenteritis infections are caused by serotypes Ad40 and Ad41, collectively known as "enteric adenoviruses" (43). In developed countries, adenoviruses account for 5–10% of pediatric hospitalizations due to diarrhea, with incidence peaking in the second year of life (42, 76). The epidemiology of the viruses are less well known in developing countries, but the proportion of disease appears to be lower, accounting for only 1–3% of hospitalization (77–79). Disease is

typically mild, but symptoms can be severe in immunosuppressed individuals (43). Compared to other causes of viral gastroenteritis, adenovirus has a notably long incubation period, 8–10 days, and long duration, lasting from 5 to 12 days (76). Up to one-third of cases persist past 14 days (43). Additionally, adenovirus is unique in that is shows little seasonal variation when compared with other viruses (43).

5.5.1.5 Measles

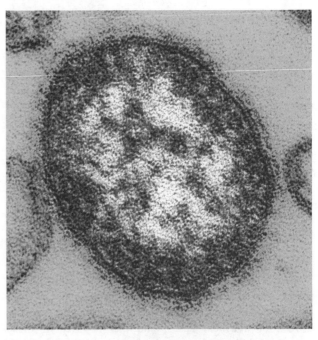

Thin-section electron micrograph transmission of a single measles. (Content provider: CDC/Cynthia S Goldsmith; William Bellini, Ph.D. *Source*: CDC, Public Health Image Library, Image ID # 10707.) (*See insert for color representation of this figure.*)

Measles virus is an enveloped, single-stranded RNA virus that is spread by respiratory secretion and most known for its characteristic rash (80). Although generally considered a mild disease in developed countries, severe complications are common in the developing world. Diarrhea is one of the most common complications of measles infections, reported in 12–63% of all measles cases identified by community surveillance (44–48). In unvaccinated populations, measles has been estimated to be a factor in 2–5% of all diarrhea cases (48). The measles virus itself is not generally considered the causative agent of diarrhea. Rather, infection with the measles virus produces a long-lasting immunosuppression, which predisposes children to secondary bacterial and parasitic infections that manifest as diarrhea (81–83). This immunosuppression can last for weeks and as a result the majority of measles associated diarrheal complications occur after the measles rash itself has cleared (48, 81, 82). Additionally measles-associated diarrhea is known to persist longer and have increased severity compared to nonmeasles-associated diarrhea, increasing the likelihood of poor outcomes and making diarrhea a leading cause of death in measles patients (48, 82, 83).

5.5.2 Bacterial Pathogens

Globally, most disease and death caused by bacterial diarrhea are the result of *E. coli*, *Shigella*, *V. cholerae*, *Salmonella*, and *Campylobacter* infections. Bacteria, much more than viruses, have a strong association with poor sanitary conditions and are exclusively transmitted by the fecal–oral route. Many bacterial pathogens have zoonotic reservoirs (34, 35). As a result, the burden of bacterial diarrhea is located almost exclusively in developing countries, where they have been found to be responsible for about half of all cases of acute diarrhea (37). Some bacteria, notably *Clostridium difficile*, cause significant disease in patients with preexisting medical conditions, which has been documented predominantly in developed countries (35).

Endemic bacterial diseases tend to predominate in warmer months, although outbreaks can occur in any season (34). Bacteria are a relatively minor cause of diarrhea in developed countries, being responsible for only 2–10% of diarrheal cases, and the majority of these cases in the United States are food-related illnesses (64). Additionally, bacterial gastroenteritis is more common in patients returning from travel to developing countries, causing an estimated 80% of traveler's diarrhea (35). When gastroenteritis develops, most bacterial diarrhea is indistinguishable from viral diarrheas, causing acute watery diarrhea, fever, abdominal pain, and vomiting (64). However, inflammatory and persistent diarrhea is common with bacterial diarrhea, which can progress to overt dysentery and sepsis (27, 64). Leukocytes in the stool can be a useful indicator of bacterial and invasive diarrhea, and etiology can be confirmed using stool cultures, alone or in combination with polymerase chain reaction (PCR) tests. However, etiological diagnosis is unnecessary in uncomplicated bacterial diarrhea, which can be treated with rehydration therapy (64). Antibiotic treatment is only indicated in cases complicated with dysentery or sepsis or the development of chronic symptoms, and culture is used in the development of regional empiric treatment guidelines, as antibiotic resistance is of growing concern across all species (34). However, it is not necessary to perform diagnostic tests for bacteria in uncomplicated acute watery diarrhea.

5.5.2.1 *Escherichia coli*

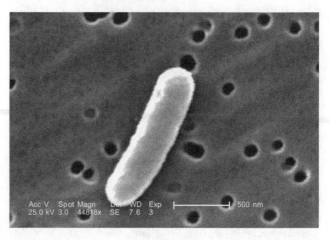

Scanning electron micrograph a single enterotoxigenic *E. coli* bacterium. (Content provider: CDC/Janice Haney Carr. *Source*: CDC, Public Health Image Library, Image ID # 10577.) (*See insert for color representation of this figure.*)

Members of the species *E. coli* group of microbes, the majority of which are nonpathogenic, are Gram-negative organisms and flora of the healthy human gastrointestinal tract (50). However, there are five major classes of enterovirulent *E. coli*: enterohemorrhagic *E. coli* (EHEC), enteropathogenic *E. coli* (EPEC), enterotoxigenic *E. coli* (ETEC), enteroinvasive *E. coli* (EIEC), and enteroaggregative *E. coli* (EAEC) (64), all with distinct virulence factors that cause diarrhea. Collectively, these pathogens are the most common bacterial cause of diarrhea in developing countries, and are implicated in 30–40% of cases of diarrhea (34). In addition they are the most common pathogen associated with persistent diarrhea, causing 23–30% of persistent diarrhea cases (27). The clear attribution of the burden of disease caused by these organisms is evolving as multiplex PCR assays make diagnosis in research settings more practical and affordable than previously used probe assays or cell culture techniques.

ETEC, EPEC, and EAEC predominantly cause watery diarrhea in children under than 2 years old (34). ETEC is the most common cause of bacterial gastroenteritis, and is alone responsible for 15–30% of diarrhea in developing countries and an estimated 380,000 deaths per year in children under age 5 (49, 50). ETEC causes disease by producing several enterotoxins, including a cholera-like toxin, but invasive and persistent ETEC disease is rare (35). In developed countries ETEC is the main cause of traveler's diarrhea (64). EPEC is the second leading cause of *E. coli* acute watery diarrhea and is responsible 4–8% of cases in developing countries (49). Disease is caused by tight attachment to the intestinal lining, causing loss of microvilli, resulting in malabsorption (35). EPEC is commonly seen in younger children, particularly in children under 2 years of age (84). EAEC is a heterogeneous group of *E. coli* subtypes that express many possible virulence factors including enterotoxins and cytotoxins.

Together, EPEC and EAEC account for the majority of persistent *E. coli* diarrheas (51). EHEC is the only *E. coli* class that is of epidemiological concern in the United States, but it is virtually unseen outside of the United States (50). EHEC is naturally found in the gastrointestinal tract of healthy cattle living in intense feed lots, and most outbreaks are the result of contamination of food during processing (64, 85). Disease with EHEC begins as acute nonbloody diarrhea that progresses to become dysenteric and can lead to hemolytic-uremic syndrome and, in up to 8% of infections, hemorrhagic colitis (38, 64). EIEC is a little-studied pathogen with a clinical course similar to shigellosis. Its pathogenic factors appeared to be originally derived from a virulence plasmid of *Shigella*, and, as a result, is often confused for *Shigella* when diagnosis is based on symptoms alone (35). It is considered of minor relevance compared to other classes of *E. coli*; there are isolated instances where EIEC appear to be more common, although its uniform distribution across all ages in such instances suggest protracted outbreaks rather than endemic transmission (53).

5.5.2.2 Shigella

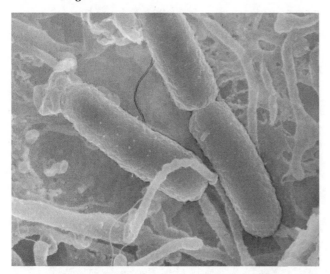

Scanning electron micrograph of a *Shigella* sp. bacteria. (Copyright Institut Pasteur.) (*See insert for color representation of this figure.*)

Shigella is a genus of nonmotile, uncapsulated, Gramnegative bacteria of the family Enterobacteriaceae (54).

The majority of disease is caused by two species: *Shigella sonnei* and *Shigella flexneri* (38). *Shigella dysenteriae* is less common but remains an important pathogen due to its propensity for high mortality outbreaks in developing countries, which effect all ages. *Shigella* bacteria are implicated in 5–15% of diarrheal cases, but diarrheal symptoms are usually mild and dehydration is rare (54). Globally there are approximately 160 million symptomatic *Shigella* infections a year, predominantly occurring as endemic disease among toddlers and older children in developing countries (38). *Shigella* is

best known for causing dysentery, responsible for 30–50% of all cases (54). Mainly due to dysentery and the resulting bacteremia, Shigella infections are associated with 10% of all diarrheal deaths globally (54). *Shigella* have a complex set of mechanisms for invading intestinal cells, multiplying intracellulary, and spreading between cells. This is achieved with the help of several toxins that have enterotoxic, cytotoxic, and neurotoxic effects (54). The combination of toxins and invasion result in a highly inflammatory disease that begins with acute diarrhea and often develops into dysentery and, occasionally, seizures (54). Bacterial shedding lasts for a few weeks, but chronic carriage is rare (37, 64). Malnutrition increases the persistence and severity of dysentery (34). Epidemic disease is common in settings of extreme poverty, natural disasters, and wars, and it affects all ages (34). Shigella disease is less common in developed countries, and occurs predominantly in travelers who have recently visited developing countries and during foodborne outbreaks (64). In the United States alone *Shigella* species are responsible for over 400,000 cases per year (57). Antibiotic treatment is indicated in dysenteric cases, but multidrug resistance in *Shigella* is a global problem, so knowledge of antibiotic susceptibility is key to treatment (34). Another chapter in this book provides a detailed discussion of *Shigella*.

5.5.2.3 Salmonella

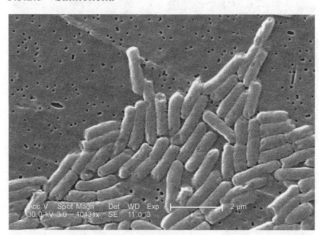

Scanning electron micrograph of a *Salmonella* sp. colony. (Content provider: CDC/Janice Haney Carr. *Source*: CDC, Public Health Image Library, Image ID # 10896.) (*See insert for color representation of this figure.*)

Salmonella is a large genus of bacteria, all of which are pathogenic to humans (38). The most important disease associated with Salmonella is typhoid fever, which is caused by *Salmonella enterica enterica*, serovar Typhi and rarely manifests with diarrheal symptoms (86). However, non-typhi *Salmonella* strains are common residents of the gastrointestinal tracts of many wild and domesticated animals and cause gastroenteritis in humans similar to other bacterial pathogens.

In developed countries, nontyphi *Salmonella* are a common cause of gastroenteritis outbreaks resulting in 1.2 million cases of the disease each year, mostly from contaminated poultry and unpasteurized dairy products (35, 64). Nontyphi *Salmonella* are less well characterized in developing countries but are associated with endemic diarrhea in children, sepsis in neonates, and sporadic outbreaks in all ages (34, 64). Dysentery is uncommon. Prolonged asymptomatic carriage in children is common. In developing countries up to 45% infected children under age 5 can transmit *Salmonella* for up to 12 weeks after primary infection (62, 64). Treatment of complicated *Salmonella* infections has been hindered in recent years due to a large increase in antimicrobial resistance caused by the practice of feeding antibiotics to poultry (52).

5.5.2.4 Campylobacter

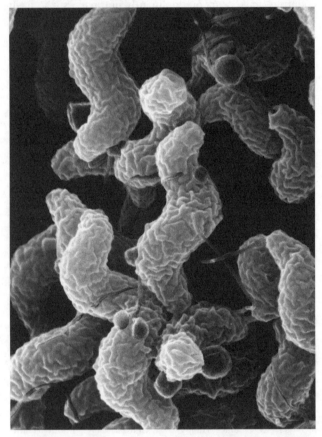

Scanning electron micrograph of *Campylobacter jejuni* bacteria. (*Source*: Agricultural Research Service (ARS) is the U.S. Department of Agriculture's Chief Scientific Research Agency.) (*See insert for color representation of this figure.*)

Campylobacter are a motile genus of Gram-negative bacteria (64). Two species of *Campylobacter*, *jejuni* and *coli*, are most commonly implicated in human disease, although *C. jejuni* alone is responsible for 90% of *Campylobacter* disease (24). *Campylobacter* are naturally found in the gastrointestinal track of animals, with poultry being the main source of bacteria for human infections (64). Gastrointestinal symptoms are generally self-limiting, although the bacteria can occasionally cause dysentery (38). Most cases self-resolve in about 1 week and only require supportive care, although 20% persist beyond a week or relapse later (64). Because of its mild nature, *Campylobacter* disease is often unreported: In Europe the incidence of campylobacteriosis is thought to be 8–30 times the reported numbers (24). An estimated 1.4–2.4 million cases per year occur in the United States, mostly associated with food contamination and undercooked poultry (35). In developing countries *Campylobacter* is responsible for 2–14% of all diarrhea disease (49), with rates peaking in younger children in rural areas compared to urban areas as a result of higher exposure to the pathogen in drinking water and living in close proximity to poultry (24, 38). Of note, *Campylobacter* infections can cause autoimmune sequelae in patients, particularly Guillain–Barré syndrome, which occurs in 1 in 1000 Campylobacter patients after resolution of infection (24). Other chapters in this book provide further discussions of Campylobacter disease.

5.5.2.5 Vibrio cholerae

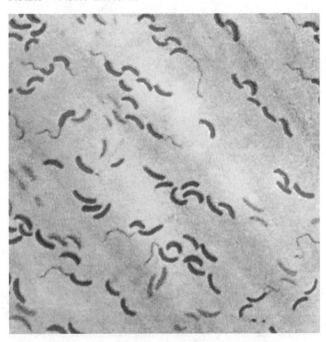

Gram-stained light micrograph of *V. cholerae*. (Content provider: CDC. *Source*: CDC, Public Health Image Library, Image ID # 5324.) (*See insert for color representation of this figure.*)

Cholera is caused by ingesting *V. cholerae*, Gram-negative bacteria that naturally reside in brackish rivers and coastal waters. All epidemic cholera disease is caused by the O1 serogroup of *V. cholerae* (55). In the past multiple serotypes of O1 have caused pandemic disease, but the current global pandemic is caused by the El Tor serotype, which spread out of Southeast Asia in the early 1960s. However, in 1993 a new serotype, O139, emerged in Bangladesh and has spread throughout Asia. Cholera is characterized by voluminous diarrhea, which can lead to rapid dehydration, hypokalemia, acidosis, and shock within as little as 4–12 h. It is believed to cause as many as 120,000 deaths each year (55). Cholera is most famous for its role in epidemic disease, which, throughout history, has occurred in settings of extreme poverty, natural disasters, and wars, where case fatality may exceed 20% (34, 55). However, cholera is minimally inflammatory, is never invasive, and is rarely fatal with appropriate rehydration therapy. Unfortunately, care is often limited in endemic areas, and case fatality rates exceeding 20% are not uncommon. Currently, South Asia and Africa are the only areas endemic for cholera; outside of these regions the disease is extremely rare (34).

5.5.3 Parasitic Pathogens

Compared to viruses and bacteria, parasites play a relatively minor role in the global burden of acute diarrhea, being responsible for only 1–8% of cases, but they are an important cause of persistent diarrhea (64). Numerous types of helminths and protozoa have been identified as causes of diarrhea, but *G. lamblia*, *Cryptosporidium*, and *E. histolytica* are the most common. All are transmitted through the fecal–oral route. Etiology is confirmed through visual detection of eggs or cysts in the stool; however, sensitivity is low even with multiple stool samples (31). Improved tests including PCR and enzyme-linked immunosorbent assay (ELISA) are available, but are not commonly used in clinical situations in impoverished areas.

5.5.3.1 *Cryptosporidium*

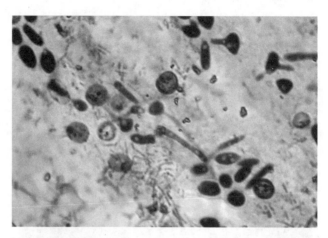

Light micrograph of *Cryptosporidium* sp. oocysts in direct fecal smear, acid-fast stained red. (Content provider: CDC/*J Infect Dis* 1983 May;147(5):824–828. *Source*: CDC, Public Health Image Library, Image ID # 5242.) (*See insert for color representation of this figure.*)

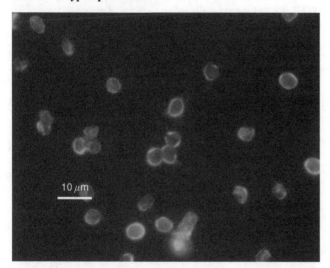

Immunofluorescence light micrograph of purified *Cryptosporidium parvum* oocysts. (Photo credit: H.D.A Lindquist, U.S. EPA. *Source*: http://www.epa.gov/nerlcwww/cpt_seq1.html.) (*See insert for color representation of this figure.*)

The genus *Cryptosporidium* is composed of numerous species of protozoa that find hosts in more than 40 mammalian species, although 97% of *Cryptosporidium* infections in humans are the result of one species: *C. parvum* (15, 87). Zoonotic reservoirs, particularly cattle, are a major source of infection in developed countries, but human feces are a more important vector of transmission in developing countries. In developed countries, drinking water contaminated with *Cryptosporidium* spores have been the cause of several major outbreaks, one, which occurred in Milwaukee, resulted in over 400,000 cases of cryptosporidiosis (15). Asymptomatic carriage is common in developing countries, with *Cryptosporidium* present in 4–5% of healthy children in Southeast Asia and Latin America (29). Most symptomatic cases are self-resolving and clinically similar to other enteric pathogens, with the exception of an increase in mild respiratory symptoms: cough occurs in 20–33% of children (87). *Cryptosporidium* may cause up to 16% of persistent diarrhea in children (29). Bloody stools are common, with 14% of acute cases and 31% of persistent cases also showing signs of blood in their stool (15). HIV-positive patients are at an increased risk of severe and persistent disease. A high prevalence of anticryptosporidium antibodies in the United States, 25–35% of children and adults, suggests exposure may be more common than previously believed, but low carriage of the protozoan in the general population suggests most *Cryptosporidium* is eliminated from the gastrointestinal tract without apparent disease (87).

5.5.3.2 *Giardia lamblia*

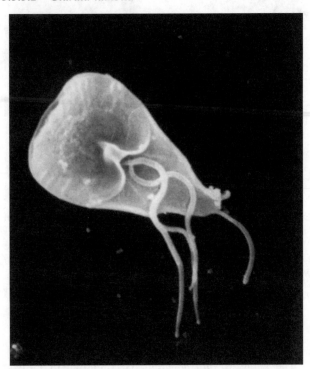

Scanning electron micrograph of a *G. lamblia* trophozoite. (Content provider: CDC/Janice Carr. *Source*: CDC, Public Health Image Library, Image ID # 8698.) (*See insert for color representation of this figure.*)

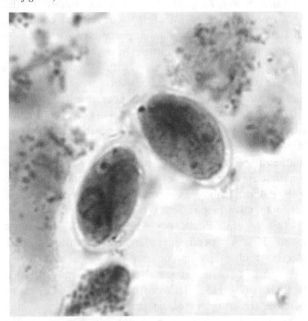

Light micrograph of *G. lamblia* stained with trichrome. (Content provider: CDC/Division of Parasitic Diseases. *Source*: http://www.dpd.cdc.gov/dpdx/HTML/ImageLibrary/Giardiasis_il.htm.) (*See insert for color representation of this figure.*)

G. lamblia is a noninvasive flagellated protozoa whose transmission occurs through ingestion of giardial cysts (27).

Humans are thought to be the principal reservoirs, although *Giardia* find hosts in other animals, and zoonosis has been implicated in some outbreaks (18, 64). The prevalence of *Giardia* in the intestinal tracts of children is low in developed countries, but can reach 20–30% in children in developing countries (38). However, most persons infected with *Giardia* do not show symptoms of disease (27). In symptomatic cases, a 1–2 week incubation period precedes symptoms that include acute diarrhea characterized by foamy, foul smelling stools, flatulence, abdominal distention, and anorexia (31, 64, 88). Most cases resolve spontaneously, although a subset becomes persistent. *Giardia* are the second most common cause of persistent parasitic infections, although they account for less than 10% of all persistent diarrhea cases in developing countries (27). Persistent diarrhea and anorexia put children at an increased risk of malnutrition (38).

5.5.3.3 *Entamoeba histolytica*

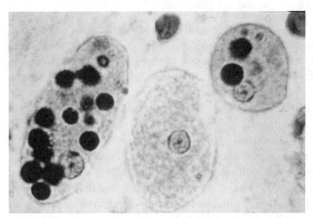

Light micrograph of *E. histolytica* trophozoites with ingested erythrocytes stained with trichrome. (Content provider: CDC/Division of Parasitic Diseases. *Source*: http://www.dpd.cdc.gov/dpdx/HTML/ImageLibrary/A-F/Amebiasis/body_Amebiasis_il6.htm.) (*See insert for color representation of this figure.*)

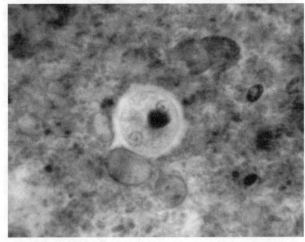

Light micrograph of *E. histolytica* cyst stained with chlorazol black. (Content provider: CDC/Dr. George Healy. *Source*: CDC, Public Health Image Library, Image ID # 1474.) (*See insert for color representation of this figure.*)

E. histolytica is a species of amoeba that infects the human gut, subsisting on intestinal bacteria and food particles. Transmission occurs through ingestion of cysts from fecal contamination of raw or undercooked food. Greater than 90% of infections are asymptomatic, however E. histolytica has the ability to invade the intestinal lining, causing highly inflammatory disease (89). Globally, E. histolytica causes an estimated 50 million illness episodes and 40,000 deaths a year (12, 64). Invasion may occur days or years after infection, manifesting as amoebic colitis, causing bloody diarrhea and abdominal pain (89). Rarely, the disease will progress, resulting in necrotizing colitis and disseminated disease, which can lead to amoebic liver abscesses.

5.6 COMPLICATIONS AND MANAGEMENT

Although infectious diarrhea is one of the most common causes of deaths globally, most cases can be easily treated with simple, inexpensive rehydration, appropriate feeding, and nutritional supplements. Key treatment strategies are summarized in Table 5.4.

5.6.1 Dehydration

Approximately 90% of deaths caused by infectious diarrhea are the result of the loss of bodily fluids (90). Mild and moderate dehydration in children is characterized by thirst, restlessness, irritability, decreased skin turgor, sunken eyes,

and, in infants, sunken fontanel (28). Signs of severe dehydration include altered consciousness, lack of urine output, cool moist extremities, quickened, weak pulse, low blood pressure, and peripheral cyanosis (107). As dehydration is the primary cause of mortality from diarrhea, the main strategy for treatment is the replacement of fluids along with appropriate feeding.

5.6.1.1 Intravenous Fluid Therapy Thomas Latta was first to document the use of intravenous fluids in 1830 to manage a young woman with severe dehydration secondary to cholera. Latta subsequently treated several other patients with severe diarrhea; however, most of his patients died either because of a lack of replacement of ongoing fluid loss or from nosocomial infections (108). As a result, IV fluids were not commonly used for the next nearly 100 years and diarrheal mortality of patients with severe dehydration continued to be close to 100% (109). In the early 1900s, an IV solution containing sodium and bicarbonate was developed which resulted in a dramatic reduction in mortality among hospitalized individuals with diarrhea. However, the case fatality rate from diarrhea even in the best hospitals was still approximately 20%. In the mid 1940s, Daniel Darrow described the importance of potassium loss with diarrhea, and the addition of potassium into IV fluids resulted in the virtual elimination of diarrheal deaths among hospitalized patients with diarrhea in the United States (110).

However, with the advent of oral rehydration solution (ORS), IV fluids are now recommended only in cases of

TABLE 5.4   Treatment and Prevention Strategies for Acute Diarrhea Aimed at Developing Countries

Treatment	
Original ORS	Effective in treating 90% of diarrhea cases (90)
Reduced osmolarity ORS	Same efficacy original ORS (91)
	20% lower stool output (91)
	30% reduction in vomiting (91)
	40% reduction in the need for unscheduled IV treatment (91)
Zinc	12 h mean reduction in duration of acute diarrhea episodes (91)
	29% lower relative risk of diarrhea continuing to day 7 (92)
	51% decrease in overall mortality (93)
Prevention	
Zinc	34% decreased prevalence of diarrhea in the 2–3 months following treatment (94)
Vitamin A	32% reduction in diarrhea associated mortality (95)
Breastfeeding	Two- to threefold reduction in diarrhea incidence (96–99)
Handwashing promotion	31% decrease in risk of diarrheal illness (100)
Installation of latrines	30% reduction in diarrheal illness (101)
Water treatment	11% reduction in diarrheal illness when treating the water source (101)
	39% reduction when treating at the household level (101)
Rotavirus vaccine	98–100% efficacy against severe rotavirus gastroenteritis (102, 103)
	74–85% protection against all rotavirus gastroenteritis (102, 103)
	45% reduction in hospitalizations due to acute gastroenteritis in across all ages (60)
Measles	18% lower incidence of diarrhea (104)
Cholera vaccine	85–86% effective against cholera (105, 106)

TABLE 5.5 Summary of Treatment Based on Degree of Dehydration

Degree of Dehydration	Rehydration Therapy	Replacement of Losses	Nutrition
Minimal or no dehydration	Not applicable	<10 kg body weight: 60–120 mL ORS for each diarrheal stool or vomiting episode; >10 kg body weight: 120–140 mL ORS for each diarrheal stool or vomiting episode	Continue breastfeeding or resume age-appropriate normal diet after initial hydration, including adequate caloric intake for maintenance
Mild-to-moderate dehydration	ORS, 50–100 mL/kg body weight over 3–4 h	Same; monitor stool output and alter maintenance appropriately	Same
Severe dehydration	Lactated Ringer's or 0.9% saline in 20 mL/kg body weight intravenous boluses until perfusion and mental status improve; then administer 100 mL/kg body weight ORS over 4 h or 5% dextrose 0.45% saline intravenously at twice the stool output rate	Same; monitor stool output and alter maintenance appropriately. If unable to drink, administer through nasogastric tube or administer 5% dextrose, 0.45% saline with 20 mequiv/L potassium chloride intravenously	Same

Adapted from Ref. 111.

severe dehydration, in which a patient may be obtunded and unable to drink. IV or intraosseous fluids should be used in the initial management if appropriate equipment is available. ORS should then be instituted as soon as the patient is able to drink fluids. In addition to delivering the appropriate quantity of ORS to replace the calculated fluid deficit, ongoing fluid losses should be replaced. Table 5.5 summarizes the recommended quantities of fluid for different levels of dehydration.

5.6.1.2 Oral Rehydration Solution Harold Harrison first demonstrated the successful use of an oral rehydration solution containing 62 mmol/L of sodium and 3.3% glucose for the management of diarrhea in Baltimore during the mid-1940s (112). A commercial company marketed a modified solution based on Harrison's formulation throughout the United States, which, as a result of both inappropriate reconstitution and the hyperosmolarity of the solution, caused many cases of hypernatremia throughout the country (113). Thus, ORS was not accepted by health care providers as standard therapy for diarrhea for decades.

In the mid-1960s ORS received renewed attention as physiologic studies demonstrated that glucose was essential to facilitate absorption of water from the gut, and this led to the development of an effective oral rehydration solution. Subsequently, in the summer of 1971, ORS received major attention for successfully saving the lives of thousands of cholera patients during the Bangladeshi refugee crisis. ORS was shown to be safe and effective even when given by untrained family members; the case fatality rate of diarrhea in the ORS intervention areas was 3%, compared with 20–30% in areas that used only intravenous fluids (114). In 1978, the

World Health Organization began recommending a standard formulation of ORS containing 111 mmol/L of glucose, 90 mmol/L of sodium, 80 mmol/L of chloride, 20 mmol/L of potassium, and 20 mmol/L of citrate as the principal strategy for diarrhea management. Multiple studies demonstrated that ORS could effectively manage over 90% of diarrhea episodes, and ORS was adopted and used by numerous national governments and international relief agencies (90, 112, 115–121). The extensive use of this solution worldwide is credited with saving millions of lives each year.

The sodium concentration of the original WHO-ORS elicited concerns that hypernatremia could occur in cases of noncholera diarrhea, where salt loss is lower. This led to the development of ORS formulations with lower osmolarity, which contained 75–90 mmol/L glucose and 60–75 mequiv/L sodium (122). In the early 1990s, laboratory studies showed that hypotonic-reduced osmolarity (245 mmol/L) ORS resulted in increased intestinal water absorption when compared with standard WHO-ORS. Several efficacy studies comparing low osmolarity and standard ORS demonstrated further reduction in stool output and a reduction in diarrhea duration (123–126). A meta-analysis of trials of low osmolarity ORS for acute diarrhea in children found that there was a 20% reduction in stool output, a 30% reduction in vomiting, and close to a 40% reduction in the need for unscheduled IV treatment in patients receiving the low osmolarity ORS compared to those receiving original ORS (91). Although data are more limited, a similar analysis of studies conducted in people with cholera concluded that there was no difference in the safety and efficacy of low osmolarity ORS and original ORS in those patients (127). The new low osmolarity ORS has been shown to be more acceptable among children (128).

In May 2002, based on the recommendations of an expert panel, the WHO changed its recommendation to the use of low osmolarity ORS containing 75 mmol/L glucose, 75 mmol/L sodium, 65 mmol/L chloride, 20 mmol/L potassium, and 10 mmol/L citrate for the treatment of children and adults with acute diarrhea. Although ORS can be prepared from home ingredients, errors in mixing can occur that may result in serious electrolyte imbalances. Therefore, commercially available ORS packets should be recommended whenever possible.

5.6.2 Malnutrition

Malnutrition and diarrhea have long been known to form a destructive cycle: Malnutrition increases the susceptibility of children to pathogens that cause diarrhea and other disease, and diarrhea decreases the gut's ability to absorb nutrients, leading to malnutrition. The malnutrition/diarrhea cycle is so strong that 61% of children who die of diarrhea have malnutrition as an underlying risk factor (2). Diarrhea in malnourished children is on average of longer duration. A case control study in India found that malnourished children have 3.25 increased odds of developing persistent diarrhea compared to nonmalnourished controls (29). Additionally, the case fatality rate of diarrhea increases progressively as malnutrition worsens, with a 24-fold increase in mortality in severely malnourished children compared to normally nourished children (30). Beyond increased rates of mortality, the malnutrition caused by diarrhea leads to significant physical and mental growth shortfalls, which negatively affect children throughout their lives (129). The number of diarrheal episodes in infancy has been shown in a Brazilian cohort study to negatively impact cognitive test scores 4–7 years later after controlling for a variety of socioeconomic indicators (130). A similar study by the same group found a strong association with increased number of diarrhea episodes in the first two years of life and decreased school performance and height and weight for age 6–9 years on (131). As a result, feeding and other nutritional interventions are critical components of treatment for amelioration of the short- and long-term consequences of diarrhea.

5.6.2.1 *Continued Feeding*
Until the 1970s the medical community endorsed the belief that feeding should be withheld during diarrhea in order to "rest the gut." In the 1980s, a series of studies demonstrated that continued feeding during a diarrheal episode is safe and improves outcomes (132–137). Based on these studies, the WHO incorporated early refeeding into their recommendations for diarrhea treatment (138): Breastfed infants should continue breastfeeding, and formula-fed infants should continue their usual formula. Commercial carbonated beverages and undiluted fruit juices should be avoided because the high carbohydrate content results in increased osmolarity of intestinal contents, which can aggravate diarrhea.

5.6.2.2 *Zinc Supplementation*
Zinc is a micronutrient that plays key roles in effective immune response and healing. In developing countries, zinc supplementation as a treatment for diarrhea, when given in addition to ORS, has been shown to be highly effective at decreasing the duration and severity of diarrheal episodes (92, 139, 140). Studies have shown that zinc decreased the duration of diarrhea by 12 h (95% CI: −23, −2), decreased use of unnecessary antibiotics, and reduced medical visits for acute diarrhea (92, 141–143). Furthermore, a 10–14 day course of zinc supplementation decreased the prevalence of diarrhea in the 2–3 months following treatment (OR 0.66; 95% CI: 0.52–0.83) (94). Introduction of zinc along with ORS led to a decrease in hospitalizations for diarrhea as well as other illnesses in two large-scale studies (93, 141). In Bangladesh, Baqui et al. reported a 24% reduction in the diarrhea hospitalization rate (95% CI: 0.59–0.96) and resulted in an overall decrease in mortality (risk ratio 0.49; 95% CI: 0.25, 0.94) (93). In India, Bhandari et al. showed a reduction in all-cause diarrhea (OR 0.56; 95% CI: 0.41, 0.75) and all-cause hospitalizations (OR 0.41; 95% CI: 0.29–0.57) (141). Based on proven safety and efficacy of zinc supplementation as an adjunct in diarrhea treatment, WHO recommends 10–14 days of oral zinc at 20 mg/day for children >6 months of age, and 10 mg/day for children <6 months of age for acute diarrhea episodes (138).

5.6.3 Antimicrobial Therapy

Antimicrobial therapy should be used only in specific cases of infectious diarrhea with complications such as dysentery, severe persistent diarrhea, or in patients at increased risk of developing severe complications, such as immunocompromised patients, although health and medical organizations in developed countries often recommend treatment for traveler's diarrhea. Due to widespread prevalence and variation in antimicrobial resistance, it is critical that therapeutic decisions be based on regional antibiotic susceptibility patterns of likely pathogens or specific pathogens if etiology is confirmed. Guidebooks such as the Academy of Pediatrics' Red Book® are available in many regions to provide regularly updated guidelines for treating specific pathogens. Table 5.6 summarizes the therapies indicated for otherwise healthy individuals.

5.6.4 Nonantimicrobial Drug Therapy

Several over-the-counter and prescription antimotility (e.g., loperamide), antisecretory (e.g., racecadotril), and toxin-binding agents (e.g., cholestyramine) are available for symptomatic relief of diarrhea (146–148). Because of the limited

TABLE 5.6 Antimicrobial Recommendations for Infectious Diarrhea Caused by Specific Pathogens

Organism	Antimicrobial Regimen
Campylobacter[a]	Erythromycin (5–7 days); or ciprofloxacin (5–7 days)
C. parvum[a]	Nitazoxanide (3 days); or paromomycin once *plus* azithromycin (28 days)
E. histolytica[b]	Metronidazole (7–10 days); or tinidazole (3 days) followed by Iodoquinol (20 days); or paromomycin (7 days)
Enteropathogenic *E. coli*[b]	Ciprofloxacin (adults, 5 days); or trimethoprim/sulfamethoxazole (5 days)
Enterotoxigenic *E. coli*[b]	Ciprofloxacin (adults, 5 days); or trimethoprim/sulfamethoxazole (5 days)
G. lamblia[b]	Metronidazole (5–7 days); or tinidazole (adults, single dose); or furazolidone (children, 7–10 days)
Shigella[a]	Azithromycin (3 days); or ceftriaxone (5 days); or ciprofloxacin (adults, 3 days); or levofloxacin (adults, 3 days)
V. cholerae[b]	Doxycycline (adults, 3 days); or trimethoprim/sulfamethoxazole (3 days); or erythromycin (3 days)

[a] Adapted from Ref. 144.
[b] Adapted from Ref. 145.

evidence and uncertain side effect profiles, most experts do not recommend the use of these agents, particularly for pediatric patients (149, 150). In fact, many experts believe that the use of antimotility agents could be harmful, especially in children. For example, antimotility agents have been implicated in prolonging fever in *Shigellosis* and in gastrointestinal complications or death when given initially prior to antibiotic treatment for *C. difficile* colitis (151, 152). In general, the use of these agents should be discouraged due to lack of efficacy in most cases, except loperamide, which may provide symptom relief in chronic diarrhea, or in addition to antibiotic therapy in traveler's diarrhea.

5.7 PREVENTION

A large proportion of morbidity and mortality due to infectious diarrhea can be prevented using well-characterized intervention, including vaccines, nutritional supplementation, breastfeeding, sanitation, and hygiene. Key treatment prevention strategies are summarized in Table 5.4.

5.7.1 Vaccines

5.7.1.1 Rotavirus Vaccines Rotateq™ (manufactured by Merck) is an oral human-bovine pentavalent reassortant vaccine, which was shown in one large clinical trial to prevent 74% of cases of rotavirus gastroenteritis and 98% of severe rotavirus gastroenteritis in a child's first rotavirus season (102). Rotarix™ (GlaxoSmithKline) is an oral live attenuated monovalent vaccine that was shown to prevent 85% of cases of rotavirus hospitalizations and 100% of severe rotavirus gastroenteritis (103). Preliminary evaluations of both vaccines indicate that vaccine immunogenicity/efficacy does not differ by breastfeeding status (153–155), and there is no reduction in immune response to either vaccine with concomitant administration of oral polio

vaccine (156, 157). Clinical trials and postmarketing surveillance for both vaccines have not found any evidence of vaccine-associated intussusceptions or other significant safety concerns (158, 159). A Phase III trial of Rotarix™ was recently completed in South Africa and Malawi, which showed a vaccine efficacy against severe gastroenteritis of 76.9% (95% CI: 56.0, 88.5) in South Africa and 49.5% (95% CI:19.2, 68.3) in Malawi (160). WHO's strategic advisory experts now recommend the inclusion of rotavirus vaccination into all national immunization programs (160).

Recently, the widespread use of rotavirus vaccine has altered the epidemiology of rotavirus infections in several countries. The United States has seen a 42–74% reduction, depending on the region, in the number of positive rotavirus tests at sentinel hospitals compared to the time before rotavirus vaccine was introduced (161). More important, rotavirus has been shown to be efficacious in developing countries. Since introduction in Nicaragua there has been a 23% reduction in acute gastroenteritis in vaccinated children compared to nonvaccinated children aged 0–11 months (162), with similar results seen in Brazil after introduction (60). A study in Mexico showed a 35% reduction (95% CI: 29, 39) in diarrhea-related mortality rate among children <5 years of age following rotavirus vaccine introduction (163). The long-term effect on serotype prevalence and replacement disease has yet to be conclusively determined (161, 164).

5.7.1.2 Cholera Vaccines There are currently two available licensed oral cholera vaccines. The first contains killed whole-cell *V. cholerae* O1 and a recombinant cholera toxin β-subunit (Dukoral®, Crucell), given in two doses to persons 6 years and over or in three doses to children 2–6 years of age. Studies in Peru and Bangladesh showed vaccine efficacies of 86% (95% CI: 37, 97) and 85% (95% CI: 62, 94), respectively, in the first 6 months (105, 106). The second is a bivalent oral vaccine containing killed whole-cell *V. cholerae* O1 and 0139. This vaccine has been shown to be safe and immunogenic in trials in Vietnam and India (165, 166).

5.7.1.3 Measles Vaccine No randomized control trials have been carried out to determine the effect of measles vaccination on diarrhea, but an observational review of Demographic and Health Surveys found that the incidence of diarrhea was 18% lower in measles-vaccinated children than in their unvaccinated counterparts (104). A case control study in Bangladesh found measles vaccination to be very protective, reducing deaths due to diarrhea by 57% in measles vaccine recipients aged 10–60 months (167).

In recent years, measles vaccination promotion programs have successfully increased global measles vaccination dramatically: from 20% in 1980 to 80% in 2007, resulting in a decrease of measles deaths from 7 to 8 million per year prior to vaccination to fewer than 200,000 in 2007 (168, 169). However, most deaths now occur in a few countries in Southeast Asia and Africa, where immunization rates remain low; the same areas that also have the highest number of diarrhea deaths (170). In these countries, measles will remain a significant factor in diarrheal disease until vaccination coverage improves.

5.7.2 Nutritional Interventions

5.7.2.1 Zinc Supplementation Prophylactic zinc supplementation has been shown to reduce the relative risk of diarrhea by 27% in studies that enrolled children with a mean age of at least 12 months (RR 0.73; 95% CI: 0.61, 0.87) (171). The additional benefits of zinc on reduction in pneumonia incidence and improvement in child growth add to the body of evidence supporting the importance of dietary zinc for overall child health (171). Additional operational research is needed to develop successful delivery strategies for routine zinc supplementation as zinc needs to be delivered daily, or at least several times weekly to be effective.

5.7.2.2 Vitamin A Vitamin A supplementation has been shown to result in a 32% (95% CI: 20, 43%) reduction in diarrhea-specific mortality and a 23% (RR 0.77; 95% CI: 0.68, 0.88) pooled overall reduction in the mortality rate of children 6–59 months of age (95). A meta-analysis showed no effect of vitamin A on diarrhea incidence (RR 1.00; 95% CI: 0.94, 1.07) (172). The effect of vitamin A on diarrhea duration or severity is less clear (173–176); there appears to be an effect on diarrhea severity, explaining the effect on mortality in spite of the lack of effect on diarrhea incidence (177). Because of the recognized overall benefits of vitamin A, the WHO recommends vitamin A supplementation every 4–6 months in children <5 years of age who live in areas where vitamin A intake is inadequate.

5.7.2.3 Breastfeeding Breastfeeding is considered one of the most basic and effective interventions in the prevention of infectious diseases in children. One meta-analysis showed that in children <6 months of age, any breastfeeding decreased diarrheal disease by over sixfold (OR 6.1; 95% CI: 4.1, 9.0) (178). Lack of breastfeeding in the first 2 months of life has also been shown to increase the risk of diarrhea mortality by over 20-fold (179). Based on the overall health and nutrition benefits, WHO recommends exclusive breastfeeding up to 6 months of age, and breastfeeding along with complementary feeds through 2 years of age or beyond.

5.7.3 Water, Sanitation, and Hygiene

The World Health Organization estimates that 88% of diarrheal diseases are attributable to unsafe water, sanitation, and hygiene, including food contamination (180). Improvement in water quality through water treatment was shown to reduce diarrheal illness by 11% when the water source was treated (RR 0.89; 95% CI: 0.42–1.90), and by 39% when treated at the household level (RR 0.61; 95% CI: 0.46, 0.81) (101). Studies have demonstrated benefits of improved personal and home-level hygiene in reducing diarrheal disease episodes. A meta-analysis showed hygiene interventions can reduce diarrheal disease incidence by a pooled relative risk of 0.55 (95% CI: 0.40, 0.75) (101, 181, 182). With the promotion of handwashing, specifically, the risk of diarrheal illness was reduced by 31% (RR 0.69; 95% CI: 0.55, 0.87) (100). However, antimicrobial soap does not impact diarrheal disease more than plain soap. A randomized trial of handwashing promotion of 36 squatter settlements in Karachi comparing regular and antimicrobial soap to controls found an overall 53% lower incidence of diarrhea in settlements that received handwashing interventions, but no difference between settlements using antimicrobial soap or regular soap (183). Although efficacy studies have been impressive, they have shown considerably fewer benefits than those seen in short-term research settings.

5.7.4 Global Control Efforts

In 1978 WHO launched its Program for the Control of Diarrheal Disease, which emphasizes home and clinic diarrheal illness management using oral rehydration therapy and continued feeding, and, to a lesser extent, on improving personal hygiene, breastfeeding, drinking water, sanitation, outbreak control, and measles vaccination (90, 184, 185). Continued feeding/early refeeding was added to WHO's recommendations in the 1980s as the importance of these became clear. The global effort to control diarrheal disease met with remarkable success in the 1980s and 1990s, decreasing the number of diarrheal deaths from 5 million deaths per year in 1976 to 1.6 million in 2004

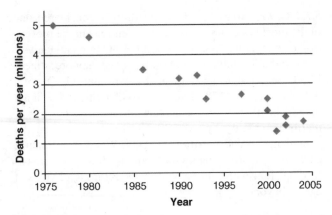

FIGURE 5.3 Mean estimates from studies of global burden of deaths due to diarrhea in children. (Adapted from Refs 1, 2, and 39.)

(Fig. 5.3), accompanied by a modest decrease in diarrheal incidence (Fig. 5.4) (38, 186).

Unfortunately, support for control of diarrhea has waned. In most countries, although caretakers and health care providers have knowledge of ORS, use rates remain poor. In addition, among those that use ORS, it is often prepared inappropriately or given below the recommended quantities (187–189). In India, which has had an active campaign focusing on raising awareness of ORS for decades, a 2007 survey showed that over 73% of Indian women were familiar with ORS. However, only 26% of children with diarrhea were given ORS (190). Similarly, Ellis and colleagues showed that in southern Mali, although parents were generally knowledgeable about ORS, they continued to seek a combination of multiple ineffective treatments, such as antidiarrheals and antibiotics, to treat the illness (191).

The promotion of ORS with zinc has been shown to simultaneously increase ORS use and acceptability among caretakers and physicians. A cluster-randomized study in

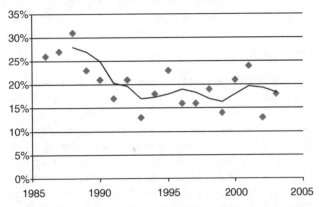

FIGURE 5.4 Proportion of children in Demographic Health Surveys reporting an episode of diarrhea in the previous 2 weeks. Line represents 3-year moving average of proportion of children reporting diarrhea. (Adapted from Ref. 186.)

India comparing ORS to ORS with zinc showed that 6 months following the intervention in the communities that received ORS with zinc, unwarranted medical care seeking was reduced by 34% (95% CI: 0.43, 0.99), oral antibiotic use was decreased by 88% (95% CI: 0.05, 0.32), intravenous fluid use rates decreased by 50% (95% CI: 0.22, 1.14), and ORS prescription rates increased over 12-fold (95% CI: 8.95, 17.35) (141). A six-site multicountry study introducing zinc for diarrhea treatment showed similar decreases in antibiotic use, and in five of six sites, ORS use rates were the same or higher in intervention communities (142). Additionally, use of smaller packets of ORS can enable caregivers to more accurately prepare ORS, thereby limiting wasted dry salts as well as the likelihood of the prepared ORS becoming contaminated through the day. For example, the National Control of Diarrheal Diseases Program in Egypt used 200 mL packets effectively in the 1980s (192).

For many countries diarrhea treatment has not been a priority in the past 10–15 years. Funding for training and special promotional activities is minimal. If the coverage of key diarrhea prevention and treatment strategies is to increase (see Supplementary Figures for coverage of key interventions), national level governments as well as donors will need to recognize the dire need for new resources to reinstitute proven treatment and prevention options to reduce mortality from diarrhea.

5.8 CONCLUSION

Infectious diarrhea caused by viral, bacterial, and parasitic pathogens is one of the leading causes of disease globally, causing billions of disease episodes a year. The dehydration and malnutrition caused by diarrhea results in millions of deaths each year, predominantly in children in developing countries. However, numerous well-characterized treatment and prevention strategies exist, which combined have the potential of eliminating the mortality associated with infectious diarrhea. However, a renewed international commitment to control diarrheal diseases is necessary if the progress made against infectious diarrhea in the last few decades is to be maintained and strengthened.

REFERENCES

1. WHO. *The Global Burden of Disease: 2004 Update*. Geneva: World Health Organization, 2008.

2. Bryce J, et al. WHO estimates of the causes of death in children. *Lancet* 2005;365(9465):1147–1152.

3. de Wit MA, Koopmans MP, van Duynhoven YT. Risk factors for norovirus, sapporo-like virus, and group A rotavirus gastroenteritis. *Emerg Infect Dis* 2003;9(12):1563–1570.

4. Malek, MA, et al. Diarrhea- and rotavirus-associated hospitalizations among children less than 5 years of age: United States, 1997 and 2000. *Pediatrics* 2006;117(6):1887–1892.

5. Black RE, et al. Incidence and etiology of infantile diarrhea and major routes of transmission in Huascar, *Peru. Am J Epidemiol* 1989;129(4):785–799.

6. Guerrant RL, et al. Prospective study of diarrheal illnesses in northeastern Brazil: patterns of disease, nutritional impact, etiologies, and risk factors. *J Infect Dis* 1983;148(6):986–997.

7. Singleton RJ, et al. Diarrhea-associated hospitalizations and outpatient visits among American Indian and Alaska Native children younger than five years of age, 2000–2004. *Pediatr Infect Dis J* 2007;26(11):1006–1013.

8. Salama P, et al. Lessons learned from complex emergencies over past decade. *Lancet* 2004;364(9447):1801–1813.

9. Schwartz BS, et al. Diarrheal epidemics in Dhaka, Bangladesh, during three consecutive floods: 1988, 1998, and 2004. *Am J Trop Med Hyg* 2006;74(6):1067–1073.

10. Cronin AA, et al. Quantifying the burden of disease associated with inadequate provision of water and sanitation in selected sub-Saharan refugee camps. *J Water Health* 2009;7(4):557–568.

11. UNICEF *Diarrhoea: Why Children Are Still Dying and What Can be Done.* Geneva: WHO/UN Children's Fund, 2009.

12. Baldi F, et al. Focus on acute diarrhoeal disease. *World J Gastroenterol* 2009;15(27):3341–3348.

13. Navaneethan U, Giannella RA. Mechanisms of infectious diarrhea. *Nat Clin Pract Gastroenterol Hepatol* 2008;5(11):637–647.

14. Greenberg HB, Estes MK. Rotaviruses: from pathogenesis to vaccination. *Gastroenterology* 2009;136(6):1939–1951.

15. Huang DB, Chappell C, Okhuysen PC. Cryptosporidiosis in children. *Semin Pediatr Infect Dis* 2004;15(4):253–259.

16. Laohachai KN, et al. The role of bacterial and non-bacterial toxins in the induction of changes in membrane transport: implications for diarrhea. *Toxicon* 2003;42(7):687–707.

17. Lapointe TK, O'Connor PM, Buret AG. The role of epithelial malfunction in the pathogenesis of enteropathogenic *E. coli*-induced diarrhea. *Lab Invest* 2009;89(9):964–970.

18. Buret AG. Pathophysiology of enteric infections with *Giardia duodenalius. Parasite* 2008;15(3):261–265.

19. Moon HW. Comparative histopathology of intestinal infections. *Adv Exp Med Biol* 1997;412:1–19.

20. Flores J, Okhuysen PC. Enteroaggregative *Escherichia coli* infection. *Curr Opin Gastroenterol* 2009;25(1):8–11.

21. Hookman P, Barkin JS. *Clostridium difficile* associated infection, diarrhea and colitis. *World J Gastroenterol* 2009;15(13):554–580.

22. Guttman JA, Finlay BB. Tight junctions as targets of infectious agents. *Biochim Biophys Acta* 2009;1788(4):832–841.

23. Schroeder GN, Hilbi H. Molecular pathogenesis of *Shigella* spp.: controlling host cell signaling, invasion, and death by type III secretion. *Clin Microbiol Rev* 2008;21(1):134–156.

24. Janssen R, et al. Host–pathogen interactions in *Campylobacter* infections: the host perspective. *Clin Microbiol Rev* 2008;21(3):505–518.

25. Haque R, et al. Amebiasis. *N Engl J Med* 2003;348(16):1565–1573.

26. Levine MM. Enteric infections and the vaccines to counter them: future directions. *Vaccine* 2006;24(18):3865–3873.

27. Abba K, et al. Pathogens associated with persistent diarrhoea in children in low and middle income countries: systematic review. *BMC Infect Dis* 2009;9:88.

28. WHO. *The Treatment of Diarrhoea. A Manual for Physicians and Other Senior Health Workers,* 1995 (WHO/CDR/95.3).

29. Lima AA, Guerrant RL. Persistent diarrhea in children: epidemiology, risk factors, pathophysiology, nutritional impact, and management. *Epidemiol Rev* 1992;14:222–242.

30. Bhandari N, Bhan MK, Sazawal S. Mortality associated with acute watery diarrhea, dysentery and persistent diarrhea in rural North India. *Acta Paediatr* 1992;81 (Suppl 381): 3–6.

31. Donowitz M, Kokke FT, Saidi R. Evaluation of patients with chronic diarrhea. *N Engl J Med* 1995;332(11):725–729.

32. Holtz LR, Neill MA, Tarr PI. Acute bloody diarrhea: a medical emergency for patients of all ages. *Gastroenterology* 2009;136(6):1887–1898.

33. Jones G, et al. How many child deaths can we prevent this year? *Lancet* 2003;362(9377):65–71.

34. O'Ryan M, Prado V, Pickering LK. A millennium update on pediatric diarrheal illness in the developing world. *Semin Pediatr Infect Dis* 2005;16(2):125–136.

35. DuPont HL. Clinical practice. Bacterial diarrhea. *N Engl J Med* 2009;361(16):1560–1569.

36. Uga S, et al. Parasites detected from diarrheal stool samples collected in Nepal. Southeast Asian *J Trop Med Public Health* 2004;35(1):19–23.

37. Huilan S, et al. Etiology of acute diarrhoea among children in developing countries: a multicentre study in five countries. *Bull World Health Organ* 1991;69(5):549–555.

38. Podewils LJ, et al. Acute, infectious diarrhea among children in developing countries. *Semin Pediatr Infect Dis* 2004;15(3):155–168.

39. Parashar UD, et al. Rotavirus and severe childhood diarrhea. *Emerg Infect Dis* 2006;12(2):304–306.

40. Glass RI, et al. The epidemiology of rotavirus diarrhea in the United States: surveillance and estimates of disease burden. *J Infect Dis* 1996;174 (Suppl 1): S5–S11.

41. Parashar UD, et al. Global mortality associated with rotavirus disease among children in 2004. *J Infect Dis* 2009;200 (Suppl 1):S9–S15.

42. Goodgame RW. Viral causes of diarrhea. *Gastroenterol Clin North Am* 2001;30(3):779–795.

43. Clark B, McKendrick M. A review of viral gastroenteritis. *Curr Opin Infect Dis* 2004;17(5):461–469.

44. Grais RF, et al. Unacceptably high mortality related to measles epidemics in Niger, Nigeria, and Chad. *PLoS Med* 2007;4(1):e16.

45. Sniadack DH, et al. Measles epidemiology and outbreak response immunization in a rural community in Peru. *Bull World Health Organ* 1999;77(7):545–552.

46. Thakur JS, et al. Measles outbreak in a Periurban area of Chandigarh: need for improving vaccine coverage and strengthening surveillance. *Indian J Pediatr* 2002;69(1):33–37.

47. Sharma MK, Bhatia V, Swami HM. Outbreak of measles amongst vaccinated children in a slum of Chandigarh. *Indian J Med Sci* 2004;58(2):47–53.

48. Feachem RG, Koblinsky MA. Interventions for the control of diarrhoeal diseases among young children: measles immunization. *Bull World Health Organ* 1983;61(4):641–652.

49. Guerrant, R.L., et al. Diarrhea in developed and developing countries: magnitude, special settings, and etiologies. *Rev Infect Dis* 1990;12 (Suppl 1):S41–S50.

50. Nataro, J.P. and J.B. Kaper, Diarrheagenic *Escherichia coli*. *Clin Microbiol Rev* 1998;11(1):142–201.

51. Pawlowski, S.W., C.A. Warren, and R. Guerrant, Diagnosis and treatment of acute or persistent diarrhea. *Gastroenterology* 2009;136(6):1874–1886.

52. DuPont, H.L., The growing threat of foodborne bacterial enteropathogens of animal origin. *Clin Infect Dis* 2007;45(10):1353–1361.

53. Vieira, N., et al. High prevalence of enteroinvasive *Escherichia coli* isolated in a remote region of northern coastal Ecuador. *Am J Trop Med Hyg* 2007;76(3):528–533.

54. Niyogi, S.K., Shigellosis. *J Microbiol* 2005;43(2):133–143.

55. Cholera vaccines. *Wkly Epidemiol Rec* 2001;76(16):117–124.

56. Wilhelmi I, Roman E, Sanchez-Fauquier A. Viruses causing gastroenteritis. *Clin Microbiol Infect* 2003;9(4):247–262.

57. Mead PS, et al. Food-related illness and death in the United States. *Emerg Infect Dis* 1999;5(5):607–625.

58. Glass RI, et al. Gastroenteritis viruses: an overview. *Novartis Found Symp* 2001;238:5–19; discussion 19–25.

59. Walter JE, Mitchell DK. Astrovirus infection in children. *Curr Opin Infect Dis* 2003;16(3):247–253.

60. Gurgel RG, et al. Incidence of rotavirus and all-cause diarrhea in northeast Brazil following the introduction of a national vaccination program. *Gastroenterology* 2009;137(6):1970–1975.

61. Walter JE, Mitchell DK. Role of astroviruses in childhood diarrhea. *Curr Opin Pediatr* 2000;12(3):275–279.

62. Cheng AC, McDonald JR, Thielman NM. Infectious diarrhea in developed and developing countries. *J Clin Gastroenterol* 2005;39(9):757–773.

63. Pang XL, Joensuu J, Vesikari T. Human calicivirus-associated sporadic gastroenteritis in Finnish children less than two years of age followed prospectively during a rotavirus vaccine trial. *Pediatr Infect Dis J* 1999;18(5):420–426.

64. Dennehy PH. Acute diarrheal disease in children: epidemiology, prevention, and treatment. *Infect Dis Clin North Am* 2005;19(3):585–602.

65. Marks PJ, et al. A school outbreak of Norwalk-like virus: evidence for airborne transmission. *Epidemiol Infect* 2003;131(1):727–736.

66. Martella V, et al. Zoonotic aspects of rotaviruses. *Vet Microbiol* 2010;140(3–4):246–255.

67. Bernstein DI. Rotavirus overview. *Pediatr Infect Dis J* 2009;28 (3 Suppl):S50–S53.

68. Dennehy PH. Rotavirus vaccines: an overview. *Clin Microbiol Rev* 2008;21(1):198–208.

69. Santosham M, et al. Detection of rotavirus in respiratory secretions of children with pneumonia. *J Pediatr* 1983;103(4):583–585.

70. Santos N, Hoshino Y. Global distribution of rotavirus serotypes/genotypes and its implication for the development and implementation of an effective rotavirus vaccine. *Rev Med Virol* 2005;15(1):29–56.

71. Gray J, et al. Rotavirus. *J Pediatr Gastroenterol Nutr* 2008;46 (Suppl 2):S24–S31.

72. Dai YC, et al. Surveillance and risk factors of norovirus gastroenteritis among children in a southern city of China in the fall-winter seasons of 2003–2006. *J Paediatr Child Health* 2010;46(1–2):45–50.

73. Yori PP, et al. Norovirus highly prevalent cause of endemic acute diarrhea in children in the Peruvian Amazon. *Pediatr Infect Dis J* 2009;28(9):844–847.

74. Guix S, et al. Molecular epidemiology of astrovirus infection in Barcelona, Spain. *J Clin Microbiol* 2002;40(1):133–139.

75. Lenaerts L, De Clercq E, Naesens L. Clinical features and treatment of adenovirus infections. *Rev Med Virol* 2008;18(6):357–374.

76. Spitzer MD. Viral causes of diarrhea. *Pediatr Rev* 2002;23(7):257–258.

77. Cunliffe NA, et al. Detection of enteric adenoviruses in children with acute gastroenteritis in Blantyre, Malawi. *Ann Trop Paediatr* 2002;22(3):267–269.

78. Liu C, et al. Identification of viral agents associated with diarrhea in young children during a winter season in Beijing, China. *J Clin Virol* 2006;35(1):69–72.

79. Qiao H, et al. Viral diarrhea in children in Beijing, China. *J Med Virol* 1999;57(4):390–396.

80. Rima BK, Duprex WP. Morbilliviruses and human disease. *J Pathol* 2006;208(2):199–214.

81. Schneider-Schaulies S, Schneider-Schaulies J. Measles virus-induced immunosuppression. *Curr Top Microbiol Immunol* 2009;330:243–269.

82. Hussey GD, CJ Clements, Clinical problems in measles case management. *Ann Trop Paediatr* 1996;16(4):307–317.

83. Carlos, C.C., et al. Enteropathogens among measles patients with diarrhea in urban Filipino children. *Phil J Microbiol Infect Dis* 1992;21(2):53–60.

84. Ochoa, T.J., et al. New insights into the epidemiology of enteropathogenic *Escherichia coli* infection. *Trans R Soc Trop Med Hyg* 2008;102(9):852–856.

85. Callaway, T.R., et al. Diet, *Escherichia coli* O157:H7, and cattle: a review after 10 years. *Curr Issues Mol Biol* 2009;11 (2):67–79.

86. Crump, J.A., S.P. Luby, and E.D. Mintz, The global burden of typhoid fever. *Bull World Health Organ* 2004;82(5):346–353.

87. Griffiths, J.K., Human cryptosporidiosis: epidemiology, transmission, clinical disease, treatment, and diagnosis. *Adv Parasitol* 1998;40:37–85.

88. Escobedo, A.A. and S. Cimerman, Giardiasis: a pharmacotherapy review. *Exp Opin Pharmacother* 2007;8(12): 1885–1902.

89. Pritt BS, Clark CG. Amebiasis. *Mayo Clin Proc* 2008;83(10): 1154–1159; quiz 1159–1160.

90. Claeson, M. and M.H. Merson, Global progress in the control of diarrheal diseases. *Pediatr Infect Dis J* 1990;9(5):345–355.

91. Hahn, S., Y. Kim, and P. Garner, Reduced osmolarity oral rehydration solution for treating dehydration caused by acute diarrhoea in children. *Cochrane Database Syst Rev* 2002;(1): CD002847.

92. Lazzerini, M. and L. Ronfani, Oral zinc for treating diarrhoea in children. *Cochrane Database Syst Rev* 2008;I (3): CD005436.

93. Baqui, A.H., et al. Effect of zinc supplementation started during diarrhoea on morbidity and mortality in Bangladeshi children: community randomised trial. *BMJ* 2002;325 (7372):1059.

94. Bhutta, Z.A., et al. Prevention of diarrhea and pneumonia by zinc supplementation in children in developing countries: pooled analysis of randomized controlled trials. *Zinc Investigators' Collaborative Group. J Pediatr* 1999;135 (6):689–697.

95. Beaton G, Martorell R, Aronson KJ.Effectiveness of Vitamin A Supplementation in the Control of Young Child Morbidity and Mortality in Developing Countries. International Nutrition Program, Toronto, 1993.

96. al-Ali, F.M., M.M. Hossain, and R.N. Pugh, The associations between feeding modes and diarrhoea among urban children in a newly developed country. *Public Health* 1997;111(4): 239–243.

97. Huffman, S.L. and C. Combest, Role of breast-feeding in the prevention and treatment of diarrhoea. *J Diarrhoeal Dis Res* 1990;8(3):68–81.

98. VanDerslice, J., B. Popkin, and J. Briscoe, Drinking-water quality, sanitation, and breast-feeding: their interactive effects on infant health. *Bull World Health Organ* 1994;72(4): 589–601.

99. Ahiadeke, C., Breast-feeding, diarrhoea and sanitation as components of infant and child health: a study of large scale survey data from Ghana and Nigeria. *J Biosoc Sci* 2000;32(1): 47–61.

100. Ejemot RI, et al. Hand washing for preventing diarrhoea. *Cochrane Database Syst Rev* 2008;(1): CD004265.

101. Fewtrell, L., et al. Water, sanitation, and hygiene interventions to reduce diarrhoea in less developed countries: a systematic review and meta-analysis. *Lancet Infect Dis* 2005;5(1):42–52.

102. Vesikari, T., et al. Safety and efficacy of a pentavalent human–bovine (WC3) reassortant rotavirus vaccine. *N Engl J Med* 2006;354(1):23–33.

103. Ruiz-Palacios, G.M., et al. Safety and efficacy of an attenuated vaccine against severe rotavirus gastroenteritis. *N Engl J Med* 2006;354(1):11–22.

104. Ryland S, Raggers H. *Childhood Morbidity and Treatment Patterns.* DHS Comparative Stodies No. 27. Claverton, Maryland: Macro International Inc., 1998.

105. Clemens, J.D., et al. Field trial of oral cholera vaccines in Bangladesh. *Lancet* 1986;2(8499):124–127.

106. Sanchez, J.L., et al. Protective efficacy of oral whole-cell/recombinant-B-subunit cholera vaccine in Peruvian military recruits. *Lancet* 1994;344(8932):1273–1276.

107. Duggan, C., M. Santosham, and R.I. Glass, The management of acute diarrhea in children: oral rehydration, maintenance, and nutritional therapy. Centers for Disease Control and Prevention. *MMWR Recomm Rep* 1992;41(RR-16): 1–20.

108. Latta T. Malignant cholera: comments contributed by the Central Board of Health, London, relative to the treatment of cholera by copious injection of aqueous and saline fluids into the veins. *Lancet* 1831–1832;2: 274–277.

109. Santosham, M., et al. Oral rehydration therapy for diarrhea: an example of reverse transfer of technology. *Pediatrics* 1997;100(5):E10.

110. Darrow, D.C., E.L. Pratt, and et al. Disturbances of water and electrolytes in infantile diarrhea. *Pediatrics* 1949;3(2): 129–156.

111. King, C.K., et al. Managing acute gastroenteritis among children: oral rehydration, maintenance, and nutritional therapy. *MMWR Recomm Rep* 2003;52 (RR-16):1–16.

112. Harrison HE. The treatment of diarrhea in infancy. *Pediatr Clin North Am* 1954: 335–348.

113. Paneth, N., Hypernatremic dehydration of infancy: an epidemiologic review. *Am J Dis Child* 1980;134(8):785–792.

114. Mahalanabis, D., et al. Oral fluid therapy of cholera among Bangladesh refugees. *Johns Hopkins Med J* 1973;132 (4):197–205.

115. Ruxin, J.N., Magic bullet: the history of oral rehydration therapy. *Med Hist* 1994;38(4):363–397.

116. al-Awqati, Q.S., M. Mekkiya, and M. Thamer, Establishment of a cholera treatment unit under epidemic conditions in a developing country. *Lancet* 1969;1(7588):252–253.

117. Chatterjee, H.N., Control of vomiting in cholera and oral replacement of fluid. *Lancet* 1953;265(6795):1063.

118. Nalin, D.R., et al. Oral maintenance therapy for cholera in adults. *Lancet* 1968;2(7564):370–373.

119. Nalin, D.R., R.A. Cash, and M. Rahman, Oral (or nasogastric) maintenance therapy for cholera patients in all age-groups. *Bull World Health Organ* 1970;43(3):361–363.

120. Cash, R.A., et al. A clinical trial of oral therapy in a rural cholera-treatment center. *Am J Trop Med Hyg* 1970;19(4): 653–656.

121. Nalin, D.R. and R.A. Cash, Oral or nasogastric maintenance therapy for diarrhoea of unknown aetiology resembling cholera. *Trans R Soc Trop Med Hyg* 1970;64(5):769–771.

122. Duggan, C., et al. Scientific rationale for a change in the composition of oral rehydration solution. *JAMA* 2004;291 (21):2628–2631.

123. Mahalanabis D, Faruque AS, Hoque SS, Faruque SM. Hypotonic oral rehydration solution in acute diarrhoea: a controlled clinical trial. *Acta Paediatrica* 1995;84(3):289–293.

124. Santosham, M., et al. A double-blind clinical trial comparing World Health Organization oral rehydration solution with a reduced osmolarity solution containing equal amounts of sodium and glucose. *J Pediatr* 1996;128(1):45–51.

125. Multicenter, randomized, double-blind clinical trial to evaluate the efficacy and safety of a reduced osmolarity oral rehydration salts solution in children with acute watery diarrhea. Pediatrics 2001;107(4):613–618.

126. Multicentre evaluation of reduced-osmolarity oral rehydration salts solution. International Study Group on Reduced-osmolarity ORS solutions. Lancet 1995;345(8945):282–285.

127. Murphy, C., S. Hahn, and J. Volmink, Reduced osmolarity oral rehydration solution for treating cholera. *Cochrane Database Syst Rev* 2004;(4): CD003754.

128. Sarker, S.A., N. Majid, and D. Mahalanabis, Alanine- and glucose-based hypo-osmolar oral rehydration solution in infants with persistent diarrhoea: a controlled trial. *Acta Paediatr* 1995;84(7):775–780.

129. Guerrant, R.L., et al. Updating the DALYs for diarrhoeal disease. *Trends Parasitol* 2002;18(5):191–193.

130. Niehaus, M.D., et al. Early childhood diarrhea is associated with diminished cognitive function 4 to 7 years later in children in a northeast Brazilian shantytown. *Am J Trop Med Hyg* 2002;66(5):590–593.

131. Lorntz, B., et al. Early childhood diarrhea predicts impaired school performance. *Pediatr Infect Dis J* 2006;25(6):513–520.

132. Chung, A.W. and B. Viscorova, The effect of early oral feeding versus early oral starvation on the course of infantile diarrhea. *J Pediatr* 1948;33(1):14–22.

133. Armitstead, J., D. Kelly, and J. Walker-Smith, Evaluation of infant feeding in acute gastroenteritis. *J Pediatr Gastroenterol Nutr* 1989;8(2):240–244.

134. Hjelt, K., et al. Rapid versus gradual refeeding in acute gastroenteritis in childhood: energy intake and weight gain. *J Pediatr Gastroenterol Nutr* 1989;8(1):75–80.

135. Sandhu, B.K., et al. A multicentre study on behalf of the European Society of Paediatric Gastroenterology and Nutrition Working Group on Acute Diarrhoea. Early feeding in childhood gastroenteritis. *J Pediatr Gastroenterol Nutr* 1997;24(5):522–527.

136. Santosham, M., K.H. Brown, and R.B. Sack, Oral rehydration therapy and dietary therapy for acute childhood diarrhea. *Pediatr Rev* 1987;8(9):273–278.

137. Santosham, M., et al. Role of soy-based, lactose-free formula during treatment of acute diarrhea. *Pediatrics* 1985;76 (2):292–298.

138. WHO and UNICEF. Clinical Management of Acute Diarrhoea, WHO/UNICEF Joint Statement. New York: WHO/ UNICEF, 2004.

139. Patro B, Golicki D, Szajewska H. Meta-analysis: zinc supplementation for acute gastroenteritis in children. *Aliment Pharmacol Ther* 2008.

140. Lukacik, M., R.L. Thomas, and J.V. Aranda, A meta-analysis of the effects of oral zinc in the treatment of acute and persistent diarrhea. *Pediatrics* 2008;121(2):326–336.

141. Bhandari, N., et al. Effectiveness of zinc supplementation plus oral rehydration salts compared with oral rehydration salts alone as a treatment for acute diarrhea in a primary care setting: a cluster randomized trial. *Pediatrics* 2008;121(5): e1279–e1285.

142. Awasthi, S., Zinc supplementation in acute diarrhea is acceptable, does not interfere with oral rehydration, and reduces the use of other medications: a randomized trial in five countries. *J Pediatr Gastroenterol Nutr* 2006;42(3): 300–305.

143. Baqui, A.H., et al. Zinc therapy for diarrhoea increased the use of oral rehydration therapy and reduced the use of antibiotics in Bangladeshi children. *J Health Popul Nutr* 2004;22 (4):440–442.

144. Mandell GL, Bennett JE, Dolin R. *Principles and Practice of Infectious Diseases*, 7th edition. Philadelphia: Churchill Livingston/Elsevier, 2010.

145. Pickering LK, editor. *Red Book: Report of the Committee on Infectious Diseases*, 27th edition. American Academy of Pediatrics, 2006.

146. Pulling, M. and C.M. Surawicz, Loperamide use for acute infectious diarrhea in children: safe and sound? *Gastroenterology* 2008;134(4):1260–1262.

147. Santos, M., et al. Use of racecadotril as outpatient treatment for acute gastroenteritis: a prospective, randomized, parallel study. *J Pediatr* 2009;155(1):62–67.

148. Isolauri, E., V. Vahasarja, and T. Vesikari, Effect of cholestyramine on acute diarrhoea in children receiving rapid oral rehydration and full feedings. *Ann Clin Res* 1986;18 (2):99–102.

149. Richards, L., M. Claeson, and N.F. Pierce, Management of acute diarrhea in children: lessons learned. *Pediatr Infect Dis J* 1993;12(1):5–9.

150. Alam, N.H. and H. Ashraf, Treatment of infectious diarrhea in children. *Paediatr Drugs* 2003;5(3):151–165.

151. Koo, H.L., et al. Antimotility agents for the treatment of *Clostridium difficile* diarrhea and colitis. *Clin Infect Dis* 2009;48(5):598–605.

152. Bhattacharya, S.K. and D. Sur, An evaluation of current shigellosis treatment. *Exp Opin Pharmacother* 2003;4 (8):1315–1320.

153. Van der Wielen, M. and P. Van Damme, Pentavalent human-bovine (WC3) reassortant rotavirus vaccine in special populations: a review of data from the Rotavirus Efficacy and Safety Trial. *Eur J Clin Microbiol Infect Dis* 2008;27 (7):495–501.

154. Goveia, M.G., et al. Efficacy of pentavalent human-bovine (WC3) reassortant rotavirus vaccine based on breastfeeding frequency. *Pediatr Infect Dis J* 2008;27(7):656–658.

155. Dennehy, P.H., et al. Comparative evaluation of safety and immunogenicity of two dosages of an oral live attenuated human rotavirus vaccine. *Pediatr Infect Dis J* 2005;24 (6):481–488.

156. Zaman, K., et al. Successful co-administration of a human rotavirus and oral poliovirus vaccines in Bangladeshi infants in a 2-dose schedule at 12 and 16 weeks of age. *Vaccine* 2009;27(9):1333–1339.

157. Ciarlet M, et al. Safety and immunogenicity of concomitant use of the pentavalent rotavirus vaccine (RotaTeq) and oral poliovirus vaccine (OPV). *European Society for Pediatric Infectious Diseases (ESPID),* Porto, Portugal, 2007.

158. Linhares, A.C., et al. Efficacy and safety of an oral live attenuated human rotavirus vaccine against rotavirus gastro-enteritis during the first 2 years of life in Latin American infants: a randomised, double-blind, placebo-controlled phase III study. *Lancet* 2008;371(9619):1181–1189.

159. Haber P, et al. Postlicensure monitoring of intussusception after RotaTeq vaccination in the United States, February 1 2006, to September 25 2007. *Pediatrics* 2008;121(6):1206–1212.

160. Rotavirus Vaccination. *Wkly Epidemiol Rec* 2009;23: 232–236.

161. Reduction in rotavirus after vaccine introduction—United States, 2000–2009. *MMWR Morb Mortal Wkly Rep* 2009;58(41):1146–1149.

162. Orozco, M., et al. Uptake of rotavirus vaccine and national trends of acute gastroenteritis among children in Nicaragua. *J Infect Dis* 2009;200 (Suppl 1): S125–S130.

163. Richardson, V., et al. Effect of rotavirus vaccination on death from childhood diarrhea in Mexico. *N Engl J Med* 2010;362 (4):299–305.

164. Payne, D.C., et al. Secular variation in United States rotavirus disease rates and serotypes: implications for assessing the rotavirus vaccination program. *Pediatr Infect Dis J* 2009;28 (11):948–953.

165. Trach, D.D., et al. Investigations into the safety and immunogenicity of a killed oral cholera vaccine developed in Viet Nam. *Bull World Health Organ* 2002;80(1):2–8.

166. Mahalanabis, D., et al. A randomized, placebo-controlled trial of the bivalent killed, whole-cell, oral cholera vaccine in adults and children in a cholera endemic area in Kolkata, India. *PLoS One* 2008;3(6):e2323.

167. Clemens, J.D., et al. Measles vaccination and childhood mortality in rural Bangladesh. *Am J Epidemiol* 1988;128 (6):1330–1339.

168. Moss, W.J., Measles control and the prospect of eradication. *Curr Top Microbiol Immunol* 2009;330:173–189.

169. Progress in global measles control and mortality reduction, 2000–2007. *MMWR Morb Mortal Wkly Rep* 2008;57(48): 1303–1306.

170. Boschi-Pinto, C., L. Velebit, and K. Shibuya, Estimating child mortality due to diarrhoea in developing countries. *Bull World Health Organ* 2008;86(9):710–717.

171. Brown, K.H., et al. Preventive zinc supplementation among infants, preschoolers, and older prepubertal children. *Food Nutr Bull* 2009;30(1 Suppl): S12–S40.

172. Grotto, I., et al. Vitamin A supplementation and childhood morbidity from diarrhea and respiratory infections: a meta-analysis. *J Pediatr* 2003;142(3):297–304.

173. Faruque, A.S., et al. Double-blind, randomized, controlled trial of zinc or vitamin A supplementation in young children with acute diarrhoea. *Acta Paediatr* 1999;88(2): 154–160.

174. Henning, B., et al. Lack of therapeutic efficacy of vitamin A for non-cholera, watery diarrhoea in Bangladeshi children. *Eur J Clin Nutr* 1992;46(6):437–443.

175. Hossain, S., et al. Single dose vitamin A treatment in acute shigellosis in Bangladesh children: randomised double blind controlled trial. *BMJ* 1998;316(7129):422–426.

176. Khatun, U.H., et al. A randomized controlled clinical trial of zinc, vitamin A or both in undernourished children with persistent diarrhea in Bangladesh. *Acta Paediatr* 2001;90 (4):376–380.

177. Barreto, M.L., et al. Effect of vitamin A supplementation on diarrhoea and acute lower-respiratory-tract infections in young children in Brazil. *Lancet* 1994;344(8917):228–231.

178. Effect of breastfeeding on infant and child mortality due to infectious diseases in less developed countries: a pooled analysis. WHO Collaborative Study Team on the Role of breastfeeding on the prevention of infant mortality. Lancet 2000;355(9202):451–455.

179. Victora, C.G., et al. Evidence for protection by breast-feeding against infant deaths from infectious diseases in Brazil. *Lancet* 1987;2(8554):319–322.

180. WHO. *Emerging Issues in Water and Infectious Disease.* Geneva: World Health Organization, 2003.

181. Haggerty, P.A., et al. Community-based hygiene education to reduce diarrhoeal disease in rural Zaire: impact of the inter-vention on diarrhoeal morbidity. *Int J Epidemiol* 1994;23 (5):1050–1059.

182. Pinfold, J.V. and N.J. Horan, Measuring the effect of a hygiene behaviour intervention by indicators of behaviour and diar-rhoeal disease. *Trans R Soc Trop Med Hyg* 1996;90 (4):366–371.

183. Luby, S.P., et al. Effect of handwashing on child health: a randomised controlled trial. *Lancet* 2005;366(9481): 225–233.

184. Control of diarrhoeal diseases: WHO's programme takes shape. *WHO Chron* 1978;32(10):369–372.

185. Feachem, R.G., R.C. Hogan, and M.H. Merson, Diarrhoeal disease control: reviews of potential interventions. *Bull World Health Organ* 1983;61(4):637–640.

186. Forsberg, B.C., et al. Diarrhoea case management in low- and middle-income countries—an unfinished agenda. *Bull World Health Organ* 2007;85(1):42–48.

187. Barros FC, V.C., Forsberg B, Maranhao AGK, Stegeman M, Gonzalez-Richmond A, Martins RM, Neuman ZA, McAuliffe J, Branco JA, Management of childhood diarrhoea at the

household level: a population-based survey in north-east Brazil. *Bull World Health Organ* 1991;69(1):59–65.

188. McDivitt, J.A., R.C. Hornik, and C.D. Carr, Quality of home use of oral rehydration solutions: results from seven HEALTHCOM sites. *Soc Sci Med* 1994;38(9):1221–1234.

189. Dua, T., R. Bahl, and M.K. Bhan, Lessons learnt from Diarrheal Diseases Control Program and implications for the future. *Indian J Pediatr* 1999;66(1):55–61.

190. *National Family Health Survey (NFHS-3), 2005–2006: India,* Vol. I. Mumbai: IIPS, 2007.

191. Ellis, A.A., et al. Home management of childhood diarrhoea in southern Mali—implications for the introduction of zinc treatment. *Soc Sci Med* 2007;64(3):701–712.

192. Miller, P. and N. Hirschhorn, The effect of a national control of diarrheal diseases program on mortality: the case of Egypt. *Soc Sci Med* 1995;40(10):S1–S30.

6

MALNUTRITION AND UNDERNUTRITION

Jeffrey K. Griffiths

6.1 INTRODUCTION

Malnutrition and undernutrition remain prominent global causes of ill health and premature mortality. These conditions contribute to high rates of mortality in children under the age of 5 and in pregnant women, as well as to morbidity in many age ranges. A recent estimate is that malnutrition contributes to the deaths of one-third of all children under the age 5 (1). Malnutrition can kill directly, as in famine, and also indirectly, by increasing both the rate and severity of infectious diseases. Above and beyond these medical outcomes, mal- and undernutrition contribute to lower educational achievement, to lower physical and mental productivity, and to lower social achievement and earning power.

Unsafe water and a lack of sanitation contribute to malnutrition by fostering the transmission of infectious diseases, which sap nutrition through nutrient loss, malabsorption, and anorexia during the illness. The classic example of such disease is recurrent diarrhea and respiratory disease in infants and children under the age of 5, which leads to impaired nutrition. Deficiencies of both calories and micronutrients critical to growth and a robust immune system may result from these infections. An impaired immune system may then predispose individuals with malnutrition to yet more infection in a vicious cycle that worsens both nutrition and resistance to infection (2). Availability of clean water decreases the rates of infections such as diarrhea and hepatitis and allows hygienic behaviors such as washing hands after defecation or food preparation. The provision of sanitation separates people from the very pathogens most adapted to humans and interrupts the cycle of transmission from person

to person or from animal to human. These fundamental preventive public health measures are among the most important ones extant.

The logic model for the relationship between water and sanitation and nutrition is thus simple at its core. Safe drinking water and sanitation result in less infectious disease and, therefore, better nutrition and nutritional reserves. The diseases most prevented are those spread by contaminated water (or food washed with such water), and those diseases that are prevented by simple hygiene measures such as the washing of hands and utensils. However, important nuances exist that will affect the logic model we ultimately use. For example, in areas where water is scarce, considerable energy is expended—principally by children and women—to carry water to households for cooking, cleaning, and hygiene. This may represent a very substantial investment of both energy and time. There are direct and inverse relationships among water availability, quantity, the distance one must travel to acquire the water, and health risks (see Table 6.1). (3). Contact with open bodies of water such as rivers or streams during water collection may place the individual at increased risk of water-related diseases such as schistosomiasis or malaria. The relationships between these diseases and nutrition may be more complex than the relationship between nutrition and recurrent diarrhea or pneumonia. Malaria, for example, is more severe when the individual is anemic.

Addressing nutrition through water and sanitation initiatives will likely also improve human health outcomes that are not classically associated with the cycle of infection and malnutrition outlined above. Measles, for example, is not usually considered a water-related disease. However, the risk

Water and Sanitation-Related Diseases and the Environment: Challenges, Interventions, and Preventive Measures, First Edition.
Edited by Janine M. H. Selendy.
© 2011 Wiley-Blackwell. Published 2011 by John Wiley & Sons, Inc.

TABLE 6.1   Water Quantity Provision, Access to Water, Health Needs Met, and Level of Health Risk

Quantity Available per Person per Day	Access (Distance and/or Time)	Health Needs Met and Unmet	Level of Health Risk (Relative Risk for Child Mortality Estimated by Gakidou et al.)
0–5 L/day	>1 km or >30 min total collection time	Consumption needs cannot be assured; hygienic needs unmet	Very high (×4.40 child mortality)
Basic access; unlikely to be more than 20 L/day	100–1000 m; 5–30 min collection time	Consumption needs should be assured; handwashing and basic food hygiene possible	High (×3.48 for child mortality)
Intermediate access; quantity ~50 L/day	Water delivered through a tap or within 100 m or within 5 min collection time	Consumption needs assured and all basic personal and food hygiene assured; laundry and bathing needs should be met	Low (×2.76 for child mortality)
Optimal access; quantity of 100 L/day or more	Water consistently supplied through multiple days	All consumption and hygiene needs should be met	Very low (×1.00 for child mortality) Gakidou et al.)

Adapted from Refs 3 and 4.

of death during a measles infection becomes as high as 50% in highly malnourished children, and incremental improvements in nutrition related to decreased diarrhea and pneumonia will also mean incrementally fewer deaths from measles. There are strong historical reasons for believing this will be the case as outlined below.

The provision of safe water and of sanitary services is historically well correlated with improved nutrition in more developed countries, and there is a growing body of evidence that the same is true in currently less developed nations. One of the challenges to prioritizing and funding improvements in safe water and sanitation is that the evidence for this linkage with nutrition in the developed countries was acquired in the past, and the current evidence for the linkage in developing countries is more piecemeal and fragmentary. To an extent we must impute the benefits that will arise from improved water and sanitation on nutrition in our logic model. This imputation is based upon historical work in developed countries, and upon reasonable assumptions that link current information on infection, nutrition, and morbidity with mortality relating to malnutrition. Improving nutrition and improving water and sanitation typically occur over decades and the benefits are reaped over a similar period, and no long-term documentary studies are currently being done or otherwise exist that explicitly and prospectively follow the relationship of these factors over time. Indeed, funding for this kind of documentation is often as short term as the attention span of busy policy makers who control the allocation of resources.

Traditionally, nutrition scientists have also worked separately from water and sanitation professionals. Both have tended to focus within their discipline rather than on the interdisciplinary boundary areas that link them. At times, amelioration of malnutrition may have been viewed as being in competition for resources with clean water and sanitation provision. Better nutrition can, however, be one of the desired outcomes of clean water and sanitation. As outlined below and discussed elsewhere in this book, malnutrition arises for a variety of reasons, only some of which are addressed by clean water and sanitation; acknowledging this, amelioration of malnutrition globally through improved water and sanitation is likely to be significant. This chapter outlines key referential points for understanding the connection between water, sanitation, the environment, and nutrition, while denoting the kinds of studies that would help us better understand the exact magnitude of the relationships.

Long-term trends in water supply relating to climate change will also directly affect food supply. As clean water and sanitation are extended to a growing global population, more thoughtful and rational use of water—and reuse of water—may be needed. Water is used for crops, for raising livestock and other food animals, for maintaining the integrity of a healthy ecosystem and environment, and for industrial purposes.

6.2   AN OVERVIEW OF MALNUTRITION AND UNDERNUTRITION: MAGNITUDE, TYPES, AND CAUSATION

Malnutrition and undernutrition remain major global contributors to poor health, disease, and death. Overt malnutrition is almost always related to poverty. Recent UNICEF reports state that 195 million, or one-third, of the world's children under age 5, are stunted, and 129 million are underweight. One-quarter of all humans, about 1.6 billion people, are anemic, principally because of iron deficiency (5). Table 6.2 outlines the magnitude of some of the principal aspects of malnutrition.

TABLE 6.2 Anemia

Population Group	Prevalence of Anemia (%, with 95% CI)	Population Affected
Preschool-age children	47.4 (45.7–49.1%)	293,000,000
School-age children	25.4 (19.9–30.9%)	305,000,000
Pregnant women	41.8 (39.9–43.8%)	56,000,000
Nonpregnant women	30.2 (28.7–31.6%)	468,000,000
Men	12.7 (8.6–16.9%)	260,000,000
Elderly	23.9 (18.3–29.4%)	164,000,000
Total population	24.8 (22.9–26.7%)	1,620,000,000

Adapted from de Benoist et al (5)

6.2.1 Forms of Malnutrition

It is useful, at the individual level, to categorize malnutrition which is deficient in caloric needs, or which is deficient in essential vitamins and micronutrients. Calorically restricted diets may lead to restricted growth, but they do not necessarily lead to increased susceptibility to infection seen in the case of specific micronutrient deficiency. However, one condition usually accompanies the another. Individuals with diets deficient in calories including protein suffer from protein–calorie malnutrition (PCM), also called protein–energy malnutrition (PEM). This may manifest in different ways, depending upon the relative deficiency of calories and protein. One specific form of PCM, termed *kwashiorkor* (Fig. 6.1), is seen with severe protein deficiency. Children

with kwashiorkor are very thin but have edema of their lower limbs, muscle wasting, protuberant abdomens, and pale hair and skin. This condition tends to occur after a child is weaned from breastfeeding and is most common where children are fed a monotonous diet of rice or other starchy foods such as cassava, plantains, or sweet potatoes. High protein foods may be reserved, on a cultural basis, for the adults. There is evidence that aflotoxins, the product of molds that grows on these starchy foods when poorly stored, may precipitate or worsen kwashiorkor by damaging the liver and its capacity to synthesize protein. Another form of PEM, termed *marasmus* (Fig. 6.2), is the result of very severe caloric deprivation, and is characterized by a very thin, stunted child with wasted muscles and tissues. These children urgently require nutritional rehabilitation as their risk of death is very high, usually from infection (6). In recent years special foodstuffs have been developed that provide a balance of calories and micronutrients for use when death from malnutrition is imminent (6).

Deficiencies relating to vitamin A, zinc, folate (vitamin B9), iodine, and iron are common. There is an evolving literature suggesting that vitamin D deficiency is both more widespread and more important than previously recognized. Iodine deficiency is usually related to a diet poor in iodine, and it leads to mental retardation and (with severe deficiency) cretinism, which includes severe brain damage, dwarfism, and deafness. Approximately 2 billion persons have insufficient iodine in their diet, particularly in Southeast Asia and

FIGURE 6.1   Children with protein–calorie malnutrition (kwashiorkor). (Photo courtesy U.S. CDC.)

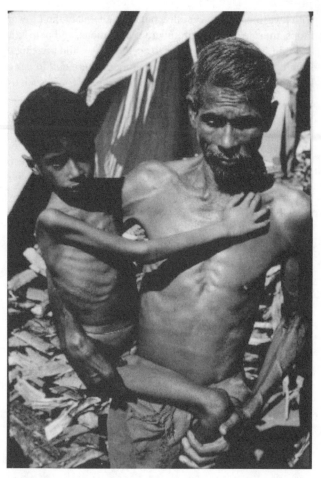

FIGURE 6.2   Marasmus (extreme emaciation). (Photo courtesy U.S. CDC.)

sub-Saharan Africa, although the number with overt disease is much smaller. Moderate iodine deficiency can reduce IQ scores by 10–15 points (7). Iron deficiency leads to anemia, an increased risk of infection, and decrements in intellectual development. Folate is required, along with iron, for the production of erythrocytes. Folate supplementation in pregnant women decreases the incidence of some congenital neurological disease, such as spina bifida, and it is often given in combination with iron to anemic pregnant women. There is a long and successful history of targeting iodine deficiency and iron-deficiency anemia by the supplementation of foodstuffs (such as by iodinization of salt and market availability of iron-fortified wheat flour), and these methods are still being used today (8). It has become clear that micronutrient deficiencies in vitamin A and zinc should be targeted as well. Vitamin A deficiency leads to an increased risk of infectious diseases such as diarrhea and pneumonia, maternal mortality, and blindness (9). Indeed, in 2008 it was estimated that an annual modest investment of 60 million USD for micronutrient supplements would yield a benefit of ~1 billion USD (10). The World Health Organization

recommends that all children at risk of vitamin A deficiency receive supplements twice a year starting at birth.

Zinc deficiency is common when diets are deficient in breast milk, animal protein, or crustacean. Most vegetables and grains have little zinc in them, and if these foodstuffs are grown in soils depleted of zinc, the levels may be very low. Zinc is an essential cofactor for growth, immunity, and normal mucosal surface. Inhibitors of zinc absorption include phytates, which are commonly found in plant foods. Zinc is soluble in water, and so it is easily lost during episodes of diarrhea. The incidence of diarrhea, and of pneumonia, is elevated in zinc-deficient children, and the additional risk is approximately 30% for diarrhea and ~70% for pneumonia in young children. Many studies have been conducted that show supplementing children's foodstuffs with low doses of zinc (10–20 mg of Zn) per day decreases the incidence of diarrhea by 20–27% and of respiratory diseases by ~15% (11).

6.2.2   Mechanisms for the Adverse Effects of Infection upon Nutrition

Infection typically leads to anorexia and decreased nutrient intake, while at the same time there is increased demand for nutrients as the metabolic rate increases with fever (2). This often leads to the breakdown of muscle protein so the body may provide itself adequate energy supply. Nitrogen may be lost in urine and feces, and borderline deficiencies may be precipitated during this period. With diseases such as diarrhea, absorption of any food ingested will decrease, further tilting the equation toward net loss of nutrients (12). In many cultures food and liquids are commonly withheld during diarrhea, with the aim of decreasing the flow of the body's fluids (while worsening nutrition and hydration). Further details on specific disease features are given in chapters on shigellosis and diarrhea. The exact magnitude of the nutritional deficits caused by respiratory infections such as pneumonia (many which can be prevented by handwashing with clean water) is far less well documented than that caused by diarrheal disease, but these deficits may, in fact, be larger.

6.2.3   Relationship between Degree of Malnutrition and Death

There is a well-documented relationship between degree of malnutrition and risk of death. In order to illustrate this point, and (later on in this chapter) to better understand some of the existing data on the benefits of clean water and sanitation, an introduction to the jargon of nutrition is provided here. In order to compare children of different ages, age and gender-specific normative values for height, weight, and other anthropometric values have been obtained. It is possible to measure a child and thereby discover how far the child deviates from a normative value for other children of their age and gender. This statistic is the standard deviation (SD)

TABLE 6.3   **Relative Risk of Mortality Associated with Low WAZ Estimated from Regression Analysis by Cause of Death**

Cause of Death	<3 SDs, WAZ $\leq$ -3	-3 to -2 SDs, $-3 \leq$ WAZ < -2	-2 to -1 SDs, $-2 \leq$ WAZ < -1	SD ≥ -1 SDs, WAZ ≥ -1
All-cause	8.72 (5.55–13.72)	4.24 (3.13–5.73)	2.06 (1.77–2.39)	1.0
Diarrhea	12.50 (7.19–21.73)	5.39 (3.73–7.79)	2.32 (1.93–2.79)	1.0
Pneumonia	8.09 (4.36–15.01)	4.03 (2.67–6.08)	2.01 (1.63–2.47)	1.0
Malaria	9.49 (3.25–27.66)	4.48 (2.20–9.15)	2.12 (1.48–3.02)	1.0
Measles	5.22 (2.29–11.88)	3.01 (1.74–5.21)	1.73 (1.32–2.28)	1.0

Adapted from Ref. 13.

from a median value. A child who is >1.96 (usually rounded to 2) standard deviations from the mean is significantly different on a statistical basis. A child is usually described as malnourished if his or her anthropometrics are significantly below (2 or more standard deviations lower) than the mean reference value for age and gender; a child is described as undernourished if these values are between 1 and 2 standard deviations below the mean. For statistical reasons each standard deviation from the mean is known as a Z score. Thus, a child whose weight (for his or her age in month and gender) is 2 standard deviations below the mean has, in the parlance of nutrition, a weight-for-age Z score (WAZ) of -2. Children with WAZ scores ≤ -2 are termed *wasted*. Similarly, a child whose height is 2 standard deviations below the mean for his or her age and gender is termed *stunted*.

Table 6.3 is taken from information collated and analyzed by Fishman et al. and published in 1994. Estimations were done by regression analysis of cause of death. The table shows that for each step a child is less than the mean for his or her age, the risk of death increases during an episode of diarrheal disease, pneumonia, malaria, or measles. The same is true when all causes of death are combined. For children with a WAZ score of < -3, the relative risk of dying was estimated to be more than 12 times that of a child with normal nutritional status. For purposes of supporting the logic model, this information is critical. With increasing malnutrition, mortality increases.

6.2.4   Why Malnutrition Exists and Critical Periods for Intervention

People must have both physical access to food and the economic capacity to purchase it. When these are present, then they are determined to have "food security." Malnutrition exists for a variety of reasons. Most are related to poverty. The majority of people living with malnutrition exist in countries where food surpluses exist, which has led prominent philosophers such as Amartya Sen to comment that the cause of famine and malnutrition is actually a problem related to the equitable distribution and purchase of food (14). Other reasons range from a simple lack of food supply to cultural and behavioral factors, some of which unintentionally promote either the acquisition of infectious

diseases or contribute to their severity. Food production may be insufficient because of drought, poor soil, or a lack of technology and the resources required for higher crop yields. The care and feeding of children may be faulty. Neonates and infants who are not exclusively breastfed are at increased risk of enteric diseases because the food and water used to supplement breast milk are often contaminated with pathogens; indeed, exclusive breastfeeding is shown clearly to decrease death rate (15, 16). Children might receive a diet deficient in protein or essential nutrients such as vitamin A because cultural norms reserve foods rich in these for adults. Maternal malnutrition is in itself common. Undernourished women are less likely to survive pregnancy, and more likely to have low birthweight infants. In some countries the most severely malnourished children are likely to have mothers with no formal education. Maternal educational attainment has repeatedly been found to be positively linked to improved child nutrition (17, 18). Foodstuffs must also be stored so that, subsequently, they are safe to eat; for example, it has already been pointed out that aflotoxins, which are by-products of fungal growth on foods, may be a precipitating factor in the development of kwashiorkor.

It is illustrative to examine data collected globally to help identify critical periods of vulnerability. Shrimpton et al. (19) published analyses of the timing of growth faltering in 2001 and in 2010 (20). Their analyses illustrate several important points. First, children born in Asia had birthweights substantially lower than did children born in either Africa or Latin America and the Caribbean, demonstrating the negative effects of intrauterine birth retardation. Low birthweight is widely recognized as a risk factor for illness and death. In terms of WAZ (weight for age) scores in the 2001 analysis, the children born in Asia were nearly 0.5 WAZ score below the children born in Africa, which, in turn, was about 0.25 WAZ score lower that of children born in Latin America and the Caribbean region. This is the result of maternal malnutrition and undernutrition preceding and during pregnancy. The second key finding is that at about 3 months of age, the WAZ score of the children began a remarkable decline, which continued until it reached at a nadir at 12–15 months. In Asia the average child ended up with a WAZ score of approximately -1.75, in Africa -1.5, and in Latin America and the Caribbean -0.75. This period corresponds to the period when infants are either weaned or no longer exclusively breastfed. In the 2010 analysis, the authors noted that in Asia this decline continued to 5 years of age, when the median WAZ score was −2.0. Height for age (HAZ) scores, an indicator of stunting, followed a pattern of continuous decline during the first 21–24 months of life, with an HAZ score of approximately −2.0 for both Asian and African children, and −1.3 for children from Latin America and the Caribbean. Analyses by other groups have led to similar and highly consistent conclusions. The implications, important for the logic model, are therefore that interventions to improve nutrition should occur before and during pregnancy and during the first 18–24 months of life. This is not to say that interventions should not occur at other points in the human life cycle; it is simply to say these are particularly critical times. Iron deficiency is also most critical during pregnancy and early childhood as well, as it is associated with an increased rate of infectious diseases, adverse events during pregnancy, and reductions in work capacity and school performance. Given this congruence between caloric and micronutrient deficiency, a life cycle approach targeting pregnancy and early childhood has arisen.

6.3 EXPECTED BENEFITS OF CLEAN WATER AND SANITATION PROVISION

There are three main bodies of evidence to be followed in describing the benefits of clean water and sanitation provision. The first relates to the burden of illness due to common infectious diseases and the reductions in illness or death that can be expected with the provision of sanitation or clean water. The second body of evidence is the at-times more fragmentary data relating to access to clean water and sanitation and nutritional status. These first two lines of evidence are increasingly combined and synthesized into estimates that inform the relationships between nutrition, water and sanitation, and health.

The third body of evidence shows a remarkable relationship that was first thoroughly documented a century ago, the Mills–Reincke phenomenon. This dealt with the unexpected outcome of additional lives saved with the provision of clean water and sanitation. In the early 1900s it was noted—by different observers in Europe and in the United States (21)—that for each death from typhoid disease prevented by water treatment (chlorination, filtration, or both), between two and five deaths due to other causes were prevented. Hiram Mills, a civil engineer working in Lawrence, Massachusetts, during the 1880s and 1890s, noted this outcome, while a contemporary, Dr. J.J. Reincke in Hamburg, Germany, independently observed the same phenomenon in Europe. The most likely explanation for this phenomenon is that persons enjoying improved water and sanitation did not acquire typhoid, and conse-

quently did not have their nutritional status impaired by this waterborne disease. Being more nutritionally robust, they would in turn survive, rather than succumb to, other conditions such as tuberculosis where nutritional status is a determinant of survival.

The conclusions drawn from the first two lines of evidence suffer from the fact that they either concluded a benefit resulted from associations found in non-perspective, cross-sectional data, because the studies were not prospective, or if the studies were prospective, the timeline for observation was relatively short, perhaps only a few years. Nonetheless the information is consistent and extensive, adding weight to the linkage between nutrition and water and sanitation. Conclusions based on the third body of evidence has the strengths of a perspective that extends over decades and addresses cumulative benefit over time to nutrition and to declines in mortality. It suffers at times from retrospective analysis of periods of time when improvements to water and sanitation were accompanied by advances in immunization, higher incomes, and expanded education. This mixture of interventions makes it difficult at times to tease apart the actual, fractional benefit derived from better water and sanitation.

6.3.1 Burden of Disease Related to a Contaminated Environment and Expected Decrease in Disease with Clean Water and Sanitation

An astonishing 94% of the burden of disability adjusted life years (DALY) due to diarrheal disease, overwhelmingly caused by bacterial and viral infections, is related to a contaminated environment, which includes unsafe water and a lack of sanitation (22). Thus, it is clear that efforts to decrease diarrheal disease must likely involve putting barriers between the groups at risk and the contaminated environment. In 1985, Esrey et al. (23) reviewed 67 studies from 28 countries and concluded that improved water supply and sanitation led to a median decrease in diarrhea morbidity of 22% and a reduction in mortality of 21%. They found that sanitation led to greater benefits than improved water quality. Twenty years after this important review, Fewtrell et al. (24) examined the data on the impact of clean water and sanitation on diarrheal disease incidence. They found an incidence reduction of between 25 and 37%, overall. The central estimates for the risk reductions they found were for hygiene (handwashing and education), 45%; for sanitation, 32%; for improved water supply, 25%; and household water treatment, 39%. The benefit of the latter in rural areas was again 39%. Multiple interventions, such as the concurrent introduction of sanitation, improved water, or hygienic measures, led to an estimated decrease in diarrheal diseases of 33%. It is notable that multiple interventions have not been found to lead to additive benefits in this and other studies.

Of interest, given the estimate by Fewtrell et al. of a 32% decline in diarrheal disease after the provision of sanitation is a study reported in 2007 relating to the decrease in diarrheal disease in Salvador, Brazil, after the implementation of a citywide sanitation system. Before (1997–1998) and after (2003–2004) the intervention, children in 24 sentinel areas were followed for up to 8 months so as to determine rates of diarrhea. Diarrhea fell by 22% overall, 43% in the areas where the baseline prevalence had been highest. Evidence within the study suggested that diarrhea disease transmission was prevented in the public, not private, domain, where household factors are largely unimportant. Furthermore, this benefit was citywide and not confined to wealthier households with greater resources, which addresses issues of environmental and social equity.

Few studies have directly examined the link between clean water and sanitation on the one hand and nutritional status on the other. In 1988, Stanton et al. (25) published work from Bangladesh suggesting that a simple educational intervention to decrease diarrhea might not lead to an improvement in nutritional status as measured in young children over a period of one year. No physical changes were made to water supplies or sanitation facilities. In contrast, more concrete evidence of the benefits of water and sanitation came from Daniels et al. in 1991, when they reported that children under the age of 5 from homes in Lesotho with latrines were on average 0.27 HAZ score larger than children from households without latrines, after adjustment for confounding factors. Magnani et al. (26) found similar results; the two greatest predictors of better nutritional status were found to be breastfeeding and adequate water supply with sanitation in the household. Of interest, breastfeeding was more common in the least wealthy households, and was protective against malnutrition until weaning occurred.

Several prospective studies linking water and sanitation and nutrition have been published. Esrey et al. (27) collected data on 119 children in Lesotho in 1982–1983, and reported that infants from households with latrines and which increased water use for hygienic purposes in the wet/warm season had significantly better growth than children from households without latrines or with latrines but without increased water use. Infants from households with both a latrine and increased water use gained more than a kilogram more in weight and 2 cm or more in length than infants whose household had only a latrine or only increased water use but did not have a latrine. Merchant et al. (28) reanalyzed data from 25,483 children enrolled in a vitamin A supplementation trial that was conducted in 1988 in the Sudan. This is one of the few prospective studies with a large sample size that has been published, which adds confidence to the results. Children coming from homes with water and sanitation facilities were significantly better nourished than children without household water or sanitation, after adjustment for confounding factors such as breastfeeding, wealth, and maternal literacy. HAZ scores for the former cohort were -1.66 and -1.55, respectively, at the beginning and end of the study, versus -2.03 and -1.94 for the latter ($p < 0.001$). For children with normal stature at the beginning of the study, the risk of developing stunting (a HAZ score of -2 or lower) was 21% lower for children whose households had both water and sanitation, compared to households without these features. Consistent with this observation, stunted children who came from households with water and sanitation had a 17% greater likelihood of reversing stunting than did children from households without water and sanitation.

The themes of infectious diseases, especially diarrheal diseases, safe water, maternal literacy, poverty and equity, and agricultural abundance, are seen to be important in another set of analyses. Moore et al. (29) analyzed World Bank and UNICEF data from 22 countries in Latin America and the Caribbean dating to the 1990s. Five measures were independent predictors for childhood mortality: vaccinations (which prevent infectious diseases, which in turn prevents nutritional losses due to the infections), the use of oral rehydration therapy during diarrhea episodes, access to safe water, female literacy, and per capita income. Consistent with this work, Milman et al. (30) examined demographic health data from 85 countries worldwide. They studied at least two demographic surveys that were taken at least 4 years apart for variables that would explain changes in stunting. They found that immunizations, the provision of safe drinking water, female literacy, government expenditures, income distribution (equity), and proportion of the economy devoted to agriculture explained the majority of changes in stunting rates. Taken together these explanatory variables relate to both stunting and to mortality, providing further linkage between clean water, nutrition, and mortality. It should be noted that evidence has been provided that strongly suggests analysis of household level data from demographic surveys underestimates the potential benefits of improving sanitation facilities (31).

Based upon a similar approach, and combining information on reductions in mortality with household information collected from national demographic surveys, Gakidou et al. (4) recently estimated that implementing measures to improve nutrition, water and sanitation, and clean biofuels (to reduce indoor air pollution and pneumonia, for example) would reduce childhood deaths by 49,700 (14%) per year in Latin America and the Caribbean, 0.8 million (24%) in South Asia, and 1.47 million (31%) in sub-Saharan Africa. They point out that this represents a substantial fraction of the Millennium Development Goal of reducing child mortality by 50%. Their approach was to estimate the benefits of reducing excess mortality due to unsafe water and sanitation and to underweight using estimates for the excess risks similar those shown in Tables 6.1 and 6.3, respectively. Again addressing social equity, they noted that if half the resources required for this benefit were first targeted to the poorest half of children, much more than half the benefit would occur.

In summary, there is a consistent finding that the provision of clean water and sanitation will result in decreased infectious disease, principally diarrhea, and related malnutrition.

6.3.2   Historical Information Regarding Connection between Nutrition and Water and Sanitation

As noted earlier, a third line of evidence relating to this relationship is analysis of the historical record. One advantage to this approach is the ability to judge cumulative benefit over time, while the major disadvantage to this approach is the difficulty in disentangling the effects of simultaneous efforts to improve the health of the population. In a practical sense, the benefits from water and sanitation improvement may be difficult to distinguish from concurrent decreases in poverty or other socioeconomic conditions.

An example of such an approach was published by Thomas Merrick in 1985 (32), in which he analyzed the effects of piped water on early childhood mortality in urban Brazil during the period from 1970 to 1976, using national demographic information. During this period the Brazilian government targeted the provision of water and sanitation to the urban population, and it followed a major improvement in raising the educational attainment of women in the 1960s. These better educated women became the mothers of the 1970s, and Merrick chose to analyze data from households with women in their twenties. Overall infant mortality declined from 98 per 1000 to 89 per 1000 during this period, which is roughly comparable to much of Africa and Southeast Asia today. Merrick found, using regression analysis, that the education of mothers and fathers had the greatest total effect on differences in child mortality (about half), while improved access to piped water (and possibly sanitation, which was not independently measured) conservatively accounted for approximately a fifth of the decline in infant mortality. In general, the educational status of the parents—a proxy for other socioeconomic factors such as income—would have existed for a longer period of time than the presence of piped water in the household. Merrick pointed out that had there been no improvement in educational achievement in the population during this period, the magnitude of the benefit of improved access to piped water would likely have been larger. In addition, he noted that Brazil had chosen to invest in education before it invested in water supply, and "the differences in the relative impact of the two variables may also reflect these differences in timing." Despite this work being over 25 years old it remains an important and seminal work.

In 2006, Sepulveda et al. (33) from Mexico reviewed the improvements in child, infant, and neonatal mortality rates that had occurred in Mexico over the years 1980–2005. Stunting had declined from just under 23% to less than 13% of children, and underweight declined from 14 to 5%. There was coincidentally a remarkable decline in diarrhea-specific mortality from 1163 per 100,000 in 1980 to 71 per 100,000 in 2005. In a retrospective analysis of the most important predictive factors to explain the decrease in childhood mortality during this period, the proximate determinants were nutritional (underweight or stunted growth), and the most important contextual determinants were maternal education, population access to sewage systems, residence in a household with a dirt floor (a proxy for poverty), and the percentage of children without an immunization against measles. When cholera was introduced in Mexico in 1991, it led to a substantial increase in the percentage of Mexican communities with potable water, increasing from 55% in 1990 to 90% in 1992 (34), it has not subsequently fallen to a floor beneath it. Again, the issue here is disentangling the benefit of water and sanitation measures from other social advances, such as increased social security and nutrition campaigns targeted to the poor. This study exemplifies a number of studies that descriptively portray simultaneous social events that include provision of water and sanitation and attempt to identify the most predictive factors for ameliorating malnutrition and improving lifespan.

Another approach to understanding the relationships between water and sanitation, nutrition, and overall health is exemplified by the econometric and demographic studies of Ferrie and Troesken (35) and Cutler and Miller (36). Ferrie and Troesken examined the expansion of water and sanitation supply in Chicago from 1850 to 1925 and related it to the decreasing incidence of typhoid fever (a waterborne disease that affects only humans) and other noninfectious diseases. They found that for every death averted from typhoid disease, approximately three other deaths were averted from other diseases such as tuberculosis, pneumonia, diphtheria, and chronic bronchitis. Before 1880, over half the deaths were in children under the age of 5, with diarrhea as the leading cause; by 1925 less than a quarter of the deaths were in children of this age. Using econometric modeling they found that the provision of water and sanitation accounted for 35–50% of the decline in deaths during the period from 1850 to 1925, when the overall mortality rate decreased by 60%. The leading causes of death in 1925 had become cardiovascular disease and cancer. When we try to understand these data, consistent with the Mills–Reincke phenomenon, it is difficult to enunciate a mechanism to explain these additional lives saved without invoking nutrition and the capacity to resist death from other infectious causes.

Cutler and Miller looked at the timing of water improvements in the United States during the period from 1900 to 1936 and found a consistent effect of combined chlorination and filtration on typhoid fever temporally related to water treatment. They examined individual cities and analyzed changes that occurred around the exact time of water supply improvements. These improvements occurred independently of other broader changes in the social landscape of the United States, which allowed them to quantify the benefits of improved water supply independently of the other changes. Their analysis suggests that in the United States, clean water led over time to decreases in child mortality that outweighed all other interventions. To quote them

> The magnitude of these estimated effects is striking. In our sample of major cities, the reduction in mortality from 1900 to 1936 was about 30%. Our results suggest that clean water technologies reduced mortality by 13% during this period, accounting for about 43% (13/30) of the total reduction. Infant and child mortality fell by 62% and 81%, respectively, in these cities during this period. Clean water appears to have been responsible for 74% (46/62) of the reduction in infant mortality and 62% (50/81) of the reduction in child mortality. Similarly, clean water led to the near eradication of typhoid fever. … The effect of clean water on total mortality is much larger than what can be attributed to typhoid fever alone … reductions in all waterborne diseases account for about 8% of the reduction in total mortality.

Of critical importance is the fact that, of the 43% decrease in mortality attributable to water treatment, only 8% was *directly* attributable to waterborne disease. The other 35% was *indirectly* attributable to water treatment, that is, an additional four deaths were averted per death from waterborne diseases. Once again the multiplier effect of the Mills–Reincke phenomenon was seen.

Cutler and Miller attempted to estimate the economic benefits that might accrue if only 1% of the 1.7 million annual deaths attributable to diarrheal disease could be prevented with water disinfection. Using a Mills–Reincke multiplier of 3, and 30 person-years lost per death with a value of person-year of $100,000, then the social rate of return would be 160 billion USD for deaths averted. Even if our current social order does not value the life of a person in the developing world as highly as this, this is a staggering return on investment.

Econometric and demographic analyses thus suggest that the benefits to the provision of clean water and sanitation are substantially greater than what can be ascribed to waterborne diseases, accrue over time, and outweigh in aggregate other social changes benefiting mortality, especially in children.

6.4 SUMMARY

Malnutrition remains a prominent global cause of death and morbidity. It exists in a number of forms, including both caloric and micronutrient deficiencies. The periods of greatest vulnerability are pregnancy and early childhood. Death from malnutrition frequently occurs during or after precipitating waterborne or water-related diseases. Malnutrition leads to a state of immunological vulnerability to infection, which is reflected in elevated death rates in malnourished groups. Mortality is directly linked to the extent of malnutrition.

Clean water and sanitation decrease the incidence of infectious diseases such as diarrhea and pneumonia, both of which can be prevented if clean water is available for ingestion and washing, and sanitation. Clean water and sanitation are strongly linked to better nutritional status. Current estimates are that about one-third of the episodes of these diseases can be prevented in the short term when clean water and/or sanitation are provided. A separate line of historical evidence suggests that short-term studies significantly underestimate the benefits that accrue over time to populations enjoying the benefits of clean water and sanitation. The most rational explanation for this is the mediation that results through improved nutrition as the cycle of infection and malnutrition is averted. Indeed, some have concluded that historically, there is no intervention more important to decreasing mortality than clean water and sanitation. Malnutrition can best be addressed by equity in access to food, improved water and sanitation, an adequate supply of foodstuffs, and education. Water and sanitation are critical keystones in ridding the world of the pernicious scourge of malnutrition.

REFERENCES

1. Black RE, et al. Maternal and child undernutrition: global and regional exposures and health consequences. *Lancet* 2008;371 (9608):243–260.

2. Scrimshaw NS, SanGiovanni JP. Synergism of nutrition, infection, and immunity: an overview. *Am J Clin Nutr* 1997;66 (2):464S–477S.

3. Howard G, Bartram J. *Domestic Water Quantity, Service Level, and Health*. Geneva, Switzerland: WHO Press, 2003. Available at http://www.who.int/water_sanitation_health/diseases/ WSH03.02.pdf.

4. Gakidou E, Oza S, Vidal Fuentes C, Li AY, Lee DK, Sousa A, Hogan MC, Vander Hoorn S, Ezzati M. Improving child survival through environmental and nutritional interventions: the importance of targeting interventions toward the poor. *JAMA* 2007;298:1876–1887.

5. de Benoist B, McLean E, Egli I, Cogswell M, editors. *Worldwide Prevalence of Anaemia 1993–2005: WHO Global Database on Anaemia*. Geneva, Switzerland/Atlanta, USA: World Health Organization/Centers for Diseases Control, 2008. Available at www.who.int/vmnis/publications/anaemia_ prevalence/en/index.html.

6. Manary MJ, Sandige HL. Management of acute moderate and severe childhood malnutrition. *BMJ* 2008;337:a2180–a2180.

7. Zimmerman MB. Iodine deficiency. *Endocr Rev* 2009;30 (4):376–408.

8. Tazhibayev S, Dolmatova O, Ganiyeva G, Khairov K, Ospanova F, Oyunchimeg D, Suleimanova D, Scrimshaw N.

Evaluation of the potential effectiveness of wheat flour and salt fortification programs in five Central Asian countries and Mongolia, 2002–2007. *Food Nutr Bull* 2008;29:255–265.

9. Sommer A. Vitamin A deficiency and clinical disease: an historical overview. *J Nutr* 2008;138:1835–1839.

10. Horton S, Alderman H, Rivera JA. *Copenhagen Consensus 2008 Challenge Paper: Hunger and Malnutrition*, Copenhagen Consensus Center, Frederiksberg, Denmark, May 2008, p. 32.

11. Brown KH, Peerson JM, Baker SK, Hess SY. Preventive zinc supplementation among infants, preschoolers, and older prepubertal children. *Food Nutr Bull* 2009;30 (1 Suppl):S12–S40.

12. Guerrant RL, Oria RB, Moore SB, Oria MO, Lima AA. Malnutrition as an enteric infectious disease with long-term effects on child development. *Nutr Rev* 2008;66:487–505.

13. Fishman SM, Caulfield LE, de Onis M, et al. Childhood and maternal underweight. In: Ezzati M,editor. *Comparative Quantification of Health Risks*, Vol. 1. WHO, 2004, pp. 39–162.

14. Sen AK. *Poverty and Famines: An Essay on Entitlement and Deprivation*. Oxford, UK: Oxford University Press, 1981.

15. Edmond K, et al. Delayed breastfeeding initiation increases risk of neonatal mortality. *Pediatrics* 2006;117:e380–e386.

16. Mullany LC, et al. Breastfeeding patterns, time to initiation and mortality risk among newborns in southern Nepal. *J Nutr* 2008;138:599–603.

17. Cleland JG, van Ginneken JK. Maternal education and child survival in developing countries: the search for pathways of influence. *Soc Sci Med* 1988;27:1357–1368.

18. Liaqat P, et al. Association between complementary feeding practice and mothers education status in Islamabad. *J Hum Nutr Diet* 2007;20:340–344.

19. Shrimpton R, Victora CG, do Onis M, et al. Worldwide timing of growth faltering: implications for nutritional interventions. *Pediatrics* 2001;107(5):e75.

20. Victora CG, de Onis M, Curi Hallal P, Blössner M, Shrimpton R. Worldwide timing of growth faltering: revisiting implications for interventions. *Pediatrics* 2010;125:e473–e480. Available at http://www.pediatrics.org/cgi/congent/full/125/3/e473.

21. Sedgwick WT, MacNutt JG. On the Mills–Reincke phenomenon and Hazen's theorem concerning the decrease in mortality from diseases other than typhoid fever following the purification of public water supplies. *J Infect Dis* 1910;7: 489–564.

22. Prüss-Üstün A, Corvalán C. *Preventing Disease Through Healthy Environments. Towards an Estimate of the Environmental Burden of Disease*. Geneva: World Health Organization, 2006.

23. Esrey SA, Feachem RG, Hughes JM. Interventions for the control of diarrhoeal diseases among young children: improving water supplies and excreta disposal facilities. *Bull World Health Organ* 1985;63(4):757–772.

24. Fewtrell L, Kaufmann RB, Kay D, Enanoria W, Haller L, Colford JM, Jr., Water, sanitation, and hygiene interventions to reduce diarrhoea in less developed countries: a systematic review and meta-analysis. *Lancet Infect Dis* 2005;5:42–52.

25. Stanton BF, Clemens JD, Khair T. Educational intervention for altering water sanitation behavior to reduce childhood diarrhea in urban Bangladesh: impact on nutritional status. *Am J Clin Nutr* 1988;48(5):1166–1172.

26. Magnani RJ, Mock NB, Bertrand WE, Clay DC. Breastfeeding, water and sanitation, and childhood malnutrition in the Philippines. *J Biosoc Sci* 1993;25:195–211.

27. Esrey SA, Habicht JP, Casella G. The complementary effect of latrines and increased water usage on the growth of infants in rural Lesotho. *Am J Epidemiol* 1992;135:659–666.

28. Merchant AT, Jones C, Kiure A, Kupka R, Fitzmaurice G, Herrera MG, Fawzi WW. Water and sanitation associated with improved child growth. *Eur J Clin Nutr* 2003;57:1562–1568.

29. Moore D, Castillo E, Richardson C, Reid RJ. Determinants of health status and the influence of primary health care services in Latin America, 1990–1998. *Int J Health Plann Manage* 2003;18:279–292.

30. Milman A, Fongillo EA, de Onis M, Hwang JY. Differential improvement among countries in child stunting is associated with long-term development and specific interventions. *J Nutr* 2005;135:1415–1422.

31. Lee L, Rosenzweig MR, Pitt MM. The effects of improved nutrition, sanitation, and water quality on child health in high-mortality populations. *J Econometr* 1997;77:209–235.

32. Merrick T. The effect of piped water on early childhood mortality in urban Brazil, 1970 to 1976. *Demography* 1985;22:1–24.

33. Sepulveda J, Bustreo F, Tapia R, Rivera J, Lozano R, Olaiz G, Partida V, Garcia-Garcia L, Valdespino JL. Improvement of child survival in Mexico: the diagonal approach. *Lancet* 2006;368(9551):2017–2027.

34. Sepulveda J, Valdespino JL, Garcia-Garcia L. Cholera in Mexico: the paradoxical benefits of the last pandemic. *Int J Infect Dis* 2006;10:4–13.

35. Ferrie JP, Troesken W. Water and Chicago's mortality transition, 1850–1925. *Explor Econ Hist* 2008;45:1–16.

36. Cutler D, Miller G. The role of public health improvements in health advances: the twentieth-century United States. *Demography* 2005;43:1–22.

7

SOIL-TRANSMITTED HELMINTHS: *ASCARIS*, *TRICHURIS*, AND HOOKWORM INFECTIONS

BRIAN G. BLACKBURN AND MICHELE BARRY

7.1 INTRODUCTION

Ascaris lumbricoides, *Trichuris trichiura*, and the human hookworms (*Necator americanus* and *Ancylostoma duodenale*) are the parasitic organisms collectively known as the soil-transmitted helminths (STHs). These parasites share the requirement that their eggs or larvae must develop on soil before they can become infectious to humans. Although other helminths are transmitted primarily through soil (e.g., *Strongyloides stercoralis*, *Toxocara* spp.), they are not grouped with the soil-transmitted helminths either because they lack this requirement or are parasites for which humans are not the definitive host. Endemic primarily in developing regions, these infections are intimately associated with poverty, not only resulting from poverty but also contributing to the perpetuation of poverty. Poor sanitation in endemic areas is a prime contributor to the transmission of these parasites. Control and elimination campaigns have focused on mass chemotherapy with albendazole or mebendazole, but environmental modification, improved sanitation, and especially economic development (which would facilitate better access to clean water and footwear) may also be integral components of the control of these diseases.

7.2 THE ORGANISMS

The soil-transmitted helminths (also known as the geohelminths) are intestinal nematodes, round nonsegmented worms with elongate, cylindrical bodies. The sexes are separate, and fertilized females produce eggs that subsequently give rise to larvae. After molting, larvae become adult worms. The life cycles of the four species are similar, although *Ascaris* and *Trichuris* are primarily transmitted orally, while hookworms are acquired primarily through penetration of skin in contact with infested soil. The soil-transmitted helminths do not reproduce in humans.

7.2.1 *Ascaris*

A. lumbricoides is the most common helminth infection of humans, with an estimated prevalence of about 1 billion persons globally (Table 7.1) (1). Ascariasis is widely distributed, but is most abundant in the tropics. Although most people infected with *Ascaris* are asymptomatic, mortality attributed to *Ascaris* is estimated at >60,000 people per year, and >15% of infected people suffer morbidity (2).

Adult *Ascaris* worms are typically creamy or pink-white, 15–45 cm long, and live in the small intestine, particularly the jejunum (Table 7.1, Fig. 7.1). Eggs are thick-shelled, oval, and about 65 μm × 45 μm (Fig. 7.2). Eggs are passed in the feces of infected persons; once they reach a favorable environment (warm, moist soil), they become infectious after developing on soil for an interval that is dependent mostly on temperature (e.g., 10–14 days at 30°C or 6 weeks at 17°C) (2). Humans usually acquire the infection by ingestion of these eggs via contaminated food (or water); eggs then hatch in the small intestine and release larvae that penetrate the intestine and migrate to the lungs a few days later. Over the week that follows, they ascend the tracheobronchial tree, are swallowed, and return to the intestines, where they mature into adult worms. Egg production begins approximately

Water and Sanitation-Related Diseases and the Environment: Challenges, Interventions, and Preventive Measures, First Edition.
Edited by Janine M. H. Selendy.
© 2011 Wiley-Blackwell. Published 2011 by John Wiley & Sons, Inc.

TABLE 7.1   **Epidemiologic, Parasitologic, and Clinical Characteristic of the Soil-Transmitted Helminths**

Characteristic	*A. lumbricoides*	*T. trichiura*	*N. americanus*	*A. duodenale*
Global prevalence (millions)	800–1,200	600–800	600–700[a]	600–700[a]
Size (mm)	150–450	30–50	7–13	8–13
Daily egg output	200,000	3,000–5,000	5,000–10,000	10,000–30,000
Daily blood loss	N/A	Undefined	0.03 mL	0.2 mL
Life expectancy of adult worms (years)	1–2	1–3	3–5	1
Larval developmental arrest in humans	No	No	No	Yes
Oral transmission	Yes	Yes	No	Yes
Cutaneous transmission	No	No	Yes	Yes

[a] Total for both hookworm species collectively.

2 months after infection, and adult worms live 1–2 years. Each adult female worm can produce over 200,000 eggs per day (3).

7.2.2 *Trichuris*

T. trichiura is also known as the whipworm because of its long, whip-like head, which embeds into the intestinal mucosa (Fig. 7.3). Adult worms are gray-to-pink and 3–5 cm long. They primarily reside in the cecum and ascending colon; with heavy infections, the distribution can extend to the entire colon. Eggs have a distinctive barrel-shaped appearance with a thick shell, bipolar plugs, and average 50 μm × 20 μm in size (Fig. 7.4). The life cycle of *Trichuris* is simple: Eggs shed in the feces become infective under appropriate (moist and shady) conditions once on soil, within

2–4 weeks. After infectious eggs have been ingested by a human, larvae emerge and move to the cecum. Here they molt, embed in the epithelium, and mature into adults. Oviposition begins 2–3 months after infection, and adult worms live 1–3 years. Female worms can produce up to 20,000 eggs per day (Table 7.1) (2, 4).

7.2.3 Hookworms

The human hookworms *A. duodenale* and *N. americanus* are small, gray-white nematodes about 0.7–1.3 cm long that live in the upper small intestine (Table 7.1). The ovoid, thin-shelled eggs of the two species are identical, and measure about 60 μm × 40 μm (Figs. 7.5 and 7.6). As with the other soil-transmitted helminths, eggs passed in the stool of infected persons must develop on soil (ideally warm, moist, and

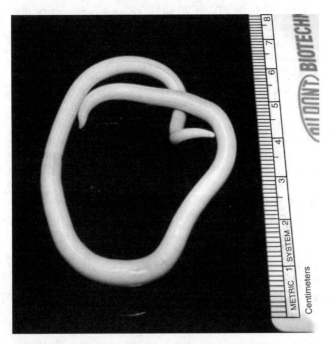

FIGURE 7.1   Adult *A. lumbricoides* worm. (Courtesy Division of Parasitic Diseases/Centers for Disease Control and Prevention (CDC).

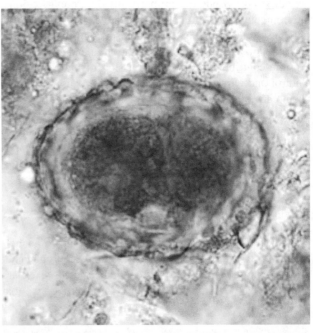

FIGURE 7.2   *A. lumbricoides* egg. (Courtesy Division of Parasitic Diseases/CDC.)

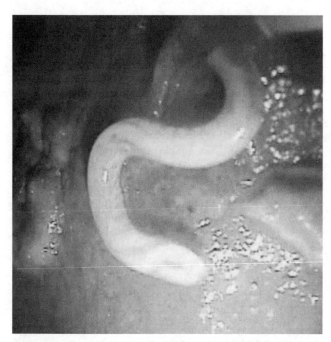

FIGURE 7.3 *T. trichiura* adult worm (colonoscopy image). (Courtesy Division of Parasitic Diseases/CDC and Duke University Medical Center.)

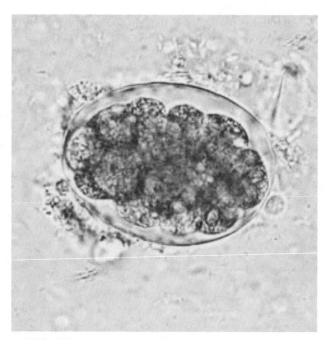

FIGURE 7.5 Hookworm egg. (Courtesy Division of Parasitic Diseases/CDC.)

shady); they hatch there within 1–2 days, becoming larvae. Larvae then molt over 5–10 days, subsequently becoming infectious. Following contact with human skin, larvae enter the body and are carried to the lungs. They then move up the trachea, are swallowed, and migrate to the small intestine.

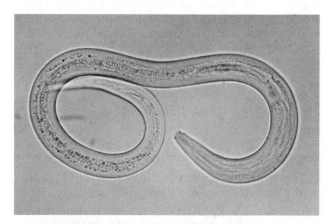

FIGURE 7.6 Infective hookworm larva. (Courtesy Division of Parasitic Diseases/CDC.)

Females start egg deposition 4–6 weeks after infection. Although *N. americanus* can only be transmitted via this percutaneous route, *A. duodenale* can be transmitted orally as well as percutaneously. The adult lifespan averages 3–5 years for *N. americanus*, and 1 year for *A. duodenale* (Table 7.1) (1).

7.3 EPIDEMIOLOGY

Soil-transmitted helminths are among the most common human parasites globally. Over a billion people are infected by at least one of these worms, and many harbor infection

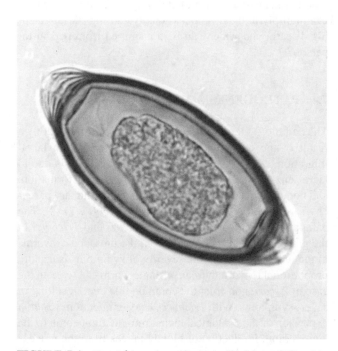

FIGURE 7.4 *T. trichiura* egg. (Courtesy Division of Parasitic Diseases/CDC.)

with multiple species. It is thought that approximately a billion people are infected with *Ascaris*, about 800 million with *Trichuris*, and about 700 million with hookworm (1–6).

The geographic distribution of these parasites is determined primarily by sanitation and climate. Eggs and larvae become infective on warm, moist soil; the tropics are thus well suited for transmission, and because eggs are spread via feces, areas with poor hygiene have the highest transmission rates. Sanitation therefore plays an important role in determining the degree of environmental contamination with eggs. Many developing countries share both the environmental and climatic conditions necessary for transmission; although most are tropical, some temperate countries sustain transmission, albeit seasonally (1–6). For both *Ascaris* and *Trichuris*, children in impoverished rural areas are particularly heavy amplifiers, as they often play on contaminated soil and more frequently are exposed by hand-to-mouth behaviors (2). The highest prevalence rates are typically seen in those under 10 years of age, and these individuals are often the most heavily infected members of a community (7–9). Even in highly endemic areas, most infected people harbor only a small number of worms; only a minority (usually children) are heavily infected (7–9). Children therefore account for the majority of the worms in a community and most of the eggs that are shed into the environment (7–9). Infection rates are also high in areas where human excrement is used to fertilize crops ("night soil"). It also appears genetic susceptibility may predispose some individuals to heavy infections (7). In endemic areas, the prevalence is commonly as high as 80% (2, 10). *Ascaris* transmission is particularly robust, given the enormous daily egg output and environmental resistance of these eggs. *Ascaris* eggs can remain viable for over 5 years in moist, loose soil, and can survive desiccation and freezing temperatures (2).

The epidemiology of hookworm infection differs slightly from that of *Ascaris* and whipworm. *N. americanus* is the predominant hookworm worldwide. Hookworm infection also occurs in tropical and subtropical areas, limited by a requirement for a warm, moist climate. Overall, the highest prevalence (and intensity) of hookworm occurs in sub-Saharan Africa, followed by China and Southeast Asia (5). Unlike *Ascaris* and *Trichuris*, hookworm prevalence is not exaggerated in urban slums, but instead occurs mostly in poor, tropical, coastal communities and agrarian communities (2, 5). Larval development for hookworm is more susceptible to climatic extremes (low rainfall or temperature) than it is for the eggs of *Ascaris*. Unlike *Ascaris* and *Trichuris*, hookworm prevalence most commonly increases throughout childhood and plateaus in young adulthood (6).

Data corroborated from geographical information systems and satellite sensors have suggested that *A. lumbricoides* and *T. trichiura* do not occur in areas where temperatures exceed 37–40°C. However, hookworm occurs throughout many such areas with these high temperatures,

suggesting that hookworm tolerates greater temperatures than *A. lumbricoides* and *T. trichiura* (11). In the tropics, hookworm is generally confined to coastal plains below 150 m elevation. Above these altitudes low temperatures (<20°C) usually limit transmission (11). Because *A. duodenale* larvae can undergo arrested development within humans, it can thrive in some areas where *N. americanus* cannot; this may allow *A. duodenale* to survive cold winter months. Hookworms are unable to survive desiccation, and a minimum amount of rainfall is also an important determinant influencing hookworm transmission. In some endemic areas where hookworm transmission is seasonal, new infections with *A. duodenale* can appear 8–10 months after the rainy season, following emergence of larvae from this arrested development stage (5, 7, 11).

Soil type is another environmental factor important for hookworm transmission. Sandy soils are well suited for this because the small particle size and well-aerated texture allows infective larvae to migrate within the soil. Clay soils inhibit migration, which is another reason hookworm is most prevalent in coastal areas, where sandy soils predominate (11).

Most cases of geohelminth infections in nonendemic regions occur among immigrants and travelers. Infections in such persons generally resolve within a few years even without treatment, when the adult worms die. Because of the requirement that geohelminth eggs or larvae develop on soil before becoming infectious, these parasites cannot be transmitted directly from person-to-person, and cannot multiply in the host. This contrasts, for example, with *S. stercoralis* (another nematode transmitted through soil contact), which can multiply in the host, occasionally causing life-threatening disease, and which can be transmitted from person-to-person.

7.4 PATHOGENESIS AND PATHOLOGY

The pathology stemming from *Ascaris* infection is mainly a result of the host response to the parasite. During larval migration, cells suffer mechanical trauma and lysis due to larval enzymes. Larvae also induce granuloma formation, and both *Ascaris* and hookworm larvae in pulmonary parenchyma cause a hypersensitivity reaction (Fig. 7.7) (2). The pathophysiologic consequences of *Ascaris* and *Trichuris* in the gastrointestinal tract result from the presence of worms in the lumen. Although the severity of symptoms is usually proportional to worm burden, a single worm can (e.g., in the case of *Ascaris*) obstruct the common bile duct, resulting in severe symptoms. With *Trichuris*, heavier infections result in expansion of the ecological niche from the right colon to the entire colon. Production of bloody mucus from the mucosa occurs, with anemia and impaired growth as possible sequelae. Hookworms, by contrast, exert their primary pathologic

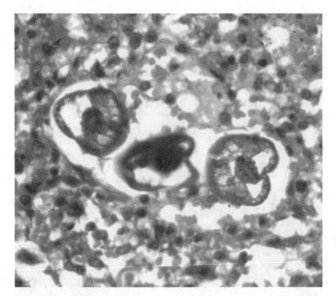

FIGURE 7.7 *A. lumbricoides* larvae in lung tissue. (Courtesy Division of Parasitic Diseases/CDC.) (*See insert for color representation of this figure.*)

effect via blood loss. They attach to the intestinal mucosa by their strong buccal capsules and cutting plates or teeth and secrete anticoagulants and anti-inflammatory factors, allowing continuous blood ingestion (Fig. 7.8a and b). Chronic iron deficiency is particularly detrimental in childhood and may directly impair cognitive and intellectual development (5, 6, 12).

Human infection with these parasites leads to a predominantly Th2 immune response. It is thought that this shift in immunological response may impact the manifestations of allergic and rheumatologic diseases as well as the response to infections typically controlled by the Th1 response (such as tuberculosis) (1). There is controversy regarding whether antibodies produced in response to geohelminth infections are protective or merely a marker of past or present infection (7).

7.5 CLINICAL MANIFESTATIONS

Most people infected with *Ascaris* are asymptomatic, although a small proportion develop pulmonary symptoms (such as cough) while larvae migrate through the lungs (1–2 weeks after infection). Eosinophilia may accompany this, as can eosinophilic pneumonia.

With most established infections, adult worms in the small bowel lumen provoke no symptoms or produce only mild abdominal pain, nausea, or anorexia. However, heavier infections can adversely affect the nutritional status of children, especially in areas where malnutrition is otherwise common (13). This can also adversely impact intellectual development and growth; treatment of heavily infected children with antihelminthics improves these outcomes (13).

Other complications of chronic ascariasis are rare, largely due to mechanical issues related to the presence of the large adult worms in the gut, and can include intestinal, biliary, or pancreatic obstruction, appendicitis, and intestinal perforation. When these complications occur, they are more common in children from endemic areas (2). Occasionally, adult worms can pass per rectum, through the nose, or tear ducts.

As with *Ascaris* infections, most people with trichuriasis are asymptomatic or have only eosinophilia. With heavy infections (particularly in children), chronic abdominal pain and diarrhea can result, as can anemia, growth retardation, and clubbing (4). Stools may have an acrid smell, and nocturnal stooling can occur (2, 4). The *Trichuris* dysentery syndrome can result, characterized by tenesmus and frequent stools containing mucus and blood. Recurrent rectal prolapse occurs, with adult worms often visible (2, 4).

Similar to the other soil-transmitted helminths, most people infected with hookworm harbor light infections and are asymptomatic. Initial infection may be characterized by "ground itch," a pruritic maculopapular rash at the site of skin penetration, seen mostly in previously sensitized individuals. Serpiginous tracks (cutaneous larva migrans) on the skin occur with infection by animal hookworms (e.g.,

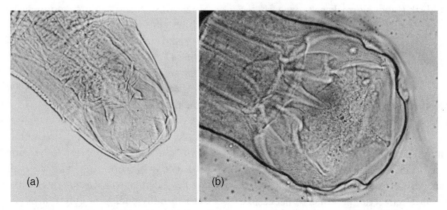

FIGURE 7.8 Adult hookworm mouthparts. (a) *A. duodenale* cutting teeth and (b) *N. americanus* cutting plates. (Courtesy Division of Parasitic Diseases/CDC.)

Ancylostoma braziliense) but not human hookworms (5). Migration of larvae through the lungs may provoke pneumonitis, which is less common and less severe than it is with *Ascaris* (2). Epigastric pain, diarrhea, and anorexia may occur about 6–12 weeks after penetration, as larvae begin attaching to the small bowel mucosa. When infection with *A. duodenale* occurs orally, the early migration of larvae can cause a syndrome known as Wakana disease, which is characterized by nausea, vomiting, pharyngeal irritation, cough, dyspnea, and hoarseness (5, 6). There are important clinical differences between the two hookworm species: *N. americanus* is shorter, removes less blood (0.03 mL/day versus 0.20 mL/day), and produces fewer eggs (5,000–10,000 versus 10,000–30,000 per day) than *A. duodenale* (Table 7.1) (5). Larvae of the latter can enter a stage of arrested development (including within humans), which may allow *A. duodenale* to persist in environmental conditions that would otherwise be too harsh (2, 5, 6). Infants can also be infected by *A. duodenale*, most likely through breast milk, and possibly transplacentally (2, 5, 6).

Some who are infected with adult hookworms experience chronic abdominal pain and eosinophilia. However, the most important manifestations of hookworm disease are iron-deficiency anemia and malnutrition. The anemia from hookworm infection develops gradually, but has especially severe consequences for children and pregnant women. It is a major health problem throughout the developing world, and can impair physical, cognitive, and intellectual growth in children, diminish productivity of workers, and threaten the outcome of pregnancy for mother and child (13). Anemia caused by hookworms is responsible for 60,000 deaths annually worldwide, and hookworms cause the greatest morbidity and mortality of the soil-transmitted helminths globally (5, 6, 12, 13). Although malaria is co-endemic to many areas with high hookworm prevalence, it is clear that increasing egg burdens are associated with lower hemoglobin levels, independent of other factors such as malaria infection (6). However, since much of the morbidity associated with both diseases results from anemia where the two are co-endemic, it is possible that hookworm disease exacerbates malarial anemia and vice versa.

7.6   DIAGNOSIS

The diagnosis of geohelminth infections is straightforward and can be made by identifying eggs in stool samples. The high daily egg output of *Ascaris* worms means that stool exams rarely miss patent infections, and concentration techniques are unnecessary (1). Trichuriasis and hookworm infections can also usually be diagnosed by simple stool smears, although concentration techniques may be necessary for light infections because the egg output is not as high as with *Ascaris* (1, 2). Stool samples may be negative until 2–3 months after infection occurs. Larvae can sometimes be

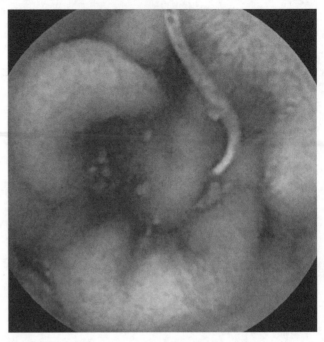

FIGURE 7.9   Hookworm in intestine (video capsule endoscopy image). (Courtesy Dr. Lauren Gerson, Stanford University School of Medicine.) (*See insert for color representation of this figure.*)

found in respiratory secretions of patients infected with *Ascaris* or hookworm during the early migratory phase of infection. At times, *Ascaris* worms may be seen radiographically (given their large size); all STHs may also be diagnosed by endoscopy (Figs. 7.3 and 7.9). Serology is rarely needed to diagnose these infections and is used more frequently in epidemiologic studies.

7.7   TREATMENT

Geohelminth infections should generally be treated with the oral benzimidazole drugs, albendazole or mebendazole. Single-dose or short-course regimens with these agents result in high cure rates for *Ascaris*; a recent meta-analysis demonstrated cure rates of 88–95% for *Ascaris* with a single dose of these agents (14). For *Trichuris* infections, single dose cure rates are low (28–36%); this parasite should therefore be treated with longer courses (at least 3–7 days) of antiparasitics to achieve higher cure rates (2, 4, 15, 16). Preliminary data suggest that mebendazole may be slightly superior to albendazole for whipworm, and that combination therapy with ivermectin may be superior to benzimidazole monotherapy (16). For hookworm, albendazole is preferred over mebendazole (72% versus 15% cure rates with single-dose regimen); even in individuals not cured, a large reduction in parasite burden and egg output occurs (2, 16). Although its use falls under pregnancy category C, the World Health Organization (WHO) has recommended the use of albendazole or mebendazole during pregnancy based on safety data, although they recommended deferring therapy during the first trimester; they also concluded that these drugs may be used in children as young as 12 months of age (17). Replacement of iron and other micronutrients is an addition to antiparasitics is an important therapeutic adjunct. Some evidence suggests that resistance to albendazole and mebendazole may be developing (18). Although alternative antiparasitics are available, none are currently as favored for use against all four geohelminth species as albendazole or mebendazole.

7.8   CONTROL AND ELIMINATION

Soil-transmitted helminth infections are prevented primarily by provision of clean food and water, good community sanitation, and adequate footwear. As provision of safe water and adequate sanitation is too expensive and logistically difficult for many developing countries to implement, WHO advocates treatment with antihelminthic drugs at regular intervals to populations at risk (17). This is done primarily to reduce individual worm burdens below those that cause significant disease and to decrease the overall community worm burden. Albendazole is safe, inexpensive, and widely available; programs employing this agent (or mebendazole) are therefore feasible for most developing countries. Several

studies have demonstrated that regular deworming of children (and pregnant/childbearing-age women) can prevent and reverse iron-deficiency anemia, impaired growth, malnutrition, poor school performance, and pregnancy complications (13).

A prominent example of successful soil-transmitted helminth elimination efforts is the southeastern United States, which was endemic for STHs through the early 1900s. An initiative (the Rockefeller Commission) was highly successful in achieving good control, and today transmission has been essentially eliminated (5, 7, 19, 20).

7.8.1   Mass Drug Administration

Current control initiatives focus largely on mass drug administration (MDA), usually with albendazole or mebendazole (Fig. 7.10). Potential population-based approaches for drug distribution include directed administration (only to those known to be infected), universal administration (to the entire population), or targeted administration (to certain subsets of the population) (9). Because of the cost and difficulty in establishing definitive diagnoses of STHs at large scale, most programs do not pursue the directed approach. Universal campaigns are often found when integrated elimination campaigns (e.g., filariasis mass drug administration campaigns) that target the entire population

FIGURE 7.10   Mass drug administration, Central Nigeria. (Photo Dr. Brian Blackburn, Stanford University School of Medicine.)

also include soil-transmitted helminth control as a goal (21). Most programs oriented solely at STHs currently follow the targeted approach, such as that of targeting schoolchildren for treatment with albendazole (9).

In 2001 the World Health Assembly passed Resolution 54.19, urging member states to control the morbidity of soil-transmitted helminths through mass drug administration programs (using albendazole or mebendazole) by targeting school-age children in developing countries; the established goal was to regularly treat >75% of school-age children at risk for infection globally by 2010 (22). Through 2006, 22 countries had reached this target (23). This program has most commonly been operationalized through school-based drug administration and has become among the largest of public health programs ever attempted. Through 2006, it was estimated that over 82 million persons globally were treated for soil-transmitted helminths, or about 10–20% of the at-risk population (23). Although not yet fully developed, the scale of this effort is beginning to approach that of more established programs, such as the lymphatic filariasis (LF) and onchocerciasis elimination and control campaigns (21).

The rationale for focusing on schools is that school-age children have the highest intensity of *Ascaris* and *Trichuris* infections (Fig. 7.11) and because schools provide a cost-effective and logistically simple way to deliver antihelminthics. However, because hookworm is frequently more prevalent in adults (including pregnant women) than children, and preschool children are vulnerable to permanent intellectual and cognitive deficits from infection as well, there is concern that school-based programs may not accurately target vulnerable hookworm-infected populations and will not effectively reduce hookworm transmission (6, 9). In addition, in areas with high-intensity transmission, hookworm infection remains endemic even when populations are regularly treated with antihelminthics. Hookworm infection routinely reoccurs, often returning to pretreatment levels within months. After one community-wide treatment, rates of hookworm infection reached 80% of pretreatment rates within 30–36 months, *A. lumbricoides* infection reached 55% of pretreatment rates within 11 months, and *T. trichiura* infection reached 44% of pretreatment rates within 17 months (24–27). In some cases, at least three treatments per year were required to improve iron status (6, 9). Another problem exists with whipworm, which requires multiple doses for a reliable cure. Logistical constraints limit most programs to single doses; whipworm control thus is achieved more slowly because of the relative resistance to treatment (9).

When implemented properly, the impact of MDA can be dramatic. In young children after 12 months of quarterly mebendazole administration in Zanzibar, moderate anemia was reduced by 59%, and growth and appetite improved (28). In India, when children 1–3 years of age received vitamin A and albendazole every 6 months, they gained 3.5 kg above

FIGURE 7.11   Typical transmission conditions for geohelminths, Central Nigeria. (Photo Dr. Brian Blackburn, Stanford University School of Medicine.)

expected in 2 years, and in Nepal twice-yearly distribution of vitamin A and albendazole to children under age 5 reduced anemia by 77% in 1 year (29, 30). Improvements in memory, language, problem solving, and attention have been seen in school-age children as a result of deworming programs (9, 31). Untreated Jamaican children with intense *T. trichiura* infections missed twice as many school days as their peers treated with albendazole (and those with lighter infection saw significantly more weight gain). In Kenya, school-based antiparasitics reduced school absenteeism by 25% (32, 33). There, a single dose of albendazole or placebo was given to schoolchildren who had very high prevalence rates of hookworm, trichuriasis, and ascariasis; at follow-up 6 months later, the treated group had gained 1.3 kg in weight and 0.6 cm in height, significantly more than the placebo group (34). The improvements were attributed more to the decrease in hookworm egg counts than to the other two parasites (34, 35). In Nepal, albendazole treatment in the second trimester of pregnancy resulted in a significant decrease in severe anemia, the mean infant birthweight rose by 59 g, and the infant mortality rate fell by 41% (36). In Uganda, two annual MDAs with albendazole resulted in a decrease in hookworm prevalence from 51% to 11% and a significant improvement in hemoglobin concentrations in just 2 years (37).

There has been some debate about the robustness of these findings. A recent meta-analysis of treatment of school-age children with antihelminthics showed a relatively consistent improvement in weight gain, but little benefit for growth, school attendance, or cognition (38). Nevertheless, because so many studies have shown impressive gains for the above indices after mass drug administration, such programs are generally considered effective and are recommended by WHO (22, 23). Another problem is the emergence of drug resistance; this has been seen already in many areas with ongoing MDA programs (39). In one example, from Zanzibar, a population treated with mebendazole every 4–6 months for the 5 years previous to the follow-up showed diminishing efficacy of mebendazole with repeated use. The cure rate (7.6%) and egg reduction rate (52.1%) for hookworm was significantly lower than in the first years of the program (22.4 and 82.4%, respectively) (18). Another problem is the lack of evidence that albendazole eradicates arrested *A. duodenale* larvae. Patients with latent *A. duodenale* infections could therefore theoretically excrete the parasite months after treatment, even without reinfection.

Several studies have demonstrated that 30–55% of moderate-to-severe anemia in pregnant women is attributable to hookworm; it is estimated that 44 million pregnancies annually are affected by hookworm (6). Severe iron-deficiency anemia during pregnancy has been linked to increased maternal mortality, prematurity, low birthweight, and impaired lactation (5, 6). Treatment of women in endemic areas with antihelminthics once or twice during pregnancy appears to substantially improve maternal hemoglobin, birthweight, and infant mortality (5, 6, 9). However, the robustness of these findings has been questioned: a meta-analysis of three studies (conducted in Peru, Sierra Leone, and Uganda) showed no effect of treatment with antihelminthics during the second or third trimester on maternal anemia, low birthweight, or perinatal mortality (40). Nevertheless, given the multiple studies that do suggest a benefit in pregnant women in hookworm-endemic areas, antihelminthic treatment is recommended during pregnancy (except during the first trimester) (22, 23).

Another question concerns the appropriate frequency of mass drug administration. Although the most common approach is annual treatment, the optimal frequency should be determined in part by the local prevalence and transmission intensity. For *A. lumbricoides* several studies have shown that in both low- and high-transmission areas, the most cost-effective option is to treat every 2 years to reduce prevalence and parasite intensity. In contrast, when prevalence reduction is the measure of effectiveness, the most cost-effective option is to treat every 4 months in high-transmission areas and every year in low-transmission areas, as prevalence reincreases more rapidly than intensity in the absence of retreatment (9, 41). Because of the nonlinear relationship between prevalence and infection intensity, when prevalence rates are high, the initial impact of mass chemotherapy is a reduction of the worm burden, rather than a reduction in prevalence (42, 43). As most complications of geohelminth infections occur in the presence of heavy worm burdens, a decline in intensity (even in the absence of a decline in prevalence) is an important goal in its own right (42, 43).

Treatment two or three times a year is effective in reducing morbidity in areas of intense transmission (prevalence >70%, with >10% of infections being moderate or heavy), such as in Zanzibar and Myanmar (9). In areas with a lower transmission intensity (prevalence 40–60% and <10% of infections moderate or heavy), such as in Oman, India, or Brazil, annual treatment was sufficient to reduce morbidity (9). WHO guidelines suggest treatment twice yearly if the STH prevalence in school-age children is >50% and once yearly if it is 20–50%. Although school-age children are the priority in such programs, WHO suggests also treating preschool children, women of child-bearing age, pregnant women (in the second and third trimester), and lactating women (17).

7.8.2 Sanitation

Sanitation is a factor closely interrelated with transmission of the soil-transmitted helminths. Ascariasis was once highly prevalent in Europe, with elimination aided by the implementation of sanitation and sewage systems (3). Given the lack of animal reservoirs for *A. duodenale* and *N. americanus*, effective sanitation should interrupt hookworm transmission (5). However, by itself, the effect of sanitation (including improved footwear) on reducing hookworm prevalence is often not realized for decades (9).

Despite the economic and public health importance of the STHs, the medical and international communities still largely fail to address their impact. They are considered among the "neglected tropical diseases," or NTDs. The neglect arises because the people most affected by these diseases are the world's most impoverished, particularly those who subsist on less than 2 USD/day (1). In addition, the infections have mostly insidious effects on health, and quantification of the impact of STHs has been difficult. When measured in disability-adjusted life years (DALYs) lost, the global burden from hookworm alone exceeds all other tropical infectious diseases except malaria, leishmaniasis, and lymphatic filariasis (12). Fortunately, over the past 5–10 years, the global community has begun to recognize the importance of these infections (1).

Poverty reduction and improved economic development have probably done more to eliminate hookworm in industrialized nations than any other single factor, including sanitation, antihelminthics, footwear, and health education. Economic development with improvement in living standards and the introduction of piped water, sewage systems, mechanized agriculture, and elimination of night soil are largely responsible for the control of hookworm infection in North America, Europe, Japan, and Korea (5). These measures would most likely be effective in many developing countries. Education regarding the proper use of sanitation facilities and the avoidance of using night soil is also essential. However, achieving these goals in resource-limited settings may prove to be more challenging than the reachable goals of mass chemotherapy programs. In addition, it is not clear that any individual component (such as education, improved sanitation, footwear) alone would achieve lasting and significant reduction in parasite burden; programs must instead integrate all of these components, and must involve a high percentage of the population, another reason such programs are difficult to successfully implement (9). An important concept is that the geohelminths not only stem from poverty but also *cause* poverty, perpetuating the cycle. In Kenya, for example, data suggest that deworming could raise per capita earning by 30% (44).

Sanitation consists of both improved personal hygiene (food preparation, handwashing, toilet use) and access to the "hardware" necessary to invoke these behaviors (latrines or toilets, wells or other safe water supply, sewage treatment) (9). Studies suggest that improved sanitary conditions can result in lower soil-transmitted helminth prevalence and reinfection rates (9). Programs have documented increased latrine number due to sanitation programs, but with little impact on STH transmission, largely because the population (especially children) was not using the latrines (45). Another major factor is the cost and sustainability of sanitation programs. Although potentially a prohibitive factor, a recent study in Vietnam assessed a market-based approach for sanitation items and found people were more willing to pay for services in an environment where they were considered customers rather than beneficiaries of a program. Proper sanitation, including decreased open defecation, increased 100% in just a year and was successful at improving STH transmission indices (46).

Such data are encouraging, as it is thought that the investment needed to provide access to adequate sanitation is currently beyond the resources of most low-income countries. In addition, coverage rates of properly built, used, and maintained sanitation likely must be >90% to decrease worm transmission (47). Recent experience in Vietnam demonstrated that the per-child cost for latrines is 7.90 USD (about 200 times the cost per dose for deworming). In addition, this particular intervention increased the latrine coverage in each community to less than 1% (9).

7.8.3 Program Integration

Integration of soil-transmitted helminth control and elimination programs with other such programs directed at parasites (such as lymphatic filariasis, onchocerciasis, schistosomiasis, and malaria) may be the way forward. Increasing public and private advocacy for this approach has been building over the last few years, and has the advantage of addressing multiple public health problems with fewer resources than each would require individually. Using existing infrastructure to deliver multiple interventions reduces costs and takes advantage of access to communities that is already in place. However, although they may lower the prevalence and intensity of STH infections, such programs are unlikely to eliminate transmission in the absence of the adjunctive factors described above, such as improved sanitation, clean water supplies, and economic development.

As an example, adding deworming medications to vitamin A distribution or immunization campaigns can significantly increase the number of preschool children reached. Integration of campaigns in this manner takes advantage of the fact that over 167 million children are reached annually by vitamin A supplementation programs globally, and more than 50 countries report >70% coverage (48). This approach is being used successfully in Cambodia, which was the first country to reach the goal of 75% coverage of school-age children with albendazole, in 2004 (9). The rationale for program integration with vitamin A supplementation seems particularly strong, given that ascariasis and trichuriasis in some studies have been associated with low serum retinol levels, as it is believed that these parasites interfere with vitamin A absorption (43, 49, 50).

Another integration example involved a single MDA with ivermectin and albendazole in Tanzania, which resulted in significant reductions in prevalence rates for whipworm and hookworm; this was done in the context of a filariasis elimination program, demonstrating the effect of a campaign active against multiple parasitic infections (51).

Combination therapy for multiple NTDs is effective and safe, including one with triple-drug combinations of antiparasitics (such as albendazole, ivermectin, and praziquantel); studies involving the addition of azithromycin for trachoma control are ongoing (52, 53). Programs that target filariasis/onchocerciasis are particularly well suited for incorporation of soil-transmitted helminth control given that the drugs used are similar (most filariasis MDA programs already include albendazole) and involve MDA as a core control measure. Some studies suggest that for trichuriasis in particular, combination therapy with albendazole or mebendazole plus ivermectin is more effective than albendazole or mebendazole alone (16, 54). In other control programs, such as for LF and onchocerciasis, community-directed distribution of drugs has been the most cost-effective means of achieving high coverage in the population (21, 55).

Cost-effectiveness is a major principle in the planning of integrated campaigns. MDA with albendazole is inexpensive, on the order of 0.03–0.04 USD per person treated per year (21, 56). It has been estimated that expanding coverage to include filariasis, onchocerciasis, schistosomiasis, vitamin A deficiency, and trachoma, in addition to the soil-transmitted helminths, could be done for as little as 0.40–0.50 USD per person per year (21, 57). Similarly, when organized as an integrated campaign, these programs would save from 25% to 50% compared to the costs that each control effort would cost individually (58, 59). Since these diseases are all best controlled presently by MDA campaigns, it is sensible to integrate them to save resources while controlling multiple public health issues simultaneously. It is even possible that combining campaigns increases compliance with all components of the program among targeted populations; the cost per DALY saved would be only 2–9 USD, indicating this approach would be highly cost-effective (58, 59).

7.9   VACCINATION

In the face of increasing drug resistance, a need for multiple retreatments, and logistical difficulties, vaccination remains the most desirable preventative measure for geohelminths by providing a simple, cost-effective single step for control or elimination. Unfortunately, the lack of good animal models and a poor understanding of how geohelminths persist in humans in the face of a potent immune response have hindered the development of an effective vaccine (60). Nevertheless, a hookworm vaccine consisting of the recombinant larval antigen ASP2 is effective in animal models and has shown a protective association in immunoepidemiology studies (60, 61). The *Na*-ASP-2 hookworm vaccine is now undergoing clinical development in humans, with Phase I data demonstrating it is safe and antigenic and Phase II trials ongoing (60, 61). This vaccine targets the larval forms of hookworms, but not adult worms. Development of a vaccine that will be effective against both the larval and adult stages of hookworms is ongoing.

7.10   CONCLUSIONS

The soil-transmitted helminths are a major global public health problem. The most important impact globally is the association between soil-transmitted helminth infection and the adverse outcomes in infected young children, who suffer anemia, impaired growth, and impaired cognitive development. Infected pregnant women also suffer higher rates of anemia, resulting in poor pregnancy outcomes. Mass drug administration, endorsed by WHO and other agencies, has been the most cost-effective and successful means of control of these parasites to date. With recent global recognition and public and private advocacy for control of these neglected tropical diseases, expansion of these successful programs is ongoing in many areas. Integration of mass treatment campaigns with interventions involving other diseases may be the best way forward to expand these programs and maintain sustainability over the long term. Improved sanitation and prevention measures are also critical long-term elements of control and elimination campaigns for these diseases. Most importantly, improved economic development in affected areas will be needed to fully break the cycle of poverty and infection associated with these parasites. Among the Millennium Development Goals put forth by the United Nations in 2000 were the eradication of extreme poverty and hunger, reduction of child and maternal mortality, and combating HIV/AIDS, malaria, other diseases. Although the recent global economic crisis has stalled progress for some of these goals, an overall trend toward fewer persons living on less than 1.25 USD/day, fewer child and maternal deaths, and improved access to treatments and prevention measures for infectious diseases over the past 15 years suggest that while much is needed, progress toward the Millennium Development Goals are moving forward (62). Breaking the cycle of poverty that is both a cause and effect of infections such as the soil-transmitted helminths is a fundamental aspect to eventual control and elimination of these diseases.

REFERENCES

1. Bethony J, Brooker S, Albonico M, et al. Soil-transmitted helminth infections: ascariasis, trichuriasis, and hookworm. *Lancet* 2006;367:1521–1532.

2. Maguire J. Intestinal nematodes (roundworms) In: Mandell G, Bennett JE, Dolin R, editors. *Principles and Practice of*

Infectious Diseases, 7th edition. Philadelphia, PA: Churchill Livingston Elsevier, 2010, pp. 3577–3586.

3. Seltzer E, Barry M, Crompton DWT. Ascariasis. In: Guerrant RL, Walker DH, Weller PF, editors. *Tropical Infectious Diseases—Principles, Pathogens, & Practice*, 2nd edition. Philadelphia: Churchill Livingstone, 2006, pp. 1257–1264.

4. Cooper ES. Trichuriasis. In: Guerrant RL, Walker DH, Weller PF, editors. *Tropical Infectious Diseases—Principles, Pathogens, & Practice*, 2nd edition. Philadelphia: Churchill Livingstone, 2006, pp. 1252–1256.

5. Hotez PJ. Hookworm infections. In: Guerrant RL, Walker DH, Weller PF, editors. *Tropical Infectious Diseases—Principles, Pathogens, & Practice*, 2nd edition. Philadelphia: Churchill Livingstone, 2006, pp. 1265–1273.

6. Hotez PJ, Brooker S, Bethony JM, et al. Hookworm infection. *N Engl J Med* 2004;351:799–807.

7. Brooker S, Bethony J, Hotez PJ. Human hookworm infection in the 21st century. *Adv Parasitol* 2004;58:197–288.

8. de Silva NR, Brooker S, Hotez PJ, Montresor A, Engels D, Savioli L. Soil-transmitted helminth infections: updating the global picture. *Trends Parasitol* 2003;19:547–551.

9. Albonico M, Montresor A, Crompton DW, Savioli L. Intervention for the control of soil-transmitted helminthiasis in the community. *Adv Parasitol* 2006;61:311–348.

10. Naish S, McCarthy J, Williams GM. Prevalence, intensity and risk factors for soil-transmitted helminth infection in a South Indian fishing village. *Acta Trop* 2004;91:177–187.

11. Brooker S, Clements AC, Bundy DA. Global epidemiology, ecology and control of soil-transmitted helminth infections. *Adv Parasitol* 2006;62:221–261.

12. Hotez PJ, Bethony J, Bottazzi ME, Brooker S, Buss P. Hookworm: "the great infection of mankind." *PLoS Med* 2005;2:e67.

13. Hall A, Hewitt G, Tuffrey V, de Silva N. A review and meta-analysis of the impact of intestinal worms on child growth and nutrition. *Matern Child Nutr* 2008;4(Suppl 1):118–236.

14. Keiser J, Utzinger J. Efficacy of current drugs against soil-transmitted helminth infections: systematic review and meta-analysis. *JAMA* 2008;299:1937–1948.

15. Smits HL. Prospects for the control of neglected tropical diseases by mass drug administration. *Expert Rev Anti Infect Ther* 2009;7:37–56.

16. Knopp S, Mohammed KA, Speich B, Hattendorf J, Khamis IS, Khamis AN, Stothard JR, Rollinson D, Marti H, Utzinger J. Albendazole and mebendazole administered alone or in combination with ivermectin against Trichuris trichiura: a randomized controlled trial. Clin Infect Dis. 2010;51:1420-8.

17. World Health Organization. Preventive chemotherapy in human helminthiasis: coordinated use of anthelminthic drugs in control interventions. *WHO Manual for Health Professionals and Programme Managers*, 2006, pp. 1–63.

18. Albonico M, Bickle Q, Ramsan M, Montresor A, Savioli L, Taylor M. Efficacy of mebendazole and levamisole alone or in combination against intestinal nematode infections after repeated targeted mebendazole treatment in Zanzibar. *Bull WHO* 2003;81:343–352.

19. Ettling J. The role of the Rockefeller Foundation in hookworm research and control. In: Schad GA, Warren KS, editors. *Hookworm Disease*. London, Philadelphia: Taylor & Francis, Ltd, 1990, pp. 3–14.

20. Rockefeller JD. The Rockefeller Commission for the Eradication of hookworm disease. *Science* 1909;30:635–636.

21. Hotez PJ. Mass drug administration and integrated control for the world's high-prevalence neglected tropical diseases. *Clin Pharmacol Ther* 2009;85:659–664.

22. Schistosomiasis and soil-transmitted helminth infections. Resolution WHA54.19. Geneva: World Health Organization, 2001. Available at http://www.who.int/wormcontrol/about_us/en/ea54r19.pdf.

23. World Health Organization. Soil-transmitted helminthiasis. Progress report on number of children treated with anthelminthic drugs: an update towards the 2010 global target. *Wkly Epidemiol Rec* 2008;82:237–252.

24. Quinnell RJ, Slater AF, Tighe P, Walsh EA, Keymer AE, Pritchard DI. Reinfection with hookworm after chemotherapy in Papua New Guinea. *Parasitology* 1993;106:379–385.

25. Elkins DB, Haswell-Elkins M, Anderson RM. The importance of host age and sex to patterns of reinfection with *Ascaris lumbricoides* following mass antihelminthic treatment in a South Indian fishing community. *Parasitology* 1988;96:171–184.

26. Chan L, Bundy DA, Kan SP. Aggregation and predisposition to *Ascaris lumbricoides* and *Trichuris trichiura* at the familial level. *Trans R Soc Trop Med Hyg* 1994;88:46–48.

27. Narain K, Medhi GK, Rajguru SK, Mahanta J. Cure and reinfection patterns of geohelminthic infections after treatment in communities inhabiting the tropical rainforest of Assam, India. *Southeast Asian J Trop Med Public Health* 2004;35: 512–517.

28. Stoltzfus RJ, Chwaya HM, Montresor A, Tielsch JM, Jape JK, Albonico M, Savioli L. Low dose daily iron supplementation improves iron status and appetite but not anemia, whereas quarterly anthelminthic treatment improves growth, appetite and anemia in Zanzibari preschool children. *J Nutr* 2004;134: 348–356.

29. Awasthi S, Pande VK, Fletcher RH. Effectiveness and cost-effectiveness of albendazole in improving nutritional status of preschool children in urban slums. *Indian J Pediatr* 2000;37: 19–29.

30. Mathema P, Pandey S, Blomquist PO.Deworming impact evaluation of preschool children deworming programme in Nepal. Abstract from International Nutritional Anaemia Consultative Group, Lima, Peru, November 2004.

31. Watkins WE, Pollit E. "Stupidity of worms": do intestinal worms impair mental performance? *Psychol Bull* 1997;121: 171–191.

32. Simeon DT, Grantham-McGregor SM, Callender JE, Wong MS. Treatment of *Trichuris trichiura* infections improves growth, spelling scores and school attendance in some children. *J Nutr* 1995;125:1875–1883.

33. Miguel E, Kremer M.Worms: Education and Health Externalities in Kenya. Working Paper no. w8481. Cambridge, MA:

National Bureau of Economic Research, 2001. Available at http://www.nber.org/papers/W8481.

34. Stephenson LS, Latham MC, Kurz KM, Kinoti SN, Brigham H. Treatment with a single dose of albendazole improves growth of Kenyan schoolchildren with hookworm, *Trichuris trichiura*, and *Ascaris lumbricoides* infections. *Am J Trop Med Hyg* 1989;41:78–87.

35. Olsen A. Experience with school-based interventions against soil-transmitted helminths and extension of coverage to non-enrolled children. *Acta Trop* 2003;86:255–266.

36. Christian P, Kathry KS, West KP. Antenatal antihelminthic treatment, birthweight, and infant survival in rural Nepal. *Lancet* 2004;364:981–983.

37. Kabatereine NB, Brooker S, Koukounari A, Kazibwe F, Tukahebwa EM, Fleming FM, Zhang Y, Webster JP, Stothard JR, Fenwick A. Impact of a national helminth control programme on infection and morbidity in Ugandan schoolchildren. *Bull WHO* 2007;85:91–99.

38. Taylor-Robinson DC, Jones AP, Garner P. Deworming drugs for treating soil-transmitted intestinal worms in children: effects on growth and school performance. *Cochrane Database Syst Rev* 2007;4:1–114.

39. Albonico M. Methods to sustain drug efficacy in helminth control programs. *Acta Trop* 2003;86:233–242.

40. Haider BA, Humayun Q, Bhutta ZA. Effect of administration of antihelminthics for soil transmitted helminths during pregnancy. *Cochrane Database Syst Rev* 2009;15:1–20.

41. Guyatt HL, Bundy DAP, Evans D. A population dynamic approach to the cost effectiveness analysis of mass antihelminthic treatment: effects of treatment frequency on *Ascaris* infection. *Trans R Soc Trop MedHyg* 1993;87:570–575.

42. Bundy DAP, Hall A, Medley GF, Savioli L. Evaluating measures to control intestinal parasitic infections. *World Health Stat Q* 1992;45:168–179.

43. de Silva NR. Impact of mass chemotherapy on the morbidity due to soil-transmitted nematodes. *Acta Trop* 2003;86:197–214.

44. Yamey G, Hotez P. Neglected tropical diseases. *BMJ* 2007;335:269–270.

45. Sow S, de Vlas SJ, Polman, Gryseels B. Hygiene practices and contamination risks of surface waters by schistosome eggs: the case of an infested village in Northern Senegal. *Bull Société Pathol Exot* 2004;97:12–14.

46. Mukherjee JFN.Harnessing Market Power for Rural Sanitation. *WSP Field Notes*. Water and Sanitation Program, Jakarta, Indonesia, 2005.

47. Esrey SA, Potash JB, Roberts L, Shiff C. Effects of improved water supply and sanitation on ascariasis, diarrhoea, dracunculiasis, hookworm infection, schistosomiasis, and trachoma. *Bull WHO* 1991;69:609–621.

48. UNICEF. *The State of the World's Children 2005*. New York: United Nation Children Fund, 2005.

49. Sivakumar B, Reddy V. Absorption of vitamin A in children with ascariasis. *J Trop Med Hyg* 1975;78:114–115.

50. Mahalanabis D, Simpson TW, Chakraborty ML, Ganguli C, Bhattacharjee AK, Mukherjee K. Malabsorption of water miscible vitamin A in children with giardiasis and ascariasis. *Am J Clin Nutr* 1979;32:313–318.

51. Massa K, Magnussen P, Sheshe A, Ntakamulenga R, Ndawi B, Olsen A. The effect of the community-directed treatment approach versus the school-based treatment approach on the prevalence and intensity of schistosomiasis and soil-transmitted helminthiasis among schoolchildren in Tanzania. *Trans R Soc Trop Med Hyg* 2009;103:31–37.

52. Reddy M, Gill SS, Kalkar SR, Wu W, Anderson PJ, Rochon PA. Oral drug therapy for multiple neglected tropical diseases: a systematic review. *JAMA* 2007;298:1911–1924.

53. Olsen A. Efficacy and safety of drug combinations in the treatment of schistosomiasis, soil-transmitted helminthiasis, lymphatic filariasis and onchocerciasis. *Trans R Soc Trop Med Hyg* 2007;101:747–758.

54. Beach MJ, Streit TG, Addiss DG, Prospere R, Roberts JM, Lammie PJ. Assessment of combined ivermectin and albendazole for treatment of intestinal helminth and *Wuchereria bancrofti* infections in Haitian schoolchildren. *Am J Trop Med Hyg* 1999;60:479–486.

55. Smits HL. Prospects for the control of neglected tropical diseases by mass drug administration. *Expert Rev Anti Infect Ther* 2009;7:37–56.

56. Kabatereine NB, Tukahebwa EM, Kazibwe F, Twa-Twa JM, Barenzi JF, Zaramba S, Stothard JR, Fenwick A, Brooker S. Soil-transmitted helminthiasis in Uganda: epidemiology and cost of control. *Trop Med Int Health* 2005;10: 1187–1189.

57. Hotez PJ, Fenwick A, Savioli L, Molyneux DH. Rescuing the bottom billion through control of neglected tropical diseases. *Lancet* 2009;373:1570–1575.

58. Hotez PJ, Molyneux DH, Fenwick A, Kumaresan J, Sachs SE, Sachs JD, Savioli L. Control of neglected tropical diseases. *N Engl J Med* 2007;357:1018–1027.

59. Brady MA, Hooper PJ, Ottesen EA. Projected benefits from integrating NTD programs in sub-Saharan Africa. *Trends Parasitol* 2006;22:285–291.

60. Diemert DJ, Bethony JM, Hotez PJ. Hookworm vaccines. *Clin Infect Dis* 2008;46:282–288.

61. Bethony JM, Simon G, Diemert DJ, Parenti D, Desrosiers A, Schuck S, Fujiwara R, Santiago H, Hotez PJ. Randomized, placebo-controlled, double-blind trial of the Na-ASP-2 hookworm vaccine in unexposed adults. *Vaccine* 2008;26: 2408–2417.

62. World Health Organization. *The Millennium Development Goals Report 2009*. New York: World Health Organization, 2009.

8

TOXIC CYANOBACTERIA

Ian Stewart, Wayne W. Carmichael, and Lorraine C. Backer

8.1 INTRODUCTION

The cyanobacteria are an ancient and remarkably successful taxonomic group of prokaryotic, single-celled microbes. They are widespread in marine, freshwater, and terrestrial environments and can colonize extreme and unusual habitats such as salt mines, alkaline "soda" lakes, hot springs, and the hollow hair shafts of polar bears (1, 2). Cyanobacteria are colloquially known as blue-green algae, from the characteristic cyan pigment many forms produce and that, together with other cyanobacteria-specific pigments, bestow a competitive advantage in allowing the organism to photosynthesize at wavelengths beyond the range that green plants and algae can utilize for growth. The term "blue-green algae" is something of a misnomer, however, as the cyanobacteria are prokaryotes, lacking membrane-bound organelles and a cell nucleus. The true algae, by contrast, are unicellular and multicellular eukaryotic organisms. So while cyanobacteria have some physiological similarities with the algae, specifically that both groups photosynthesize and release oxygen as a metabolic product, the cyanobacteria and algae are on different sides of arguably the most fundamental dividing line that helps us classify and categorize the diversity of life on earth: that separating prokaryotic from eukaryotic cells.

While cyanobacteria are essentially ubiquitous in freshwater, coastal, and marine waters, they come to the attention of public health workers when environmental conditions are such that these organisms undergo prodigious proliferation and outcompete other components of the phytoplankton to form mass developments known as blooms or waterblooms. Cyanobacterial blooms can occur in coastal and inland waters, and both kinds present their own public health dilemmas. This chapter focuses on the public health implications of cyanobacteria in freshwater systems.

8.2 CYANOBACTERIAL TOXINS ("CYANOTOXINS")

Cyanobacteria are able to produce a very broad and diverse array of unusual chemical compounds, many of which are unique to the phylum. Toxic strains and species across at least 20 cyanobacterial genera (3) can produce a suite of potent toxins that injure tissues and impair the function of various organs and organ systems. The cyanotoxins that have caused mass mortality and morbidity in humans, livestock, and wild animals are understandably those that have been the principal focus of multidisciplinary research efforts by several groups across the globe over the past 50 years. The research specialties contributing to the modern understanding of toxic cyanobacteria include structural and analytical chemistry, biochemistry, environmental toxicology and ecotoxicology, phycology, medical and veterinary epidemiology, microbiology, and microbial genetics.

The chemical structures and basic functional toxicology of the various freshwater cyanotoxins and cyanotoxin groups known to cause illness and death in vertebrates have been well characterized. A cursory overview of the medically and economically important cyanotoxins follows; interested readers will find comprehensive reviews of their chemistry

Water and Sanitation-Related Diseases and the Environment: Challenges, Interventions, and Preventive Measures, First Edition.
Edited by Janine M. H. Selendy.
© 2011 Wiley-Blackwell. Published 2011 by John Wiley & Sons, Inc.

and the health and environmental impacts of toxic blooms in the literature (4–8).

Microcystins: a group of some 80 structurally related cyclic heptapeptides, with more congeners likely to be discovered. The most toxic forms are potent hepatotoxins and tumor promoters; toxicity at the molecular level is due to their inhibition of protein phosphatases. More than 10 cyanobacterial genera can produce microcystins.

Nodularin: a cyclic pentapeptide with a similar acute toxicological profile to that of the more potent microcystins. Made by the brackish-water cyanobacterium *Nodularia spumigena.*

Cylindrospermopsin: a highly water-soluble alkaloid with a tricyclic ring structure linked to a uracil moiety. Also described as a hepatotoxin, but other tissues and organs are affected: kidney, spleen, gut, heart, thymus, skin. Cylindrospermopsin is a protein synthesis inhibitor, but this may not be the primary mechanism of toxicity. Six cyanobacterial genera are known to be cylindrospermopsin producers.

Saxitoxins: a family of tricyclic alkaloids produced by five cyanobacterial genera as well as some eukaryotic microalgae (dinoflagellates). The saxitoxins are potent neurotoxins, acting as antagonists of voltage-gated sodium channels on motor nerves, initiating a conduction defect that causes respiratory paralysis.

Anatoxin-a, homoanatoxin-a: neurotoxic alkaloids, these cyanotoxins bind nonreversibly to nicotinic acetylcholine receptors at motor nerve end plates, causing a depolarizing nerve block manifesting as acute respiratory failure and hypoxia/anoxia.

Anatoxin-a(s): structurally and functionally unrelated to anatoxin-a (the "s" refers to hypersalivation due to excessive muscarinic stimulation in experimental rodents noted by researchers who did the early work on this cyanotoxin). Anatoxin-a(s) is a phosphate ester neurotoxin, acting as a cholinesterase inhibitor resulting in a conduction deficit and death by asphyxiation. *Anabaena* is the only genus known to manufacture anatoxin-a(s).

Lyngbyatoxins, aplysiatoxins: alkaloid toxins—these are potent inflammatory agents, tumor promoters, and protein kinase C activators. Produced by marine cyanobacteria, mainly the genus *Lyngbya*, these compounds initiate irritant reactions on the skin of exposed bathers; mass outbreaks of acute dermatitis have been reported. Toxins of this group are implicated in some poisoning incidents associated with consumption of edible seaweed on which *Lyngbya* may grow epiphytically.

BMAA: a nonprotein amino acid reportedly produced by a wide range of aquatic and terrestrial cyanobacteria.

BMAA has been posited as the causative agent of a devastating degenerative neurological illness that affected a specific ethnic group on the Pacific island of Guam, with a dietary exposure route implicated. Some research groups suggest that BMAA, which is known to be neurotoxic under specific exposure conditions in experimental *in vivo* models, may be associated with chronic dementing illnesses such as Alzheimer's disease. However, the BMAA and chronic neuropathology hypothesis is controversial, with other researchers challenging most components of the theory, including that of the widespread distribution of BMAA in aquatic cyanobacteria. The body of literature on the BMAA/cyanobacteria/neurodegeneration theory is expanding rapidly. For examples from workers developing the hypothesis, see Refs 9–11. Examples of some papers that challenge various aspects of the hypothesis, including the analytical detection of BMAA in cyanobacteria and animal tissues, are Refs 12–14.

The above are examples of the main cyanotoxins and cyanotoxin groups currently known to pose a threat to public health (except BMAA, with claims of attendant public health risks currently under dispute). Cyanobacteria can produce a broad array of biologically active compounds; it is possible that as-yet undiscovered cyanobacterial metabolites may also have toxic properties.

Harmful exposures to freshwater cyanobacterial toxins can be acute or chronic. Chronic exposures may occur where drinking water supplies are untreated or processes to remove or detoxify cyanotoxins are inadequate. Feasible, but poorly researched, chronic exposures may arise where cyanotoxins are present in aquatic food products. Acute exposure and illness may be seen in situations where drinking water supplies suffer gross contamination by toxic cyanobacteria, or similarly affected waters are used for bathing or recreational purposes. These acute and chronic exposure routes for cyanotoxins and their human health consequences will now be discussed in more detail.

8.3 CHRONIC EXPOSURE TO CYANOTOXINS IN DRINKING WATER

Experimental studies using laboratory animal models have demonstrated that the microcystins and nodularin are tumor-promoting compounds (15–17). The related concern for public health is whether these toxins are carcinogens in their own right. The International Agency for Research on Cancer (IARC) has listed microcystin-LR as a Group 2B agent ("possibly carcinogenic to humans"), but has insufficient information at present to classify any other cyanotoxins (18). However, many toxicologists with specific expertise in cyanobacterial toxins suspect that nodularin, cylindrospermopsin, and the more toxic microcystin conge-ners will be shown to be carcinogenic in the future. These suspicions are based on the collective body of experimental evidence to date, which has been comprehensively reviewed by Falconer (7). Chronic (2 year) exposure studies in labo-ratory animal models and compelling human epidemiological evidence are necessary requirements for agencies such as the IARC to make definitive statements regarding the carcino-genicity of the hepatotoxic cyanotoxins. One reason why long-term feeding studies have not as yet been conducted in rodents pertains to the difficulty of securing sufficient quan-tities of purified toxins. The hepatotoxic cyanotoxins are complex molecules with numerous chiral centers; therefore, they are not able to be efficiently and cost-effectively syn-thesized. To isolate and purify the several hundreds of grams of each toxin needed for a chronic feeding study at a range of doses would require large-volume mass culture and/or wild harvesting of toxic cyanobacteria (19). Such endeavors have not as yet been attempted at the requisite scale, though prioritization of this work has been called for (20). Despite the current absence of definitive evidence for carcinogenicity from the cyanobacterial hepatotoxins, many national and international authorities consider the preliminary data on the carcinogenic potential of these compounds sufficiently con-cerning to warrant precautionary approaches. Exposure guidelines for cyanotoxins in drinking water encompassing tolerable daily intake (TDI) estimates and various safety factors have been recommended by the World Health Orga-nization and national public health agencies. The majority of countries that have adopted drinking water guideline values to date have done so for microcystin-LR (or total microcystins as MC-LR equivalents) at 1 μg/L (21), MC-LR being the cya-notoxin that has so far generated the most comprehensive body of toxicological information. Australia has embraced a "health alert" value of 1 μg/L for cylindrospermopsin and cylindrospermopsin-producing cyanobacteria in drinking water supplies, calculated by a similar risk assessment method used to derive the MC-LR guideline but using toxicological data specific to cylindrospermopsin (22). New Zealand has provisionally adopted maximum acceptable values for several of the major cyanotoxins in drinking waters (21).

Guideline values for the presence of specific cyanotoxins in drinking water supplies have meaningful benefits only when the communities they serve also have access to the infrastructure, expertise, and standardized procedures to detect, routinely monitor for, and effectively remove or detoxify cyanotoxins, and then to distribute safe drinking water to all members of those communities. Such resources are ordinarily associated with developed countries that can command centralized water treatment facilities and retic-ulated supplies. The processes used in modern water treat-ment plants for removing and decontaminating common bacterial and viral contaminants, such as coagulation, flocculation, sedimentation, or flotation, filtration, and

disinfection—chlorination, chloramination, ozonation, H_2O_2, and UV radiation—can generally remove cyano-bacterial cells and remove or safely transform most cyanotoxins (23). Specific considerations are necessary for effectively rendering reticulated drinking water supplies safe from cyanotoxins; some processes are more effective than others, and treatment failures have occurred in the past (see below). But as a general approach, drinking waters subjected to modern water treatment techniques can be considered likely to contain cyanobacterial toxins at con-centrations sufficiently low to be considered safe to con-sume, given the current level of understanding.

8.3.1 Exposure to Cyanotoxins in Untreated or Minimally Treated Drinking Water

A contrasting situation to that of developed countries with access to treated drinking water is seen in some reports that discuss a possible association between liver and colon can-cers and consumption of untreated surface water in some areas of China. Reports from the 1970s have suggested that untreated water sourced primarily from ponds or ditches was associated with hyperendemic rates of hepatocellular carci-noma (HCC) (24). China has a high average incidence rate of HCC, at 24–40 per 100,000 compared to less than 5 per 100,000 for most populations in the developed Americas, northern Europe, and Oceania (age-standardized incidence rates) (25, 26). But there is significant regional variability within China, with highest prevalence along the east coast and low rates in western regions (26). Parts of Jiangsu and Guangxi Provinces reportedly have the highest mortality rates from HCC in China at 47 per 100,000 compared to the national average mortality of 10–15 per 100,000 (26, 27). Some seriously affected villages have mortalities up to 185 per 100,000 (27).

Earlier reports drew suspicion that high rates of HCC were related to stagnant drinking water sourced from ditches and ponds that were dug specifically as domestic water supplies in areas of reclaimed land unsuitable for sinking wells (24). Later reports focused on the possibility that exposure to cyanobacterial toxins, particularly microcystins, may be an attributable risk factor for HCC. The biological plausibility for this hypothesis would appear strong, given the under-standing that microcystins are potent tumor promoters and possible carcinogens. However, the epidemiological investi-gations conducted so far do not provide convincing evidence to support the premise. Some (28–31) but not all studies (32) appear to show a slightly increased risk for HCC associated with consumption of untreated surface water. A recent report (33) described chronic exposure to microcystins in professional fishers working on Lake Chaohu, Anhui Prov-ince in southeast China. The lake is regularly affected by blooms of *Microcystis* and *Anabaena*. These workers live on their fishing boats most of the time, and use minimally treated lake water as their drinking source and aquatic food products (fish, prawns, snails) from the lake as their principal dietary intake. Microcystins were detected in serum from all individuals in the study cohort ($n = 35$). Other serum biochemistry parameters related to liver and renal function were examined, as well as hepatitis B virus (HBV) surface antigen and core antibody, and alpha-fetoprotein (AFP) as an outcome biomarker for HCC. No study subjects were AFP-positive; the test has a sensitivity of 65% (34). Three subjects were HBV-positive, of which two had raised ALT and AST levels. Otherwise, the proportion of subjects with abnormally high hepatic enzymes was low: 3% ($n = 1$) for GGT to 14% with raised ALP ($n = 5$, of which $n = 2$ were HBV-positive) (33). Similar exposure patterns appear to be replicated elsewhere in China, with artisanal fishers living on Lake Taihu (Jiansu Province), which was affected by a severe microcystin-producing bloom in 2007 (35, 36).

An epidemiological investigation (retrospective cohort design) conducted in eight townships in the Haining City area, Zhejiang Province, reported increased rates of colorectal cancer linked to untreated surface drinking water supplies. An earlier case-control study reported in a Chinese-language journal was cited by the authors (37). However, a cohort study undertaken in nearby Jiashan County did not detect a statistically significant relationship between colorectal cancer and drinking from surface waters, whereas a moderately increased risk was associated with consumption of water from wells (38). But an earlier nested case-control study, presumably largely drawn from the same cohort in Jiashan County, showed a protective effect of well-water consumption on colon cancer that disappeared after adjusting for covariates; the final multivariable model revealed a significant association between mixed drinking water sources (primarily river water and ditch water) and both colon and rectal cancers (39).

China has one of the world's highest incidence rates of HCC, with chronic viral hepatitis being the principal risk factor; aflatoxin exposure and alcohol intake are also regional risk factors (25, 26, 40). Epidemiological studies examining associations between HCC and consumption of untreated surface waters would need to incorporate reliable exposure measures as well as consider the impact of risk factors such as hepatitis B and C, aflatoxins, alcohol, and tobacco. Such studies may be difficult to realize in China, given the dynamic social and economic conditions there: HCC caused by hepatitis B can be expected to decline in coming decades as HBV vaccination rates increase; improved grain storage and food safety logistics might similarly be expected to reduce exposure to aflatoxins. Increased alcohol and tobacco consumption (41, 42) may influence HCC rates in the opposite direction. Qidong County, in Jiangsu Province, is the site of one of two population-based trials of universal infant vaccination against HBV (the other is in The Gambia, West Africa)

commenced in 1984 (43). Qidong was chosen for this trial because it has the highest HCC mortality rates in China (44). Interestingly, Qidong County was also targeted in earlier HCC prevention research, which sought to influence incidence rates by encouraging alternate sources of drinking water, with a statistically significant decrease reportedly associated with declining proportions of the community drinking from house ditch and field ditch supplies (from 79% to 54% within a 5-year period from 1972) (24).

The difficulty of teasing out causal relationships between exposure to cyanotoxins in drinking water and complex, multifactorial diseases such as primary liver cancer and colorectal cancer is implied by the inconclusive epidemiology, as discussed above. Hepatotoxic cyanotoxins may be co-carcinogens; very potent synergistic effects of dietary exposure to aflatoxins and HBV infection on the risks of HCC have been identified through epidemiologic studies (45, 46); heavy alcohol consumption and diabetes have also been shown to be important complementary risk factors in HBV-positive HCC (47). Experimental rodent studies featuring sequential exposure to aflatoxin B_1 and cyanobacterial hepatotoxins indicate increased development of hepatic tumors and dysplastic lesions as a result of such co-exposures (48, 49).

Drinking water sources have been and continue to be a priority focus of epidemiological studies into several endemic diseases in China, including HCC, Kashin–Beck disease, and skeletal fluorosis (27). Cyanotoxins in drinking water, particularly the microcystins, are the subject of detailed investigations in China from the perspective of environmental risk factors for HCC and other cancers of the digestive tract; there is a significant body of literature, mainly epidemiological investigations, on this specific topic in Chinese-language biomedical publications. Assuming that there are populations in rural China with continuing and/or recent reliance on untreated surface drinking water supplies sufficient to warrant abiding public health consideration of such exposures (again, this assumption may be somewhat redundant given the rapid pace of development within China), the question of the environmental chemistry of these waters would seem to invite further inquiry. Water from village ponds and domestic water ditches could be systematically analyzed for cyanobacteria and cyanotoxins at high-risk times (summer/autumn); other suspect carcinogens likely to be associated with intensive agricultural activities that might find their way into standing surface waters should also be considered, for example, nitrates, pesticides, and toxic metals such as arsenic.

Analytical capabilities for reliably determining microcystin concentrations in raw and treated drinking water have become routinely available only over the last two decades. So population-level exposure assessment of microcystins in household- and/or village-level drinking water supplies is the most significant piece of missing information limiting the ability of epidemiological investigations to retrospectively assess the question of whether and to what degree MCs have contributed to the development of HCC in certain rural Chinese communities. We suggest, therefore, that this question remains open.

8.3.2 Epidemiological Investigation of Cyanotoxins in Drinking Water in Other Countries

The only epidemiological investigations in countries outside of China to examine associations between HCC and exposure to cyanotoxins in drinking water would appear to be ecologic studies conducted in Florida (50) and Serbia (51). The Florida study found a slightly increased rate of HCC in populations supplied with drinking water from treatment plants that process surface water sources when compared to communities immediately outside those service areas. That finding did not hold when compared to rates of HCC across Florida generally, nor to sampled reference populations supplied with drinking water from groundwater sources. However, the results were interesting insofar as the demographic features of communities from the areas surrounding surface water supply zones might be considered similar to those serviced by the surface water treatment plants, whereas those living outside the surface water service area were thought to have significantly less exposure to treated surface drinking waters than the nearby residents serviced by those plants. An ecologic design has the advantage of being a relatively low-cost epidemiological enquiry, and can be viewed as a useful initial foray here. The principal limitation of the design is that exposures (in this case, to microcystins in surface water drinking supplies) are inferred from aggregate data, that is, the drinking water habits of individuals within study groups is not known. For example, we do not know what proportion of the community supplied with tap water actually drank it, nor do we know which households used additional point-of-use drinking water filtration. The inherent bias associated with an ecologic design is termed "ecological inference fallacy," where exposures are assumed to be homogeneous across sampled geographic areas. Yet the findings by Fleming et al. (50) would seem to invite population-based analytical epidemiology studies to explore this matter in more detail as drinking water supplies in developed countries are drawn increasingly from surface waters. The availability of modern water treatment plants in developed countries does not by definition imply that the communities they serve will always be protected from cyanotoxins. Breakthroughs can occur when cyanobacterial blooms overload the capacity of water treatment processes to remove cells and free toxins. The costs of monitoring and removing cyanobacteria and cyanotoxins are not borne equally across the developed world; likewise, regulatory and management regimes may be inconsistent within and between developed countries. Awareness of the problem by water treatment professionals and communication

with public health workers may be inadequate in regional or national agencies without a coordinated approach to address toxic cyanobacteria in drinking water.

The ecologic study by Svirčev et al. [51] reported differential rates of primary liver cancer (PLC) incidence and mortality across regions of Serbia; the authors suggest that higher rates of PLC occur in regions prone to severe cyanobacteria blooms. Apart from the inherent limitations of ecologic study designs, as discussed above, the Serbian study appears to have some specific methodological weaknesses. Criteria for determining cyanobacteria-affected and unaffected surface water reservoirs were not clearly defined, and exposure to microcystins was essentially implied, as only a single bloom event was subject to investigation for detection and quantification of microcystins. Statistical inference tests and significance levels were not described. The authors did, however, suggest that their work should be viewed as a hypothesis-generating study. Reported microcystin-LR concentrations of 650µg/L in raw water and 2.5µg/L in finished drinking water during the bloom that was investigated suggests that a cyanotoxin monitoring program and refined drinking water treatment processes are needed in Serbia, and the epidemiology of PLC in relation to microcystin exposure could be explored there with a more robust study design.

8.4 ACUTE EXPOSURES TO TOXIC CYANOBACTERIA

Water treatment failures have occurred in several countries, resulting in contamination of source waters and potable supplies by cyanotoxins. Mass poisoning outbreaks, including fatalities, have occurred as a consequence. Some examples are:

- *Queensland, Australia: Palm Island Outbreak.* In the late spring of 1979, residents of this tropical island community suffered an outbreak of hepatoenteritis. One hundred forty-eight individuals, mostly children, presented with loss of appetite, constipation, vomiting, and painful hepatomegaly. Over 100 of the children became acidotic and hypokalemic over ensuing days and the majority were treated with intravenous fluids and electrolytes. Fifty children were affected by bloody diarrhea that lasted for up to 3 weeks. Subsequent investigations identified the reticulated drinking water supply as the common exposure factor. The town dam had been treated with copper sulfate to lyse a cyanobacteria bloom. This mass outbreak, dubbed "Palm Island Mystery Disease" at the time, was retrospectively attributed to intoxication by cylindrospermopsin, produced by *Cylindrospermopsis raciborskii*, which is a common contaminant of freshwater waterbodies in Queensland. All affected individuals recovered after illnesses that lasted from 4 to 26 days (52, 53). That there were no fatalities resulting from this severe common-source outbreak is probably in the main due to it having occurred in a developed country with excellent emergency health care facilities and the capacity to rapidly airlift the more seriously ill children to a well-equipped hospital on the mainland (7).

• *Pernambuco, Brazil: Caruaru Hemodialysis Outbreak.* In February 1996 (i.e., late summer), 116 of 131 acute and chronic renal failure patients undergoing routine treatment at a hemodialysis clinic in Caruaru experienced nausea, vomiting, headaches, and ocular symptoms (visual field deficits, blurred vision). One hundred and one patients subsequently developed acute liver failure, and 76 of these died. Epidemiological investigation of this incident identified 52 of these fatalities that met the case definition for a suite of signs, symptoms, and pathological markers characterized as "Caruaru Syndrome." A water treatment failure was quickly identified as the antecedent factor that resulted in contamination of the dialysate with microcystins and cylindrospermopsin. Microcystins were detected in the source water and clinic water treatment filters, and in serum and liver samples from affected patients. Cylindrospermopsin was also found in the clinic's filtration system (54, 55). While this tragedy resulted from exposure to cyanotoxins by an unusual and very direct route (intravenous exposure during hemodialysis), it serves as a reminder that these widely distributed natural toxins can be extremely hazardous. Hemodialysis patients may be at a particular risk of exposure to water-soluble cyanotoxins, as the large volumes of water in contact with the dialyzer during treatment may provide an opportunity for some of these low molecular weight toxins to diffuse across the membrane. The possibility of low-dose chronic exposure to cyanotoxins should be considered here. An incident involving subclinical exposure to microcystins was investigated in Rio de Janeiro, Brazil, in 2001. Following detection of a microcystin-producing bloom of *Microcystis* and *Anabaena* in a water supply reservoir, water samples were obtained from 45 hemodialysis clinics across the city. Microcystins were detected by ELISA from the finished dialysate of four clinics. Serum was collected from 39 patients treated at one of these clinics; 35 (90%) returned positive for MCs. Monitoring of 12 patients continued for a further two months, with MCs detectable in serum throughout that period (56).

• *Bahia, Brazil: Itaparica Dam Outbreak.* An epidemic outbreak of diarrhea occurred through the autumn months of 1988, affecting both adults and children. Almost 2400 cases were reported over a 6 week period, with 88 fatalities. The outbreak was coincident with the flooding of the new Itaparica Dam, and epidemiological investigations indicated a water-borne source, with communities supplied by treated water from the reservoir predominantly affected. Raw and treated water met WHO standards for drinking water; common enteric pathogens did not appear to be involved. Public health recommendations to filter and boil drinking water did not prevent illness in people who followed such advice. A bloom of *Microcystis* and *Anabaena* was present in the dam at the time (57). We discuss the implications of large-scale manipulation of hydrological systems, and in particular dam building and cultural eutrophication, on the changing patterns of cyanobacterial mass developments in the next section.

The two mass poisoning incidents discussed above that were presumably associated with inadequately treated water predated the availability of routine methods for detecting and quantifying specific cyanotoxins in drinking water supply systems.

So the attribution of cyanobacterial toxin poisoning in each case was arrived at by a process of exclusion and consideration of clinical presentation, diagnostic laboratory investigations, and epidemiological surveillance. Therefore, those events cannot be definitively confirmed as being caused by exposure to cyanotoxins. By contrast, modern analytical methods and expertise were able to be applied to resolve the Caruaru hemodialysis emergency.

Acute exposure to toxic cyanobacteria can occur in other spheres of human activity. Recreational exposures are seen across extended periods of the year in tropical, subtropical, and warm temperate regions, and over more limited timelines in higher latitudes. Peak demand for access to recreational water occurs during warmer months and on sunny days, such conditions being the most conducive for the proliferation of cyanobacterial blooms. There is a high demand for recreational water in inland Australia; demand exceeds availability of choice in some locations such that recreational users will tolerate conditions of degraded water quality in order to pursue their chosen leisure activities. The public health concerns regarding exposure to cyanobacteria in recreational waters center on the potential for contact with hazardous concentrations of cyanotoxins in untreated waters in natural and artificial waterbodies. The topic of recreational and occupational field exposure to cyanobacteria was reviewed in 2006 (3); epidemiological investigations continue to be conducted, and subsequent publications are available (58, 59).

Figure 8.1 demonstrates the very high densities of cyanobacteria that can be found in freshwater waterbodies accessible for recreation. Supplementary Figures 1–10 depict a range of cyanobacterial surface scums and blooms in various natural and artificial waterbodies. Blooms of the same species can be quite diverse in their macroscopic presentation; some

FIGURE 8.1 Mixed bloom of *Anabaena circinalis*, *Microcystis flos-aquae*, and *Cylindrospermopsis raciborskii*; Murray River upstream of Torrumbarry Weir, New South Wales, Australia. Image courtesy NSW Office of Water. (*See insert for color representation of this figure.*)

expertise (and ideally a microscope) is required to differentiate cyanobacteria from eukaryotic bloom-forming microalgae. More advanced training and skills are needed to accurately determine the genus and species composition of individual blooms. Supplementary Figure 11 shows a bloom of the nontoxic protist *Euglena*; these organisms can also form green-colored mass developments that can be difficult for the inexperienced observer to identify.

Supplementary Figures of the recent epidemiological study of recreational exposure to cyanobacteria conducted by author LCB's group in California (59) are also on the accompanying DVD.

Research attention is increasingly being directed to the topic of dietary transfer of cyanotoxins in foods of freshwater and brackish water origin. While no cases of acute intoxication in humans from such exposure routes have been reported, a growing body of literature describes analytical detection and quantification of cyanobacterial toxins in a broad range of edible fish, shellfish, and crustaceans. Recent reviews discuss the human health risk potential for acute, subacute, and chronic exposures to cyanotoxins, presenting provisional exposure guideline calculations (60), and implications for the aquaculture industry (61). Another potential dietary exposure route for cyanotoxins relates to contamination of market garden produce by irrigation or rinse water affected by cyanobacteria (62).

8.5 DAM BUILDING AND PROBLEM CYANOBACTERIA

The construction and maintenance of dams to impound water for irrigation, agriculture, drinking water supplies, flood mitigation, and electricity generation has undoubtedly benefited human welfare, whether measured in economic terms or exemplified by the growth of cities and industries, or the health benefits of safe, secure water availability. At the middle of the last century about 5,000 large dams were in existence, and by the end of the century more than 45,000 large dams were in over 140 countries, built at an estimated cost of 2 trillion USD (63). Another important secondary benefit of water reservoirs is that of recreational amenity; inland reservoirs are highly valued for leisure activities such as swimming, sailing, waterskiing, and fishing. These leisure- and tourism-related functions also generate significant economic benefits for nearby communities.

Probably unsurprisingly, dam building also comes at a cost. The downsides of water impoundment are many and varied, and include social and economic disruption, for example, displacement of people and communities, changes to land use, and ecological effects such as altered flow regimes, disruption to anadromous fish, siltation, saline encroachment, and loss of sediments to downstream reaches. Dams can also result in deleterious effects on public health, especially in tropical areas, with increased incidence of vector-borne diseases such as schistosomiasis, malaria, and Rift Valley Fever associated with reservoir development. High population densities and typically substandard sanitation in construction and resettlement camps can facilitate the spread of dysentery, tuberculosis, HIV/AIDS, and other sexually transmitted diseases (63, 64). The broader public health impacts of large dam development have not been systematically studied (65). Cyanobacterial blooms are one of many adverse and interconnecting ecological effects that result from anthropogenic manipulation of hydrological systems. Some two-thirds of the global flow of freshwater is impeded on its path to the oceans by around 40,000 large dams (>15 m from foundation to crest) and over 800,000 smaller dams (66). There are many examples of disrupted phytoplankton dynamics and succession toward cyanobacterial dominance when the flow regimes, nutrient cycling, biodiversity, sediment accumulation, and thermal contours of rivers are fundamentally altered by damming them, for example, see Refs 67–70.

Australia's large-scale inland irrigation schemes in the Murray–Darling basin present a cautionary tale of the kinds of problems that can accrue in the long term from mismanagement of the water resource and fragmentation of regulatory oversight. Over-allocation of water rights to irrigators and overextraction from rivers have delivered a very complex societal, political, economic, and environmental legacy that is proving difficult to resolve. Interrelated manifestations of this dilemma are loss of productivity from dryland salinity, degraded drinking water quality (particularly with regard to rising salt concentrations), loss of biodiversity, and a looming environmental catastrophe in the Coorong lakes—a Ramsar-listed freshwater lagoon system at the lower reaches of the Murray River (71, 72). Four large dams and numerous small weirs, as well as a series of locks built to assist river transportation, contribute to the highly manipulated flow regimes in the Murray. Facilitated by nutrient run-off—most importantly from broad-acre application of phosphate fertilizers across inland Australia's high-input/high-output agricultural strategy (73), these changes have delivered ideal conditions for the proliferation of massive cyanobacterial blooms. A saxitoxin-producing bloom of *Anabaena* affected over 1,000 km of the Darling-Barwon system in 1991, resulting in considerable livestock mortality and emergency provision of portable water treatment equipment to riverside townships by the Australian army (74, 75). More recently, another 1,000 km-long bloom of *Anabaena circinalis*, *Microcystis flos-aquae*, and *C. raciborskii* impinged on the Murray River and several tributaries over a 3-month period (76). These events confer on Australia the dubious distinction of hosting the world's longest riverine cyanobacteria blooms. Less extensive blooms are regular—indeed annual—occurrences along various sections of the Murray–Darling and related river systems. Figure 8.2 shows a reach of the Murray River affected by this bloom in 2009.

Many more examples of anthropogenically enhanced cyanobacterial overgrowth in natural and artificial waterbodies can be found in the literature—a particularly telling case is that of Venezuela's Lake Valencia. The catchment area of the lake is densely populated, with intensive agricultural and industrial activities. High nutrient inputs from untreated domestic, agricultural, and industrial waste are likely to be the main cause of the extraordinarily high primary productivity in the lake, which is subject to chronic cyanobacterial overgrowth, mainly *Microcystis aeruginosa* and *Synechocystis aquatilis* (77). Apparent temporal relationships between periods of increased human activity in the Lake Valencia region and cyanobacterial dominance have been described by geochemical analysis of sediment cores (78). A massive microcystin-producing *Microcystis* bloom in 2007 on Lake Taihu, Jiangsu Province, China, initiated a critical event for the city of Wuxi. An estimated 10 million people rely on Lake Taihu as a drinking water source, but this eutrophic lake is also a sink for urban, agricultural, and industrial wastes. Some 2 million people living in Wuxi drank bottled water for a week at the peak of the emergency (36). The detailed history of the Yahara River chain of lakes near Madison, Wisconsin (USA) provides another illustration of the relationships between cultural eutrophication and severe cyanobacterial infestation (79).

Cyanobacteriologists are careful to note that cyanobacterial blooms are naturally occurring events. Convincing descriptions of these phenomena from antiquity and from more recent periods that predate significant population-dependent environmental impacts can be found in the literature; see, for example, Codd et al. (80). Yet chronic cyanobacterial blooms affecting managed river systems that in their pristine state prior to impoundment and cultural eutrophication would have had consistent flows through catchments with relatively low concentrations of soluble nutrients can reasonably be viewed as symptomatic of anthropogenic degradation of aquatic systems. The same is true for bloom-prone artificial lakes, reservoirs, and natural lakes suffering cultural eutrophication.

Control and stabilization of water resources is an understandable human aspiration; security of water supply is a concern at all strata of community organization, from the individual through to levels of household, village, town, city, region, and nation. Transnational cooperation on water security is required where national boundaries transect watersheds; such cooperation is not always successful, and conflict has occurred in the past and can be anticipated in the future where and when increased demand driven by growing populations outstrips the supply of freshwater for safe drinking water. Manipulation of river flows by impoundment in order to secure water supply can deliver very significant and immediate economic, political, and social benefits. But disadvantages that may accrue over the long term from poor decision making and mismanagement are often not viewed as costs (i.e., losses) that should be incorporated into social and economic balance

FIGURE 8.2   *Nodularia spumigena*; recreational lake, Logan Shire, Queensland, Australia. (*See insert for color representation of this figure.*)

sheets. The Australian example of the Murray River points toward immediate and medium-term direct economic benefits, sometimes disproportionately advantaging regional and private interests, contrasted with the ecological and economic debts that are spread more indirectly and borne by the broader community over longer time-scales.

Cyanobacteria blooms arising from manipulation of hydrologic regimes are but one symptom of a stressed and malfunctioning system, albeit one that is often highly visible (see Fig. 8.2) and sometimes accompanied by adverse publicity. Loss of productivity due to over-allocation of water resources or encroaching salinity can be quantified and reflected in appropriate accounting systems. Less tangible costs (in the sense of such costs generally being externalized or otherwise omitted from routine accounting) are those resulting from loss of ecosystem services such as flushing of estuaries, loss of biodiversity, decreased tourism-related income, and costs related to higher-level drinking water disinfection and treatment requirements.

A taskforce established to assess the feasibility of surface water impoundment to facilitate the expansion of agricultural development in tropical northern Australia recently recommended that a large-scale dam-building program should not proceed (81). This conclusion was made on expert scientific opinion that despite very large annual precipitation coverage—an estimated petalitre (10^{15} L) over the north of Australia—high seasonal rainfall, high evaporation rates,

interannual variability of rainfall, and lack of suitable locations for large, deep reservoirs make the capture and storage of surface water an unsuitable primary option for the region. The conclusions of the taskforce generated criticism from some political and industry groups in Australia, see, for example Refs 82, 83. Similar scenarios can be observed elsewhere in the world: Gleick (84), in his critique of the so-called "hard path" to water resource management, suggested that the belief that water demands can only be met by construction of more large-scale infrastructure is alive and well within some constituencies (84).

One supposition that might be drawn from the Australian experience is that political imperatives to build dams can be very insistent, even in locations that are unsuitable for the purpose and despite the prospect of diminishing returns due to deferred environmental costs. Yet community opposition to large-scale reservoirs and irrigation schemes is growing around the world, particularly where significant social dislocation and environmental costs accompany the proposal (84). Innovative, community-scale alternatives to large-scale manipulation of hydrologic regimes may be economically competitive in many regions, for example, micro-dams, run-of-river hydro schemes, and agricultural techniques that focus on water conservation and soil improvement, such as use of contour bunds and organic fertilizer supplements (85–89).

Large-scale dam projects will be subject to more sophisticated greenhouse gas accounting in future. Reservoirs are sources of atmospheric carbon dioxide and methane during and after construction, because flooded trees cease to absorb CO_2 and heterotrophic bacteria then decompose organic carbon stored in plants and soil, liberating CO_2 and CH_4 (90). Thus, carbon sinks are converted to sources. Eutrophic conditions in reservoirs promote further greenhouse gas production (91). Carbon accounting for hydroelectric power generation has revealed that some reservoirs have a higher greenhouse gas footprint than that of equivalent electricity production by coal-fired power. This occurs in reservoirs with low energy density, that is, broad, shallow dams (91, 92). Some of the anticipated effects of a warming climate on the problem of nuisance cyanobacteria are discussed below.

8.6   FUTURE PROBLEMS

Nuisance cyanobacteria are expected to become an increasing problem on a warming planet. Changes in the habitable range of many plant and animal species to higher latitudes and altitudes can also be anticipated for cyanobacteria. This may be occurring already in the case of the freshwater cyanobacterium *C. raciborskii*, which produces potent alkaloid toxins from the saxitoxin and cylindrospermopsin families. *C. raciborskii* was historically recognized as an inhabitant of tropical and subtropical waters but has over the past three decades been found in northern Europe, Canada, southern Australia, and New Zealand (93–95). This organism is increasingly recognized as a burgeoning public health exigency for drinking water supplies across all inhabited continents, and is described as an invasive cyanobacterium, with links to global warming and eutrophication posited as explanations for these changing bloom dynamics (96–98). Increased temperatures may favor the growth of cyanobacteria blooms because of enhanced and prolonged stratification of lakes and reservoirs, resulting in longer growth periods; intensified precipitation patterns may increase the transport of nutrients into surface waters. Incursions of seawater into freshwater systems caused by rising sea levels, as well as increased abstraction rates and drought-related inflow reductions are expected to broaden the range of brackish-water cyanobacteria (97). Paul (99) and Castle and Rodgers (100) discuss temporal associations between paleoextinctions and increased cyanobacterial abundance, suggesting that the geological record supports a hypothesis that prehistoric warmer climatic conditions, sea-level changes, elevated nutrient supplies, low faunal diversity, and reduced grazing pressure selected for cyanobacterial dominance of the oceans after mass extinction events. Cyanobacteria are well adapted to harsh environmental conditions, and can be expected to proliferate in both freshwater and marine systems under conditions of increased temperatures and ultraviolet flux (99, 100).

8.7   TOXIC CYANOBACTERIA AND SAFE DRINKING WATER SUPPLIES

Safe drinking water for all is an as-yet unrealized goal, and the burden of this failure is carried overwhelmingly by developing countries. Separating drinking water from sewage has been one of the most important and enlightened public health innovations to arise from the Industrial Revolution (101, 102). This insight is still not universally applied in the twenty-first century, due to poverty, social dislocation, and dispossession. For populations in developing countries that do not have access to safe drinking water, the immediate public health priority should always be directed toward disinfection to protect against pathogens of enteric origin, as discussed elsewhere in this volume. In such circumstances, protecting populations without access to treated water in reticulated supplies from cyanotoxins in their drinking water might be viewed as a secondary consideration. But there are small-scale, household- and community-level strategies to minimize the risks of exposure to cyanobacterial toxins that also address the most urgent daily priorities to treat bacterial and viral pathogens that may be present in water. Probably the most important factor to account for is prevention; managing source water free of cyanotoxins will generally be easier and more effective than approaches to remove the toxins that may be present in untreated water. To this end, rainwater harvesting or provision of drinking water sourced from groundwater should be considered superior to potable supplies drawn from standing surface water. Rainwater stored in underground tanks or opaque above-ground tanks will not be subject to significant proliferation of cyanobacteria because the sparse cells that may be present will be unable to photosynthesize. There are many innovative approaches to harvesting rainwater in developing countries; the 2002 report by the Third World Academy of Sciences (103) presents some fascinating case studies of its adoption in different communities, including at village scale and for apartment blocks. Rainwater harvesting is an attractive option "not only because it is economical but because of the rainwater's incomparable quality" (103).

Some micro-level water treatment techniques are being evaluated for their ability to remove cyanobacterial cells and cyanotoxins. Combined flocculant–disinfectant treatment of significantly contaminated water appears to be an effective point-of-use intervention for reducing the incidence of diarrheal disease in emergency settings and villages without centralized water treatment capacities (104, 105). Experimental investigation of flocculant–disinfectant treatment was shown to rapidly bind and precipitate cyanobacterial cells and decrease turbidity; the process did not cause a significant degree of cell lysis with liberation of free cyanotoxins (106).

Solar disinfection (SODIS) is a very low-cost water disinfection strategy in which solids are removed by sedimentation, after which pathogens are inactivated by heat and solar radiation by use of translucent containers (103).

Laboratory-based investigations have shown SODIS to be highly effective, though equivocal uptake and compliance appears to have limited its performance in some communities (107, 108). SODIS does not appear to have been studied from the perspective of managing toxigenic cyanobacteria or cyanotoxins in drinking water. Solar disinfection alone may be ineffective for treatment of surface water contaminated by cyanobacteria; the potential for cells to either proliferate within the treatment container or to lyse and release heat-stable toxins would require research investigation.

Point-of-use membrane filtration holds promise for household water disinfection. Production costs of membrane filters have decreased in recent years and household systems that operate under hydrostatic pressure (thus, independent of the requirement for powered pumps) are available. Widespread application for developing countries is not yet apparent, however. Further research and development efforts will be required; pretreatment in situations of high organic load will be necessary (109).

8.8 CONCLUDING REMARKS

Toxic cyanobacteria pose risks to human health from both acute and chronic exposures. Guidelines for exposure to cyanobacteria and cyanotoxins from drinking water and through contact in recreational waters have been developed by state and national agencies in many countries (110). Regulatory guidelines for toxic cyanobacteria are not universally adopted, however, even within developed countries. Yet there appears to be a North/South divide for protection of public health from these toxins that will be familiar to public health workers in developing countries. A disproportionate research effort has gone into techniques for removal and detoxification of cyanobacteria and their toxins from centralized, advanced water treatment systems. There is a clear requirement to research and develop strategies for both preventing contamination by cyanobacteria and to remove contaminating cyanotoxins from the wide range of microscale, decentralized drinking water purification techniques that may be utilized in developing countries.

ACKNOWLEDGMENTS

IS was supported by the Queensland Government Cabinet-approved QHFSS Research Fund, project RSS10-001.

REFERENCES

1. Fogg GE, Stewart WDP, Fay P, Walsby AE. *The Blue-Green Algae*. London: Academic Press, 1973.

2. Barsanti L, Coltelli P, Evangelista V, Frassanito AM, Passarelli V, Vesentini N, Gualtieri P. Oddities and curiosities in the algal world. In: Evangelista V, Barsanti L, Frassanito AM, Passarelli V, Gualtieri P, editors. *Algal Toxins: Nature, Occurrence, Effect and Detection*. Dordrecht: Springer, 2008, pp. 353–391.

3. Stewart I, Webb PM, Schluter PJ, Shaw GR. Recreational and occupational field exposure to freshwater cyanobacteria—a review of anecdotal and case reports, epidemiological studies and the challenges for epidemiologic assessment. *Environ Health* 2006;5:6.

4. Carmichael WW. Cyanobacteria secondary metabolites—the cyanotoxins. *J Appl Bacteriol* 1992;72:445–459.

5. Chorus I, Bartram J, editors. *Toxic Cyanobacteria in Water: A Guide to Their Public Health Consequences, Monitoring and Management*. London: E & FN Spon on behalf of the World Health Organization, 1999. Available at http://www.who.int/water_sanitation_health/resourcesquality/toxcyanobacteria.pdf

6. Codd GA, Ward CJ, Bell SG. Cyanobacterial toxins: occurrence, modes of action, health effects and exposure routes. *Arch Toxicol Suppl* 1997;19:399–410.

7. Falconer IR. *Cyanobacterial Toxins of Drinking Water Supplies: Cylindrospermopsins and Microcystins*. Boca Raton: CRC Press, 2005.

8. Hudnell HK, editor. *Cyanobacterial Harmful Algal Blooms: State of the Science and Research Needs*. New York: Springer, 2008. (Individual chapters published as the journal series *Adv Exp Med Biol*, vol. 619.).

9. Bradley WG, Mash DC. Beyond Guam: the cyanobacteria/BMAA hypothesis of the cause of ALS and other neurodegenerative diseases. *Amyotroph Lateral Scler* 2009;10 (Suppl 2):7–20.

10. Cox PA, Banack SA, Murch SJ, Rasmussen U, Tien G, Bidigare RR, Metcalf JS, Morrison LF, Codd GA, Bergman B. Diverse taxa of cyanobacteria produce beta-*N*-methylamino-L-alanine, a neurotoxic amino acid. *Proc Natl Acad Sci USA* 2005;102:5074–5078.

11. Metcalf JS, Banack SA, Lindsay J, Morrison LF, Cox PA, Codd GA. Co-occurrence of beta-*N*-methylamino-L-alanine, a neurotoxic amino acid with other cyanobacterial toxins in British waterbodies, 1990–2004. *Environ Microbiol* 2008;10:702–708.

12. Krüger T, Mönch B, Oppenhäuser S, Luckas B. LC-MS/MS determination of the isomeric neurotoxins BMAA (beta-*N*-methylamino-L-alanine) and DAB (2,4-diaminobutyric acid) in cyanobacteria and seeds of *Cycas revoluta* and *Lathyrus latifolius*. *Toxicon* 2010;55:547–557.

13. Snyder LR, Cruz-Aguado R, Sadilek M, Galasko D, Shaw CA, Montine TJ. Parkinson-dementia complex and development of a new stable isotope dilution assay for BMAA detection in tissue. *Toxicol Appl Pharmacol* 2009;240:180–188.

14. Steele JC, McGeer PL. The ALS/PDC syndrome of Guam and the cycad hypothesis. *Neurology* 2008;70:1984–1990.

15. Nishiwaki-Matsushima R, Nishiwaki S, Ohta T, Yoshizawa S, Suganuma M, Harada K, Watanabe MF, Fujiki H. Structure–function relationships of microcystins, liver tumor promoters, in interaction with protein phosphatase. *Jpn J Cancer Res* 1991;82:993–996.

16. Nishiwaki-Matsushima R, Ohta T, Nishiwaki S, Suganuma M, Kohyama K, Ishikawa T, Carmichael WW, Fujiki H. Liver tumor promotion by the cyanobacterial cyclic peptide toxin microcystin-LR. *J Cancer Res Clin Oncol* 1992; 118:420–424.

17. Ohta T, Sueoka E, Iida N, Komori A, Suganuma M, Nishiwaki R, Tatematsu M, Kim SJ, Carmichael WW, Fujiki H. Nodularin, a potent inhibitor of protein phosphatases 1 and 2A, is a new environmental carcinogen in male F344 rat liver. *Cancer Res* 1994;54:6402–6406.

18. International Agency for Research on Cancer (IARC). *IARC monographs on the evaluation of carcinogenic risks to humans - Volume 94: ingested nitrate and nitrite, and cyanobacterial peptide toxins*. Lyon: IARC for the World Health Organization, 2010.

19. Stewart I, Carmichael WW, Sadler R, McGregor GB, Reardon K, Eaglesham GK, Wickramasinghe WA, Seawright AA, Shaw GR. Occupational and environmental hazard assessments for the isolation, purification and toxicity testing of cyanobacterial toxins. *Environ Health* 2009;8:52.

20. Azevedo SM, Chernoff N, Falconer IR, Gage M, Hilborn ED, Hooth MJ, Jensen K, MacPhail R, Rogers E, Shaw GR, Stewart I, Fournie JW. Human Health Effects Workgroup report. *Adv Exp Med Biol* 2008;619:579–606.

21. Burch MD. Effective doses, guidelines & regulations. *Adv Exp Med Biol* 2008;619:831–853.

22. National Health and Medical Research Council. *Australian drinking water guidelines*. Canberra: Commonwealth of Australia, 2010.

23. Hrudey S, Burch M, Drikas M, Gregory R. Remedial measures. In: Chorus I, Bartram J, editors. *Toxic Cyanobacteria in Water: A Guide to Their Public Health Consequences, Monitoring and Management*. London: E & FN Spon on behalf of the World Health Organization, 1999, pp. 275–312. Available at http://www.who.int/water_sanitation_health/resourcesquality/toxcyanchap9.pdf

24. Su D. Drinking water and liver cell cancer: an epidemiologic approach to the etiology of this disease in China. *Chin Med J (Engl)* 1979;92:748–756.

25. El-Serag HB, Rudolph KL. Hepatocellular carcinoma: epidemiology and molecular carcinogenesis. *Gastroenterology* 2007;132:2557–2576.

26. Yuen MF, Hou JL, Chutaputti A, on behalf of the Asia Pacific Working Party on Prevention of Hepatocellular Carcinoma. Hepatocellular carcinoma in the Asia Pacific region. *J Gastroenterol Hepatol* 2009;24:346–353.

27. Lin NF, Tang J, Bian JM. Geochemical environment and health problems in China. *Environ Geochem Health* 2004;26:81–88.

28. Ueno Y, Nagata S, Tsutsumi T, Hasegawa A, Watanabe MF, Park HD, Chen GC, Chen G, Yu SZ. Detection of microcystins, a blue-green algal hepatotoxin, in drinking water sampled in Haimen and Fusui, endemic areas of primary liver cancer in China, by highly sensitive immunoassay. *Carcinogenesis* 1996;17:1317–1321.

29. Yu SZ. Primary prevention of hepatocellular carcinoma. *J Gastroenterol Hepatol* 1995;10:674–682.

30. Luo RH, Zhao ZX, Zhou XY, Gao ZL, Yao JL. Risk factors for primary liver carcinoma in Chinese population. *World J Gastroenterol* 2005;11:4431–4434.

31. Yu S, Zhao N, Zi X. The relationship between cyanotoxin (microcystin, MC) in pond-ditch water and primary liver cancer in China. *Zhonghua Zhong Liu Za Zhi* 2001;23: 96–99 (in Chinese).

32. Yu SZ, Huang XE, Koide T, Cheng G, Chen GC, Harada K, Ueno Y, Sueoka E, Oda H, Tashiro F, Mizokami M, Ohno T, Xiang J, Tokudome S. Hepatitis B and C viruses infection, lifestyle and genetic polymorphisms as risk factors for hepatocellular carcinoma in Haimen, China. *Jpn J Cancer Res* 2002;93:1287–1292.

33. Chen J, Xie P, Li L, Xu J. First identification of the hepatotoxic microcystins in the serum of a chronically exposed human population together with indication of hepatocellular damage. *Toxicol Sci* 2009;108:81–89.

34. Marrero JA, Feng Z, Wang Y, Nguyen MH, Befeler AS, Roberts LR, Reddy KR, Harnois D, Llovet JM, Normolle D, Dalhgren J, Chia D, Lok AS, Wagner PD, Srivastava S, Schwartz M. α-Fetoprotein, des-γ carboxyprothrombin, and lectin-bound α-fetoprotein in early hepatocellular carcinoma. *Gastroenterology* 2009;137:110–118.

35. Zhang H, Zhang J, Zhu Y. Identification of microcystins in waters used for daily life by people who live on Tai Lake during a serious cyanobacteria dominated bloom with risk analysis to human health. *Environ Toxicol* 2009;24:82–86.

36. Qin B, Zhu G, Gao G, Zhang Y, Li W, Paerl HW, Carmichael WW. A drinking water crisis in Lake Taihu, China: linkage to climatic variability and lake management. *Environ Manage* 2010;45:105–112.

37. Zhou L, Yu H, Chen K. Relationship between microcystin in drinking water and colorectal cancer. *Biomed Environ Sci* 2002;15:166–171.

38. Chen K, Yu W, Ma X, Yao K, Jiang Q. The association between drinking water source and colorectal cancer incidence in Jiashan County of China: a prospective cohort study. *Eur J Public Health* 2005;15:652–656.

39. Chen K, Cai J, Liu XY, Ma XY, Yao KY, Zheng S. Nested case-control study on the risk factors of colorectal cancer. *World J Gastroenterol* 2003;9:99–103.

40. Yu MC, Yuan JM. Environmental factors and risk for hepatocellular carcinoma. *Gastroenterology* 2004;127:S72–S78.

41. Hao W, Young D. Drinking patterns and problems in China. *J Subst Use* 2000;5:71–78.

42. Peto R, Chen Z-M, Boreham J. Tobacco: the growing epidemic in China. *CVD Prev Control* 2009;4:61–70.

43. Plymoth A, Viviani S, Hainaut P. Control of hepatocellular carcinoma through hepatitis B vaccination in areas of high endemicity: perspectives for global liver cancer prevention. *Cancer Lett* 2009;286:15–21.

44. Sun Z, Zhu Y, Stjernsward J, Hilleman M, Collins R, Zhen Y, Hsia CC, Lu J, Huang F, Ni Z, Ni T, Liu GT, Yu Z, Liu Y, Chen JM, Peto R. Design and compliance of HBV vaccination trial

on newborns to prevent hepatocellular carcinoma and 5-year results of its pilot study. *Cancer Detect Prev* 1991;15: 313–318.

45. Wogan GN, Hecht SS, Felton JS, Conney AH, Loeb LA. Environmental and chemical carcinogenesis. *Semin Cancer Biol* 2004;14:473–486.

46. Wild CP, Montesano R. A model of interaction: aflatoxins and hepatitis viruses in liver cancer aetiology and prevention. *Cancer Lett* 2009;286:22–28.

47. Yu MC, Yuan JM, Lu SC. Alcohol, cofactors and the genetics of hepatocellular carcinoma. *J Gastroenterol Hepatol* 2008;23 (Suppl 1):S92–S97.

48. Lian M, Liu Y, Yu SZ, Qian GS, Wan SG, Dixon KR. Hepatitis B virus x gene and cyanobacterial toxins promote aflatoxin B1-induced hepatotumorigenesis in mice. *World J Gastroenterol* 2006;12:3065–3072.

49. Sekijima M, Tsutsumi T, Yoshida T, Harada T, Tashiro F, Chen G, Yu SZ, Ueno Y. Enhancement of glutathione *S*-transferase placental-form positive liver cell foci development by microcystin-LR in aflatoxin B1-initiated rats. *Carcinogenesis* 1999;20:161–165.

50. Fleming LE, Rivero C, Burns J, Williams C, Bean JA, Shea KA, Stinn J. Blue green algal (cyanobacterial) toxins, surface drinking water, and liver cancer in Florida. *Harmful Algae* 2002;1:157–168.

51. Svirčev Z, Krstić S, Miladinov-Mikov M, Baltić V, Vidović M. Freshwater cyanobacterial blooms and primary liver cancer epidemiological studies in Serbia. *J Environ Sci Health C Environ Carcinog Ecotoxicol Rev* 2009;27:36–55.

52. Byth S. Palm Island mystery disease. *Med J Aust* 1980;2:40, 42.

53. Hawkins PR, Runnegar MTC, Jackson ARB, Falconer IR. Severe hepatotoxicity caused by the tropical cyanobacterium (blue-green alga) *Cylindrospermopsis raciborskii* (Woloszynska) Seenaya and Subba Raju isolated from a domestic water supply reservoir. *Appl Environ Microbiol* 1985;50:1292–1295.

54. Carmichael WW, Azevedo SMFO, An JS, Molica RJR, Jochimsen EM, Lau S, Rinehart KL, Shaw GR, Eaglesham GK. Human fatalities from cyanobacteria: chemical and biological evidence for cyanotoxins. *Environ Health Perspect* 2001;109:663–668.

55. Jochimsen EM, Carmichael WW, An JS, Cardo DM, Cookson ST, Holmes CEM, Antunes MBdC, de Melo Filho DA, Lyra TM, Barreto VST, Azevedo SMFO, Jarvis WR. Liver failure and death after exposure to microcystins at a hemodialysis center in Brazil. *N Engl J Med* 1998;338:873–878.

56. Soares RM, Yuan M, Servaites JC, Delgado A, Magalhães VF, Hilborn ED, Carmichael WW, Azevedo SMFO. Sublethal exposure from microcystins to renal insufficiency patients in Rio de Janeiro, Brazil. *Environ Toxicol* 2006;21:95–103.

57. Teixeira MdGLC, Costa MdCN, de Carvalho VLP, Pereira MdS, Hage E. Gastroenteritis epidemic in the area of the Itaparica Dam, Bahia, Brazil. *Bull Pan Am Health Organ* 1993;27:244–253.

58. Backer LC, Carmichael W, Kirkpatrick B, Williams C, Irvin M, Zhou Y, Johnson TB, Nierenberg K, Hill VR, Kieszak SM, Cheng YS. Recreational exposure to low concentrations of microcystins during an algal bloom in a small lake. *Mar Drugs* 2008;6:389–406.

59. Backer LC, McNeel SV, Barber T, Kirkpatrick B, Williams C, Irvin M, Zhou Y, Johnson TB, Nierenberg K, Aubel M, LePrell R, Chapman A, Foss A, Corum S, Hill VR, Kieszak SM, Cheng YS. Recreational exposure to microcystins during algal blooms in two California lakes. *Toxicon* 2010;55:909–921.

60. Ibelings BW, Chorus I. Accumulation of cyanobacterial toxins in freshwater "seafood" and its consequences for public health: a review. *Environ Pollut* 2007;150:177–192.

61. Smith JL, Boyer GL, Zimba PV. A review of cyanobacterial odorous and bioactive metabolites: impacts and management alternatives in aquaculture. *Aquaculture* 2008;280:5–20.

62. Codd GA, Metcalf JS, Beattie KA. Retention of *Microcystis aeruginosa* and microcystin by salad lettuce (*Lactuca sativa*) after spray irrigation with water containing cyanobacteria. *Toxicon* 1999;37:1181–1185.

63. World Commission on Dams. *Dams and Development: A New Framework for Decision-Making*. London: Earthscan, 2000.

64. Goldman CR. Ecological aspects of water impoundment in the tropics. *Rev Biol Trop* 1976;24:87–112.

65. Sleigh AC, Jackson S. Dams, development, and health: a missed opportunity. *Lancet* 2001;357:570–571.

66. Nilsson C, Berggren K. Alterations of riparian ecosystems caused by river regulation. *Bioscience* 2000;50:783–792.

67. Jeong KS, Kim DK, Joo GJ. Delayed influence of dam storage and discharge on the determination of seasonal proliferations of *Microcystis aeruginosa* and *Stephanodiscus hantzschii* in a regulated river system of the lower Nakdong River (South Korea). *Water Res* 2007;41:1269–1279.

68. Morais P, Chícharo MA, Chícharo L. Changes in a temperate estuary during the filling of the biggest European dam. *Sci Total Environ* 2009;407:2245–2259.

69. Sobrino C, Matthiensen A, Vidal S, Galvão H. Occurrence of microcystins along the Guadiana estuary. *Limnetica* 2004; 23:133–144.

70. Li Z, Fang F, Guo J, Chen J, Zhang C, Tian G. Spring algal bloom and nutrients characteristics in Xiaojiang River backwater area, Three Gorge Reservoir, 2007. *J Lake Sci* 2009;21:36–44.

71. Beeton RJS, Buckley KI, Jones GJ, Morgan D, Reichert RE, Trewin D, (2006 Australian State of the Environment Committee). Australia State of the Environment 2006. Independent report to the Australian Government Minister for the Environment and Heritage. Canberra: Department of the Environment and Heritage, 2006, pp. i–viii;1–132. Available at http://www.environment.gov.au/soe/2006/publications/report/pubs/soe-2006-report.pdf

72. Environment Protection Authority. The state of our environment: State of the Environment Report for South Australia 2008. EPA South Australia, 2008, p. 301. Available at http://www.epa.sa.gov.au/soe/soe_2008

73. Wallbrink PJ, Wilson CJ, Martin CE. Sources and delivery of suspended sediment and phosphorus to Australian rivers:

National Eutrophication Management Program (NEMP). Canberra: Land & Water Australia, 2009, p. 15. Available at http://lwa.gov.au/files/products/river-landscapes/pn30295/sources-and-delivery-suspended-sediment-and-phosph.pdf

74. NSW Blue-Green Algae Task Force. Final report of the NSW Blue-Green Algae Task Force. Parramatta: NSW Department of Water Resources, 1992.

75. Stewart I, Seawright AA, Shaw GR. Cyanobacterial poisoning in livestock, wild mammals and birds—an overview. *Adv Exp Med Biol* 2008;619:613–637.

76. Ryan NJ, Dabovic J, Bowling LD, Driver B, Barnes B. The Murray River algal bloom: evaluation and recommendations for the future management of major outbreaks. Sydney: NSW Office of Water, 2009. pp. i–iii;1–37. Available at http://www.water.nsw.gov.au/ArticleDocuments/34/algal_murray_evaluation.pdf.aspx.

77. Infante AG. Primary production of phytoplankton in Lake Valencia (Venezuela). *Int Rev Gesamten Hydrobiol* 1997;82:469–477.

78. Xu YP, Jaffe R. Geochemical record of anthropogenic impacts on Lake Valencia, Venezuela. *Appl Geochem* 2009;24: 411–418.

79. Lathrop RC. Perspectives on the eutrophication of the Yahara lakes. *Lake Reserv Manage* 2007;23:345–365.

80. Codd GA, Lindsay J, Young FM, Morrison LF, Metcalf JS. Harmful cyanobacteria: from mass mortalities to management measures. In: Huisman J, Matthijs HCP, Visser PM, editors. *Harmful Cyanobacteria*. Dordrecht: Springer, 2005, pp. 1–23.

81. Northern Australia Land and Water Taskforce. Sustainable development of northern Australia: a report to Government from the Northern Australia Land and Water Taskforce. Canberra: Department of Infrastructure, Transport, Regional Development and Local Government, 2009;pp. i–x, 1–38. Available at http://www.nalwt.gov.au/files/337281_NLAW.pdf

82. Knight L. Northern Taskforce told not to investigate new dams. Perth: *Farm Weekly* (*Rural Press*), 2010;Farmonline website. Available at http://fw.farmonline.com.au/news/nationalrural/agribusiness-and-general/general/northern-taskforce-told-not-to-investigate-new-dams/1747388.aspx

83. Taffa V. Canberra says no to northern Australia agriculture. Sydney: *The Southern Thunderer* (website), 2010. Available at http://www.southernthunderer.com.au/2010/02/canberra-says-no-to-northern-australia-agriculture/

84. Gleick PH. Global freshwater resources: soft-path solutions for the 21st century. *Science* 2003;302:1524–1528.

85. Gleick PH. The changing water paradigm: a look at twenty-first century water resources development. *Water Int* 2000;25:127–138.

86. Al-Seekh SH, Mohammad AG. The effect of water harvesting techniques on runoff, sedimentation, and soil properties. *Environ Manage* 2009;44:37–45.

87. Ouedraogo E, Stroosnijder L, Mando A, Brussaard L, Zougmore R. Agroecological analysis and economic benefit of organic resources and fertiliser in till and no-till sorghum production after a 6-year fallow in semi-arid West Africa. *Nutr Cycl Agroecosys* 2007;77:245–256.

88. Zougmore R, Kabore D, Lowenberg-DeBoer J. Optimal spacing of soil conservation barriers: example of rock bunds in Burkina Faso. *Agron J* 2000;92:361–368.

89. Zougmore R, Mando A, Stroosnijder L, Ouedraogo E. Economic benefits of combining soil and water conservation measures with nutrient management in semiarid Burkina Faso. *Nutr Cycl Agroecosys* 2004;70:261–269.

90. St Louis VL, Kelly CA, Duchemin E, Rudd JWM, Rosenberg DM. Reservoir surfaces as sources of greenhouse gases to the atmosphere: a global estimate. *Bioscience* 2000;50: 766–775.

91. Gunkel G. Hydropower—a green energy? Tropical reservoirs and greenhouse gas emissions. *Clean-Soil Air Water* 2009;37:726–734.

92. dos Santos MA, Rosa LP, Sikar B, Sikar E, dos Santos EO. Gross greenhouse gas fluxes from hydro-power reservoir compared to thermo-power plants. *Energy Policy* 2006; 34:481–488.

93. Kling HJ. Cylindrospermopsis raciborskii (Nostocales, Cyanobacteria): a brief historic overview and recent discovery in the Assiniboine River (Canada). *Fottea* 2009; 9:45–47.

94. Fastner J, Rucker J, Stuken A, Preussel K, Nixdorf B, Chorus I, Kohler A, Wiedner C. Occurrence of the cyanobacterial toxin cylindrospermopsin in northeast Germany. *Environ Toxicol* 2007;22:26–32.

95. Wood SA, Stirling DJ. First identification of the cylindrospermopsin-producing cyanobacterium *Cylindrospermopsis raciborskii* in New Zealand. *NZ J Mar Freshwater Res* 2003;37:821–828.

96. Haande S, Rohrlack T, Ballot A, Røberg K, Skulberg R, Beck M, Wiedner C. Genetic characterisation of *Cylindrospermopsis raciborskii* (Nostocales, Cyanobacteria) isolates from Africa and Europe. *Harmful Algae* 2008;7:692–701.

97. Paerl HW, Huisman J. Climate. Blooms like it hot. *Science* 2008;320:57–58.

98. Briand JF, Leboulanger C, Humbert JF, Bernard C, Dufour P. *Cylindrospermopsis raciborskii* (Cyanobacteria) invasion at mid-latitudes: selection, wide physiological tolerance, or global warming? *J Phycol* 2004;40:231–238.

99. Paul VJ. Global warming and cyanobacterial harmful algal blooms. *Adv Exp Med Biol* 2008;619:239–257.

100. Castle JW, Rodgers JH, Jr., Hypothesis for the role of toxin-producing algae in Phanerozoic mass extinctions based on evidence from the geologic record and modern environments. *Environ Geosci* 2009;16:1–23.

101. Dawson DJ, Sartory DP. Microbiological safety of water. *Br Med Bull* 2000;56:74–83.

102. McMichael AJ. *Planetary Overload: Global Environmental Change and the Health of the Human Species*. Cambridge: Cambridge University Press, 1993.

103. Third World Academy of Sciences. *Safe Drinking Water: The Need, the Problem, Solutions and an Action Plan*. Trieste: Third World Academy of Sciences, 2002, p. 23. Available at http://twas.ictp.it/publications/twas-reports/safedrinkingwater.pdf

104. Colindres RE, Jain S, Bowen A, Mintz E, Domond P. After the flood: an evaluation of in-home drinking water treatment with combined flocculent–disinfectant following Tropical Storm Jeanne—Gonaives, Haiti, 2004. *J Water Health* 2007;5:367–374.

105. Crump JA, Otieno PO, Slutsker L, Keswick BH, Rosen DH, Hoekstra RM, Vulule JM, Luby SP. Household based treatment of drinking water with flocculant–disinfectant for preventing diarrhoea in areas with turbid source water in rural western Kenya: cluster randomised controlled trial. *BMJ* 2005;331:478.

106. Allen EAD, Carmichael WW, Keswick B.The evaluation of a flocculating agent (PuR) for the removal of cyanobacteria and their toxins from potable water supplies [poster]. Sixth International Conference on Toxic Cyanobacteria. Bergen, 2004.

107. Altherr AM, Mosler HJ, Tobias R, Butera F. Attitudinal and relational factors predicting the use of solar water disinfection: a field study in Nicaragua. *Health Educ Behav* 2008;35: 207–220.

108. Mäusezahl D, Christen A, Pacheco GD, Tellez FA, Iriarte M, Zapata ME, Cevallos M, Hattendorf J, Cattaneo MD, Arnold B, Smith TA, Colford JM, Jr., Solar drinking water disinfection (SODIS) to reduce childhood diarrhoea in rural Bolivia: a cluster-randomized, controlled trial. *PLoS Med* 2009;6: e1000125.

109. Peter-Varbanets M, Zurbrügg C, Swartz C, Pronk W. Decentralized systems for potable water and the potential of membrane technology. *Water Res* 2009;43:245–265.

110. Chorus I, editor. Current approaches to cyanotoxin risk assessment, risk management and regulations in different countries. Berlin: Federal Environmental Agency (Umweltbundesamt), 2005. Available at http://www.umweltdaten.de/publikationen/fpdf-l/2910.pdf

9

DENGUE: A WATER-BORNE DISEASE

SCOTT B. HALSTEAD

9.1 INTRODUCTION

Dengue viruses form a group of three important human pathogens, the four dengues, yellow fever and chikungunya viruses, each transmitted in an urban cycle by the bite of *Aedes aegypti,* or in some cases, by *Aedes albopictus*. The former vector mosquito is strictly anthropophilic and breeds mainly in water stored in and around homes for drinking, cooking, or washing. The absence of reliable, affordable, and safe piped water is a major risk factor contributing to support of vector mosquito populations and the acquisition of dengue infections. The word "dengue" is a Spanish homonym of the Swahili "ki dinga pepo," a term used in Africa to describe chikungunya during the nineteenth century (1). Chikungunya virus was imported to the Caribbean producing the epidemic of 1827–1828, but then disappeared entirely from the New World, the diagnostic term being transposed to modern dengue illnesses (2). Dengue viruses cause two main syndromes: *Dengue fever* is a benign illness caused by each of the four dengue viruses whose classical features include biphasic fever, myalgia or arthralgia, rash, leukopenia, and lymphadenopathy. A more severe form, *dengue hemorrhagic fever(DHF)* is an acute febrile disease characterized by capillary permeability, abnormalities of hemostasis and the severe instance, a protein-losing shock syndrome—*dengue shock syndrome(DSS)*.

9.2 ETIOLOGY

There are four distinct antigenic types of dengue virus— named dengue (DENV) 1, 2, 3, and 4—members of the family Flaviviridae. They are enveloped viruses enclosing a single positive RNA strand composed of approximately 11,000 nucleotides. The genome codes for three structural proteins—C, prM, and E—and seven nonstructural proteins—NS1, NS2a, NS2b, NS3, NS4, NS5a, and NS5b. Dengue viruses evolved, probably in Asia, from a common ancestor adapted to subhuman primates (3). After prolonged survival in isolated primate populations four separate viruses evolved with genomes that differ from one another by around 35–40% (4). Each of these viruses emerged separately into the urban transmission cycle some 300–400 years ago, around the same time that anthropophilic *A. aegypti* may have been established in tropical Asian ports by explorers and traders from Africa (4–6).

9.3 EPIDEMIOLOGY

9.3.1 Vector Bionomics

A. aegypti, a crepuscular daytime-biting mosquito, is the principal vector. All four virus types have been recovered from naturally infected *A. aegypti*. Although the species originated from feral African ancestors, in most tropical areas, *A. aegypti* is highly domesticated and breeds in water stored for drinking, washing, or bathing or in any container collecting freshwater. Dengue viruses also have been recovered from naturally infected *A. albopictus*, which breeds outdoors in vegetation, as in Hawaii in 2001 (7). Outbreaks in the Pacific area have been attributed to *Aedes scutellaris* and *Aedes polynesiensis*. Dengue viruses replicate in the gut, brain, and salivary glands of infected mosquitoes without apparent harm to adult mosquitoes (8). Mosquitoes are infectious for a lifetime and as long as 70 days in experimental

Water and Sanitation-Related Diseases and the Environment: Challenges, Interventions, and Preventive Measures, First Edition.
Edited by Janine M. H. Selendy.

circumstances. Because female mosquitoes take repeated blood meals, long-lived female mosquitoes have great potency as vectors. Several species of *Stegomyia* and *Toxorhynchites* are infected readily orally and by intrathoracic inoculation, although the threshold of infection by oral feeding is higher than the intrathoracic route (8, 9).

A. aegypti preferentially feeds on people and, hence, is most abundant in and around human habitations. The mosquito breeds in clean water. Breeding sites may be provided by humans through living habits, as in Thailand, where water commonly is stored in and around homes in large earthenware jars (10). In contrast, *A. aegypti* is not abundant in some parts of India because only small amounts of water are brought to homes for immediate use from village wells. Water in sinks, flower vases, household offerings, ant traps, coconut husks, tin cans, plastic or glass bottles, oil drums, and rubber tires may supply breeding sites for *A. aegypti* (11). With the beginning of monsoon rains, a large number of eggs laid outdoors are hatched. Indoor populations do not show seasonal change (12).

9.3.2 Transmission

Temperature is important in controlling dengue viral transmission. Evidence indicates that the extrinsic incubation period shortens with increasing mean temperatures while mosquito-biting rates increase with seasonal increases in temperature and relative humidity (13, 14). In the tropics, outbreaks of dengue generally coincide with the monsoon season. Eggs, which resist desiccation, are deposited inside water containers above the water line (15). However, biting activity is reduced at temperatures below a wet bulb temperature of 14°C (16). During the Little Ice Age, which extended into the colonial era, *A. aegypti* became established in cities along the North American Atlantic and Gulf coasts and in urban areas in the Mississippi Basin. These vector populations supported substantial summertime outbreaks of dengue and yellow fever up through the first half of the nineteenth century (17, 18). Summertime dengue epidemics were common in temperate areas in other parts of the Americas as well as in Europe, Australia, and Asia until early in the twentieth century (10, 19, 20). Transmission of dengue in temperate countries was interrupted during winter weather and dengue has not established itself endemically at latitudes above 25 degrees north or south.

In urban areas, transmission of dengue may be explosive and involve as much as 70–80% of the population (17). Because *A. aegypti* has a limited flight range, spread of the virus is mainly via mobile viremic human beings (21, 22). *A. aegypti* and *Culex quinquefasciatus* can transmit dengue mechanically by interrupted feeding (23). Because of the "skittishness" of *A. aegypti* and its habit of feeding during the day when its intended victim is awake and often moving, interrupted feeding resulting in near simultaneous

transmission to multiple hosts within a household must commonly occur (24, 25).

9.3.3 Maintenance Cycles

Simmons and colleagues were the first to note that wild-caught *Macaca philippinensis* resisted dengue infection whereas *Macaca fuscatus* (Japanese macaque) was susceptible (17). Work by Rudnick (26) in Malaysia has revealed a jungle cycle of dengue transmission involving canopy-feeding monkeys and *Aedes niveus*, a species that feeds on both monkeys and humans. Although the existence of a jungle dengue cycle in the Malaysian rain forest has been documented, the full geographic range of the subhuman primate zoonotic reservoir in Asia is not known. In the 1980s and 1990s, extensive epizootics of dengue virus type 2 occurred in subhuman primates over wide areas of West Africa (5). Genetic and epidemiologic studies have shown that urban human dengue and jungle monkey dengue are relatively compartmentalized (27, 28). There is some evidence, however, that sylvatic dengue strains retain the ability to infect humans and cause disease (5, 29). *A. aegypti*, dengue viruses, and dengue-susceptible humans are so abundant and so widespread that if dengue viruses are exchanged between humans and monkeys, detection will be extremely difficult. If, in the future, urban dengue is eliminated by mosquito abatement but vector mosquito populations become reestablished, dengue viruses could be reintroduced from jungle cycles. Subhuman primates are generally susceptible to infection by human strains of dengue viruses. Numerous species belonging to *Macaca, Cercopithecus, Cercocebus, Papio, Hylobates*, and *Pan* can be infected by the bite of virus-infected mosquitoes or by injection of infectious virus preparations. Infection is essentially asymptomatic. Viremia occurs at levels sufficient to infect mosquitoes.

There is experimental evidence of the occurrence of transovarial transmission (30). The contribution to maintenance of viruses by transovarial transmission in a natural habitat is unknown (31).

9.3.4 Geographic Distribution

The probable spread of *A. aegypti* during historical times out of Africa throughout the world provided an ecologic niche quickly occupied by several human viral pathogens: yellow fever in the Americas, chikungunya repeatedly introduced into Asia, and the dengue viruses emerged from the Asian sylvatic cycle and spread to tropical countries throughout the world. During the eighteenth and nineteenth centuries, dengue epidemics occurred in newly settled lands, largely because of the necessity for storage of domestic water in frontier areas. Isolated shipboard or garrison outbreaks often confined to nonindigenous settlers or visitors were reported in Africa, the Indian subcontinent, and Southeast

Asia (2, 10, 17). During World War II, dengue virus infections occurred commonly in combatants of the Pacific War and spread to staging areas not normally infected: Japan, Hawaii, and Polynesia (32). DHF-like disease was described clinically in Thailand beginning in 1950 and in the Philippines from 1953. Cases were confirmed etiologically as dengue in 1958 and 1956, respectively (33). DHF first was described in Singapore and Malaysia in 1962, Vietnam in 1963, India in 1963, Ceylon (Sri Lanka) in 1965, Indonesia in 1969, Burma (Myanmar) in 1970, China in 1985, and Kampuchea and Laos from about 1985, and major outbreaks have occurred in Sri Lanka and India since 1988, in French Polynesia since 1990, in Pakistan since 1998, and in Bangladesh since 1999. DHF/DSS has occurred at consistently high endemicity in Thailand, Burma, Vietnam, Indonesia, and more recently in Sri Lanka and India (34, 35). In Thailand, it is the third ranking cause of hospitalization and death in children. Intermittent epidemics have involved Malaysia and the Philippines. The largest DHF/DSS epidemic in history occurred in Southeast Asia in 1998, when more than 490,000 hospitalizations and 4000 deaths were reported to the WHO.

During the past 50 years, major epidemics of all four dengue serotypes have occurred on many Pacific islands (34, 36–38).

After dengue viruses were imported from Africa during the slave trade era, 150 years later, they were again imported into the Americas, this time from Asia likely by viremic humans on jet airplanes. In 1977, an Asian DENV 1 (genotype III) was introduced, producing massive epidemics in Jamaica, Cuba, Puerto Rico, and Venezuela and quickly spreading throughout the Caribbean Islands, Mexico, Texas, Central America, northern South America, and then to Brazil (39). During the 1977 epidemic in Puerto Rico, in addition to DENV 1 many strains of DENV 2 and 3 were isolated. Secondary infections in the sequence DENV 2–1 or DENV 3–1 must have occurred in great numbers, but no DHF occurred; the only syndrome observed was DF in adults. While secondary infections by DENV 1 result in DHF/DSS in SE Asia, the exact frequency of severe disease is not known for specific serotypes (40). In 2001, a large DHF/DSS outbreak occurred in French Polynesia in the sequence DENV 2–DENV 1 (38). Dengue 1 spread rapidly and now is endemic on the larger Caribbean islands, in Mexico, coastal Central America, and the tropical areas of Guyana, Venezuela, Colombia, Ecuador, Peru, and Brazil. Following the island-wide transmission of DENV 1 in 1977–1979, a 3-month DENV 2 epidemic occurred in Cuba in 1981, resulting in 116,000 hospitalizations, 10,000 with DHF/DSS. This epidemic led to island-wide *A. aegypti* control and apparent eradication of the virus (41). In 1981, an Asian DENV 4 was introduced into the Caribbean, spread widely, and now is endemic throughout the Caribbean basin. In 1986 and 1987, DENV 1 spread through most of coastal Brazil and from there to Paraguay and to Peru and Ecuador. In 1990,

more than 9000 dengue cases were reported from Venezuela; 2600 of them were classified as DHF, and 74 deaths were associated with the epidemic (42). DENV 1, 2, and 4 were isolated. Shortly thereafter, DHF/DSS caused by DENV 2 was reported from Brazil and French Guiana (43). In 1994, DENV 3 was introduced into the region (44). In 1997, a DENV 2 Southeast Asian genotype was introduced into Santiago de Cuba and caused a sharp DHF/DSS outbreak observed only in individuals 20 years and older (45).

In 1981, both DENV 4 (genotype I) and DENV 2 (genotype III) were introduced into the Caribbean islands. DENV 4 rapidly spread westward to other Caribbean islands, then to Mexico, Central America, and northern South America (Caribbean basin) (46). DENV 2 was introduced into Cuba from Southeast Asia, possibly from Vietnam and from there spread to the Caribbean basin and all of tropical America more slowly than did DENV 4 (46, 47).

In the decade of the 1970s Brazil witnessed the reinvasion of *A. aegypti*. This permitted the intrusion in 1981 of DENV 1 and DENV 4 into Roraima State in the north (48, 49). But it was not until 1986, that DENV 1 caused a sharp outbreak of DF in Rio de Janeiro, then spread toward the northeast and midwest regions where it remains endemic (50). In 1990, DENV 2 (genotype III) was reported in Rio de Janeiro spreading across the country and also becoming endemic in some areas (48, 51). After dropping from sight for nearly 20 years, American genotype DENV 2 virus remerged in Iquitos, Peru, in 1995 (52).

In 1994, DENV 3 (genotype III) was first detected in Central America, then spreading to other areas in the Caribbean basin and beyond (46). In 2001–2002, DENV 3 caused a sharp outbreak in Havana and environs, producing many severe cases of DHF in adults who had experienced DENV 1 infections in 1997–1999 (53). In 2002, DENV 3 produced a major epidemic initially in Rio de Janeiro, then throughout Brazil (48, 54). In Rio de Janeiro, 62 deaths and many cases of DHF were reported. Yet, the predominant clinical expression throughout Brazil was still DF in adults (48). DHF/DSS in children caused by second infections with DENV 2 and 3 is a growing problem in several Central American and northern tier South American countries (55–60). DENV 3 is now widespread in Brazil and responsible for millions of cases and for the transition from DF to DHF, particularly in children (61–63). Remarkably, although endemic in Puerto Rico and Central American countries, DENV 4 has not yet penetrated most of Brazil.

DENV 1 and 2 have been recovered from humans with mild clinical illness in Nigeria in the absence of epidemic disease (64). In 1983, DENV 3 was isolated in Mozambique (34). DHF/DSS has not been reported, and even dengue fever outbreaks are rare. In this respect, Africa resembles the situation in Haiti, where multiple dengue serotypes are transmitted at high rates among a predominantly black population, but severe disease is not recognized (65).

9.4   CLINICAL FINDINGS

Primary infections with DENV 2 and 4 are thought to be largely inapparent, particularly in children (45, 66–68). Primary infections with dengue virus types 1 and 3 in adults produce classical dengue fever (see the next section). Infections with these same viruses in infants and young children may result in undifferentiated fever or a 1- to 5-day fever, pharyngeal inflammation, rhinitis, and mild cough or, in some instances, a febrile illness with thrombocytopenia and mild vascular permeability, which in dengue endemic countries is identified as DHF (67, 69).

9.4.1   Dengue Fever

In classic dengue fever (seen most frequently in adults), after an incubation period of 2–7 days, patients experience a sudden onset of fever, which rapidly rises to 39.5–41.4°C (103–106°F) and is usually accompanied by frontal or retro-orbital headache. Occasionally, back pain precedes the fever. A transient, macular, generalized rash that blanches under pressure may be seen during the first 24–48 h of fever. The pulse rate may be slow in proportion to the degree of fever. Myalgia or bone pain occurs soon after onset and increases in severity. During the second to sixth day of fever, nausea and vomiting are apt to occur, and generalized lymphadenopathy, cutaneous hyperesthesia or hyperalgesia, aberrations in taste, and pronounced anorexia may develop. Coincident with, 1 or 2 days after defervescence, a generalized, morbilliform, maculopapular rash appears, with sparing of the palms and soles. It disappears in 1–5 days. In some cases, edema of the palms and soles may be noted, and desquamation may occur. About the time of appearance of this second rash, the body temperature, which has fallen to normal, may become elevated slightly and establish the biphasic temperature curve.

Epistaxis, petechiae, and purpuric lesions, though uncommon, may occur at any stage of the disease. Swallowed blood from epistaxis may be passed per rectum or be vomited and could be interpreted as bleeding of gastrointestinal origin. Gastrointestinal bleeding, menorrhagia, and bleeding from other organs have been observed in some dengue fever outbreaks (70–72). Peptic ulcer disease predisposes to gastrointestinal hemorrhage; in some cases, patients may exsanguinate during an otherwise normal dengue fever (71). This syndrome may be confused with dengue shock syndrome (see below) and contributes to a misunderstanding of the pathogenesis of severe dengue disease. The mechanism of the hemorrhagic diathesis that commonly occurs with dengue virus infection is not known, but speculation centers on direct disruption of endothelial glycocalyx, possibly by viral antigens (35). The dengue hemorrhagic fever syndrome is differentiated from dengue fever by the degree of thrombocytopenia and capillary leakage (73). Laboratory findings accompanying dengue fever are an early leucopenia, at times rather profound. Mature polymorphonuclear leukocytes become necrotic and disappear and this is accompanied by a lymphopenia. Careful observations on experimental dengue infections observed a greater number of juvenile than mature forms of PMNs, the so-called "cross" of dengue, occurring on day 2–3 of fever (23). About this same time platelet counts may drop, although seldom below 100,000 mm^3. Modest elevations in serum glutamic oxaloacetic transaminase (SGOT) and serum glutamic pyruvic transaminase (SGPT) may be observed.

After the febrile stage, prolonged asthenia, mental depression, bradycardia, and ventricular extrasystoles are noted commonly in adults (20).

9.4.2   Dengue hemorrhagic fever/dengue shock syndrome (DHF/DSS)

Dengue syndromes occur on a severity continuum; for example, DF cases without clear-cut evidence of vascular permeability are often accompanied by thrombocytopenia, liver enzyme elevations, and subthreshold vascular permeability (66, 74, 75). Use of the term "hemorrhagic fever" instead of the more accurate "dengue vasculopathy" has led physicians and patients alike to anticipate bleeding as the greatest threat for fatal outcome. This has led to promiscuous use of whole blood or blood products (76). Fluid that is lost rapidly from circulation into tissue spaces, when *not replaced promptly*, may lead to shock, which if *prolonged, leads to complications* such as gastrointestinal bleeding (76–78).

DHF/DSS is an acute vascular permeability syndrome accompanied by abnormal hemostasis observed at the time of defervescence. The incubation period of DHF/DSS is unknown, but presumed to be the same as that of dengue fever. In children, progression of the illness is characteristic (73, 79–81). A relatively mild first phase with an abrupt onset of fever, malaise, vomiting, headache, anorexia, and cough may be followed after 2–5 days by rapid deterioration and physical collapse. In Thailand, the median day of admission to the hospital after the onset of fever is day 4. In this second phase, the patient usually has cold, clammy extremities, a warm trunk, a flushed face, and diaphoresis. Patients are restless and irritable and complain of midepigastric pain. Frequently, scattered petechiae appear on the forehead and extremities, spontaneous ecchymoses may develop, and easy bruisability and bleeding at sites of venipuncture are common findings. Prolonged bleeding times are usual. Mild elevations in prothrombin time, moderate elevations in partial prothrombin time with low levels of small coagulation proteins are observed. The latter phenomenon is due to loss of proteins from circulation through open vascular pores. Circumoral and peripheral cyanosis may occur. Respirations are rapid and often labored. The pulse is weak, rapid, and thready, and the heart sounds are faint. The pulse pressure is frequently narrow (<20 mm Hg); systolic and diastolic pressure may be low or unobtainable. The liver may become palpable two or three fingerbreadths below the costal margin and is usually firm and nontender. Liver enzymes, SGOT and SGPT are usually elevated, often to high levels. Chest radiographs show unilateral (right) or bilateral pleural effusions, but often only after defervescence. Approximately 10% of patients have gross ecchymosis or gastrointestinal bleeding. Bleeding in patients with dengue infections may result from a combination of thrombocytopenia, dysfunctional surviving platelets, and increased fibrinolysis, rather than from classical disseminated intravascular coagulation. After a 24- or 36-h period of crisis, convalescence is fairly rapid in children who recover. The temperature may return to normal before or during the stage of shock.

Dengue shock may be subtle, occurring in patients who are fully alert, and is accompanied by increased peripheral vascular resistance and an elevated diastolic pressure (35). Bleeding, sometimes severe, may accompany DF, most commonly in adults (71). These subtleties and complications have given rise to diagnostic problems encountered by epidemiologists or clinicians, particularly in attempting to apply WHO case definitions from community hospital records or attempting to identify dengue syndromes in the absence of an outbreak (82–86). Diagnostic problems may be avoided by early use of ultrasound. Gallbladder wall thickening may be a precursor to clinically significant vascular permeability in children with dengue (87). Furthermore, ultrasound examinations of the abdomen and thorax can reliably detect fluid accumulation in serosal cavities (88). This low-cost diagnostic tool coupled with careful examination for physical signs of vascular collapse should greatly improve the reliability of the diagnosis of DHF/DSS (89). The diagnostic tool of greatest importance is the microhematocrit centrifuge, but only if used properly. Throughout Asia, microhematocrit centrifuges are located in outpatient departments and on clinical treatment wards, where they are accessible for use by both laboratory technicians and trained nurses. The availability of these instruments makes it possible to measure hematocrits at very frequent intervals at all times of day or night. A rising hematocrit signals the need for fluid administration, while a falling hematocrit should result in reducing fluid delivery and as a caution for possible gastrointestinal bleeding.

Using WHO criteria, DHF is a dengue illness accompanied by thrombocytopenia (<100,000/mm^3) and hemoconcentration. This is accurately defined as a hematocrit that is $>/=20\%$ than recovery value (or this can be stated as a $>/=20\%$ fall in hematocrit with fluid replacement). Less accurately, individual hematocrit values have been compared with community norms. Early detection of vascular permeability remains a diagnostic problem. Visualization of significant fluid in serosal cavities constitutes evidence of increased vascular permeability. While sonograms

performed prior to defervescence often reveal gallbladder wall thickening, upright chest X-rays seldom detect pleural or pericardial effusions until after beginning fluid replacement. The use of strain-gauge plethysmography documents up to 50% higher microvascular permeability in DHF/DSS patients than controls (77). DSS is diagnosed when these manifestations are accompanied by hypotension or narrow pulse pressure ($\leq$20 mm Hg). In areas endemic for dengue, hemorrhagic fever should be suspected in children with a febrile illness who exhibit shock and hemoconcentration with thrombocytopenia. Hypoproteinemia, hemorrhagic manifestations, and hepatic enlargement are frequent accompanying findings. Because many rickettsial diseases, meningococcemia, and other severe illnesses caused by a variety of agents may produce a similar clinical picture, the diagnosis should be made only when epidemiologic or serologic evidence suggests the possibility of dengue. Hemorrhagic manifestations have been described in other diseases of viral origin, including the arenavirus hemorrhagic fevers of Argentina, Bolivia, and West Africa (Lassa fever); the tick-borne hemorrhagic fevers of India and the former Soviet Union; hemorrhagic fever with renal syndrome, which occurs across northern Eurasia, specifically, from Scandinavia to Korea; and Marburg and Ebola virus infections in central Africa.

9.5 DIAGNOSIS

Specific etiological diagnosis is obtained by isolating virus or viral RNA from blood sample obtained during the acute illness phase. Alternatively, an acute blood sample can be tested for the presence of NS1, a DENV protein that circulates in blood during infection (90, 91). At least two commercial companies, Bio-Rad and Pan-Bio, market NS1 antigen detection kits in ELISA format while rapid tests are under development. When the detection of DENV NS1 is paired with ELISA tests on a serum taken five or more days after onset of fever, it is possible to confirm an acute DENV illness and by interpreting the partition of IgM and IgG antibodies, determine if patient has experienced a primary or secondary infection. Further details are published in a WHO report (73).

9.6 TREATMENT

9.6.1 Dengue Fever

Treatment is supportive. Bed rest is advised during the febrile period. Antipyretics or cold sponging should be used to keep the body temperature below 40°C (104°F). Paracetamol (10–15 mg/kg every 4–6 h) is the preferred antipyretic agent. Analgesics or mild sedation may be required to control pain. Fluid and electrolyte replacement therapy is required when deficits caused by sweating, fasting, thirsting, vomiting, or diarrhea are present. Because of the risk of Reye's syndrome and the dengue hemorrhagic diathesis, aspirin should not be given to reduce fever or control pain.

9.6.2   Dengue Hemorrhagic Fever

Detailed instructions for clinical diagnosis and management of DHF/DSS can be found in a WHO Technical Guide and in the Textbook of Pediatric Infectious Diseases (73, 92). Instructions are also available online at: http://www.searo. who.int/EN/Section10/Section332/Section1631.htm; http:// www.searo.who.int/LinkFiles/Dengue_Guideline-dengue.pdf. For more information, see the earlier recommendations by Cohen and Halstead (79) and recent studies by Dung and colleagues (93) and Wills and colleagues (94).

No specific antiviral treatment exists, but in DHF/DSS, symptomatic and supportive measures are effective and life saving. The major pathophysiologic abnormality seen in DHF/DSS is an acute increase in vascular permeability that leads to leakage of plasma. Plasma volume studies revealed a reduction of more than 20% in severe cases. Supporting evidence of plasma leakage (and consequent hypovolemia) includes a rapid, weak pulse; diaphoresis; cool, pale skin of the extremities; decreased urine output; and direct measurement by strain-gauge plethysmography (77), pleural effusion on chest radiography or ultrasound, hemoconcentration, and hypoproteinemia. Pleural effusion may not be evident until after fluid resuscitation is started. In the absence of increased vascular permeability, clinically significant hemoconcentration may result from thirst, dehydration, fever, anorexia, and vomiting. Fluid intake by mouth should be as ample as tolerated. Electrolyte and dextrose solution (as used in diarrheal disease), fruit juice, or both are preferable to plain water. With high fever, a risk of convulsions exists, so antipyretic drugs may be indicated. Salicylates should be avoided because they are known to cause bleeding and acidosis. Acetaminophen is preferable at the following doses: younger than 1 year, 60 mg/dose; 1–3 years of age, 60–120 mg/dose; 3–6 years of age, 120 mg/dose; and 6–12 years of age, 240 mg/dose.

Children should be observed closely for early signs of shock. The critical period is the transition from the febrile to the afebrile phase. Frequent hematocrit determinations are essential because they reflect the degree of plasma leakage and the need for administration of intravenous fluid. Hemoconcentration usually precedes changes in blood pressure and pulse. The hematocrit should be determined daily from the third day until the temperature becomes normal for 1 or 2 days. Oral or parenteral fluid therapy can be administered in an outpatient rehydration unit for correction of dehydration or acidosis or when signs of hemoconcentration are present. The volume of fluid and its composition are similar to the fluids used for the treatment of diarrhea with moderate dehydration. It is important to avoid overhydration, which may lead to hypervolemia, pulmonary edema, and cardiac failure. Fluid administration must be regulated carefully by continuously monitoring microhematocrit values, pulse rate, and urine output. A full bounding rapid pulse is a danger signal.

9.7   PATHOGENESIS

Epidemiologic, clinical, and virologic studies of DHF/DSS in humans have shown a significant association between severe illness and infection in the presence of circulating dengue antibody, whether passively acquired from the mother or actively acquired from previous infection (45, 95–102). This circulating antibody has two biologic activities: neutralization of virus and enhancement of infection (102). In Thailand, DHF/DSS developed in infants during dengue virus type 2 infection only when maternal neutralizing antibody had catabolized to low titer and infection-enhancing antibodies were left in circulation (99, 103). Similarly, in a prospective study of dengue virus infection in Thai children, DHF/DSS occurred in children who had circulating enhancing antibodies from a previous single dengue virus infection, but it did not occur in children whose first infection left them with low levels of cross-reactive dengue virus type 2 neutralizing antibody at the time of the second dengue virus infection (104). A similar mechanism explains the failure of secondary infections to produce DHF/DSS with the American genotype dengue type 2 (52, 105). American genotype dengue 2 viruses are significantly neutralized by human antidengue 1, whereas Southeast Asian dengue 2 viruses are not (105). The full-length sequences of the American and Southeast Asian dengue 2 genomes reveal limited amino acid differences (106).

In vitro studies of dengue virus type 2 demonstrated enhanced growth in cultures of human mononuclear phagocytes that were supplemented with very small quantities of dengue antibodies (107). Investigators have proposed that the infected cell mass—the number of infected mononuclear phagocytes in humans with naturally or passively acquired antibody may exceed that in nonimmune individuals (108). In serial blood samples early in the illness in children experiencing secondary dengue infections, enhanced viremia levels or enhanced levels of dengue NS1 predicted severe disease (67, 109). Vascular permeability is thought to result when infected cells are attacked by activated T lymphocytes, with the subsequent release of vasoactive cytokines proportionate to infected cell mass (108, 110–114). There is a reduced risk for DHF/DSS in protein-calorie malnourished children and the increased risk for DHF/DSS in girls versus boys are consistent with the hypothesis that a competent immune elimination system must be available to generate the cytokines that produce DHF/DSS (34, 108, 115).

The genetic heterogeneity of humans may place many persons at low risk to severe dengue occurring in individuals circulating dengue antibodies. There appears to be an important human dengue resistance gene based upon studies of the 1981 Cuban outbreak where whites were hospitalized more frequently than blacks with DHF/DSS (41, 96). A search for DHF/DSS in black children in Haiti revealed no cases despite high dengue type 1, 2, and 4 infection rates and circulation of the Southeast Asian genotype dengue 2 viruses (65). Several HLA antigens have shown differing frequencies in DHF/DSS cases and controls (116–118). Early in the acute stage of secondary dengue virus infection, rapid activation of the complement system occurs (119). During shock, blood levels of C1q, C3, C4, C5, C6, C7, C8, and C3 proactivator are depressed and C3 catabolic rates are elevated. The blood clotting and fibrinolytic systems are activated (120, 121). As yet, neither the mediator of vascular permeability nor the complete mechanism of bleeding has been unequivocally identified. Recent studies suggest a role for tumor necrosis factor, interleukin-2, and interferon-γ (122). Capillary damage allows fluid, electrolytes, protein, and in some instances, red blood cells to leak into intravascular spaces (78). This internal redistribution of fluid, together with deficits caused by fasting, thirsting, and vomiting, results in hemoconcentration, hypovolemia, increased cardiac work, tissue hypoxia, metabolic acidosis, and hyponatremia. A mild degree of disseminated intravascular coagulation, plus liver damage and thrombocytopenia, could contribute additively to produce hemorrhage (121).

9.8   PREVENTION

Globally, the increased intensity of dengue virus transmission can be attributed to a combination of factors: the huge increase in human population post–World War II, migration of populations from rural to urban areas, increased vector population due in part to the growth of substandard urban residences without adequate sources of potable water and screening, increased mosquito breeding sites afforded by disposable industrial products, the vast growth of modern air transport, and finally, the low state of government-organized mosquito control efforts manifested in systematic failure to recruit competent professional staff and chronic underfunding.

9.8.1   Vaccines

Tissue culture-based vaccines for dengue virus types 1, 2, 3, and 4 are immunogenic but not yet licensed for use. Numerous multivalent dengue vaccines using a variety of approaches are in various stages of development (123–129). A phase IIb clinical trial of a live chimeric tetravalent dengue-yellow fever 17 D vaccine was begun in 4000 Thai children in 2010 (Jean Lang, September 2010, personal communication). For reasons given below, there are virtually no alternatives to vaccines to bring the dengue pandemic under control. Vaccines must be used efficiently to stop transmission to protect the unvaccinated. This will require a greatly expanded effort to understand and measure dengue endemicity.

9.8.2   Vector Control

Current programs for the control of dengue epidemics directed against populations of vector mosquitoes depend on access to reliable potable water, breeding source reduction, insecticides, repellents, and screens on houses to avoid mosquito bites (73). Source reduction requires public support either by legal sanctions or by voluntary actions. Source reduction campaigns should be well organized, supervised, and evaluated. Proper disposal of discarded cans, bottles, tires, and other potential breeding sites not used for storage of drinking or bathing water should be performed. Smaller or portable water storage containers for drinking and bathing and flower vases should be emptied completely and scrubbed once weekly. If water storage is mandatory, a tight-fitting lid or a thin layer of oil may prevent eggs from being deposited or hatching. Water storage containers too large to be emptied should be scrubbed to remove eggs from sides and treated with Abate 1% sand granules at a dosage of 1 ppm (e.g., 10 g of sand to 100 L of water). Treatments should be repeated at intervals of 2–3 months.

The public health history of the first half of the twentieth century in the Americas is one of mounting pressure against *A. aegypti*, as the urban vector of yellow fever virus. Mosquito control campaigns were invented in Cuba during the Spanish-American war, then applied in Panama during construction of the Canal, in Brazil with the work of Oswaldo Cruz, to key cities by a strategy of the Rockefeller Foundation and culminating in the 1950s with eradication campaigns mounted by Pan American Health Organization (PAHO) (130). By 1960, 13 large countries in the Americas had eradicated *A. aegypti*. After that date, one by one, each of these successful mosquito control programs was abandoned and *A. aegypti* returned. An eradication campaign in the United States was replaced by a program of disease surveillance and containment of introduced virus. *A. aegypti* eggs were exported from the United States largely through the trade in used tires (11). Discouragingly, by 1994, the species was as widely distributed in the Americas as it had been at the turn of the twentieth century.

9.8.3   Failure of Vector Control

The failure of modern societies to mount and sustain programs to control *A. aegypti* is a notable indictment of the contemporary public health establishment. This may be due to one or a combination of the six "failures" (131).

Failure to staff vector control programs adequately are the result of:

1. Failure to recruit and retain appropriate professionals.
2. Failure to train high-quality professionals in academic centers of entomologic research excellence.
3. Failure to apply sound managerial and scientific practices.
4. Failure to provide adequate funding.
5. Failure to gain public support.
6. Failure of political will.

9.8.4   Epidemic Measures

The WHO recommendations are as follows. On the basis of epidemiologic and entomologic information, the size of the area that requires adult mosquito abatement should be determined. With technical-grade malathion or fenitrothion at 438 mL/hectare, two adulticide treatments at a 10 day interval should be made with the use of a vehicle-mounted or portable ultra-low-volume aerosol generator or mist blower (73, 132). Cities of moderate size should stockpile at least 1 vehicle-mounted aerosol generator, 5 mist blowers, 10 swing fog machines, and 1000 L of ultra-low-volume insecticide to be prepared to perform adulticide operations over a $20\,km^2$ area rapidly. With limited funds, such equipment and insecticides can be stockpiled centrally for rapid transportation when required. Priority areas for launching ground applications are those with a concentration of cases. Special attention should be focused on areas where people congregate during daylight hours, such as hospitals and schools. If necessary, ultra-low-volume insecticides may be applied from aircraft. C47 or similar aircraft, smaller agricultural spray planes, and helicopters have been used to make aerial applications.

During the early stages of epidemics, an ultra-low-volume spray of 4% malathion in diesel oil or kerosene may be used to spray all houses within a 100 m radius of the residence of DHF patients.

9.8.5   Health Education

Control of *A. aegypti* has been maintained effectively in some tropical areas through the simple expedient of emptying water containers once a week. During the yellow fever campaigns, strong sanitary laws made the breeding of mosquitoes on premises a crime punishable by fine or jail (130). In the modern era, Singapore and Cuba have adopted these measures successfully. Health education through mass media or through the schools has been attempted in Burma, Thailand, Malaysia, and Indonesia, but without spectacular success (133). Community participation and mass education campaigns in Puerto Rico and Mexico have fared little

better (134, 135). The goals of health education and community participation approaches are to make the population aware of the identity of the vector of DHF, describe its biting habits (daytime feeding) and its breeding habits (containers holding clean water), and motivate people to reduce breeding sources by emptying water from containers on a regular basis. The use of piped water rather than water storage should be encouraged. Studies in Malaysia after the 1973 epidemic of DHF indicated a very low level of functional knowledge among the inhabitants of Kuala Lumpur, Malaysia, about the vector of DHF (133). Discouragingly, persons who were informed correctly, in most instances, took no action to protect themselves against mosquito breeding in their homes. This reaction is in contrast to the present situation in Singapore, where stiff fines and frequent inspections have reduced infestation by *A. aegypti* drastically. Extensive effort is being made to apply social science methods to gain the voluntary participation of the population in sustained mosquito control programs (136).

9.8.6   Future Programs

In the future, the efficient control of dengue will require an entirely new kind of intersectoral and interdisciplinary mosquito abatement program with the following features:

1. Assure that minimal housing standards include household distribution of reliable potable water and window screening.
2. Incorporate building and urban designs that promote *A. aegypti* control into urban planning (137).
3. Promote interagency cooperation. At a minimum, *A. aegypti* control requires cooperation between the industries of health, environment, municipal health departments, and urban solid waste disposal agencies.
4. Science and education. A strong university-based research program and graduate education should be developed in dengue endemic regions designed to study and understand the unique bionomics of this species and develop reliable methods to control *A. aegypti* (138).
5. Public and private sector collaboration. *A. aegypti* abatement poses an especially interesting challenge to the private sector, which could be assigned frontline abatement responsibilities with the public sector providing for quality control. A highly successful mosquito abatement program has been implemented in the Central Valley of California in which contracts for vector control are competitively awarded to private companies with government establishing standards and measuring performance (139, 140).
6. Use new "Gold Standard" to monitor dengue control. In the past, control of dengue has been measured by

monitoring mosquito population indices. Instead, control efforts should be monitored by success at controlling dengue virus transmission using appropriate population-based sero-surveillance methods.

REFERENCES

1. Christie J. Remarks on "kidinga Pepo" a peculiar form of exantematous disease. *B M J* 1872;1:577–579.

2. Carey DE. Chikungunya and dengue: a case of mistaken identity? *J Hist Med Allied Sci* 1971;26(3):243–262.

3. Vasilakis N, Weaver SC. The history and evolution of human dengue emergence. *Adv Virus Res* 2008;72:1–76.

4. Wang E, Ni H, Xu R, et al. Evolutionary relationships of endemic/epidemic and sylvatic dengue viruses. *J Virol* 2000;74(7):3227–3234.

5. Diallo M, Sall A, Moncayo AC, et al. Potential role of sylvatic and domestic African mosquito species in dengue emergence. *Am J Trop Med Hyg* 2005;73:445–449.

6. Vasilakis N, Tesh RB, Weaver SC. Sylvatic dengue virus type 2 activity in humans, Nigeria, 1966. *Emerg Infect Dis* 2008; 14(3):502–504.

7. Effler PV, Pang L, Kitsutani P, et al. Dengue fever, Hawaii, 2001–2002. *Emerg Infect Dis* 2005;11:742–749.

8. Rosen L, Gubler DJ. The use of mosquitoes to detect and propagate dengue viruses. *Am J Trop Med Hyg* 1974;23(6): 1153–1160.

9. Rosen L. The use of toxorhynchites mosquitoes to detect and propagate dengue and other arboviruses. *Am J Trop Med Hyg* 1981;30(10):177–183.

10. Halstead SB. Mosquito-borne haemorrhagic fevers of South and South-East Asia. *Bull World Health Organ* 1966;35:3–15.

11. Gubler DJ. *Aedes aegypti* and *Aedes aegypti*-borne disease control in the 1990s: top down or bottom up. Charles Franklin Craig Lecture. *Am J Trop Med Hyg* 1989;40(6):571–578.

12. Tonn RJ, Sheppard PM, Macdonald WW, Bang YH. Replicate surveys of larval habitats of *Aedes aegypti* in relation to dengue haemorrhagic fever in Bangkok, Thailand. *Bull WHO* 1969;40:819–829.

13. Sheppard PM, MacDonald WW, Tonn RJ, Grab B. The dynamics of an adult population of *Aedes aegypti* in relation to dengue hemorrhagic fever in Bangkok. *J Anim Ecol* 1969;38:661–702.

14. Watts DM, Burke DS, Harrison BA, Whitmire RE, Nisalak A. Effect of temperature on the vector efficiency of *Aedes aegypti* for dengue 2 virus. *Am J Trop Med Hyg* 1987;36(1):143–152.

15. Halstead SB. Selective primary health care: strategies for control of disease in the developing world. XI. Dengue. *Rev Infect Dis* 1984;6:251–264.

16. Derrick EH, Bicks VA. The limiting temperature for the transmission of dengue. *Proceedings of the Ninth Pacific Science Congress of the Pacific Science Association*, 1962, pp. 40–41.

17. Siler JF, Hall MW, Hitchens AP. Dengue: its history, epidemiology, mechanism of transmission, etiology, clinical manifestations, immunity, and prevention. *Philippine J Sci* 1926;29:1–304.

18. Rush B. An account of the bilious remitting fever, as it appeared in Philadelphia, in the summer and autumn of the year 1780. In: *Medical Inquiries and Observations*, 1st edition. Philadelphia: Prichard and Hall, 1789, pp. 89–100.

19. Halstead SB, Papaevangelou G. Transmission of dengue 1 and 2 viruses in Greece in 1928. *Am J Trop Med Hyg* 1980;29(4): 635–637.

20. Lumley GF, Taylor FH. Dengue. Glebe, NSW, Australia, 1943.

21. Shu PY, Chien LJ, Chang SF, et al. Fever screening at airports and imported dengue. *Emerg Infect Dis* 2005;11:460–462.

22. Shu PY, Su CL, Liao TL, et al. Molecular characterization of dengue viruses imported into Taiwan during 2003–2007: geographic distribution and genotype shift. *Am J Trop Med Hyg* 2009;80(6):1039–1046.

23. Simmons JS, St John JH, Reynolds FHK. Experimental studies of dengue. *Philippine J Sci* 1931;44:1–252.

24. Halstead SB, Scanlon J, Umpaivit P, Udomsakdi S. Dengue and chikungunya virus infection in man in Thailand, 1962–1964: IV. Epidemiologic studies in the Bangkok metropolitan area. *Am J Trop Med Hyg* 1969;18(6):997–1021.

25. Gubler DJ, Kuno G, Sather GE, Waterman SH. A case of natural concurrent human infection with two dengue viruses. *Am J Trop Med Hyg* 1985;34:170–173.

26. Rudnick A. Ecology of dengue virus. *Asian J Infect Dis* 1978;2:156–160.

27. Rico-Hesse R. Molecular evolution and distribution of dengue viruses type 1 and 2 in nature. *Virology* 1990;174:479–493.

28. Diallo M, Ba Y, Sall AA, et al. Amplification of the sylvatic cycle of dengue virus type 2, Senegal, 1999–2000: entomologic findings and epidemiologic considerations. *Emerg Infect Dis* 2003;9:362–367.

29. Cardosa J, Ooi MH, Tio PH, et al. Dengue virus serotype 2 from a sylvatic lineage isolated from a patient with dengue hemorrhagic fever. *PLoS Negl Trop Dis* 2009;3(4):e423.

30. Rosen L, Shroyer DA, Tesh RB, Freier JE, Lien JC. Transovarial transmission of dengue viruses by mosquitoes: *Aedes albopictus* and *Aedes aegypti*. *Am J Trop Med Hyg* 1983;32:1108–1119.

31. Watts DM, Harrison BA, Pantuwatana S, Klein TA, Burke DS. Failure to detect natural transovarial transmission of dengue viruses by *Aedes aegypti* and *Aedes albopictus* (Diptera: Culicidae). *J Med Entomol* 1985;22(3):261–265.

32. Sabin AB. Research on dengue during World War II. *Am J Trop Med Hyg* 1952;1:30–50.

33. Hammon WM, Rudnick A, Sather GE. Viruses associated with epidemic hemorrhagic fevers of the Philippines and Thailand. *Science* 1960;131:1102–1103.

34. Halstead SB. Epidemiology of dengue and dengue hemorrhagic fever. In: Gubler DJ, Kuno, G., editors. *Dengue and Dengue Hemorrhagic Fever*. Wallingford, UK: CAB, 1997, pp. 23–44.

35. Halstead SB. Dengue. *Lancet* 2007;370(9599):1644–1652.

36. Gubler DJ, Reed D, Rosen L, Hitchcock JC, Jr., Epidemiological, clinical, and virologic observations on dengue in The Kingdom of Tonga. *Am J Trop Med Hyg* 1978;27(3):581–589.

37. Rosen L. A recent outbreak of dengue in French Polynesia. *Jpn J Med Sci Biol* 1967;20:67–69.

38. Hubert B, Halstead SB. Dengue 1 virus and dengue hemorrhagic fever, French Polynesia, 2001. *Emerg Infect Dis* 2009;15:1265–1270.

39. PAHO. *Dengue in the Caribbean, 1977; 1978.* Montego Bay, Jamaica: PAHO, 1978, pp. 1–186.

40. Nisalak A, Endy TP, Nimmannitya S, et al. Serotype-specific dengue virus circulation and dengue disease in Bangkok, Thailand form 1973 to 1999. *Am J Trop Med Hyg* 2003;68:191–202.

41. Kouri GP, Guzman MG, Bravo JR, Triana C. Dengue haemorrhagic fever/dengue shock syndrome: lessons from the Cuban epidemic, 1981. *Bull World Health Organ* 1989;67:375–380.

42. Ramirez Ronda CH. Dengue in Puerto Rico: clinical manifestations and management from 1960s to 1987. *PR Health Sci J* 1987;6:113–118.

43. Nogueira RM, Zagner SM, Martins IS, Lampe E, Miagostovich MP, Schatzmayr HG. Dengue haemorrhagic fever/dengue shock syndrome (DHF/DSS) caused by serotype 2 in Brazil. *Mem Inst Oswaldo Cruz* 1991;86(2):269.

44. Halstead SB. Dengue in the Americas and Southeast Asia: do they differ? *Rev Panam Salud Publica* 2006;20(6):407–415.

45. Guzman MG, Kouri G, Valdes L, et al. Epidemiologic studies on dengue in Santiago de Cuba, 1997. *Am J Epidemiol* 2000;152(9):793–799.

46. Gubler DJ. Dengue and dengue hemorrhagic fever: its history and resurgence as a global public health problem. In: Gubler DJ, Kuno G, editors. *Dengue and Dengue Hemorrhagic Fever.* New York: CAB International, 1997, pp. 1–22.

47. Carrington CV, Foster JE, Pybus OG, Bennett SN, Holmes EC. The invasion and maintenance of dengue virus type 2 and type 4 in the Americas. *J Virol* 2005;79:14680–14687.

48. Siqueira JBJ, Turchi Martelli CM, Evalim Coelho G, da Rocha Simplicio AC, Hatch DL. Dengue and dengue hemorrhagic fever, Brazil, 1981–2002. *Emerg Infect Dis* 2005;11:48–53.

49. Neto RJP, Lima DM, de Paula SO, Lima CM, Rocco IM, Fonseca BAL. Molecular epidemiology of type 1 and 2 dengue viruses in Brazil from 1988 to 2001. *Braz J Med Biol Rsch* 2005;38:843–852.

50. Nogueira RM, Schatzmayr HG, Miagostovich MP, Farias MF, Farias Filho JD. Virological study of a dengue type 1 epidemic at Rio de Janeiro. *Mem Inst Oswaldo Cruz (Rio de Janeiro)* 1989;83:219–225.

51. Nogueira RM, Miagostovich MP, Schatzmayr HG, et al. Dengue type 2 outbreak in the south of the state of Bahia, Brazil: laboratorial and epidemiological studies. *Rev Inst Med Trop Sao Paulo* 1995;37(6):507–510.

52. Watts DM, Porter KR, Putvatana P, et al. Failure of secondary infection with American genotype dengue 2 to cause dengue haemorrhagic fever [see comments]. *Lancet* 1999;354(9188): 1431–1434.

53. Alvarez M, Rodriguez R, Bernardo L, et al. Dengue hemorrhagic fever caused by sequential dengue 1–3 infections at a long interval: Havana epidemic, 2001–2002. *Am J Trop Med Hyg* 2006;75:1113–1117.

54. Nogueira RM, Schatzmayr HG, de Filippis AM, et al. Dengue virus type 3 in Brazil, 2002. *Emerg Infect Dis* 2005;11: 1376–1381.

55. Mendez A, Gonzalez G. Dengue haemorrhagic fever in children: ten years of clinical experience. *Biomedica* 2003;23: 180–193.

56. Hammond SN, Balmaseda A, Perez L, et al. Differences in dengue severity in infants, children, and adults in a 3-year hospital-based study in Nicaragua. *Am J Trop Med Hyg* 2005;73:1063–1070.

57. Balmaseda A, Hammond SN, Perez L, et al. Serotype-specific differences in clinical manifestations of dengue. *Am J Trop Med Hyg* 2006;74:449–456.

58. Larru Martinez B, Quiroz E, Bellon JM, Esquivel R, Nieto Guevara J, Saez-Llorens X. Dengue pediatrico en Panama. *An Pediatr (Barc)* 2006;64:517–522.

59. Ocazionez RE, Gomez SY, Cortes FM. Dengue hemorrhagic fever serotype and infection pattern in a Colombian endemic area. *Rev Salud Publica (Bogota)* 2007;9(2):262–274.

60. Comach G, Blair PJ, Sierra G, et al. Dengue virus infections in a cohort of schoolchildren from Maracay, Venezuela: a 2-year prospective study. *Vector Borne Zoonotic Dis* 2008;9:87–92.

61. Teixeira MG, Costa MC, Coelho G, Barreto ML. Recent shift in age pattern of dengue hemorrhagic fever, Brazil. *Emerg Infect Dis* 2008;14(10):1663.

62. Barreto FR, Teixeira MG, Costa Mda C, Carvalho MS, Barreto ML. Spread pattern of the first dengue epidemic in the city of Salvador, Brazil. *BMC Public Health* 2008;8:51.

63. Barreto ML, Teixeira MG. Dengue fever: a call for local, national, and international action. *Lancet* 2008;372(9634): 205.

64. Halstead SB. Dengue haemorrhagic fever—a public health problem and a field for research. *Bull World Health Organ* 1980;58(1):1–21.

65. Halstead SB, Streit TG, Lafontant JG, et al. Haiti: absence of dengue hemorrhagic fever despite hyperendemic dengue virus transmission. *Am J Trop Med Hyg* 2001;65:180–183.

66. Kalayanarooj S, Vaughn DW, Nimmannitya S, et al. Early clinical and laboratory indicators of acute dengue illness. *J Infect Dis* 1997;176(2):313–321.

67. Vaughn DW, Green S, Kalayanarooj S, et al. Dengue viremia titer, antibody response pattern, and virus serotype correlate with disease severity. *J Infect Dis* 2000;181(1):2–9.

68. Vaughn DW. Invited commentary: dengue lessons from Cuba. *Am J Epidemiol* 2000;152(9):800–803.

69. Scott RM, Nimmannitya S, Bancroft WH, Mansuwan P. Shock syndrome in primary dengue infections. *Am J Trop Med Hyg* 1976;25(6):866–874.

70. Rice L. A clinical report of the Galveston epidemic of 1922. *Am J Trop Med* 1923;3:73–90.

71. Tsai CJ, Kuo CH, Chen PC, Changcheng CS. Upper gastrointestinal bleeding in dengue fever. *Am J Gastroenterol* 1991;86:33–35.

72. Wen KH, Sheu MM, Chung CB, Wang HZ, Chen CW. The ocular fundus findings in dengue fever. *Kao-Hsiung I Hsueh Ko Hsueh Tsa Chih (Kaohsiung Journal of Medical Sciences)* 1989;5:24–30.

73. World Health, Organization. *Dengue Haemorrhagic Fever: Diagnosis, Treatment, Prevention and Control*, 2nd edition. Geneva: World Health Organization, 1997, pp. 1–84.

74. Kuo CH, Tai DI, Chang Chien CS, Lan CK, Chiou SS, Liaw YF. Liver biochemical tests and dengue fever. *Am J Trop Med Hyg* 1992;47(3):265–270.

75. Gamble J, Bethell D, Day NP, et al. Age-related changes in microvascular permeability: a significant factor in the susceptibility of children to shock? *Clin Sci (Colch)* 2000;98(2):211–216.

76. Lum LC, Abdel-Latif Mel A, Goh AY, Chan PW, Lam SK. Preventive transfusion in Dengue shock syndrome—is it necessary? *J Pediatr* 2003;143(5):682–684.

77. Bethell DB, Gamble J, Pham PL, et al. Noninvasive measurement of microvascular leakage in patients with dengue hemorrhagic fever. *Clin Infect Dis* 2001;32(2):243–253.

78. Wills BA, Oragui EE, Dung NM, et al. Size and charge characteristics of the protein leak in dengue shock syndrome. *J Infect Dis* 2004;190:810–818.

79. Cohen SN, Halstead SB. Shock associated with dengue infection. I. Clinical and physiologic manifestations of dengue hemorrhagic fever in Thailand, 1964. *J Pediatrics* 1966;68:448–456.

80. Nimmannitya S. Clinical spectrum and management of dengue haemorrhagic fever. *Southeast Asian J Trop Med Public Health* 1987;18(3):392–397.

81. Nimmannitya S, Halstead SB, Cohen S, Margiotta MR. Dengue and chikungunya virus infection in man in Thailand, 1962–1964. I. Observations on hospitalized patients with hemorrhagic fever. *Am J Trop Med Hyg* 1969;18(6):954–971.

82. Deen JL, Harris E, Wills B, et al. The WHO dengue classification and case definitions: time for a reassessment. *Lancet* 2006;368:170–173.

83. Rigau-Perez JG. Severe dengue: the need for new case definitions. *Lancet Infect Dis* 2006;6:297–302.

84. Balasubramanian S, Jankiraman L, Kumar SS, Muralinath S, Shivbalen S. A reappraisal of the criteria to diagnose plasma leakage in dengue hemorrhagic fever. *Indian Pediatr* 2006;43:334–339.

85. Balmaseda A, Hammond SN, Perez MA, et al. Short report: assessment of the World Health Organization scheme for classification of dengue severity in Nicaragua. *Am J Trop Med Hyg* 2005;73:1059–1062.

86. Bandyopadhyay S, Lum LCS, Kroeger A. Classifying dengue: a review of the difficulties in using the WHO case classification for dengue haemorrhagic fever. *Trop Med Int Health* 2006;11:1238–1255.

87. Colbert JA, Gordon A, Roxelin R, et al. Ultrasound measurement of gallbladder wall thickening as a diagnostic test and prognostic indicator for severe dengue in pediatric patients. *Pediatr Infect Dis J* 2007;26:850–852.

88. Srikiatkhachorn A, Krautrachue A, Ratanaprakarn W, et al. Natural history of plasma leakage in dengue hemorrhagic fever: a serial ultrasonographic study. *Pediatr Infect Dis J* 2007;26:283–290.

89. Halstead SB. The dengue case definition dilemma. *Pediatr Infect Dis J* 2007;26:291–292.

90. Alcon S, Talarmin A, Debruyne M, Falconar A, Deubel V, Flamand M. Enzyme-linked immunosorbent assay reveals high levels of the dengue virus protein NS1 in the sera of infected patients. *J Clin Microbiol* 2002;40:376–381.

91. Young PR, Hilditch PA, Bletchly C, Halloran W. An antigen capture enzyme-linked immunosorbent assay reveals high levels of the dengue virus protein NS1 in the sera of infected patients. *J Clin Microbiol* 2000;38(3):1053–1057.

92. Halstead SB. Dengue and dengue hemorrhagic fever. In: Feigin RD, Cherry JD, Demmler-Harrison DJ, Kaplan SL, editors. *Textbook of Pediatric Infectious Diseases*, 6 edition. Philadelphia: Saunders Elsevier, 2009, pp. 2347–2356.

93. Dung NM, Day, NPJ, Tam, DTH, Loan, HT, Chau, HTT, Minh, LN, Diet, TV, Bethell, DB, Kneen, R, Hien, TT, White, NJ, Farrar, JJ. Fluid replacement in dengue shock syndrome: a randomized, double-blind comparison of four intravenous-fluid regimens. *Clin Infect Dis* 1999;29:787–794.

94. Wills BA, Dung NM, Loan HT, et al. Comparison of three fluid solutions for resuscitation in dengue shock syndrome. *New Engl J Med* 2005;353:877–889.

95. Burke DS, Nisalak A, Johnson DE, Scott RM. A prospective study of dengue infections in Bangkok. *Am J Trop Med Hyg* 1988;38(1):172–180.

96. Guzman MG, Kouri GP, Bravo J, Soler M, Vazquez S, Morier L. Dengue hemorrhagic fever in Cuba, 1981: a retrospective seroepidemiologic study. *Am J Trop Med Hyg* 1990;42:179–184.

97. Halstead SB, Nimmannitya S, Yamarat C, Russell PK. Hemorrhagic fever in Thailand; recent knowledge regarding etiology. *Jpn J Med Sci Biol* 1967;20s:96–103.

98. Halstead SB. Observations related to pathogenesis of dengue hemorrhagic fever. VI. Hypotheses and discussion. *Yale J Biol Med* 1970;42:350–362.

99. Kliks SC, Nimmannitya S, Nisalak A, Burke DS. Evidence that maternal dengue antibodies are important in the development of dengue hemorrhagic fever in infants. *Am J Trop Med Hyg* 1988;38(2):411–419.

100. Russell PK, Yuill TM, Nisalak A, Udomsakdi S, Gould D, Winter PE. An insular outbreak of dengue hemorrhagic fever. II. Virologic and serologic studies. *Am J Trop Med Hyg* 1968;17(4):600–608.

101. Sangkawibha N, Rojanasuphot S, Ahandrik S, et al. Risk factors in dengue shock syndrome: a prospective epidemio-

logic study in Rayong, Thailand. I. The 1980 outbreak. *Am J Epidemiol* 1984;120:653–669.

102. Halstead SB. Neutralization and antibody-dependent enhancement of dengue viruses. In: Chambers TJ, Monath, T.P, editors. *The Flaviviruses: Pathogenesis and Immunity.* New York: Elsevier Academic Press, 2003, pp. 422–467.

103. Halstead SB, Lan NT, Myint TT, et al. Infant dengue hemorrhagic fever: research opportunities ignored. *Emerg Infect Dis* 2002;12:1474–1479.

104. Kliks SC, Nisalak A, Brandt WE, Wahl L, Burke DS. Antibody-dependent enhancement of dengue virus growth in human monocytes as a risk factor for dengue hemorrhagic fever. *Am J Trop Med Hyg* 1989;40(4):444–451.

105. Kochel TJ, Watts, DM, Halstead, SB, Hayes, CG, Espinosa, A, Felices, V, Caceda, R, Bautista, T, Montoya, Y, Douglas, S, Russell, KL. Effect of dengue-1 antibodies on American dengue-2 viral infection and dengue haemorrhagic fever. *Lancet* 2002;360:310–312.

106. Leitmeyer KC, Vaughn DW, Watts DM, et al. Dengue virus structural differences that correlate with pathogenesis. *J Virol* 1999;73(6):4738–4747.

107. Halstead SB, O'Rourke EJ. Dengue viruses and mononuclear phagocytes. I. Infection enhancement by non-neutralizing antibody. *J Exp Med* 1977;146:201–217.

108. Halstead SB. Immunological parameters of Togavirus disease syndromes. In: Schlesinger RW, editor. *The Togaviruses, Biology, Structure, Replication.* New York: Academic Press, 1980, pp. 107–173.

109. Libraty DH, Young PR, Pickering D, et al. High circulating levels of the dengue virus nonstructural protein NS1 early in dengue illness correlate with the development of dengue hemorrhagic fever. *J Infect Dis* 2002;186:1165–1168.

110. Bethell DB, Flobbe K, Cao XT, et al. Pathophysiologic and prognostic role of cytokines in dengue hemorrhagic fever. *J Infect Dis* 1998;177(3):778–782.

111. Green S, Vaughn DW, Kalayanarooj S, et al. Elevated plasma interleukin-10 levels in acute dengue correlate with disease severity. *J Med Virol* 1999;59(3):329–334.

112. Green S, Vaughn DW, Kalayanarooj S, et al. Early immune activation in acute dengue illness is related to development of plasma leakage and disease severity. *J Infect Dis* 1999;179 (4):755–762.

113. Kurane I, Ennis FA. Cytokines in dengue virus infections: role of cytokines in the pathogenesis of dengue hemorrhagic fever. *Semin Virol* 1994;5:443–448.

114. Rothman AL, Ennis FA. Immunopathogenesis of dengue hemorrhagic fever. *Virology* 1999;257(1):1–6.

115. Thisyakorn U, Nimmannitya S. Nutritional status of children with dengue hemorrhagic fever. *Clin Infect Dis* 1993;16: 295–297.

116. Coffey LL, Mertens E, Brehin AC, et al. Human genetic determinants of dengue virus susceptibility. *Microbes Infect* 2009;11(2):143–156.

117. Hibberd ML, Ling L, Tolfvenstam T, et al. A genomics approach to understanding host response during dengue

infection. *Novartis Found Symp* 2006;277:206–214; discussion 214–217, 251–253.

118. Mathew A, Rothman AL. Understanding the contribution of cellular immunity to dengue disease pathogenesis. *Immunol Rev* 2008;225:300–313.

119. Bokisch VA, Top FH, Jr., Russell PK, Dixon FJ, Muller-Eberhard HJ. The potential pathogenic role of complement in dengue hemorrhagic shock syndrome. *N Engl J Med* 1973;289:996–1000.

120. Halstead SB. Dengue: hematologic aspects. *Semin Hematol* 1982;19(2):116–131.

121. Wills BA, Oragui EE, Stephens AC, et al. Coagulation abnormalities in dengue hemorrhagic fever: serial investigations in 167 Vietnamese children with dengue shock syndrome. *Clin Infect Dis* 2002;35:277–285.

122. Rothman AL. Immunology and immunopathogenesis of dengue disease. *Adv Virus Res* 2003;60:397–419.

123. Apt D, Raviprakash K, Brinkman A, et al. Tetravalent neutralizing antibody response against four dengue serotypes by a single chimeric dengue envelope antigen. *Vaccine* 2006; 24(3):335–344.

124. Putnak RJ, Coller BA, Voss G, et al. An evaluation of dengue type-2 inactivated, recombinant subunit, and live-attenuated vaccine candidates in the rhesus macaque model. *Vaccine* 2005;23(35):4442–4452.

125. Whitehead SS, Blaney JE, Durbin AP, Murphy BR. Prospects for a dengue virus vaccine. *Nat Rev Microbiol* 2007;5(7): 518–528.

126. Blaney JE, Jr., Matro JM, Murphy BR, Whitehead SS. Recombinant, live-attenuated tetravalent dengue virus vaccine formulations induce a balanced, broad, and protective neutralizing antibody response against each of the four serotypes in rhesus monkeys. *J Virol* 2005;79(9):5516–5528.

127. Butrapet S, Kinney RM, Huang CY. Determining genetic stabilities of chimeric dengue vaccine candidates based on dengue 2 PDK-53 virus by sequencing and quantitative Taq-MAMA. *J Virol Methods* 2006;131(1):1–9.

128. Guirakhoo F, Kitchener S, Morrison D, et al. Live attenuated chimeric yellow fever dengue type 2 (ChimeriVax-DEN2) vaccine: Phase I clinical trial for safety and immunogenicity: effect of yellow fever pre-immunity in induction of cross neutralizing antibody responses to all 4 dengue serotypes. *Hum Vaccin* 2006;2(2):60–67.

129. Guirakhoo F, Pugachev KV. Viremia and immunogenicity in nonhuman primates of a tetravalent yellow fever-dengue chimeric vaccine: genetic reconstructions, dose adjustment, and antibody responses against wild-type dengue virus isolates. *Virology* 2002;298:146–159.

130. Strode GK, editor. *Yellow Fever*. New York: McGraw-Hill Book Company, 1951.

131. Halstead SB. *Aedes aegypti:* why can't we control it? *Bull Soc Vector Ecol* 1988;13:304–311.

132. Lofgren CS, Ford HR, Tonn RJ, Jatanasen S. The effectiveness of ultra-low volume applications of malathion at a rate of 6 US fluid ounces per acre in controlling *Aedes aegypti* in a

large-scale field test at Nakhon Sawan. Thailand. *Bull World Health Organ* 1970;42:15–25.

133. Dobbins JG, Else JG. Knowledge, attitudes and practices related to control of dengue hemorrhagic fever in an urban Malay kampung. *South East Asian J Trop Med Pub Health* 1975;6:120–126.

134. Lloyd LS, Winch P, Ortega-Canto J, Kendall C. Results of a community-based *Aedes aegypti* control program in Merida, Yucatan, Mexico. *Am J Trop Med Hyg* 1992;46(6):635–642.

135. Winch PJ, Leontsini E, Rigau-Perez JG, Ruiz-Perez M, Clark GG, Gubler DJ. Community-based dengue prevention programs in Puerto Rico: impact on knowledge, behavior, and residential mosquito infestation. *Am J Trop Med Hyg* 2002; 67(4):363–370.

136. Lloyd LS, Winch P, Ortega-Canto J, Kendall C. The design of a community-based health education intervention for the control of *Aedes aegypti*. *Am J Trop Med Hyg* 1994;50(4): 401–411.

137. Lee YF. Urban planning and vector control in Southeast Asian cities. *Kaohsiung J Med Sci* 1994;10:S39–S51.

138. Gratz NG. Education and employment of medical entomologists in *Aedes aegypti* control programmes. *Kaohsiung J Med Sci* 1994;10:S19–S27.

139. Challet GL. Mosquito abatement district programs in the United States. *Kaohsiung J Med Sci* 1994;10:S67–S73.

140. Clarke JL. Privatization of mosquito control services in urban areas. *Kaohsiung J Med Sci* 1994;10:S74–S77.

10

DRACUNCULIASIS (GUINEA WORM DISEASE): CASE STUDY OF THE EFFORT TO ERADICATE GUINEA WORM

DONALD R. HOPKINS AND ERNESTO RUIZ-TIBEN

10.1 INTRODUCTION

Dracunculiasis (Guinea worm disease) is a water-borne parasitic infection that is intimately tied to the environment and to human behavior. The campaign to eradicate this disease lasted nearly 30 years and provides many lessons that may be adapted for attacking other problems successfully, particularly where change in human habits is required in order to mitigate or interrupt transmission.

The adult parasite *Dracunculus medinensis* occurs only in humans, who become infected after drinking stagnant surface water from sources such as ponds where microscopic water fleas (copepods) have been infected by ingesting immature forms of the parasite. The parasite matures in humans over a period of 1 year, during which the infection evokes no symptoms or outward signs. After a year, the thin white adult worm, which measures up to 3 feet (1 m) long, emerges slowly and painfully through a ruptured blister that it raises on the skin (Fig. 10.1). Most worms emerge on the leg, ankle, or foot, but they may emerge from anywhere on the body, and a patient may have up to a dozen or more worms emerge around the same time. The emerging adult worms are all females, which spew hundreds of thousands of immature larvae into the water when the infected person enters a surface source such as a pond for any reason. Thus, the life cycle of the worm is perpetuated by humans who drink untreated water from such contaminated sources, as well as by those who enter sources of drinking water while a Guinea worm is emerging or about to emerge from their body.

Since only a few centimeters of a worm can usually be pulled out manually by rolling it on a small stick or rolled

gauze in a day, the painful incapacitation associated with this infection normally lasts up to 1 or 2 months or more. The site of worm emergence often becomes infected secondarily with various bacteria, increasing the local inflammation and pain that are the hallmarks of this disease. Infection with tetanus bacilli at the site of the ulcer, which is the base of the ruptured blister caused by the emerging worm, is the most dangerous secondary complication of Guinea worm disease, which otherwise is not usually fatal. In about one-half of 1% of cases, a worm that emerges in or near a major joint such as the elbow or knee can cause permanent freezing of that joint, with consequences very similar to crippling damage from paralytic poliomyelitis. Past infections do not confer immunity, so people can be reinfected year after year.

There is no animal or any other environmental reservoir of *D. medinensis* outside of human beings and the infection has been recognized in ancient sources such as Egyptian mummies and the Old Testament. The immature larvae last only a few weeks outside of the human body, and only if they are ingested by a copepod. That is why this disease is vulnerable to complete eradication, even though copepods are ubiquitous in stagnant surface sources of freshwater.

The impact of Guinea worm disease on affected communities can be devastating, even though the infection is not usually fatal. This is a disease of remote, neglected rural populations that do not have easy access to clean drinking water. Before the eradication campaign began, up to 50% or more of a village's population could be infected at the same time, for periods averaging 1–2 months. Working age adults of both sexes are affected most of all because they tend to drink larger volumes of water. And the worms emerge

Water and Sanitation-Related Diseases and the Environment: Challenges, Interventions, and Preventive Measures, First Edition.
Edited by Janine M. H. Selendy.

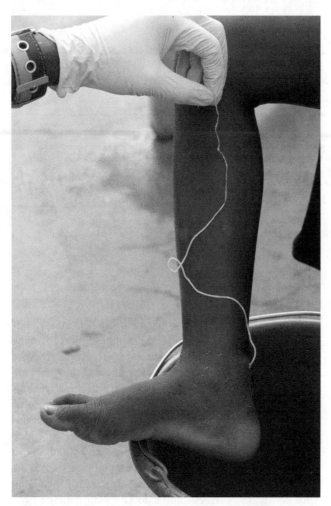

FIGURE 10.1 Emerging Guinea worm. Savelugu, northern Ghana: Patient with a Guinea worm emerging, at the Savelugu Case Containment Center. (Photo credit: The Carter Center/Louise Gubb.) (*See insert for color representation of this figure.*)

seasonally, precisely during the time of year of peak agricultural labor, when villagers need to harvest or plant their crops. Hence, Guinea worm disease negatively impacts not only people's health but also their agricultural productivity and school attendance (1).

In drier areas such as the Sahel just below the Sahara Desert, the "Guinea worm season" coincides with the brief rainy season, because that is the time of year when sources of surface water are abundant, and available to receive and transmit new parasites. In better watered climates nearer the Atlantic coast in West Africa, for example, the dry season is the optimal period for the disease because that is when stagnant water sources are shrinking and most contaminated; the abundant rainy season causing rivers to flow (nonstagnant) and diluting any contamination. Areas where the period of peak prevalence of Guinea worm disease coincides with the rainy season are doubly handicapped, since that is also the time of year when travel into remote

areas, and hence, provision and supervision of control measures, is most difficult.

There is no treatment, cure, or vaccine for Guinea worm disease, except to provide analgesics to relieve the pain and antibiotics to mitigate secondary infections. The ancient practice of carefully winding the emerging worm around a small stick is still the best way to try to shorten the period of disability. If a worm is broken while being coaxed to emerge, however, the remainder of the worm retracts into the body and spills larvae into the tissues, thereby causing severe inflammation and abscess formation, which is very painful and requires incisions and drainage by medically trained personnel, ideally at a clinic facility.

Although Guinea worm disease is not treatable, fortunately transmission of the infection can be prevented in several ways, by protecting drinking water from contamination by infected humans, and/or by protecting humans from drinking contaminated water. The most basic intervention is to teach people in endemic areas to avoid entering sources of drinking water when a worm is emerging or about to emerge from their body, and to always filter any water they drink from an open pond or other stagnant source through a finely woven cloth. Convincing conservative villagers that this disease is caused by water they drank a year ago, and that it is not because of any number of long and strongly held traditional beliefs, is easier said than done. One of the most effective ways to promote such understanding is to filter water from the usual local pond through a cloth and backwash the filtered material into a clear glass or jar, which is then held up to the light (with or without a magnifying glass) so villagers can see the tiny organisms moving around in the water they are drinking. That revelation is instinctively repulsive, and is persuasive to many. Distributing cloth filters free of charge for villagers to use to protect themselves and their families, and demonstrating how to use them properly, also serves as a tangible reminder to persons at risk of what they should do. Special attention is paid to educating women and girls, who are most often the ones who collect water for household use. As of 2008, about 40% of the workforce in the national Guinea Worm Eradication Programs (GWEPs) was female. Specially woven nylon material, which originally was donated by the E. I. DuPont Company, and which lasts longer and filters faster than cotton cloth, is most commonly used in the eradication campaign. Portable "pipe filters," a type of large plastic straw with a filter inside, can be hung on a string around the neck, and are provided to allow persons to filter the water they drink when traveling or farming away from home (Fig. 10.2).

Another effective intervention is to treat potentially contaminated ponds with ABATE® Larvicide, which at the recommended concentration is colorless, odorless, tasteless, and harmless to humans, fish, and plants, but lethal to copepods and the parasites they convey. This larvicide, which is donated to The Carter Center by BASF

FIGURE 10.2   (a) Cloth filter. Savelugu, northern Ghana: Technical assistant Eugene Yeng and a young community-based surveillance volunteer, Abdulhai Idiris. (Photo credit: The Carter Center/ Louise Gubb.) (b) Pipe filters . (Photo credit: The Carter Center/Emily Staub.) (*See insert for color representation of this figure.*)

Corporation for use in the eradication campaign, is applied at a concentration of one part per million every 4 weeks during the transmission season. Providing clean water from underground sources by borehole wells or similar means is the ideal intervention, since such clean water prevents many other infections in addition to Guinea worm disease, and is often much more accessible than the traditional pond, which may be a mile or more away from the village. Unfortunately, this is also the slowest and most expensive of the available preventive measures, and assuring priority to specific villages for any reason, including Guinea worm disease, is difficult because of political influences, corruption, and powerlessness of residents in the remote villages concerned.

Once village-based interventions are in place in as many affected communities as possible, an additional barrier to transmission can be "case containment," designed to prevent transmission from each patient. This is done by individual health education to urge the patient not to enter any water source, by manual extraction of the worm (slowly wrapping around a stick) and daily bandaging and treatment of the wound, and by caring for the patient in a home, clinic, special "Case Containment Center" or other facility. This provides less disability and early return to normalcy for the patient, and discourages further contamination of water sources by ensuring voluntary isolation of infected individuals. Specific criteria, including detection of the patient before or within 24 h of emergence of the worm, have been developed for classifying cases as contained or uncontained (2).

10.2   IMPLEMENTING THE ERADICATION CAMPAIGN

The global Guinea Worm Eradication Program was conceived and initiated at the Centers for Disease Control and Prevention in Atlanta, Georgia, in October 1980, originally as an adjunct to the International Drinking Water Supply and Sanitation Decade (1981–1990), one of the goals of which was to provide safe drinking water to all who did not yet have safe water. Early in the twentieth century, dracunculiasis was widely prevalent in many Asian and African countries, from southern parts of the Soviet Union to British India, Persia, and Saudi Arabia, and much of North, East, and West Africa. By the time The Carter Center began leading the global campaign in 1986, an estimated 3.5 million persons were being infected annually, in 3 Asian and 17 African countries (3) (Fig. 10.3). Because of its remote habitat, marginalized affected population and lack of available treatment, dracunculiasis was vastly

underreported before the campaign began. Although Nigeria and Ghana each officially reported about 3000–5000 cases of the disease to the World Health Organization annually in the 1980s, for example, when the two countries conducted nationwide village-by-village searches for the disease in 1989, they enumerated over 650,000 and almost 180,000 cases, respectively.

Village volunteers (VVs) are the bedrock of the GWEP. Sometimes selected by village chiefs or government health workers, but often chosen by fellow villagers, the VVs are trained by the program to do four things: educate their neighbors about how to prevent Guinea worm disease, distribute cloth filters, record and report cases monthly to a supervisor, and provide first aid care to villagers with Guinea worm disease. The immediate supervisors are peripheral governmental health workers or are national and international technical advisors who are specially hired and supported by The Carter Center for the national GWEP. This much-envied system comprised more than 6000 VVs each in Ghana and Nigeria in the early 1990s, for example, pioneering unprecedented monthly reporting and surveillance for cases of Guinea worm disease from every known endemic village, including some of the most remote villages imaginable. A few such villages were previously unknown to government authorities. The key to sustaining this system is and was assiduous attention to training, retraining, supply, constructive supervision, encouragement, and feedback of the VVs and their supervisors. Compilation, display, and regular monitoring of monthly surveillance data and of operational indices are other vital ingredients that are particularly important to motivating workers and to mobilizing financial support for the program. This combination of dedicated volunteers, their support, supply and supervision is what has brought this ancient loathsome disease to the brink of eradication.

Villagers are bombarded with simple messages via as many channels as possible (radio, videos, discussions

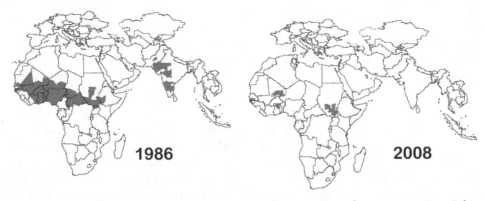

FIGURE 10.3   Guinea worm reduction over time. (*See insert for color representation of this figure.*)

illustrated with flip charts, drama, songs, posters, etc.), encouraging them to always filter unsafe drinking water, to avoid entering or allowing others to enter sources of drinking water when a Guinea worm is emerging or about to emerge, and to report any cases of the disease to a health worker. By these means, the GWEP seeks to engender a climate whereby undesirable behavior is communally regarded as an unneighborly act, without stigmatizing those infected, and where repeated messaging provides constant and involuntary reminders of what villagers should and should not do when they encounter a source of drinking water. More recently, behavior change communication techniques have been taught to more effectively engage communities in seeking agreement on what to do on their own to prevent Guinea worm disease from affecting their community. Political mobilization is equally important as community mobilization. Here, one seeks to motivate political, traditional, and religious leaders to actively support eradication efforts, by using data from the campaign to inspire, encourage, and if necessary, make the right people uncomfortable. The fact that Guinea worm disease is so visible, has important economic impact, and can be prevented, even *eradicated*, provides powerful ammunition for such political advocacy (4).

As mentioned earlier, the ecology and epidemiology of dracunculiasis is intimately bound to local environments where the infection thrives. Copepods are ubiquitous, but vary in their ability to transmit Guinea worms to humans. By definition, all humans must have some source of drinking water, but these also vary in their susceptibility to harbor copepods, which require stagnant, not flowing freshwater. Streams and small rivers that contain flowing water during the rainy season but become almost dry riverbeds from which humans collect drinking water from stagnant intermittent pools and excavated holes during the dry season are simultaneously transformed into effective potential sources of infection. In northern Ghana, small dams constructed in the 1950s and 1960s for watering cattle and irrigating crops inadvertently provided ideal habitat for undesirable contact between Guinea worms, receptive copepods, and many thirsty people (5). Moreover, many such sources of impounded water are too large to reasonably treat with ABATE. This well-intended ecological alteration caused much of northern Ghana to stand out most unfavorably as far as Guinea worm disease was concerned, from otherwise similar neighboring areas of Burkina Faso, Cote d'Ivoire, and Ghana itself. And just as surface sources of drinking water are determined by the environment and rainfall, the availability of safer underground sources, which are not at risk of contamination by people with emerging Guinea worms, is conditioned by geology. Some populations still manage to subsist in areas underlain by rock that prohibits or severely limits the drilling of wells, by collecting rainwater and/or traveling long distances to fetch surface water from far away. A result of these many factors

is that Guinea worm disease is distributed sporadically, often being highly endemic or absent altogether, in adjacent villages and districts.

Despite the many environmental determinants of Guinea worm disease's occurrence and intensity, human agency has been a major factor in the eradication campaign in other ways besides health-related behavior, primarily because of insecurity and sporadic violence engendered neither by Guinea worms nor by copepods, but by people. Such disruptive insecurity has been most prominent in the GWEP during the civil war in Sudan, but has also been a significant impediment to control measures in many other affected countries in Africa at one time or another. In Ghana, an outbreak of ethnic violence in the Northern Region during the peak transmission season in 1994 and 1995 disrupted interventions, led to loss of supplies and equipment, and caused villagers and health workers to flee to district capitals, where some fleeing villagers contaminated the drinking water sources at local dams, causing the number of Guinea worm cases to double between January and April 1994 and the same period of 1995. Violence involving nomadic Tuaregs in northern Mali delayed the reporting and deployment of remedial measures to control an outbreak in 2006 and 2007. Between January and August 2009, 27 incidents of sporadic violence or petty banditry disrupted operations of the GWEP in southern Sudan, causing Guinea worm workers to be confined to their dwellings or evacuated for various periods, often in highly endemic areas. In addition to civil war in Cote d'Ivoire, other ethnic, political, or religious conflicts have also impeded Guinea worm eradication activities in Chad, Ethiopia, Niger, Nigeria, Togo, and Uganda.

Addressing the political issues effectively has been a critical component of the campaign that requires engagement of persons outside of the traditional public health arena. Involvement of former U.S. President Jimmy Carter has been key to mitigating or removing political barriers in the Guinea Worm Eradication Program. In 1995, as the global campaign sought to help Sudan launch its program despite the ongoing civil war between the north and south of that country, by drawing on his prior contacts with both sides, President Carter was able to negotiate a cease-fire. By then, Sudan was known to be among the most highly endemic countries, and it seemed wise to try to start working in accessible areas of the country, which included most of the north and some of the more highly affected southern states where most of the fighting was occurring. Eventually known as the "Guinea Worm Cease-Fire," the hiatus in the civil war lasted nearly 6 months, and was even more successful than we expected. Sudanese health workers and their foreign Non-Governmental Organization (NGO) partners on both sides surprised themselves by what they were able to accomplish, visiting almost 2000 endemic villages to inquire whether Guinea worm disease occurred there and beginning to educate local villagers where it did, and distributing over 200,000 cloth filters (6). And despite resumption of the fighting, in 1996 Sudan's GWEP distributed even more cloth filters without a cease-fire than it did the year before.

Another masterstroke was to engage two prominent African leaders in the campaign. In 1992 President Carter persuaded then former Malian head of state General Amadou Toumani Toure to help energize his country's listless eradication effort. General Toure eventually toured all the major endemic areas of Mali, and visited all nine other endemic francophone countries to lobby their heads of state and ministers on behalf of the campaign as well. His popularity enhanced, Toure was later elected president of Mali (he had ousted a military dictator in a *coup d'etat* in 1991 and turned the government over to an elected civilian a year later) and is now serving his second term. Popular former Nigerian head of state General (Dr.) Yakubu Gowon was recruited in 1998 to help reverse 5 years of stagnation due partly to sporadic insecurity and ineffectiveness of the military government in Nigeria, which had begun its program in 1988 with more reported cases than any other country. By the end of 2005, General Gowon had personally visited over a hundred endemic villages in 18 Nigerian states, urging local medical, traditional, and political leaders to intensify control measures, including giving priority to endemic villages for provision of safe drinking water (l). Nigeria's reported cases began declining again dramatically in 2000, and it is now on the verge of eliminating dracunculiasis entirely.

Marginalization and migration are two other human factors affecting Guinea worm disease and its eradication. Many of the most resistant foci of disease are thus characterized because the poor rural populations in them are disenfranchised and long neglected by others in the same country. Salient examples are the Nyangatom in southern Ethiopia, Konkomba in northern Ghana, and black Tuareg in Mali and Niger. Overcoming decades or centuries of such enmity in order to do the necessary in such circumstances, by paying careful attention to ethnicity of health workers, to language used for written and verbal messages, and to placement of case containment centers and borehole wells, for example, can be a challenge. The frequency and unpredictability of human travel to attend weddings and funeral ceremonies (often in distant neighboring villages or even other districts), and to farm or trade pose other challenges, in addition to more predictable transhumance and annual movements of some pastoral groups. In Mali, an infected Koranic student triggered an explosion of 85 cases in one village in a previously non-endemic area after having walked over 300 km from his home district. Similar exceptional treks by infected individuals have been reported in persons moving from southern Sudan into Ethiopia.

For all its potential benefits and seemingly irresistible logic, the global GWEP got underway very slowly, because of insufficient funding during the 1980s and early 1990s to assist national eradication programs. Although it was conceived in 1980 just before the launch of the International Drinking Water and Sanitation Decade (1981–1990), by 1990 only 4 of the 20 endemic countries (India, Pakistan, Ghana, and Nigeria) had actually launched nationwide Guinea Worm Eradication Programs. Twelve countries began their national programs in 1992–1994, and all but one (Central African Republic) got underway by the end of 1995 (1).

10.3 IMPACT AND FINAL CHALLENGES

As already noted, the numbers of reported cases of dracunculiasis increased rapidly as individual countries conducted their first nationwide surveys to count cases of the disease, starting in 1988. And as the same countries began implementing control measures successively, their numbers of reported cases began to decline significantly, but not as abruptly as they had risen. The relatively rapid initial reductions in cases reflected impact of control measures in the most receptive populations, gradually revealing the most resistant and difficult clusters of endemic villages, families, and individuals as the campaign progressed at the frustratingly stately pace dictated by the infection's 1 year long incubation period and aided by human mistakes, neglect, bureaucratic and financial shortcomings, and by the sporadic insecurity described above.

Although the global GWEP missed the original target date of 1995 for interrupting transmission of dracunculiasis, by 2003 the campaign had reduced the number of cases to 32,193 cases reported from 4659 villages in 12 countries, of which 20,299 cases were reported from Sudan, 8285 from Ghana, and 1459 from Nigeria. Heartened by the recent results, representatives of the 12 remaining endemic countries resolved during the World Health Assembly in May 2004 to intensify implementation of control measures so as "to free the world of *dracunculiasis* by the end of 2009" (7). Another major advance occurred in January 2005 when the Comprehensive Peace Agreement was signed to end the two-decade long civil war in Sudan, thus removing the single most important obstacle to complete eradication. Sustained and generous financial support provided to The Carter Center for the campaign by the Bill & Melinda Gates Foundation and other donors provided the wherewithal to capitalize on the favorable political developments. After a brief upsurge in the number of cases reported from Sudan in 2006 as that program gained access to previously inaccessible endemic areas, the newly constituted South Sudan Guinea Worm Eradication Program (SSGWEP) quickly began building on the momentum and experience gained since the Guinea Worm Cease-Fire in 1995 (8).

By the end of 2008, dracunculiasis had been eliminated from Asia altogether, the number of endemic countries had been reduced from 20 to 6, the number of endemic villages was down from over 23,000 in 1991 to 1025, and the number

Number of cases

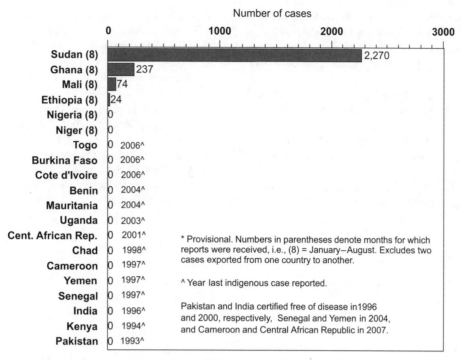

FIGURE 10.4 Distribution of indigenous 2604 cases of dracunculiasis reported during 2009.

of cases reported was reduced from an estimated 3.5 million in 1986 (and over 900,000 counted in 1989) to 4619. Most of the cases reported in 2008 were from Sudan (3618 cases), with Ghana (501 cases) and Mali (417 cases) reporting most of the remainder. The number of cases exported from one country to another declined from 154 in 2002 to 6 in 2008 (9).

A total of 2604 indigenous cases have been reported during January–August 2009, which is a reduction of 33% from the 3902 indigenous cases that were reported during the same period of 2008. Sudan reported 2270 (87%) of the cases so far in 2009, while Ghana reported 236 cases, Mali reported 74 cases, and Ethiopia reported 24 cases. Nigeria and Niger, which reported 38 and 2 indigenous cases in 2008, respectively, have reported no indigenous cases in January–August 2009 (Fig. 10.4). So far only two cases are known to have been imported from one country to another in 2009, one having traveled from Ghana to Niger and the other from Mali to Niger. At this stage of the campaign, the proportion of cases that were discovered before or within 24 h of emergence of the Guinea worm and are successfully contained is the single most important indicator of the effectiveness of national programs. As of the end of August 2009, Sudan had reportedly contained 83% of its cases in 2009, Ghana had contained 94%, Mali 76%, and Ethiopia 96% (10).

The World Health Organization bears sole responsibility for certifying interruption of transmission of dracunculiasis.

To do so, it requires adequate surveillance having detected no indigenous case of dracunculiasis in a recently endemic country for at least three consecutive years after the last known indigenous case. An International Commission for the Certification of Dracunculiasis Eradication reviews data submitted by countries and the reports of teams sent to verify the status of surveillance in selected countries, and makes recommendation to the director-general of WHO. As of September 2009, WHO had officially certified the absence of dracunculiasis from 180 countries, and 21 countries remain to be certified (9).

It is likely that only four endemic countries will remain as of January 2010, with the overwhelming majority of cases being in southern Sudan. The global Guinea Worm Eradication Program is thus poised to achieve its ultimate goal soon, despite many challenges described above, provided sporadic insecurity in the main currently endemic area, southern Sudan, is contained. The Southern Sudan Guinea Worm Eradication Program is quite able to contain the remaining Guinea worms, if the human insecurity is contained.

REFERENCES

1. Ruiz-Tiben E, Hopkins DR. Helminthic diseases: dracunculiasis. In: *International Encyclopedia of Public Health*, Vol. 3, 2008, pp. 294–311.

2. Hopkins DR, Ruiz-Tiben E. Strategies for eradication of dracunculiasis. *Bull World Health Organ* 1991;69:533–540.

3. Watts SJ. Dracunculiasis in Africa: its geographic extent, incidence, and at-risk population. *Am J Trop Med Hyg* 1987;37:119–125.

4. Edungbola LD, et al. Mobilization strategy for Guinea worm eradication in Nigeria. *Am J Trop Med Hyg* 1992;47: 529–538.

5. Hunter JM. Bore holes and the vanishing of Guinea worm disease in Ghana's Upper Region. *Soc Sci Med* 1997;45: 71–89.

6. Hopkins DR, Withers PC, Jr., Sudan's war and eradication of dracunculiasis. *Lancet* 2002;(Suppl 360): s21–s22.

7. World Health Organization. Dracunculiasis eradication: Geneva declaration on Guinea-worm eradication, Geneva, 2004. *Wkly Epidemiol Rec* 2004;79:234–235.

8. Hopkins DR, Ruiz-Tiben E, Downs P, Withers PC, Jr., Roy S. Dracunculiasis eradication: neglected no longer. *Am J Trop Med Hyg* 2008;79:474–479.

9. World Health Organization. Dracunculiasis eradication-global surveillance summary, 2008. *Wkly Epidemiol Rec* 2009;84: 162–171.

10. Centers for Disease Control and Prevention. Guinea Worm Wrap-Up #192. September 30, 2009. Available online at http://www.cdc.gov/ncidod/dpd/parasites/dracunculiasis/moreinfo_dracunculiasis.htm#wrap.

11

ONCHOCERCIASIS

Adrian Hopkins and Boakye A. Boatin

11.1 INTRODUCTION

Onchocerciasis (or river blindness) is a parasitic disease cause by the filarial worm, *Onchocerca volvulus*. Man is the only known animal reservoir. The vector is a small black fly of the *Simulium* species. The black fly breeds in well-oxygenated water and is therefore mostly associated with rivers where there is fast-flowing water, broken up by cataracts or vegetation. All populations are exposed if they live near the breeding sites and the clinical signs of the disease are related to the amount of exposure and the length of time the population is exposed. In areas of high prevalence first signs are in the skin, with chronic itching leading to infection and chronic skin changes. Blindness begins slowly with increasingly impaired vision often leading to total loss of vision in young adults, in their early thirties, when they should be at their most productive and when, as a result, their children are forced to be permanently at their side to accompany them. Other effects include epilepsy and growth retardation and these are most evident again in communities with high prevalence.

The disease is found mostly in Africa but is also found in Latin America and in Yemen as shown in Table 11.1, which gives the global estimates of the population at risk, infected, and blind in 1995 (1).

In Africa more than 100 million people are at risk of getting the infection in 30 countries south of the Sahara. However, further mapping of the disease in non-Onchocerciasis Control Programme (OCP) areas has shown that more people are infected in these areas than was thought; the total number of people infected is now estimated as 37 million (2). The control measures are now having an impact; the risk of the infection is actually much reduced and elimination of transmission in some areas has been achieved. Differences in the vectors in different regions of Africa, and differences in the parasite between its savannah and forest forms led to different presentations of the disease in different areas.

It is probable that the disease in the Americas was brought across from Africa by infected people during the slave trade and found different *Simulium* flies, but ones still able to transmit the disease (3). Around 500,000 people were at risk in the Americas in 13 different foci, although the disease has recently been eliminated from some of these foci, and there is an ambitious target of eliminating the transmission of the disease in the Americas by 2012.

Host factors may also play a major role in the severe skin form of the disease called Sowda, which is found mostly in northern Sudan and in Yemen.

The disability-adjusted life years (DALYs) lost due to onchocerciasis is 1.49 million as of 2003 of which 60% is attributable to "troublesome itching." Although the disease is not normally associated with mortality, a few studies have shown that life expectancy of the blind due to onchocerciasis was greatly reduced and that mortality in blind adults on the average was three to four times greater than in the fully sighted population.

The unprecedented donation of Mectizan® (ivermectin) by Merck & Co Inc. to as many people as needed it, for as long as it was needed was a watershed in public health (4). This donation, which began in 1987, has not only revolutionized the strategies for control of onchocerciasis but has also led to other major donations by the pharmaceutical industry for the elimination of trachoma, lymphatic filariasis, and other neglected tropical diseases of poverty (5).

Water and Sanitation-Related Diseases and the Environment: Challenges, Interventions, and Preventive Measures, First Edition.
Edited by Janine M. H. Selendy.

TABLE 11.1 Global Estimates of the Population at Risk, Infected and Blind from Onchocercisis in 1995 (1)

Region	Population at Risk of Infection (Millions)	Population Infected	Number Blind as a Result of Onchocerciasis
Africa			
OCP area:			
Original area	17.6[a]	10,032	17,650
Extensions	6.0	2,230,000	31,700
Non-OCP area	94.5	15,246,800	217,850
African subtotal	118.1	17,486,832	267,200
Arabian peninsula	0.1	30,000	0
Americas	4.7	140,455	750
Total	122.9	17,657,287	267,950

[a] The population given is that which would have been at risk had the OCP not existed.

11.2 THE PARASITE, THE VECTOR, AND THE RELATED LIFE CYCLES

11.2.1 The Parasite

The adult forms of *O. volvulus* are found in the human host in fibrous nodules, or onchocercomata, many of which are found in the subcutaneous tissues especially over bony prominences, although others are found deeper in the tissues. Female worms, which are considerably larger than the male (30–80 cm and 3–5 cm long, respectively), are found entwined around each other in the nodules and each nodule may contain from one to two male worms and two to three female worms, although larger nodules may exceptionally contain up to 50 worms. Whereas the females remain in the nodules, the males are more mobile and may go between nodules. Females release 700–1,500 larvae or microfilariae (MF) each day. The vast majority of these MF are found in the skin. Some find their way to the eye and probably other tissues of the body. If they are not ingested by the vector, they live in the host from 6 to 24 months.

In order to develop further the MF must be ingested by the vector, a small black fly of the *Simulium* species where, after penetrating the wall of the mid gut, they migrate to the thoracic muscles (6). Here initially they undergo a big increase in size into a sausage-shaped larva. After a few days these sausage-shaped larvae elongate and become much more mobile becoming the preinfective form and eventually migrating into the hemocoel and toward the head where they wait in the proboscis for the next time the *Simulium* requires a blood meal. This part of the life cycle in the fly takes around 10 days (6–12 days). When the *Simulium* bites again the infective larvae (L3 stage) migrate actively into the wound made by the bite and live in the tissues for a few days before going through a further larval stage (L4). After about a week they develop into juvenile worms (L5). It takes a further 7–15 months before these juvenile becomes mature and moves toward the nodules where mating takes place and the cycle begins again. The adult worms live for 10–14 years.

The life cycle of the parasite and its relationship to the human host and vector are shown in Fig. 11.1.

The pathogenicity of *O. volvulus* varies with different strains of the parasite. Savannah strains of the parasite provoke much more blindness than the forest strains (7–10). Both provoke skin changes. The different strains of *O. volvulus* are particularly well adapted to the vectors, forest strains being transmitted by the vectors adapted to the forest areas and similarly with savannah strains (11–14). The strains of *O. volvulus* in the Americas are also quite pathogenic to the eyes adding to the theory that the disease was introduced into the Americas with the slave trade.

11.2.2 The Vector

The most common vector is the very aptly named *Simulium damnosum*, "the damned black fly." This small fly, about the size of a mosquito, requires well-oxygenated water to lay its eggs, and where the larvae can develop. The female fly requires a blood meal to provide the necessary nutrition to produce each batch of eggs, and the time between blood meals varies from 6 to 12 days.

S. damnosum is actually a complex of sibling species. There are minor variations between the species but they show adaptations to local circumstances; the *S. damnosum* subcomplex contains species mostly responsible for transmission in savannah areas whereas the *S. sanctipauli* and *S. squamosum* subspecies are found in more forest areas. One intriguing group is the *S. neavei* complex. *S. neavei* larvae develop on the back of crabs, which are found in forest areas in eastern Africa. Another interesting species is *S. albivirgulatum*, which is found in the "Cuvette Central" of the River Congo. The larvae of this species are found on the underside of leaves floating down the River Congo and its many tributaries, thus, finding enough oxygen and nutrition for larval development. In the Americas the most common complexes are *S. ochraceum*, *S. metallicum*, and *S. exiguum*.

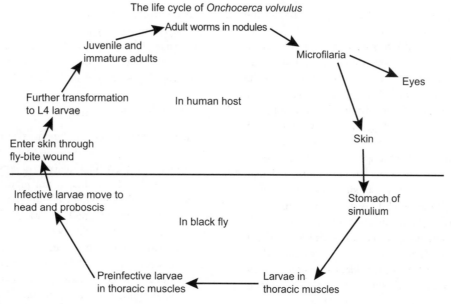

FIGURE 11.1   The life cycle of the parasite and its relationship to the human host and vector.

The importance of the different species relates to their capacity for transmission of the disease, and also the ease of control. Where vector control/elimination has been the main tool in the control of onchocerciasis, it has been important to understand the detailed biology of each species, to establish the best method of control. Some species of *Simulium* such as *S. neavei* do not fly far from breeding sites. *S. damnosum* however can fly long distances (400–500 km) by flying with the prevailing winds in West Africa. Generally, the *Simulium* try to feed as close to the breeding sites as possible. Populations living within 5 km of the breeding site are the most at risk although communities living up to 12–15 km may be affected. Communities living on the river's edge are more at risk than those a little further away. In fact, these communities very close to the breeding site to a certain extent "protect" the population living a little further away. This is one reason why blindness is relatively rare in towns, even if they are riverside communities.

The *Simulium* rarely enters houses to bite. Its preferred biting times are in the early morning and late afternoon. These are also times when populations who depend on the rivers for water, washing, and food are most likely to be at the river for these various activities. Those working in fields nearby or fishing in the river itself also tend to be less active during the midday heat.

11.3   RELATIONSHIPS BETWEEN PARASITE, VECTOR, AND THE HOST POPULATION

Onchocerciasis has had a significant impact on the populations in the savannah areas of West Africa. The close contact of man and the black fly poses not only medical problems for the communities that live near the river basin but also the bite of the flies causes intense irritation and a huge nuisance to the population. Communities are therefore compelled to abandon their villages and flee from the river basin for fear of acquiring the disease, notwithstanding the nuisance from the bite of the flies. Such movements have dire consequences on socioeconomic development. Moving to areas removed from the rivers has likewise led to poverty because the soil is much poorer away from the fertile valleys and water is more difficult to find so agriculture is gravely affected. Productivity is also affected negatively, leading to diminished earnings and adversely affecting the supply of labor. A high prevalence of blindness in young adults leads to further deterioration and deepens the cycle of poverty. These young adults are inflicted at the time when they should be their most productive and when they have young families to care for and educate. Often the children are forced to lead their blind parents around rather than go to school. Civil conflict and insecurity in Africa (particularly in Central Africa and the Sudan) has led to large population movements, often out of infected areas. However, when peace returns, populations may return to their traditional homelands and once again become exposed to the disease, although the chronic skin changes and blindness may take some time to develop.

The biting nuisance even where flies have been well controlled for a period has had a very positive effect on development, but the return of the flies after the control measures have been withdrawn lead to a lot of complaints from the population, even if the Simulium are not infected and not transmitting any disease (15). In the area of Inga on the Congo River during peak breeding times, the biting rate can reach to 13,000 bites per day and the population is often forced to wear protective clothing, long-sleeved shirts, even gloves and hats, in spite of tropical weather just to reduce the irritation from these bites.

Migration of the fly can also be significant. Movements of the *Simulium* have been mentioned above. It is possible that infected flies can be carried by winds into areas where the disease has already been controlled. *S. neavei* is found in forest areas in central and eastern Africa. The Potamonautes crabs, on which the larva develop, like shady areas. With deforestation these crabs have disappeared in some areas and transmission of the disease has stopped as a result. However, unfortunately *S. damnosum*, which prefers more savannah areas, has invaded some of these sites and once again transmission of the disease could become a problem.

Because flies can travel long distances in Africa, care must be taken when development projects include the construction of dams. The fast runoff of water around a hydro turbine produces highly oxygenated water, ideal to establish new breeding sites for *Simulium*. This must be considered during all construction and maintenance of new dams.

11.4 THE DISEASE

11.4.1 Adult Worms

The onchocercomata produce little in the way of symptoms. Those that are found in subcutaneous tissues are easily palpable as a mobile mass in the subcutaneous tissues with a firm consistency at the center. With experience these masses can easily be differentiated from other subcutaneous masses, such as small lipomas, sebaceous cysts, lymph nodes, and even subcutaneous cysticerci. However, as the disease is

more controlled and nodules become less numerous, the accuracy of diagnosis also declines.

One use of nodules has been in establishing a community diagnosis for decisions on mass treatment and this is used in rapid mapping strategies described later.

11.4.2 Microfilaria (MF)

The principal symptoms of the onchocerciasis are due to the MF. Each female can produce up to 1,500 MF per day. Live MF do not really create a major problem. However, the MF start to die after 6 months, if not ingested by a *Simulium*. When MF die, they produce an inflammatory reaction. Even if these reactions are small and localized, due to the large number of MF dying each day (particularly in the skin), they provoke widespread and chronic reactions. At first the reactions are reversible, but as the infection continues permanent changes occur. Recent evidence suggests *Wolbachia* endobacteria (symbionts of arthropods and filarial nematodes) may contribute to the inflammatory pathology associated with the disease (16–18).

11.5 CLINICAL SYMPTOMS

11.5.1 Skin

The first changes are due to itching and the lesions produced by energetic scratching. This is officially described as acute papular onchodermatitis (APOD), but the original description from West Africa of craw-craw somehow describes it more effectively (19). Sometimes patients use sticks or stones to scratch more vigorously. Figure 11.2 shows the typical

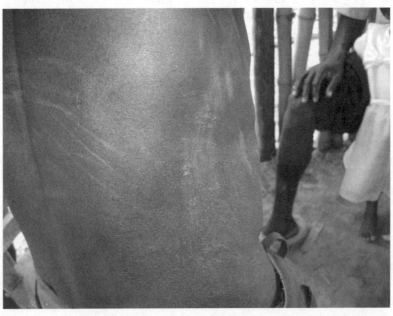

FIGURE 11.2 The typical changes of early onchocerciasis infection. (*See insert for color representation of this figure.*)

changes of early onchocerciasis infection. Trauma to the skin often leads to secondary infection, with localized edema and local lymphadenopathy. "Troublesome itching" has been shown to be one of the most important symptoms of onchocerciasis (20) and has recently been recognized as affecting more than 50% of the populations in some communities in the rain forest belt where blinding onchocerciasis is relatively rare.

Chronic onchocercal skin disease (OSD) leads to two skin changes. A scaly, often itchy thickening of the skin, called chronic papular onchodermatitis (CPOD) or lizard skin dermatitis is often seen over the back and buttocks. The more classic changes are changes in pigmentation of the skin, classically described on the lower limbs although sometimes seen elsewhere on the body. Patches of depigmentation give a classic appearance of "leopard skin" (depigmentation, DPM), as shown in Fig. 11.3.

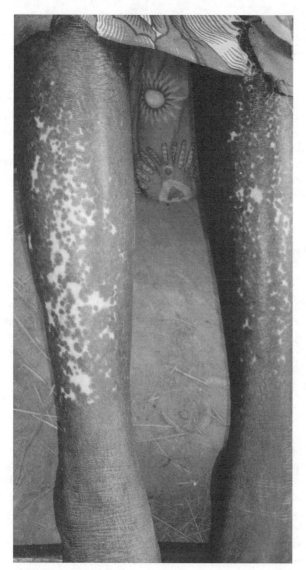

FIGURE 11.3 Patches of depigmentation give a classic appearance of "leopard skin" (DPM).

A further change that happens to the skin is atrophy (ATR) manifested by loss of elasticity. The skin of someone in their thirties or forties already looks like that of an old person. Swollen lymph nodes are found just under the skin in the inguinal region and the weight of these nodes drags down on the skin, if it has lost its elasticity, creating the so-called "hanging groin."

In Yemen and the north of Sudan and in a small proportion of patients elsewhere there is a lichenified onchodermatitis (LOD), which typically affects young adults and is also called Sowda (21). The skin is itchy, thickened, and hyperpigmented. This gradually spreads and may cover large areas, particularly the lower limbs, but is often asymmetrical. Draining lymph nodes become inflamed, which usually results in lymphedema or thickening of the tissues.

11.5.2 Eyes

MF are found in all the tissues of the eye from the conjunctiva anteriorly, to the optic nerve posteriorly (22, 23). MF in the conjunctiva provoke itching.

Active MF can invade the cornea, where if they die, they provoke a punctuate keratitis, or fluffy, white inflammatory reaction, in the clear cornea. This is reversible initially, but severe chronic infections lead to permanent changes. The cornea gradually loses its transparency and becomes white and hard (sclerosing keratitis). This begins at the 3 o'clock and 9 o'clock positions and then gradually fills in from the bottom, creating a semilunar keratitis until the whole cornea is affected with so-called sclerosing keratitis, as shown in Fig. 11.4.

MF are also found in the anterior chamber of the eye. In this space between the back of the cornea and the pupil, MF can be seen with the slit lamp, swimming in the aqueous humor, sticking to the inner surface of the cornea, the endothelium, or in severe cases the microfilaria sink to the bottom of the anterior chamber and can be seen as a mass, a so-called pseudo-hypopyon.

MF also penetrates the iris and can cause inflammation (a chronic anterior uveitis). This chronic inflammation can also lead to blindness due to secondary glaucoma or secondary cataract. The pupil often becomes small and deformed and sticks to the lens (posterior synechiae) and therefore nonreactive. The patient from Sudan in Fig. 11.5 shows a typical chronic uveitis with secondary cataract in her right eye and sclerosing keratitis in her left eye.

In the posterior segment of the eye MF can be seen in the vitreous. The most serious aspects of posterior segment disease, however, are chronic chorioretinitis and optic nerve atrophy. The chronic chorioretinitis is an inflammation that often begins around the periphery of the retina and causes gradual loss of visual field starting in the periphery. The macula (for central vision) is often spared initially, but with

FIGURE 11.4 The evolution of sclerosing keratits.

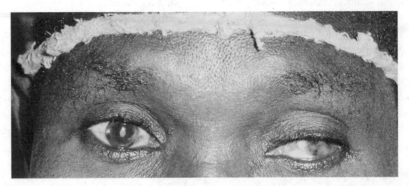

FIGURE 11.5 A patient from Sudan showing typical chronic uveitis with secondary cataract in her right eye and sclerosing keratits in her left eye. (*See insert for color representation of this figure.*)

advancing disease central vision is also lost. Optic nerve atrophy also provokes loss of vision (24, 25).

11.5.3 Systemic Effects

Systemic effects such as low body weight, general debility, and diffuse musculoskeletal pain have been described as other features of onchocerciasis. Although bleeding and ulceration of skin, secondary infections, bone pain, headache, and fatigue have also been suggested, these are relatively minor. Evidence suggests that onchocerciasis is a risk factor for epilepsy (26, 27) and may be responsible for a type of dwarfism in certain areas (28). With a heavy microfilarial load there may be generalized lymphadenopathy and some dilatation of lymph vessels, leading to tissue swelling and mild elephantiasis, particularly in areas of Sowda.

11.5.4 Social Effects

Onchocercal skin disease (OSD) has been associated with a variety of psychosocial and economic effects. The disease also leads to stigmatization of affected persons and their families. Unsightly acute and chronic skin lesions and thickened and irritated skin limit the chances of young adolescent girls finding marriage partners. Likewise negative sociocultural aspects of the skin disease (people worried that skin disease would affect their ability to interact socially, fear of being ostracized, a feeling of low esteem, and children more likely to be distracted in school due to constant itching) have now been recognized.

At the community level, studies have shown a reduced productivity due to the incessant itching, leading to increased poverty, increased expenditure on health in spite of a reduced income, all adding to the vicious cycle of poverty that this disease creates; this is apart from the effects of blindness. Many young people leave the villages for the towns and areas free of *Simulium*, increasing the level of poverty and destitution among the older people left behind and further contributing to the shortened life span of the blind (29).

11.6 CONTROL MEASURES

11.6.1 Control of the Vector

The most effective way to control the vector is to attack the larval forms. They are easier to control because of their relatively easy-to-reach known habitats in fast flowing rivers, with a flow rate of 5 m/s. The fly has a wind-assisted flight range of about 400 km, making the adult fly more difficult to control directly (30).

Various methods have been used to control the vector, including attempts at environmental management. Environmental management involves clearing the vegetation and applying agents like Paris Green and Creosote to the breeding sites of the vector. This approach was tried in the Chiapas focus of Mexico without success. *S. neavei* was, however, successfully eradicated from the Raina focus—a rather small focus—in Kenya through simply clearing the riverine forest. Deforestation in some parts of Uganda has also seen the disappearance of *S. neavei*, although there is now the risk of invasion by *S. damnosum*. The application of dichlorodiphenyltrichloroethane (DDT) was instrumental in the eradication of *S. neavei* from the Kodera valley in Kenya (31).

The mainstay for vector control has been the use of insecticides with the aim to interrupt transmission of the parasite. In the Onchocerciasis Control Programme (OCP) of West Africa the objective was to continue larviciding for a sufficiently long period during which time the human reservoir of the parasite would die out.

For vector control to be effective larviciding needs to be carried out preferably weekly. The larval stages rarely take more than 7 days to about 12 days to change to the next stage. Ground and aerial larviciding are both used for control. Ground larviciding is best applied where this is feasible, such as in small foci, and where the type of vegetation allows it. Aerial larviciding on the other hand has been used more extensively to cover large tracts of river basins, which would otherwise be either impossible or very difficult to cover by ground larviciding.

The formulation of insecticides used for large-scale treatment of rivers must meet a range of requirements. They must be highly effective against the vectors, safe to the rest of the environment, including human and other life forms, with guaranteed supply for a long period and the cost should be low. They must be biodegradable but with a good carry downstream.

Various insecticides are available. Temephos (AgrEvo France) is a cheap and efficient organophosphorus with insignificant impact on the nontarget fauna. Its use, however, needs careful attention given that insecticide resistance emerges very quickly with its application. Other environmentally friendly larvicides that have been used are the organophosphates, carbamates, pyrethroids, and the biological *Bacillus thuringiensis* serotype H14-B.tH14. These are used in rotation in accordance with river flow rate, to help avoid the emergence of insecticide resistance in the vector; to minimize adverse impact on nontarget organisms, riverine flora and fauna; and to increase cost effectiveness (32). For the OCP an expert independent ecological advisory group was set up to help monitor the impact of insecticides on the nontarget organisms in the rivers under larviciding. The Ecological Group also advised on the best possible insecticides to use for the maximum impact on the vector, while still having the minimal impact on the environment.

11.6.2 Surgical Removal of Nodules

Nodulectomy can remove some of the adult worms and has been used in some circumstances to try to control symptoms of the disease. Because they are often multiple, and many are also impalpable in the deep tissues, it is impossible to eliminate all adult worms by nodulectomy. Nodules may be removed for cosmetic reasons or because of annoyance due to their position, for example, around the waist where a belt might irritate. The logistical exercise of removing all palpable nodules from all patients in an endemic area would also be overwhelming, so this is not a strategy used for control.

11.6.3 Control of the Disease with Medication

The ideal drug would be one that kills the adult worm causing no side effects and would be safe for mass distribution. Unfortunately Suramin, which does kill adults (a macrofilaricide), is also toxic and unsafe to give as a mass treatment. One drug is currently undergoing trials, Moxidectin, but it is unlikely to be cleared for mass usage before 2014 or 2015 even if it does prove to be effective. Doxycycline also has macrofilaricidal effects by killing *Wolbachia*, obligatory endosymbiotic bacteria in some species of filaria including onchocerciasis. If *Wolbachia* are killed, the female cannot reproduce and will eventually die. The problems with Doxycycline are that it has to be given for a prolonged period, at least a daily dose for 4 weeks, and it cannot be given to children under 12, which is a very significant proportion of the population in most endemic areas (33).

Ivermectin is an effective microfilaricide (killing microfilaria). It also has an effect on reproduction preventing the release of microfilaria by the adult female for approximately 4 months. When microfilariae are released the repopulation of the skin is very slow and after 12 months it reaches around 20% of the original level. There is also a macrofilaricidal effect. This effect was not thought to be particularly important but recent research shows that some of the early studies were not done in a closed system and some of the live adult worms found were possible worms from new infections rather than adult worms resistant to ivermectin. Ivermectin is contraindicated in children under 5 (less than 90 cm height or 20 kg weight), pregnancy and during the first week after delivery and also patients with chronic disease, or central nervous system diseases.

There is another microfilaricide, Diethylcarbamazine (DEC). DEC causes much greater inflammatory reactions, or Mazzoti reactions, due to the massive destruction of microfilaria, than is the case with ivermectin. This is especially important in the eye where the inflammation provoked can cause significant loss of vision and even blindness. This does not occur with ivermectin. DEC is therefore contraindicated for use in onchocerciasis. Flubendazole is a macrofilaricide but the complications of administering the drug have prevented its use (34).

11.6.4 The Role of Ivermectin in Disease Control and Elimination

Ivermectin has proved to be such a safe and effective drug that most onchocerciasis control programmes are now using mass drug administration (MDA) as the principal strategy for onchocerciasis control. MDA with ivermectin is done once or twice yearly. In well-defined endemic areas where treatment is given every 6 months, transmission of the disease can be virtually eliminated in 6–8 years provided there is a high enough coverage (around 80% of the total population, or over 95% of those eligible to take the drug). In Africa the foci of the disease are not so small and well defined. In most areas ivermectin MDA is done annually, which is sufficient to control the main symptoms of disease but will require 16 or more years to achieve elimination depending on the initial prevalence and the coverage of treatments during the MDA.

11.7   ELIMINATION PROGRAMS

11.7.1   Onchocerciasis Control Programme in West Africa

Figure 11.6 (1) shows the geographical distribution of onchocerciasis in Africa and the Arabian Peninsula. The Onchocerciasis Control Programme (OCP) in West Africa was established as a regional program in the early 1970s. Its prime aim was to eliminate onchocerciasis as a disease of public health importance in West Africa and to place participating countries in a position to control any recrudescence of infection should it occur. The program was executed by the World Health Organization (WHO) on behalf of the Committee of Sponsoring Agencies (CSA) (Food and Agricultural Organization, United Nations Development Programme, the World Bank with a rich donor participation, and the WHO) and the 11 participating countries—Benin, Burkina Faso, Cote d'Ivoire, Ghana, Guinea Bissau, Guinea Conakry, Mali, Niger, Senegal, Sierra Leone, and Togo (35, 36).

The OCP started as a vertical control program with its vector operations in 1975. At the close of the program in 2002, the operations covered an area of 1,200,000 km² of which 764,000 km² benefited directly from vector control. Over 50,000 km of rivers was surveyed and appropriately larvicided. The objective of vector control was to interrupt transmission of the parasite for long enough periods to allow the human reservoir to die out. Initially planned to last 20 years, the period for larviciding was reduced to 14 years upon new evidence, which suggested a shorter reproductive life span (12 years) for the worm. This period of vector control was further reduced to 12 years when larviciding was combined with ivermectin treatment in light of additional information from model predictions. Vector control did not benefit the population already infected, or those who were already suffering the morbidity associated with the disease before larviciding started. However, as from 1987, when ivermectin treatment was added to larviciding in some areas, direct morbidity control became part of the program strategy.

The OCP is cited as one of the most successful large-scale control program ever undertaken in the developing world. Several reasons may be responsible for this success. First, there was a sense of common purpose by all concerned. There was a single objective that was developed by all, agreed upon, accepted, and was stuck to by the group concerned. It was as a "regional program" that the participating countries, which bore the brunt of the disease burden, wanted. It is clear that given the long range of the vector and the complex nature and heavy infrastructure involved in the operations only a regional approach was likely to succeed. All concerned devoted their attention in terms of time, funds, and in kind contribution to the OCP. Donor participation and funding was unflinching and the contribution through the free drug donation by Merck was remarkable. Furthermore, the program was driven by motivated and competent staff and by operational research whose findings underpinned the strategies that were used.

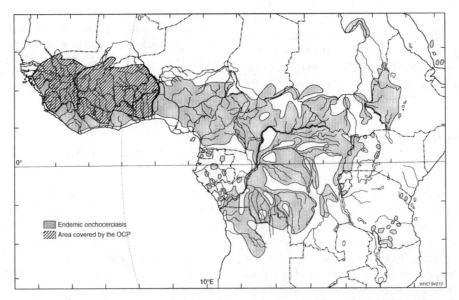

FIGURE 11.6   The geographical distribution of onchocerciasis in Africa and the Arabian Peninsula.

As the flies in the river basins were controlled, the communities returned to the previously abandoned villages. This, in several situations, was not done in a controlled fashion, creating a new challenge of intense pressure on the river basins and the land, although in some countries organizations were set up to coordinate this rehabilitation of abandoned land. Since the program has ended, flies have begun to reinvade the areas. Although these flies are no longer infected, and there is no risk of river blindness, the population in some areas is complaining of the "failure of the program," as they suffer once again from the biting nuisance.

11.7.2 African Programme for Onchocerciasis Control (APOC)

The African Programme for Onchocerciasis Control (APOC) was established at the end of 1995 to help control onchocerciasis in 19 countries that were thought to be endemic for onchocerciasis in Africa outside of the OCP area. The principal strategy was mass distribution of ivermectin and the chief objective of the program was to set up sustainable ivermectin distribution that would continue when the program finished. A secondary objective was to eliminate the vector completely in the few sites where the conditions existed to do so.

One of the first activities of APOC was to map out not only where the disease existed but where the prevalence of the disease was sufficiently high to warrant mass treatment (37).

It was found that there is a fairly consistent relationship between the presence of nodules in the population and the overall prevalence of the disease, nodules being present in one third to one half of the population who are positive by skin snip (a small bloodless skin biopsy examined for skin microfilaria). Fifty people resident in a community for at least 10 years can be examined and an estimation of the prevalence can be made dividing populations into hyperendemic where mass treatment is urgent, mesoendemic where mass treatment is desirable, and hypoendemic where only individual cases with symptoms need treatment. Skin and eye disease is infrequent in hypoendemic communities. This assessment by village called Rapid Epidemiological Assessment (REA) was further adapted using knowledge of the terrain and the population into a broader based mapping system based on river valleys where a biased sample of villages were evaluated, according to strict criteria. Although this Rapid Epidemiological Mapping of Onchocerciasis (REMO) is not a detailed epidemiological study, it enables decisions to be made where mass treatment should be carried out (38). Although some refinement is still needed in difficult areas, most of the 19 countries have been mapped and red areas (needing mass treatment) and green areas (not needing mass treatment) have been defined. Three of the 19 countries—Kenya, Mozambique, and Rwanda—only have hypoendemic disease and are not therefore receiving MDA. The current REMO map can be seen on the APOC website (see Fig. 11.7).

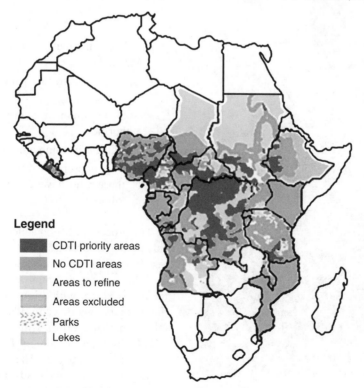

FIGURE 11.7 The current REMO map seen on the APOC website. http://www.who.int/blindness/partnerships/APOC/en/ (*See insert for color representation of this figure.*)

Because of this need for a sustainable method of MDA, a low-cost distribution system was needed. Early trials using mobile teams were not sustainable and community-based approaches were required. Research showed that if communities were empowered to make their own plans of actions once they had been fully sensitized and had been trained in the necessary technical details, they were able to organize their own MDA with minimal input from the health services, which are often nonexistent in remote communities. This led to a refinement of community-based approaches called Community-Directed Treatment with Ivermectin (CDTI) (39). Once communities are informed, they undertake the following tasks:

- Choose a distributor for training
- Do a census to calculate Mectizan requirements
- Organize the collection of Mectizan from a health center or other distribution point
- Organize a distribution method (house to house, fixed point in village, etc.)

FIGURE 11.9   Registration of residents in a Burundi village.

- Help the distributor calculate the dose and distribute the Mectizan
- Organize transfer of patients with adverse events, if required
- Note the treatment statistics and report to the health authorities
- Participate in community supervision
- Arrange appropriate recognition at the community level of those who have given their time to work for the distribution.

The process of health education, registration of families, calculating the dose using a dose pole, and giving out tablets in a Burundi village are shown in Figs. 11.8–11.11.

After 5 years communities should be ready to continue treatment alone, with minimal help and supervision from the primary health care (PHC) services, so although it takes time and effort with each community to initiate CDTI, once the system is functional, it becomes more and more sustainable. The development of CDTI has been one of the major advances in PHC over the last decade and has been the foundation for other community activities (40).

FIGURE 11.8   Health education about onchocerciasis in a Burundi village.

FIGURE 11.10   Dose pole used to estimate the dose of ivermectin based on height.

11.7.3   Ivermectin: The Stimulation of a New Public Health Approach—Community-Directed Intervention (CDI)

The strategy for onchocerciasis control until 1987 was entirely by controlling the vector. The community was hardly involved apart for the employment of some local people as fly catchers and participation in parasitological and epidemiological studies. With the introduction of ivermectin and mass drug administration (MDA), different strategies were needed. Many onchocerciasis MDA programs in Africa were begun initially by nongovernmental development organizations (NGDOs) working in eye care; these MDA were often a part of an integrated eye care program, and vitamin A distribution was added on to the MDA. In some areas MDA with Praziquantel for Schistosomiasis (Bilharzia), and/or MDA including Albendazole for Lymphatic Filariasis elimination were added on as well. Repeated MDA requires much more community participation and research has shown that if communities are empowered to organize their own treatment, they are fully able to do so, not just with ivermectin but also with other MDA. This co-implementation has now been fully tested and found to achieve better results than traditional methods using the health services in an approach called community-directed intervention (CDI). It has also been extended to other non-MDA interventions such as distribution of bed nets, and home treatment of malaria (41).

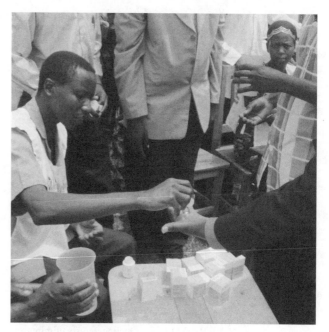

FIGURE 11.11   Giving out tablets.

11.7.4   A Paradigm Shift in Africa

APOC was conceived as a control program with two main objectives. First, the main objective was to set up sustainable program for MDA, with the eventual aim of eliminating onchocerciasis as a "public health problem." Second, another less important objective was to eliminate the vector of the disease where possible, which included a few small foci in non-OPC areas in Uganda, Tanzania, and the Island of Bioko (Equatorial Guinea) in the Gulf of Guinea. It was considered that onchocerciasis could not be eliminated in Africa using current tools (42).

Studies in Senegal and Mali have now showed that ivermectin could be stopped after 16 years, at least in some areas of savannah onchocerciasis (43). In some areas of Uganda, where there has been a combined vector and ivermectin approach treatment is on the point being discontinued (44). Skin snip research in other foci of infection have indicated similar results after 14–18 years, but further studies are required to confirm the interruption of transmission. Results have not been uniform and are probably related to the intensity of the original infection as well as levels of treatment coverage.

These studies have stimulated a major change in thinking, that onchocerciasis could be eliminated at least from some areas of Africa using ivermectin distribution. It is now conceivable to think of eliminating the transmission of onchocerciasis in western and eastern Africa, even though the central African region will remain a problem due to the problems caused by conflict and the presence of another filarial infection Loa-Loa (45).

Geographical distribution of endemic onchocerciasis in the Americas

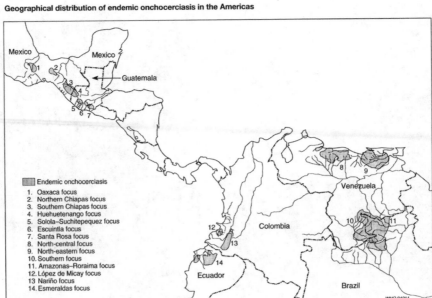

FIGURE 11.12   Onchocerciasis elimination program in the Americas is limited to small foci in six countries: Mexico, Guatemala, Columbia, Ecuador, Venezuela, and Brazil.

11.7.5   The Onchocerciasis Elimination Program for the Americas (OEPA)

Onchocerciasis in the Americas is limited to small foci in six countries—Mexico, Guatemala, Columbia, Ecuador, Venezuela, and Brazil (see Fig. 11.12) (1). The total number of people infected was half a million. In 1995 the program changed the policy of treatment to twice a year and from 2000 began a process of pushing for maximum coverage. Unlike APOC there is an attempt to treat everyone infected, even in hypoendemic areas. Because these foci are small and well localized, it has been possible to create closed systems and with the twice yearly treatments for 7 years the program has already managed to stop treatment in some areas where transmission has been interrupted (46). The program has been so successful that it is hoped that treatment can be stopped everywhere in 2012, although the focus on the Venezuelan–Brazil border with the Yanomami Indians deep in the Amazon forest remains a challenge. Intensive treatment is ongoing in the remaining foci with treatment cycles every 3 months and intensive epidemiological and entomological surveys both to follow progress and also for surveillance in areas where treatment has stopped. Figure 11.13 shows the treatment with ivermectin since the beginning of the program in the Americas and the projections into 2015 (47).

11.7.6   Yemen

The disease in Yemen was probably imported from Africa and is found in the river valleys (Wadis) draining into the Red Sea. The strategy has been based on individual treatment of symptomatic cases of Sowda, although this has also been extended to family members in some cases. Yemen has now planned an elimination campaign. This will have two parts. The first, which is vector control, should be reasonably easy because some of the Wadis dry up in the dry season. However, some breeding sites will be difficult to locate due to the difficult terrain. The second strategy is mass treatment with ivermectin in all the areas affected. This will be for approximately 300,000 people and will be four times a year. Individual patients already on treatment will be followed up regularly. Once the plan is fully implemented, it is hoped that the program will take around 7 years to complete.

11.8   CRITERIA FOR ELIMINATION OF TRANSMISSION OF ONCHOCERCIASIS

The World Health Organization has produced criteria for certifying elimination of the transmission of the disease. These are not applicable in all circumstances and will have to be adapted to different epidemiological situations.

Figure 11.14 shows a schematic representation of the phases in programs for elimination of onchocerciasis transmission, in relation to the theoretical fall-off of the adult worm population and annual transmission potential (ATP). Arrows mark major achievements, which indicate the transition between phases and changes in required interventions or surveillance activities as described. The four phases shown are critical periods in progress to elimination.

In Phase 1 transmission of the disease continues but the annual transmission potential is gradually reduced due to an

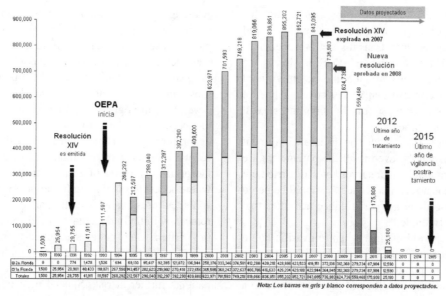

FIGURE 11.13 Graph of treatment with ivermectin since the beginning of the program in the Americas and the projections into 2015 (47).

appropriate intervention (ivermectin treatment, vector control, or both). The ATP is the number of infective larvae that could be transmitted to an individual if all the flies biting him during 1 year were able to transfer all the load of L3s. This figure varies with vector differences and the strains of *O. volvulus*. An ATP of less than 100 is considered to be safe to protect from eye manifestations of the disease but must be brought down much lower to eliminate transmission. This is considered the "break point" in modeling of the disease (48). As the treatment strategy continues, there will be no more transmission, either due to absence of flies using vector control or absence of infected bites if ivermectin treatment is used.

In Phase 2, transmission has ceased and the adult population will age and will die off. It should be noted, however, that while some adults are alive, there is always the possibility of transmission, if control activities are suspended. It is in fact vital during this phase that control efforts continue to the end of the phase, if not all that has been gained will be lost (49). OCP using vector control, used 14 years as the most appropriate length of Phase 2. A similar period is proposed for ivermectin, but if multiple treatments per year are used,

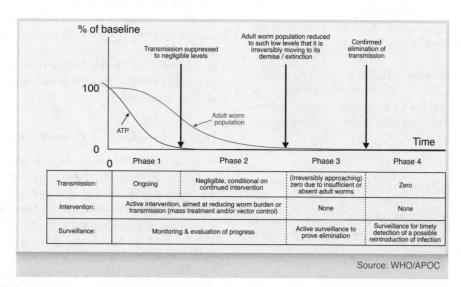

FIGURE 11.14 A schematic representation of the phases in programs for elimination of onchocerciasis transmission, in relation to the theoretical fall-off of the adult worm population and ATP. http://www.who.int/blindness/partnerships/APOC/en/ (*See insert for color representation of this figure.*)

the adult worms will age and die much more quickly and this period could be shortened to 6 or 7 years. It should be noted that Phase 2 only begins when transmission has been suppressed because until that is achieved, there is always the possibility of introducing new filaria into the system. Annual treatments with ivermectin, especially when coverage has not been adequate, will not suppress transmission and so will delay the beginning of Phase 2.

In Phase 3, treatment is stopped. The disease will die out under normal circumstances, even if all the parameters are not right down to zero, but regular surveillance is carried out to confirm that there is no recrudescence. It should be noted that during this phase there are surveillance activities that need to be continually funded!

Phase 4 covers a period of time when reinvasion could occur from another focus of infection, or alternatively due to massive movements of populations, for example, in a war situation, infected patients in sufficient numbers move to a new area where there are flies, so that transmission could recommence.

11.9 GUIDELINES FOR ELIMINATION

The World Health Organization has produced guidelines to certify elimination (50).

Elimination of Morbidity:
   Prevalence of microfilariae (MF) in the cornea or anterior eye chamber <1%.
Elimination of Transmission:
   L3 in flies <0.05% (0.1% in parous flies).
   ATP lower than 5–20 L3 per season.
   Absence of detectable infection in school children and antibody prevalence of <0.1%.

To certify elimination by WHO, this has to be country wide, for all foci of infection. Some modifications of these criteria are needed to adapt to different situations, for example, where flies have been eliminated by vector control methods.

11.10 CHALLENGES

There is a mixture of programmatic challenges as well as technical issues to be resolved (51) for the ongoing control or elimination of onchocerciasis.

11.10.1 Co-infection with Loa-Loa

In areas of tropical rain forest in Africa there is a common infection of patients with another filarial parasite, Loa-Loa.

Although the mobile adult worm is found in the tissues, often not far under the skin, the MF are present in the blood. When patients with high levels of MF in the blood are treated with ivermectin, they can provoke serious adverse events. Although the pathology of these events is still poorly understood, the most significant event is an "encephalopathy." The patient develops neurological symptoms such as drowsiness, slurring of speech, walking difficulties, and eventually coma may occur. If properly managed, patients with coma will normally recover after a few days, although some sequelae may remain. Unfortunately many of these patients live in remote areas where access to suitable health services is non-existent, also in some cultures, patients with coma are treated with local remedies, initially delaying transfer to suitable health facilities. Often when patients eventually arrive at these health facilities, there are already major complications due to pressure sores and other infections, putting the lives of these patients further at risk. Special precautions must be put in place to manage these patients quickly and effectively where treatment is required for onchocerciasis.

11.10.2 Transmission Zones

This is an important issue for the APOC elimination program. Mass treatment is ongoing in mesoendemic and hyperendemic regions only. These are areas where skin disease and eye complications are present. In hypoendemic areas only patients with symptoms are treated. Transmission is certainly taking place in some of these hypoendemic zones, although it may not be important in some areas where the mesoendemic and hyperendemic foci have been well controlled. These areas need to be fully investigated to see where transmission is occurring and what would be an appropriate treatment strategy. In many of the foci in APOC countries there is co-infection with lymphatic filariasis, another filarial disease that uses ivermectin together with albendazole for an elimination program. Distribution of ivermectin may already be ongoing in some of these hypoendemic areas and programs must be coordinated.

11.10.3 Frequency of Treatment with Ivermectin

A single dose of ivermectin will reduce MF in the skin to negligible levels in a day or so. The adult females start to reproduce after 3–4 months and microfilaria can appear once again in the skin. However, as already mentioned, even after a year the worm load is around 20% of the pre-treatment levels. This has been the basis for suggesting annual treatment as a control measure. After a few rounds of annual treatment, levels of MF in the skin are so low as to already have an effect on transmission. If the objective is to change to the elimination of transmission, this may or may not be sufficient. Annual treatment has certainly proved effective in some circumstances and has eliminated the disease in some foci in West Africa. To move from Phase 1 to Phase 2 of elimination twice yearly treatment would be quicker. However, in programs already well established with annual treatment and already moving toward elimination, the logistical complications of changing strategy is probably not warranted. This needs further investigation.

11.10.4 Conflict Areas

Civil conflict has unfortunately been a reality for several African countries, particularly in the central African region. Conflict not only causes significant population movements but also destroys infrastructure and leads to a significant brain drain when competent health workers look for jobs elsewhere where their skills can be better utilized. Annual treatment coverage in many areas in central Africa remains far too low to have any impact on transmission of the disease and efforts at scaling up are also difficult for the reasons mentioned above (52, 53).

11.10.5 Diagnostics

Skin snipping and fly dissection have been the mainstay of control diagnostics. Fly dissection is tedious, since many flies need to be dissected. However, for the flies, pool screening using polymerase chain reaction (PCR) is proving very effective and is very practical. Skin snipping becomes increasingly less sensitive as numbers of MF diminish. Also skin snipping is a painful procedure; to confirm the absence of the disease, it is recommended to check 3,000 children in the area. Two methods are useful, but both have drawbacks. First, the DEC patch test is noninvasive, but the patch must remain on the skin until read the following day and health staff must be available to read the result. Serological tests such as the OV16 test are very useful but require laboratories some distance from the field to analyze the results. If OV16 could be developed into a simple card test carried out in the field and read in the field, it could become an ideal test.

11.10.6 Sustainability

Whatever strategy is used for ivermectin distribution, it is clear that distribution is an ongoing activity and must be maintained at high levels of coverage if transmission is to be interrupted. It is difficult to maintain both donor interest and patient interest once the initial impact of the treatment is no longer evident. Programs must be fully integrated into the primary health care system and become part of the normal package of activities at this level in order to be maintained.

11.10.7 Integration with MDA for Other Health Activities

Several of the so-called neglected tropical diseases (NTDs) also use MDA as their primary strategy for control or

elimination (54). It is logical to integrate MDA where it is safe to do so and where there is an obvious fit (55). For some of the diseases such as soil-transmitted helminths (intestinal worms) or schistosomiasis (bilharzia), the target population is school children, and treatment strategies involve school-based distribution. Other programs use child and maternal health weeks. For onchocerciasis and lymphatic filariasis the total population (except those excluded from treatment for medical reasons) are treated and some sort of CDTI is necessary. Programs should be integrated where it is a natural fit and safe to do so, but should not be forced at the risk of losing coverage or the losing the specificity related to any one of the diseases targeted in a control or elimination program.

11.11 CONCLUSION

Various public health measures and different programs for over 30 years have led to onchocerciasis, once the scourge of many areas of Africa and the Americas, gradually becoming controlled and even eliminated. This has been effective through different methods but all with effective partnership. Onchocerciasis is now being controlled with mass drug administration, which would not have happened without the donation of Mectizan (ivermectin) by Merck in 1987. This donation has been the inspiration for donations for the control of other neglected tropical diseases. The objective of eliminating onchocerciasis, however, should not be forgotten. If disease control efforts should falter during the next few years, there is a grave risk or recrudescence and populations once freed of the disease will once again be exposed. All that has been gained must not be lost through short-sighted policies or lack of funding for the last few years, needed to complete the elimination of this debilitating disease.

REFERENCES

1. WHO. Onchocerciasis and its Control. Report of a WHO Expert Committee on Onchocerciasis Control. WHO Technical Report, Series 852. WHO, Geneva, 1995.
2. Noma M, Nwoke BEB, Nutall I, Tambala PA, Enyong P, Namsenmo A, Remme J, Amazigou UV, Kale OO, Seketeli A. Rapid epidemiological mapping of onchocerciasis (REMO) its application by the African Programme for Onchocerciasis Control. *Ann Tropical Med Parasitol* 2002;96(Suppl 1):29–39.
3. Zimmerman PA, Katholi CR, Wooten MC, Lang-Unnasch N, Unnasch TR. Recent evolutionary history of American *Onchocerca volvulus,* based on analysis of a tandemly repeated DNA sequence family. *Mol Biol Evol* 1994;11(3):384–392.
4. Colatrella B. The Mectizan donation program: 20 years of successful collaboration. *Ann Trop Med Parasitol* 2008; 102(Suppl 1):S7–S11.

5. Gustavsen KM, Bradley MH, Wright AL. GlaxoSmithKline and Merck: private-sector collaboration for the elimination of lymphatic filariasis. *Ann Trop Med Parasitol* 2009;103 (Suppl 1):S11–S15.

6. Blacklock DB. The development of *Onchocerca volvulus* in *Simulium damnosum*. *Ann Trop Med Parasitol* 1926;20: 1–48.

7. Duke BOL, Anderson J. A comparison of the lesions produced in the cornea of the rabbit eye by microfilaria of the forest and Sudan-Savannah strains of *Onchocerca volvulus* from Cameroon. *Tropenmed Parasitol* 1972;23:354–368.

8. Duke BOL, Garner A. Fundus lesions in the rabbit eye following inoculation of *Onchocerca volvulus* in the posterior segment. *Tropenmed Parasitol* 1976;27:3–17.

9. Dadzie KY, Remme J, Rolland A, Thyelfors B. Ocular onchocerciasis and intensity of infection in the community II West African Savannah. *Ann Trop Med Parasitol* 1989;40: 348–354.

10. Remme J, Dadzie KY, Rolland A, Thyelfors B. Ocular onchocerciasis and intensity of infection in the community I West African Savannah. *Ann Trop Med Parasitol* 1989;40: 340–347.

11. Duke BOL. The population dynamics of *Onchocerca volvulus* in the human host. *Trop Med Parasitol* 1993;44:61–68.

12. Duke BOL, Lewis DJ, Moore PJ. *Onchocerca: Simulium* complexes I. Transmission of forest and Sudan-savannah strains of *Onchocerca volvulus* from Cameroon, by *Simulium damnosum* from various West Africa bioclimatic zones. *Ann Trop Med Parasitol* 1996;60:318–336.

13. Lewis DJ, Duke BOL. *Onchocerca: Simulium* complexes II. Variation in West African *Simulium damnosum*. *Ann Trop Med Parasitol* 1996;60:318–326.

14. Anderson J, Fuglsang H, Hamilton PJS, Marshall TFdeC. Studies on onchocerciasis in the United Cameroon Republic II. Comparison of onchocerciasis in rain forest and Sudan Savannah. *Tran R Soc Trop Med Hyg* 1974;68:209–222.

15. Hougard J-M, Sékétéli A 1998. Combatting onchocerciasis in Africa after 2002: the place of vector control. *Ann Trop Med Parasitol, Suppl*; 92;165–166.

16. Taylor MJ, Hoerauf A. *Wolbachia* bacteria of filarial nematodes. *Parasitol Today* 1999;15:437–442.

17. Hise AG, Gillette-Ferguson I, Pearlman E. Immunopathogenesis of *Onchocerca volvulus keratitis* (river blindness): a novel role for TLR4 and endosymbiotic *Wolbachia* bacteria. *J Endotoxin Res* 2003;9(6):390–394.

18. Saint André A, Blackwell NM, Hall LR, Hoerauf A, Brattig NW, Volkmann L, Taylor MJ, Ford L, Hise AG, Lass JH, Diaconu E, Pearlman P. The role of endosymbiotic *Wolbachia* bacteria in the pathogenesis of river blindness. *Science* 2002;295(5561):1809–1811.

19. O'Neil J. On the presence of a filaria in "craw-craw." *Lancet* 1875;105(2686):265–266.

20. Murdoch ME, Hay RJ, Mackenzie CD, et al. A clinical classification and grading system of the cutaneous changes in onchocerciasis. *Br J Dermatol* 1993;129:260–269.

21. Anderson J, Fuglsang H, al-Zubaidy A. Onchocerciasis in Yemen with special reference to Sowda. *Trans R Soc Trop Med Hyg* 1973;67(1):30–31.

22. Hissette J. Mémoire on *Onchocerca volvulus* and ocular manifestations in the Belgian Congo. *Ann Soc Belge Méd Trop* 1932;12:433–529.

23. Ridley H. Ocular onchocerciasis, including an investigation in the Gold Coast. *Br J Ophthalmol* 1945;10(Suppl):1–58.

24. Abiose A, Jones BR, Murdoch I, et al. Reduction in incidence of optic nerve disease with annual ivermectin to control onchocerciasis. *Lancet* 1993;341:130–134.

25. Cousens SN, Yahaya H, Murdoch I, Samaila E, Evans J, Babalola OE, et al. Risk factors for optic nerve disease in communities mesoendemic for savannah onchocerciasis, Kaduna State, Nigeria. *Trop Med Int Health* 1997;2(1):89–98.

26. Druet-Cabanac M, Preux PM, Bouteille B, Bernet-Bernady P, Dunand J, Hopkins AD, et al. Onchocerciasis and epilepsy: a matched case-control study in the Central African Republic. *Am J Epidemiol* 1999;149(6):565–570.

27. Pion SDS, Kaiser C, Boutros-Toni F, Cournil A, Taylor MM, et al. Epilepsy in onchocerciasis endemic areas: systematic review and meta-analysis of population-based surveys. *PLoS Negl Trop Dis* 2009;3(6):e461.

28. Kipp W, Burnham G, Bamuhiiga J, Leichsenring M. The Nakalanga Syndrome in Kabarole District, Western Uganda. *Am J Trop Med Hyg* 1996;54(1):80–83.

29. Prost A, Vaugelade J. La surmortalite des avegles en zone de savanne ouest-africaine. *Bull World Health Org* 1981;59: 773–776.

30. Thompson BH. Studies on the flight rage and dispersal of *Simulium damnsosum* (Diptera Simuliidae) in the rain forest of Cameroon. *Ann Trop Med Parasitol* 1976;70: 343–354.

31. Roberts JMD, Neumann E, Guckel CW, Highton RB. Onchocerciasis in Kenya, 9, 11, and 18 years after elimination of the vector. *Bull World Health Org* 1986;64:667–681.

32. Hougard J-M, Poudiougo P, Guillet P, Back C, Akpoboua L, Quioevere D. Criteria for the selsction of larvicides by the onchocerciasis control programme in West Africa. *Ann Trop Med Parasitol* 1993;85:435–442.

33. Wanji S, Tendongfor N, Nji T, Esum ME, Ngwa JC, Nkwescheu A, et al. Community-directed delivery of doxycycline for the treatment of onchocerciasis in areas of co-endemicity with loiasis in Cameroon. *Parasit Vectors* 2009;2:39.

34. Dominguez-Vazquez A, Taylor HR, Greene BM, Ruvalcaba-Macias AM, Rivas-Alcala AR, Murphy RP, Beltran-Hernandez F. Comparison of flubendazole and diethylcarbamazine in treatment of onchocerciasis. *Lancet* 1983; 1(8317):139–143.

35. Samba EM. The Onchocerciasis Control Programme in West Africa. An Example of Effective Public Health Management. Geneva: World Health Organization, 1994.

36. Boatin B. The Onchocerciasis Control Programme in West Africa (OCP). *Ann Trop Med Parasitol* 2008;102(Suppl 1): 13–17.

37. Taylor HR, Duke BOL, Munoz B. The selection of communities for treatment of onchocerciasis with ivermectin. *Trop Med Parasitol* 1992;43:267–270.

38. Ngoumou P, Wash JF. A manual for rapid epidemiological mapping of onchocerciasis. Doc No TDR/TDE/ONCHO/93. 4 World Health Organisation Geneva, 1993.

39. Amazigo U. The African Programme for Onchocerciasis Control (APOC). *Ann Trop Med Parasitol* 2008;102(Suppl 1): S19–S22.

40. WHO/APOC. Community-directed treatment with ivermectin (CDTI). 2009. Available at www.who.int/apoc/cdti/en.

41. WHO/TDR. Community-directed interventions for major health problems in Africa—a multi-country study: final report 2008. Available at www.who.int/tdr/publications WHO/APOC 2009. Informal consultation on elimination of onchocerciasis transmission using current tools in Africa "Shrinking the Map," WHO/APOC/2009.

42. Dadzie KY, Neira M, Hopkins D. Final report on the conference on the eradicability of onchocerciasis. *Filaria J* 2003;2:2.

43. Diawara L, Traoré MO, Badji A, Bissan Y, Doumbia K, et al. Feasibility of onchocerciasis elimination with ivermectin treatment in endemic foci in Africa: first evidence from studies in Mali and Senegal. *PLoS Negl Trop Dis* 2009;3(7):e497.

44. Ndyomugyenyi R, Lakwo T, Habomugisha P, Male B. Progress towards the elimination of onchocerciasis as a public-health problem in Uganda: opportunities, challenges and the way forward. *Ann Trop Med Parasitol* 2007;101(4): 323–333.

45. WHO/TDR. Press release. http://apps.who.int/tdr/svc/news-events/news/phase3-trial-moxidectin 2009.

46. Sauerbrey M. The Onchocerciasis Elimination Program for the Americas. *Ann Trop Med Parasitol* 2008;102(Suppl 1): S25–S29.

47. PAHO. Pan American Health Organisation 48th Directing Council (CD48-10-e), 2008. Available at www.paho.org/English/GOV/CD/cd48-10-e.pdf.

48. Plaisier AP, van Oortmarssen GJ, Habbema JD, Remme J, Alley ES. ONCHOSIM: a model and computer simulation program for the transmission and control of onchocerciasis. *Comput Methods Programs Biomed* 1990;31(1):43–56.

49. Hopkins AD. Onchocerciasis control: impressive achievements not to be wasted. *Can J Ophthalmol* 2007;42:13–15.

50. WHO. Certification of elimination of human onchocerciasis criteria and procedures, 2001. Available at WHO/CDS/CPE/CEE/2001.18a Accessed at http://whqlibdoc.who.int/hq/2001/WHO_CDS_CPE_CEE_2001.18b.pdf.

51. Hopkins AD. Ivermectin and onchocerciasis: is it all solved? *Eye* 2005;19:1057–1066.

52. Hopkins AD. Distribution d'ivermectine dans les pays en conflit. *Cahiers Santé* 1998;8(1):72–74.

53. Homeida MM, Goepp I, Ali M, et al. Medical achievements under civil war conditions. *Lancet* 1999;354:601.

54. Molyneux DH, Hotez PJ, Fenwick A. "Rapid impact interventions" how a policy of integrated control for Africa's neglected tropical diseases could benefit the poor. *PLoS Med* 2005;2(11):e336. Doi:10.1371/journal.pmed.002033.

55. Hopkins AD. Challenges for the integration of mass drug administration against multiple "neglected tropical diseases." *Ann Trop Med Parasitol* 2009;103(Suppl 1):S23–S31.

12

REASSESSING MULTIPLE-INTERVENTION MALARIA CONTROL PROGRAMS OF THE PAST: LESSONS FOR THE DESIGN OF CONTEMPORARY INTERVENTIONS

Burton H. Singer and Marcia C. Castro

12.1 INTRODUCTION

Malaria is no novelty in the historical medical and public health literature. Sumerians, Assyrians, Chinese, Indians, Egyptians, Greeks, and Romans left a series of records of deadly fevers (1–3).[1] One of the best descriptions of malaria was provided by Hippocrates (about 460 BC to about 377 BC), considered the Father of Medicine, who observed a relationship between stagnant waters and enlarged spleens on the nearby population, and highlighted the importance of the season of the year and place of residence for one's health (1–3, 5, 6). Although naive in their formulation, hypotheses about relations between malaria and the environment (particularly stagnant waters) led to the adoption of engineering works aimed at reducing marshes and swamps. Probably one of the earliest records of such initiatives dates back to 490–430 BC, and refers to the work of Empedocles of Agrigentum, a citizen of Sicily, who altered the course of rivers to avoid the formation of stagnant and bad smelling water in the city of Selinus (7). This was followed by similar initiatives over the centuries, including the construction of surface drainage, removal of swamps, and drainage of

coastal and wetland areas (8). All these initiatives were based on the miasmatic theory of malaria transmission,[2] and continued after the etiology and life cycle of the disease were elucidated. Countries such as Italy conquered malaria through massive drainage and land reclamation by hydraulic work (9, 10).

Gradually accumulating evidence over centuries was augmented dramatically by discoveries from the late nineteenth century to the present day. They revealed the nature of the disease, its form of transmission, the characteristics of the vector, the life cycle of the parasite, a diversity of strategies for prevention and control, alternatives for treatment, and the genome sequencing of malaria vectors and parasites (11–17). Yet, apart from all the accumulated knowledge, there were an estimated 243 million malaria cases worldwide in 2008, the vast majority (85%) in Africa. The disease accounted for an estimated 863,000 deaths in 2008, 89% of which were recorded in Africa (18). Children younger than 5 years are often under higher risk, and Africa accounts for 9 out of every 10 child deaths due to malaria (19).

Geographically, malaria is currently restricted to some tropical areas in Central and South America, Africa, and

---

[1] The earliest genetic evidence of *Plasmodium falciparum* malaria was established after the examination of 16 mummies from the eighteenth dynasty (circa 1550–1295 BC). The malaria parasite was found in four of the mummies, including king Tutankhamen (4).

[2] Malaria was believed to be caused by breathing in the foul vapors emanating from stagnant and fetid waters. Indeed, the word malaria (of Roman origin) means bad air (*mala aria*).

Water and Sanitation-Related Diseases and the Environment: Challenges, Interventions, and Preventive Measures, First Edition.
Edited by Janine M. H. Selendy.
© 2011 Wiley-Blackwell. Published 2011 by John Wiley & Sons, Inc.

Central and Southeast Asia (20). This is a much different picture than that observed in the eighteenth century, when malaria was present in all continents (1, 21), a pattern that continued well through the mid-twentieth century (22–26). The changing spatial distribution of malaria is a result of organized and successful efforts to reduce the burden of malaria. The nature, focus, and coverage of these efforts changed over time (1, 6, 27–32). Funding for malaria control in the public sector has increased fourfold in the past two decades, and strategies focus on prevention (42%), treatment (31%), health systems strengthening (14%), and program support (13%) (33). A recent Global Fund report estimates that in 12 African countries (Democratic Republic of the Congo, Ethiopia, Ghana, Kenya, Mauritania, Mozambique, Nigeria, Rwanda, Senegal, Sierra Leone, United Republic of Tanzania, and Zambia), approximately 384,000 child lives were saved between 2000 and 2009 through scaling up the use of insecticide-treated mosquito nets (ITNs) and intermittent preventive treatment for pregnant women (IPTp) (33).

In 2007, a call for malaria elimination by 2015 and long-term eradication was made during the Gates Malaria Forum (34, 35). Recommended interventions to achieve this goal include long-lasting insecticidal nets (LLINs), indoor residual spraying (IRS), and IPTp; adoption of other vector control strategies, however, were awaiting what would be viewed as "sound scientific evidence" for their efficacy (36). As a result, environmental management for malaria control is not a priority, despite its long history of success in reducing malaria transmission (37), and its inclusion in the Chinese National Malaria Control Program from 1955 to the present time (38, 39). Of particular interest is the use of intermittent irrigation strategies for control of rice-field malaria in China (37, 38).

Current challenges for malaria elimination are immense, and could be augmented by the consequences of climatic change (e.g., favorable conditions for vector and parasite development in areas previously inhospitable for malaria; increased occurrence of extreme climatic events, such as flooding and drought; and climate-induced population displacements), and by the construction of large-scale infrastructure and development projects—some in response to climatic changes (e.g., dams, water transfer schemes, irrigation, and sea walls). All these factors involve significant environmental modification, commonly associated with increases in malaria transmission, which could be mitigated through environmental management strategies for vector control (40).

In this paper, we review a variety of multiple-intervention malaria control programs where vector control strategies played a decisive role. We present examples of integrated malaria control programs in which the choice of control strategies, as well as any needed fine-tuning and modification of the program components over time, were based on regular surveillance and performance evaluation. We selected five case studies for consideration: Federated Malay States, Northern Rhodesia, Panama Canal Zone, Borneo, and China. These examples have a particular focus on the relationships between local environmental modification and local hydrology.

We organize the discussion in five sections. We start (Section 12.2) with a brief summary of the cycle of malaria transmission, types of vectors and parasites, and common symptoms of the disease. We also discuss a systemic view of the disease based on a multidisciplinary framework of analysis. In Section 12.3, we review the types of control interventions currently available. In Section 12.4, we discuss the components and structural organization and management of past successful malaria control interventions, illustrated by five case studies. In Section 12.5, we discuss the need to revisit the successful interventions (as illustrated by the case studies) in order to set up dynamic decision systems for current control programs. Finally, we conclude with some remarks on lessons learned.

12.2   MALARIA: THE DISEASE

Human malaria is a disease of the blood caused by parasites of the genus *Plasmodium*. Except for the zoonotic form, *Plasmodium knowlesi*, transmission involves two hosts: humans and mosquitoes. Humans are the intermediate host, where the asexual phase of the parasite life cycle takes place and where a sexual phase (gametocytes) is initiated and is infective to mosquitoes. Hybridization of parasites occurs in the mosquito. The life cycle of the disease has three basic stages. The first is called the human liver stage, and starts when an infected mosquito feeds on a human. *Plasmodium* cells in the form of sporozoites enter the human bloodstream reaching and infecting the liver, where they mature and multiply asexually producing merozoites. The second, named human blood stage, starts when merozoites infect red blood cells, and additional asexual reproduction occurs until male and female gametocytes are formed. Following emergence from the liver stage, the first clinical manifestations of malaria can be detected, and the disease can be diagnosed by a blood test. The final stage starts when a mosquito bites an infected human, ingesting blood that contains gametocytes. Inside the mosquito's stomach, gametocytes are transformed into gametes. When fertilized, gametes generate zygotes, which develop into sporozoites. The cycle of the disease is perpetuated when this infected mosquito bites another human, introducing the sporozoites into the human bloodstream (5, 41).

Five types of parasites are responsible for human malaria: *P. falciparum*, *P. vivax*, *P. malariae*, *P. knowlesi*, and *P.*

ovale. The first, *P. falciparum* is the most severe and lethal parasite, responsible for the majority of deaths attributed to malaria; *P. ovale* is the least prevalent parasite. Infections with the monkey parasite *P. knowlesi* (until recently commonly misdiagnosed as *P. malariae*) have been reported across Malaysian Borneo and Peninsular Malaysia, Vietnam, and Thailand (5, 41–45). Mixed infections of two or more parasites also occur.

Common symptoms of malaria include intermittent fevers and chill, usually accompanied by headache, myalgias, arthralgias, weakness, vomiting, nausea, abdominal pain, anorexia, and diarrhea. Other clinical manifestations include splenomegaly (enlarged spleen), anemia, thrombocytopenia, hypoglycemia, pulmonary or renal dysfunction, and neurological changes. *P. falciparum* cases, if not well treated, can progress to cerebral malaria, acute renal and liver failure, icterus, severe anemia and coagulation defects, pulmonary and cerebral edema, adult respiratory distress syndrome, coma, and death. Complications of *P. vivax* cases can result in splenomegaly, while those of *P. malariae* include nephritic syndrome (41, 46).

Female mosquitoes of the genus *Anopheles* are the vectors responsible for the transmission of malaria. There are approximately 400 species of *Anopheles* mosquitoes, but only 60 or so are able to carry malaria, and roughly half of them are important for malaria transmission. The female *Anopheles* is able to release eggs every 2–4 days, and it takes between 7 and 20 days for the egg to turn into an adult mosquito, depending on temperature and humidity conditions (47, 48). Each species may have different characteristics regarding biting pattern (unimodal versus bimodal biting hours, and indoors versus outdoors), feeding preferences (humans versus animal blood), flight range, and choice of breeding sites (5).

One episode of malaria does not offer immunity to the individual, and a person can be infected as many times as he or she is bitten by infected mosquitoes. However, exposure to malaria over several years can provide acquired immunity, attenuating the most severe symptoms of the disease, and resulting in asymptomatic infections (5, 42, 49, 50). Recently, it has been shown that immunity to the most severe effects of malaria occurs after one or two prior infections (51). Acquired immunity is lost when exposure to a malaria infection ceases (e.g., migration to a malaria-free area). Two genetic mutations confer some protection against a malaria infection. The first, red blood cell Duffy antigen negativity, is widespread in Africa and renders protection against *P. vivax* without producing any negative consequences for the carrier (52, 53). However, a few cases of *P. vivax* malaria infection among Duffy-negative individuals have been reported recently (54–57). The second is called sickle-cell hemoglobin, and offers some protection against *P. falciparum*. Children may inherit a sickle-cell

gene from one parent and a normal gene from the other, in which case there is protection against the most severe consequences of *P. falciparum*, especially against cerebral malaria (58). The negative consequence of this mutation, however, is that children who inherit the sickle-cell gene from both parents develop sickle-cell anemia and exhibit significant morbidity and mortality (59).

12.2.1   A Systemic View of Malaria

Understanding of the relationships in the triangle human–vector–parasite demands a multidisciplinary approach, which combines epidemiology with environmental sciences, demography, anthropology, entomology, economics, politics, geography, hydrology, and molecular biology (60). Figure 12.1 presents a systemic view of malaria, where the triangle human–vector–parasite exists within (and interacts with) a local environment (61, 62). The human vertex of the triangle includes variables that characterize individuals (e.g., their susceptibility to infection, their knowledge and behavior, and ways through which they can transform the local environment) and the context in which individuals interact (e.g., political and social). The vector vertex includes characteristics that are crucial for choosing the ideal package of interventions to launch an integrated vector management program (63, 64), and for measuring local exposure to malaria infection. The parasite vertex refers to issues that determine the effectiveness of drug-based intervention, and that indicate the levels of disease severity. Finally, the local environment includes both the natural environment and the built-habitat. The three vertices are not static, but rather interact among each other and with the local environment, producing unique local profiles of malaria risk.

12.3   MALARIA CONTROL

The discoveries of Charles Louis Alphonse Laveran (1845–1922) in 1880 (who saw the malaria parasite under a microscope), of Ronald Ross (1857–1932) in 1898 (who discovered that malaria was transmitted by female *Anopheles* mosquitoes), of Carlos Chagas (1879–1934) in 1905 (who proposed the concept of malaria as a household infection), of Malcolm Watson (1873–1955) and N. H. Swellengrebel (1885–1970) in the 1910s (who found that only some species of *Anopheles* were able to transmit malaria, coining the term "species sanitation"), and of Paul H. Müller (1899–1965) in 1948 (who discovered the insecticidal properties of DDT) brought about significant changes in the strategies utilized for malaria control (3, 31, 65). Moreover, the experiences from past control efforts and the knowledge accumulated over the years indicate that no single package of interventions

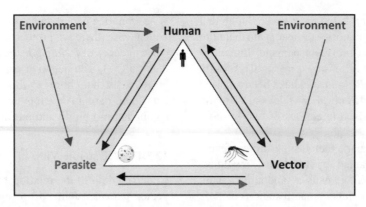

Human

Genetic immunity; acquired immunity; knowledge (transmission, prevention, treatment); migratory pattern; age; activity; education; cultural beliefs; population density; access to health care; environmental transformation (man-made environment); political context (support and infrastructure, construction of major development projects, implementation of public health interventions, foreign assistance, governance, and political stability); globalization (e.g., market pressures, crop selection, exploitation of natural resources); behavior.

Parasite	**Vector**
Type and strain, resistance to antimalarials, duration of infection	Species; feeding, resting, biting, host, and breeding preference; flying range; vectorial capacity; life span; reproduction rate; mosquito resistance to insecticides; and larval resistance to chemicals.

Environment

Natural: Temperature, humidity, and rainfall; soil quality; elevation/slope; vegetation; hydrography; presence of natural enemies of mosquitoes and larvae; natural disasters.
Man-made:land use and land change; deforestation; housing; local infrastructure (water, sanitation, waste collection); urbanization; development projects (e.g., roads, railways, dams, irrigation, mining, resettlement projects, and oil pipelines); "Natural" disasters (facilitated by man-made changes).

FIGURE 12.1   Systemic view of malaria.

can be adopted in all malaria endemic areas. Instead, efforts need to address local idiosyncrasies in malaria transmission resulting from the dynamic interaction between humans, vector, parasite, and the environment (Fig. 12.1) (66).

12.3.1   Types of Control Interventions

Traditional measures for malaria control, which focus on the mechanisms of disease transmission and on the local environment, can be grouped according to their main objectives (41), as shown in Table 12.1.

Regarding antimalarials, until the 1930s quinine was the only effective drug against malaria available (67). Since then, a handful of drugs have been produced, but widespread drug resistance developed at a faster pace than drug

production (68). At present, the World Health Organization (WHO) recommends the use of artemisinin-based combination therapy (ACT)[3] as the first-line drug treatment (69). However, artemisinin-resistant *P. falciparum* malaria has been reported in the Cambodia–Thailand border (70), which raises much concern since there is no new drug available to replace artemisinin.

[3] Four ACT combinations are recommended: artemether plus lumefantrine (AL), artesunate plus amodiaquine (AS + AQ), artesunate plus mefloquine (AS + MQ), and artesunate plus sulfadoxine-pyrimethamine (AS + SP). The choice depends on the level of drug resistance of the partner medicine in the combination. In addition, dihydroartemsinin plus piperaquine (DHA + PPQ) has been recommended for the treatment of uncomplicated *P. falciparum* malaria (69).

TABLE 12.1   Types of Control Interventions

Objective of the Intervention	Type of the Intervention
1. Reduce vector–human contact	Site selection, improvement of house conditions (e.g., closing eaves), house screening, mosquito nets, indoor residual spraying (IRS), repellents, and protective clothing
2. Reduce breeding habitats	Environmental management (i) Environmental modifications—irrigation systems, filling of water bodies sanitation, and drainage (ii) Environmental manipulations—water-level fluctuations, regulation and improvement of natural watercourses, intermittent irrigation, sluicing, salinity, vegetation clearing, and aforestation/deforestation
3. Reduce *Anopheles* larvae	Chemical (e.g., oil) and biological larviciding (e.g., *Bacillus thuringiensis* and *B. sphericus*), predators
4. Reduce adult *Anopheles*	IRS, insecticide-treated nets (ITNs), long-lasting insecticide-treated nets (LLITNs), genetic control[a]
5. Eliminate parasite in humans	Antimalarial drugs (for both prophylactic and therapeutic purposes), intermittent preventive treatment for malaria in infants (IPT$_i$), intermittent preventive treatment of malaria during pregnancy (IPTp), vaccine[a]

[a] Indicates that the intervention is under research, and therefore not yet available.

Control efforts may also address other issues highlighted in Fig. 12.1, such as individual behavior and knowledge about malaria transmission and prevention. Community-based approaches and sensitization campaigns are ideal strategies to address these issues, but their incorporation in malaria control program is still limited (71).

12.4   MALARIA AND HYDROLOGY

Next we present five case studies that illustrate successful integrated malaria control efforts, constrained by local eco-logical and cultural characteristics. None of these examples fully resemble current National Malaria Control Programs implemented in malaria-endemic countries. Yet, their lessons have much to contribute to the ongoing discourse about malaria eradication.

12.4.1   Case Study 1—Malaria Control in the Federated Malay States (FMS)

The control program began in 1901 with initiation of planning and implementation in the town of Klang and the nearby seaport, Port Swettenham. In qualitative terms, the baseline conditions were (a) a large percentage—by far the majority—of hospital patients at Klang had malaria; (b) the town of Klang and the people along the entire coastline were suffering from a major outbreak of malaria; (c) there was a proposal to eliminate the town of Jugra—20 miles south of Klang—because of high malaria rates; and (d) only a small fraction of the sick ever came to a hospital. It was anticipated that if curative treatment was the response to the malaria problem, the hospital system, however extended, could not accommodate the demand. Furthermore, focusing on treatment would provide no benefit to the overall community. Every well, ditch, and swamp teemed with *Anopheles* larvae.

12.4.1.1   Control at Klang Town   Based on the above assessment, Malcolm Watson, who had been appointed surgeon general and who was responsible for mitigation of the malaria problem, made the decision that prevention should be attempted. Malaria was to be studied outside the hospital ward. He assembled baseline data having two com-ponents: (i) a survey conducted to ascertain where the documented malaria cases came from, and the case rates (measured by spleen rates) at selected communities; and (ii) a survey of breeding sites followed by preparation of a map indicating their location.

With this information at hand, an initial intervention plan for malaria control had to be delineated. A combination of preventive and curative tools was available, including the extant hospital system for treatment. The options that, in principle, could be utilized were quinine (for treatment and prophylaxis, whether in hospital or outside), bed nets, house screening, spreading of petroleum on swamps, continuous sweeping out and elimination of individual collections of water, and provision of permanent drains and filling of water bodies. There were also a priori constraints on implementa-tion: (a) it was impossible to force use of quinine dispensed to Chinese workers, and it was difficult to get consistent use of it from Indian workers; (b) it was impossible to enforce use of bed nets, however persuasive the advertising about their benefits, as night visitation for surveillance would be re-quired; (c) compulsory screening of all houses was impos-sible because of financial limitations; (d) use of petroleum on swamps was not practical due to cost and temporary efficacy; (e) the vast area at Klang prevented sweeping out or dealing with individual collections of water on any continuous, ongoing basis; and (f) it would have been necessary to constantly supervise coolies on measures that would suppress mosquitoes only as long as they were applied.

Thus, measures such as quinine distribution, house screening, and bed nets could be anticipated to be used sporadically, and with low population coverage. The limited experience prior to 1901 supported this contention. This led to a proposed trial policy and intervention; namely, any expenditure should be on works of a permanent nature (e.g., draining and filling of standing water bodies). The decision was made and implemented. Klang town should be drained.

Surveillance of malaria cases (spleen rates) and mortality was carried out from the time drain construction was initiated. Within 3 years malaria ceased to be a problem of any practical consequence at Klang. In particular, it no longer inhibited work associated with production and shipping of rubber to Port Swettenham and out of the FMS to international markets. Ongoing maintenance of the drains was the only follow-up at Klang. There were no *a priori* targets of spleen rates or mortality rates that were designated as representing "success" of the program. There was also no control, or comparison, community designated as part of the evaluation of the drainage and filling intervention at Klang. Comparisons with preintervention levels—that is, historical controls—were the basis for evaluation. Achieved spleen rates below 5% were pronounced to be acceptable—a decline from anywhere between 30% and 80 + %, depending on the specific locality. Total mortality counts dropped from above 250 per year to 50 per year by 1903, with these levels sustained and/or improved upon through 1921. The cost of installation of the drainage system and filling of swamps, covering 300 acres, was USD 30,000 in 1901 dollars.

12.4.1.2 Control at Port Swettenham
The opening of the port on September 15, 1901 was followed by an abrupt explosion of malaria transmission, which prevented shipping by both boat and railway. A proposal was made to close the port by the end of November 1901 until it could be made safe. Although the port was not formally closed, a package of antimalarial interventions was immediately put in place. These were drainage, oiling of all pools of water, and universal offering of quinine. By 1904, malaria ceased to be a problem at the port. This continued to be the case until 1906 when the first signs of recrudescence appeared. Monitoring of drains revealed that some malaria cases coincided with the placing of cement inverts into the road drains at a level that blocked the flow of water into all lateral drains that had previously been open to them. This led to standing water that was the site of new *Anopheles* larval development. The situation was made worse by the extension of a railroad embankment. With the rapid flare-up of malaria cases, the government authorized new expenditures to repair and enhance the drainage system. This was immediately followed by a sharp reduction in the malaria case rate.

The ongoing surveillance and adaptive tuning of interventions exemplified by the Port Swettenham drain modifications was a central feature of the FMS program as it continued to develop in urban areas and rural rubber estates. Of particular interest was the fact that not only the malaria death rate, but the all-cause death rate dropped precipitously at Port Swettenham following the initial introduction of a drainage system. The high all-cause mortality rate prior to the interventions was a consequence of many malaria asymptomatic cases becoming vulnerable to other diseases (e.g., dysentery, diarrhea). Eliminating malaria resulted in reduced case rates and mortality from the other causes. The cost of drainage and filling of swamps, covering 100 acres, was USD 50,000 in 1901 dollars.

12.4.1.3 Control at Flat Land Coastal Rubber Estates
The baseline information consisted of knowledge that *Anopheles umbrosis* was the primary malaria vector and that it bred in stagnant jungle pools. Spleen rate and mortality rate were high, and the constraints on enforcing bed net usage and proper use of quinine were the same as listed above for Klang. The following interventions were initiated: (i) drainage work, consisting of 37 miles of drains, 20 feet wide, which allowed rubber planters to drain 24,000 acres; (ii) drainage of stagnant jungle pools at estate site and cultivation were designed to eliminate pools locally; and (iii) surveillance of children via blood samples and of the general population via spleen rates. Death rates were also monitored. Annual surveillance of malaria cases was designed to cover 500 square miles. Regular monitoring of breeding sites at estates was initiated.

Initial findings indicated that there was no malaria among individuals with houses and work sites located at more than 0.5 miles from breeding sites, and that death rates were correlated with spleen rates. Recommended actions based on 1–2 years of monitoring and interventions were to (i) remove houses to distances exceeding 0.5 miles from jungle pools; or to (ii) fell jungle and drain pools for at least 0.5 miles from extant houses and work areas. Implementation of these two options at flat land estates resulted in sharp decrease in spleen and mortality rates. Subsequently, new flat land estates were opened and designed based on these two recommendations, and malaria was not a problem. However, at one of the new estates, the surveillance revealed a sudden fresh outbreak of malaria cases. Close checking of these cases revealed that the coolies had been working near some new lines on the estate that were close to a point where the estate boundaries bent inward toward the rubber plants. Jungle clearance and pool drainage had not taken place outside the bent-in boundary, thereby putting the coolies less than 0.5 miles from *An. umbrosis* breeding sites. Jungle clearance and shifting of the boundaries for coolie lines solved the problem. This particular step in the program of flat land estate malaria control represents an instance of adaptive tuning of interventions over time.

12.4.1.4   Control at Hilly Estates   In 1909, 99% of hillside rubber estates had high malaria rates. There had been no improvement since the estates opened several years earlier. The only water was from crystal springs and brooks. There was a very high hospital admission rate and high mortality rates. Further, quinine did not stop or significantly reduce transmission, although it did serve to reduce hospital admission rates and the mortality rate. The principal vector was *Anopheles maculatus*. Its larvae can survive even strong currents and streams filled with grass, as long as the water is fresh. Indeed, a preliminary experiment at the Seafield estate showed that thoroughness of quinine dispensing and weeding of the edges of streams had not been effective at sharply reducing malaria rates.

With this background information and the experience from lowland estates at hand, a new intervention experiment was implemented at the Seafield estate. The coolie population was placed—housing and working—in selected "sanitary areas," and all water less than 0.375 miles from them was to be carried off by underground pipes. The idea was to create a mosquito-free circle on hillsides, with drainage up to 0.375 miles from the coolie line. The smaller drainage zone than that used at the lowland estates (0.375 miles rather than 0.5 miles) was motivated by the fact that *An. maculatus* was a much more delicate vector than *An. umbrosis*, and thus would be likely to have a shorter dispersal range. This intervention led to a short-term (within 2 years) reduction in malaria cases, but the reduction was decidedly inadequate for maintaining a healthy workforce. The drainage was then extended to 0.5 miles, but it needed larger pipes to carry away rainy season water. The result, although costly, was a major reduction in malaria cases.

The above experimental experience led to the following general specifications for malaria control at hilly estates: (i) rebuild coolie lines on flat land at 0.5 miles from the hills; (ii) if a low-lying swamp is present, open it and move coolie lines proximal to it (recall that *An. maculatus* requires very clean water, and swamps are not hospitable to their larvae); and (iii) if there are hills behind and flat land in front of coolie lines, pipe drain hills where *An. maculatus* is breeding. Following this protocol, there was a decline in the sickness rate for workers from 13.5% in November 1911 to 2.2% in December 1912. The spleen rate declined in the same period from 91% to 14%.

12.4.2   Case Study 2—Malaria Control at Copper Mines in Northern Rhodesia, 1928–1950

Economic exploitation of copper in the former Northern Rhodesia (now Zambia) began in 1909, following the completion of a railway into the copper belt. Until the 1920s, Northern Rhodesia's copper production (entirely from outcrop copper deposits) was insignificant compared to the neighboring Belgian Congo. However, starting in 1923 there was a sharp increase in the world demand for copper, and an accompanying rise in copper prices on the world market. Four copper mining companies were started with the initiation of major production during the years 1929–1936.

A primary prior obstacle to copper production was difficulty in recruiting an adequate labor force, as a result of intense malaria transmission at the mining sites (28). Health records from the medical department of the Roan Antelope mine, the first of the four mines in the area, indicate that when Europeans arrived there were 105 malaria attacks per thousand people in a single month. As with most data from health clinics, this was an underestimate of the extent of the malaria problem. Indeed, parasite rates and spleen rates among children in villages proximal to the Roan Antelope mine ranged between 36% and 45% (72). *Anopheles gambiae* and *Anopheles funestus* were the vectors responsible for transmission at the mining communities.

Members of the mining boards quickly realized that sustaining a healthy labor force and recruiting new workers would require effective disease control measures. An initial malaria control effort consisting of house screening, use of mosquito nets, and administration of quinine for prophylaxis and treatment was not sufficiently effective in reducing illness rates, and the malaria incidence rates remained high. There is clear documentation that this unsuccessful attempt at malaria control prior to 1929 resulted in migrant workers abandoning the Roan Antelope site. The initial interventions were supplemented by an environmental management program, initiated in 1929, and consisted of vegetation clearance, swamp drainage, river boundary modification (to enhance flowing water), and regular application of oil to open water bodies. This package of interventions at the Roan Antelope mine was adaptively tuned and maintained, with the addition of dichlorodiphenyltrichloroethane DDT in 1946, for two decades. Comparable packages of interventions were implemented at Mufulira, Nkana-Kitwe, and Nchanga mines between 1930 and 1936, and maintained until 1950. The baseline annual malaria incidence rate at the Roan Antelope mine was 514 per thousand. During the first 3 years of the full intervention program, the annual incidence rate was reduced to 135 per thousand and maintained below this level until 1946, when DDT was introduced. This led to a further reduction in annual incidence to 21 per thousand. The net result of the malaria control initiative was a dramatic increase in population at the mining sites from an estimated 10,964 persons in 1930 to 140,368 by 1949. The workforce at the mining sites grew by a factor of 5 during this period. To ensure steady revenue flows, the management boards of the Northern Rhodesian mines closely monitored the work shifts lost due to malaria, and regularly tuned the package of control measures to minimize this outcome measure. Ultimately, the copper production also increased dramatically, and Northern Rhodesia, which contributed about 0.3% of annual worldwide copper extraction between 1925 and 1930, became the third largest producer of copper ore in the world by 1940, with a global share just exceeding 11% (73).

It is important to observe that the FMS and Northern Rhodesian malaria control programs were carried out under corporate sponsorship. Their objectives were to ensure a healthy workforce. They did virtually nothing to improve public health on a broader scale than communities proximal to the work sites. The large up-front expenses needed to install drainage systems, and the use of packages of interventions for the full programs were not at issue due to the large financial returns. Discourse about cost-effectiveness of single interventions such as bed nets, insecticidal spraying, and drugs that pervade the contemporary tropical public health arena was absent from the colonial corporate environment of the FMS and Northern Rhodesian programs. In addition, stable management of the programs facilitated their successful operation for 20 + years. There is no counterpart to this in the public health arena with the exception of the Chinese national malaria control program that we discuss in Section 12.4.5.

12.4.3 Case Study 3—Malaria Control in the Panama Canal Zone

The control of malaria was a necessity for the construction of the Panama Canal. The initial groundwork for malaria control in Panama was an important effort directed by Surgeon Major William Gorgas (1854–1920), initially focused on yellow fever, during the American occupation of Havana, Cuba. Here the U.S. Army imposed a strategy of screening houses and extensive drainage to reduce, if not eliminate, breeding sites of mosquitoes. Yellow fever was eradicated from Havana, and as a by-product malaria transmission was also greatly reduced.

The Isthmus of Panama was (in the early 1900s), and still is, an ideal environment for many species of mosquitoes. High temperatures persist throughout the year. The rainy season lasts for 9 months, and the interior of the Isthmus is tropical jungle. Antimalarial work was primarily carried out in rural areas along 47 miles of railroad track between Panama and Colon. Approximately 80,000 people lived within one-half mile of the railroad, occupying 30 villages and camps. It was estimated that roughly 17% of the population in Colon experienced at least one malarial attack each week.

Starting in 1904, a multiple-intervention malaria control program was initiated. It consisted of seven interventions: (i) drainage, (ii) brush and grass cutting, (iii) oiling, (iv) larviciding, (v) quinine distribution, (vi) house screening, and (vii) killing adult mosquitoes. Each of these interventions was monitored, evaluated analogous to the performance-based ratings used in the FMS and Northern Rhodesian programs, and adaptively modified/tuned to facilitate reduction in the *Anopheles* mosquitoes and, correlatively, reduction in malaria incidence.

The drainage component consisted of elimination of all pools of water within 200 yards of village boundaries and within 100 yards of individual houses. Subsoil drainage was preferred, followed by concrete ditches and open ditches. Paid inspectors were responsible for maintaining drains free of obstructions. All bush and grass was kept at a height of less than one foot over the same distances as used for pool drainage. When drainage was impossible, oil was added along the edges of ponds and swamps to kill mosquito larvae. Since there were no commercially available insecticides in the early 1900s, a mixture of carbolic acid, resin, and caustic soda was used as a larvicide. Finally, the killing of adult mosquitoes was carried out by feeding collectors who gathered adult mosquitoes that remained in houses and tents during the day. This was an era long before DDT and other modern adulticides.

Because of the control program, the death rate due to malaria for canal construction employees dropped from 11.6 per thousand in November 1906 to 1.23 per thousand in December 1909. Malaria-specific deaths in the general population proximal to the railroad fell from 16.2 per thousand in July 1906 to 2.6 per thousand in December 1909. The percentage of employees hospitalized for malaria dropped steadily from 9.6% in December 1905 to 1.6% in 1909. Although malaria was never eliminated from the Canal Zone, it was controlled at a sufficiently low level that construction could be completed.

This brief summary of the Panama Canal project does not begin to do justice to the complex day-to-day decision making, performance-based rating evaluations, and intervention adjustments that had to be made throughout the program. These details, from various perspectives, have been extensively documented by the program managers (74–78). However, a particularly useful comparative analysis was assembled by Watson (12), where he contrasts his own FMS program with the Panama Canal efforts, and provides unparalleled insight—valuable to the present day—about the nuances of adaptively working toward an overall effective malaria control program.

12.4.4 Case Study 4—Malaria Control in Borneo

Beginning with his arrival in Borneo in 1938, John McArthur (1901–1996) initiated surveys to determine the distribution of malaria in areas proximal to where transmission was already known to occur. His overall goal was to establish a program of reduction, and even eradication, if possible, of malaria from the area (79). *An. maculatus*, a well-known malaria vector in other parts of Asia, was present in abundance in villages on open sunlit plains, but there was no human malaria at these localities. Surprisingly, villages on hillsides adjacent to dense jungle, where there was no *An. maculatus*, were found to be highly malarious, with spleen rates above 80%.

Extensive experience with *An. maculatus* in Malaya (28), where transmission was promoted whenever jungle was disturbed, was automatically translated into expensive efforts with oiling, drainage, and other antimosquito measures in Borneo by the local malaria service. These efforts, however, were totally ineffective in Borneo. Here, following a remarkable piece of detective work by McArthur (80), *Anopheles leucosphyrus* was incriminated as the primary malaria vector in Borneo. *An. leucosphyrus* is a purely jungle breeding mosquito, operating only in seepages under jungle shade and very difficult to detect. With this knowledge in hand, it was easily controlled by localized jungle clearance around the breeding-places.

An important experiment was conducted on the Tambunan plain in North Borneo from 1949 to 1952 to determine whether the admission of sunlight to seepages, followed where possible by land cultivation, was able to eradicate *An. leucosphyrus* and malaria cases from jungle villages. An area of approximately 10 square miles of jungle-covered hills, where spleen rates had been maintained at above 80% for at least 14 years, was selected as the experimental site. Jungle was cleared from approximately 200 seepages along 32 miles of waterway. This was followed by a 95% reduction in vector breeding that was maintained for 3 years with only minimal maintenance. Spleen rates declined during the first year from 86% to 56%. The decline continued to 53% and then 45% by the third year. There was some transmission remaining, as demonstrated by parasites in the blood of infants. This was attributed to infected mosquitoes from surrounding uncontrolled villages with very high malaria rates. Through the first 2 years of the experimental period, the spleen rates in a control area remained unchanged at nearly 100%. In the third year, the village chief in the control sites decided to imitate the program from experimental villages. This led to a precipitous drop in spleen rates, as had occurred in the first year at the experimental sites. While the experimental protocol (originally intended to run several years) was destroyed by this action, the drop in spleen rates actually added further evidence to support the control strategy for *An. leucosphyrus*, as well as to demonstrate that community-based jungle clearance was feasible.

In further follow-up, McArthur suggested the following as a continuation to the Tambunan study: "The clearing in the experimental area lasted much longer than was at first expected, and such clearing can be maintained permanently if the people cultivate, or graze their cattle in, the moist fertile areas thus exposed. Newer methods of clearing by spraying with selective herbicides—one spraying of which is claimed to be effective for five years—may prove an economical means of malaria control which is worthy of further study." (79, p. 656).

Following his 1950s journal articles (79–82), the publication of a book (83), and a presentation to the World Health Organization (84), McArthur's work—and its potential—regrettably seems to have vanished from the contemporary international discourse on malaria control. This is unfortunate in light of the recent documentation of transmission of *P. knowlesi* from monkeys to humans via *An. leucosphyrus* in Borneo (43). *P. knowlesi* could easily have been misclassified as *P. malariae* by McArthur, using the microscope diagnostic technology of the 1940s and 1950s. However, contemporary PCR analyses (43) document *P. knowlesi* infections in humans in Borneo, thereby clarifying that five species of plasmodia, rather than the long-established four species, infect humans and that one of them is a zoonosis with *An. leucosphyrus* as the primary transmitter.

12.4.5 Case Study 5—Malaria Control in China

China has had the longest running successful public health initiative focused on malaria of any country in the world. Starting gradually in 1950, systematically organized with a National Malaria Control Program in 1955 (39), and continuing to the present day, this adaptive, multiple-intervention, locally tuned effort at malaria suppression warrants in-depth examination and much more attention in contemporary discourse than it is receiving. Using 1949 as a starting point for baseline statistics, there were more than 30 million malaria cases in the country, and the mortality rate was approximately 1% per annum. Malaria was epidemic in 1,829 counties/cities, accounting for 70–80% of all counties in the country. Further, malaria represented 61.8% of the total recorded cases of acute infectious diseases in China in 1949.

When the national control program was initiated in 1955, it relied on primary health care networks as an organizational base, and made extensive use of community participation to respond to local needs. An intensive educational program was put in place that featured advertising of integrated sets of interventions, giving balanced emphasis to both prevention and curative medicine. This balance has persisted to the present day. Indeed, successful suppression of malaria in the diverse Chinese ecosystems owes much to this holistic philosophy. Figure 12.2 shows one of the many posters that were used for education at the local level as part of the broad-based Chinese antimalarial campaign. It emphasizes community work toward the elimination of breeding habitats, and the use of preventive measures.

By the year 2000, there were 1.202 billion people in 2,787 counties/cities living in areas where malaria incidence was less than 0.1 per thousand (including original malaria-free areas as well as endemic areas now free from the disease). There were 30.62 million people in 72 counties/cities with an incidence of 0.1–1.0 per thousand, and 3.94 million in 17 counties/cities with an incidence of 1.1–10 per thousand. There was no county or city in the entire country with an incidence above 10 per thousand. Figure 12.3 shows the

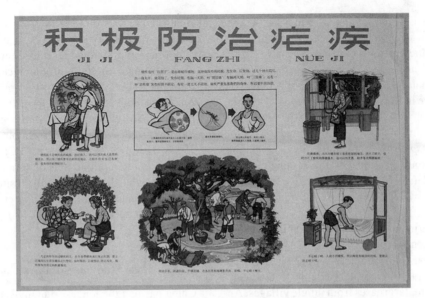

FIGURE 12.2 Poster from the Chinese National Malaria Control Program. *Source*: Available at the U.S. National Library of Medicine, National Institutes of Health website: http://www.nlm.nih.gov/exhibition/chineseantimalaria/gallery.html. Accessed March 2010. *Note*: The poster emphasizes malaria prevention and timely treatment. The title is *Actively Prevent and Treat Malaria*. Text was written by Wang Liancheng, and the art work was performed by Zhao Shuqi and Mo Gong. The images have the aesthetic features of Chinese New Year Pictures and paper-cutting, two popular media of Chinese folk art. The poster was published by Science Popularization Press, and printed by People's Education Printing Factory, May 1963.

national level trends in incidence and mortality rates from 1955 through 1998 (85). This coarse level of description masks considerable variation from community to community. It also masks the many examples of recrudescence of malaria that were detected in diverse communities, but followed by adjustment and/or intensification of interventions leading to reduced transmission and malaria case rates (86–88).

The guiding framework for malaria control in China is the adaptive tuning of multiple interventions guided by performance-based ratings carried out over time. However, an additional form of monitoring needs to be included here, namely, assessment of drug resistance. Not surprisingly, this phenomenon was, and continues to be, a challenging feature of antimalarial drug distribution in China. However, a research program focused on drug resistance and the development of new drugs was initiated in response to this problem (88) and plays an important role in the current version of the national program.

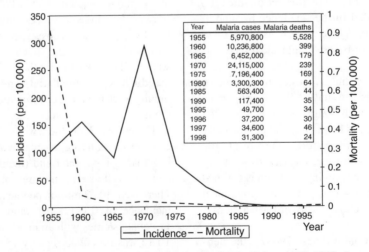

Year	Malaria cases	Malaria deaths
1955	5,970,800	5,528
1960	10,236,800	399
1965	6,452,000	179
1970	24,115,000	239
1975	7,196,400	169
1980	3,300,300	64
1985	563,400	44
1990	117,400	35
1995	49,700	34
1996	37,200	30
1997	34,600	46
1998	31,300	24

FIGURE 12.3 Trends in malaria incidence and mortality in China, 1955–1998. *Source*: Ref. 85.

12.5 DISCUSSION

Critiques of vector control strategies suggesting inadequate "scientific evidence" for their performance indicate the need to reformulate Watson's intervention strategies in contemporary statistical terms. This exercise has the potential to contribute to the ongoing efforts of national malaria control programs implemented in malaria-endemic countries. Here, we offer a few suggestions and an outline, without attempting to fully develop the model.

We begin by focusing on a single outcome measure (e.g., spleen rate or mortality rate) and view it as functionally dependent on the intensity of application of a set of K interventions (e.g., drugs, larvicides, drainage, bed nets, etc.), ecological conditions, and calendar time. Thus, if $p(t)$ is the outcome measure, we can write it formally as

$$p(t) = f(x_1, x_2, \ldots, x_K; y, t), \qquad (12.1)$$

where x_i is a measure of the intensity of application of the ith intervention (e.g., percentage of the population using bed nets, percentage of the potential larval habitats with no *Anopheles* larvae, etc.), and y is a vector of ecological conditions. The independent time variation can represent seasonal effects such as rainfall. The x_i values and y can also vary with time. Thus, equation (12.1) represents a response surface representing the consequences of application of any one, or a combination of K interventions. The objective of a control program is to reduce and maintain $p(t)$ at a low level for all times t beyond some initial start-up period. If the target is eradication, then obviously we want $p(t) = 0$ for all t after some prescribed intervention period.

Operating with a fixed set of K interventions, we assume that $\partial f / \partial x_i < 0$ for all i. This simply means that we anticipate that each of the interventions will reduce p, even if only by a very small amount. Changes in one or more x-variables can influence ecological conditions (one or more components of y)—and it is not necessarily the case that changes in components of y lead to a decrease in p. Indeed, a striking example of initiation of interventions leading to deleterious ecological conditions that increase an outcome, p, is the sharp rise in malaria mortality at Klang town between 1901 and 1902, consequential to the initiation of drain construction. This activity simultaneously created new *An. umbrosis* breeding sites and brought in a fresh cadre of exposed workers.

Within the context of malaria control, we do not necessarily know all of the x variables that should be included in equation (12.1). For example, if vaccines or the utilization of transgenic mosquitoes were to become viable forms of intervention, they could be included in a future specification of equation (12.1). In addition, we may not know the region R $(x_1, \ldots, x_K)$, which leads to the lowest level of p consistent with possible constraints on application of some of the interventions. Third, we do not know the functional form of f, and observations on the x variables and p are subject to moderate, and sometimes large, errors. This is precisely the set of conditions that are the setting for Evolutionary Operation (EVOP) put forth in the chemical engineering and statistics literatures in the 1960s (89).

In terms of Watson's FMS program, we begin with baseline values of p, y, and some of the x variables. Then one or more additional interventions were applied and we can examine the extent to which p is reduced. Further, associated with each of the interventions is a separate measure for rating its performance on each round of application. These performance-based ratings serve as a guide for which interventions to intensify at the next round of testing. An example of a table of performance-based ratings tuned to malaria control is shown in Table 12.2 (90). The values in the table are qualitatively consistent with judgments in Watson's decision making. However, the central point is that such a rating system was utilized in the FMS program, and with greater detail in Watson's malaria control programs at copper mines in Northern Rhodesia (29), as discussed in Section 12.4.4. Chemical sprays, intervention type 8 in Table 12.2, were not utilized in the FMS programs. However, they did appear with DDT in 1946, and this was a late-stage part of the Rhodesian copper mine programs.

In EVOP, as formulated by Box and Draper (89), decisions about new interventions to put in place are based on factorial experiments that involve two or more interventions. This allows the investigator to determine more points on the response surface represented in equation (12.1) than is possible when experimenting with one or a combination of interventions applied in a single run, as in the malaria control context. In the malaria context there is also no randomization of experimental protocols across communities, because identification of ecologically matched control communities is a virtual impossibility. Interestingly, this is also an issue in the chemical process control context: *"If runs can be randomized without too much difficulty, then this should be done. Our experience has been, however, that a random sequence is difficult to organize under actual production conditions and for the 2^2 design with added center points"* (89, p. 75).

Decisions about new intervention protocols in the malaria context are also based on regular surveillance with updating of the performance-based ratings chart. The analog of this is the information board in the EVOP set-up (89, p. 13). Beyond the routine assessments in performance-based ratings—and information boards—there is also an inevitable component of subjective judgment in deciding on new runs of interventions. Watson's explication of his own judgments in the FMS program and in the copper mine programs (12, 28, 29) provides considerable insight about the limits of automating decision processes in malaria control. This has an exact counterpart in chemical process control, as described in Box and Draper (89). Yet, Watson's programs ran successfully for

TABLE 12.2   Performance-Based Ratings of Malaria Control Programs According to Potential Problems Associated with Different Types of Interventions

Possible Problem with Intervention	Type of Intervention	Rating Index[a] Categories for Program Outcome			Factors Associated with Poor Performance
		Good	Fair	Poor	
Noncompliance	1. Antimalarial drugs	$a > 80$	20–80	<20	Inadequate community education; inadequate training of medical staff
Noncompliance	2. Bed nets	$b > 60$	30–60	<30	Inadequate community education
Noncompliance	3. House screening	$b > 85$	30–85	<30	Inadequate education and training
Inadequate coverage	4. Education	$c > 85$	50–85	<50	Poor contact strategy
Inadequate coverage	5. Drainage	$d < 10$	10–70	>70	Improper drainage; lack of trained field staff
Inadequate coverage	6. Vegetation clearance	$d < 10$	10–70	>70	Previously unsuspected breeding sites; incomplete cutting
Inadequate coverage	7. Larvicides	$d < 10$	10–70	>70	Low oil supply; insufficient effort by staff
Noncompliance	8. Chemical sprays	$b > 70$	30–70	<30	Inadequate education; poor relationship between program staff and household heads

Source: Adapted from Ref. 90.
[a] The Rating Index is defined as: a = percentage of compliance among those self-selecting for treatment; b = percentage of households in compliance with intervention; c = percentage of those with symptoms volunteering for treatment; d = maximum value of the ratio: [(mean mosquito catch in current year/mean mosquito catch in baseline year) × 100] observed over surveillance sites.

periods of 20 years, prior to its formalization by Box and Draper (89).

An interesting open question in the malaria context concerns the extent to which you can have an "Expert Computer System" that could be widely employed for ongoing decision making in the adaptive tuning and ongoing operation of a multiple intervention control program. The point of raising this is that if effective malaria control programs are to be dependent on a manager with Watson's remarkable acuity and insight, then this is a very tall order for the tropical disease control community. Answers here lie in the future, and will require a great deal of local experimentation.

12.6   CONCLUSION

An understanding of the mechanism of malaria transmission has been at hand since the late nineteenth century. Combining this information with a large literature on *Anopheles* vector ecology and bionomics, multiple-intervention malaria control programs were successfully implemented more than a century ago. There has also been dramatic progress in our understanding of the biology of vectors and parasites, much of it deriving from molecular level studies, as well as considerable advances in pharmacological and larvicide/adulticide interventions. These more recent scientific advances have been accompanied by a decline in use and emphasis on vector control strategies that played such a central role during the first 60 years of the twentieth century.

With an eye toward restoring a balance between preventive and curative strategies as components of malaria control programs, each finely tuned to local ecological conditions, we have presented five case examples of successful integrative programs from the past. One example, the Chinese National Malaria Control Program is a public sector initiative that continues to operate to the present day. Three of the examples derive from the British colonial experience in Asia and Africa, and one of them—malaria control to facilitate construction of the Panama Canal—is an unusual exception in terms of money invested and intensity of effort expended. Nevertheless, all of these programs had adaptive tuning of a multiplicity of interventions, together with ongoing performance-based ratings of the component tools, and measurement of outcomes as a basic mode of operation that led to successful suppression of malaria.

As a preliminary response to the contemporary criticism of vector control strategies as lacking a base of "scientific evidence," we have outlined a framework (building on EVOP concepts) for providing rigorous evaluation of the multiple-intervention programs of the past, as well as contemporary initiatives. A full development of EVOP tuned to malaria control would go beyond the scope of the present paper. However, our brief outline does at least open a new avenue for providing what has recently been regarded as the missing base of "scientific evidence" for a diverse set of vector control methods, integrated with drugs, bed nets, IRS, and IPTp. Expanding on this outline and linking it with cost-effectiveness analyses could be an important step toward bringing back the many positive features of integrative malaria control programs of the past, and redressing them in modern clothes.

REFERENCES

1. Bruce-Chwatt LJ. History of malaria from prehistory to eradication. In: Wernsdorfer WH, McGregor I, editors. *Malaria: Principles and Practice of Malariology*. Edinburgh; New York: Churchill Livingstone, 1988, pp. 1–59.

2. Bruce-Chwatt LJ, Zulueta J. *The Rise and Fall of Malaria in Europe: A Historic-Epidemiological Study*. Oxford; New York: Oxford University Press, 1980, p. 240

3. Desowitz RS. *The Malaria Capers: More Tales of Parasites and People, Research and Reality*. New York: W.W. Norton, 1991, p. 288

4. Hawass Z, Gad YZ, Ismail S, Khairat R, Fathalla D, et al. Ancestry and pathology in King Tutankhamen's family. *JAMA* 2010;303:638–647.

5. Bailey NTJ. *The Biomathematics of Malaria*. London: C. Griffin, 1982, p. 210.

6. Boyd MF. Historical review. In: Boyd MF, editor, *Malariology: A Comprehensive Survey of All Aspects of This Group of Diseases from a Global Standpoint*. Philadelphia: W. B. Saunders, 1949, pp. 3–25.

7. Slater LB. *War and Disease: Biomedical Research on Malaria in the Twentieth Century*. New Brunswick, NJ: Rutgers University Press, 2009, p. 249.

8. Celli A, Celli-Fraentzel A. *The History of Malaria in the Roman Campagna from Ancient Times*. London: Bale & Danielsson, 1933, p. 226.

9. Carter ED. Development narratives and the uses of ecology: malaria control in Northwest Argentina, 1890–1940. *J Hist Geog* 2007;33:619–650.

10. Snowden FM. *The Conquest of Malaria: Italy, 1900–1962*. New Haven: Yale University Press, 2006, p. 296.

11. Ross R. *The Prevention of Malaria*. New York: E.P. Dutton & Company, 1910, p. 669.

12. Watson MS. *Rural Sanitation in the Tropics. Being Notes and Observations in the Malay Archipelago, Panama and Other Lands*. London: J. Murray, 1915, p. 320.

13. Warrell DA, Gilles HM. *Essential Malariology*. London; New York: Arnold, 2002, p. 348.

14. Wernsdorfer WH, McGregor I. *Malaria: Principles and Practice of Malariology*. Edinburgh; New York: Churchill Livingstone, 1988, p. 1818.

15. Gardner MJ, et al. Genome sequence of the human malaria parasite *Plasmodium falciparum*. *Nature* 2002;419:498–511.

16. Holt RA, et al. The genome sequence of the malaria mosquito *Anopheles gambiae*. *Science* 2002;298:129–149.

17. Carlton JM, Adams JH, Silva JC, Bidwell SL, Lorenzi H, et al. Comparative genomics of the neglected human malaria parasite *Plasmodium vivax*. *Nature* 2008;455:757–763. Doi: 10.1038/nature07327.

18. World Health, Organization. *World Malaria Report 2009*. Geneva: World Health Organization, 2009, p. 77.

19. World Health, Organization. *Global Health Risks: Mortality and Burden of Disease Attributable to Selected Major Risks*. Geneva: World Health Organization, 2009, p. 62.

20. Hay SI, Guerra CA, Gething PW, Patil AP, Tatem AJ, et al. A world malaria map: *Plasmodium falciparum* endemicity in 2007. *PLoS Med* 2009;6(3):e1000048. Doi: 1000010.1001371/journal.pmed.1000048.

21. Bruce-Chwatt LJ. *Essential Malariology*. New York: Wiley, 1985, p. 452.

22. Costa DAM. *A Malária e Suas Diversas Modalidades Clínicas*. Rio de Janeiro: Imprensa a Vapor Lombaerts & Comp, 1885.

23. Hirsch A. *Geographical and Historical Pathology*. Creighton C, translator. London: The New Sydenham Society, 1883.

24. Watson M. The geographical aspects of malaria. *Geogr J* 1942;99:161–172.

25. Lysenko AJ, Semashko IN. Geography of malaria. A medical-geographical study of an ancient disease, 1968, p. 85. Moscow: Available through the Roll Back Malaria website: http://www.rollbackmalaria.org/docs/lysenko/lysenko_toc.htm. Accessed March 2010.

26. Hay SI, Guerra CA, Tatem AJ, Noor AM, Snow RW. The global distribution and population at risk of malaria: past, present, and future. *Lancet Infect Dis* 2004;4:327–336.

27. Konradsen F, van der Hoek W, Amerasinghe FP, Mutero C, Boelee E. Engineering and malaria control: learning from the past 100 years. *Acta Trop* 2004;89:99–108.

28. Watson M. *The Prevention of Malaria in the Federated Malay States: A Record of Twenty Years' Progress*. London: J. Murray, 1921, p. 381.

29. Watson SM. *African Highway: The Battle for Health in Central Africa*. London: J. Murray, 1953, p. 294.

30. Sallares R. *Malaria and Rome: A History of Malaria in Ancient Italy*. Oxford: Oxford University Press, 2002, p. 341.

31. Bradley DJ. Watson, Swellengrebel and species sanitation: environmental and ecological aspects. *Parassitologia* 1994;36:137–147.

32. Hamoudi A, Sachs JD. The changing global distribution of malaria: a review. Cambridge, MA: CID Working Paper No. 2. Center for International Development, 1999, p. 32.

33. UNICEF, World Health Organization, PATH. *Malaria Funding and Resource Utilization: The First Decade of Roll Back Malaria*. Geneva: World Health Organization, 2010, p. 98.

34. Roberts L, Enserink M. Did they really say ... eradication? *Science* 2007;318:1544–1545.

35. Tanner M, de Savigny D. Malaria eradication back on the table. *Bull WHO* 2008;86:82–83.

36. Roll Back Malaria, Partnership. *The Global Malaria Action Plan. For a Malaria-Free World*. Geneva: World Health Organization, 2008, p. 274.

37. Keiser J, Singer BH, Utzinger J. Reducing the burden of malaria in different eco-epidemiological settings with environmental management: a systematic review. *Lancet Infect Dis* 2005;5:695–708.

38. Baolin L.Environmental management for the control of rice-field-breeding mosquitoes in China. In: IRRI, editor. *Vector-Borne Disease Control in Humans through Rice Agroecosystem Management Proceedings of the Workshop on Research and Training Needs in the Field of integrated Vector-borne Disease Control in Riceland Agroecosystems of Developing Countries.* Philippines: International Rice Research Institute, in collaboration with the WHO/FAO/UNEP, 1988, pp. 111–121.

39. Tang LH, Qian HL, Xu SH. Malaria and its control in the People's Republic of China. *Southeast Asian J Trop Med Pub Health* 1991;22:467–476.

40. WHO. *Manual on Environmental Management for Mosquito Control: With Special Emphasis on Malaria Vectors.* Geneva: World Health Organization, 1982, p. 283.

41. Bruce-Chwatt LJ. Malaria and its control: present situation and future prospects. *Ann Rev Public Health* 1987;8:75–110.

42. Institute of Medicine, (U.S.). *Malaria: Obstacles and Opportunities. A Report of the Committee for the Study on Malaria Prevention and Control: Status Review and Alternative Strategies, Division of International Health, Institute of Medicine.* In: Oaks SC, Mitchell VS, Pearson GW, Carpenter CJ, editors, Washington, DC: National Academy Press, 1991, p. 309.

43. Cox-Singh J, Davis T, Lee K, Shamsul S, Matusop A, et al. *Plasmodium knowlesi* malaria in humans is widely distributed and potentially life threatening. *Clin Infect Dis* 2008;46: 165–171.

44. Eede PVd, Vythilingam I, Duc TN, Van HN, Hung LX, et al. *Plasmodium knowlesi* malaria in Vietnam: some clarifications. *Malaria J* 2010;9:20. Doi: 10.1186/1475-2875-9-20.

45. Jongwutiwes S, Putaporntip C, Iwasaki T, Sata T, Kanbara H. Naturally acquired *Plasmodium knowlesi* malaria in human, Thailand. *Emerg Infect Dis* 2004;10:2211–2213.

46. Benenson AS. *Control of Communicable Diseases in Man: An Official Report of the American Public Health Association.* Washington DC: American Public Health Association, 1990, p. 532.

47. Deane LM, Causey OR, Deane MP. *Studies on Brazilian Anophelines from the Northeast and Amazon Regions.* Baltimore: The Johns Hopkins Press, 1946, p. 50.

48. Consoli RAGB, Oliveira RL. *Principais Mosquitos de Importância Sanitária no Brasil.* Rio de Janeiro: Editora Fiocruz, 1998, p. 228.

49. McGregor IA, Wilson RJM. Specific immunity: acquired in man. In: Wernsdorfer WH, McGregor I, editors, *Malaria: Principles and Practice of Malariology.* Edinburgh; New York: Churchill Livingstone, 1988, pp. 559–619.

50. Molineaux L. The epidemiology of human malaria as an explanation of its distribution, including some implications for its control. In: Wernsdorfer WH, McGregor I editors, *Malaria: Principles and Practice of Malariology.* Edinburgh; New York: Churchill Livingstone, 1988, pp. 913–998.

51. Gupta S, Snow RW, Donnelly CA, Marsh K, Newbold C. Immunity to non-cerebral severe malaria is acquired after one or two infections. *Nat Med* 1999;5:340–343.

52. Webb JLA. *Humanity's Burden: A Global History of Malaria.* Cambridge; New York: Cambridge University Press, 2009, p. 236.

53. Cavalli-Sforza LL, Menozzi P, Piazza A. *The History and Geography of Human Genes.* Princeton, NJ: Princeton University Press, 1994, p. 518.

54. Ménard D, Barnadas C, Bouchier C, Henry-Halldin C, Gray LR, et al. *Plasmodium vivax* clinical malaria is commonly observed in Duffy-negative Malagasy people. *Proceedings of the National Academy of Sciences.* Published online before print March 15, 2010. Doi: 10.1073/pnas.0912496107.

55. Ryan JR, Stoute JA, Amon J, Dunton RF, Mtalib R, et al. Evidence for transmission of *Plasmodium vivax* among a duffy antigen negative population in Western Kenya. *Am J Trop Med Hyg* 2006;75:575–581.

56. Cavasini CE, Mattos LC, Couto AA, Bonini-Domingos CR, Valencia SH, et al. *Plasmodium vivax* infection among Duffy antigen-negative individuals from the Brazilian Amazon region: an exception? *Trans R Soc Trop Med Hyg* 2007;101: 1042–1044.

57. Cavasini CE, Mattos LCd, Couto AADA, Couto VSDA, Gollino Y, et al. Duffy blood group gene polymorphisms among malaria vivax patients in four areas of the Brazilian Amazon region. *Malaria J* 2007;6:167. Doi: 10.1186/1475-2875-6-167.

58. Molineaux L. The epidemiology of human malaria as an explanation of its distribution, including some implications for its control. In: Wernsdorfer WH, McGregor I, editors, *Malaria: Principles and Practice of Malariology.* Edinburgh; New York: Churchill Livingstone, 1988, pp. 913–997.

59. Ashley-Koch A, Yang Q, Olney RS. Sickle hemoglobin (*Hb S*) allele and sickle cell disease: a huge review. *Am J Epidemiol* 2000;151:839–845.

60. Castro MC, Monte-Mór RL, Sawyer DO, Singer BH. Malaria risk on the Amazon frontier. *Proc Natl Acad Sci* 2006;103: 2452–2457.

61. Ault SK. Effect of demographic patterns, social, structure, and human behavior on malaria. In: Service MW, editor, *Demography and Vector-Borne Diseases.* Boca Raton, FL: CRC Press, 1989, pp. 283–301.

62. Ault SK. Effect of malaria on demographic patterns, social structure, and human behavior. In: Service MW, editor, *Demography and Vector-Borne Diseases.* Boca Raton, FL: CRC Press, 1989, pp. 271–282.

63. Beier JC, Keating J, Githure JI, Macdonald MB, Impoinvil DE, et al. Integrated vector management for malaria control. *Malaria J* 2008;7:S4.

64. WHO. *Global Strategic Framework for Integrated Vector Management.* Geneva: World Health Organization. WHO/CDS/CPE/PVC, 2004, p. 15.

65. Cruz O. Prophylaxis of malaria in central and southern Brazil. In: Ross R, editor, *The Prevention of Malaria.* New York: E.P. Dutton & Company, 1910, pp. 390–399.

66. Carter R, Mendis KN, Roberts D. Spatial targeting of interventions against malaria. *Bull WHO* 2000;78:1401–1411.

67. Davis R, Icke G. Malaria: an on-line resource, 2002. Available at http://www.rph.wa.gov.au/malaria.html. Accessed February 2010.

68. Peters W. *Chemotherapy and Drug Resistance in Malaria*, Vol. 2 London; Orlando: Academic Press, 1987, p. 1085.

69. World Health, Organization. *Guidelines for the Treatment of Malaria*. Geneva: World Health Organization, 2010, p. 194.

70. Dondorp AM, Yeung S, White L, Nguon C, Day NPJ, et al. Artemisinin resistance: current status and scenarios for containment. *Nat Rev Microbiol* 2010;8:272–280. Doi: 210.1038/nrmicro2331.

71. Bjørndal A. Improving social policy and practice: knowledge matters. *Lancet* 2009;373:1829–1831.

72. Utzinger J, Tozan Y, Singer BH. Efficacy and cost-effectiveness of environmental management for malaria control. *Trop Med Int Health* 2001;6:677–687.

73. Utzinger J, Tozan Y, Doumani F, Singer BH. The economic payoffs of integrated malaria control in the Zambian copperbelt between 1930 and 1950. *Trop Med Int Health* 2002;7:657–677.

74. Haskin FJ. *The Panama Canal*. Garden City, NY: Doubleday Page & Company, 1913, p. 386.

75. Darling ST. Factors in the transmission and prevention of malaria in the Canal Zone. *Ann Trop Med Parasitol* 1910;4:179–223.

76. Darling ST. *Studies in Relation to Malaria*. Mount Hope: Isthmian Canal Commission, Laboratory of the Board of Health. Quartermaster's department, 1910, p. 42.

77. Gorgas WC. Malaria prevention on the Isthmus of Panama. In: Ross R editor, *The Prevention of Malaria*. New York: E.P. Dutton & Company, 1910, pp. 346–352.

78. Le Prince JA. Anti-malarial work on the Isthmus of Panama. In: Ross R, editor, *The Prevention of Malaria*. New York: E.P. Dutton & Company, 1910, pp. 353–368.

79. McArthur J. Malaria control in Borneo. *Lancet* 1953;262: 655–656.

80. McArthur J. The transmission of malaria in Borneo. *Trans R Soc Trop Med Hyg* 1947;40:537–558.

81. McArthur J. The control of malaria in Borneo (an account of the Tambunan experiment). *Trans R Soc Trop Med Hyg* 1954;48:234–241.

82. McArthur J. The importance of *Anopheles leucosphyrus*. *Trans R Soc Trop Med Hyg* 1951;44:683–694.

83. McArthur J. Malaria in Borneo: an account of the work of the Malaria Research Department, North Borneo, 1939–42. Labuan: Malaria Research Headquarters, 1948, p. 250.

84. McArthur J. Malaria control by methods other than insecticides. Malaria carried by *Anopheles leucosphyrus.* The World Health Organization, 1953, p. 4. WHO/MAL/84. Available at whqlibdoc.who.int/malaria/WHO_MAL_84.pdf.   Accessed March 2010.

85. Tang LH. Progress in malaria control in China. *Chinese Med J* 2000;113:89–92.

86. Sleigh AC, Lin X-L, Jackson S, Shang L. Resurgence of *vivax* malaria in Henan Province, *China. Bull WHO* 1998;76: 265–270.

87. Jackson S, Sleigh AC, Lin X-L. Cost of malaria control in China: Henan's consolidation programme from community and government perspectives. *Bull WHO* 2002;80:653–659.

88. Yang H, Yang Y, Yang P, Li X, Gao B, et al. Monitoring *Plasmodium falciparum* chloroquine resistance in Yunnan Province, China, 1981–2006. *Acta Trop* 2008;108:44–49.

89. Box GEP, Draper NR. *Evolutionary Operation: A Statistical Method for Process Improvement*. New York: Wiley, 1969, p. 237.

90. Singer BH. Self-selection and performance-based ratings: a case study in program evaluation. In: Wainer H, editor. *Drawing Inferences from Self-Selected Samples*. New York: Springer-Verlag, 1986, pp. 29–49.

13

SCHISTOSOMIASIS

Pascal Magnussen, Birgitte Jyding Vennervald, and Jens Aagaard-Hansen

13.1 INTRODUCTION

Schistosomiasis, a freshwater-related, helminth infection, is considered the second most important parasitic infection after malaria in terms of public health impact (1). It is endemic in almost 80 countries in the Americas, Africa, and Asia, with an especially high burden of infection and disease on the African continent especially in sub-Saharan countries. Approximately 600–700 million people are estimated to be at risk of infection and approximately 160–200 million individuals to be infected. Mortality figures for schistosomiasis are difficult to estimate because death due to schistosomiasis rarely takes place in health facilities, but in sub-Saharan Africa figures between 20,000 and 200,000 per year have been suggested (2). Morbidity figures have recently been reassessed and disability estimates are that the global cost to a *Schistosoma*-infected individual is at least 2–15% chronic disability compared to previous figures of around 0.5% (1). The main schistosome species causing human infection are *Schistosoma mansoni*, *S. haematobium*, and *S. intercalatum* in Africa and the Middle East, the Americas, and the Caribbean, and *S. japonicum* and *S. mekongi* in China, the Philippines, Laos, Kampuchea, and South Thailand. For the African schistosome species, humans are the most important definitive host. *S. japonicum* and *S. mekongi* are both zoonotic parasites, with a wide range of definitive hosts such as pigs, water buffaloes, and various rodents.

In the following sections various aspects of schistosomiasis transmission, pathogenesis, epidemiology, clinical manifestations, diagnosis, treatment, social dynamics influencing transmission, and various control measures will be described.

13.2 SCHISTOSOMIASIS TRANSMISSION CYCLE

The schistosome life cycle is shown in diagrammatic form in Fig. 13.1. The parasite alternates between two hosts: the definitive human host, in whom pairing of adult worms and sexual replication occurs; and the intermediate snail host, in which asexual replication occurs (3).

Parasite eggs are excreted via the feces or urine (point 1). The eggs hatch in freshwater to release miracidia (point 2), which locate vector snails (point 3), where asexual multiplication with formation of sporocysts occurs (point 4). *S. mansoni* is transmitted by snails of the genus *Biomphalaria*, whereas *S. haematobium* and *S. intercalatum* by the genus *Bulinus*. *S. japonicum* and *S. mekongi* are transmitted by amphibious snails of the species *Oncomelania* and *Tricula*, respectively. After approximately 5 weeks, large numbers of cercariae (point 5), the infectious larval stage, are shed into the water and people are infected when the cercariae penetrate intact skin (point 6). People come into contact with water containing infectious larvae for a wide range of reasons, including washing themselves or their clothes and collecting water for drinking or cooking or for irrigation. Among children, playing or swimming in water is an important activity, while some adults, such as fishermen, car washers, canal cleaners, or people working in irrigation schemes or rice fields, may have a high occupational exposure (4). When the cercariae have penetrated the skin they transform into other larval forms, the schistosomula (point 7), which are either male or female and migrate in the body via the blood vessels (point 8). They pass through the lungs via the capillaries and reach the portal venous system where maturation takes place (point 9), and after mating, the worm pair travels to the small blood vessels around the intestines (point 10A, point 10B), or

Water and Sanitation-Related Diseases and the Environment: Challenges, Interventions, and Preventive Measures, First Edition.
Edited by Janine M. H. Selendy.
© 2011 Wiley-Blackwell. Published 2011 by John Wiley & Sons, Inc.

Schistosomiasis

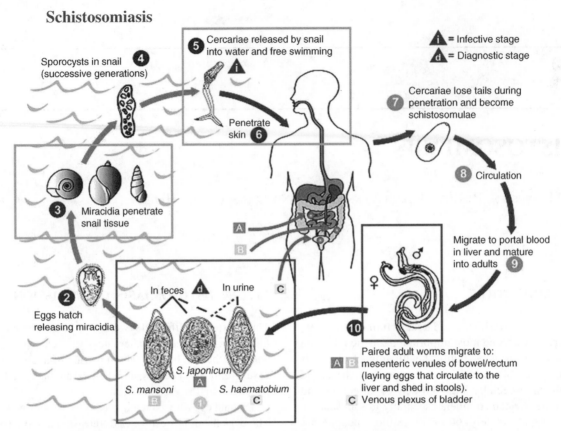

FIGURE 13.1 The Transmission Cycle of Schistosomiasis. Courtesy of www.cdc.gov. The numbers in the text correspond to numbers indicated on the drawing. (*See insert for color representation of this figure.*)

urinary bladder (point 10C). Once they have established themselves, adult worms have an estimated mean lifespan of 3 years (*S. haematobium*) to 7 years (*S. mansoni*). The female worms produce between 300 and 3000 eggs per day depending on the species. To reach the lumen of the intestine or urinary bladder and the exterior via the feces (point 1A, point 1B) or urine (point 1C), eggs have to pass through the wall of the blood vessel as well as the intestinal or bladder wall respectively (5).

13.3 EPIDEMIOLOGY

In endemic countries children become infected when they start having contact with infected water. In many instances, young children begin to acquire the infection from the age of 3 or 4 years, but sometimes infections are found in even younger children. These very young children may sit in the shallow water where they are exposed to infection while the mother does her domestic chores such as washing clothes or utensils (6). Both prevalence (the proportion of infected individuals) and intensity (the mean egg output) then rise sharply and in most endemic areas the proportion of infected

individuals will be highest in the group aged 10–15 years, after which it reaches a plateau or declines among adults (7). This pattern is caused by various factors. First, children generally have more frequent, prolonged, and extensive contact with water than adults. However, water contact patterns vary markedly between different locations, and some communities show very high levels of occupational exposure among adults. Second, several studies have indicated that acquired immunity to schistosomiasis gradually develops among adults, so that they are reinfected to a much lesser extent than children (7). Third, a progressive spontaneous death of adult worms from early infections acquired during childhood takes place over the years.

Local variations of infection patterns are pronounced and strongly influenced by gender roles (8). Thus, in communities where women are secluded to the domestic area, for instance for religious reasons, they are relatively protected; as opposed to communities where they play an active role in agricultural work and fetching of water. Also local perceptions of "appropriate behavior" for boys and girls may influence the likelihood that they swim in water bodies thereby exposing themselves to the infection. Numerous anthropological studies have shown the many local varieties and adaptation patterns, which are important to know in order to plan appropriate control programs (8–10).

13.4 PATHOGENESIS

The eggs are the major cause of pathology in schistosomiasis. Approximately half of the eggs are not excreted with feces or urine but are retained in the tissue close to the adult worm pairs or swept into the portal circulation, where they lodge in the capillaries of the liver. A granulomatous inflammatory response is formed around the egg in the tissue. The inflammation is induced by antigens secreted by the miracidium inside the egg, and this response is followed by fibrosis. Granuloma formation is a crucial feature in schistosomiasis pathology. Depending on the localization of the adult worms, granulomas will form in various organs, primarily the liver, large intestines including the rectum, and the genitourinary tract (11).

13.5 CLINICAL MANIFESTATIONS

Most infected people living in endemic areas have few overt symptoms but even a low intensity infection may be associated with anemia and have an impact on growth in children (12). The clinical manifestations depend on the stage of infection. Cercarial penetration of the skin can provoke an immediate hypersensitivity reaction with intense itching and a rash, caused by an inflammatory reaction to cercaria. This is known as cercarial dermatitis or "swimmer's itch." A special and often severe type of cercarial dermatitis is seen after exposure to cercariae from bird schistosomes, for example, *Trichobilharzia*, and this condition is regarded as a public health problem in areas where freshwater lakes are used for recreational purposes (13).

The migration of the young schistosomes does not normally give rise to symptoms but when egg production starts, an acute condition known as Katayama fever can be seen. It presents with high fever, eosinophilia, headache, and gastrointestinal symptoms, especially in people coming from nonendemic areas. It can resemble acute malaria, typhoid fever, or other febrile illnesses, but eosinophilia indicates Katayama fever (5).

It is the accumulation of eggs in the tissue and the resulting tissue inflammation and granuloma formation that causes the organ specific morbidity in schistosomiasis. The signs and symptoms seen in S. *haematobium* infection is the result of deposition of eggs in the urinary bladder wall, ureters, and genital organs. When the infection is established, the following two stages can be recognized:

- An active stage with egg deposition in the genitourinary tract and egg excretion in urine. This is accompanied by blood (hematuria) and protein (proteinuria) in the urine. The blood in the urine is often clearly visible with the naked eye. This is mainly seen in children, adolescents, and younger adults.
- A chronic stage with sparse or absent urinary egg excretion but with urinary tract pathology and sometimes genital lesions.

Eggs are deposited in large conglomerates in the bladder wall from where they are sloughed off into the urine with ulceration and bleeding as a consequence. Hence, hematuria is often seen in S. *haematobium* infection, especially in the terminal portion of the urine voided. Eggs retained in the bladder wall cause inflammation and thickening of the wall. When the granulomas are situated in the ureters or near the ureteric openings into the urinary bladder, these can be constricted or even blocked, leading to obstructive uropathy with development of hydroureter and hydronephrosis. In some cases, eggs can be found in other organs as well, such as the male and female genital organs, causing genital schistosomiasis, a condition that has received increasing attention because it has been hypothesized that genital lesions and inflammation caused by schistosome eggs may increase the transmission of HIV (14). In rare cases, eggs can lodge in the central nervous system or the spinal cord and give rise to severe neurological symptoms as a result of inflammation and edema. Infection with S. *haematobium* is associated with development of a specific type of cancer in the urinary bladder (squamous cell carcinoma) (15).

Intestinal schistosomiasis is associated with granuloma formation, inflammation, and fibrosis, primarily in the large intestine and abdominal pain and bloody diarrhea can be seen especially in high-intensity infections. Eggs that fail to traverse the wall of the blood vessel will be carried by the venous blood to the liver, and end in the small venuoles of the portal tract. Here they give rise to granulomas, which may result in enlargement of liver and spleen. Granulomas can, in due course, lead to replacement by fibrous tissue (collagen) and result in periportal fibrosis (Symmers' fibrosis), a typical lesion in schistosomiasis. Hepatosplenic schistosomiasis is a chronic manifestation of S. *mansoni* and S. *japonicum* infection. The term covers two clinical entities:

- Early inflammatory hepatosplenic disease with enlargement of liver and spleen, but generally no fibrosis. This is mainly seen in children and adolescents and the severity of organ enlargement is related to intensity of infection. This type of hepatosplenic disease may also be associated with concomitant chronic exposure to malaria.
- Hepatosplenic disease with periportal fibrosis, which is mainly seen later in life, generally in young and middle-aged adults with long-standing intense exposure to

infection. If organ enlargement and fibrosis is pronounced, portal hypertension may develop, and in severe cases esophageal varices can occur. These can give rise to severe, life-threatening bleeding, where the infected person vomits large quantities of fresh blood (hematemesis). This is the most common cause of death in late-stage intestinal schistosomiasis. The frequency varies from area to area.

Beside the specific pathology and morbidity related to the species of schistosomes infecting humans (*S. haematobium* in the urinary tract, *S. mansoni*, *S. japonicum*, *S. mekongi*, and *S. intercalatum* in the gastrointestinal tract), there are more general and subtle effects on health such as anemia and growth retardation, which may lead to reduced school performance among children as well as impaired physical performance among adults (16).

13.6 DIAGNOSIS

Diagnosis of schistosomiasis is still made primarily by detection of eggs in feces or urine. The most widely used methods are urine filtration where 10 mL of urine is filtered where after the eggs trapped on the filter are counted or the Kato–Katz method for quantification of eggs in 25–50 mg of stool. The eggs are easily recognized and both methods are highly specific but they have a low sensitivity—especially the Kato–Katz method, which may easily underdiagnose low-intensity infections (17). Serological testing with detection of specific antibodies can be used in travelers from nonendemic areas returning with an infection, whereas antibody determination is of limited value in endemic areas with continued transmission. Another diagnostic approach is detection of circulating worm antigens in serum or urine and a simple reagent strip test for detection of worm antigens in urine has been developed (18). Diagnosis is another issue directly influenced by sociocultural and economic factors (8). As the diagnosis is based on either urine or stool samples, local perceptions of privacy and embarrassment may in some cases lead to underdiagnosis, simply because certain groups (e.g., women) shy away. To the extent that the health care providers charge for the diagnostic procedures, or the clinics are situated far away, this may also inhibit diagnosis due to lack of funds. Also with respect to diagnosis, the local variations in contextual factors should be taken into consideration.

Detection of blood in urine can be used as an indirect indicator of *S. haematobium* infection, either by direct visual inspection or by use of urine analysis reagent strips. This has been employed successfully, as a means of cost-effective community diagnosis—especially in control programs directed toward school children (19). No similar diagnostic procedure exists for intestinal schistosomiasis.

13.7 SOCIAL DYNAMICS INFLUENCING SCHISTOSOMIASIS TRANSMISSION

Population movements are relevant for the spread of schistosomiasis in many ways. The main mechanism is that movement leads to fresh exposure to the infectious agents—either because the migrating population moves into new areas and gets in touch with infective populations or environments, or because the migrating population carries the infection and brings it to hitherto unexposed populations at their point of arrival. People move for a multitude of different reasons, for example, pilgrimage, labor migration, pastoralism, trade, or tourism. It may also be caused by forced resettlement or refuge due to either war or natural disasters. All these aspects of population movements are directly linked to the spread of schistosomiasis as well as other infectious diseases. In some cases, snails containing the parasites may also be transported from one place to another, thereby transferring the infection.

Urbanization is rapidly increasing also in schistosomiasis endemic countries. This phenomenon usually entails a combination of poor infrastructure in terms of water supply, sewage, and sanitation, and people often migrate to and from rural disease endemic areas. This combination of factors may create local environments suitable for transmission of schistosomiasis.

Large-scale irrigation and hydroelectric power schemes play an important role for new establishment of, or increase in existing, transmission (20). To mitigate the negative effects of such water management projects, it has become increasingly accepted that health impact assessments are needed as part of the project development process. Major health problems may ensue when people migrate to newly developed irrigation schemes, especially when coming from non-schistosomiasis endemic areas. Schistosomiasis with epidemic levels of transmission, high morbidity, and difficulties in treatment due to high infection intensities and rapid reinfection has occurred following such influx of previously non-exposed populations (21, 22). Sustaining control in such situations has also proven difficult due to inadequate health infrastructure and public health resources. However, once established, the transmission and continued exposure leads to fast development of partial immunity, thereby reducing the level of reinfection and intensity (23). Another environmental aspect of schistosomiasis relates to climate change. The distribution of *Oncomelania hupensis*, the intermediate snail host for *S. japonicum*, is limited to areas where the mean January temperature is above 0°C. Furthermore, the development of the schistosome stages in the snail is arrested at temperatures below 15°C. Thus, the expected global warming may lead to an increase in the distribution of host snails and favor parasite development and result in significant increases in populations at risk especially in China (24).

On a more general level, poverty constitutes an important social determinant for schistosomiasis in a multitude of ways. Poor people are likely to be more exposed due to more risky work and inadequate access to piped water, sanitary facilities, and medical services (25).

13.8 CONTROL

With the discovery of the complete life cycle of schistosomiasis during the period 1902–1915 (3) and the subsequent description of the various species of intermediate vector snails involved in the life cycle as well as the epidemiology and the pathogenesis, it became possible to put forward strategies for schistosomiasis control. The transmission cycle (Fig. 13.1) illustrates the various places where the transmission can be interrupted and the disease thereby controlled.

Transmission of schistosomiasis is closely linked to human practices related to contact with freshwater bodies and sanitation practices. The following parameters are prerequisites for the transmission of schistosomiasis:

- Freshwater bodies harboring efficient snail intermediate hosts. It is the distribution of the snails that determines the distribution of schistosomiasis in any given area.
- A human population susceptible to infections with schistosomes.
- Poor sanitation leading to contamination of snail-infested water with feces and urine containing schistosome eggs.
- Existence of an infectious water source leading the population to have direct contact with contaminated water for work, washing, and leisure purposes.

In principle, schistosomiasis should be a totally preventable infection. However, poor access to safe water in most rural areas of sub-Saharan Africa and elsewhere in developing countries in South America and East Asia, combined with lack of adequate and appropriate sanitation keep transmission at a high level in many endemic areas. As can be seen in Fig. 13.1, measures preventing transmission can be aimed basically at preventing or reducing contact with infested water (Fig. 13.1, green boxes) and/or preventing or reducing contamination of water with schistosome eggs (Fig. 13.1, blue boxes). Both aims can be obtained using disease-specific and nondisease-specific control strategies and tools, as shown in Table 13.1.

Initial attempts to control schistosomiasis focused on snail control, health education, and in some instances provision of safe water supply and sanitation as part of more extensive development projects. Especially in Egypt and Sudan, snail control in irrigation canals remained the first-line strategy for decades. Major problems encountered were high cost of

TABLE 13.1 Classification of Control Strategies

Strategy	Disease Specific	Nondisease Specific
Prevent or reduce contact with infected water	Control of intermediate vector	Adequate safe water supply for all domestic and leisure purposes and health education
Prevent or reduce contamination of water with schistosome eggs	Morbidity reduction by chemotherapy	Improve sanitation (latrines) and health education

Source: Adapted from Ref 3.

molluscicides and the need of very regular retreatment of water bodies due to reinvasion of snails. Furthermore molluscicides were environmentally problematic because they also affected other species such as fish.

With the arrival during 1970–1980 of safe and effective anti-schistosomicidal drugs (metrifonate for *S. haematobium* and oxamniquine for *S. mansoni* and later praziquantel, which is effective against all species of schistosomes (26)), rational chemotherapy became possible at the individual level. With praziquantel, which can be given as a single dose, the concept of large-scale treatment targeted to at risk populations or even whole populations developed. Due to the epidemiology of schistosomiasis, the greatest risk of high-intensity infection and pathology is found in school-age children and this group has thus routinely been targeted for mass drug administration programs. However, there are high-risk occupational groups among adults such as fishermen, car washers, and irrigation workers who should also be targeted for mass chemotherapy. It should be borne in mind that a control strategy, which depends mostly on one drug, praziquantel, is very vulnerable, because there is a risk that the parasites develop resistance toward the drug.

Several studies have shown that repeated single-dose treatments with praziquantel once a year have an impact on morbidity (27). Due to reinfection, prevalence of infection remains high but intensity of infection is significantly reduced and as morbidity is related to infection intensity, a prolonged effect on morbidity can be sustained with repeated annual treatments. Special considerations are needed regarding the zoonotic varieties of schistosomiasis caused by *S. japonicum* and *S. mekongi* where other mammals such as pigs and water buffaloes are also definitive hosts. In those endemic areas, pharmaceutical treatment of animals or use of tractors instead of buffaloes to pull ploughs can be an integrated part of the control programs (28).

In the short term, morbidity control will remain the cornerstone of schistosomiasis control often disseminated by mass drug administration programs. Though usually effective in the short term, they may entail long-term challenges. Partly, such programs depend on continued external funding. Changing priorities or financial difficulties of the donors may interrupt the program activities, thereby threatening their sustainability. Partly, there can be significant problems in securing high coverage if a strategy of mass drug administration is adopted unless the involved populations take ownership of the strategy and get involved in its implementation. Poor communication and insufficient sensitization of target populations and lack of understanding of local perceptions, attitudes, and practices may lead to non-acceptance and resistance against a program (29–31). In control programs with a more community directed approach, involving village health workers (32), it will be possible to improve existing strategies, taking factors into account such as timing of treatment according to transmission season and including short interval (weeks) spaced treatments to ensure that juvenile worms get treated, thereby reducing morbidity (33, 34).

In the long term, schistosomiasis control should be based on the combination of regular treatment of high-risk population segments combined with an increasing effort to provide safe water and sanitation as well as appropriate health education approaches. Health impact assessments should be undertaken before major water management projects are implemented in order to provide information predicting changes in the epidemiology of existing schistosomiasis or provide bases for interventions to prevent introduction of schistosomiasis transmission (35).

Health education is a cross-cutting control activity, which underpins most of the other interventions. It is meant to motivate and guide the patients and targeted population segments in safe practices (avoiding risky water contact and using latrines) and the most appropriate ways of utilizing health care services with respect to diagnosis and treatment. However, numerous anthropological studies have shown that lay people's perceptions of illness are often quite different from those promoted by the biomedical establishment and treatment-seeking practices often lead to self-treatment or take the patients to traditional healers. For instance, in many disease endemic countries the symptom, hematuria, which is among teenagers seen as a very certain sign of infection, is locally perceived as a normal sign of coming to age. Consequently, it is mandatory that health education interventions be based on a thorough understanding of local perspectives, so that they can be formulated accordingly.

In order to ensure sustainability, schistosomiasis control should be undertaken within the framework of existing health services and integrated with other disease control programs. The importance of the way in which health care systems are organized is clearly illustrated by a case from China (36). Whereas China achieved remarkable results in schistosomiasis control based on prevention and mass campaigns during the communist economic system, the neoliberal reorganization has introduced an incentive structure for the schistosomiasis control centers that encourages the health staff to emphasize curative services—often with unnecessarily expensive drugs for extended periods.

Several combinations of environmental and snail control, water, sanitation, and health education approaches with or without treatment has been tested in small scale with varying results. However, sustainability has been difficult to ensure on the long term in areas of high transmission.

Historically, schistosomiasis control has thus gone through a variety of different strategies. Initially it focused on vector snail control combined with health education and provision of safe water and sanitation. Later focus turned to morbidity reduction using regular administration of safe single-dose drugs (37–40). Throughout the history of schistosomiasis control and up till now, the implemented strategies have been focused on biomedical and technical aspects (vector control and morbidity control), while the social context of schistosomiasis has played a minor role for strategy and tools development (8).

13.9   CONCLUSION

Schistosomiasis is considered to be the second most important parasitic infection after malaria in terms of public health impact and poor access to safe water combined with lack of adequate and appropriate sanitation keep transmission at a high level in many endemic areas. Based on recommendations from WHO, population-based treatment with praziquantel is now the main component in most national control programs. However, in the absence of ecological and behavioral changes, this strategy has minimal effect on transmission. This means that regular retreatment of affected populations has to be continued for an unknown period and sustainability becomes a key element in treatment-based control. The transmission pattern and control of schistosomiasis is complex due to the intricate interaction of biological, environmental, and sociocultural factors. Consequently, control measures should adapt to local conditions rather than applying simplistic, universal strategies. Controlling schistosomiasis is not just a matter of distributing drugs; strengthening the local health systems so they are able to take care of patients and to integrate sustainable control measures are equally important and the social context of schistosomiasis should not be ignored. Schistosomiasis can in principle be controlled by behavioral changes, sanitation, and safe water supply, but these measures will solve the problem only if poverty is eliminated.

REFERENCES

1. King HC, Dickmann K, Tisch D. Reassessment of the cost of chronic helminth infection: a meta-analysis of disability-relat-

ed outcomes in endemic schistosomiasis. *Lancet* 2005;365:1561–1569.

2. van der Werf MJ, de Vlas SJ, Brooker S, Looman CW, Nagelkerke NJ, Habbema JD, Engels D. Quantification of clinical morbidity associated with schistosome infection in sub-Saharan Africa. *Acta Trop* 2003;86:125–139.

3. Jordan P, Webbe G, Sturrock RF, editors. *Human Schistosomiasis*. Wallingford, UK: CAB International, 1993.

4. Tameim O, Abdu KM, el Gaddal AA, Jobin WR. Protection of Sudanese irrigation workers from schistosome infections by a shift to earlier working hours. *J Trop Med Hyg* 1985;88:125–130.

5. Gryseels B, Polman K, Clerinx J, Kestens L. Human schistosomiasis. *Lancet* 2006;368:1106–1118.

6. Odogwu SE, Ramamurthy NK, Kabatereine NB, Kazibwe F, Tukahebwa E, Webster JP, Fenwick A, Stothard JR, *Schistosoma mansoni* in infants (aged<3 years) along the Ugandan shoreline of Lake Victoria. *Ann Trop Med Parasitol* 2006;100:315–326.

7. Kabatereine NB, Vennervald BJ, Ouma JH, Kemijumbi J, Butterworth AE, Dunne DW, Fulford AJ. Adult resistance to *Schistosomiasis mansoni*: age-dependence of reinfection remains constant in communities with diverse exposure patterns. *Parasitology* 1999;118:101–105.

8. Bruun B, Aagaard-Hansen J. *The Social Context of Schistosomiasis and Its Control. An Introduction and Annotated Bibliography*. Geneva: UNICEF/UNDP/World Bank/WHO Special Programme for Research & Training in Tropical Diseases (TDR), 2008.

9. Parker M. Bilharzia and the boys: questioning common assumptions. *Soc Sci Med* 1993;37(4):481–492.

10. El Katsha S, Watts S. *Gender, Behavior, and Health: Schistosomiasis Transmission and Control in Rural Egypt*. Cairo: American University in Cairo Press, 2002.

11. Mahmoud AAF, Pasvol G, Hoffman SL, editors. Schistosomiasis. In: *Tropical Medicine. Science and Practice*, Vol. 3. London: Imperial College Press, 2001.

12. King CH, Dangerfield-Cha M. The unacknowledged impact of chronic schistosomiasis. *Chronic Illness* 2008;4:65–79.

13. Schets FM, Lodder WJ, van Duynhoven YT, de Roda Husman AM. Cercarial dermatitis in the Netherlands caused by *Trichobilharzia* spp. *J Water Health* 2008;6(2):187–195.

14. Kjetland EF, Ndhlovu PD, Gomo E, Mduluza T, Midzi N, Gwanzura L, Mason PR, Sandvik L, Friis H, Gundersen SG. Association between genital schistosomiasis and HIV in rural Zimbabwean women. *AIDS* 2006;20(4):593–600.

15. Bouvard V, Baan R, Straif K, Grosse Y, Secretan B, El Ghissassi F, Benbrahim-Tallaa L, Guha N, Freeman C, Galichet L, Cogliano V. WHO International Agency for Research on Cancer Monograph Working Group. A review of human carcinogens–Part B: biological agents. *Lancet Oncol.* 2009;10(4):321–322.

16. World, Bank. *Disease Control Priorities in Developing Countries*, Chapter 24, 2nd edition. New York: A co-publication of the World Bank and Oxford University Press, 2006.

17. Enk MJ, Lima AC, Drummond SC, Schall VT, Coelho PM. The effect of the number of stool samples on the observed prevalence and the infection intensity with *Schistosoma mansoni* among a population in an area of low transmission. *Acta Trop* 2008;108(2–3):222–228.

18. van Dam GJ, Wichers JH, Ferreira TM, Ghati D, van Amerongen A, Deelder AM. Diagnosis of schistosomiasis by reagent strip test for detection of circulating cathodic antigen. *J Clin Microbiol* 2004;42:5458–5461.

19. Red Urine Study Group. Identification of high-risk communities for schistosomiasis in Africa: a multicountry study. Geneva: World Health Organization, Special Programme for Research & Training in Tropical Diseases (Social and Economic Research Projects Report, No. 15; TDR/SER/PRS/15), 1995.

20. Steinmann P, Keiser J, Bos R, Tanner M, Utzinger J. Schistosomiasis and water resources development: systematic review, meta-analysis, and estimates of people at risk. *Lancet Infect Dis* 2006;6:411–425.

21. Gryseels B, Stelma FF, Talla I, van Dam GJ, Polman K, Sow S, Diaw M, Sturrock RF, Doehring-Schwerdtfeger E, Kardorff R, et al. Epidemiology, immunology and chemotherapy of *Schistosoma mansoni* infections in a recently exposed community in Senegal. *Trop Geogr Med* 1994;46 (4 Spec No.):209–219.

22. Stelma FF, Talla I, Sow S, Kongs A, Niang M, Polman K, Deelder AM, Gryseels B. Efficacy and side effects of praziquantel in an epidemic focus of *Schistosoma mansoni*. *Am J Trop Med Hyg* 1995;53(2):167–170.

23. Polman K, Stelma FF, Le Cessie S, De Vlas SJ, Falcão Ferreira STM, Talla I, Deelder AM, Gryseels B. Evaluation of the patterns of *Schistosoma mansoni* infection and re-infection in Senegal, from faecal egg counts and serum concentrations of circulating anodic antigen. *Ann Trop Med Parasitol* 2002;96(7):679–689.

24. Zhou X-N, Yang G-J, Yang K, Wang X-H, Hong Q-B, Sun L-P, Malone JB, Kristensen TK, Bergquist NR, Utzinger J. Potential impact of climate change on schistosomiasis transmission in China. *Am J Trop Med Hyg* 2008;78(2):188–194.

25. Commission on Social Determinants of Health. Closing the Gap in a Generation: Health Equity through Action on the Social Determinants of Health. Final Report. Geneva: World Health Organization, 2008.

26. WHO Model Prescribing, Information. *Drugs Used in Parasitic Diseases*, 2nd edition. Geneva: World Health Organization, 1995.

27. Magnussen P. Treatment and re-treatment strategies for schistosomiasis control in different epidemiological settings: a review of 10 years' experiences. *Acta Trop* 2003;86:243–254.

28. Zheng J, Zheng Q, Wang X, Hua Z. Influence of livestock husbandry on schistosomiasis transmission in mountainous regions of Yunnan province. *SE Asian J Trop Med* 1997;28(2):291–295.

29. Mwanga JR, Magnussen P, Mugashe CL, Gabone RM, Aagaard-Hansen J. Schistosomiasis related perceptions, attitudes and treatment seeking practices in Magu district, Tanzania: public health implications. *J Biosoc Sci* 2004;36:63–81.

30. Parker M, Allen T, Hastings J. Resisting control of neglected tropical diseases: dilemmas in the mass treatment of schistosomiasis and soil-transmitted helminths in north-west Uganda. *J Biosoc Sci* 2008;40:161–181.

31. Fleming FM, Fenwick A, Tukahebwa EM, Lubanga RGN, Namwangye H, Zaramba S, Kabatereine NB. Process evaluation of schistosomiasis control in Uganda, 2003–2007: perceptions, attitudes and constraints of a national programme. *Parasitology* 2009;136(13):1759–1769.

32. Massa K, Olsen A, Sheshe A, Ntakamulenga R, Ndawi B, Magnussen P. Can coverage of schistosomiasis and soil-transmitted helminthiasis control programmes targeting school-aged children be improved? *New approaches. Parasitology* 2009;136:1781–1788.

33. Augusto G, Magnussen P, Kristensen T, Appleton C, Vennervald BJ. The influence of transmission season on parasitological cure rates and intensity of infection after praziquantel treatment of Schistosoma haematobium in Mozambique. *Parasitology* 2009;136:1771–1779.

34. Sacko M, Magnussen P, Traoré M, Landouré A, Doucouré A, Reimert CM, Vennervald BJ. The effect of single dose versus two doses of praziquantel on Schistosomiasis haematobium infection and pathology among school-aged children in Mali. *Parasitology* 2009;136:1851–1857.

35. Lock K. Health impact assessment. *BMJ* 2000;320:1395–1398.

36. Bian Y, Sun Q, Zhao Z, Blas E. Market reform: a challenge to public health—the case of schistosomiasis control in China. *Int J Health Planning and Management* 2004;19: 79–94.

37. Jordan P, Rosenfield PL. Schistosomiasis control: past, present, and future. *Annu Rev Public Health* 1983;4:311–334.

38. WHO Technical Report Series. Epidemiology and Control of Schistosomiasis. Technical Report, Series No. 643. Geneva: World Health Organization, 1980.

39. Report of a WHO Expert Committee. The Control of Schistosomiasis. Technical Report, Series No. 728. Geneva: World Health Organization, 1985.

40. Report of a WHO Expert Committee. Prevention and Control of Schistosomiasis and Soil-Transmitted Helminthiasis. Technical Report, Series No. 912. Geneva: World Health Organization, 2001.

14

TRACHOMA

JOSEPH A. COOK AND SILVIO P. MARIOTTI

14.1 INTRODUCTION

Trachoma, a chronic bacterial keratoconjunctivitis, is the most common infectious cause of blindness. One hundred years ago trachoma was the most important cause of blindness of any etiology and the cause of the founding of many eye hospitals. Trachoma has always been linked to poverty, poor hygiene, and availability of water. Fortunately, with improving socioeconomic conditions, including especially access to water and improved hygiene, trachoma as a cause of blindness is absent from industrialized countries. It remains a significant cause of blindness in the poor, dry developing countries and the productivity costs of trachoma-related reduced vision and blindness have been estimated USD 5.3 billion annually in 2003 dollars (1). Population-based surveys provided recent information for 42 out of 57 endemic countries; 40.6 million people are estimated to be suffering from active trachoma and 8.2 million are estimated to have trichiasis, the trachomatous condition that leads to blindness (2). The current estimate of prevalence of trachoma is lower than the previous World Health Organization (WHO) estimates: This can be explained by the success in implementing a control strategy, by more accurate data, as well as by socioeconomic development in endemic countries. In the past 12 years, great changes in trachoma control have taken place as a result of the WHO's leadership in developing a comprehensive strategy for control (*SAFE*, described below) and the founding of the Alliance for Global Elimination of Trachoma by the year 2020 (*GET2020*). As a result, the chance for reducing or perhaps eliminating this debilitating cause of blindness is within reach (3, 4).

14.2 CAUSATIVE ORGANISM AND NATURAL HISTORY

Trachoma is caused by four serovars of *Chlamydia trachomatis* (A, B, Ba, and C); serovars D through K cause genital tract infection. This obligate intracellular gram-negative bacterium prefers the epithelial surfaces of the eye; the other serovars affect the epithelial surfaces of the genital tract, causing the most common form of sexually transmitted disease. Crossover of these serovars to the alternative site is rare. *C. trachomatis* is an obligate intracellular bacterium and cannot replicate outside eukaryotic host cells. The metabolically active *Chlamydia* reticulate body matures, enlarges, and finally erupts causing cell death and releasing spore-like elementary bodies, which are the infecting agent. The elementary body attaches to epithelial cells on contact through its major outer membrane protein (MOMP). The presence of these elementary bodies in the secretions from the eyes and noses of infected persons (children especially) facilitate further transmission to family or contacts.

The intracellular location of the organism leads to protection from antibody and complement and there is a down-regulation of the host major histocompatibility complex or MHC class I molecules by infected cells, thereby reducing killing by cytotoxic T cells. Episodes of infection seem to fall in severity as the child ages, suggesting that there is some development of immunity. On the other hand, this may reflect a decreasing of frequency of reinfection that occurs with age. While the immunopathology of trachoma is not fully defined, it is accepted that more severe trachoma, scarring, and corneal damage occurs after repeated infections and that this is related to a delayed hypersensitivity reaction. Detailed

Water and Sanitation-Related Diseases and the Environment: Challenges, Interventions, and Preventive Measures, First Edition.
Edited by Janine M. H. Selendy.
© 2011 Wiley-Blackwell. Published 2011 by John Wiley & Sons, Inc.

reviews of immunity and immunopathology are available in two *Lancet* seminars on trachoma (5, 6).

14.3   CLINICAL MANIFESTATIONS

After a brief incubation period of 5–10 days, the initial infection will result in a mild conjunctivitis that heals without permanent damage. As little immunity exists, repeated infections result in an exaggerated response: intense inflammation and scarring of the upper subtarsal conjunctiva, distortion of the lid margin that results in a shortened upper lid pulling the eye lashes inward (trichiasis). The early signs of trachoma (follicular disease and intense inflammation) are seen in children; scarring and trichiasis is observed in the older population. The resulting abrasion of the cornea by inturned lashes causes pain and eventually leads to corneal opacity and blindness. The World Health Organization has devised a simplified grading scheme for assessment of trachoma in communities (Fig. 14.1), which demonstrates the progression of disease that may take place over years of infection and reinfection. The early signs of trachoma are detected by everting the upper eyelid and examining with a 2.5× loupe (preferably) for the follicular stage (TF): characteristic white or yellow follicles of 0.5–2.0 mm. As disease worsens, the intense inflammatory stage (TI) may be seen. Trachomatous scarring (TS) begins as small stellate scars that with time coalesce to form the dense scar tissue that distorts normal lid architecture. This is followed by frank trachomatous trichiasis (TT) and corneal opacity (CO). The WHO scheme shown in Fig. 14.1 has been taught to eye workers in

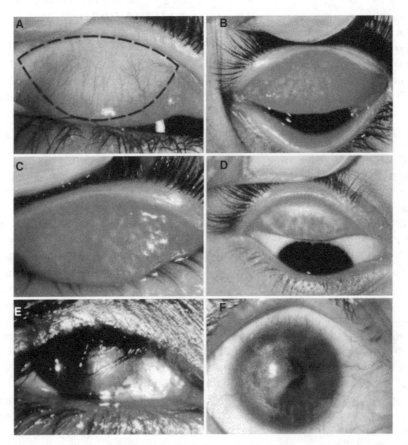

FIGURE 14.1 Clinical appearance of trachoma and the WHO trachoma grading scheme. (a) Normal everted upper tarsal conjunctiva; pink, smooth, thin, and transparent. Large deep-lying blood vessels that run vertically are present over the tarsal conjunctiva. The dotted line shows the area to be examined. (b) Trachomatous inflammation-follicular (TF). Five or more follicles of >0.5 mm. (c) Trachomatous inflammation-follicular and intense (TF + TI). Inflammatory thickening obscuring >1/2 the normal deep tarsal vessels. (d) Trachomatous scarring (TS). (e) Trachomatous trichiasis (TT). (f) Corneal opacity (CO). These photographs are reproduced with permission from the WHO Programme for the Prevention of Blindness and Deafness. (*See insert for color representation of this figure.*)

many trachoma endemic countries as a means of assessing the burden of disease within communities.

14.4 DIAGNOSIS

Diagnosis is generally a clinical diagnosis made by the examination described above. Giemsa stain of the intracytoplasmic inclusions and or culture of the organism are difficult and laborious and generally unavailable in endemic areas. Although both are specific, they are not sensitive. Four nuclei acid amplification tests are commercially available for the diagnosis of *C. trachomatis* infection. All were developed primarily for diagnosis of urogenital chlamydial infections and only two (Amplicor (Roche) and LCx (Abbott)) have been tested in ocular chlamydial infections. It is clear that the positive predictive value of clinical exams falls with falling prevalence and it is postulated that in persons who have had trachoma, any ocular infection or irritation may stimulate the typical follicular response seen in early trachoma. In at least one study, while clinical exams continued to find a few active cases, no infection could be detected by PCR tests done simultaneously with clinical exams in a control program in Tanzania (7, 8). Therefore, for practical public health purposes, the available and cost-effective method is the clinical examination. Should the cost and ease of application of the PCR tests be reduced and their role with respect to prevention of blindness be clarified, these tests could add an increased measure of accuracy to the current public health control programs—possibly allowing the discontinuation of antibiotic treatment sooner than reliance on clinical grading would dictate.

14.5 EPIDEMIOLOGY

Trachoma remains highly endemic in many parts of Africa and continues to persist in a number of countries in the Middle East, Asia, and Latin America (9). As noted above, 57 countries are known or considered endemic for the disease and more than 40 million people have active trachoma. Another 8.2 million have trichiasis and therefore at high risk of irreversible and severe visual impairment. Population-based surveys provided recent information for 42 out of 57 endemic countries. Globally 1.2 billion people live in endemic areas, 40.6 million people are suffering from active trachoma, and 8.2 million have trichiasis. In addition, 48.5% of the global burden of active trachoma is concentrated in 5 countries: Ethiopia, India, Nigeria, Sudan, and Guinea. On the other hand, 50% of the global burden of trichiasis is concentrated in only three countries: China, Ethiopia, and Sudan. Overall, Africa is the most affected continent—27.8 million cases of active trachoma (68.5% of all) and 3.8 million cases of trichiasis (46.6% of all) are located in 28 of the 46 countries in the WHO African Region, with an estimated population of 279 million living in endemic areas (2).

Children are the main reservoir of infection with *C. trachomatis* and trachoma is a family-based disease with clustering in households and communities as a result of the ease of transmission of infected ocular secretions between people, especially other family members as noted above and in Fig. 14.2 (6). While children below 10 years of age account for the preponderance of active infections, the severe stages are seen in adults. Women are three times as likely to be

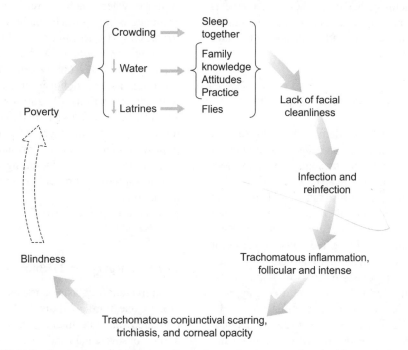

FIGURE 14.2   Interaction of risk factors for trachoma. Lancet reproduced with permission (6).

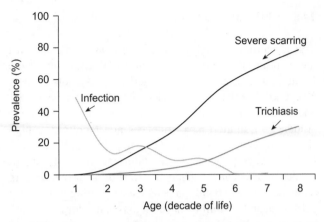

FIGURE 14.3 Schematic presentation of trachoma infection and disease based on community control programs in Ethiopia (courtesy of Dr. Bruce Gaynor, University of California San Francisco).

blinded by trachoma as are men; the cause of this is thought to be the frequent reinfection they acquire in child care. Figure 14.3 is a schematic demonstration of the sequence of early infection in childhood, but trichiasis and corneal opacity occurring in the older ages drawn from data collected in control programs in Ethiopia. This pattern to a greater or lesser level has been seen in all areas where trachoma is endemic.

Infection is easily transmitted from eye to eye by fingers, flies, and fomites (shared cloths, clothes, towels). Households with young children with eye discharge and external nasal exudates, posterior pharyngeal infection or drainage through the nasolacrimal duct are the greatest source of infection (10). Risk factor studies for trachoma have repeatedly shown the association of young children, poor sanitation, poor facial cleanliness, and inadequate access to water with trachoma (11, 12).

14.6 TRACHOMA: WATER AND SANITATION

The importance of water in the epidemiology of this disease is intuitively obvious when one considers the modes of transmission. But, water's link to trachoma is perhaps more complex than it would seem. In a detailed review of available studies of water and trachoma, Prost and Negrel concluded that there was a general reduction in trachoma rates as water access improved (13). However, it is also not clear whether water availability/access should be measured in terms of distance to source or time to fetch water. Cairncross has noted that a "water use plateau" exists in areas of scarce water supplies throughout the world. The amount used may vary, but the plateau of use exists from a few minutes from the house to 30 min away (14). After 30 min, water use declines from the plateau and hence, those beyond 30 min of a water source are likely to benefit most from a new and closer water

source. In a rigorous study of children ages 1–9 years in Tanzania involving 99 randomly selected households, active trachoma increased with increasing water collection time but was unrelated to amount of water collected. In a smaller sub-study in the same area, active trachoma prevalence was substantially lower in children from households where more water was used for personal hygiene independent of the total amount of water used (15). The allocation of water to hygiene was directly associated to shorter water collection time. Similarly in Gambia trachoma-free households used more water for washing children than households with trachoma cases (16). The key issue affecting trachoma in households is then the use of water in personal hygiene, especially in children.

14.7 PREVENTION AND CONTROL

It is water's link to poverty and poor hygiene that makes its role crucial in halting trachoma transmission. Water is essential but not sufficient to control trachoma. That is, access to water is usually less among the poor and likewise, poor hygiene cannot improve without water. This was the clear intention of the WHO in developing a strategy that incorporated the three major elements of public health improvement: primary, secondary, and tertiary preventive measures. In 1997, WHO founded an alliance (of ministries of health as well as nongovernmental organizations) for the global elimination of trachoma by 2020 (*GET2020*), and in 1998, a World Health Assembly resolution called for the 57 countries where trachoma remains endemic to take steps to eliminate blinding trachoma by implementing the "*SAFE*" strategy: WHA Resolution 51.11, *Global Elimination of Blinding Trachoma* (17). This strategy was devised by WHO and includes four proven elements leading to control. *SAFE* entails: (*S*) trichiasis surgery to halt corneal damage (tertiary prevention); (*A*) antibiotic treatment (single-dose azithromycin, 20 mg/kg of body weight for children and 1 g for adults (secondary prevention); (*F*) face washing or improved facial hygiene (primary prevention); and (*E*) environmental change including access to water and improved sanitation (primary prevention) was first described in the WHO manual, *Achieving Community Support for Trachoma* (see Section 14.7.2).

14.7.1 Surgery for Trichiasis—S

In untreated trichiasis, corneal opacity may develop in around one third of individuals whose trichiasis remains untreated for more than a year (18) and trichiasis itself is a cause of severe pain and reduced vision. Surgical treatment (tertiary prevention) is the only means of halting further damage. The WHO recommends the use of the bilammellar tarsal rotation procedure or one of the two other similar procedures in use in endemic countries, depending on the experience of the eye care community in the country and ensuring that quality of service is carefully respected. Reacher and colleagues undertook a randomized controlled trial of this procedure in Oman and obtained at least a 70% success rate 6–24 months after surgery (19). The upper-eyelid tarsal plate is incised with external rotation of the distal margin by insertion of 3–4 sutures from the external surface to the incision to the internal surface of the upper part of the tarsal plate. This rotates the lashes up and away from the cornea. Recurrence of trichiasis has been a major concern and no doubt affects the acceptability of patients to have this procedure carried out. These rates have varied from as low as 8% at 1 year to 47% in a 17-year follow-up in Oman (20). The continuing trachomatous process of lid shortening around scarring, coupled with the possibility of reinfection, may explain the wide differences in recurrence rates. The other possibility could be less attentive eye surgeons with less experience with the procedure. In a trial under field conditions lid surgery by well-trained and well-equipped eye nurses has been shown to be as good as surgery provided by ophthalmologists (21) and has been successfully carried out in villages in national control programs in many countries. Moving this needed surgery to the villages improves acceptability, reduces costs, and is more direct in dealing with the need to halt further vision impairment. To assure improved performance of surgery, the WHO has made available two manuals for teaching and evaluating performance (quality control) of this procedure (see Section 14.A.1).

14.7.2 Antibiotics—A

From 1952 until the early 1990s, the recommended treatment for trachoma had been topical tetracycline ointment four times a day at first, later reduced to twice a day—for 6–8 weeks. While results could be good in the case of programs where therapy was observed or administered, in fact, when tubes of ointment were simply given to children or patients, the results were often poor because the ointment disturbed vision and because it is also irritable, it was discontinued as soon as any improvement in symptoms was obtained. Later doxycycline, an oral tetracycline with long half-life, was found to be effective if given once daily for 2 or 3 weeks. However, since the target population is young children, it was not suitable for wide use (or use at all) because of the dental damage in children as well as photosensitivity caused by doxycycline.

In the mid-1990s, three randomized trials of azithromycin showed that azithromycin in a single oral dose was at least as effective as tetracycline ointment (22–24). Community use of azithromycin in trachoma control was successfully carried out using 3 weekly doses in three countries: Egypt, Tanzania, and Gambia (25). Having an oral antibiotic that is effective in a single dose has revolutionized the control programs for trachoma. Additionally, this systemic antibiotic has the benefit of treating extraocular sites of chlamydial infection, especially the posterior pharynx not reached by ocular ointments. Azithromycin, a macrolide antibiotic with few side effects, has the unusual quality of concentration in macrophages, polymorphonucleocytes, and epithelial cells—where the intracellular *C. trachomatis* resides. Studies in conjunctival biopsies of uninflamed eyes showed that the inhibitory concentrations were maintained in these tissues for at least 15 days after a single treatment (26).

This extraordinary means of treating trachoma was enthusiastically supported by the trachoma control community and at least four large cohort studies have been carried out in Tanzania (8, 27), Ethiopia (28), and Gambia (29), showing reduced prevalence of chlamydial infection or signs of active trachoma after community-wide treatment with single-dose oral azithromycin. In a region of Gambia where the disease is hypoendemic a sustained reduction in the prevalence of chlamydial infection was achieved in 12 of 14 villages receiving oral azithromycin (29). The results of a single treatment in a village hyperendemic to trachoma in the Kongwa district of Tanzania showed a reduction in the prevalence of chlamydial infection from 57% to 12% 2 months after treatment. In another Tanzania district, Rombo, trachoma prevalence showed a sustained reduction after a single treatment; however, in this case, all children with active disease at 6, 12, and 18 months were offered tropical tetracycline treatment. In this district of low prevalence (pretreatment prevalence = 9.5%) no evidence of *C. trachomatis* DNA was detected in the community at 24 months (7, 8).

Many areas in Ethiopia are hyperendemic for trachoma and prevalence of infection, although initially decreased, was found to increase after treatment, in absence of the other transmission control measures included in the SAFE strategy. Theoretical models suggested that biannual treatments may be required when the initial prevalence of trachoma is >50%. In a study of 16 adjacent villages 14,897 of 16,403 eligible individuals (90.8%) received their scheduled treatment. In the eight villages in which residents were treated annually, the prevalence of infection in preschool children was reduced from a mean of 42.6% (range, 14.7–56.4%) to 6.8% (range, 0.0–22.0%) at 24 months. In the eight villages in which residents were treated biannually, infection was reduced from 31.6% pretreatment (range, 6.1–48.6%) to 0.9% (range, 0.0–4.8%) at 24 months. Biannual treatment was associated with a lower prevalence at 24 months ($P = 0.03$, adjusting for baseline prevalence). At 24 months, no infection could be identified in six of eight of those treated biannually and in only one of eight of those treated annually ($P = 0.049$, adjusting for baseline prevalence) (28). Following the same

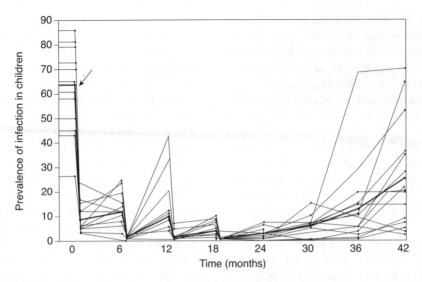

FIGURE 14.4 The mean prevalence of infection (black) and the prevalence in each of the individual 16 study villages (gray) in Ethiopia based on estimates of the prevalence of infection in 1–5 year-old children at a single village visit. After the fourth treatment at 18 months after the study began, infection returned to the 16 villages as shown (30).

16 villages, the same group of investigators found in a much larger study but with only four mass treatments 6 months apart, mean prevalence in 1–5 year-old children began a steady rise in the 24 months following the last treatment. The initial prevalence was 63.5%, falling to 2.6% 6 months after the last treatment. But in the next 18 months prevalence had increased to 25.2% (Fig. 14.4) (30). This was a steady increase after the last treatment but still was 40% less than the original prevalence. Also in Ethiopia, these investigators conducted a smaller study of two villages with a total population of 768 who were offered six biannual mass treatments with oral azithromycin. At 12 months after the final treatment, each village member was examined clinically and swabs were taken for testing for evidence of *C. trachomatis* RNA. The average antibiotic coverage was high, >90% in both villages. Clinical activity in children aged 1–5 years decreased from 78% and 83% in the two villages before treatment to 17% and 24% at 42 months. But, PCR evidence of infection in the same age group decreased from 48% to 0% in both villages at 42 months (31). These important studies are indicative of both the power and limitations of effective antibiotic treatment. That is, if mass antibiotic treatment is offered at 6-month intervals, prevalence will remain low even in areas of the highest transmission. However, this is untenable in terms of public health delivery, costs, and bacterial resistance issues. This is especially true in light of what is known about the importance of water and behavioral change for facial cleanliness.

A topical azithromycin topical eye drop has been prepared, but is in preliminary trials and not yet available under a drug donation framework. In a trial in Cameroon in children between 1 and 10 years, prevalence of active infection (TF and TI) fell from 31.5% before treatment to 6.3% at 12 months after the treatment and the drops were reported to be well accepted (32). This preliminary report will require comparison with oral and therefore systemic treatment with azithromycin, or define its role in trachoma elimination campaigns that use the MDA approach.

While oral azithromycin has proved effective for trachoma, there are additional community benefits observed in those receiving mass treatment. Beneficial effects on mortality rates have been observed in a large study in Ethiopia (33). Mortality rates in children 1–9 years of age in two large groups of villages of 15,902 and 14,716 participants. The control group of villages received biannual treatment with oral azithromycin 1 year after the first group. In the treated communities, the estimated overall mortality rate during this period for children in the untreated group was 8.3 per 1000 person-years, while among the treated communities, the estimated overall mortality rate was 4.1 per 1000 person-years. It is likely that this mass treatment had beneficial effects on gastrointestinal and pulmonary disease reducing mortality in the treated group. With the broad use of any antibiotic, the specter of antibiotic resistance must be considered. No evidence of azithromycin resistance has been reported in *Chlamydia* species. However, azithromycin resistance in other pathogens, particularly pneumococci has been shown to develop after mass treatment. Fortunately the prevalence of macrolide-resistant strains generally returns to pretreatment levels between 6 and 12 months (34, 35).

In 1996 at its Global Trachoma Scientific Meeting, the World Health Organization endorsed the use of azithromycin in trachoma control programs (36). In the following year, the WHO founded the Global Alliance for the elimination of blinding trachoma by the year 2020—GET2020. WHO's endorsement of the use of azithromycin for trachoma led to a generous decision by Pfizer Inc. to donate azithromycin for this purpose. Pfizer joined the Edna McConnell Clark Foundation to found the International Trachoma Initiative, a nongovernmental organization, to advance this elimination program. Since 1998, 18 countries have begun national trachoma control programs using the SAFE strategy and donated mass treatment with azithromycin. The WHO recommends mass treatment for at least 3 years if the prevalence of trachoma in children aged 1–9 years in greater than 10%, and then reassessment of prevalence. Recent experience suggests that trachoma is unlikely to be eliminated in a single year and a delay of 2 or 3 years for reassessment would be saving in public health resources. While the impact of this long-acting antibiotic is truly remarkable, it is clear that long-term elimination requires the full SAFE strategy.

14.7.3   F and E—Facial Cleanliness and Environmental Change

The F and E of this strategy translate into the role of water and sanitation in the transmission and control of blinding trachoma. Mass antibiotic treatment alone is unlikely to eliminate ocular chlamydial infection especially in areas such as Ethiopia where infection is clearly hyperendemic (Emerson and Ngondi, 2009) (37). And, the efficacy of access to water and improved hygiene has been shown even in the absence of antibiotics in western Europe and North America by the mid-twentieth century. On the other hand, Rabiu and colleagues assessing evidence for the effectiveness of environmental sanitary measure on the prevalence of active trachoma in endemic area for the Cochrane Database concluded that there was a dearth of acceptable data to determine the effectiveness of all aspects of F and E (38). In this case, the absence of randomized trials in this area is not surprising as it involves the difficult areas of cultural/behavioral change and whole communities assigned to an intervention. Also, in the face of the ease of providing a single dose drug, behavioral change/ health education and/or water supplies/latrine building is much more labor intensive and expensive. Still, it is obvious that these measures have benefits far beyond trachoma control, both in terms of disease control as well as better quality of life and expenditure of time and energy in obtaining water. Water, and therefore the link to sanitation, affects trachoma transmission by access and use of water (behavioral factors), sanitation/latrine availability, and fly control.

There is, however, evidence outside of rigid strictures of randomized controlled trials that F and E are effective. A randomized control trial on facial cleanliness was undertaken by West and her colleagues in six villages of central Tanzania (West, 1995) (27). In this case, all members of the villages received topical tetracycline daily for 30 days.

Three villages also received an additional program focusing on facial cleanliness in children. At the end of 1 year, clean faces in children in the intervention villages had increased by 60% and the rate of active trachoma was reduced by 42%; the rate of intense trachoma (TI) fell by 65%. That is, there was greater reduction in severe disease before reduction in the milder first stage of infection. These results were obtained in addition to a benefit seen in both arms of the study as a result of the use of the tetracycline ointment. Using an adult village participatory learning program, the same investigators were able to increase the rate of clean faces in children from 9% to 33% in 1 year (39, 40).

Water is clearly linked to two other risk factors for trachoma: sanitation (latrines) and flies. A case-control study in almost 1000 households in 8 Tanzanian villages examined the association between use and quality of latrines and the risk of trachoma. Use of latrines was significantly associated with a decreased risk of trachoma. The condition of the latrine or quality was not a determining factor. Latrines then need not be expensive or elaborate to provide health benefits (41). In a community-based study of 507 children 1–9 years of age in 232 Ethiopian households, lack of access to a latrine increased the risk of having active trachoma by 4.36-fold but perhaps more important, absence of a clean face (defined simply as only absence of discharge from eyes or nose) increased the odds of having trachoma by 7.59-fold (42). The prevalence of trachoma in this community was 53%.

It is clear from studies in Gambia (43) and Ethiopia (44) that *Chlamydia* DNA can be found by polymerase chain reaction (PCR) on flies caught on children's faces. These flies are water-seeking flies: largely, *Musca sorbens*, a species that preferentially breeds in human feces, hence, the link to latrines. However, trachoma transmission has been found to exist in areas where flies are *not* a problem. This was observed in Kentucky in the United States in 1911 where the basic lack of hygiene and sanitation was felt to be the major cause of continuing transmission (45). It has been shown that in a trial in Gambia of use of fogging with insecticides that the fly population was reduced and fly populations remained so long as insecticides were continued. Similarly, the expected rate of new cases of trachoma rates was shown to fall concomitantly (46). In a larger study in Gambia, villages were randomized to receive fly control, the provision of new latrines, or no additional activities. Persons with intense trachoma were treated with either tetracycline or azithromycin. After 6 months the number of *M. sorbens* on children's faces had decreased by 88% with fly control and 30% in the villages with new latrines. At 6 months, the prevalence of trachoma in children in the fly control villages fell from 14% to 7%, in the latrine villages from 11% to 8%, and increased in the control villages from 9% to 10% (47). However, studies in Tanzania using the same fly control techniques did not show any benefit from fly control when used in conjunction with antibiotic treatment in this hyper-endemic area (48). Similarly and perhaps more definitively, studies in Ethiopia where transmission is extraordinarily high, studies have shown that in the presence of mass treatment with azithromycin, fly control may not add to reducing transmission of trachoma through fly–eye contact. Flies were collected in a village that had received mass oral azithromycin distribution and were compared with flies in an untreated village. Polymerase chain reaction (PCR) was performed to detect chlamydial DNA on the flies. Conjunctival swabs were also taken to assay for chlamydial prevalence in the children. Chlamydia was found on 23% of the flies in the untreated villages but only 0.33% in the treated villages. Therefore, mass treatment of children with antibiotics would seem to drastically reduce the role of flies as a disease vector without controlling the fly population (44).

In Vietnam a public health intervention study was conducted between 2002 and 2005 to assess the impact of improved water and sanitation compared with villages that received only the S and A components of SAFE (49). Two villages of 1300–1500 population, separated by 10 km and five other villages were studied. The F and E components applied to the village receiving the full SAFE strategy included health education (through the Women's Union and school authorities), as well as 284 latrines, 241 bathrooms, 273 dug wells, and 252 water tanks. The comparison village received only SA. The SA village prevalence fell from 10.2% to 6.7%, while the SAFE village fell from 13.3% to 1.4%. The additional decline of active trachoma due to the addition of health education and water/sanitation was responsible for 58.7% of the decline at all ages and 37.4% in children under the age of 15 years.

14.8   THE PATH TOWARD ELIMINATION BY 2020

Since the founding of the WHO's GET 2020 alliance and the Pfizer donation of azithromycin though ITI, 18 countries have embarked on national trachoma control programs using mass treatment with azithromycin in the SAFE strategy (Box 14.1). Of these, Morocco and Ghana have informed the WHO that they have reached their target for elimination of blinding trachoma. In addition, Mexico and Saudi Arabia are in the process of assessing the situation and inform the WHO of having achieved the elimination of blindness from trachoma.

In April 2007, WHO launched its new approach to the control of neglected tropical diseases (NTDs). These diseases have a number of common features that are all associated with impoverished settings, and industry donations of the antibiotics needed for mass treatment are available. Control of these diseases is now considered to be part of the global drive to reduce poverty and to attain the United Nations Millennium Development Goals. The NTDs affect

---

BOX 14.1 COUNTRIES WITH NATIONAL TRACHOMA CONTROL USING THE SAFE STRATEGY

- Burkina Faso
- Ethiopia
- Eritrea
- Gambia
- Ghana
- Guinea-Bissau
- Kenya
- Mali
- Mauritania
- Morocco
- Nepal
- Niger
- Nigeria
- Senegal
- Sudan
- Tanzania
- Uganda
- Vietnam

---

one third of the global population, especially poor populations living in remote rural areas. Trachoma is one of the "major five" NTDs; the others are onchocerciasis, lymphatic filariasis (LF), three soil-transmitted helminths (STH), and schistosomiasis. For trachoma and the first three of these, there are drug donation programs by major pharmaceutical companies: ivermectin (Merck) for onchoceriasis; albendazole (GlaxoSmithKline) along with an inexpensive generic, diethyl carbamazine for LF; and mebendazole (Johnson & Johnson) for STH. For schistosomiasis, the major generic drug, praziquantel, has fallen in price so that adults can be treated for as little as USD 0.25—children much less. These five diseases kill large numbers of people: 20% of deaths and 24% of disability-adjusted life years (DALYs) lost as a result of communicable diseases are due to these NTDs. Because of the availability of cheap or donated effective drugs for these diseases, they can be treated for as little as USD 0.40–0.79 per person (50). A major thrust for countries where there is overlap of these diseases is to integrate control programs wherever possible. This includes opportunistic synergy with programs focused on the "big three" (HIV, malaria, and tuberculosis) (51). Such a program has been reported in the Amhara region of Ethiopia where treatment and health education were combined in a joint trachoma and malaria effort under a Carter Center/Lions Club program with the Ministry of Health (52).

14.9 CONCLUSION

The evidence gathered from operational research studies cited here and from the experience of national control programs clearly supports the effectiveness of the SAFE strategy and the implementation of it in full as the path to secure elimination of blinding trachoma. The impact of mass treatment with azithromycin is remarkable; but alone it will not eliminate the irreversible consequences of the disease, with the potential exception of the areas of lowest transmission. The added benefits of access to water and hygiene (including availability of latrines, education on their use, and behavioral changes for personal hygiene) for general health and quality of life make this a needed inclusion in all NTD elimination programs. All programs that result in the behavioral change needed to increase clean faces in children will have a profound effect on transmission of trachoma. The SAFE strategy adds another benefit beyond simply mass drug administration (MDA). Since it is a complete public health strategy, it provides a linkage with chronic disease control programs and encourages development of a stable health care system. When and where trachoma elimination is/will be achieved, the people of those endemic areas will be left with a functional health care system, an increased education on the real benefits of public health and clinical care, and with first-hand evidence that prevention works and induces a higher quality of life, making the paradigm of primary health care a living truth.

14.A.1 APPENDIX: RESOURCES AVAILABLE FOR ELIMINATION OF TRACHOMA

The World Health Organization Program in Prevention of Blindness makes available the following invaluable resources for trachoma elimination programs.

These manuals can be obtained by writing: The World Health Organization, Prevention of Blindness Program, 1211 Geneva 27 Switzerland or they may be downloaded by accessing the WHO website: www.who.int/pdb/publications/trachoma/en.

14.A.1.1 Guides and Manuals

- *The SAFE Strategy: Preventing Trachoma*
- *Trachoma Control: A Guide for Program Managers*
- *Guidelines for Rapid Assessment for Blinding Trachoma*
- *Final Assessment of Trichiasis Surgeons*
- *Trichiasis Surgery for Trachoma*
- *Primary Health Care Level Management of Trachoma*
- *Achieving Community Support for Trachoma Control: A Guide for District Health Work*

- *Zithromax—Program Manager's Guide*
- *A Guide: Trachoma Prevention through School Health Curriculum Development*

14.A.1.2 Definitive Textbook on Trachoma

Hugh R. Taylor. *Trachoma: A Blinding Scourge from the Bronze Age to the Twenty-First Century*, Illustrated. East Melbourne, Australia: Centre for Eye Research Australia/Haddington Press, 2008, 282 pp. ($112 ISBN 978-0-9757695-9-1).

14.A.1.3 Information on Availability of Azithromycin for National Control Programs

The International Trachoma Initiative (ITI)
The Task Force for Global Health
325 Swanton Way
Decatur, GA 30030
Phone: 1 800 765 7173
Web: www.trachoma.org

14.A.1.4 Trachoma Information Service

In partnership with the ITI, the Kilimanjaro Centre for Community Ophthalmology (KCCO) in Tanzania manages a Trachoma Information Service—a bimonthly e-mailing of trachoma-related information including the latest trachoma research articles. Summaries provided by the authors pay particular attention to what the findings meant for people involved in trachoma control in developing countries.

Persons who wish to receive this material via e-mail may send name and e-mail address to KCCO, Dr. Paul Courtright at pcourtright@kcco.net.

14.A.1.5 Women and Trachoma

The Kilimanjaro Centre for Community Ophthalmology and the Carter Center make available a useful manual on *Women and Trachoma*. It can be downloaded from www.kcco.net or www.cartercenter.org; or, by phone to Carter Center 404-420-3830.

REFERENCES

1. Frick KD, Hanson CL, Jacobson GA. Global burden of trachoma and economics of the disease. *Am J Trop Med Hyg* 2003;69 (Suppl 5):1–10.
2. Mariotti SP, Pascolini D, Rose-Nussbaumer J. Trachoma: global magnitude of a preventable cause of blindness. *Br J Ophthalmol* 2009;93(5):563–568.
3. Mariotti SP. New steps toward eliminating blinding trachoma. *New Engl J Med* 2004;351(19):2004–2007.

4. Cook JA. Eliminating blinding trachoma. *New Engl J Med* 2008;358(17):1777–1779.

5. Mabey DC, Solomon AW, *Foster A. Trachoma. Lancet* 2003;362(9379):223–229.

6. Wright HR, Turner A, *Taylor HR. Trachoma. Lancet* 2008;371 (9628):1945–1954.

7. Solomon AW, Harding-Esch E, Alexander ND, Aguirre A, Holland MJ, Bailey RL, Foster A, Mabey DC, Massae PA, Courtright P, Shao JF. Two doses of azithromycin to eliminate trachoma in a Tanzanian community. *New Engl J Med* 2008;358 (17):1870–1871.

8. Solomon AW, Holland MJ, Alexander ND, Massae PA, Aguirre A, Natividad-Sancho A, Molina S, Safari S, Shao JF, Courtright P, Peeling RW, West SK, Bailey RL, Foster A, Mabey DC. Mass treatment with single-dose azithromycin for trachoma. *New Engl J Med* 2004;351(19):1962–1971.

9. Polack S, Brooker S, Kuper H, Mariotti S, Mabey D, Foster A. Mapping the global distribution of trachoma. *Bull World Health Organ* 2005;83(12):913–919.

10. Malaty R, Zaki S, Said ME, Vastine DW, Dawson DW, Schachter J. Extraocular infections in children in areas with endemic trachoma. *J Infect Dis* 1981;143(6):853.

11. West S, Lynch M, Turner V, Munoz B, Rapoza P, Mmbaga BB, Taylor HR. Water availability and trachoma. *Bull World Health Organ* 1989;67(1):71–75.

12. Schemann JF, Sacko D, Malvy D, Momo G, Traore L, Bore O, Coulibaly S, Banou A. Risk factors for trachoma in Mali. *Int J Epidemiol* 2002;31(1):194–201.

13. Prost A, Negrel AD. Water, trachoma and conjunctivitis. *Bull World Health Organ* 1989;67(1):9–18.

14. Cairncross S. Trachoma and water. *Community Eye Health/Int Centre Eye Health* 1999;12(32):58–59.

15. Polack S, Kuper H, Solomon AW, Massae PA, Abuelo C, Cameron E, Valdmanis V, Mahande M, Foster A, Mabey D. The relationship between prevalence of active trachoma, water availability and its use in a Tanzanian village. *Trans R Soc Trop Med H* 2006;100(11):1075–1083.

16. Bailey R, Downes B, Downes R, Mabey D. Trachoma and water use: a case control study in a Gambian village. *Trans R Soc Trop Med Hyg* 1991;85(6):824–828.

17. World Health Organization. World Health Assembly Resolution 51.11 Global Elimination of blinding trachoma, 1998. Available at www.who.int/ncd/vision2020_actionplan/. Accessed 30 March 2010.

18. Bowman RJ, Faal H, Myatt M, Adegbola R, Foster A, Johnson GJ, Bailey RL. Longitudinal study of trachomatous trichiasis in the Gambia. *Br J Ophthalmol* 2002;86(3):339–343.

19. Reacher MH, Munoz B, Alghassany A, Daar AS, Elbualy M, Taylor HR. A controlled trial of surgery for trachomatous trichiasis of the upper lid. *Arch Ophthalmol* 1992;110 (5):667–674.

20. Khandekar R, Al-Hadrami K, Sarvanan N, Al Harby S, Mohammed AJ. Recurrence of trachomatous trichiasis 17 years after bilamellar tarsal rotation procedure. *Am J Ophthalmol* 2006;141(6):1087–1091.

21. Alemayehu W, Melese M, Bejiga A, Worku A, Kebede W, Fantaye, D. Surgery for trichiasis by ophthalmologists versus integrated eye care workers: a randomized trial. *Ophthalmology* 2004;111(3):578–584.

22. Dawson CR, Schachter J, Sallam S, Sheta A, Rubinstein RA, Washton H. A comparison of oral azithromycin with topical oxytetracycline/polymyxin for the treatment of trachoma in children. *Clin Infect Dis* 1997;24(3):363–368.

23. Tabbara KF, Abu-el-Asrar A, al-Omar O, Choudhury AH, al-Faisal Z. Single-dose azithromycin in the treatment of trachoma. A randomized, controlled study. *Ophthalmology* 1996;103(5):842–846.

24. Bailey RL, Arullendran P, Whittle HC, Mabey DC. Randomised controlled trial of single-dose azithromycin in treatment of trachoma. *Lancet* 1993;342(8869):453–456.

25. Schachter J, West SK, Mabey D, Dawson CR, Bobo L, Bailey R, Vitale S, Quinn TC, Sheta A, Sallam S, Mkocha H, Mabey D, Faal H. Azithromycin in control of trachoma. *Lancet* 1999;354 (9179):630–635.

26. Tabbara KF, al-Kharashi SA, al-Mansouri SM, al-Omar OM, Cooper H, el-Asrar AM, Foulds G. Ocular levels of azithromycin. *Arch Ophthalmol* 1998;116(12):1625–1628.

27. West SK, Munoz B, Mkocha H, Holland MJ, Aguirre A, Solomon AW, Foster A, Bailey RL, Mabey DC. Infection with *Chlamydia trachomatis* after mass treatment of a trachoma hyperendemic community in Tanzania: a longitudinal study. *Lancet* 2005;366(9493):1296–1300.

28. Melese M, Alemayehu W, Lakew T, Yi E, House J, Chidambaram JD, Zhou Z, Cevallos V, Ray K, Hong KC, Porco TC, Phan I, Zaidi A, Gaynor BD, Whitcher JP, Lietman TM. Comparison of annual and biannual mass antibiotic administration for elimination of infectious trachoma. *JAMA* 2008;299 (7):778–784.

29. Burton MJ, Holland MJ, Makalo P, Aryee EA, Alexander ND, Sillah A, Faal H, West SK, Foster A, Johnson GJ, Mabey DC, Bailey RL. Re-emergence of *Chlamydia trachomatis* infection after mass antibiotic treatment of a trachoma-endemic Gambian community: a longitudinal study. *Lancet* 2005;365 (9467):1321–1328.

30. Lakew T, House J, Hong KC, Yi E, Alemayehu W, Melese M, Zhou Z, Ray K, Chin S, Romero E, Keenan J, Whitcher JP, Gaynor BD, Lietman TM. Reduction and return of infectious trachoma in severely affected communities in Ethiopia. *PLoS Negl Trop Dis* 2009;3(2):e376.

31. Biebesheimer JB, House J, Hong KC, Lakew T, Alemayehu W, Zhou Z, Moncada J, Roger A, Keenan J, Gaynor BD, Schachter J, Lietman TM. Complete local elimination of infectious trachoma from severely affected communities after six biannual mass azithromycin distributions. *Ophthalmology* 2009;116(11):2047–2050.

32. Huguet P, Bella L, Einterz E.M., Goldschmidt P, Bensaid P. Mass treatment of trachoma with azithromycin 1.5% eye drops in the Republic of Cameroon: feasibility, tolerance and effectiveness. *Br J Ophthalmol* 2010;94(2):157–160.

33. Porco TC, Gebre T, Ayele B, House J, Keenan J, Zhou Z, Hong KC, Stoller N, Ray KJ, Emerson P, Gaynor BD, Lietman TM.

Effect of mass distribution of azithromycin for trachoma control on overall mortality in Ethiopian children: a randomized trial. *JAMA* 2009;302(9):962–968.

34. Leach AJ, Shelby-James TM, Mayo M, Gratten M, Laming AC, Currie BJ, Mathews JD. A prospective study of the impact of community-based azithromycin treatment of trachoma on carriage and resistance of *Streptococcus pneumoniae*. *Clin Infect Dis* 1997;24(3):356–362.

35. Chern KC, Shrestha SK, Cevallos V, Dhami HL, Tiwari P, Chern L, Whitcher JP, Lietman TM. Alterations in the conjunctival bacterial flora following a single dose of azithromycin in a trachoma endemic area. *Br J Ophthalmol* 1999;83(12): 1332–1335.

36. World Health Organization. Planning for the Global Elimination of Trachoma, 1996. Available at www.who.int/pdb/publications/trachoma/en. Accessed 30 March 2010.

37. Emerson PM, Ngondi J. Mass antibiotic treatment alone does not eliminate ocular chlamydial infection. *PLoS Negl Trop Dis* 1995;3(3):e394–1.

38. Rabiu M, Alhassan M, Ejere H. Environmental sanitary interventions for preventing active trachoma. *Cochrane Database Syst Rev (Online)* 2007;4(4):CD004003.

39. Lynch M, West SK, Munoz B, Kayongoya A, Taylor HR, Mmbaga BB. Testing a participatory strategy to change hygiene behaviour: face washing in central Tanzania. *Trans R Soc Trop Med Hyg* 1994;88(5):513–517.

40. McCauley AP, Lynch M, Pounds MB, West S. Changing water-use patterns in a water-poor area: lessons for a trachoma intervention project. *Soc Sci Med (1982)* 1990;31(11): 1233–1238.

41. Montgomery MA, Desai MM, Elimelech M. Assessment of latrine use and quality and association with risk of trachoma in rural Tanzania. *Trans R Soc Trop Med Hyg* 2010;104(4): 283–289.

42. Golovaty I, Jones L, Gelaye B, Tilahun M, Belete H, Kumie A, Berhane Y, Williams MA. Access to water source, latrine facilities and other risk factors of active trachoma in ankober, Ethiopia. *PloS One* 2009;4(8):e6702.

43. Emerson PM, Bailey RL, Mahdi OS, Walraven GE, Lindsay SW. Transmission ecology of the fly *Musca sorbens,* a putative vector of trachoma. *Trans R Soc Trop Med Hyg* 2000;94(1): 28–32.

44. Lee S, Alemayehu W, Melese M, Lakew T, Lee D, Yi E, Cevallos V, Donnellan C, Zhou Z, Chidambaram JD, Gaynor BD, Whitcher JP, Lietman TM. Chlamydia on children and flies after mass antibiotic treatment for trachoma. *Am J Trop Med Hyg* 2007;76(1):129–131.

45. Stucky JA. Ophthalmia and trachoma in the mountains of Kentucky. *Trans Am Ophthal Soc* 1911:321–328.

46. Emerson PM, Lindsay SW, Walraven GE, Faal H, Bøgh C, Lowe K, Bailey RL. Effect of fly control on trachoma and diarrhoea. *Lancet* 1999 Apr 24;353(9162):1401–3.

47. Emerson PM, Lindsay SW, Alexander N, Bah M, Dibba SM, Faal HB, Lowe KO, McAdam KP, Ratcliffe AA, Walraven GE, Bailey RL. Role of flies and provision of latrines in trachoma control: cluster-randomised controlled trial. *Lancet* 2004;363 (9415):1093–1098.

48. West SK, Emerson PM, Mkocha H, McHiwa W, Munoz B, Bailey R, Mabey D. Intensive insecticide spraying for fly control after mass antibiotic treatment for trachoma in a hyperendemic setting: a randomised trial. *Lancet* 2006;368 (9535):596–600.

49. Khandekar R, Ton TK, Do Thi P. Impact of face washing and environmental improvement on reduction of active trachoma in Vietnam-a public health intervention study. *Ophthal Epidemiol* 2006;13(1):43–52.

50. Hotez PJ, Molyneux DH, Fenwick A, Kumaresan J, Sachs SE, Sachs JD, Savioli L. Control of neglected tropical diseases. *New Engl J Med* 2007;357(10):1018–1027.

51. Gyapong JO, Gyapong M, Yellu N, Anakwah K, Amofah G, Bockarie M, Adjei S. Integration of control of neglected tropical diseases into health-care systems: challenges and opportunities. *Lancet* 2010;375(9709):160–165.

52. International Trachoma Initiative 2010. *Trachoma Matters, Winter 2010.* Available at www.trachoma.org/core/resources/ publications. Accessed 12 February 2010.

15

SHIGELLOSIS

Michael L. Bennish and M. John Albert

15.1 INTRODUCTION

Shigellosis is the archetype of a disease resulting from poor personal and environmental hygiene and sanitation. With humans as its only host, and with the gut as its place of residence within humans, infections with *Shigella*—the causative agent of shigellosis—are primarily the result of fecal–oral spread via contaminated food or fomites (1). Because *shigellae* are not hardy in the environment, infections usually occur among persons living in close quarters or lacking facilities (or habits) for hand washing and the isolation of stool from the general environment—as occurs with the poor in developing countries, or young children in day-care facilities.

Shigella infections continue to cause an unacceptably large number of deaths worldwide—estimated at more than 1,000,000 annually (2)—primarily among malnourished children (3). Increasing affluence is associated with a diminution in the incidence and severity of *Shigella* infections (3).

Although millions of persons have in recent decades benefited from improved economic conditions and a lessened risk of shigellosis, continued (and in some cases increasing) economic and social disparities have left many in developing countries at continued high risk of infection with *Shigella*.

In this chapter, we review the biology and epidemiology of *Shigella* infections, their clinical manifestations, and the environmental and social conditions that make this disease so emblematic of our failure to resolve the social inequities that lead to ill health for so many.

15.2 PATHOGEN AND PATHOGENESIS

15.2.1 Pathogen

Shigella species are gram-negative, nonsporulating, nonmotile, rod-shaped bacteria that belong to the family *Enterobacteriaceae*. Four species of *Shigella* are recognized based on differences in biochemical and serological characteristics. These are *S. dysenteriae* (15 established and 2 provisional serotypes), *S. flexneri* (15 serotypes and subtypes), *S. boydii* (20 serotypes), and *S. sonnei* (a single serotype with 5 biotypes) (4). *S. dysenteriae* type 1 is the causative agent of epidemic dysentery and serotypes belonging to the other species cause mainly endemic diarrheal illnesses. *S. dysenteriae* type 1 produces Shiga toxin, which is closely related to Shiga toxin 1 produced by enterohemorrhagic *Escherichia coli*. Of all the *Shigella* species and serotypes, *S. dysenteriae* type 1 produces the most severe form of illness, presumably attributable to Shiga toxin. Some *Shigella* strains produce enterotoxins known as *Shigella* enterotoxin 1 (ShET1) and ShET2, which may mediate the watery phase of *Shigella* diarrhea (5, 6). All *virulent shigellae* possess a 120–140 MDa virulence plasmid, which contains genes for invasion and intracellular survival.

15.2.2 Pathogenesis

The infectious dose is very small. Volunteer studies have shown that ingestion of as few as 100–200 viable organisms in milk is able to cause disease without the need to neutralize

Water and Sanitation-Related Diseases and the Environment: Challenges, Interventions, and Preventive Measures, First Edition.
Edited by Janine M. H. Selendy.
© 2011 Wiley-Blackwell. Published 2011 by John Wiley & Sons, Inc.

gastric acidity with bicarbonate (7). *Shigella* is able to survive the low acidity of the stomach by upregulating the acid-resistance genes (8). Shigellosis is characterized by a severe inflammatory response at the colonic mucosa and destruction of colonic epithelial cells. There are excellent reviews on the pathogenesis of the disease process (9–11). Pathogenesis can be divided into five stages: (i) entry of the bacterium into the host colonic cell; (ii) lysis of phagosomes (the vacuole that forms from the fusion of the cell membrane around the phagocytized *Shigella*) and bacterial multiplication; (iii) intra- and intercellular spread of the bacterium; (iv) death of the host cell and ulceration of the mucosa; and (v) resultant inflammatory response. This model for pathogenesis has been developed largely on studies with *S. flexneri* strains and it is assumed that disease caused by other *Shigella* (and *E. coli* possessing an invasion plasmid similar to that found in *Shigella*) has a similar pathogenesis. Bacterial pathogenesis is largely mediated by genes on the invasive plasmid while genes regulating expression of virulence are mostly on the bacterial chromosome.

Initially, *shigellae* are phagocytized by specialized antigen processing "M cells," which are located in the gut epithelium overlying the mucosa-associated lymph nodes. Once internalized in an endocytic vacuole, the organism is transcytosed across the epithelial layer into an intraepithelial pocket where it encounters resident macrophages. The organism is able to evade the killing mechanisms of the macrophage by inducing lysis of the phagocytic vacuole. Once the bacterium is in the cytoplasm, apoptosis (programmed cell death) of the macrophage is induced by *Shigella*, with the release from the macrophage of proinflammatory cytokines, which are chemotactic for polymorphonuclear cells. The polymorphonuclear cells then reach the infected subepithelial area where they transmigrate through the epithelial cell lining to reach luminal bacteria. The influx of polymorphonuclear cells across the epithelial layer disrupts the integrity of the epithelium allowing luminal bacteria to cross into other cells of the submucosa in an M cell-independent manner. Another possible pathway of invasion is by manipulating the cellular tight-junction-associated proteins allowing bacterial paracellular movement.

15.3 EPIDEMIOLOGY

15.3.1 Modes of Transmission and Risk Groups

Shigellosis is overwhelmingly a disease of the poor, the young, and persons in developing countries. All three groups have a salient characteristic in common—enhanced likelihood of exposure to and ingestion of feces or fecal material or organisms. This increased exposure results from living in what is a "fecally promiscuous" environment—an environment in which feces is not contained and isolated—resulting in the fecal–oral spread of *Shigella* from food or fomites contaminated by persons in the immediate environment. This occurs because of an absence of water and soap to wash hands, a lack of latrines or toilets for disposal and isolation of feces, and crowding, or a failure to embrace sanitary measures when they are available.

Daycare centers are often the industrialized country equivalent of these fecally promiscuous environments. They often have large number of infants or young children defecating under crowded conditions, cared for by too few common staff, with inadequate facilities for hand washing (12). Day-care center-related infections (infections of children in day care or their caregivers and family members) account for a substantial (up to 35% in some estimates) proportion of *Shigella* infections in industrialized countries (12, 13).

Other groups are at risk—including persons whose sexual practices (primarily men who have sex with men) include oral–anal contact (14), persons in custodial institutions (15), and displaced persons in refugee camps, which commonly have inadequate sanitation and access to water (16). Travelers from rich to poor countries are at high risk of being infected with *Shigella*, presumably because of a lack of preexisting immunity and exposure to unhygienically prepared foods. Some of these travelers may simply be emigrants returning for visits to their home countries. A substantial proportion of *Shigella* infections in industrialized countries, up to one-quarter in a recent survey in the United States (13), may be the result of foreign travel. Indeed, because of the lack of functioning microbiologic laboratories and surveillance in many developing countries, the identification of travel-related infections in industrialized countries—including *Shigella*—has been proposed as a means of monitoring patterns of infections in developing countries, including changes in the pattern of organisms causing enteric infection and antimicrobial resistance patterns. (17).

Bacillary dysentery has—not unexpectedly given the crowding and lack of sanitation in field conditions—been a major affliction of armies during armed conflicts (18), and at times has been severe enough to affect the outcome of battles. Despite the high-tech nature of many modern armies, shigellosis continues to be a major problem, being the second most common cause of diarrhea among U.S. military personnel during the recent Iraq war (19).

In societies where improved sanitation exists, infections with *Shigella*—though decreasing in number—are more likely to occur from commercial food (20) or recreational water outbreaks (21) in which swimmers ingest water that others have defecated into (22, 23) than they are in poor countries with inadequate sanitation. Drinking water is rarely if ever a cause of outbreaks of *shigellosis* (24). Compared to *Salmonella*, *Shigella* remains an uncommon source of outbreaks (20) and the majority of infections—80% by some estimates (25)—still occur from person-to-person transmission.

15.3.2 Burden of Disease

Statements on the global burden of shigellosis are hampered by a lack of data (3). Most surveillance relies on health facility or laboratory-based reporting (20, 26, 27), which is known to underestimate the true incidence of infection by a factor of 20–100 (2, 28). Because persons coming to health facilities are self-selected, they may misrepresent the age and gender of persons infected in the community, along with underestimating the total numbers of those infected. Active surveillance at the household level is expensive, usually involves small numbers of individuals, and cannot be undertaken on a regular basis. Thus, active surveillance, unless sustained, will miss changing patterns of infection, as well as the heterogeneity that can exist between communities even within the same country. And for some areas of the world—especially Africa—very little information—either through passive or active surveillance—exists (3).

Additionally, *Shigella* is not a hardy organism, and fresh stool specimens need to be plated directly onto media for cultivation of the organism—something not always possible in community-based surveillance programs (29). The use of transport media inevitably results in a diminishment of the rate of isolation (30)—a loss, based on recent studies using molecular techniques for identification, that may be as high as one-third of all infections (27).

Thus, any estimates of the worldwide impact of shigellosis are at best order of magnitude estimates. The most widely quoted estimate of disease burden is that of Kotloff (2), which, based upon previously published reports, estimated that in 1999 there were 165 million *Shigella* infections annually causing diarrhea, with 99% occurring in developing countries, and 1.1 million deaths. That review, as it acknowledged, suffered from the limitations of the data at hand—only one study from Africa (Nigeria) was included, and estimates of the number of infections that result in care at treatment centers and hospitals was extrapolated from a single study in Chile.

Although this and other estimates of the burden of disease from *Shigella* have limitations, they all serve to emphasize that *Shigella* infections remain an important health problem, primarily in developing countries, but also present an important health burden among selected populations in industrialized countries.

15.3.3 Age-Specific Incidence

Shigellosis is predominantly a disease of childhood, with the highest incidence in children 1–5 years of age, and within that age group the highest incidence is in children 12–23 months of age. That age distribution can be explained by childhood behavior and development. Infants younger than 12 months retain some passive immunity, and are less likely to be exposed because of breast-feeding and their immobility and diminished exposure to environmental objects. Toddlers (children 1–3 years of age) toddle, thus having greater exposure to fomites, and remain in an oral stage—with many environmental discoveries going into their mouths. And they lack passive immunity.

Incidence rates among children have been reported to be as high as one infection per year for malnourished children 2 years of age (31). A recent study from rural Peru found an incidence rate of 0.43 *Shigella* infections per year in children 12–23 months (28). But there is a wide incidence range reported, even from community-based studies (3), with one study in Thailand finding an improbably low incidence rate of only 0.04 episodes per year in children (32).

15.3.4 Geographic Distribution

Shigella is distributed worldwide wherever humans reside. In addition to differences in the burden of infection, there are pronounced differences in the distribution of species and serotypes. Infection with *S. sonnei* predominates in richer, industrialized nations, and infections with *S. flexneri*—which causes more severe illness—predominates in poorer countries (Fig. 15.1). Indeed, the proportion of *Shigella*

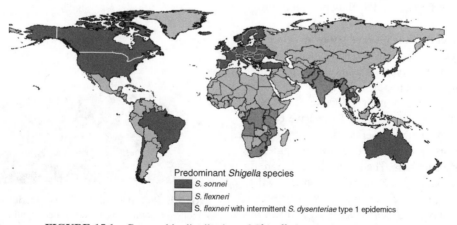

Predominant *Shigella* species
- S. sonnei
- S. flexneri
- S. flexneri with intermittent S. dysenteriae type 1 epidemics

FIGURE 15.1   Geographic distribution of *Shigella* by species and serotype.

infections caused by *S. sonnei* is a reasonable indicator of per capita gross domestic product (GDP)—having a regression coefficient of 0.55 in 56 studies when GDP and proportion of infections caused by *S. sonnei* were compared (3).

As countries become wealthier, their proportion of infections caused by *S. sonnei* and *S. flexneri* flip—with *S. sonnei* becoming the more common species. This happened in the South of the United States, where until the 1960s (but no longer) *S. flexneri* predominated (33), while in the more industrialized north *S. sonnei* was the predominant. And it has occurred more recently in Thailand, Vietnam, and Brazil—all countries that have shown rapid economic development—and where *S. sonnei* has either increased in frequency or become the predominant *Shigella* species (27, 34, 35).

S. dysenteriae type 1—the serotype of *Shigella* causing the most severe disease—was associated with the most impoverished conditions (3), causing epidemic disease with high fatality rates in Latin America (36), South Asia (37), and Central Africa (38). Other serotypes of *S. dysenteriae* are less common (3, 27), as is the fourth species of *Shigella*, *S. boydii*. These account for small proportions of *Shigella* infections wherever they occur (3, 27).

The reasons for these geographic differences in distribution of *Shigella* species and serotypes are unclear. *S. sonnei* consists of only one serotype—whereas the three other species consist of multiple serotypes. In industrialized countries, infection with *S. sonnei* may occur early in life, with subsequent exposure to other *Shigella* less common, whereas in developing countries exposure may continue through a lifetime. But this would not explain the difference in infection patterns that already occur in childhood.

An intriguing, but speculative hypothesis for geographic differences in *S. sonnei* and *S. flexneri* prevalence is that lack of access to clean water in poor countries results in greater exposure to an environmental organism—*Plesiomonas shigelloides*—one serotype of which has an LPS that cross-reacts with the *S. sonnei* LPS, thus providing cross-protective immunity (39).

But in general, there is little understanding of differences in serotype and species distribution, and changes in that distribution. *S. dysenteriae* type 1 traditionally had caused epidemics with a high mortality rate in Bangladesh and other south Asian countries, along with epidemics in Central America (40) and Central Africa (38). There have been no reported epidemics in Central America in almost 40 years, and in the last 10 years, few reported infections or epidemics from Africa (16) or South Asia. The International Centre for Diarrhoeal Disease Research, Bangladesh, which used to report repeated epidemic and endemic *S. dysenteriae* type 1, has not identified any *S. dysenteriae* type 1 infections since 2002.

15.4   CLINICAL MANIFESTATIONS AND TREATMENT

15.4.1   Clinical Manifestations

The pathognomonic feature of shigellosis is dysentery—hence, the synonym bacillary (for the structure of the organism and to distinguish it from amoebiasis) dysentery.

Dysentery is characterized by tenesmus—urgent, painful, repeated straining at attempted defecation resulting from inflammation of the recto-sigmoid—and production of stool consisting of a mixture of mucus, blood, and fecal matter (Fig. 15.2).

The majority of infections with *Shigella* do not, however, result in dysentery (27, 28, 41). Watery or mucoid diarrhea (diarrhea in which mucus is prominent in the stool) are more common than frank dysentery. Hospital and clinic-based surveillance—because of the self-selection for severity—will almost inevitably report a higher proportion of *Shigella*-infected patients having dysentery than community-based studies using active surveillance.

The proportion of *Shigella*-infected patients having dysentery varies by the infecting species or serotype of *Shigella*. Dysentery is most common with *S. dysenteriae* type 1 infections, next most common with *S. flexneri* infection, and least common with *S. sonnei* infection. Infections caused by *S. dysenteriae* other than type 1, and by *S. boydii*, probably result in dysentery at rates intermediate between *S. flexneri* and *S. sonnei*—though there is little active surveillance data on these less common causes of shigellosis.

Because *Shigella* invades the gut mucosa and elicits a pronounced local and systemic inflammatory response patients often appear toxic—with chills, a high fever, anorexia, and malaise (42). The intestinal inflammation often results in cramps and tenderness on palpation of the abdomen.

FIGURE 15.2   Characteristic dysenteric stool of a patient with shigellosis, containing blood, mucus, and small amounts of fecal matter. (*See insert for color representation of this figure.*)

Because it is an infection of the colon and rectum, vomiting is less common than with enteric infections of the small bowel (27).

Shigellosis can be accompanied by intestinal, systemic, or metabolic complications (43, 44). Intestinal complications include rectal prolapse—an extrusion of the rectum through the anus resulting from the straining of tenesmus, and intestinal obstruction and toxic megacolon. The latter resembles the toxic megacolon seen with ulcerative colitis, and presumably results from a similar pathology of extensive ulceration and disruption of neural and muscular function leading to functional obstruction. Intestinal perforation can occur even in the absence of toxic megacolon and result in peritonitis and septicemia.

Because of the extensive damage to the colonic epithelium, patients—especially children—can develop a protein-losing enteropathy (45), which varies in severity but can contribute to the malnutrition and diminished growth that often results from *Shigella* infection.

Systemic complications include the hemolytic-uremic syndrome, seen almost exclusively with *S. dysenteriae* type 1 infections (46, 47) and related in uncertain fashion to the production of Shiga toxin by *S. dysenteriae* type 1 (48). The hemolytic-uremic syndrome consists of the triad of microangiopathic hemolytic anemia, thrombocytopenia, and renal failure. Profound leucocytosis is also an accompanying feature of HUS, but may also occur in isolation. Onset of HUS is usually 5–7 days after the initiation of illness.

Septicaemia—with *Shigella* or with other organisms—can occur even in the absence of frank gut perforation—presumably either because of leakage of gut flora across microabscesses or ulcerations, or nosocomial infections from intravenous needles and fluids (49). *Shigella* may also, uncommonly, infect other epithelial surfaces, including vaginal and conjunctival mucosa (50). Indeed, inoculation of *Shigella* into guinea pig conjunctiva with subsequent inflammation (the Sereny test) had been a classical means of identifying *Shigella* or *E. coli* capable of invading mucosal cells (51).

Conjunctivitis may also occur as a result of the post-infectious arthritis syndrome—a complication rarely reported from developing countries (3, 52), perhaps because this complication is uncommon, or perhaps because follow-up of patients is difficult.

Altered consciousness and seizures occur with all species of *Shigella* (53). Most seizures are presumably febrile in origin—though they may also occur with the metabolic derangements that occur with shigellosis. Encephalopathy—manifested as altered consciousness and coma—may simply be a result of fever and systemic toxicity, be a premorbid condition, or be related to metabolic abnormalities. Although Shiga toxin was first described as a neurotoxin, there is no evidence that it plays a role in altered consciousness during shigellosis, and no *Shigella*-specific virulence factor has been described that could account for the central nervous

manifestations of shigellosis. Ekiri—shigellosis accompanied by a frequently lethal encephalopathy—was commonly reported in pre- and post-Second World war Japan (54). It is now rare there, and is a term that generally has fallen into disuse.

Metabolic complications include dehydration, hyponatremia (55), and hypoglycemia (56). Because the site of infection with shigellosis is the large bowel, the voluminous watery diarrhea that characterizes the toxin-mediated enteric infections of the small bowel (enterotoxigenic *E. coli*, *Vibrio cholerae*) is not a part of shigellosis. In most clinical settings, severe dehydration during shigellosis is rare. Dehydration, if it does occur, is presumably a result of some water loss in the stool, increased evaporative loss from fever, and diminished fluid intake.

Hypoglycemia does not appear to be specific for shigellosis, but may occur with any enteric infection, and is likely due to failed gluconeogenesis in marginally nourished children (56). Severe hyponatremia—with serum sodium concentration below 120 mmol/L—seems to more specifically occur with *S. dysenteriae* type 1 infection, and is associated with inappropriate secretion of antidiuretic hormone.

Adverse nutritional consequences may be the most important of all complications from shigellosis. Effects of shigellosis on nutrition occur from multiple insults. These include protein loss from the gut as discussed above (the colonic epithelium in severe *Shigella* colitis resembles a burn surface), diminished intake because of anorexia, and increased metabolism because of inflammation and fever. A number of studies have shown that shigellosis has, when compared to other enteric infections, a disproportionately adverse effect on nutritional status.

15.4.2 Diagnosis

Definitive diagnosis of *Shigella* infection is made by isolation of the organism from stool. Stool samples are plated upon media selective for gram-negative organisms. Such selective agars include MacConkey, *Salmonella–Shigella*, Xylose–Lysine–Deoxy cholate, or Hektoen Enteric agars. Usually two agar plates are used—one that is less selective (such as MacConkey) and one that is more selective (such as *Salmonella–Shigella*) to give the highest yield. *Shigella* does not ferment lactose and thus appears white or colorless on selective media containing acid-sensitive indicator dyes. Nonlactose fermenting colonies are then further analyzed using biochemical tests for x presumptive identification and then confirmed as *Shigella* by slide agglutination test with specific antiserum.

In most health facilities in developing countries where shigellosis is endemic or epidemic, diagnostic microbiologic services are not available. Rapid bedside tests for diagnosis of *Shigella* are not currently commercially available (57)—though an immunochromatographic dipstick for identification of *S. flexneri* 2a has been found to have good specificity and sensitivity in preliminary field testing (58), and rapid tests for detection of Shiga toxin may be useful in diagnosis of *S. dysenteriae* type 1 infection. Microscopic examination of stool—using either a wet mount or a Gram stain—can be supportive of the diagnosis of shigellosis by showing large numbers of stool leukocytes and erythrocytes (59).

But for the most part, the diagnosis of shigellosis remains a clinical one, and treatment is empiric based upon the clinical suspicion of shigellosis. In regions where *Shigella* is known to be endemic (virtually all of the developing world) or epidemic, children and adults with dysentery—bloody mucoid diarrhea—should be assumed to have shigellosis and treated accordingly (29). Even when diagnostic microbiologic facilities are available, treatment needs to be initiated early in the course of disease if complications are to be prevented (60).

Entamoeba histolytica infection is the other common cause of acute bloody diarrhea in developing countries. Tenesmus is generally less severe in amoebiasis than in shigellosis. Symptomatic amoebiasis has generally thought to be rare in children (61), but some recent studies suggest it might be more common than previously suspected (62). Microscopic examination of the stool can be helpful in differentiating the two—patients with *Entamoeba* trophozoites showing erythrophagocytosis are diagnostic of amoebiasis (63), and patients with shigellosis have more erythrocytes and leukocytes (59).

15.4.3 Treatment

Antimicrobial treatment is indicated for all persons with dysenteric shigellosis. Initiation of antimicrobial therapy early in the course of illness with an effective agent shortens the duration of illness (64), reduces its severity, and prevents complications, including death (60, 64). For those with primarily watery or mucoid diarrhea, no systemic toxicity, and without underlying illnesses (malnutrition, HIV-infection), antimicrobial therapy is generally not indicated.

Antimicrobials used for the treatment of shigellosis need to be active *in vitro* against the majority of circulating strains of *Shigella*, systemically absorbed rather than simply intraluminal (65), available in an oral formulation for use for children in the poor and often rural settings in developing countries where *Shigella* morbidity is greatest, inexpensive, and shown effective in randomized, controlled clinical trials (66).

Because treatment should be initiated early in the course of disease, treatment cannot await isolation of the organism and identification of its antimicrobial susceptibility, which usually takes 72 h. Empiric therapy thus depends on regular surveillance of antimicrobial susceptibility patterns of *Shigella* strains circulating in the community. Because *Shigella*, unlike *V. cholerae* and enterotoxigenic *E. coli*, invades the gut mucosa, agents that are not absorbed are ineffective. But even agents that meet most of the above criteria—such as

oral cephalosporins (67) and amoxicillin (68)—can be ineffective when evaluated in controlled trials.

Finding agents that currently meet all of these criteria is increasingly difficult. Preeminent among current challenges is the widespread resistance to agents that had historically been used for treatment of *Shigella* infections (12, 27, 28, 35, 38, 69–76), and the dearth of new antimicrobial agents active against *Shigella* (77). Ampicillin and cotrimoxazole had been the mainstay of treatment of shigellosis in the 1970s and 1980s, but widespread resistance beginning in the 1980s (78) made their use increasingly problematic. Though resistance can wax and wane (ampicillin resistance among *S. sonnei* isolates in Belgium decreased from 31.8% of all isolates in 2000 to 9.6% in 2007, for instance), it mostly waxes. Thus, once resistance to an agent occurs, it puts more pressure on the few remaining agents that are useful, increasing the likelihood that they will develop resistance. It also limits opportunities to rotate the use of agents.

Currently, azithromycin (79) and the fluoroquinolones (80) are the only oral agents generally useful in the treatment of shigellosis (Table 15.1). Ceftriaxone is also effective (81), but must be administered parenterally, a limiting factor in health facilities in most developing countries. Resistance to all of these agents has been reported (72, 74, 82)—and if it becomes widespread, there are no current additional options for treatment of *Shigella* infections.

Because shigellosis can have such a profound effect on nutritional status, there is a need to assure continued and enhanced feeding during illness and recovery. The use of a high-protein diet, if available, and even if used for a short period, can enhance recovery, increase serum protein concentrations, and improve weight gain and growth following illness (83–86). Rectal prolapse should be treated with compresses and will abate if the primary infection is treated.

Metabolic complications—hypoglycemia and hyponatremia—should be identified with the use of biochemical tests. Hypoglycemia can be corrected with intravenous dextrose—2.0 mL/kg of 25% glucose. Patients with severe hyponatremia (Na < 120 mmol/L) should receive 3% NaCl (12 mL/kg over a 4-h period); less severe hyponatremia (Na; 120–130 mmol/L) can have infusion of 0.9% NaCl and restriction of water intake. Severe hemolytic-uremic syndrome (renal failure) requires hemodialysis.

Zinc supplementation improves immunity and mucosal integrity. Zinc treatment given for 10–14 days during and after a diarrheal episode, as now recommended by the WHO and UNICEF (87) is associated with a reduction in severity and duration of the diarrhea episode and reduction in the incidence of diarrhea in the three months following zinc therapy. In at least one study conducted in shigellosis, zinc-supplemented children gained significantly more weight at recovery and experienced significantly fewer episodes of diarrhea compared to the control children during a 6-month follow-up period. Zinc supplementation also improved immune response to *Shigella* (88).

TABLE 15.1 Current Options for the Treatment of *Shigella* Infections

Antimicrobial Agent		Pediatric			Adult			Comment
		Dose/Route of Administration	Frequency	Duration	Dose/Route of Administration	Frequency	Duration	
Drugs of Choice								
Ciprofloxacin	Multiple dose	15 mg/kg orally to maximum adult dose	Every 12 h	3 days	500 mg	Every 12 h	3 days	Single-dose therapy is not effective in treatment of *S. dysenteriae* type 1 infections, and may not be effective in the treatment of infections caused by other *Shigella* species that are not highly susceptible to ciprofloxacin (MIC $\leq 0.004\,\mu g/mL$)
	Single dose	Not evaluated			1 g	Upon diagnosis	Single dose	
Azithromycin		15 mg/kg orally to maximum adult dose as initial dose, followed by 10 mg/kg day to maximum adult dose on subsequent days	Every 24 h	5 days	500 mg day 1 followed by 250 mg on subsequent days	Every 24 h	5 days	
Alternative Drug Therapies								
Pivmecillinam (registered as amdinocillin pivoxcil in the United States, where it currently is not available)		20 mg/kg orally to maximum adult dose	Every 6 h	5 days	400 mg	Every 6 h	5 days	In contrast to ampicillin, pivmecillinam selectively binds to penicillin-binding protein 2, and is relatively resistant to many common β-lactamases. Thus, may be effective in treatment of strains resistant to ampicillin
Ceftriaxone		50 mg/kg intravenously or intramuscularly to maximum individual dose of 1.5 g	Every 24 h	2–5 days	Not evaluated			Because this agent must be administered parenterally it is reserved for use in patients who are infected with strains of *Shigella* resistant to first-line oral drugs or are severely ill and need parenteral drugs in hospital. It has not been evaluated in controlled studies for treatment of *S. dysenteriae* type 1. For infection with other species of *Shigella* both two-dose 2 day and five-dose 5 day courses of therapy have proven successful. Although not evaluated in adults, there is no reason to believe it would not be effective in adults. In controlled trials oral cephalosporins have been ineffective in the treatment of shigellosis and should not be used

Vitamin A is essential for the functioning of the immune system and growth and development of children. Vitamin A supplementation of children is a national policy in developing countries where deficiency is prevalent. It reduces childhood mortality from diarrheal diseases and measles-related diarrhea. It also reduces disease severity in patients with diarrhea. Although one study found that vitamin A supplementation resulted in higher rates of clinical but not bacteriological cure in children with shigellosis (89), the study has a number of limitations (90), and vitamin A is not routinely recommended in the treatment of shigellosis.

The problem with most of these treatment recommendations is that they presuppose a functioning, equipped health care system. Because shigellosis is a disease of poverty, where the majority of shigellosis cases occur the ability to treat these complications—even in referral hospitals—is limited. Thus, the focus in management should be on early antimicrobial therapy using well-defined algorithms such as the Integrated Management of Childhood Illness (91), accompanied by knowledge of the current antimicrobial susceptibility patterns of *Shigella*. Early treatment, along with efforts at prevention and improved nutrition, will be much more cost effective than treating severe complications of shigellosis.

15.4.4   Outcome

Death in shigellosis usually occurs in two circumstances: epidemic or endemic *S. dysenteriae* type 1 infections, which can cause appreciable mortality in those who were previously healthy, especially children and the elderly (37), and infections with other *Shigella* species in debilitated and malnourished infants and children (42). Case fatality rates have ranged from 0% to 2.6% in community-based studies, and from 0% to 21% in health-facility-based studies (3).

More recent surveillance studies—both community- and facility-based—have found no mortality resulting from shigellosis (27, 28), perhaps because of generally improving nutritional status, enhanced immunization rates for measles and other childhood infections, and the absence of *S. dysenteriae* type 1 infections. These secular changes are reminiscent of what happened in the West and Japan, where bacillary dysentery was an important cause of death in the first half of the twentieth century (92, 93), but rarely if ever results in death or serious complications now (26). That law firms specializing in medical malpractice still troll for patients with shigellosis (http://www.pritzkerlaw.com/section-foodborne-illness/shigella/shigella-complications.html, http://www.marlerblog.com/legal-cases/shigella-foodpoisoning-outbreak-at-an-illinois-subway—shigella-has-a-long-history-of-foodborne-ill/ suggest that the occasional, potentially lucrative (if overstated) complication from shigellosis must still occur, even in industrialized countries.

Among the selected group of patients admitted to hospital in poor countries, death rates can be as high with *S. sonnei*—

thought to be the least virulent species of *Shigella*—as with the much more virulent *S. dysenteriae* type 1 (42). This is presumably because *S. sonnei* is more common among infants (41), among whom most infections are more likely to be fatal, and because *S. dysenteriae* type 1 infections, because of the more severe dysentery, are likely to prompt hospital admission at all ages. The majority of deaths, however, are caused by *S. flexneri* (94), the most common cause of *Shigella* infections in developing countries.

Among risk factors for death identified in different studies are hypoproteinemia, which presumably reflects the intensity of colitis and underlying malnutrition, being young or being elderly, initial ineffective antimicrobial therapy, thrombocytopenia, hyponatremia, septicemia, and altered consciousness or seizures (41, 42, 53, 94–97).

The other common deleterious outcome of shigellosis is diminished nutritional status (84, 98). Shigellosis has a strong negative effect on linear growth of children (99). The increased death rate of patients infected with *Shigella*, compared to those with watery diarrhea, after hospital discharge suggest the important consequences of the nutritional impact of shigellosis (94).

Persistent diarrhea is also more frequent in children with *Shigella* dysentery (100) and undoubtedly worsens the effect of shigellosis on nutritional status. Moreover, the duration of diarrhea increases progressively as nutritional status indicators worsen (99) and severely stunted children are most at risk of having prolonged episodes of dysentery (100).

There are other long-term sequelae, most of which are uncommon. Long-term renal impairment can follow the hemolytic-uremic syndrome. Patients may develop reactive arthritis following shigellosis, but this has been rarely reported from the tropics (52)—either because it is truly rare (most likely) or because ascertainment is poor (also probably likely). Metabolic abnormalities, if not recognized and treated promptly, can have long-term neurologic consequences.

15.5   PREVENTION AND CONTROL

Shigellosis is a water-washed (in contrast to a waterborne) disease. Unlike cholera (a waterborne infection), which is transmitted via contaminated water, *Shigella* infections occur because of inadequate availability and use of water for hand washing and sanitary disposal of feces. Thus, the focus for prevention and control of shigellosis is improving access to clean water and increasing its use, along with soap, for washing of hands whenever they come into contact with feces and before preparing or eating foods. Other prevention strategies can complement the effort to improve hand washing (101). Here we briefly discuss control strategies known or likely to be effective in shigellosis, some of which are discussed in more detail elsewhere in this book, and are summarized in Table 15.2.

TABLE 15.2   Options for Prevention of Shigellosis

Option	Efficacy	Cost	Practicality	Challenges
Hand washing	High—one study found a sevenfold decrease in secondary attack rate when hand washing was implemented	Low if the other elements for hand-washing—soap and water—are in place	Extremely practical if persons can be persuaded to make it a habit	Major challenges include sufficient provision of water to households, lack of money to purchase soap, and eliciting changes in behavior so that hand washing—especially after defecation—is seen as a normative behavior. That latter also requires some conception of the dangers posed by feces, and contact with feces
Provision of soap	A critical element for effective hand washing	Still an expensive item for the poorest persons in society—and those most at risk of *Shigella* infection	In poor countries access to soap is still limited. A free market distribution system—in contrast to government provision—is likely to be most effective, and avoid the inevitable corruption that occurs with government provision of commodities	As with hand washing, important for persons to understand the health benefits of soap, and not just their cosmetic benefit. Other option is for the mechanical cleansing of hands—using ash for instance. How culturally acceptable the use of ash is remains uncertain
Provision of water	A crucial element in the hand washing strategy	The infrastructure for provision of water access can be costly—especially in rural areas and in urban slums. For prevention of shigellosis, provision of clean water is not as crucial as having water available for hand washing	Once water is available, using it for hand washing is not difficult	There remain immense challenges in providing the poor of the world (and increasingly richer citizens) ready availability to water. This is discussed in detail in other chapters of this book
Fly control	One study in military camps reported an 85% decrease in shigellosis with the use of fly traps. How effective this would be in other contexts—where the majority of x *Shigella* transmission is likely to be person-to-person—is uncertain	Likely to be high	Most areas require x fundamental improvements in sanitation—isolation of both human and animal excrement and trash	Fly traps (as used in the one study showing an impact on reduction of shigellosis) and pesticides are not likely to be effective sustained measures for fly control in developing countries. Improved sanitation—both in urban and rural areas—is likely to have more pronounced and sustainable effects. In developing countries containment and isolation of animal feces as a site of fly breeding is a problem as great as that of isolation of human feces

(continued)

TABLE 15.2   (*Continued*)

Option	Efficacy	Cost	Practicality	Challenges
Improved sanitation	If combined with hand washing, likely to have large impact on *Shigella* transmission	Very considerable	An inevitable accompaniment of socioeconomic development, but innovations exist—in latrine design and public sanitation—that work in low-income areas	The challenge is whether it can be accomplished absent fundamental changes in socioeconomic and political conditions. Lots of options are available if the political will and community cohesion exist
Breast-feeding	Effect most dramatic in those less than 1 year, but benefit can extend through age 3	Very low—actually saves money over formula feeding	Extremely	Commercial formulas still heavily promoted, and breast-feeding challenging for the increasing number of women working outside the household
Improved socioeconomic conditions	Industrialized (developed countries) have rates of *Shigella* infection 1/10 or less than those of poor countries. The elements of development that contribute to the diminishment of shigellosis presumably are some combination of most of the interventions listed above	Not a health intervention per se—and the cost of stimulation of socioeconomic development presumably recovered many times over in subsequent years	It has happened to many countries, so possible for it to happen in others	The determinants of socioeconomic development and ways to enhance the rate at which it happens (or to ensure it happens at all) remain a subject of debate among development economists, political leaders, and aid agencies
Vaccine	No vaccine currently available, so cannot estimate efficacy	Most vaccines, especially those given in childhood, have proved to be cost-effective	Depending on vaccine, if immunity is long-lived, and vaccine can be given in infancy, very practical	Has been the Holy Grail of *Shigella* control for more than 100 years. Efforts in part frustrated by the multiple species and serotypes

15.5.1   Hand Washing

The effect of hand washing on transmission of *Shigella* is dramatic. In a study conducted in 1982 by M. U. Khan in Bangladesh, simple provision of soap, and vessels for storing water, to households in which someone had been identified as having shigellosis reduced the secondary attack rate of shigellosis by sevenfold—from 14% to 2% (104). Many other studies have confirmed the effectiveness of water availability and hand washing on reducing the incidence of diarrhea (102), and a systematic review of all published studies concluded that risk of *Shigella* infection could be reduced by 59% with hand washing (103). Hand washing should be done after defecation, after cleaning a child who has defecated, after disposal of a child's feces, and before handling or eating food. If soap is not available, ash or earth can be used as an alternative to scrub the hands (104). Soiled cloths should not be used to wipe the washed hands.

15.5.2   Water Supply

Effective hand washing requires water, and too many households in developing countries lack adequate access to water—even for use in hand washing. The impact that limited access to water has on the incidence of shigellosis is reflected in the experience of daycare centers. In one study, comparing daycare centers with a sink in every room versus those with fewer sinks, day care centers with a sink in every room was had a significantly lower secondary attack rates of shigellosis (12). Patterns of water use and storage are also associated with shigellosis attack rates in developing countries (28).

Providing water to households in developing countries remains a major challenge (Fig. 15.3). Piped, chlorinated water from a central supply is optimal, but is currently available to only a minority of the world's population. The use of surface water for drinking and cooking should be avoided. If this is not possible, water should be chlorinated or boiled before use. The source of water for culinary use should be different from the one used for general purposes such as bathing and washing.

In the household, water should be stored in narrow-mouthed containers to prevent dipping of contaminated hands. If water is stored in wide-mouthed containers, a long-handled dipper should used to draw water. Water containers should be covered and kept away from children and animals. The water containers should be cleaned regularly.

15.5.3   Safe Disposal of Human Waste

There should be health messages on the dangers of defecation outside of a toilet—that is, defecating on the ground, in water or near a water supply. It is desirable for every household to have a toilet for defecation. For rural communities, a low-cost and effective toilet is the ventilated improved pit (VIP) latrine. The feces of children unable to use a toilet should be disposed of in the toilet. In the absence of a toilet, burying excreta in the ground reduces transmission of pathogens.

15.5.4   Fly Control

The common housefly, *Musca domestica,* feeds on human excreta, then mechanically carries organisms and contaminates exposed food items by depositing the organisms on the food. Fly density in the kitchen area is associated with an increased risk of shigellosis (105). Reducing the density of flies in military camps has been shown to significantly reduce the transmission of *Shigella.* Baited fly-traps have been shown to be effective in reducing fly density (106).

15.5.5   Breast-Feeding

Breast-milk is a source of nutrition, and it possesses specific and nonspecific antibacterial factors. Moreover, breast-feeding reduces the exposure of children to contaminated food. Breast-fed children have fewer episodes of diarrhea due to *Shigella* (107). When infections do occur, they are less severe compared to infections in children who are not breast-fed. This effect on severity of infection is evident

(a)

(b)

FIGURE 15.3   (a) Children playing in pond outside their village in Bangladesh. (b) Mother washing dishes in pond where children were playing, and possibly defecating. Defecating in water used for recreation has been an increasingly frequent cause of *Shigella* transmission in the United States. It is exacerbated in developing countries by using the same water for cooking and cleaning purposes.

particularly in malnourished children (107). Protection against shigellosis is greatest in exclusively breast-fed infants up to 6 months of age, but remains significant when breast-milk is supplemented with other foods in children into the third year of life (107).

15.5.6 Food Safety

Food items can be contaminated at any stage from "farm to table." To avoid contamination of food, food safety practices should be emphasized. These include hygienic washing of hands before handling food, cooking food until hot throughout, eating food while hot, storing food at a safe temperature (4°C), reheating stored food thoroughly, keeping cooked food and clean utensils separate from uncooked food and unclean utensils, not eating raw food except fruits that are peeled and eaten immediately, and protecting food from flies by means of fly screens.

15.5.7 Measles Immunization

There are conflicting reports on the role of *Shigella* in the etiology of measles-associated dysentery. A study in Bangladesh found an association between measles-associated dysentery and *Shigella* (108), but such an association was not found in studies in Peru (109), Kenya (110), and Rwanda (111). Nevertheless, measles immunization of children, which is now increasingly common, should protect against measles and hence, measles-associated diarrhea.

15.5.8 Vaccines against *Shigella*

The definitive prevention strategy is to develop an effective vaccine. An effective vaccine could theoretically lead (as with efforts with smallpox) to the eradication of the disease, since humans are the only reservoir of the organism, and humans are rarely long-term carriers of the organism.

Although efforts to develop an effective vaccine have been going on since soon after the identification of the causative agent more than 100 years ago, there are still no licensed vaccines for use in humans. Immunity in shigellosis is largely serotype-specific, thus complicating the development of a vaccine. Therefore, vaccine development strategies are to make vaccines against the species and serotypes that cause the majority of serious infections. These include *S. dysenteriae* type 1; *S. flexneri* types 2a, 3a, and 6; and *S. sonnei*. There are approaches to make either live, attenuated strains for oral vaccination or subunit vaccines for parenteral immunization. The attenuated strains currently under investigation are either metabolic mutants that maintain immunogenicity but lack sustained virulence or strains with multiple mutations affecting virulence and metabolism. The challenge remains

achieving attenuation in virulence while maintaining immunogenicity.

A subunit vaccine containing detoxified lipopolysaccharide (LPS) antigen from *S. sonnei* conjugated to a carrier protein and administered as a parenteral vaccine induced significant protection during an outbreak of *S. sonnei* dysentery. A similar *S. flexneri* 2a vaccine is being tested. Another way to making a subunit vaccine is the use of glycoconjugates that are synthetic mimics of protective epitopes of LPS antigens that can be conjugated to common virulence antigens as carriers (112). A 34 kDa outer membrane protein common to all species of *Shigella* appears to be a promising antigen for broad protection. Immunization with this protein afforded protection against dysentery due to *S. flexneri* 2a in a rabbit model of dysentery (113).

15.6 CONCLUSION

Shigellosis, as much as any disease, results from lack of access to sufficient water for personal hygiene, and the behavioral changes necessary to utilize water for hand washing when it is available, and for safe disposal of feces.

It is now more than 160 years since Semmelweis showed—before the germ theory of disease had been elaborated—that hand washing can be crucial to interrupting transmission of disease (114). Despite the evidence—continued studies showing the benefit of hand washing in community and in health care settings (102–104, 115, 116)—hand washing is often more honored in the breach than in the observance, even among the highly educated and professionals who know its importance (117). And while it is clear that demonstration projects can have a marked effect on improving hand washing and reducing diarrhea transmission (103), the challenge remains how to imbed these practices within society.

The problem is in part cultural beliefs and practices—either among health care professionals or among the poor in developing countries—that do not reinforce practices that would be effective in controlling transmission of infections like *Shigella*. Profound changes in personal and community hygiene occurred in Europe and the United States in the nineteenth and early twentieth century—and in the United States was an important part of the acculturation process for immigrants (118). Those efforts in the United States stretch back to the homilies of Benjamin Franklin and the early church leaders. Indeed, one of Franklin's "13 virtues" was "tolerate no uncleanliness in body, clothes, or habitation." That the interpretation of what constitutes cleanliness coincided with what is necessary to control infections might have been in part serendipity—after all these efforts at defining what is clean predated the germ theory of disease (118). Obsessive cleanliness in the United States—even before the current fad for hand sanitizers accentuated by the recent

FIGURE 15.4 Advertisement for soap in Bangladesh, which appeals to its cosmetic, rather than its hygienic, properties.

H1N1 influenza epidemic—has nonetheless been effective at diminishing much of the burden of shigellosis.

Persons in developing countries often are as strongly committed to cleanliness—especially ritual cleanliness—as those in industrialized countries (119). But the conceptualization of what constitutes cleanliness may not coincide with what is required for control of shigellosis. Hands may be washed before prayer but not before eating, and a common bowl of water used for hand washing. Soap may be viewed (enhanced by advertisements and promotions) more as a beauty product than one for killing microorganisms and interrupting transmission of infections (Fig. 15.4).

But no change in perceptions can affect behavior if the elements required for that changed behavior—ready access to clean water, income to buy soap, diminished household crowding, support for decent child care and day care—are lacking.

Although book chapters are supposed to be impersonal and constrained, rather than personal and discursive, we cannot help reflecting about our personal situation, water in the contemporary world, and how it affects control of shigellosis.

One of us (MLB) is living in rural South Africa (the economic powerhouse of Africa) on a farm previously owned by a white farmer. This farm—unlike the homesteads where blacks lived—had a connection to the water supply of the town 10 km distant, and its own borehole for supplemental water. Because of drought over the last 6 years, the river that is the source for the municipal water is virtually dry for 3 or more months every year forcing severe water rationing to the more affluent who are on the municipal supply, and the bore hole is also becoming undependable as more and more water is drawn from it.

The impoverished majority of the population that historically had to rely on contaminated surface water (and suffered

period epidemics of *S. dysenteriae* type 1 and *V. cholerae* as a result) (120, 121) has seen, since the advent of democratic government, extensive efforts to enhance their access to water. That remains difficult in rural settings, where homesteads are widely scattered, and persons trudging long distances to water sources with an array of water containers heaped in a wheelbarrow remains an all too common sight. In urban townships and slums, efforts to ensure that costs of water supply are recouped have limited access to water for many (122).

The other of us (JA) works in a country that historically had poor sanitation, where flies were ubiquitous, and dysentery common. It now depends on desalination for much of the water it uses—an expensive process funded by the sale of a nonrenewable resource—oil.

All of this points out the continuing challenges of increasing provision of water to households for use in hand washing—and the challenges of maintaining what we now have given the widespread environmental challenges to water accessibility.

The effect of global warming on *Shigella* transmission is unknown. Some (28, 44), but not all (3), studies have found that shigellosis is greater during the summer—possibly because of enhanced survival of *Shigella* on food, fomites, and in the environment. Global warming would exacerbate this problem.

There is also much to be encouraged by. Death rates from shigellosis appear to be declining—presumably in part because of the apparent disappearance of *S. dysenteriae* type 1 epidemics from the world during the first decade of the twenty-first century, and also because, despite the challenges, of the general improvement in child nutrition and general health worldwide. The latter can be attributed in part to widespread micronutrient supplementation, enhanced immunization coverage, improvements in maternal education, and enhanced economic circumstances for many. But much remains to be done in controlling shigellosis, and crucial among these are making available to all the water access and sanitary facilities that should be a basic human right.

REFERENCES

1. Islam MS, Hossain MA, Khan SI, Khan MN, Sack RB, Albert MJ, Huq A, Colwell RR. Survival of *Shigella dysenteriae* type 1 on fomites. *J Health Popul Nutr* 2001;19:177–182.

2. Kotloff KL, Winickoff JP, Ivanoff B, Clemens JD, Swerdlow DL, Sansonetti PJ, Adak GK, Levine MM. Global burden of *Shigella* infections: implications for vaccine development and implementation of control strategies. *Bull World Health Organ* 1999;77:651–666.

3. Ram PK, Crump JA, Gupta SK, Miller MA, Mintz ED. Part II. Analysis of data gaps pertaining to *Shigella* infections in low and medium human development index countries, 1984–2005. *Epidemiol Infect* 2008;136:577–603.

4. Grimont F, Lejay-Collin M, Talukder KA, Carle I, Issenhuth S, Le Roux K, Grimont PA. Identification of a group of x *Shigella*-like isolates as *Shigella boydii* 20. *J Med Microbiol* 2007;56:749–754.

5. Nataro JP, Seriwatana J, Fasano A, Maneval DR, Guers LD, Noriega F, Dubovsky F, Levine MM, Morris JG, Jr., Identification and cloning of a novel plasmid-encoded enterotoxin of enteroinvasive *Escherichia coli* and *Shigella* strains. *Infect Immun* 1995;63:4721–4728.

6. Fasano A, Noriega FR, Liao FM, Wang W, Levine MM. Effect of shigella enterotoxin 1 (ShET1) on rabbit intestine *in vitro* and *in vivo*. *Gut* 1997;40:505–511.

7. DuPont HL, Levine MM, Hornick RB, Formal SB. Inoculum size in shigellosis and implications for expected mode of transmission. *J Infect Dis* 1989;159:1126–1128.

8. Small P, Blankenhorn D, Welty D, Zinser E, Slonczewski JL. Acid and base resistance in *Escherichia coli* and *Shigella flexneri*: role of rpoS and growth pH. *J Bacteriol* 1994;176:1729–1737.

9. Jennison AV, Verma NK. *Shigella flexneri* infection: pathogenesis and vaccine development. *FEMS Microbiol Rev* 2004;28:43–58.

10. Sansonetti PJ. The bacterial weaponry: lessons from *Shigella*. *Ann NY Acad Sci* 2006;1072:307–312.

11. Schroeder GN, Hilbi H. Molecular pathogenesis of *Shigella spp.*: controlling host cell signaling, invasion, and death by type III secretion. *Clin Microbiol Rev* 2008;21:134–156.

12. Arvelo W, Hinkle CJ, Nguyen TA, Weiser T, Steinmuller N, Khan F, Gladbach S, Parsons M, Jennings D, Zhu BP, Mintz E, Bowen A. Transmission risk factors and treatment of pediatric shigellosis during a large daycare center-associated outbreak of multidrug resistant *Shigella sonnei*: implications for the management of shigellosis outbreaks among children. *Pediatr Infect Dis J* 2009;28:976–980.

13. Haley CC, Ong KL, Hedberg K, Cieslak PR, Scallan E, Marcus R, Shin S, Cronquist A, Gillespie J, Jones TF, Shiferaw B, Fuller C, Edge K, Zansky SM, Ryan PA, Hoekstra RM, Mintz E. Risk factors for sporadic shigellosis, FoodNet 2005. *Foodborne Pathog Dis* 2010;7:741–747.

14. Aragon TJ, Vugia DJ, Shallow S, Samuel MC, Reingold A, Angulo FJ, Bradford WZ. Case-control study of shigellosis in San Francisco: the role of sexual transmission and HIV infection. *Clin Infect Dis* 2007;44:327–334.

15. DuPont HL, Gangarosa EJ, Reller LB, Woodward WE, Armstrong RW, Hammond J, Glaser K, Morris GK. Shigellosis in custodial institutions. *Am J Epidemiol* 1970;92:172–179.

16. Kerneis S, Guerin PJ, von Seidlein L, Legros D, Grais RF. A look back at an ongoing problem: *Shigella dysenteriae* type 1 epidemics in refugee settings in Central Africa (1993–1995). *PLoS One* 2009;4:e4494.

17. Guerin P, Grais R, Rottingen J, Valleron A, Group SS. Using European travellers as an early alert to detect emerging pathogens in countries with limited laboratory resources. *BMC Public Health* 2007;7:8.

18. Hardy AV, Mason RP, Martin GA. The dysenteries in the armed forces. *Am J Trop Med Hyg* 1952;1:171–175.

19. Thornton SA, Sherman SS, Farkas T, Zhong W, Torres P, Jiang X. Gastroenteritis in US Marines during Operation Iraqi Freedom. *Clin Infect Dis* 2005;40:519–525.

20. CDC. Surveillance for foodborne disease outbreaks—United States 2006. *Morb Mortal Wkly Rep* 2009;58:609–615.

21. Yoder JS, Hlavsa MC, Craun GF, Hill V, Roberts V, Yu PA, Hicks LA, Alexander NT, Calderon RL, Roy SL, Beach MJ. Surveillance for waterborne disease and outbreaks associated with recreational water use and other aquatic facility-associated health events—United States, 2005–2006. *MMWR Surveill Summ* 2008;57:1–29.

22. Sorvillo FJ, Waterman SH, Vogt JK, England B. Shigellosis associated with recreational water contact in Los Angeles County. *Am J Trop Med Hyg* 1988;38:613–617.

23. Keene WE, McAnulty JM, Hoesly FC, Williams LP, Hedberg K, Oxman GL, Barrett TJ, Pfaller MA, Fleming DW. A swimming-associated outbreak of hemorrhagic colitis caused by *Escherichia coli* O157:H7 and *Shigella sonnei*. *N Engl J Med* 1994;331:579–584.

24. Yoder J, Roberts V, Craun GF, Hill V, Hicks LA, Alexander NT, Radke V, Calderon RL, Hlavsa MC, Beach MJ, Roy SL. Surveillance for waterborne disease and outbreaks associated with drinking water and water not intended for drinking—United States, 2005–2006. *MMWR Surveill Summ* 2008;57:39–62.

25. Mead PS, Slutsker L, Dietz V, McCaig LF, Bresee JS, Shapiro C, Griffin PM, Tauxe RV. Food-related illness and death in the United States. *Emerg Infect Dis* 1999;5:607–625.

26. Gupta A, Polyak CS, Bishop RD, Sobel J, Mintz ED. Laboratory-confirmed shigellosis in the United States, 1989–2002: epidemiologic trends and patterns. *Clin Infect Dis* 2004;38:1372–1377.

27. von Seidlein L, Kim DR, Ali M, Lee H, Wang X, Thiem VD, Canh do G, Chaicumpa W, Agtini MD, Hossain A, Bhutta ZA,

Mason C, Sethabutr O, Talukder K, Nair GB, Deen JL, Kotloff K, Clemens J. A multicentre study of *Shigella* diarrhoea in six Asian countries: disease burden, clinical manifestations, and microbiology. *PLoS Med* 2006;3:e353.

28. Kosek M, Yori PP, Pan WK, Olortegui MP, Gilman RH, Perez J, Chavez CB, Sanchez GM, Burga R, Hall E. Epidemiology of highly endemic multiply antibiotic-resistant shigellosis in children in the Peruvian Amazon. *Pediatrics* 2008;122:e541–e549.

29. Ronsmans C, Bennish ML, Wierzba T. Diagnosis and management of dysentery by community health workers. *Lancet* 1988;2:552–555.

30. Wells JG, Morris GK. Evaluation of transport methods for isolating *Shigella* spp. *J Clin Microbiol* 1981;13:789–790.

31. Black RE, Brown KH, Becker S, Alim AR, Huq I. Longitudinal studies of infectious diseases and physical growth of children in rural Bangladesh. II. Incidence of diarrhea and association with known pathogens. *Am J Epidemiol* 1982;115:315–324.

32. Punyaratabandhu P, Vathanophas K, Varavithya W, Sangchai R, Athipanyakom S, Echeverria P, Wasi C. Childhood diarrhoea in a low-income urban community in Bangkok: incidence, clinical features, and child caretaker's behaviours. *J Diarrhoeal Dis Res* 1991;9:244–249.

33. Eichner ER, Gangarosa EJ, Goldsby JB. The current status of shigellosis in the United States. *Am J Public Health Nations Health* 1968;58:753–763.

34. Peirano G, Souza FS, Rodrigues DP. Frequency of serovars and antimicrobial resistance in *Shigella* spp. from Brazil. *Mem Inst Oswaldo Cruz* 2006;101:245–250.

35. Vinh H, Nhu N, Nga T, Duy P, Campbell J, Hoang N, Boni M, My P, Parry C, Van Minh P, Thuy C, Diep T, Phuong L, Chinh M, Loan H, Tham N, Lanh M, Mong B, Anh V, Bay P, Chau N, Farrar J, Baker S. A changing picture of shigellosis in southern Vietnam: shifting species dominance, antimicrobial susceptibility and clinical presentation. *BMC Infect Dis* 2009;9:204.

36. Gangarosa EJ, Perera DR, Mata LJ, Mendizabal-Morris C, Guzman G, Reller LB. Epidemic *Shiga bacillus* dysentery in Central America. II. Epidemiologic studies in 1969. *J Infect Dis* 1970;122:181–190.

37. Rahaman MM, Khan MM, Aziz KM, Islam MS, Kibriya AK. An outbreak of dysentery caused by *Shigella dysenteriae* type 1 on a coral island in the Bay of Bengal. *J Infect Dis* 1975;132:15–19.

38. Ries AA, Wells JG, Olivola D, Ntakibirora M, Nyandwi S, Ntibakivayo M, Ivey CB, Greene KD, Tenover FC, Wahlquist SP, et al. Epidemic *Shigella dysenteriae* type 1 in Burundi: panresistance and implications for prevention. *J Infect Dis* 1994;169:1035–1041.

39. Sack DA, Hoque AT, Huq A, Etheridge M. Is protection against shigellosis induced by natural infection with *Plesiomonas shigelloides*? *Lancet* 1994;343:1413–1415.

40. Mendizabal-Morris CA, Mata LJ, Gangarosa EJ, Guzman G. Epidemic Shiga-Bacillus dysentery in Central America: derivation of the epidemic and its progression in Guatemala, 1968–69. *Am J Trop Med Hyg* 1971; 20:927–933.

41. Huskins WC, Griffiths JK, Faruque AS, Bennish ML. Shigellosis in neonates and young infants. *J Pediatr* 1994;125:14–22.

42. Bennish ML, Harris JR, Wojtyniak BJ, Struelens M. Death in shigellosis: incidence and risk factors in hospitalized patients. *J Infect Dis* 1990;161:500–506.

43. Bennish ML. Potentially lethal complications of shigellosis. *Rev Infect Dis* 1991;13 (Suppl 4):S319–S324.

44. Abu-Elyazeed RR, Wierzba TF, Frenck RW, Putnam SD, Rao MR, Savarino SJ, Kamal KA, Peruski LF, Jr., Abd-El Messih IA, El-Alkamy SA, Naficy AB, Clemens JD. Epidemiology of *Shigella*-associated diarrhea in rural Egyptian children. *Am J Trop Med Hyg* 2004;71:367–372.

45. Bennish ML, Salam MA, Wahed MA. Enteric protein loss during shigellosis. *Am J Gastroenterol* 1993;88:53–57.

46. Rahaman MM, JamiulAlam AK, Islam MR, Greenough III WB. Shiga bacillus dysentery associated with marked leukocytosis and erythrocyte fragmentation. *Johns Hopkins Med J* 1975;136:65–70.

47. Bhimma R, Rollins NC, Coovadia HM, Adhikari M. Post-dysenteric hemolytic uremic syndrome in children during an epidemic of *Shigella* dysentery in Kwazulu/Natal. *Pediatr Nephrol* 1997;11:560–564.

48. Palermo MS, Exeni RA, Fernandez GC. Hemolytic uremic syndrome: pathogenesis and update of interventions. *Expert Rev Anti Infect Ther* 2009;7:697–707.

49. Struelens MJ, Patte D, Kabir I, Salam A, Nath SK, Butler T. *Shigella* septicemia: prevalence, presentation, risk factors, and outcome. *J Infect Dis* 1985;152:784–790.

50. Baiulescu M, Hannon PR, Marcinak JF, Janda WM, Schreckenberger PC. Chronic vulvovaginitis caused by antibiotic-resistant *Shigella flexneri* in a prepubertal child. *Pediatr Infect Dis J* 2002;21:170–172.

51. Cristea D, Ceciu S, Chitoiu DT, Bleotu C, Lazar V, Chifiriuc MC. Comparative study of pathogenicity tests for *Shigella* spp. and enteroinvasive *Escherichia coli* strains. *Roum Arch Microbiol Immunol* 2009;68:44–49.

52. Mazumder RN, Salam MA, Ali M, Bhattacharya MK. Reactive arthritis associated with *Shigella dysenteriae* type 1 infection. *J Diarrhoeal Dis Res* 1997;15:21–24.

53. Khan WA, Dhar U, Salam MA, Griffiths JK, Rand W, Bennish ML. Central nervous system manifestations of childhood shigellosis: prevalence, risk factors, and outcome. *Pediatrics* 1999;103:E18.

54. Sakamoto A, Kamo S. Clinical, statistical observations on ekiri and bacillary dysentery; a study of 785 cases. *Ann Paediatr* 1956;186:1–18.

55. Rahaman MM, Alam AK, Islam MR. Letter: leukaemoid reaction, haemolytic anaemia, and hyponatraemia in severe *Shigella dysenteriae* type-1 infection. *Lancet* 1974;1:1004.

56. Bennish ML, Azad AK, Rahman O, Phillips RE. Hypoglycemia during diarrhea in childhood. Prevalence, pathophysiology, and outcome. *N Engl J Med* 1990;322:1357–1363.

57. Abubakar I, Irvine L, Aldus CF, Wyatt GM, Fordham R, Schelenz S, Shepstone L, Howe A, Peck M, Hunter PR. A systematic review of the clinical, public health and cost-effectiveness of rapid diagnostic tests for the detection and identification of bacterial intestinal pathogens in faeces and food. *Health Technol Assess* 2007;11:1–216.

58. Nato F, Phalipon A, Nguyen TL, Diep TT, Sansonetti P, Germani Y. Dipstick for rapid diagnosis of *Shigella flexneri* 2a in stool. *PLoS One* 2007;2:e361.

59. Speelman P, McGlaughlin R, Kabir I, Butler T. Differential clinical features and stool findings in shigellosis and amoebic dysentery. *Trans R Soc Trop Med Hyg* 1987;81:549–551.

60. Bennish ML, Khan WA, Begum M, Bridges EA, Ahmed S, Saha D, Salam MA, Acheson D, Ryan ET. Low risk of hemolytic uremic syndrome after early effective antimicrobial therapy for *Shigella dysenteriae* type 1 infection in Bangladesh. *Clin Infect Dis* 2006;42:356–362.

61. Albert MJ, Faruque AS, Faruque SM, Sack RB, Mahalanabis D. Case-control study of enteropathogens associated with childhood diarrhea in Dhaka, Bangladesh. *J Clin Microbiol* 1999;37:3458–3464.

62. Haque R, Mondal D, Kirkpatrick BD, Akther S, Farr BM, Sack RB, Petri WA, Jr., Epidemiologic and clinical characteristics of acute diarrhea with emphasis on *Entamoeba histolytica* infections in preschool children in an urban slum of Dhaka, Bangladesh. *Am J Trop Med Hyg* 2003;69:398–405.

63. Tanyuksel M, Petri WA, Jr., Laboratory diagnosis of amebiasis. *Clin Microbiol Rev* 2003;16:713–729.

64. Haltalin KC, Nelson JD, Ring III R, Sladoje M, Hinton LV. Double-blind treatment study of shigellosis comparing ampicillin, sulfadiazine, and placebo. *J Pediatr* 1967;70:970–981.

65. Haltalin KC, Nelson JD, Hinton LV, Kusmiesz HT, Sladoje M. Comparison of orally absorbable and nonabsorbable antibiotics in shigellosis. A double-blind study with ampicillin and neomycin. *J Pediatr* 1968;72:708–720.

66. Salam MA, Bennish ML. Antimicrobial therapy for shigellosis. *Rev Infect Dis* 1991;13:S332–S341.

67. Salam MA, Seas C, Khan WA, Bennish ML. Treatment of shigellosis: IV. Cefixime is ineffective in shigellosis in adults. *Ann Intern Med* 1995;123:505–508.

68. Nelson JA, Haltalin KC. Amoxicillin less effective than ampicillin against *Shigella in vitro* and *in vivo*: relationship of efficacy to activity in serum. *J Infect Dis* 1974;129 (Suppl):S222–S227.

69. Bennish ML, Salam MA, Hossain MA, Myaux J, Khan EH, Chakraborty J, Henry F, Ronsmans C. Antimicrobial resistance of *Shigella* isolates in Bangladesh, 1983–1990: increasing frequency of strains multiply resistant to ampicillin, trimethoprim-sulfamethoxazole, and nalidixic acid. *Clin Infect Dis* 1992;14:1055–1060.

70. Anh NT, Cam PD, Dalsgaard A. Antimicrobial resistance of *Shigella* spp. isolated from diarrheal patients between 1989 and 1998 in Vietnam. *SE Asian J Trop Med Public Health* 2001;32:856–862.

71. Guerin PJ, Brasher C, Baron E, Mic D, Grimont F, Ryan M, Aavitsland P, Legros D. Case management of a multidrug-resistant *Shigella dysenteriae* serotype 1 outbreak in a crisis context in Sierra Leone, 1999–2000. *Trans R Soc Trop Med Hyg* 2004;98:635–643.

72. Huang IF, Chiu CH, Wang MH, Wu CY, Hsieh KS, Chiou CC. Outbreak of dysentery associated with ceftriaxone-resistant *Shigella sonnei*: first report of plasmid-mediated CMY-2-type AmpC beta-lactamase resistance in *S. sonnei. J Clin Microbiol* 2005;43:2608–2612.

73. Jain SK, Gupta A, Glanz B, Dick J, Siberry GK. Antimicrobial-resistant *Shigella sonnei*: limited antimicrobial treatment options for children and challenges of interpreting in vitro azithromycin susceptibility. *Pediatr Infect Dis J* 2005;24:494–497.

74. Taneja N, Lyngdoh V, Vermani A, Mohan B, Rao P, Singh M, Dogra A, Singh MP, Sharma M. Re-emergence of multi-drug resistant *Shigella dysenteriae* with added resistance to ciprofloxacin in north India & their plasmid profiles. *Indian J Med Res* 2005;122:348–354.

75. Sivapalasingam S, Nelson JM, Joyce K, Hoekstra M, Angulo FJ, Mintz ED. High prevalence of antimicrobial resistance among *Shigella* isolates in the United States tested by the National Antimicrobial Resistance Monitoring System from 1999 to 2002. *Antimicrob Agents Chemother* 2006;50:49–54.

76. Vrints M, Mairiaux E, Van Meervenne E, Collard J-M, Bertrand S. Surveillance of antibiotic susceptibility patterns among *Shigella sonnei* strains isolated in Belgium during the 18-year period 1990 to 2007. *J Clin Microbiol* 2009;47:1379–1385.

77. Wenzel RP. The antibiotic pipeline—challenges, costs, and values. *N Engl J Med* 2004;351:523–526.

78. Bennish M, Eusof A, Kay B, Wierzba T. Multiresistant *Shigella* infections in Bangladesh. *Lancet* 1985;2:441.

79. Khan WA, Seas C, Dhar U, Salam MA, Bennish ML. Treatment of shigellosis: V. Comparison of azithromycin and ciprofloxacin. A double-blind, randomized, controlled trial. *Ann Intern Med* 1997;126:697–703.

80. Bennish ML, Salam MA, Haider R, Barza M. Therapy for shigellosis. II. Randomized, double-blind comparison of ciprofloxacin and ampicillin. *J Infect Dis* 1990;162:711–716.

81. Kabir I, Butler T, Khanam A. Comparative efficacies of single intravenous doses of ceftriaxone and ampicillin for shigellosis in a placebo-controlled trial. *Antimicrob Agents Chemother* 1986;29:645–648.

82. Boumghar-Bourtchai L, Mariani-Kurkdjian P, Bingen E, Filliol I, Dhalluin A, Ifrane SA, Weill FX, Leclercq R. Macrolide-resistant *Shigella sonnei. Emerg Infect Dis* 2008;14:1297–1299.

83. Mazumder RN, Hoque SS, Ashraf H, Kabir I, Wahed MA. Early feeding of an energy dense diet during acute shigellosis enhances growth in malnourished children. *J Nutr* 1997;127:51–54.

84. Kabir I, Rahman MM, Haider R, Mazumder RN, Khaled MA, Mahalanabis D. Increased height gain of children fed a high-protein diet during convalescence from shigellosis: a six-month follow-up study. *J Nutr* 1998;128:1688–1691.

85. Mazumder RN, Ashraf H, Hoque SS, Kabir I, Majid N, Wahed MA, Fuchs GJ, Mahalanabis D. Effect of an energy-dense diet on the clinical course of acute shigellosis in undernourished children. *Br J Nutr* 2000;84:775–779.

86. Kabir I, Butler T, Underwood LE, Rahman MM. Effects of a protein-rich diet during convalescence from shigellosis on catch-up growth, serum proteins, and insulin-like growth factor-I. *Pediatr Res* 1992;32:689–692.

87. World Health Organization and UNICEF. *Clinical Management of Acute Diarrhoea*. WHO/UNICEF Joint Statement. Geneva: The United Nations Children's Fund and World Health Organization, 2004.

88. Roy SK, Raqib R, Khatun W, Azim T, Chowdhury R, Fuchs GJ, Sack DA. Zinc supplementation in the management of shigellosis in malnourished children in Bangladesh. *Eur J Clin Nutr* 2008;62:849–855.

89. Hossain S, Biswas R, Kabir I, Sarker S, Dibley M, Fuchs G, Mahalanabis D. Single dose vitamin A treatment in acute shigellosis in Bangladesh children: randomised double blind controlled trial. *BMJ* 1998;316:422–426.

90. Salam MA, Khan WA, Dhar U, Ronan A, Rollins NC, Bennish ML. Vitamin A for treating shigellosis. Study did not prove benefit. *BMJ* 1999;318:939–940.

91. World Health Organization and UNICEF. *Handbook: IMCI Integrated Management of Childhood Illness*. Geneva: World Health Organization, 2005.

92. Dodd K, Buddingh CJ, Rapoport S. Causation of ekiri, a highly fatal disease of Japanese children. *Am J Dis Child* 1950;79:949–951.

93. Trofa AF, Ueno-Olsen H, Oiwa R, Yoshikawa M. Dr. Kiyoshi Shiga: discoverer of the dysentery bacillus. *Clin Infect Dis* 1999;29:1303–1306.

94. Bennish ML, Wojtyniak BJ. Mortality due to shigellosis: community and hospital data. *Rev Infect Dis* 1991;13 (Suppl 4):S245–S251.

95. van den Broek JM, Roy SK, Khan WA, Ara G, Chakraborty B, Islam S, Banu B. Risk factors for mortality due to shigellosis: a case-control study among severely-malnourished children in Bangladesh. *J Health Popul Nutr* 2005;23:259–265.

96. Legros D, Paquet C, Dorlencourt F, Saoult E. Risk factors for death in hospitalized dysentery patients in Rwanda. *Trop Med Int Health* 1999;4:428–432.

97. Guerin P, Brasher C, Baron E, Mic D, Grimont F, Ryan M, Aavitsland P, Legros D. *Shigella dysenteriae* serotype 1 in west Africa: intervention strategy for an outbreak in Sierra Leone. *Lancet* 2003;362:705–706.

98. Kabir I, Malek MA, Mazumder RN, Rahman MM, Mahalanabis D. Rapid catch-up growth of children fed a high-protein diet during convalescence from shigellosis. *Am J Clin Nutr* 1993;57:441–445.

99. Black RE, Brown KH, Becker S. Effects of diarrhea associated with specific enteropathogens on the growth of children in rural Bangladesh. *Pediatrics* 1984;73:799–805.

100. Henry FJ. The epidemiologic importance of dysentery in communities. *Rev Infect Dis* 1991;13 (Suppl 4):S238–S244.

101. World Health Organization. *Guidelines for the Control of Shigellosis, Including Epidemics Due to Shigella dysenteriae 1*. Geneva: World Health Organization, 2005.

102. Luby SP, Agboatwalla M, Painter J, Altaf A, Billhimer WL, Hoekstra RM. Effect of intensive handwashing promotion on childhood diarrhea in high-risk communities in Pakistan: a randomized controlled trial. *JAMA* 2004;291:2547–2554.

103. Curtis V, Cairncross S. Effect of washing hands with soap on diarrhoea risk in the community: a systematic review. *Lancet Infect Dis* 2003;3:275–281.

104. Khan MU, Khan MR, Hossain B, Ahmed QS. Alum potash in water to prevent cholera. *Lancet* 1984;2:1032.

105. Chompook P, Todd J, Wheeler JG, von Seidlein L, Clemens J, Chaicumpa W. Risk factors for shigellosis in Thailand. *Int J Infect Dis* 2006;10:425–433.

106. Cohen D, Green M, Block C, Slepon R, Ambar R, Wasserman SS, Levine MM. Reduction of transmission of shigellosis by control of houseflies (*Musca domestica*). *Lancet* 1991;337:993–997.

107. Ahmed F, Clemens JD, Rao MR, Sack DA, Khan MR, Haque E. Community-based evaluation of the effect of breast-feeding on the risk of microbiologically confirmed or clinically presumptive shigellosis in Bangladeshi children. *Pediatrics* 1992;90:406–411.

108. Koster FT, Curlin GC, Aziz KM, Haque A. Synergistic impact of measles and diarrhoea on nutrition and mortality in Bangladesh. *Bull World Health Organ* 1981;59:901–908.

109. Greenberg BL, Sack RB, Salazar-Lindo E, Budge E, Gutierrez M, Campos M, Visberg A, Leon-Barua R, Yi A, Maurutia D, et al. Measles-associated diarrhea in hospitalized children in Lima, Peru: pathogenic agents and impact on growth. *J Infect Dis* 1991;163:495–502.

110. O'Donovan C. Measles in Kenyan children. *East Afr Med J* 1971;48:526–532.

111. De Mol P, Bosmans E. *Campylobacter* enteritis in Central Africa. *Lancet* 1978;1:604.

112. Phalipon A, Mulard LA, Sansonetti PJ. Vaccination against shigellosis: is it the path that is difficult or is it the difficult that is the path? *Microbes Infect* 2008;10:1057–1062.

113. Mukhopadhaya A, Mahalanabis D, Chakrabarti MK. Role of *Shigella flexneri* 2a 34 kDa outer membrane protein in induction of protective immune response. *Vaccine* 2006;24:6028–6036.

114. Noakes TD, Borresen J, Hew-Butler T, Lambert MI, Jordaan E. Semmelweis and the aetiology of puerperal sepsis 160 years on: an historical review. *Epidemiol Infect* 2008;136:1–9.

115. Khan MU. Interruption of shigellosis by hand washing. *Trans R Soc Trop Med Hyg* 1982;76:164–168.

116. Luby SP, Agboatwalla M, Feikin DR, Painter J, Billhimer W, Altaf A, Hoekstra RM. Effect of handwashing on child health: a randomised controlled trial. *Lancet* 2005;366:225–233.

117. Pittet D, Simon A, Hugonnet S, Pessoa-Silva CL, Sauvan V, Perneger TV. Hand hygiene among physicians: performance, beliefs, and perceptions. *Ann Intern Med* 2004;141:1–8.

118. Hoy S. *Chasing Dirt: The American Pursuit of Cleanliness*. New York City: Oxford University Press, 1995.

119. Zeitlyn S, Islam F. The use of soap and water in two Bangladeshi communities: implications for the transmission of diarrhea. *Rev Infect Dis* 1991;13 (Suppl 4):S259–S264.

120. Pillay D, Karas A, Sturm A (1997). An outbreak of *Shiga bacillus* dysentery in KwaZulu/Natal, South Africa. *J Infect* 34:107–111.

121. Keddy KH, Nadan S, Govind C, Sturm AW. Evidence for a clonally different origin of the two cholera epidemics of 2001–2002 and 1980–1987 in South Africa. *J Med Microbiol* 2007;56:1644–1650.

122. Pauw J. The politics of underdevelopment: metered to death-how a water experiment caused riots and a cholera epidemic. *Int J Health Serv* 2003;33:819–830.

16

THE ZIMBABWE CHOLERA EPIDEMIC OF 2008–2009

Edward Dodge

16.1 BRIEF OVERVIEW OF CHOLERA

Cholera has been known as a major health threat for centuries. Hippocrates described a disease similar to cholera, and Portuguese explorers depicted a disease outbreak in the Indian subcontinent that was clearly consistent with cholera in the late 1400s.

Robert Koch identified the comma-shaped cholera bacillus in 1884 and established it as the cause of cholera. For many years, up to the 1960s, the scientific community credited him with the discovery of the cholera bacillus.

Unbeknownst to Koch, Filippo Pacini had discovered a comma-shaped bacillus that he called a *Vibrio* in the course of investigating the cholera epidemic that came to Florence, Italy, in 1854. He published a paper in 1854 entitled, "Microscopical Observations and Pathological Deductions on Cholera," in which he described the bacillus and its possible relationship to the disease of cholera. Most scientists of that era believed in the miasmic theory of contagion (disease spread by "bad air") and Pacini's work was totally ignored. It was not rediscovered until many decades later.

Even so, 1854 was a pivotal year in advancing our understanding of cholera, for that was also when John Snow made his famous association between the Broad Street pump handle and the cholera epidemic that hit London. Most scientists did not believe in the germ theory of contagion, but Snow's research made an indelible impression, and he became known as one of the fathers of epidemiology. He was the first to clearly show the importance of good water management in the control of disease (1).

The first known cholera pandemic originated in the Ganges River Delta of India in 1817 and spread over much of Asia and the Middle East over the next 5 years. A second pandemic wave began in 1829, reaching Europe and the Americas for the first time in 1832. By the 1950s, six major pandemic waves had occurred. The seventh pandemic is said to have begun in the early 1960s, and is still ongoing (2).

Several varieties of *Vibrio cholerae* are known. The *El Tor* variant has been responsible for most epidemics since 1961, though classic *V. cholerae* remains a threat in some areas. *El Tor cholera* spread around the world after 1961, erupted in Peru in 1991, and then spread throughout Latin America. Largely controlled in South America now, cholera has nevertheless become endemic in large areas of Asia, Africa, and Latin America (2).

The clinical course of cholera lasts from 2 to 7 days. It is rapidly progressive in many cases. The time from the first liquid stool to clinical shock may occur in less than 12 h. If untreated, death can follow within less than 24 h from the onset of symptoms.

The classic "rice-water" stools are the result of the cholera toxin triggering voluminous small intestinal secretions of salt and water. Such large volumes are secreted that fluid loss can measure up to 60 L in a span of 5 days. This large, rapid fluid loss is responsible for the early onset of shock and the rapid progression of the illness.

Surprisingly, in spite of the strong intestinal reaction to the cholera toxin, no pathological changes are found in the small intestine, and the cholera organism does not invade the blood stream as a rule. Perhaps because of this, there is usually no fever and little or no abdominal pain with cholera. In severe cases, pulse and blood pressure may become extremely faint or even unobtainable because of the rapid dehydration that develops.

Water and Sanitation-Related Diseases and the Environment: Challenges, Interventions, and Preventive Measures, First Edition.
Edited by Janine M. H. Selendy.
© 2011 Wiley-Blackwell. Published 2011 by John Wiley & Sons, Inc.

Treatment is relatively simple and straightforward, but must be started quickly because of the rapid development of the illness. If begun soon enough, oral rehydration solution is very effective. In severe cases, intravenous lactated Ringer's solution may be needed. Rehydration is the key to effective treatment. When it is adequate, recovery is the rule, with less than 1% of cases proceeding to death. Without such treatment, over 50% of serious cases may die.

Antibiotic treatment is secondary to rehydration, but it may help shorten the course of the illness and may help prevent carrier status from developing in recovering patients. Several antibiotics are effective, including tetracycline, doxycycline, and several others.

In endemic areas, some degree of immunity seems to develop, and many cases may be mild or asymptomatic. Carriers can carry the organism for months or even years without any symptoms. That may be one reason why the microbe persists during interepidemic periods. Outside of the human host, the organism does not survive for more than a few days in most environments. An exception is in brackish seawater, where many *Vibrio* species are abundant, and where *V. cholerae* variants have been proved to survive (3).

Cholera became endemic in sub-Saharan Africa in the 1970s. In 2005, 31 of the 40 countries (78%) that reported indigenous cases of cholera to WHO were in Africa, where the reported incidence was far higher than in Asia or Latin America (2).

It was a special concern for Mozambique during its civil war and on up to 2009. In many of those years, notified cholera cases in Mozambique represented a fifth to a third of all cases reported in Africa. Fernanda Teixeira, secretary-general of the Mozambican Red Cross, said that 2009 was worse than previous years, with over 12,000 cases and 157 deaths recorded in the first 3 months of the year. Teixeira said this outbreak was not necessarily related to the situation in Zimbabwe, even though they share a long border (4).

Some Zimbabweans believe that cholera originally came into their country with migrant workers from Mozambique, where it had been endemic since 1973 (5). Regardless of how it came, cholera has also been endemic in rural areas of Zimbabwe for several years.

16.2 THE ZIMBABWE CHOLERA EPIDEMIC OF 2008–2009

In August 2008, a cholera outbreak is reported to have begun in Chitungwiza, near the capital city of Harare. Rapid population growth is said to have overwhelmed the sewage and water supply systems of the town. From that area, the disease is reported to have spread to most of the other provinces in the country (6).

A number of factors converging in Zimbabwe in 2008 contributed to the risk of a cholera epidemic. A deteriorating economy over several preceding years seems to have been a major underlying factor. The severity of the economic decline is cited as a reason for the deterioration of the country's infrastructure, including breakdown of basic municipal services such as sewage treatment and water supply in many areas (7).

The difficult economy combined with runaway inflation led to major disruption of health services in 2008. Some health workers migrated to neighboring countries. Food supplies were scarce, leading to nutritional problems for many people. Political turmoil in the wake of elections early in 2008 also contributed to social unrest. All of these factors seemed to play a role in setting the stage for the cholera epidemic (7).

In previous years, cholera had been mainly an episodic rural phenomenon, but in 2008 it became primarily an urban problem. The capital city of Harare became a major focus of the epidemic, especially in impoverished, high-density neighborhoods. As increasing numbers of cases began turning up in late 2008, the health system was not prepared to cope with them adequately. Cholera patients, requiring rapid treatment, did not fare well.

Budiriro was one of the high-density suburbs of Harare at the epicenter of the cholera epidemic. With a working-class population of about 130,000, the suburb had "seven primary schools, two secondary schools, and six health facilities." It also had five business centers with stores ranging from "retail supermarkets" to small shops.

Like other high-density suburbs, Budiriro was scarred by the crises that have gripped Zimbabwe in recent years. According to one well-researched report, "it is characterized by poor municipal water supply, untreated sewage, collapsed medical infrastructure, and over-crowding. . . . Burst sewage pipes left puddles and permanent stench, while months of uncollected refuse littered the streets" (7).

A sample survey of residents done part way through the epidemic revealed that a majority saw the unavailability of clean water as the major cause of the outbreak, even though most of them did not know the specific bacterial cause of cholera. The broken-down sewage system was also identified as a main cause (7).

Lake Chivero is the major source of water for Harare. Paradoxically, however, "the city of Harare [is] located in Lake Chivero's catchment area, which means that all the city's effluent flows into the lake to be pumped back into the city. The quality of the water pumped at the Morton Jeffrey waterworks was below the acceptable standards due to the acute shortage of water purification chemicals" (7).

Constance Manika, a Zimbabwean reporter who lived in one of Harare's high-density suburbs with her family, wrote about the water quality there: "Every weekend for the past eight months, my husband and I have been forced to make the

20-kilometer trip by road from our home in the high density suburbs of Harare to the affluent suburb of Belvedere to fetch clean water."

"We have tap water where we live, but it can hardly be said to be safe for human consumption. When you pour the water into a clear cup or container and let it sit for a few minutes, a green, sewage-like substance settles to the bottom" (8).

Much of the high-density residential area did not have access to even poor-quality water. According to Dube's report, "the suburb [Budiriro] is located at an altitude which makes it difficult for water to reach all parts of the suburb. Harare water is gravity driven and high altitudes naturally militate against efficient water supply. The situation is aggravated by the fact that the water pipes . . . were not the right size and frequent pipe bursts were a sign of aging and stress" (7).

According to Dube, "At least forty percent of the water pumped from Morton Jeffrey is lost through leakages" (7).

In his review of the water problems in Budiriro, Dube concludes: "The responses of key respondents and the residents of Budiriro concurred on the fact that the water crisis was the major cause of the outbreak" (7).

While Harare suffered the worst of the cholera epidemic, cases were reported from all of the country's provinces and cities. A report by OCHA (the United Nations Office for the Coordination of Humanitarian Affairs) in May 2009 showed that cholera had infected 98,429 people and claimed 4276 lives between August 2008 and May 31, 2009 (9).

Bulawayo, Zimbabwe's second largest city with a population of 1.5 million, did not fare as badly as Harare. By May 2009 Bulawayo had recorded fewer than 20 deaths. Analysts concluded that better control of the water supply was a factor in letting Bulawayo escape the cholera epidemic relatively unscathed. The city's water supply, though inadequate, was at relatively low risk of industrial and sewage contamination (10).

Even so, the city's sewage system was severely defective due to years of neglect. An article in *The Standard*, dated August 15, 2009, states that "human waste from burst sewer pipes are now a common sight in the city." The Bulawayo Sewerage Task Force estimates that it will take USD 20 million to replace the aged sewer systems. Another USD 10,000 is needed "for equipment to rehabilitate the water treatment plants" (11).

Mutare (Fig. 16.1) and Gweru, smaller cities in Zimbabwe, also have water sources that are better protected than Harare's, but their sewerage treatment systems have been deficient for years. Both cities are reported to have discharged untreated sewage into nearby rivers, with obvious deleterious results (12).

Beitbridge, a town of about 40,000 people on the border between Zimbabwe and South Africa, suffered disproportionately from the Zimbabwe cholera epidemic. Excerpts from a report by *Doctors Without Borders* outline what the desperate situation was like when the cholera epidemic overwhelmed the town and its hospital.

On Friday, November 14, [2008] when the Zimbabwean Health Authorities in Beitbridge first reported cholera to MSF, there were five cases. Two days later, there were already more than 500; by the end of the week, there were more than 1,500. . . . In one week, fifty-four people died.

FIGURE 16.1   An distant view of the city of Mutare. This gives an idea of the city size and the kind of terrain typical of eastern Zimbabwe. (Photo: E. Dodge.)

Patients were first placed inside Beitbridge's main hospital, most lying on cement floors, in very poor hygienic conditions. On Sunday morning, the hospital had to make the decision to put all the patients out behind the buildings, on the dirt, so that body excretions could be absorbed into the ground. The sight was appalling: patients lying in the dust in the scorching heat; all asking for the life-saving drip (Ringer's lactate IV fluid). There wasn't even any water to give them, since the hospital, as everywhere in town, has its water supply cut on most days.

With the current crises in Zimbabwe, basic services are lacking. ... There is trash everywhere, and open sewage runs through most of Beitbridge's streets. Almost everyday there are cuts in the water and power supplies.

At the beginning of the crisis, the Beitbridge hospital did not have any IV fluid or oral rehydration salts (ORS) tablets in stock. MSF shipped over 800 liters of the Ringer's fluid the first day of the intervention. ... Shipments of medical and logistical supplies arrived over ten days. A team of 16 expatriates, comprised of doctors, nurses, logisticians, and administrators were sent to Beitbridge. And more than 100 additional health workers, cleaners, and day workers [were] hired locally. (13)

Matthew Cochrane from the International Federation of Red Cross and Red Crescent Societies tells about the problems in a clinic in the town of Kwekwe when it was hit with cholera. In his report of February 2, 2009, he says the clinic was overwhelmed with 131 cholera patients in the space of 5 days. The ward was "chaotic ..., and in the midst of it all lay three bodies wrapped in plastic." Most of the cases came from Tiger Reef, a small mining town that had been without water for months.

The Zimbabwe Red Cross set up a tent in the clinic's yard to provide treatment for the more serious cases. It also decided to focus its public education efforts in Tiger Reef, the main source of this particular outbreak. Residents there expressed their frustration: "How can we deal with this cholera when we don't have toilets and have to walk three kilometers to the river to get water?" (14).

Water sources throughout Zimbabwe were suspect. The International Centre for Diarrheal Diseases Research (ICDDR), based in Bangladesh, sent investigators to Zimbabwe. They identified 30 strains of cholera from 16 test sites across the country. In Budiriro, all available water sources were contaminated with cholera (15).

As the epidemic worsened all across the country, the Zimbabwean government realized that it needed assistance. It requested help through the United Nations and international nongovernmental organizations (NGOs). They responded and teamed together, along with the government, to strengthen cholera containment efforts where most needed.

"By early 2009 the Ministry of Health and Child Welfare (MoHCW) was receiving support from ... the World Health Organization (WHO), United Nations Children's Education Fund (UNICEF), Doctors Without Borders (MSF), and Oxfam."

Cholera Control Command Centers were set up in appropriate areas to assess and respond to critical local needs. New bore-holes began to be drilled. Treatment centers were provided with critically needed supplies (7).

After the unity government was formed in February 2009, hospitals reopened and many health personnel returned to work. Programs to educate the public about cholera, its transmission, and its treatment were strengthened. As all these programs took effect, the epidemic gradually began to ebb.

By late July 2009, Zimbabwe's health minister, Dr. Henry Madzorera, announced that the cholera epidemic had ended. The International Federation of Red Cross and Red Crescent Societies (IFRC) said the epidemic was the worst to hit Africa in 15 years. In the 10 months between August 2008 and June 2009, nearly 100,000 cases and over 4200 deaths due to cholera had been reported in Zimbabwe.

Dr. Madzorera emphasized that future outbreaks were a risk, and that the country could not relax its vigilance regarding cholera. The health minister called on all provinces and cities to be prepared for the likelihood of recurring outbreaks because of continuing problems with sewage and water lines (6).

Leaders of several NGOs expressed similar concerns. IFRC spokesman Matthew Cochrane was quoted as saying, "Our concern is that the fundamental issues—access to sanitation and access to clean water—haven't been meaningfully addressed" (6).

Rian van de Braak, spokesperson for *Medecins Sans Frontieres*, stated that the root causes of the initial outbreak—including dilapidated drinking water systems and leaky sewage systems—have not been addressed adequately. She added that with the rainy season fast approaching, "everyone expects cholera to be back" (16).

Agostinho Zacarias, U.N. Development Program representative in Zimbabwe, said at a World Humanitarian Day ceremony in Harare: "... humanitarian threats such as food shortages and outbreak of diseases such as cholera pose a significant challenge." The United Nations has said 6 million people in Zimbabwe—more than half the population—have limited or no access to safe water and sanitation in rural and urban areas (17).

An editorial in *The Herald*, published by the government of Zimbabwe, stated on August 28, 2009: "... the problems that helped cholera get a hold on our cities, and spread so fast among the large crowded populations, are still with us. Water supplies in Harare are still problematical. The north and the east of the city are still without water almost all the time. The outer suburbs in the south and southwest have very interrupted supplies. ... The city has also been running inadequate and, at times, quite unsafe rubbish removal services. Garbage is piling up in suburbs, and Budiriro is one of the worst affected" (18).

16.3 LESSONS FROM THE ZIMBABWE CHOLERA EPIDEMIC OF 2008–2009

Although many factors were involved in Zimbabwe's cholera epidemic of 2008–2009, the major lesson is that good sanitation and safe adequate water supplies are critically important to protect people and communities from water-borne diseases like cholera. As one cholera expert has noted: "If there were no opportunities for food or water to be contaminated with cholera *vibrios*, there would be no spread of disease" (19).

Since Zimbabwe has major continuing problems with adequate sanitation and safe water supplies, it continues to be at serious risk of recurring cholera outbreaks. Serious water and sanitation problems exist in most communities throughout Zimbabwe. The magnitude of these problems make it clear that there is no easy fix close at hand.

It is estimated that Zimbabwe needs an infusion of over 8 billion dollars to rebuild its damaged and inadequate sewage and water lines. Even with such funding, the repair and rebuilding process could take several years, underlining the minister of health's call for widespread preparedness to deal with the probability of future cholera outbreaks (6).

The fact that the Ministry of Health is calling for plans and preparations to deal with such risk at all levels of government and community organization suggests that any future outbreaks may be contained more quickly and effectively than was the case in the 2008–2009 epidemic.

The readiness of NGOs to step up assistance rapidly should also help stop any future outbreaks more quickly than was the case in the 2008–2009 epidemic. As reported on August 23, 2009 in the online edition of the *Christian Science Monitor*:

> Aid agencies as diverse as UNICEF and Doctors Without Borders are preparing short-term solutions, drilling new wells in cholera hotspots, and contacting nurses and doctors who had been on the front-lines during the previous outbreak.
>
> Doctors Without Borders is also distributing 11,000 cholera kits—complete with water purification tablets, rehydration salts, soap, and other basic sanitation items—in fifty of the most remote clinics in the countryside and has trained local health staff on how to intervene when the first cases appear. (16)

However, it clearly would be much better to prevent any future outbreaks than to cope with them after they have erupted. The chaos caused by cholera epidemics, from the heart-rending collapse of individuals, families, and communities, to the tremendous drain on health facilities and personnel, is unnecessary. All that is needed for prevention is the provision of good sanitation and a sufficient supply of clean safe water.

From a humanitarian perspective, it would be desirable for the international community to assist Zimbabwe in the rebuilding of its public services infrastructure as quickly as possible. It would certainly be more desirable to prevent future cholera outbreaks than to continue to deal with them after they occur. From both humanitarian and economic standpoints, prevention is more effective than treatment.

REFERENCES

1. Frerichs RR. Dept. of Epidemiology, UCLA; Who First Discovered *Vibrio Cholera*? Posted 5 August 2001. Available at http://www.ph.ucla.edu/EPI/snow/first/discoveredcholera.html. Accessed August 2009.
2. Gaffga NH, Tauxe RV, Mintz ED. Cholera: a new homeland in Africa? *Am J Trop Med Hyg*, 2007;77(4):705–713.
3. Barua D, Greenough III WB, editors. *Cholera (Current Topics in Infectious Disease)*. New York: Springer, 1992.
4. IRIN. Mozambique: worst cholera outbreak in a long time. WASH new Africa. March 30, 2009. Available at http://washafrica.wordpress.com/2009/03/30/mozambique-worst-cholera-outbreak-in-a-long-time/. Accessed August 2009.
5. DeWolfe. Africa University, Murtare, Zimbabwe. Personal communication. August 2009.
6. Jongwe F. Zimbabwe says cholera epidemic "has ended." AFP. July 30, 2009. Available at http://www.google.com/hosted-news/afp/article/ALeqM5gBFe6MjuxYG89JSjY94jNTf_vV-VA. Accessed August 2009.
7. Dube D. *An Assessment of the Nature and Context of Aid Agency Intervention in the Budiriro Cholera Outbreak* [unpublished Master's Thesis]. Mutare, Zimbabwe: Africa University, 2009, pp. 3, 7, 54, 55–57.
8. Manika C. *The Battle to Stay Alive: Surviving in Zimbabwe by the Mercy of God*. May 18, 2009. Available at http://thewip.net/contributors/2009/05/the_battle_to_stay_alive_survi.html. Accessed July 2009.
9. OCHA Monthly Humanitarian Update, Zimbabwe. May 2009. Available at http://www.reliefweb.int/rw/rwb.nsf/db900SID/MUMA-7SR4CA?OpenDocument. Accessed August 2009.
10. Tearfund. Tide is turning in Zimbabwe cholera outbreak. May 11, 2009. Available at http://www.tearfund.org/News/Zimbabwe/Cholera + outbreak + latest.htm. Accessed August 2009.
11. Ndlovu N. Bulawayo's ticking health time bomb. *Zimbabwe Standard*, August 15, 2009. Available at http://www.thezimbabwestandard.com/index.php?option=com_content&view=article&id=21055:bulawayos-ticking-health-time-bomb&catid=31:zimbabwe-stories&Itemid=66. Accessed August 25, 2009.
12. Mtisis S. Promoting water quality laws enforcement and implementation in Zimbabwe's urban areas. *Eighth International Conference on Environmental Compliance and Enforcement*

2008, pp. 505, 506. Available at http://www.inece.org/conference/8/proceedings/68_Mtisis.pdf. Accessed August 22, 2009.

13. Stavropoulous J. *Zimbabwe: Cholera Hits Beitbridge, Exposes Major Health Risks. Doctors Without Borders, Field News, December 1, 2008.* Available at http://edition.cnn.com/2009/WORLD/africa/08/19/zimbabwe.humanitarian.day/index.html?eref=edition_africa. Accessed August 8, 2009.

14. Cochrane M. Zimbabwe Aid Diary. *BBC News*, February 2, 2009. Available at http://news.bbc.co.uk/2/hi/africa/7864575.stm. Accessed July 5, 2009.

15. Chinnock P. Cholera report from Zimbabwe: "The whole aquatic environment seems to be heavily contaminated." TropIKA.net, March 9, 2009. Available at http://www.tropika.net/svc/news/20090309/Chinnock-20090309-News-Zim-Cholera. Accessed August 22, 2009.

16. Maromo J. Aid groups warn cholera could return in Zimbabwe. *The Christian Science Monitor*, August 23, 2009. Available at http://www.csmonitor.com/2009/0823/p06s13-woaf.html. Accessed August 25, 2009.

17. CNN.com. *U.N. Official: Zimbabwe's woes pose significant challenge*, August 19, 2009. Available at http://edition.cnn.com/2009/WORLD/africa/08/19/zimbabwe.humanitarian.day/index.html?eref=edition_africa. Accessed August 25, 2009.

18. The Herald (Editorial), Published by the government of Zimbabwe. August 28, 2009. Available at http://allafrica.com/stories/200908280078.html. Accessed August 29, 2009.

19. Mandell GL, Bennet JE, Dolin R, editors. *Mandell, Douglas and Bennett's Principles and Practices of Infectious Diseases*, 4th edition. Philadelphia: Churchill Livingstone, 1995.

17

INFECTIOUS DISEASE CONTROL IN GHANA: GOVERNMENT'S INTERVENTIONS AND CHALLENGES TO MALARIA ERADICATION

Julius N. Fobil, Juergen May, and Alexander Kraemer

17.1 INTRODUCTION

Ghana is generally touted as a great example of a democratic success story in Africa following two decades of smooth political transitions. While the national macroeconomic and microeconomic climate appears to be imprecisely reported, the country is at the crossroads of an epidemiologic transition (1–4). Whereas cases of many noncommunicable diseases are fast growing in the population, many infectious and preventable diseases still constitute a huge public health threat in the communities (3, 5–8).

Within continental Africa, while malaria is largely eliminated in North African and Mediterranean countries, it remains on the top of the list of most important public health problems in countries south of the Sahara including Ghana (9, 10). It is also the primary reason for outpatient visitation and the number one childhood killer which is only closely followed by acute lower respiratory infections (ALRIs) and diarrhea in that order (11). In Ghana in particular, despite overwhelming efforts in control campaigns, malaria, diarrhea, and ALRIs remain the three leading causes of morbidity and mortality in the population.

This raises doubts about the understanding of the range of risk factors for many of the infectious diseases afflicting the population. Fundamental questions that need to be answered are: (i) Do these infections have common risk factors or will the occurrence of one of them increase the risk of the other in infected persons? (ii) At what level are these factors operating and what will be the most effective control strategies to be deployed to contain the persistence of these common infections in the population? (iii) How do we prioritize resource allocation for national health programs in order to have the most impact or, in other words, why are the previous and current disease control initiatives not able to eliminate the target diseases and what needs to be done differently? (iv) How should the intervention strategies be packaged and delivered to guarantee acceptability and use in the communities in order to achieve optimum effects? Given the range of unresolved questions, there is an important need to look for new evidence beyond the etiology of the infections, especially to find out how the risk factors operate at behavioral and lifestyle levels so that disease control programs can be best focused in order to be most effective.

For instance, the evidence that children in low-income settings suffering from frequent diarrhea episodes were also at a high risk of ALRI was explored using data from Brazil and Ghana (11). The investigation by Schmidt and co-workers aimed to determine whether the observed association was due to common risk factors for both conditions or whether diarrhea could potentially directly increase the risk of ALRI. In that study, while a roughly linear relationship was observed between the number of diarrhea days over the previous 28 days and the risk of ALRI in the Brazil data, they found no evidence of an association between diarrhea and ALRI. In the Ghana data, however, it was reported that every additional day of diarrhea within 2 weeks increased the risk of ALRI by a factor of 1.08 (95% CI 1.00–1.15) and that 26% of ALRI episodes were due to recent exposure to diarrhea (11). The authors of the study therefore cautiously concluded that diarrhea could contribute to the burden of

Water and Sanitation-Related Diseases and the Environment: Challenges, Interventions, and Preventive Measures, First Edition.
Edited by Janine M. H. Selendy.
© 2011 Wiley-Blackwell. Published 2011 by John Wiley & Sons, Inc.

ALRI in malnourished child populations (11–13). This study and many others on how risk factors for infectious diseases may be interrelated provide evidence that infectious disease control programs would benefit from further studies that seek to understand the transmission dynamics of co-infections and convergence of these infections in populations of low income settings in particular (7, 11–15).

In this chapter, we discuss Ghana National Malaria Control Program (NMCP), highlighting the successes, the challenges, and failures, and attempt to offer suggestions that could improve program performance.

Most malaria infections in Africa south of the Sahara are caused by *Plasmodium falciparum*, the severest and the deadliest as well as the most persevering form of the parasite (8, 16–24). It is the main cause of severe clinical malaria. Africa south of the Sahara is home to the most efficient and well-adapted species of the mosquitoes that transmit the disease.

In Ghana, malaria continues to be the most important public health problem (7, 15, 25, 26), with its main effects on children aged under 5 years and pregnant women especially among primigravidae (7, 15, 26). Malaria is responsible for 40% of all admissions at hospitals and accounts for 22% of mortality under age 5 years (2, 15, 22). It is also estimated that an average loss of 3.7 days of male and 4.7 days of female output is experienced by the Ghanaian worker (14, 16, 27). The major focus of control of malaria in the absence of an affordable and effective vaccine is the promotion of the use of insecticide-treated nets, which currently stands at 9.1% for children under 5 and 7.8% for pregnant women (16, 25, 27). Only 11.7% of children under 5 years have access to prompt and correct treatment within 24 h of illness (16).

17.2   GHANA MALARIA CONTROL INITIATIVES—PAST AND PRESENT

The immediate past and current Ghana national malaria control initiatives have relied on effective case management and a large-scale deployment of insecticide-treated bed nets (ITNs), while awaiting the discovery of an effective and affordable vaccine (28). At the personal level, an important control strategy has focused on the manipulation of behavioral factors so as to reduce the transmission rates. Other strategies have concentrated strictly on ecological factors in an effort to disrupt the vector breeding potential, while also focusing on treatment for groups at highest risk of the disease and those who are prone to death from malarial attacks (2, 20, 23, 24, 29–31).

17.2.1   Key Milestones

The first major intervention for malaria control in Ghana was proposed and supported by the WHO which saw indoor residual spraying using DDT in many households in Volta and northern regions benefited from 1950 to 1960 (28). Covered in the intervention were aerial spraying in Accra and surrounding areas and injection of chloroquine into salt sold to the public at post offices. The period spanning 1970–1980 saw intensified prescription and use of chloroquine at all health facilities in Ghana. From 1996 to 1997, Ghana approved a WHO/AFRO program of accelerated malaria control in 30 pilot districts that emphasized chloroquine use and building the capacity of health workers to be able to recognize malaria signs and symptoms and to properly treat. In 1999, the Roll Back Malaria (RBM) Partnership began in Ghana with adoption of the following response strategies: improvement in case management through capacity building for health practitioners and caregivers at home (home-based care), multiprevention strategies such as ITNs, intermittent preventive treatment (IPT), and environmental management, which made malaria control effort a collective priority for all sectors in Ghana. In 2000 Ghana adopted the Abuja Targets and removed import duties on bed nets (32) and received USD 8.8 million from the Global Fund to Fight AIDS, TB, and Malaria in order to sustain the control efforts. The year 2005 saw a policy switched from a monotherapy strategy to artemisinin combination therapies (ACTs) as a result of increased parasite resistance to chloroquine with additional funding of USD 38.8 million from the Global Fund for this purpose (28, 32). Under this initiative, the implementing agencies ensured that the new drug policy was enforced at the dispensing units to replace chloroquine with ACTs for the management of uncomplicated malaria and sulfadoxine–pyrimethamine (SP) for prevention of malaria in pregnancy. This initiative undertaken in partnership with local community-based organizations (CBOs) and nongovernmental organization (NGOs) led to the extension of intermittent preventive treatment (IPT) from 20 initial districts to all 138 districts nationwide greatly boosting ITN use among children under 5 and pregnant women. In 2006, an augmentation strategy (a child health campaign) to provide all children under 2 with nets was introduced.

With support from the U.S. President's Malaria Initiative (PMI), the National Malaria Control Program (NMCP) embarked upon a plan to update and revise national malaria strategy in 2007. A 5-year, USD 1.2 billion program administered by the U.S. Agency for International Development (USAID) and implemented together with the Centers for Disease Control and Prevention (CDC) was designed to benefit 15 countries in total. The main goal of the PMI has been to reduce malaria deaths by 50% in 15 countries in Africa by reaching 85% of the most vulnerable groups—principally, pregnant women and children under 5 years of age—with lifesaving prevention and treatment measures. The PMI presently partners with National Malaria Control Program (NMCP) and other interest groups, including the

World Health Organization (WHO); the Global Fund to Fight AIDS, TB, and Malaria; the World Bank Malaria Booster Program; the Roll Back Malaria partnership; the Bill & Melinda Gates Foundation; and local CBOs and NGOs— including faith-based and community groups as well as those in the private sector. The program execution plans spanning 2008–2013 in Ghana are largely to build upon and expand the previous control efforts in the following strategic intervention areas:

- *Insecticide-Treated Mosquito Nets (ITNs)*. Sleeping under a long-lasting ITN provides protection from malaria-carrying mosquitoes. The nets are nontoxic to humans, last up to 3 years, and do not need retreatment. PMI supports Ghana's mixed model for the distribution of long-lasting ITNs, which includes mass free distribution; targeted subsidy programs, including voucher schemes in some regions; subsidized sales at public clinics; commercial sales; distribution at the community level through NGOs and CBOs; and workplace distributions. In 2007, PMI worked with its partners to distribute more than 236,000 nets through a voucher program and another 750,000 ITNs, purchased by UNICEF and the World Bank, during a nationwide campaign to distribute 1.5 million ITNs. Since beginning work in Ghana, PMI has procured 410,000 ITNs, most of which were distributed in early 2009.

- *Indoor Residual Spraying with Insecticides (IRS)*. IRS acts to kill or shorten the lives of adult female malaria-carrying mosquitoes when they rest on the sprayed inside walls of homes after feeding, cutting transmission of malaria. During 2008, PMI in collaboration with the Ghana Health Service supported the first large-scale implementation of IRS in the public sector in Ghana. PMI also worked with the mining company AngloGold Ashanti, which provided technical assistance and collaboration, particularly in the areas of community sensitization, training of trainers, and training of spray operators. During 2008, PMI supported the spraying of more than 254,000 houses in Ghana, protecting more than 600,000 people.

- *Intermittent Preventive Treatment for Pregnant Women (IPTp)*. Pregnant women are particularly vulnerable to malaria, as pregnancy tends to reduce levels of immunity to malaria. Unborn children may then suffer the consequences of maternal malaria infections, which can result in low birth weight and a higher risk of death early in infancy. During 2008, PMI provided technical and financial support to update the health care worker training curriculum on malaria in pregnant women and supported training of 464 health workers on IPTp.

- *Diagnosis of Malaria and Treatment with Artemesinin-Based Combination Therapy (ACT)*. ACTs are extremely effective against malaria parasites and have few or no side effects. In 2008, PMI provided technical assistance to strengthen the supply chain for ACTs and to promote their rational use. PMI also supported the development of the national policies on treatment, including the expansion of home management of malaria, update of training manuals, and the use of both microscopy and rapid diagnostic tests for malaria diagnosis. To avoid a nationwide shortage of ACTs, PMI procured an emergency supply of more than 1.1 million ACT treatments, which arrived in Ghana in November 2008 for distribution to health facilities. During the second year of its implementation in Ghana, the PMI allocated USD 17.3 million in funding for malaria prevention and treatment (10). Of this amount, 33% supported malaria diagnosis and procurement and treatment with ACTs, 28% for ITNs, 21% for IRS, 3% for IPTp, and 2% for monitoring and evaluation activities.

17.3 SUCCESSES, CHALLENGES, AND WEAKNESSES OF CONTROL STRATEGIES

A review of reports revealed that the performance of the control efforts has been measured based on the levels of program implementation targets met and the coverage of control services. Although the success of the control programs should be based upon the level of reduction in malaria transmission rate, this has not been possible due to lack of reliable and complete population-based data on malaria morbidity and mortality. However, the analysis of administrative data shows that malaria fatalities in children under 5 dropped by 38% from 3.26/100 in 2003 to 2/100 in 2005. The Ministry of Health records show that malaria-attributable deaths in pregnancy were reduced to zero in the 20 districts where IPT was implemented commencing in 2003. In 2005, 48.6% of mothers and child caretakers responded correctly to malaria in their children, compared to 22% in 2003. Other significant successes include improvements in the availability and coverage of malaria preventive services as follows:

- ITN use in children under 5 increased from 3.5% in 2003 to 38.7% in 2005.
- ITN use in pregnant women increased from a low 3.3% in 2003 to 46.5% in 2005.
- IPT coverage increased from 0% at the beginning of program in 2003 to 71.9%.

Additionally, the proportion of households owning one or more ITN increased from 3.2% in the 2003 Demographic Health Survey (DHS) to 18.7% in the 2006 Multi-Indicator

Cluster Survey (MICS). The proportion of children reported to have slept under an ITN the night before the survey was 3.5% in the 2003 DHS, increasing to 21.8% in the 2006 MICS. In the 2003 DHS, ITN ownership was found to vary significantly by region, with the highest rate in the Upper West Region (32%) and the lowest in the Western Region (8%). The proportion of women reported to have slept under an ITN the night before the survey was 2.2% in 2003 DHS data, and there were no comparable data from the 2006 MICS data. Implementation of IPTp with SP began in 2003. IPTp coverage rose from 2% in the 2003 DHS to 28% for at least two doses of IPTp with SP in the 2006 MICS.

In the area of case management, the proportion of children receiving antimalaria treatment within 24 h of fever onset increased from 44% in 2003 DHS to 48% in 2006 MICS. ACTs were introduced in 2004, and according to the 2006 MICS, among children with fever only 3.4% received an ACT. However, several positive trends in malaria management at the health facility level were documented by the USAID-supported Quality Health Partners (QHP) project in 30 focus districts nationwide. Almost all facilities, 96.7%, had adopted the use of artesunate/amodiaquine (AS/AQ) combination as the first-line outpatient treatment for malaria, and the percentage of providers who treated malaria appropriately had increased from 1.0% at baseline to 56.7% 3 years later.

17.3.1 Challenges and Weaknesses

In spite of the steady progress made over the years, malaria continues to negatively impact Ghana (10, 28, 32). It remains a major cause of poverty and low productivity. It accounts for 44.5% of all outpatient attendances and 12% of under 5 mortality (33, 34). Malaria is also a major cause of absenteeism from schools and from work (7, 9, 35–37).

The persistence of the disease in Ghana appears to be a consequence of a combination of social, economic, and technical factors operating across a broad spectrum of ecological and environmental factors. In all the 135 districts, efforts to eradicate malaria have increased and some progress has been achieved in reducing the case fatality rate (28, 32). However, analysis of routine outpatient data shows that the disease has consistently remained the topmost cause of ill health and death in both rural and urban areas (28, 32, 38–40) (Fig. 17.1).

Despite all the efforts by the national malaria control program to reduce the transmission of malaria through intensive mass campaigns, it does not seem to decline and still contributed nearly 50% of the reason for outpatient visitation among related infectious diseases in Ghana as of 2007 (see Fig. 17.1). Although there seems to be a consensus that implementation and behavioral factors largely account for the observed trends, other challenges including development of resistance by the vector (mosquito) itself to the active ingredient in the indoor residual sprays and

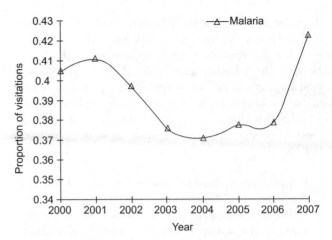

FIGURE 17.1 Proportion of annual outpatient visitations due to malaria, 2000–2007.

parasite resistance to antimalarials also play a very important role (15, 22).

This review identified several obstacles that militate against the eradication of malaria in Ghana. These obstacles are both technical and non-technical. Notable among these obstacles are the methods of delivery of the intervention programs. For example, most of the communities in which the interventions are deployed lack personnel with the required advanced skills for the implementation and management of the intervention programs. The success and sustainability of community-level interventions are not simply a matter of availability and access to the interventions on a sustained basis, but primarily relate to the complete understanding of program objectives and the intended purpose by members of the beneficiary communities (34, 41). Community members have been reported to assume ownership of these programs if they understood the program objectives and intentions right at commencement (28, 39, 40). The review observed a lack of coordination or overlap of roles among state agencies responsible for malaria control, especially among the Ministry of Health (MoH), the Ghana Health Service (GHS), the District Assemblies, the complementary and subsidiary institutions such as disease control and surveillance units as well as other implementing arms (10, 28, 32, 38, 40). This institutional weakness gives rise to a conflict of roles, a lack of synergies, and, in some instances, gaps in roles, thus reducing the efficiency of the collaborating departments and agencies involved in national malaria control programs (38–40).

17.4 CONCLUSION AND RECOMMENDATIONS

This review found evidence of a strong national will for eradicating malaria and all other forms of infectious diseases by way of several intervention programs. We noted a wide adoption of global malaria control strategy including treatment with ACTs, pre-packs, IPTp, IPTi, ITNs, and vector control initiatives such as LLINS, IRS, and so on, in Ghana. However, some strategic focal areas within the overall program structure have remained weak. These include lack of or inadequate laboratory facilities for accurate malaria diagnoses and lack of empirical data on intervention outcomes, the absence of which does not permit accurate evaluation of interventions.

17.4.1 Recommendations

It was noted that technical areas of malaria control (e.g., treatment, chemotherapy, human behavior, vector life-cycle interruption, attack on the parasite itself, etc.) were well covered in the national programs. Additionally, the area of malaria case management has seen significant improvement. Once accurately diagnosed, there is now a strong possibility of effective treatment. However, accurate diagnoses and proper recordkeeping in health facilities needs to be strengthened so that accurate determination of malaria can be periodically undertaken in order to generate continuous program-specific data that could offer opportunities for the evaluation of program effectiveness. Provision of laboratory equipment (e.g., microscopy or rapid diagnostic test—RDT) would reduce the chances of misdiagnoses and offer new opportunities for improving diagnostic accuracy and the quality of routine surveillance data from the health facilities.

REFERENCES

1. Adjei S. Key issues in capacity strengthening and management of reforms in Africa: Ghana's experience. *Ghana National Healthcare Conference and Exhibition*, Accra, 2004.

2. Appawu M, et al. Malaria transmission dynamics at a site in northern Ghana proposed for testing malaria vaccines. *Trop Med Int Health* 2004;9(1):164–170.

3. Awusabo-Asare K, Annim SK. Wealth status and risky sexual behaviour in Ghana and Kenya. *Appl Health Econ Health Policy* 2008;6(1):27–39.

4. Robine J-M. *Determining Health Expectancies.* Chichester; Hoboken, NJ: Wiley, 2003, p. xiii, 428 p.

5. Accra Sanitation Workshop. Report of the Discussion Group on Financing and Cost Recovery Options, 1988.

6. Afrane YA, et al. Does irrigated urban agriculture influence the transmission of malaria in the city of Kumasi, Ghana? *Acta Trop* 2004;89(2):125–134.

7. Ehrhardt S, et al. Malaria, anemia, and malnutrition in African children—defining intervention priorities. *J Infect Dis* 2006; 194(1):108–114.

8. Ofosu-Okyere A, et al. Novel *Plasmodium falciparum* clones and rising clone multiplicities are associated with the increase in malaria morbidity in Ghanaian children during the transition into the high transmission season. *Parasitology* 2001;123 (Pt 2): 113–123.

9. Africa Malaria Report. Roll Back Malaria Report, 2003. [Cited 20 October 2009.] Available at http://www.rollbackmalaria. org/amd2003/amr2003/ch1.htm.

10. GPMI. The Burden of Malaria in Africa, 2009. Available at http://www.fightingmalaria.gov/countries/profiles/ghana_ profile.pdf.

11. Schmidt WP, et al. Recent diarrhoeal illness and risk of lower respiratory infections in children under the age of 5 years. *Int J Epidemiol* 2009;38(3):766–772.

12. Cairncross S. Editorial: water supply and sanitation: some misconceptions. *Trop Med Int Health* 2003;8(3):193–195.

13. Cairncross S, et al. The public and domestic domains in the transmission of disease. *Trop Med Int Health* 1995;39: 173–176.

14. Donnell MJ, et al. Malaria and urbanization in sub-Saharan Africa. *Malaria J* 2005;4(12):1–5.

15. Kobbe R, et al. Seasonal variation and high multiplicity of first *Plasmodium falciparum* infections in children from a holoendemic area in Ghana, West Africa: *Trop Med Int Health* 2006; 11(5):613–619.

16. Binka FN, et al. Patterns of malaria morbidity and mortality in children in northern Ghana. *Trans R Soc Trop Med Hyg* 1994;88 (4):381–385.

17. Schandorf A. Problems in practice in a developing country. *Ugeskr Laeger* 1972;134(43):2283–2287.

18. Wagner G, et al. High incidence of asymptomatic malaria infections in a birth cohort of children less than one year of age in Ghana, detected by multicopy gene polymerase chain reaction. *Am J Trop Med Hyg* 1998;59(1):115–123.

19. Brady J. Calculation of anopheline man-biting densities from concurrent indoor and outdoor resting samples. *Ann Trop Med Parasitol* 1974;68(3):359–361.

20. Baird JK, et al. Seasonal malaria attack rates in infants and young children in northern Ghana. *Am J Trop Med Hyg* 2002; 66(3):280–286.

21. McGuinness D, et al. Clinical case definitions for malaria: clinical malaria associated with very low parasite densities in African infants. *Trans R Soc Trop Med Hyg* 1998;92(5): 527–531.

22. Ofori MF, et al. Malaria-induced acquisition of antibodies to *Plasmodium falciparum* variant surface antigens. *Infect Immun* 2002;70(6):2982–2988.

23. Chinery WA. Effects of ecological changes on the malaria vectors *Anopheles funestus* and the *Anopheles gambiae* complex of mosquitoes in Accra, Ghana. *J Trop Med Hyg* 1984;87(2):75–81.

24. Birley MH, Lock K. Health and peri-urban natural resource production. *Environ Urban* 1998;10(1):89–106.

25. Koram KA, et al. Seasonal profiles of malaria infection, anaemia, and bednet use among age groups and communities in northern Ghana. *Trop Med Int Health* 2003;8(9): 793–802.

26. Koram KA, et al. Severe anemia in young children after high and low malaria transmission seasons in the Kassena-Nankana district of northern Ghana. *Am J Trop Med Hyg* 2000;62(6): 670–674.

27. Afari EA, et al. Seasonal characteristics of malaria infection in under-five children of a rural community in southern Ghana. *West Afr J Med* 1993;12(1):39–42.

28. NMCP. Ghana's golden jubilee celebrates independence, but freedom from malaria offers real progress. *Ghana Malaria Action Alert* 2007;1(1):1–2.

29. Yoon PW, et al. Effect of not breastfeeding on the risk of diarrheal and respiratory mortality in children under 2 years of age in Metro Cebu, The Philippines *Am J Epidemiol* 1996; 143(11):1142–1148.

30. Cavanagh DR, et al. Antibodies to the N-terminal block 2 of *Plasmodium falciparum* merozoite surface protein 1 are associated with protection against clinical malaria. *Infect Immun* 2004;72(11):6492–6502.

31. Kristan M, et al. Pyrethroid resistance/susceptibility and differential urban/rural distribution of *Anopheles arabiensis* and *An. gambiae s.s.* malaria vectors in Nigeria and Ghana. *Med Vet Entomol* 2003;17(3):326–332.

32. NMCP. The road to malaria eradication—getting involved and making it happen! *Ghana Malaria Action Alert* 2008;1(7):1–2.

33. Klinkenberg E, et al. Urban malaria and anaemia in children: a cross-sectional survey in two cities of Ghana. *Trop Med Int Health* 2006;11(5):578–588.

34. Ng'ang'a PN, et al. Malaria vector control practices in an irrigated rice agro-ecosystem in central Kenya and implications for malaria control. *Malar J* 2008;7:146.

35. Vogel G. Infectious diseases. Will a preemptive strike against malaria pay off? *Science* 2005;310(5754):1606–1607.

36. Carter R, Mendis KN, Roberts D. Spatial targeting of interventions against malaria. *Bull World Health Organ* 2000; 78(12):1401–1411.

37. Phillips RS. Current status of malaria and potential for control. *Clin Microbiol Rev* 2001;14(1):208–226.

38. WHO. World Malaria Report 2009. Geneva: World Health Organization, 2009, pp. 1–25.

39. MoH. *Policy and Strategies for Improving the Health of Children Under-Five in Ghana*. Accra, Ghana: Ministry of Health, 1999.

40. NPC. Ghana Country Report. 5 Years Implementation of ICPD Programme of Action, 2000.

41. Zhou G, et al. Spatial relationship between adult malaria vector abundance and environmental factors in western Kenya highlands. *Am J Trop Med Hyg* 2007;77(1): 29–35.

SECTION III

WATER RESOURCES

This section addresses means of accessing clean water and preventing contamination of water. It includes case studies on innovative and successful solutions that include traditional and innovative means of achieving clean water and ways of protecting water supplies.

18

HOUSEHOLD WATER TREATMENT AND SAFE STORAGE IN LOW-INCOME COUNTRIES

Thomas F. Clasen

18.1 INTRODUCTION

The World Health Organization (WHO) estimates that improving water, sanitation, and hygiene could prevent at least 9.1% of the global burden of disease and 6.3% of all deaths (1). Diarrhea represents a significant share of this burden, resulting in an estimated 4 billion cases and 1.9 million deaths each year of children under 5 years, or 19% of all such deaths in developing countries (2). Diarrheal disease also contributes to decreased food intake and nutrient absorption leading to malnutrition, reduced resistance to infection, and impaired physical growth and cognitive development (3, 4).

As part of its Millennium Development Goals (MDGs), the United Nations (UN) expressed its commitment by 2015 to reduce by half the proportion of people without "sustainable access to safe drinking water" (5). As of 2006, an estimated 884 million people worldwide lacked access to improved water sources (6). While progress is being made, current trends will leave hundreds of millions unserved by the target date. Three-quarters of these will live in rural areas where poverty is often most severe and where the cost and challenge of delivering safe water are greatest. Even improved water supplies, however, such as protected wells and communal stand posts often fail to deliver safe drinking water in settings with poor sanitation due to infusion of fecal contamination (7). Moreover, water that is microbiologically safe at the source or other point of distribution is subject to frequent and extensive fecal contamination during collection, transport, and storage in the home (8). Thus, the health benefits of safe drinking water will remain elusive for vast populations for years to come.

Providing safe, reliable, piped-in water to every household is an essential goal, yielding optimal health gains, while contributing to the MDG targets for poverty reduction, nutrition, childhood survival, school attendance, gender equity, and environmental sustainability. While committed strongly to this goal, and to incremental improvements in water supplies wherever possible, the WHO and others have called for targeted, interim approaches that will accelerate the heath gains associated with safe drinking water for those whose water supplies are unsafe (9). While careful not to encourage diversion of resources away from connected taps, public health officials have called for other approaches that will provide some of the health benefits of safe drinking water, while progress is being made in improving infrastructure (10, 11).

One such alternative is household water treatment and safe storage (HWTS) (12). In many settings, both rural and urban, populations have access to sufficient quantities of water, but that water is unsafe for consumption as a result of microbial or chemical contamination. This is increasingly true even for piped-in water, since supplies are rarely provided on a 24–7 basis, forcing householders to store more water in the home in ways subject to recontamination and leading to microbial infiltration of poorly maintained systems. Effective treatment at the household level—often using the same basic approaches of filtration, disinfection, and assisted sedimentation, as characterize conventional water treatment—can remove, kill, or inactivate most microbial

Water and Sanitation-Related Diseases and the Environment: Challenges, Interventions, and Preventive Measures, First Edition.
Edited by Janine M. H. Selendy.
© 2011 Wiley-Blackwell. Published 2011 by John Wiley & Sons, Inc.

pathogens (13–16). Moreover, by focusing at the point of use rather than the point of delivery, treating water at the household level minimizes the risk of recontamination that even improved water supplies can present (8).

Although HWTS is not new, its potential as a focused public health intervention strategy is just emerging. For centuries, householders have used a variety of methods for improving the appearance and taste of drinking water, including filtering it through porous rock, sand, and other media or using natural coagulants and flocculants to reduce suspended solids. Even before germ theory was well established, successive generations were taught to boil water, expose it to the sun, or store it in metal containers with biocidal properties, all in an effort to make it safer to drink. In 2008, the WHO acknowledge the potential contribution of HWTS in its *Guidelines for Drinking-Water Quality* (GDWQ): "HWT technology has the potential to have rapid and significant positive health impacts in situations where piped water systems are not possible and where people rely on source water that may be contaminated, or where stored water becomes contaminated because of unhygienic handling during transport or in the home" (17). The WHO also recognized the contribution that HWTS can make among people living with HIV/AIDS, both in providing safe drinking water for those on drug therapy and for the preparation of replacement feeds for mothers who are HIV + and choose not to breast-feed in order to prevent transmission of the virus in breast-milk (18). In 2009, UNICEF announced a seven-point strategy for the treatment and prevention of diarrhea among children that includes HWTS, calling for the adoption of "household water treatment and safe storage systems, such as chlorination and filtration, in both development and emergency situations to support reductions in the number of diarrhea cases" (19).

Despite this support for HWTS as a targeted intervention for preventing diarrhea and other waterborne diseases among vulnerable populations, the promise of the intervention has yet to be fully realized. This is due in part not only to suboptimal methods for treating water at the household level but also to the challenge of reaching the target population with affordable HWTS solutions and securing their correct, consistent, and sustained use.

The objective of this chapter is to review the evidence around HWTS as a health intervention in low-income settings. The chapter begins by reviewing some of the leading methods for treating water at the household level and the research on their microbiological performance. It then summarizes the evidence concerning the effectiveness and cost-effectiveness of HWTS to prevent diarrheal disease. The chapter ends by noting the major challenges in scaling up HWTS by achieving coverage and sustained uptake among populations that might benefit most from the intervention.

18.2 HWTS METHODS

In 2000, the WHO commissioned a comprehensive study to review HWTS methods and practices. The review identified 37 different options for treating and safely storing water at the household level and assessed the available evidence on their microbiological effectiveness, health impact, acceptability, affordability, sustainability, and scalability (9). This section summarizes those options for which there is significant research on microbiological effectiveness and health impact. For additional details regarding these and other HWTS interventions, readers are referred to the websites on the intervention maintained by the WHO (http://www.who.int/household_water/en/) and the U.S. Centers for Disease Control and Prevention (CDC) (www.cdc.gov/safewater), which both contain other useful links.

18.2.1 Boiling

Boiling or heating with fuel is perhaps the oldest means of effectively disinfecting water at the household level. It is certainly the most common, with an estimated 600 million people in low- and middle-income countries reporting that they usually boil their water before drinking it (20). If practiced correctly, boiling is also one of the most effective, killing or deactivating all classes of waterborne pathogens, including bacterial spores and protozoan cysts that have shown resistance to chemical disinfection and viruses that are too small to be mechanically removed by microfiltration. Moreover, while chemical disinfectants and filters are challenged by turbidity and certain dissolved constituents, boiling can be used effectively across a wide range of waters. While some authorities recommend boiling water for 10, 20, and even 25 min, the WHO GDWQ simply recommend bringing water to a rolling boil of 1 min as an indication that a disinfection temperature has been achieved (21).

For decades, governments, NGOs, and others have promoted boiling, both in developing countries where water is routinely of uncertain microbial quality and in developed countries when conventional water treatment fails or water supplies are interrupted due to disasters or other emergencies (22). Among householders who report that they "always" or "almost always" treat their water by boiling, the practice has been associated with reductions of fecal bacteria of 99% in Vietnam (23), 97% in India (24), and 86.2% in Guatemala (20). In rural Kenya, pasteurization of water using a simple wax indicator to show householders when water reached 70°C increased the number of households whose drinking water was free of coliforms from 10.7% to 43.1% and significantly reduced the incidence of severe diarrhea compared to a control group (25). Given this evidence of microbiological effectiveness and the widespread use of the practice, Clasen and colleagues (23, 24) have argued that boiling is the benchmark against which all other HWTS methods should be measured.

Nevertheless, boiling water presents certain disadvantages that may limit its scalability as a means of routinely treating drinking water. First, the evidence does show that as actually practiced in the home, boiling and storing water often does not always yield microbiologically safe drinking water. Among 137 households in Pakistan who reported boiling as their only method for treating water, only 24 (17.5%) of samples from stored water were free of fecal coliforms (26). In hygiene-challenged emergency settings, householders who reported boiling their drinking water were no more likely than others to have water that was free of fecal contamination (27). Outside the emergency context, boiling has been shown to be more effective, though a quarter of water samples in field studies in Vietnam and India still contained medium or high levels of fecal contamination (23, 24).

Second, more than half of the world's population relies chiefly on wood, charcoal, and other biomass for their energy supplies (28). In sub-Saharan Africa, Southeast Asia, and the Western Pacific regions, the figures are 77%, 74%, and 74%, respectively. The procurement of these fuels represents a substantial commitment of time and energy, primarily for women and girls, and may detract from other productive and potentially health-promoting activities (29). An alternative means of treating water that does not require the use of such fuel may reduce the time spent collecting the same. Third, because many of those who must treat their own water rely on high-emission biofuels and frequently cook indoors on low-efficiency stoves in poorly ventilated rooms, boiling can aggravate what is often already poor indoor air quality (28). The use of these fuels is also environmentally unsustainable in many countries and contributes to greenhouse gases. Fourth, boiling water at home has also been associated with higher levels of burn accidents, especially among young children (30). Finally, the long-term cost of boiling is greater than some alternatives (23, 24), and may be the principal reason why the practice is not even more widespread. Research on the affordability of boiling in a village in Bangladesh found that families in the lowest income quartile would have had to spend 22% of their yearly income on fuel; even those in the highest income bracket would have spent 10% (22). For a typical family in the lowest income quartiles, boiling of drinking water would require an 11% increase in household budget.

18.2.2 Chlorination

Chlorination is the most widely practiced means of treating water at the community level; it has been used continuously in the Europe and North America for more than 100 years (9). Combined in most cases with some level of filtration to remove solids and reduce chlorine demand, community-based chlorination of drinking water has been credited with nearly half the total mortality reduction, three-quarters of the infant mortality reduction, and nearly two-thirds of the child mortality reduction in major cities in the early twentieth century, the period that recorded the largest gains in childhood health (31).

Chlorination is also used increasingly at the household level. An estimated 5.6% of households in middle- and low-income countries, or 67 million people, report that they usually treat their water with bleach (sodium hypochlorite) or another source of chlorine at home before drinking it (20). Tablets formed from dichloroisocyanurate (NaDCC), a leading emergency treatment of drinking water, and novel systems for on-site generation of oxidants such as chlorine dioxide, also have a role in household water treatment. At doses of a few mg/L and contact time of about 30 min, free chlorine inactivates more than 99.99% of enteric pathogens, the notable exceptions being *Cryptosporidium* and *Mycobacterium* species.

The impact of chlorine interventions in improving water quality and reducing diarrheal diseases in field studies has been documented. In a systematic review of 21 intervention studies of point-of-use water treatment with chlorine, Arnold and Colford (32) reported that the intervention significantly reduced the risk of stored water contamination with *Escherichia coli* (pooled relative risk: 0.20, 95%CI: 0.13–0.30). They also reported that the intervention reduced the risk of child diarrhea (pooled relative risk: 0.71, 95%CI: 0.58–0.87). A major finding of their review, however, was an attenuation of the intervention's reduction of child diarrhea in longer trials. The authors noted that this could be attributable to seasonal variations in microbiological performance or to user fatigue that may be more acute for interventions such as chlorine that adversely impact water aesthetics (taste and odor). The diminished effect of chlorine and other disinfection interventions over time has been contrasted with filter interventions, leading some investigators to conclude that this absence of sustained effect, after adjusting for possible reporting bias, raises questions about the suitability of scaling up chlorine interventions at the household level until further efforts to secure long-term uptake can be shown (33).

The Safe Water System (SWS) is a programmatic chlorination intervention developed by the U.S. Centers for Disease Control and Prevention in response to a cholera outbreak in Latin America. It combines bottles of dilute sodium hypochlorite with safe storage and behavior change techniques (www.cdc.gov/safewater). The microbiological efficacy of the SWS has been demonstrated (13, 15), and the effectiveness of the intervention in reducing diarrheal disease has been reported in a variety of settings (34–39).

One chlorine alternative to sodium hypochlorite is sodium dichloroisocyanurate (NaDCC), also known as sodium dichloro-s-triazinetrione or sodium troclosene. For more than

30 years, an effervescent tablet version of NaDCC (Aquatabs® sold by Ireland-based Medentech, Ltd.) has been used for the emergency treatment of water. Its widespread use in nonemergency household water treatment applications began after review of the chlorinated isocyanurates by the U.S. Environmental Protection Agency (USEPA), WHO, and the Food and Agriculture Organization of the United Nations for the routine treatment of drinking water. NaDCC may also offer certain other advantages over sodium hypochlorite in terms of safety, shelf life, up-front cost and convenience, although these have not yet been shown to impact coverage or uptake (40). A blinded field trial has shown the product to be effective in treating drinking water (41). However, a blinded health impact trial has in Ghana reported no impact from the intervention on diarrhea, though the authors suggested that this might have been attributable to low levels of fecal contamination in the water or to the use of a control (improved storage vessel) that may have also been protective (42).

Despite impressive gains in coverage, the adoption and long-term use of chlorine-based HWTS options has been challenging. Follow-up studies on populations exposed to the intervention on a programmatic basis have reported mixed but often low levels of adoption despite extensive programmatic campaigns by implementation partners (43). As discussed in Section 18.2.5, achieving correct, consistent use of the intervention by the target population is a significant challenge for all HWTS options. In an effort to address the deficiencies with respect to chlorine, investigators have begun experimenting with the placement of chlorine dispensers at protected springs and other water sources (44). After collecting a vessel of water, users turn a valve to deliver a measured dose of a stock solution of sodium hypochlorite directly into their water, starting the 30 min contact time even before they reach home. There is some initial evidence that these chlorine dispensers have higher levels of consistent uptake than chlorine products that are purchased for household use (44). Research is underway to explore this alternative in greater scale and to investigate strategies for ensuring that the dispensers are filled when necessary. As most of the cost of the SWS and similar sodium hypochlorite solutions is in the packaging, bulk chlorine dispensers offer an alternative that is significantly lower in cost per liter or cost per household. Whether there is a way to recover the cost of treatment is not yet clear.

18.2.3 Solar Disinfection

Solar disinfection has been subject to rigorous efficacy testing, both in the laboratory and under field conditions, and evaluated for effectiveness in preventing diarrheal disease. While such testing has included permanently mounted panels and other configurations (45), the "Sodis" bottle system, which simply involves filling 1–2 L plastic PET bottles with water and exposing them to the sun for 6 or more hours, has been particularly well documented. Testing in the laboratory has demonstrated the microbiological efficacy of the method against a variety of pathogens (46, 47).

The Sodis method has been shown to be effective in reducing diarrhea (48–50) and cholera (51). However, in one of the largest and most rigorous (though not yet published) studies of the Sodis method, Mausezahl and colleagues (52) found no statistically significant reduction in diarrhea among <5 s in 22 villages in Bolivia (RR: 0.81, 95%CI: 0.59–1.12). The cluster-randomized trial followed 725 children from 425 households for 1 year after a local NGO implemented the intervention to half the villages and allowed the other half to follow their existing water management practices. The investigation was designed to assess the effectiveness of the intervention as actually deployed by a program implementer; accordingly, the local NGO conducted a standardized interactive Sodis promotion and hygiene education campaign. The investigators found, however, that mean compliance with the intervention was only 32.1%, and ascribed the insignificant result to poor uptake.

Unfortunately, none of these health impact studies included assessments of the microbiological performance of the intervention. Accordingly, they do not shed additional light on the contribution of Sodis in improving water quality. To date, only Rainey and Harding (53) have reported on the microbiological performance of the Sodis method in a field study. This was an efficacy study in 40 households in Nepal. The investigators reported that Sodis reduced the level of fecal contamination under household conditions, with 9 of 10 household water samples free of fecal bacteria despite high levels of contamination in source water. However, the study was small, follow-up limited (three times over 5 months) and the intervention was delivered in a research rather than programmatic context. The investigators also reported that only 10% of study households actually adopted the intervention.

18.2.4 Filtration

Unlike boiling, chlorination, and solar disinfection, household filtration encompasses a variety of different processes and products for improving the microbiological quality of drinking water. Mechanical filtration involves the physical removal of suspended solids (including microbes) from water by employing a porous media whose pore size is smaller than the target contaminant. Common media include cloth, sand, porous rock, unglazed ceramics. Advanced membranes for microfiltration, ultrafiltration, nanofiltration, and reverse osmosis are also used, but must be specially configured for gravity-pressure applications common in low-income settings. Adsorption involves the retention of contaminants within the medium itself, much like a sponge retains water. Activated carbon, either in block or granulated form, is the most common adsorption media for treating water, but is rarely used in low-income settings because of the inability to determine when its adsorption capacity has been exhausted.

Slow-sand filters, which remove suspended solids and microbes by means of a slime layer (*schmutzdecke*) that develops within the top few centimeters of sand, are capable of removing 99% or more of enteric pathogens if properly constructed, operated, and maintained (54). A simpler but more advanced version, known as the "bio-sand" filter, was specifically designed for intermittent use and is more suitable for household applications. It has been tested both in the laboratory and the field (55, 56) and is being deployed widely in development settings by the Centre for Alternative Water and Sanitation Technologies (www.cawst.org) and others (43).

Ceramic filters have been used for treating drinking water for more than a century (9). Higher-quality ceramic filters treated with bacteriostatic silver have been shown effective in the lab at reducing waterborne protozoa by more than 99.9% and bacteria by more than 99.9999%; to date, however, they have not met targets for reducing waterborne viruses, though viral sorption and inactivation can be enhanced by special processing (57). Commercial gravity filters are typically formed into hollow cylindrical "candles," which are mounted into the top of a two-compartment vessel; locally fabricated filters usually consist of a ceramic pot that serves as the upper chamber and is designed to fit directly into a lower plastic vessel. Pathogens are removed as contaminated water passes through the ceramic in the top compartment to the lower holding compartment. Because the filtered water can only be accessed from this lower compartment by a tap or spigot, it is protected from another significant risk—recontamination prior to consumption. Their potential usefulness as a public health intervention has been reported in development and emergency settings (40, 58, 59). The improving quality of locally fabricated silver-coated ceramics is particularly promising as a sustainable and low-cost alternative (59).

18.2.5 Combination of Flocculation and Disinfection

Sachets combining flocculation and disinfection agents were developed in South Africa more than two decades ago. The original products, still sold under the Watermaker™ and Chlor-floc™ brands, employ alum to reduce turbidity and chlorine-resistant protozoan cysts and dichloroisocyanurate to inactivate bacteria and viruses. In 2002, Procter & Gamble Company began field testing its own flocculation/disinfectant sachets. Marketed under the PUR® brand, the product uses ferric sulfate as the flocculant and calcium hypochlorite as the disinfectant, and was designed to address perceived deficiencies in other combination products. Users open the sachet, pour the contents into 10 L of water, stir it repeatedly for several minutes until the floc settles out in the bottom of the vessel, pour the supernatant through a clean cloth into another vessel, then allow it to stand for 30 min.

Few HWTS technologies have been tested as extensively as PUR sachets, both in the laboratory and the field. Laboratory tests demonstrated that the product is highly efficacious, not only against bacteria (>99.99999% reduction), virus (>99.99%), and cysts (>99.95%), but also in reducing levels of arsenic, a significant chemical health hazard in many South Asian water supplies (15). Aside from boiling and certain commercial filters that have not yet been deployed widely among low-income populations (e.g., Hindustan Unilever "Pureit®" and the Vestergaard-Frandsen "LifeStraw Family®"), PUR sachets are the only household water treatment option designed for use in low-income settings that would appear to satisfy the requirements of a microbiological water purifier established under the EPA Protocol and NSF P-231 (16). Significantly, the sachets have also shown efficacy against arsenic, an important waterborne risk in many Asian countries (16, 60).

A series of rigorous field trials have been undertaken, mainly to assess the efficacy of the intervention in reducing diarrheal disease. Some of these studies also assessed the microbiological performance of the intervention (37–39, 61, 62).

18.2.6 Safe Storage

Households without access to piped-in water are normally required to store water in the home. Research has consistently found that even water that is safe at the point of collection is subject to frequent and extensive microbial contamination during collection, transport, and storage in the home, usually as a result of poor hand hygiene and the practices of drawing drinking water by dipping into an large-mouth vessel (8). The microbiological quality of water from contaminated surface or other sources will be aggravated as a result of poor storage practices. While there is some debate over the pathogenicity of microbes circulating in the household (domestic domain) as opposed to those outside the home (public domain) (63), rigorous field trials have reported substantial reductions in diarrheal disease from the distribution and use of improved vessels for storing water (64). Some of the key factors influencing the impact of storage vessels and conditions on household water quality are: (i) portability and ease of use, based on capacity, size, shape, weight, presence of handles; (ii) durability, weight, and other properties related to resistance and longevity; (iii) presence of a coverable (preferably screw-cap) opening for filling and cleaning access but small enough to reduce the potential for introducing contaminants by contaminated hands, dipping utensils, and other vehicles (e.g., airborne dust), vectors, or other sources; (iv) ability to withdraw water in a sanitary manner, such as via a tap, spigot, spout, or other narrow orifice; and (v) presence and accessibility of documentation describing how to properly use the container for water treatment and sanitary storage (65).

18.3 EFFECTIVENESS AND COST-EFFECTIVENESS OF HWTS TO PREVENT DIARRHEA

18.3.1 Effectiveness

Two decades ago, Esrey and colleagues reviewed previous studies on the impact of environmental interventions on diarrhea, and found improvements in water quality to be considerably less effective than those aimed at water quantity, water availability, and sanitation (66, 67). Ubiquitously cited in both professional journals and practical guides, the reviews have led to the dominant paradigm respecting water supply and sanitation interventions: that to achieve broad health impact, greater attention should be given to safe excreta disposal and proper use of water for personal and domestic hygiene rather than to drinking-water quality. The corollary has become equally established: that interventions aimed solely at improving drinking-water quality would have relatively little impact in reducing diarrheal disease.

Recently, however, an increasing body of evidence has suggested the need to refine the dominant paradigm. Esrey's conclusions that water quality improvements could reduce diarrheal disease by 15%–17% were based exclusively on studies involving interventions at the point of distribution, such as protected wells and springs. Thus, it did not capture the potential additional health gains that could be achieved by ensuring water quality to the point of use by treating it at the household level or by preventing recontamination after collection, transport, and storage in the home. An analysis of the actual review, however, suggests that only when the water supply is delivered on plot are the health gains realized. Waddington and colleagues reached the same conclusion in a recent systematic review (68) that showed that water quality interventions at the source were protective against diarrhea

among young children if implemented through the point of use (RR: 0.79, 95% CI: 0.63–0.98), but only marginally so if implemented to the point of delivery (RR: 0.95, 95% CI: 0.90–1.00).

Table 18.1 summarizes the results of three more recent systematic reviews of water interventions to prevent diarrheal diseases. They suggest that improvements in water quality make substantial contributions to the prevention of diarrheal diseases. They also suggest, however, that interventions at the household level may be more effective in preventing diarrhea than conventional interventions at the point of distribution. Among household-based interventions, filtration was associated with the largest reductions in diarrheal disease, perhaps because it also improves water aesthetics, which may increase use (compliance) with the intervention.

There is increasing evidence, however, that the results from HWTS interventions may be exaggerated due to placebo effect and reporting bias. Clasen and colleagues observed that while more than two dozen studies of HWTS interventions reported a protective effect, none of the four studies that attempted to blind the intervention with a placebo found the effect to be statistically significant (40). A recent field trial of pipe-style filters in Ethiopia also found significant evidence of reporting bias (71). This and other evidence of the lack of successful strategies for achieving adoption of HWTS interventions at scale has led some investigators to conclude that efforts to scale up the intervention are premature (72). Other investigators have noted, however, that if the estimates of effect of HWTS interventions are discounted by the exaggerated effect associated with open trials of reported outcomes (like self-reported diarrhea), the interventions are still protective, particularly for those HWTS methods whose sustainability can be maintained (33, 43).

There is a need for additional assessments of HWTS interventions using placebos and objective outcomes (such as anthropometrics) in order to determine the actual protective effect of HWTS interventions to prevent diarrhea (40, 43). Even so, the size of the effect, if any, is likely to depend largely on the prevailing conditions. The major factor influencing the effect is more likely to be the extent to

TABLE 18.1 Pooled Estimate of Risk (and Number of Studies) of Systematic Reviews of Water Supply and Water Quality Interventions to Prevent Diarrheal Disease

Intervention (Improvement)	Fewtrell et al. (69) (95% CI)	Clasen et al. (70) (95% CI)	Waddington et al. (68) (95% CI)
Water supply	0.75 (0.62–0.91) (6)		0.98 (0.89–1.06) (8)
Water quality	0.69 (0.53–0.89) (15)	0.57 (0.46–0.70) (38)	0.58 (0.50–0.67) (31)
Source	0.89 (0.42–1.90) (3)	0.73 (0.53–1.01) (6)	0.79 (0.62–1.02) (3)
Household	0.65 (0.48–0.88) (12)	0.53 (0.39–0.73) (32)	0.56 (0.45–0.65) (28)
Chlorination		0.63 (0.52–0.75) (16)	
Filtration		0.37 (0.28–0.49) (6)	
Solar disinfection		0.69 (0.63–0.74) (2)	
Floc-disinfection		0.69 (0.58–0.82) (6)	

which water is the dominant pathway for transmission of diarrheagenic agents. Thus, any given trial, blinded or open, is unlikely to deliver an estimate of effect that is fully generalizable, since the underlying effect will depend on transmission dynamics that vary according to settings, seasons, and other factors. This is consistent with the large range of effect sizes and heterogeneity reported in the meta-analyses that accompanied each of these systematic reviews.

18.3.2   Cost-Effectiveness

While HWTS offers the potential for superior health gains, the economic advantages over conventional improvements in water supplies are also compelling. The cost of implementing water quality interventions varies, from a low of USD 0.63 per person per year (solar disinfection) and USD 0.66 (chlorination) to USD 3.03 (ceramic filters) and USD 4.95 (combined flocculation/disinfection); this compares to an average USD 1.88 per person per year for installing and maintaining wells, boreholes, and communal tap stands in Africa (73). The combination of lower cost and higher effectiveness renders household-based chlorination the most cost effective of water quality interventions to prevent diarrhea, with a cost-effectiveness ratio in Africa of USD 53 per disability-adjusted life year (DALY) averted, compared to USD 123 for conventional source-based interventions. When health cost savings are included in the analysis, implementing low-cost HWTS interventions actually results in net savings to the public sector (73). A recent WHO-sponsored analysis also concluded that household-based chlorination was among the most cost-beneficial of the various options for pursuing the MDG water and sanitation targets, yielding high returns on every dollar invested mainly from lower health care costs but also increased productivity and value of school attendance (74). Finally, there is considerable evidence that the target population is willing and able to pay for some or all of the cost of household-based water treatment products (75), leveraging public sector and donor funding and allowing it to be more focused on the base of the economic pyramid.

In summary, (i) the up-front cost of providing low-cost household water treatment is about half that of conventional source-based interventions; (ii) most or all of that cost can be borne directly by the beneficiary, not the public sector; and (iii) the public sector will nevertheless recover more than the full cost of implementation from reduced health costs for disease treatment. On the other hand, HWTS does not improve access to water or the quantity of water available, both of which are usually higher priorities for low-income populations and can deliver other important gains in productivity and health. Moreover, there is evidence that HWTS provides little benefit in settings without sufficient quantities of water (40). Accordingly, the intervention is only appropriate in situations where basic levels of access and quantity have been met.

18.4   OPTIMIZING THE POTENTIAL OF HWTS: ACHIEVING TARGETED COVERAGE AND UPTAKE AT SCALE

Despite the long history of HWTS, the practice is still at a nascent stage in terms of public health interventions. There is considerable research to support the microbiological effectiveness of certain approaches, and a growing body of research about its health impact on diarrheal disease. In order to realize the full potential of the intervention, however, it is essential that effective and affordable HWTS options reach the most vulnerable populations at scale (coverage) and that those populations use the intervention correctly and consistently over the long term (uptake). To date, there is mixed evidence on both of these fronts.

18.4.1   Scaling Up Coverage among Target Populations

Using data from household surveys from 67 middle- and lower-income countries, Rosa and Clasen (76) estimate that 33% of households—or more than 1.1 billion people—report treating their water at the household level before drinking it. However, more than 20% of this figure is using practices such as filtering their water through a cloth or letting it stand and settle that are unlikely to render the water safe. Most of the balance consists of those that reported boiling, which evidence has shown to be practiced less frequently and more inconsistently than the JMP household surveys reflect (77). Moreover, the practice of boiling is still largely regional and probably does not correspond with risk. Rural populations, for example, are much less likely to have piped water supplies or other improved water sources; in 2006, rural dwellers represented 84% of the population worldwide using unimproved sources of drinking water (6). However, in all geographical regions, rural populations are less likely to be boiling their water (20). With only 58% of its population having access to improved water sources, sub-Saharan Africa is one of the regions most likely to rely on unsafe water supplies. Nevertheless, only 4.9% of the population in 17 African region countries report boiling their water before drinking it. This compares to 20.4% in Latin America and the Caribbean, 21.2% in Southeast Asia, 34.8% in Central Europe, and 68.3% in the Western Pacific region. Like other HWTS methods, householders in the richest quintile are more than twice as likely to practice boiling than those of the poorest quintile (20). There is also evidence that children under 5 years, who are most vulnerable to diarrheal diseases, may not be benefiting from the HWTS even if it may be practiced by other members of the household (20).

A report commissioned by the WHO used data from manufacturers and program implementers to estimate coverage of HWTS in developing countries (43). In 2007, the combined efforts of the above-described HWTS products—exclusive of boiling and emergency applications—produced approximately 15.5 billion L of treated water. This represents an average annual growth of 25.5% over 2005 and 2006 levels of 9.9 billion and 12.2 billion L treated, respectively. Manufacturers and program implementers reported 18.8 million users of these HWTS products as of the end of 2007, compared with 12.0 million in 2005 and 15.5 million in 2005—an average annual growth rate of 25.1%. It is noted that these figures do not include boiling, which Rosa and Clasen (76) estimated to be practiced by nearly 600 million people. They also do not include emergency use of HWTS products—an important role for point-of-use water treatment, but one that does not contribute to the overall levels of coverage under routine conditions. The largest single contribution to the figures supplied by the manufacturers and program implementers was socially marketed sodium hypochlorite, representing approximately 57% of the 18.8 million users in 2007. While social marketing is expressly targeted to low-income populations, however, there is evidence that that these initiatives also result in inequitable uptake that favors urban and higher-income households (78), much like HWTS practices generally.

These figures are impressive, especially given that most of these HWTS programs have been under way for less than 10 (and some less than 5) years. However, given 884 million people relying on unimproved water supplies and many more whose water is not consistently safe for drinking, the results to date for HWTS methods other than boiling provide a perspective that is more sobering. Using the figures from the WHO report, these HWTS methods will not cover even 100 million users until 2015. Except for the SWS in Zambia and, arguably, Madagascar, and some promising ceramic filter coverage in Cambodia, no HWTS method other than boiling can be said to have achieved scale in coverage (43).

Coverage must also focus on the most vulnerable populations, especially children under 5 years, persons living with HIV/AIDS who with compromised immune systems, and people at the base of the economic pyramid who are most likely to rely on untreated water supplies and least likely to be practicing effective HWT practices or to have access to effective health care. Commercial and quasi-commercial strategies such as social marketing have sought to target some of these markets, using strategies such as microfinance linked to women's self-help groups and bicycle vendors to reach the "last mile" of the distribution chain (http://www.path.org/projects/safe_water.php; http://www.psi.org/our-work/healthy-lives/diarrheal-disease). However, it is likely that the public sector and donors will need to play a leading role in securing coverage among these target populations (43).

18.4.2 Ensuring Uptake (Correct, Consistent Use)

Coverage is an important metric for measuring progress toward scaling up. Like most health interventions, HWTS

promoters must actually reach the target population with safe, effective, appropriate, and affordable solutions. This can be a daunting challenge, even for an intervention such as a vaccine. However, unlike vaccines and certain other interventions, water treatment in the home requires householders to embrace and routinely use the intervention in order to provide protection; even occasional consumption of untreated water may neutralize the potential health benefits of the intervention (79). Thus, the real potential for the benefits of HWTS depends not only on the extent to which it can be made available to the target population but also on the extent to which it is adopted by that population and used correctly and consistently (uptake) on a sustained basis.

Most of the evidence to date on uptake of the HWTS comes from research-driven efficacy trials. These typically involve intensive campaigns at the outset of the study to ensure compliance with the intervention, an effort that cannot always be duplicated programmatically. Because these studies typically involve continuous water sampling and health surveillance, the study population is also visited frequently, which may itself increase compliance (Hawthorne effect) in ways that will not characterize the programmatic delivery of the intervention by governments, NGOs, and entrepreneurs. Even so, these efficacy studies report a wide range of compliance. A review by Arnold and Colford (32) reported that the percentage of households with detectable chlorine in their water supplies ranged from 36% to 100%. There is some evidence of higher uptake of chlorine from dispensers placed at water collection points (44). There is also some evidence of longer-term uptake of ceramic and biosand filters (43). Procter & Gamble ultimately withdrew its flocculant/disinfectant product from the market when penetration did not meet its expectations, reaching about 15% in the Philippines and only 5% in Guatemala (80). Even among participants in a prior intervention study, an aggressive 6-month marketing campaign did not increase adoption beyond such 5%, leading investigators to conclude that demonstrated effectiveness of the intervention was not enough to increase uptake (81).

Two recent assessments of programs by NGOs to introduce HWT among vulnerable populations demonstrate the challenges of achieving uptake and the corresponding lack of any measurable health impact. Arnold and colleagues (82) assessed the uptake and health impact of a 3-year program by NGOs in Guatemala to promote HWT (boiling, Sodis, and chlorination) and hand washing with soap (HWWS). The 6-month study used propensity scoring to match and compare 600 households in 30 villages (15 intervention/15 control). They found no statistically meaningful difference in adoption of intervention and control households for HWT (9% vs. 3%) or HWWS. Consistent with the low-sustained behavior adoption, the investigators found no difference between intervention and control villages in child diarrhea, respiratory infections, or growth. Mausezahl and colleagues used a cluster-randomized controlled trial design in 22 rural communities in Bolivia to evaluate the effect of Sodis in reducing diarrhea among children under the age of 5 (52). A local NGO conducted a standardized interactive Sodis-promotion campaign in 11 villages targeting households, communities, and primary schools. Mothers completed a daily child health diary for 1 year. Despite this extensive promotion campaign, investigators found only 32% compliance with the intervention and no strong evidence for a substantive reduction in diarrhea among children (RR: 0.81, 95% CI: 0.59–1.12).

Finally, as noted above, there is evidence that the health impact of some HWTS interventions diminishes over time (32, 33, 68). Sustained uptake of healthful interventions involving behavior change is a common problem; recent evidence suggests that adoption of oral rehydration solution (ORS) is falling from levels achieved a generation ago (Bowen, unpublished data, 2006). There are also issues concerning the sustainability of common HWTS options themselves, due to affordability and the need to ensure continued supplies of consumables and replacement filters (83). However, there is evidence that HWWS has been able to achieve sustained adoption in some settings (68). Promoters of HWTS have begun to follow successful hygiene promotion strategies, such as delivery in schools where children become change agents for influencing behaviors at home (84). Although there is increasing research on behavior change and communication in the context of environmental interventions such as HWTS (85), the lessons learned have rarely been incorporated effectively into HWTS campaigns. HWTS may actually benefit by being delivered synergistically with hygiene promotion, both to increase sustained adoption and to reduce costs.

18.5 CONCLUSION

Drinking water that is free from fecal contamination is not a sufficient condition to human health, but it is a necessary one that is far from the reach of billions of people worldwide. Combined with improved sanitation and hygiene, HWTS can prevent disease and save lives. Importantly, HWTS provides a tool for the most vulnerable populations to take charge of their own water security, while they patiently wait for the pipe to finally reach them.

Additional work is necessary to develop affordable HWTS solutions that operate effectively and sustainably in the challenging environments facing low-income populations and that will be embraced and used correctly and consistently. Research is also required to identify and target the populations that could benefit most from HWTS and to develop delivery strategies that can be implemented at scale.

REFERENCES

1. Pruss-Üstün A, Bos R, Gore F, Bartram J. *Safe Water, Better Health: Costs, Benefits and Sustainability of Interventions to Protect and Promote Health*. Geneva: World Health Organization, 2008.

2. Boschi-Pinto C, Velebit L, Shibuya K. Estimating child mortality due to diarrhoea in developing countries. *Bull WHO* 2008;86(9):710–717.

3. Baqui AH, Black RE, Sack RB, Chowdhury HR, Yunus M, Siddique AK. Malnutrition, cell-mediated immune deficiency and diarrhoea: a community-based longitudinal study in rural Bangladeshi children. *Am J Epidemiol* 1993;137:355–365.

4. Guerrant DI, Moore SR, Lima AAM, Patrick P, Schorling JB, Guerrant RL. Association of early childhood diarrhoea and cryptosporidiosis with impaired physical fitness and cognitive function four–seven years later in a poor urban community in Northeast Brazil. *Am J Trop Med Hyg* 1999;61:707–713.

5. United Nations. United Nations Millennial Declaration. General Assembly Res. 55/2 (18 September 2000).

6. WHO/UNICEF. *Progress in Drinking-Water and Sanitation: Special Focus on Sanitation*. Geneva: World Health Organization and United Nations Children's Fund, 2008.

7. RADWQ. *Summary of Results of Rapid Drinking Water Quality Analysis*. Geneva: World Health Organization, 2008.

8. Wright J, Gundry S, Conroy R. Household drinking water in developing countries: a systematic review of microbiological contamination between source and point-of-use. *Trop Med Intl Health* 2003;9(1):106–117.

9. Sobsey MD. *Managing Water in the Home: Accelerated Health Gains from Improved Water Supply*. Geneva: The World Health Organization (WHO/SDE/WSH/02.07), 2002.

10. Mintz ED, Reiff FM, Tauxe RV. Safe water treatment and storage in the home. A practical new strategy to prevent waterborne disease. *JAMA* 1995;273(12):948–953.

11. WHO/UNICEF. *Progress Toward the Millennium Development Goals, 1990–2005*. Geneva: World Health Organization, 2005.

12. WHO. *Combating Waterborne Disease at the Household Level*. Geneva: World Health Organization, 2007.

13. Quick R, Venczel L, Gonzalez O, Mintz E, Highsmith A, Espada A, Damiani E, Bean N, De Hannover R, Tauxe R. Narrow-mouthed water storage vessels and *in situ* chlorination in a Bolivian community: a simple method to improve drinking water quality. *Am J Trop Med Hyg* 1996;54:511–516.

14. Luby S, Agboatwalla M, Razz A, Sobel J. A low-cost intervention for cleaner drinking water in Karachi, Pakistan. *Intl J Infect Dis* 2001;5(3):144–150.

15. Rangel JM, et al. A novel technology to improve drinking water quality: a microbiological evaluation of in-home flocculation and chlorination in rural Guatemala. *J Water Health* 2003;1:15–22.

16. Souter PF, Cruickshank GD, Tankerville MZ, Keswick BH, Ellis BD, Langworthy DE, Metz KA, Appleby MR, Hamilton N, Jones AL, Perry JD. Evaluation of a new water treatment

for point-of-use household applications to remove micro-organisms and arsenic from drinking water. *J Water Health* 2003;1(2):73–84.

17. WHO. *Guidelines for Drinking-Water Quality, Vol. 1. Recommendations*. Second Addendum to 3rd edition. Geneva: World Health Organization, 2008.

18. WHO. *Essential Prevention and Care Interventions for Adults and Adolescents Living with HIV in Resource-Limited Settings.* Geneva: World Health Organization, 2008.

19. UNICEF. *Diarrhea: Why Children Are Still Dying and What Can Be Done*. New York: United Nations Children's Fund, 2009.

20. Rosa G, Miller L, Clasen T. Microbiological effectiveness of disinfecting water by boiling in rural Guatemala. *Am J Trop Med Hyg* 2010;82(2):289–300.

21. WHO. *Drinking-Water Quality Standards*, 3rd edition. Geneva: The World Health Organization, 2004.

22. Gilman RH, Skillicorn P. Boiling of drinking-water: can a fuel-scarce community afford it? *Bull WHO* 1985;63:157–163.

23. Clasen T, Do Hoang T, Boisson S, Shippin O. Microbiological effectiveness and cost of boiling to disinfect water in rural Vietnam. *Environ Sci Tech* 2008; doi 10.1021/es7024802.

24. Clasen T, McLaughlin C, Nayaar N, Boisson S, Gupta R, Desai D, Shah N. Microbiological effectiveness and cost of disinfecting water by boiling in semi-urban India. *Am J Trop Med Hyg* 2009;79(3):407–413.

25. Iijima Y, Karama M, Oundo JO, Honda T. Prevention of bacterial diarrhea by pasteurization of drinking water in Kenya. *Microbiol Immunol* 2001;45(6):413–416.

26. Luby SP, Syed AH, Atiullah N, Faizan MK, Fisher-Hoch S. Limited effectiveness of home drinking water purification efforts in Karachi, Pakistan. *Int J Infect Dis* 1999;4:3–7.

27. Gupta SK, Suantio A, Gray A, Widyastuti E, Jain N, Rolos R, Hoekstra RM, Quick R. Factors associated with E. coli contamination of household drinking water among tsunami and earthquake survivors, Indonesia. *Am J Trop Med Hyg* 2007;76:1158–1162.

28. Rehfuess E, Mehta S, Pruss-Ustun A. Assessing household solid fuel use: multiple implications for the Millennium Development Goals. *Environ Health Perspect* 2006;114(3):373–378.

29. Biran A, Abbot J, Mace R. Families and firewood: the costs and benefits of children in firewood collection and use in two rural communities in sub-Saharan Africa. *Hum Ecol* 2004;32(1):1–25.

30. Houangbevi A. Burns in children in Africa (experience of the Burns Department at the Lome University Hospital Center). *Annales de l'anesthésiologie française*, 1981;22(4):346–350 (in French).

31. Cutler D, Miller G. The role of public health improvements in heath advances: the twentieth-century United States. *Demography* 2005;42:1–22.

32. Arnold B, Colford J. Treating water with chlorine at point-of-use to improve water quality and reduce child diarrhea in developing countries: a systematic review and meta-analysis. *Am J Trop Med Hyg* 2007;76(2):354–364.

33. Hunter P. Household water treatment in developing countries: comparing different intervention types using meta-regression. *Environ Sci Technol* 2009;43(23):8991–8997.

34. Semenza JC, Roberts L, Henderson A, Bogan J, Rubin CH. Water distribution system and diarrhoeal disease transmission: a case study in Uzbekistan. *Am J Trop Med Hyg* 1998;59(6):941–946.

35. Quick RE, Venczel LV, Mintz ED, Soleto L, Aparicio J, Gironaz M, Hutwagner L, Greene K, Bopp C, Maloney K, Chavez D, Sobsey M, Tauxe RV. Diarrhoea prevention in Bolivia through point-of-use water treatment and safe storage: a promising new strategy. *Epidemiol Infect* 1999;122(1):83–90.

36. Quick RE, Kimura A, Thevos A, Tembo M, Shamputa I, Hutwagner L, Mintz E. Diarrhoea prevention through household-level water disinfection and safe storage in Zambia. *Am J Trop Med Hyg* 2002;66(5):584–589.

37. Reller ME, Mendoza CE, Lopez MB, Alvarez M, Hoekstra RM, Olson CA, Baier KG, Keswick BH, Luby SP. A randomized controlled trial of household-based flocculant-disinfectant drinking water treatment for diarrhoea prevention in rural Guatemala. *Am J Trop Med Hyg* 2002;69:411–419.

38. Luby SP, Agboatwalla M, Hoekstra RM, Rahbar MH, Billhimer W, Keswick B. Delayed effectiveness of home-based interventions in reducing childhood diarrhea, Karachi, Pakistan. *Am J Trop Med Hyg* 2004;71(4):420–427.

39. Chiller TM, Mendoza CE, Lopez MB, Alvarez M, Hoekstra RM, Keswick BH, Luby SP. Reducing diarrhea in Guatemalan children: a randomized controlled trial of a flocculant-disinfectant for drinking water. *Bull WHO* 2006;84(1):28–35.

40. Clasen T, Edmondson P. Sodium dichloroisocyanurate (NaDCC) tablets as an alternative to sodium hypochlorite for the routine treatment of drinking water at the household level. *Intl J Hyg Environ Health* 2006;209:173–181.

41. Clasen T, Saeed T, Boisson S, Edmondson P, Shipin O. Household-based chlorination of drinking water using sodium dichloroisocyanurate (NaDCC) tablets: a randomized, controlled trial to assess microbiological effectiveness in Bangladesh. *Am J Trop Med Hyg* 2007;76(1):187–192.

42. Jain S, Sahanoon OK, Blanton E, Schmitz A, Wannemuehler KA, Hoekstra RM, Quick RE. Sodium dichloroisocyanurate tablets for routine treatment of household drinking water in periurban Ghana: a randomized controlled trial. *Am J Trop Med Hyg* 2010;82(1):16–22.

43. Clasen T. *Scaling Up Household Water Treatment Among Low-Income Populations*. Geneva: World Health Organization, 2009.

44. Kremer M, Miguel E, Mullainathan S, Null C, Zwane A. Trickle down: chlorine dispensers and household water treatment. Workshop on Scaling Up the Distribution of Water Treatment Technologies, Harvard University, Cambridge, MA, 12 December 2008.

45. Kang G, Roy S, Balraj V. Appropriate technology for rural India—solar decontamination of water for emergency settings and small communities. *Trans R Soc Trop Med Hyg* 2006;100(9):863–866.

46. Wegelin M, Canonica S, Mechsner K, Fleischmann T, Pesaro F, Metzler A. Solar water disinfection: scope of the process and analysis of radiation experiments. *J Water SRT-Aqua* 1994;43(4):154–169.

47. McGuigan KG, Méndez-Hermida F, Castro-Hermida JA, Ares-Mazás E, Kehoe SC, Boyle M, Sichel C, Fernández-Ibáñez P, Meyer BP, Ramalingham S, Meyer EA. Batch solar disinfection (SODIS) inactivates oocysts of Cryptosporidium parvum and cysts of Giardia muris in drinking water. *J Appl Microbiol* 2006;101(2):453–463.

48. Conroy RM, Elmore-Meegan M, Joyce T, McGuigan KG, Barnes J. Solar disinfection of drinking water and diarrhoea in Maasai children: a controlled field trial. *Lancet* 1996;348(9043):1695–1697.

49. Conroy RM, Meegan ME, Joyce T, McGuigan K, Barnes J. Solar disinfection of water reduce diarrhoeal disease: an update. *Arch Dis Child* 1999;81(4):337–338.

50. Rose A, Roy S, Abraham V, Holmgren G, George K, Balraj V, Abraham S, Muliyil J, Joseph A, Kang G. Solar disinfection of water for diarrhoeal prevention in southern India. *Arch Dis Child* 2006;91(2):139–141.

51. Conroy RM, Meegan ME, Joyce T, McGuigan K, Barnes J. Solar disinfection of drinking water protects against cholera in children under 6 years of age. *Arch Dis Child* 2001;85(4):293–295.

52. Mausezahl D, Christen A, Duran Pacheco G, Tellez FA, Iriarte M, et al. Solar drinking water disinfection (SODIS) to reduce childhood diarrhoea in rural Bolivia: a cluster-randomized, controlled trial. *PLoS Med* 2009;6(8):e1000125. doi: 10.1371/journal.pmed.1000125.

53. Rainey RC, Harding AK. Acceptability of solar disinfection of drinking water treatment in Kathmandu Valley, Nepal. *Int J Environ Health Res* 2005;15(5):361–372.

54. Hijnen WAM, et al. Elimination of viruses, bacteria and protozoan oocysts by slow sand filtration. *Water Sci Technol* 2004;50(1):147–154.

55. Stauber CE, Elliott MA, Koksai F, Ortiz GM, DiGiano FA Sobsey MD. Characterisation of the biosand filter for E. coli reductions from household drinking water under controlled laboratory and field use conditions. *Water Sci Technol* 2006;54:1–7.

56. Tiwari SS, Schmidt WP, Darby J, Kariuki ZG, Jenkins MW. Intermittent slow sand filtration for preventing diarrhoea among children in Kenyan households using unimproved water sources: randomized controlled trial. *Trop Med Int Health* 2009 Nov;14(11):1374–1382.

57. Brown J, Sobsey MD. Ceramic media amended with metal oxide for the capture of viruses in drinking water. *Environ Technol* 2009;30(4):379–391.

58. Clasen T, Brown J, Suntura O, Collin S, Cairncross S. Reducing diarrhoea through household-based ceramic filtration of drinking water: a randomized, controlled trial in Bolivia. *Am J Trop Med Hyg* 2004;70(6):651–657.

59. Brown J, Sobsey MD, Proum S. *Use of Ceramic Water Filters in Cambodia*. Washington, DC: World Bank, Water, Sanitation and Hygiene, 2007.

60. Norton DM, Rahman M, Shane AL, Hossain Z, Kulick RM, Bhuiyan MI, Wahed MA, Yunus M, Islam MS, Breiman RF, Henderson A, Keswick BH, Luby SP. Flocculant-disinfectant point-of-use water treatment for reducing arsenic exposure in rural Bangladesh. *Int J Environ Health Res* 2009 Feb; 19(1):17–29.

61. Crump JA, Otieno PO, Slutsker L, Keswick BH, Rosen DH, Hoekstra RM, Vulule JM, Luby SP. Household based treatment of drinking water with flocculant-disinfectant for preventing diarrhea in areas with turbid source water in rural western Kenya: cluster randomized controlled trial. *BMJ* 2005;331 (7515):478–484.

62. Doocy S, Burnham G. Point-of-use water treatment and diarrhoea reduction in the emergency context: an effectiveness trial in Liberia. *Trop Med Int Health* 2006;11(10): 1542–1552.

63. Cairncross S, Blumenthal U, Kolsky P, Moraes L, Tayeh A. The public and domestic domains in the transmission of disease. *Trop Med Int Health* 1996;1(1):27–34.

64. Roberts L, et al. Keeping clean water clean in a Malawi refugee camp: a randomized intervention trial. *Bull WHO* 2001;79:280–287.

65. Mintz E, Bartram J, Lochery P, Wegelin M. Not just a drop in the bucket: expanding access to point-of-use water treatment systems. *Am J Pub Health* 2001;91(10):1565–1570.

66. Esrey SA, Feachem RG, Hughes JM. Interventions for the control of diarrhoeal diseases among young children: improving water supplies and excreta disposal facilities. *Bull WHO* 1985;64:776–772.

67. Esrey SA, Potash JB, Roberts L, Shiff C. Effects of improved water supply and sanitation on ascariasis, diarrhoea, dracunculiasis, hookworm infection, schistosomiasis, and trachoma. *Bull WHO* 1991;69:609–621.

68. Waddington H, Snilstveit, White H, Fewtrell L. *Water, Sanitation and Hygiene Interventions to Combat Childhood Diarrhoea in Developing Countries*. Delhi: International Initiative for Impact Evaluation, 2009.

69. Fewtrell L, Kaufmann R, Kay D, Enanoria W, Haller L, Colford J. Water, sanitation, and hygiene interventions to reduce diarrhoea in less developed countries: a systematic review and meta-analysis. *Lancet Infect Dis* 2005;5:42–52.

70. Clasen T, Roberts I, Rabie T, Schmidt W-P, Cairncross S. Interventions to improve water quality for preventing diarrhoea (Cochrane Review). In: The Cochrane Library, Issue 3, 2006. Oxford: Update Software.

71. Boisson S, Schmidt W-P, Berhanu T, Gezahegn H, Clasen T. Randomized controlled trial in rural Ethiopia to access a portable water treatment device. *Environ Sci & Tech*, 2009; 43(15):5934–5999.

72. Schmidt WP, Cairncross S. Household water treatment in poor populations: is there enough evidence for scaling up now? *Environ Sci Technol* 2009;43(4):986–992.

73. Clasen T, Haller L, Walker D, Bartram J, Cairncross S. Cost-effectiveness analysis of water quality interventions for preventing diarrhoeal disease in developing countries. *J Water Health* 2007;5(4):599–608.

74. Hutton G, Haller L. *Evaluation of the Costs and Benefits of Water and Sanitation Improvements at the Global Level.* Geneva: World Health Organization, 2004.

75. Ashraf N, Berry J, Shapiro J. Can Higher Prices Stimulate Product Use? Evidence from a Field Experiment on Clorin in Zambia. Presentation at the Berkeley Water Center Workshop on water sanitation and hygiene, Berkeley, California, 6 November 2006.

76. Rosa G, Clasen T. Estimating the scope of household water treatment in low-and medium-income countries. *Am J Trop Med Hyg* 2010;82(2):289–300.

77. Makutsa P, Nzaku K, Ogutu P, Barasa P, Ombeki S, Mwaki A, Quick R. Challenges in implementing a point-of-use water quality intervention in rural Kenya. *Am J Pub Health* 2001;91(10):1571–1573.

78. Rheingans R, Dreibelbis R. Disparities in Sûr'Eau use and awareness: results from the 2006 PSI TraC survey (Preliminary Results), 2007.

79. Hunter PR, Zmirou-Navier D, Hartemann P. Estimating the impact on health of poor reliability of drinking water interventions in developing countries. *Sci Total Environ* 2009;407(8):2621–2624.

80. Hanson M, Powell K. *Procter & Gamble and Population Services International (PSI): Social Marketing for Safe Water.* Paris: Institut Européen d'Administration des Affaires, 2006.

81. Luby SP, Mendoza C, Keswick BH, Chiller TM, Hoekstra RM. Difficulties in bringing point-of-use water treatment to scale in rural Guatemala. *Am J Trop Med Hyg* 2008;78(3): 382–387.

82. Arnold B, Arana B, Mausezahl D, Hubbard A, Colford J. Evaluation of a pre-existing, 3-year household water treatment and handwashing. *Intl J Epidemiol* 2009;1–11.

83. Sobsey MD, Stauber CE, Casanova LM, Brown JM, Elliott MA. Point of use household drinking water filtration: a practical, effective solution for providing sustained access to safe drinking water in the developing world. *Environ Sci Technol* 2008;42(12):4261–4267.

84. O'Reilly CE, et al. The impact of a school-based safe water and hygiene programme on knowledge and practices of student and their parents: Nyanza Province, western Kenya, 2006. *Epidemiol Infect* 2008;136(1):80–91.

85. Figueroa ME, Kincaid DL. *Social, Cultural and Behavioral Correlates of Household Water Treatment and Storage.* Geneva: World Health Organization, 2009.

19

INFORMAL SMALL-SCALE WATER SERVICES IN DEVELOPING COUNTRIES: THE BUSINESS OF WATER FOR THOSE WITHOUT FORMAL MUNICIPAL CONNECTIONS

Laura Sima and Menachem Elimelech

19.1 INTRODUCTION

In many parts of the world, access to high-quality drinking water and proper sanitation remains a luxury. Conventional centralized systems are unable to cope with rapid population growth in urban areas and high per-connection costs in rural areas. Where formal, centralized municipal services are lacking, the informal private sector has filled service gaps. This is especially the case in rapidly urbanizing centers, peri-urban areas, and urban slums. Indeed, studies suggest that over 25% of the population in Latin American cities depends on the informal sector, while this number may be as high as 50% in Africa (1). Furthermore, the informal sector appears to be growing more quickly than the formal sector in rapidly urbanizing zones. As of 1999, there were over 200 independent service providers in Guatemala servicing over half of the population (1).

Although the importance of the informal sector, in terms of its ability to either prevent or facilitate the spread of water-borne disease in urban areas, is undeniable, relatively few studies to date have addressed this sector. Over 10,000 water businesses, servicing 49 countries, have been documented thus far, but very few published studies are quantitative or comparative in nature. Rather, most studies to date are anec-dotal accounts of the sector in certain zones. This chapter will summarize current knowledge of this sector, while making generalizations and arguing for a health-focused, quantitative approach to such studies whenever possible.

We begin by broadly discussing the current state of water-services provisions in urban areas, especially with regards to changing demographics and urban growth. We then discuss how and why the informal sector has emerged, and why it has remained relatively understudied. We then reference anecdotal studies to further define and clarify this sector, including examples of the extensive variety of services around the world. Using these anecdotal studies, we make some descriptive generalizations about the sector, discussing both the problems associated with and benefits offered by the sector. We conclude with a discussion of different attitudes toward the sector and anticipate future trends in interacting with and, possibly, regulating, informal water services. It is our goal to compile studies from various disciplines to provide a coherent and complete understanding of the sector. As informal providers are often the *major* water provider in many urban areas, we must understand and work with their industry if we seek to improve water access and reduce waterborne disease in these areas.

19.2 ACCESS TO IMPROVED WATER SOURCES IN URBAN AREAS

In general, informal small-scale service providers "appear most prevalent in countries with low coverage levels, inef-fective public utilities that provide inadequate or partial

Water and Sanitation-Related Diseases and the Environment: Challenges, Interventions, and Preventive Measures, First Edition.
Edited by Janine M. H. Selendy.

services, and remote, difficult-to-access regions" (2). Unfortunately, areas without efficient public utilities, where the majority of urban or peri-urban residents are not connected to municipal water supplies, are highly prevalent throughout the developing world. The possibilities for rapidly increasing access to treated water in the developing world have diminished because government and private-sector investment in this area has decreased (2). The problem is especially acute in poor peri-urban and urban areas. In Pakistan, for example, household access to safe water has actually fallen in the past 10 years by 20%, partially due to increased pressures from rapid urbanization (3).

Municipal infrastructure design, based on the Western model of distribution, was developed to serve the well-defined, slowly growing, urban boundaries of nineteenth-century London. Investing in pipes and calculating flow rates, for example, require the ability to estimate population settlements and project for future population changes. This is nearly impossible in rapidly growing cities, where current populations are often unaccounted for in municipal records. Although alternative water treatment has focused on rural areas, by 2015, more than half of the world's population will reside in urban areas (4). The problem is especially acute in rapidly urbanizing regions of sub-Saharan Africa. A recent survey of the three largest cities in Kenya showed that, on average, 39% of urban households do not have access to piped water in their house or yard. Water coverage levels for a number of sub-Saharan capital cities are shown in Fig. 19.1.

The number of households that have access on a daily level is much lower, however, due to interruption of services as a result of unreliable electricity, overutilized systems, and leaks (5). The same is true of many South Asian megacities. For example, only slightly over 20% of the residents living in the urban zone of Jakarta have access to treated water (6, 7).

Often, infrastructure projects in poor urban communities are viewed as "risky investments" (8), so minimal funds are directed to the neediest areas. This is because the poor are unable to pay connection charges or monthly bills. Even when funds are available, legal hurdles prevent connecting new households. This is especially the case in urban slums, where most residents lack land tenure or property rights. Without property rights, it is impossible to apply for individual tap connections. Even when legal hurdles are removed, most slum residents expect to live in their current homes for a short term, and are thus unlikely to invest in household improvements, such as the installation of a tap (8). Currently, 79% of the urban population in less-developed nations live in "slums" (areas characterized by insecurity of tenure and severe overcrowding) (9); this population is expected to double in the next 30 years (9, 10). These populations are expected to grow most drastically in areas of the world where access to clean water is already low (e.g., sub-Saharan Africa and South Asia). Meeting the needs of these populations will be a significant challenge for municipal providers, both in terms of the sheer increase in water demand and systematic

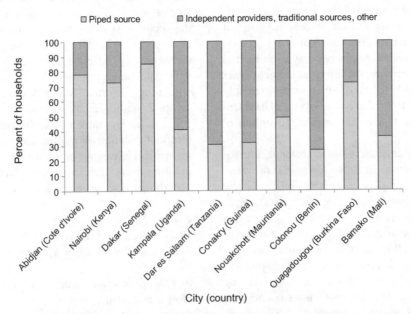

FIGURE 19.1   Percent of households with direct access to piped water in sub-Saharan African cities. Households without tap connections use water from a variety of sources, including informal water providers, traditional sources (e.g., rivers or ponds, though these are often very polluted and inaccessible in large urban zones), or a combination of these. The graph is adapted from a table published by Kariuki and Schwartz (2).

difficulties associated with connecting those living in slums. As a result, increasing numbers of the urban poor are expected to suffer hardships over the next few decades (11).

The traditional alternatives to municipal water services are naturally occurring water sources—water from rivers, ponds, or wells. However, for those living in densely populated urban zones, these free, traditional sources of water are inaccessible. Even if traditional water sources are available, high population density contributes to extreme pollution of lakes, rivers, and groundwater, making traditional water sources unsafe in urban zones. As a result, those residing in poor urban areas are left with few options but to purchase water from the increasingly sophisticated "informal water sector" discussed here.

As mentioned in the introduction of this chapter, this informal system accounts for a significant percentage of water services in urban areas of developing countries. A few examples may give a sense of the importance of this sector. In Guatemala City (Guatemala) and Lima (Peru), over 30% of the freshwater need is met by water trucks; in Bamako (Mali), 25% of the population purchases water from households with private tap connections; privately owned water points managed by individuals make up the largest growing water sector in Dhaka (Bangladesh); and, in Asunción (Paraguay), 400 privately managed aqueducts run from individual ground wells, providing services for nearly 2,000 families (1). Surveys conducted in the mid-1980s show that the informal water vending system has been a critical part of urban water acquisition for some time. Based on a comparison of five African cities, the percentage of households that primarily purchased water from vendors ranged from 15% (Garkida, Nigeria) to 90% (Diourbel, Senegal and Mandera, Kenya), with an average of 60% (2).

19.3 TYPES OF INFORMAL BUSINESSES

In general, small-scale informal providers occupy the role of: *gap filler* in areas where services exist but may be intermittent; *pioneer* in areas where no formal services exist; or *subconcessionaire*, buying water from formal utilities and reselling in areas where services are unavailable to large portions of the urban population (2). An illustrative example may best express the complexity of these systems. In Onitsha, Nigeria, 275 tanker trucks collect water from private boreholes and sell it to households and businesses equipped with storage facilities (12). These households then serve as small-scale private businesses, which in turn sell the water by the bucketful to other households that lack storage capacity. To further complicate matters, some persons in the community work as water distributors—they purchase water from either the households that have storage capacity or directly from deep boreholes, and deliver water by the bucketful to

persons who would not otherwise have the time to wait in line and purchase water.

Businesses in the informal sector are incredibly diverse, capable of responding appropriately to local needs. Nonetheless, several types of informal water businesses have been identified in a variety of locations. These types of businesses are described below to give a better idea of what is meant by private water providers.

19.3.1 Private Tankers

Private tankers are a common sight in peri-urban areas of South America, as well as in urban areas of sub-Saharan Africa (1). With a capacity of around 1 000 L, these tankers tend to collect water from public utilities or from deep wells and resell it to households. They charge by the volume of water purchased, and maintain regular contracts with consumers, as well as selling to one-time purchasers. The case of peri-urban Mumbai, India, is interesting. Private tankers have been present in the area since the mid-1980s. By 2005, 320 water trucks (belonging to 220 companies) were supplying water in the Vasai-Virar Sub-Region of peri-urban Mumbai. Business has grown rapidly in the past few years, as there has been an increase in the number of middle- and high-income households in the area, which are increasingly unsatisfied with irregular municipal services. Sales tend to be seasonal, peaking in the summer, when the municipality is unable to provide the necessary levels of service. Although there is no formal government registry for these businesses, they have organized their own associations, the goal of which is to organize the water supply within different areas of the town, control prices, and organize routes among themselves to service new consumers in the area (13).

19.3.2 Water Kiosks

Water kiosks are by far the most important alternatives for poorer consumers, especially in sub-Saharan Africa. Water kiosks can be highly advanced public access points, with multiple taps and full-time employees, or simply a small hose running from the tap of a connected household. A survey of 700 households in Kiberia (Africa's largest slum in Nairobi) found that over 19% of the population depends on water kiosks for their primary water needs. The same survey localized 650 independent kiosks in this area (5). The phenomenon is also common in peri-urban Mumbai, India, where it is not uncommon to see a queue of 20 people waiting to fill up from the tap of a neighbor (13). The volume of water extracted from the tap does not increase a family's monthly bill, since households often pay flat connection fees, rather than metered rates. Thus, many households opt to sell water from their own tap to service unconnected neighbors. Even where metered rates are charged, the per-volume, subsidized municipal rates are insignificant, considering the rates charged per volume to purchasers.

19.3.3 Water Cart Vendors

These mobile vendors tend to travel in areas where few households are connected to municipal water services, and sell water by the bucketful or jerrycan-full. The source water may be traditional (e.g., river, well) or purchased from the municipality (e.g., from public kiosks). Water cart networks can be thought of as above-ground pipes, which take water from cheaper, more accessible areas to areas with lower coverage and less access. In sub-Saharan Africa, water cart vendors tend to be most commonly utilized by double-income households, where the lost-work opportunity cost to collect water is far greater than the cost of purchasing water, which may be delivered directly to the household. A survey of residents (both middle- and low-income) found that about 5% of the population in the cities of Nairobi, Mombasa, and Kakamega in Kenya depend on water cart vendors (5).

19.3.4 Water Refill Stations

Water refill stations, which are most active in the Philippines and Indonesia, provide consumers with treated water at neighborhood stands for household consumption. These stands are generally small businesses or franchises with several small locations, operating in urban slums and peri-urban areas. Most stations use combined filtration/ultraviolet disinfection or filtration/reverse osmosis systems imported from South Korea or Singapore (14, 15). In South Asia, WaterHealth International, a nonprofit organization, uses the same concept—profit-generating, community-scale water treatment kiosks—to provide water in rural areas. The organization received a significant amount of funding to scale up efforts in India and Ghana (16). Similar, community-scale decentralized systems have been proposed in South Africa (17). In peri-urban Kisumu, Kenya, a water treatment business owner has been making a significant profit, while improving clean water access in the area for nearly 8 years (18). Acumen Fund, a revolving fund that invests in nonprofit groups with high health value, recently invested several hundred thousand dollars in two public–private Indian decentralized water treatment schemes: reverse osmosis-based systems and solar-powered desalination (19, 20).

19.3.5 Private Piped Networks

Small-scale private piped networks tend to be rarer than the aforementioned alternatives. There are several reasons for this trend, including the high startup costs (e.g., purchase of piping, digging), high market volatility (it is difficult to anticipate future needs for these services), and the illegal status of informal providers. Additionally, as there are already many alternative water access methods, which do not require such a high initial investment, it is difficult for network services to compete (2). Nonetheless, several case studies demonstrate that such networks are profitable, and do exist, in specific instances. In Ghana, for example, small private companies may sign contracts with local community government bodies that enable them to provide services for small communities and charge each household at a profit (21). In the areas where this has happened, operators have improved dilapidated networks, which the government was unable to repair due to lack of funding. In Mauritania, operators provide water to small towns (ranging from 500 to 2,000 residents), via formal agreements with the state. Since 1993, over 300 independent operators have significantly increased access to tap water in small towns around Mauritania (21).

The Informal Services Network in Kisumu, Kenya

With an official population of 700,000, Kisumu is easily Kenya's third largest city (22). Like most rapidly developing cities around the world, population growth far exceeds the capacity of the municipality to increase water services (in terms of both quantity and quality of service) (23). This is especially the case in the poorest neighborhoods, where there are few taps, and a high percentage of family income is spent on water. A rising middle class is moving from the unsafe sprawling urban slums to peri-urban communities, which are unsupplied by the municipality. As a result, most middle-class families now rely on water purchased from water cart vendors. An increasingly wealthy upper class is dissatisfied with the service interruptions and low-quality available from the municipal water provider. As a result, in Kisumu, water bottle distribution, private water treatment, and household water delivery are becoming increasingly profitable sectors. Although the municipality treats potable water, private deep wells, boreholes, and independent water treatment provide alternative sources. For households without tap connections, water can be purchased from any of the over 300 legal and illegal kiosks in town. In middle-income peri-urban areas, mobile water cart vendors distribute water by the jerrycan load, while in the wealthier area of town, water trucks are available to refill water tanks around the clock. Figure 19.2 provides estimates of water sources and their contribution to household water supply in Kisumu.

19.4 GENERAL CHARACTERISTICS OF THE INFORMAL SECTOR

Comparative studies of the informal sector have identified several characteristics common to the diverse body of

Total Volume of Water to Households

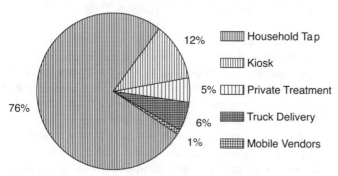

- 12% Household Tap
- Kiosk
- 5% Private Treatment
- 76%
- 6% Truck Delivery
- 1% Mobile Vendors

FIGURE 19.2 Approximate contribution of different water sources to household water supply in Kisumu. Water collected by households directly from traditional sources (e.g., rivers) is not included in the total presented above. Data was collected by the authors based on interviews with vendors representative of every informal water industry in Kisumu, as well as members of the municipal utility.

informal businesses operating internationally. These are discussed briefly below.

19.4.1 Adaptable

One primary advantage of the informal sector is its ability to respond quickly and specifically to changing conditions and needs. When population needs change, it may take between 5–10 years for a municipal system to respond with proper infrastructure, because pipeline projects must be proposed, designed, funded, and constructed. By the time infrastructure projects are finally ready for use, they are often already insufficient, because population growth rates can be higher than anticipated in rapidly growing cities. The private sector, on the other hand, is able to respond quickly, since its operation is often independent of permanent structures, such as piped distribution systems. A case study in the Vasai-Virar Sub-Region of Mumbai is an excellent example of this trait. During the heavy rains of the late summer (July and August) of 2005, several kilometers of pipeline were destroyed. It took nearly 6 months for repairs to be made, during which hundreds of thousands were without proper water services. Local water vendors saw a unique business opportunity and responded rapidly, rerouting tankers full of purified water to the area. The tankers became the primary water source for residents in this area. Without the rapid response of the private sector, it is difficult to imagine how the residents would have been able to survive until the piped system was fully reconstructed in early 2006 (13). Another example is that of Lima, Peru in the 1980s, when peri-urban populations were growing at unprecedented rates in the mountainsides. The municipality was simply unable to expand infrastructure networks quickly enough to keep up with this rapid growth in

such difficult terrain; however, given their mobility, water trucks and water carts easily navigated the difficult terrain, and thus, were able to supply peri-urban slums with consistent water (1).

Water entrepreneurs are consistent innovators, often being the first to introduce novel, water-delivery technologies to an area. An interesting example comes from Buenos Aires, Argentina, where IIED-América Latina introduced the first small-bore sewer systems. These systems first appeared in Malang, Indonesia, where they were developed by a small-scale entrepreneur. Another entrepreneur in Columbia found out about the system, and informed IIED-América Latina about it. Thus, small-scale entrepreneurs, each eager to make a profit, spread a technology that is highly adapted to developing regions, from Southeast Asia to Latin America, without any direct government interaction (2). Large-scale companies, especially those managing municipal-scale monopolistic networks, tend to be sluggish and hesitant to invest in new technologies, especially those that require significant financial risk. Smaller companies tend to be more nimble, often investing in technology that provides a competitive edge over others. Thus, small-scale private water providers speed up the adoption of technology that offers increased efficiency and is better suited to market conditions. In the past 10 years, for example, the traditional water vendors in Malawi (women who carried jars on their heads and sold by the quart or cup) have been displaced by water cart vendors, who can transport much higher volumes of water in oil drums. During the same time period, water cart vendors from Ciudad del Este in Paraguay ceased the use of donkey-drawn carts, and began using trucks, and even installed house-to-house piped connections (1).

Post-Conflict Areas

The emergence of unregulated, small-scale private infrastructure providers (in both the electricity and water sectors) has been documented in post-conflict areas, where the state fails to bring services to poor, peri-urban, and rural communities. An in-depth journal review in 2004 found that, of 46 countries known to have an informal water market, nearly half were defined as post-conflict countries (24). It was argued that "while private water vendors are common in the fringes of cities throughout the developing world, the role they play in post-conflict countries can be crucial to the subsistence of large numbers of consumers—both in urban and in rural settings" (24). This can be explained by the capacity of the private sector to mobilize resources quickly, taking advantage of irregular regulations and pent-up demand. In Cambodia, for example, private companies that treat water through metered, piped systems that they manage and construct can be found throughout the countryside (2). These companies depended entirely on local financing to offer desperately needed services in the war-torn countryside.

19.4.2 Responsive to Consumers' Needs

Since the private sector is dependent on direct purchases from consumers, and competition is often fierce among service providers, the informal sector is highly responsive to consumer needs, much more so than municipal bodies. Indeed, the continued operation of a small-scale provider depends greatly on consumer loyalty, especially in light of competition with public operators, who can offer services at government-subsidized rates. Contracts among the informal sector are made with consumers, rather than governments, and thus, awareness of consumer needs is inherently higher (1). The informal system has evolved independently in each city, as individual entrepreneurs pursue personal business opportunities. Thus, the system is responsive to local needs and conditions. Indeed, a review of hundreds of individual providers concludes that "the wide range of forms which the small-scale providers have evolved in response to local conditions suggests something close to endemic speciation" (1).

19.4.3 Financial Viability and Efficiency

Unlike the municipal water system, which is subsidized by local governments, small-scale providers are forced to operate efficiently. It has been estimated that there is not enough funding to build a public water system in Onitsha (Nigeria) (12), as the population is too poor to support the high, estimated costs of such a system. However, a highly advanced private vending system is able to collect as much as 24 times the cost of constructing a public system from this same population by operating efficiently.

In the informal sector, there is virtually no "unaccounted-for water," a universal occurrence in municipal systems. This is because, rather than seeing themselves as managing a large system, informal water providers see themselves as business people, who value the goods they sell and account for them. A private water treatment plant in Kenya, for example, treats only as much water as is necessitated by demand, minimizing waste of treatment chemicals or power for pumping, in the same way that a restaurant would minimize the amount of food prepared during non-meal hours (18). The same plant is also much more reliable when it comes to paying its electricity bills on time, and therefore, unlike the municipal provider, never suffers from a disruption in its electricity. This is done to prevent the loss of profit associated with a disruption in services. By being more cautious about resource use, this private water treatment plant, which operates in parallel with municipal treatment, is able to make over 80% profit, while the municipality is unable to operate without subsidy (18).

19.5 CONSEQUENCES AND TRENDS OF THE INFORMAL WATER SECTOR

Small-scale informal water services vary widely, based on local needs and conditions. However, common to all small-scale informal industries is an ability to fully recover costs (and make profits), often in areas that have been deemed unprofitable by formal municipal companies. In order to do so, such enterprises must be extremely efficient, with low amounts of "unaccounted-for" water. This is especially impressive, considering that, unlike public utilities, such businesses are unable to benefit from public subsidies, debt, or cost of scale (1).

Why Informal?

Many water providers remain informal due to a number of barriers to becoming formal, including high entry costs to legality (such as excessive license fees or registration requirements that small-scale businesses are unable to meet) and expensive tax structures (the payment of which may undermine any hope for profitability). In most cases, however, the cause for informality is simpler. In many cities with municipal networks, the municipal provider (whether a public or private company) is often granted sole water rights, even in areas to which the piped network has not yet been extended. Thus, alternative, informal small-scale operators generally operate illegally (1). As small-scale providers are often forced to remain informal, they encounter many costs, including high risk (activities may be subject to stiff penalties and/or capital confiscation) and lack of access to public services (e.g., protection of property through the use of law enforcement or of rights during arbitration of contracts).

19.5.1 Financial Burden of Informal Water Access

Consumers of the informal sector incur two types of costs, both monetary and nonmonetary, in order to acquire water. Often, the type of service utilized, and the price a family is willing to pay for water depends on a subtle balance of these costs, in light of a family's income level and status.

The primary reason that small-scale water providers have been understudied is that they are viewed as profiteers. This is an easy conclusion considering the per-unit cost of water purchased from the informal sector in light of the cost charged by most public operators. Specifically, in Port-au-Prince, Haiti, based on estimates of 1988 water prices, the poorest citizens were purchasing water at prices ranging from USD 1.1 to USD 5.5 per cubic meter (25). The same trend was noted in Jakarta, Indonesia, when early studies found that some vendors were charging USD 2.60 per cubic meter, over 30 times the price charged by the public utility (6). In Kenya, consumers often pay 18 times the cost of public utilities to purchase water at private kiosks (5). Households in several African cities, including those surveyed in Guidan Rouondji (Niger), Mandera (Kenya), and Ali Matan (Somalia), spent over 25% of their disposable income on water (26). In other words, although the municipality (since it subsidizes water treatment and provides water at a lower per-volume cost than it costs to produce) is losing money by supplying to kiosks, none of these subsidies reach the poor (5). Similar anecdotal examples can be found throughout the developing world.

A more generalized picture of private-sector water prices can be gained by comparing the water prices of informal distributors from around the world. A recent study based on data from 47 countries analyzed the price variability for water around the world based on water sources (2). This study showed that water purchased from mobile water cart vendors can come at the highest price and variability, while water purchased from piped networks, both public and private, can be up to 100 times less expensive. Interestingly, although price variability is high, based on the source of purchased water, variability is minimal in terms of the region where water is purchased (2).

In addition, collecting water from informal distributors can be associated with high labor costs. In Kenya, households with private connections may spend 5 min on water collection (or less) daily, while those with yard or community taps may spend 15 min, and those purchasing from kiosks may spend over 55 min waiting for water daily. Households using a variety of alternatives (e.g., stream water for washing or kiosk water for drinking) spend an average of 37 min on water collection daily (5). The amount of time spent collecting water varies somewhat according to location and season, but the comparison between private- and public-sector water collection in terms of time spent and water access, universally and consistently shows that public-sector water collection is often more time efficient. Based on household income level and the value of time within different income groups, households may opt to use different water sources. In Kenyan studies (5), middle-income households were found to bear the largest and lowest costs for water. This is because middle-income households, in comparison to poor households, were most likely to have private household tap connections (e.g., the cheapest water source), as well as purchase from the most expensive water source (e.g., opting for services from water tankers, water vendors, or bottled water rather than waiting in line at public kiosks).

A variety of water sources is often used, depending upon season, household needs, prices, and other factors. In Kenya, on average, households reported that the primary source of water for a household constitutes an average of 66% of the total volume of water used within that household. Nearly 42% of the studied population reported that they use two or more sources of water on a regular basis to meet daily needs (5).

19.5.2 Volume of Water Used by Those Who Purchase Water from Informal Providers

As with any other economic good, water, when it is purchased on a volumetric basis from informal providers, tends to be rationed, depending on a household's economic status. This is not often the case in households that have household tap connections, because volumetric charges for water retrieval tend to be marginal. Since a certain quantity of water is necessary for good household hygiene and health, this outcome, particularity in the private sector, is a major reason that the sector is often viewed in a less-than-positive light (1).

An economic analysis in Kenya found that the quantity of water consumed within a household is negatively correlated with price and positively correlated with income and household size (21). Indeed, partially as a result of a growing private provision of water, average consumption of water in Kenya fell from 105 L per capita per day (Lcd) in 1967 to 45 Lcd in 1997 (27). Similar trends can be seen in Port-au-Prince, Haiti, where it is estimated that households with tap connections use 156 Lcd, compared to 11 Lcd by the poor who rely on the vendor system (25). The same is true in Jakarta, where households with private connections use an average of 62.2 Lcd, households with access to public hydrants use 27.5 Lcd, and those that rely on vendors use only 14.6 Lcd (6).

19.5.3 Water Quality

There are few studies of water quality within the informal sector. Thus, it is difficult to come to any clear conclusions regarding the quality of water provided by informal services. Nonetheless, poor water quality is often cited as a problem inherent to informal services. It will be difficult to draw conclusions before further tests have been made, but it is undeniable that, since they are unregulated, private informal suppliers are unlikely to be very concerned with achieving World Health Organization standards for pollutants and other measurements of quality that consumers from a high-income country would expect (24). This is especially the case since water quality does not have a direct impact on the providers' bottom line. Although some regard may be given to the turbidity of water (physical appearance may have a bearing on the price a consumer is willing to pay for water), purchasers have no means of assessing (and thus paying more for) the microbiological quality of their water. Furthermore, unsanitary water handling practices may infect water that was initially safe, thereby spreading disease within cities.

19.6 GENERAL VIEWS OF THE SECTOR

19.6.1 Traditional Attitudes

Most commonly, small-scale operators are regarded as "'bad guys' who charge usurious rates" (1). The most positive traditional view of the sector is that they are "temporary," though they may have been in business for over 20 years (2). Similar is the view of the sector in post-conflict areas, where the development of the sector is regarded as a "public nuisance," though these operators may be providing essential public goods (24).

Often informal water services are not considered in policy discussions because, at the outset, water development funds are specifically allocated to improve in-house networks, whether by improving water treatment capabilities or extending network coverage (12, 28). As a result of privatization efforts in the 1990s, administrative and regulatory reforms are increasingly advocated as a means of improving water services (2). Since attempts to service the poor through nontraditional means are relatively recent and decisions have been made in a top-down fashion, little consumer-level information has been collected. Indeed, information regarding water demand by urban poor, including the amount currently paid, the most common, current sources of water, and opinions regarding possible changes to water methods, are largely unknown and undocumented (12). As a result, consumer demand is not often considered when new policies and administrative reforms are enacted.

19.6.2 Recent Attitudes

Views of the sector have changed recently, mostly as a growing number of studies are demonstrating the scale to which the sector is providing access in the public realm. This is happening as those who study the sector are coming to realize that the alternative to informal service providers in poor rural, peri-urban, or slum areas is not service from a struggling state-owned utility, but no service whatsoever. In recognition of this fact, several post-conflict countries, such as Mozambique, Sri Lanka, El Salvador, and Cambodia, are introducing regulations that recognize private providers, while encouraging the development of associations to introduce minimal levels of oversight (24). Even in non-post-conflict areas, practitioners are increasingly recognizing the potential role of small-scale providers in increasing the coverage of services. As such, the United Kingdom's Department for International Development and the Inter-American and Asian Development Banks have undertaken the work of researching this sector with a more organized and less anecdotal approach (2). This study found that the small-scale private sector has been estimated to account for 85% of all private-sector investment in water security (2).

The concept of regulating the water-vending industry is traditional practice in some regions of the world. In the Dominican Republic, for example, large-scale private trucking has been regulated since the early 1980s. Since then, the government has regulated the 10 largest active water firms by inspecting the product weekly, and specifying a legal maximum charge to protect consumers (29). Supporting the growth of informal industries, by lowering the investment risk and "formalizing the informal" may be the best solution to lowering costs, by preventing monopolistic price gouging. In Nairobi and Mombasa, the establishment of kiosks run by the government and NGOs (which charge fair, fixed prices) has significantly reduced water prices in the areas where they operate (5). Nonprofit donor bodies have also recognized the benefit of private enterprises in providing unmet services, with a high benefit from relatively minimal aid monies. In Cambodia, for example, targeted subsidies and contracts that reduce investment risks have been put in place by donors to encourage the spread of private piped networks throughout the countryside (24). In Kenya, regional water boards have been working with small providers to help them gain legal status. By formally applying for legal status as companies or "associations," independent service providers can operate legally under Kenyan law. Such agreements could be a relatively simple method to impose cost and quality regulations (30).

In Nairobi, the picture is a little clearer. Community management of water schemes was introduced in both rural and urban areas in the 1980s, and today, community-based organizations (CBOs) and small companies serve at least 60% of Nairobi households. Indeed, in Nairobi's large urban slums, CBOs or informal neighborhood groups are typically the only providers, because residents have no legal tenure and therefore, cannot be served by formal utilities. Many well-documented weaknesses of the private sector are however, also evident in Kenya. These providers typically lack both access to finance and the technical and managerial capacity to use finance effectively. They also face resistance from formal public utilities, lack legal status or tenure, and provide services that are largely unregulated in terms of both price and quality. Kenya has already made progress on some of these fronts. To deal with the unclear legal status, resistance from utilities, and lack of regulation, regional water boards—which generally own water service assets—are working with small providers to help them gain legal status as companies or associations. They can then operate as independent service providers under Kenyan law and enter into service provision agreements with the boards. These agreements regulate prices and service quality (30).

In the early 1990s, less than 20% of the population in Jakarta, Indonesia, had piped water connections. As it was becoming increasingly evident that traditional water sources (wells, rivers, and lakes) were increasingly polluted and inaccessible, the local government undertook a number of measures to address this matter. For those without private tap connections in the 1990s, water was not a public service, but a private good, purchased at exorbitant rates from water vendors and standpipes. It was in this environment that, in April of 1990, the government undertook a deregulatory measure, permitting private homes with private water connections to resell municipal water legally. A study after this regulation was implemented concluded that the advantages were both direct, since they benefited those who began purchasing water from these metered households, as well as indirect, because they forced water vendors to decrease water tariffs by increasing market competition. The most notable effect of legalizing private household sales was a drastic decrease in water pricing and associated increase in water consumption by the poor (6).

Legalization and regulation may be the best means of taking advantage of the benefits of a competitive small-scale water services sector, while minimizing the negative health consequences (those associated with the use of low-quality water in small volumes, see above). Attempts to do so are relatively recent, and have been met with varied response. As governments increase interaction with this sector, the best means of legalizing and regulating may soon be discovered. Although it is possible to continue demonizing informal providers, a more balanced view, one that recognizes their importance in light of unmet urban water demand without ignoring their profit-driven motives, may be the best means to improve services to the poor and reduce mortality and morbidity caused by diseases related to a lack of safe water.

19.7 CONCLUDING REMARKS

The informal water services sector is critical to those who would not otherwise have access to potable water in urban areas. This has been documented in metropolitan regions of Latin America, sub-Saharan Africa, and Southeast Asia. This sector is highly adaptable and responsive. Unfortunately, a lack of hygiene and sanitary water handling may undermine water quality and spread disease. Additionally, as a result of high water tariffs in the informal sector, poor households purchase suboptimal amounts of clean water. Disease is spread when water is insufficient for basic hygiene tasks in the household, and thus, high water tariffs may indirectly increase the incidence of disease in poor areas. Regulation and formalization may aid in recognizing the important role of this sector, while minimizing negative outcomes of its operation. Yet, information on the sector and its activities is necessary before any appropriate regulation or legalization steps may be taken. Most studies referenced in this paper are anecdotal in nature, since little academic or institutional work has been undertaken to understand and analyze this sector. Such studies will be critical for better understanding

the informal water services sector and its impact on public health and local economies.

REFERENCES

1. Solo TM. Small-scale entrepreneurs in the urban water and sanitation market. *Environ Urban* 1999;11(1):117–132.
2. Kariuki M, Schwartz J. Small-scale private service providers of water supply and electricity. In: *A Review of Incidence, Structure, Pricing and Operating Characteristics.* World Bank Policy Research Working Paper, 2005.
3. Hussain M, et al. *Enterococci* vs. *coliforms* as a possible fecal contamination indicator: baseline data from Karachi. *Pak J Pharm Sci* 2007;20(2):107–111.
4. United Nations. *World Urbanization Prospects: The 2001 Revisions.* New York: ESA/P/WP, 2002.
5. Gulyani S, Talukdar D, Kariuki RM. Universal (non)service? Water markets, household demand and the poor in urban Kenya. *Urban Stud* 2005;42(8):1247–1274.
6. Crane R. Water markets, market reform and the urban poor: results form Jakarta, Indonesia. *World Dev* 1994;22(1): 71–83.
7. SUSENAS. Indonesia's Socio-Economic Survey. 2004.
8. Budds J, McGranaham G. Are the debates on water privatization missing the point? Experiences from Africa, Asia, and Latin America. *Environ Urban* 2003;15(2):87–113.
9. Carolini G, Garau P, Sclar E. The 21st century health challenge of slums and cities. *Lancet* 2005;365(9462):901–903.
10. United Nations HABITAT. *Guide to Monitoring Target 11: Improving the Lives of 100 Million Slum Dwellers.* Nairobi, Kenya, 2003.
11. Allen A, Davila JD, Hofmann P. The peri-urban poor: citizens or consumers. *Environ Urban* 2006;18(2):333–352.
12. Whittington D, Lauria D, Mu X. A study of water vending and willingness to pay for water in Onitsha, Nigeria. *World Dev* 1991;19(2/3):179–198.
13. Angueletou A. Informal water suppliers meeting water needs in peri-urban areas of Mumbai. *Development* 2007;51(1):2–5.
14. Darmawan B. Industry Review—Association of Water Refill Stations, Equipment Vendors, and Tank Distributors. In: *Small-Scale Water Purification Businesses in East Africa: Entrepreneurial Strategies for Providing Clean Drinking Water.* Nairobi, Kenya, 2009.
15. Asian Development Bank. The Water Supply & Sanitation Sector: Current Challenges and Opportunities. In: *2nd National Sanitation Summit: Better Quality and Safety through Improved Sanitation.* Manila, Philippines, 2008.
16. Minakshi S. *IFC, WaterHealth to Help India's Rural Poor Access Affordable Drinking Water.* IF Press, 2009.
17. Pryor MJ, et al. A low pressure ultrafiltration system for water supply to developing communities in South Africa. *Desalination* 1998;119:103–111.
18. Guidthai A. Niamsaria water treatment plant: from protecting chicken to providing for a community. In: *Small-Scale Water*

Purification Businesses in East Africa: Entrepreneurial Strategies for Providing Clean Drinking Water. Nairobi, Kenya, 2008.

19. Acumen Fund. *Investments: Water Portfolio*. 2009.

20. Acumen Fund. *Aqua-Aero Water Systems: Drinking Water for Rural Rajasthan*. 2009.

21. Visser BV, et al. Access through innovation: expanding water service delivery through independent network providers. In: *Building Partnership for Development in Water and Sanitation*. London: BPD Water and Sanitation, 2006.

22. Kenya Bureau of Statistics. *Population Figures for Towns and Municipalities*. Republic of Kenya, 1999.

23. Sima L, Elimelech, M.A survey of the informal water services sector in Kisumu, Kenya. In: *OU Water Conference*. Norman, OK, 2009.

24. Schwartz J, Hahn S, Bannon I. The private sector's role in the provision of infrastructure in post-conflict countries: patterns and policy options. *Social Development Papers: Conflict Prevention & Reconstruction* 2004, p. 16.

25. Fass S. *Political Economy in Haiti: The Drama of Survival*. New Brunswick, NJ: Transaction Publishers, 1988.

26. Zaroff B, Okun DA. Water vending in developing countries. *Aqua* 1984;5:289–295.

27. World Bank. *Kenya: Review of the Water Supply and Sanitation Sector*. Bank TW, editor. Washington, DC, 2001.

28. Munasinghe M. Water supply policies and issues in developing countries. *Natural Resources Forum* 1990;14(1):33–48.

29. Lewis MA, Miller TR. Public–private partnership in water and supply sanitation in Sub-Saharan Africa. *Health Policy Plan* 1987;2(1):70–79.

30. Mehta M, Virjee K, Njoroge S. Helping a new breed of private water operators access infrastructure finance. *Gridlines* 2007;25.

SECTION IV

SANITATION AND HYGIENE

According to the May 2003 issue of *Lancet Infectious Diseases*

Human excreta is the source of most diarrheal pathogens ... hands should be washed with soap after contact with human excreta and before handling food. ... A recent review of all the available evidence suggests that handwashing with soap could reduce diarrhea incidence by 47% and save at least one million lives.

This section addresses the range of sanitation and hygiene measures available to reduce disease. Presentation of sanitation and hygiene coverage includes needs and goals of sanitation systems and types of effective sanitation systems. Social, economic, technological, and cultural factors that need to be taken into account when designing and implementing successful systems are discussed. Case studies of successful approaches to meeting sanitation needs and educational campaigns to increase awareness of the importance of and application of adequate hygiene are presented.

20

THE SANITATION CHALLENGE IN INDIA

BINDESHWAR PATHAK

20.1 THE SANITATION CHALLENGE IN INDIA

Sanitation has been declared as the most important medical advance since 1840, ahead of antibiotics and vaccines. In developed countries of the West, generally urban and even rural communities are served by water carriage systems or sewerage systems for the disposal of wastewater and human excreta from the households. Household toilets and bathrooms connected to sewerage systems are the symbol of Western civilization.

However, in India after the decline of the Mohen-jo-daro and Harrapan civilization, which had developed an elaborate system of excreta disposal, the practice of defecation in the open became common. During the Pauranic period (CA 600–323 BCE), it found support in the religious text like the Devi Puran, which advocated that people should not defecate near human habitation but at a distance—as far as where an arrow would fall. It suggested that one should dig a small pit—put some grass and leaves, defecate, and again put some grass and leaves and fill it with earth. Such a sanction in and of a religious text had a great influence on the people. This method was safe and hygienic. Also the geographical factors of existence of bushes and trees and presence of undulating land and other natural prominences providing cover prompted people to defecate in the open, which in a way became an integral part of the sanitation-oriented cultural mode prevalent at that time and which also continued in later times. The tropical climate in India also helped in adoption of the practice of defecation in the open. It was not too cold to prevent people from going out. The houses were made of bricks, or mud and thatch. But, what was common in all the shelters was that there were no toilets.

In the Mauryan period (322–185 BCE), when Patliputra was the capital of Magadh empire, Chanakya, Prime Minister of Chandragupta Maurya, wrote in Arthashastra, that a kitchen, a toilet, and a bathroom should be made in the house. Those who defecate outside should be penalized 100 "puns" (currency of that time) and those who urinate in the open should pay 10 "puns." If a person was sick and under the influence of drugs, no penalty should be imposed on him. Despite admonitions in nature of "Thou shall not ...," with defiance met with the pain of penalty the practice of open defecation continued.

Even during the Mughal period (1526–1772), the system continued. By the time the British came and ruled India (1772–1947), John Harington had developed the flush toilet in 1596 in Britain. Sewerage was first introduced in London in 1850, in New York in 1860, and in India in Calcutta in 1870; yet after 138 years, of the 5,161 towns/cities only 232 towns are covered by sewage system and that, too, partially because of the heavy cost of construction and maintenance and enormous quantity of water required to flush human excreta. Septic tanks also need a lot water to flush (10 L per flush) and a gas pipe is needed to allow exit and intake of gases to help bacteria survive to degrade excreta. When septic tank fills up, it has to be cleaned. Decomposed sludge, raw human excreta, and water have to be manually cleaned, needing the services of scavengers. After 2 years, another trenching ground is required for further decomposition before it can be used as fertilizer. Thus, only 15% of the urban population adopted the septic tank system and hardly by anyone in rural India. The reason was that the construction and maintenance were too expensive and enormous quantity of water was required. Developing countries

Water and Sanitation-Related Diseases and the Environment: Challenges, Interventions, and Preventive Measures, First Edition.
Edited by Janine M. H. Selendy.
© 2011 Wiley-Blackwell. Published 2011 by John Wiley & Sons, Inc.

FIGURE 20.1   Scavenger women set out in the morning with broom and pan to clean bucket toilets.

could not afford it. So culture and lack of affordable technology played a major role in the continuance of the practice of defecation in the open.

Because these technologies were not affordable in terms of both construction and maintenance and required enormous quantity of water to flush, the problem remained in Asia, Africa, and Latin America and 2.6 billion people in these continents even by the turn of the millennia do not enjoy access to safe and hygienic toilets (Fig. 20.1).

In India, the sanitation scenario, till the late 1960s was dismal and in rural areas no house had a toilet. Everybody in the village used to go for defecation in the open. Because of lack of toilets, women had to suffer the most; they had to go for defecation before sunrise or after sunset. Sometimes they had to face criminal assaults or snake and scorpion bites while going after dark for defecation. Girls generally did not go to schools, since no school in rural areas had toilets. In the villages many children used to die because of diarrhea and dehydration.

In urban areas only few towns had the facilities of the sewerage system. As observed earlier, 15% of the urban population used septic tanks and the rest used to go outside for defecation in parks, lanes, and on both sides of railway tracks and the remaining population used bucket toilets cleaned by the human scavengers who carried human excreta as a head-load (Fig. 20.2).

Had scavengers not cleaned bucket toilets, there would have been epidemics of cholera, diarrhea, dysentery, and so on, and people would have died in large numbers. Yet society kept them at the lowest ladder of the social

structure of the caste system and gave them the stigma of being "untouchable." They were ostracized and had to live on the outskirts of the city/town, lest they touched anyone by mistake. Women would not give them food hand to hand but dropped it in their palms. If they were thirsty, water was poured from a safe distance into their cupped palms.

Even the traders would accept their money only after cleaning the coins. There was no question of their going to school, entering temples, or of their children playing with children of other communities. They played among themselves or with pigs.

FIGURE 20.2   An untouchable human scavenger taking out human excreta manually from a bucket toilet to carry as a head load for its further disposal.

FIGURE 20.3   A life of degradation, humiliation, and discrimination in store for the new bride of a scavenger family.

The human scavengers were treated as untouchables and they were hated, humiliated, and insulted by the people for whom they used to work. In the Indian society, before the Independence of India in 1947, a person born in the "untouchable" caste died as an "untouchable." There was no chance of any change in the caste structure (Fig. 20.3).

There are two types of prisons—one is the physical prison run by the government, wherefrom a prisoner can be released after days, months, or years, except for heinous crimes, because there is provision for remission on account of good behavior but in India there is the social prison without walls called the caste system, where before independence of India a person born an untouchable would die as an untouchable.

Another problem was that Indians were not accustomed to making payments for use of any public facility much less for use of toilets. The British government passed an act in 1878 to maintain public toilets on "pay and use" basis, but it did not work. Toilets built then by local bodies did not succeed because they were ill maintained, hence, they were considered as a veritable hell on earth. Nobody liked to go inside the public toilets and tried to avoid even passing by in their vicinity because of the terrible stink. Therefore, the absence of public toilets in public places was a great problem in India. Another critical problem was that there were no community toilets and urinals in public places near railway stations, bus stands, market yards, and places of religious and tourist interest. It was very difficult for the people to manage when they used to feel the call of nature, and had to go to a nearby pond or bush or any dirty place just to defecate. Because of lack of public conveniences, people, especially foreigners, used to get discouraged to visit India.

Lack of safe water and facilities of human excreta disposal are the two key factors behind the huge burden of infectious diseases such as diarrhea, dysentery, cholera, typhoid, hepatitis, worm-infections, and so on, particularly in the developing countries.

Mahatma Gandhi was the first person whose attention was drawn toward the plight of scavengers. He wanted scavengers to be relieved from their subhuman occupation of cleaning human excreta manually and wished to restore their human rights and dignity, to bring them on a par with others in society.

I wanted to be a teacher of sociology in the university but somehow could not become one. I did sundry jobs, ranging from being a part-time teacher, a clerk at a power plant to selling medicines, and finally joined the Gandhi Birth Centenary Celebration Committee in Bihar in the year 1968. There I got the idea of fulfilling the dream of Mahatma Gandhi. It was difficult for me to solve the above problem of sanitation, which India was facing, as I belonged to an orthodox Brahmin family, which raised cultural and social barriers between me and solutions, and because I was not an engineer to enable me find solutions to technical problems. Thus, once when I touched an untouchable Dome lady, my grandmother made me swallow cow dung, cow urine, and Ganges water to purify me.

But somehow or the other, I decided to solve the problem unlike many others in the world who are experts in collection of information but unable to apply it to solve problems. So I applied my mind, which alone makes imbibing knowledge meaningful, and prepared myself to shed my own prejudices against untouchables, by going to and living in a colony of scavengers and thereby putting to practice what in sociology was taught to us that if one wanted to work for a community, one must build a rapport with the people of that community so that one comes to know about them. I did so with the help of a scavenger when I went and lived in the colony of scavengers for 3 months and came to know about their origins, culture, values, mores, and so on. While I was coming to live in the scavengers' colony named after Mr. Jagjivan Ram, a freedom fighter and the former Deputy Prime Minister of India, I was not quite sure whether to continue in the profession, as my father was very upset because Brahmins and toilets did not go together. By that time I was also married and my father-in-law was distressed and berated me in a language I am loathe to repeat. The people of the Brahmin community also ridiculed and humiliated me occasionally. The situation was totally disconcerting, and nobody appreciated my initiative to change the lives of the untouchable scavengers.

While living in Bettiah in the scavengers' colony, one particular morning, it came to my notice that a newly married girl was being forced by her in-law's family to go to clean bucket toilets and that she was crying bitterly as she was most unwilling to do so. On hearing her cries I went and intervened, trying to persuade the family members not to force her, if she was unwilling to go and clean toilets. They heard me but did not agree, and countered by asking me what would she do from the morrow if she did not do the work of scavenging and earn some money and even if she sold vegetables who would buy them from her, she being an

FIGURE 20.4 No help was extended to the boy, attacked by a bull, as he was an untouchable.

FIGURE 20.5 Dr. Pathak with colleagues during the Bihar Gandhi Birth Celebration.

untouchable. Finally, despite my protests, they sent her to clean bucket toilets (Fig. 20.4).

After a few days, as I was going to the market, with a colleague of mine of that colony, we saw a bull attacking a boy of 10–12 years, who was wearing a red shirt. When people rushed to save him, on hearing someone shouting that he belonged to the "untouchable" scavengers' colony, everyone left him. We took him to the hospital, but the boy died. After this incident I took a vow to fulfill one of the dreams of Mahatma Gandhi viz. to get the scavengers relieved from their subhuman and health-hazardous occupation of cleaning and carrying human excreta manually.

While living in the colony, I studied carefully the books written by Mr. Rajendra Lal Das and the book *Excreta Disposal for Rural Areas and Small Communities*, published by the World Health Organization in 1958. The following sentence from the WHO book left a deep mark in my memory: "Suffice it to say here that out of the heterogeneous mass of latrine designs produced all over the world, the sanitary pit privy emerges as the most practical and universally applicable type."

This book was about disposal of human waste in rural areas, but the problem of scavenging was mainly an urban one because most scavengers used to work in urban areas. Here I applied my mind and thought that if the soil condition in a rural and an urban area is the same, then there should be no reason why a technology recommended for rural areas would not be applicable in an urban area. So here was a design, practical and with universal applicability except for its implementation. It was my reasoning and application of mind that led to implementation. It was demonstration of the fact that application of mind is more important than knowledge. Knowledge can be borrowed but application has to be your own (Figs. 20.5 and 20.6).

Being a follower of Mahatma Gandhi, I decided to start a silent revolution of nonviolence and peace for the removal of untouchability through liberation of human scavengers with the help of a toilet technology whereby services of scavengers would not be required. To get them relieved from this occupation, I had to find out technology/ies that would be appropriate, affordable, indigenous, and culturally acceptable.

To overcome the problem of safe disposal of human wastes I developed two major technologies—one for household individual toilets and another for community and public toilets. I developed the design of eco-friendly two-pit, pour-flush compost toilet (popularly known as Sulabh Shauchalaya) that is technically appropriate, economically affordable, and culturally acceptable. For the safe reuse of human wastes from public toilets, housing colonies, high-rise buildings, hostels, hospitals, and so on. I developed a technology for recycling and reuse of excreta through biogas generation and on-site treatment of effluent through a simple and convenient technology for its safe reuse without health or environmental risk.

After I requested a member of the legislative assembly of Bihar to write to Mrs. Indira Gandhi, the then prime minister of India, about the poor sanitation status of Bihar, my file moved ahead in the department, and I was granted funds. Mrs. Gandhi replied in March 1973 that she was writing to the chief minister and asked him to give his personal attention to this matter. This letter fortuitously helped the program a lot, and finally in 1974, the government of Bihar named the Sulabh technology and the organization for implementation of the program.

Once the program started and became successful, the government public health engineers raised objections on the contamination of water from these toilets. The government of Bihar asked the Environmental Engineering Research Institute, Nagpur, to give its opinion on the matter. Later, the institute opined that there were no chances of contamination of soil, if these toilets were constructed with due care.

FIGURE 20.6 Bhangi Mukti Andolan—Liberation of Scavengers program being discussed by Dr. Pathak with a group of people in Patna.

The government of Bihar used to give grants to the Bihar State Gandhi Centenary Committee every year for making the scheme a success, and the Bihar Gandhi Centenary Committee somehow kept the program going, but there was no visible impact.

I had learnt the hard way that the people wanted result-oriented work and not preaching. I suggested to Bhangi-Mukti Cell to undertake actual conversions. This was, however, opposed on the ground that taking up of actual conversions would be beyond the scope of the cell. But I felt strongly about the desirability of taking up practical work to some extent. Such opposite views brought about parting of ways. I resigned from the Bhangi-Mukti Cell. The Bihar Gandhi Centenary Committee was wound up in the year 1970, and there was no voluntary agency to implement the scheme. The government of Bihar continued to give financial assistance to the local bodies from 1967 to 1974. The implementation of the scheme was left to the municipalities, but the results were not visible. On being directed by the then prime minister to take personal interest in this matter, the then chief minister of Bihar immediately directed the then minister of Urban Development to find out some ways for the proper utilization of the funds and to achieve the target of "Bhangi Mukti" (liberation of scavengers), who, in turn, directed the department to take concrete steps for the implementation of the scheme with the assistance of the Sulabh Shauchalaya Sansthan.

The Urban Development department remained undecided for a year on how to utilize the services of the Sansthan. But after a year Urban Development department formulated the methodology for the implementation of this scheme with the help of the Sulabh Shauchalaya Sansthan, which was agreed upon by the then secretary of Urban Development and the minister, the methodology became so successful that this scheme of conversion of service latrines into Sulabh Shauchalayas received a boost. This methodology was thrice scanned by the same department, but it finally agreed to keep up this methodology.

In 1974, the government of Bihar recognized the Sulabh Shauchalaya Sansthan as a catalytic agency between the government, local bodies, and the house owners for the conversion of bucket privies into Sulabh Shauchalayas. Later the chief minister of Bihar ordered conversion of all the existing service latrines into Sulabh Shauchalayas. He said paucity of funds would not stand in the way. The scheme gathered momentum, and it made a great impact on the people of the state.

In the year 1980–1981, at my initiative, the Ministry of Home Affairs, government of India, included the program of conversion of bucket toilets in the centrally sponsored scheme of "Implementation of Protection of Civil Rights Act (PCR) PCR Act" of 1955 for the liberation of scavengers. Under the scheme, there was a provision of 50% loan and 50% subsidy to the beneficiaries, and it worked very well in the states where the program was taken up.

Funds were allotted to the government of Bihar for conversion of dry latrines into Sulabh toilets first in two towns and then three more towns, which were made scavenging-free. This program of conversion was a great success in Bihar and the Ministry of Home Affairs allotted funds to the other state governments as well from 1982 to 1983 onwards. Since then, conversion of hundreds of thousands of dry latrines to pour-flush has been carried out by Sulabh International in various states of the country.

Earlier there was a social stigma and psychological taboo against handling of human excreta. It was also partially due to the fact that only people of lowest economic strata were supposed to be associated with this work. Since human excreta was considered as the most hated object in the society, it was difficult for anyone to visualize or accept that a project related to its disposal could be financially viable. However, Sulabh made it financially viable under an arrangement where the cost of construction was (or in some cases, even now) is borne by the local body and the cost of maintenance of toilet blocks and day-to-day expenses is met from out of user fee. This arrangement has changed in recent years to build-operate-transfer (BOT) system. The organization builds the toilet complex, with its own financial resources, operates it, and eventually transfers it to the local body after a certain period of time, in some cases.

Sulabh does not depend on external agencies for finances and meets all the financial obligations through internal resources. All the toilet complexes are not self-sustaining, particularly those located in slums and less-developed areas. The maintenance of such toilet complexes is cross-subsidized from the surplus income-generating toilet complexes in busy and developed areas.

After my intervention in the year 1968, the situation has changed, but even today 630 million people in India out of more than a billion people defecate in the open and nearly 0.64 million bucket toilets are still cleaned by approximately 128,000 number of scavengers daily.

In the 1970s, when people heard about my idea of running public toilets on a "pay and use" basis, they initially were skeptical about the success of the model. They used to ridicule me, saying when people did not pay bus or rail fare, there was hardly any reason to believe that they would pay for use of toilets. However, in 1974, when I set up the first public toilet in Patna, on the very first day, 500 people came to use the toilets and paid for the use. As a result, the first day's collection was USD 5. Now this concept has been accepted not only in India but also in Bhutan and Afghanistan. In the last three decades and more, Indians have become habituated to making payment for the use of public toilets throughout the country. This has helped reduce the burden on public exchequer for the maintenance of public toilets and bath facilities. This is a fine example of public–private partnership.

Women now go to toilets with dignity and in privacy and safety, and girls go to schools because there they now have toilets to be used. The mortality rate of children has reduced from 129 per thousand in the 1970s to 66 per thousand in 2009. Millions of "untouchable" scavengers have been relieved from their subhuman occupation and their human rights and dignity have been restored.

Two or three significant changes in this area have taken place in India. The subject of toilet earlier was a taboo in Indian culture. Nobody talked about it while having meals, but now they talk about Sulabh without much hesitation. We have been able to change the thoughts, attitudes, and behavior of the Indian people toward toilets and "untouchable" human scavengers.

As part of its educational campaign, Sulabh International established a Museum of Toilets with "facts, pictures and objects detailing the historic evolution of toilets from 2500 BC to date" (Fig. 20.7).

The museum was established with the following objectives:

1. To educate students about the historical trends in the development of toilets;
2. To provide information to researchers about the design, materials, and technologies adopted in the past and those in use in the contemporary world;
3. To help policy makers to understand the efforts made by predecessors in this field throughout the world;
4. To help the manufacturers of toilet equipment and accessories in improving their products by functioning as a technology storehouse; and
5. To help sanitation experts learn from the past and solve problems in the sanitation sector.

Source: www.sulabhtoiletmuseum.org/.

FIGURE 20.7 Sulabh International's Museum of Toilets.

FIGURE 20.8 One of the many Sulabh Toilets spread all across India. Details about the Sulabh International sanitation systems are in Chapter 22, "Household Centered Environmental Sanitation Systems," also by author Bindeshwar Pathak, and in the book's DVD.

Second, the concept of untouchability is no longer attached to the profession of construction and maintenance of the toilets. Third, in India, in interactions in day-to-day social life, there are frequent communal clashes. But in the use of public toilets by people irrespective of religion and caste, we have not come across any case of communal tension so far. Anyone can use the toilet and nobody discriminates against anyone else. There is a sea-change in the attitude and thinking of society and now scavengers merge with and in the mainstream of society.

Recently a Brahmin invited a woman scavenger to the marriage of his daughter, accepted a gift from her, and gave food to her, along with serving her family members. This was unheard of in the social history of India, but it has now happened. The pages of history have been turned. Scavengers socially discriminated till now have been accepted by the society.

The other advantages of the aforementioned Sulabh technologies are that (1) their working reduces global warming because of elimination of emission of gases into the atmosphere because the gases are absorbed in the soil in the two-pit toilets and methane, a component of the biogas produced from biogas digester, is burnt on-site. (2) Less water is required in Sulabh twin-pit toilets, because only 1–1.5 L of water is sufficient for ablution as well as cleaning the toilet pan compared to 10–12 L of water required in conventional sewer and septic tank systems, thus saving an enormous quantity of water.

The task of total sanitation coverage in any country can be fulfilled only through close cooperation between local government and the community or local NGOs involved with such work. Neither government nor NGOs/CBOs can fulfill the task alone. The problem requires both technical as well as social aspect to be tackled. Therefore, a holistic effort is needed to overcome the problem. The strategy developed by Sulabh can easily be replicated in other developing countries to improve the sanitation status and quality of life (Fig. 20.8).

Thus, the Sulabh technologies are suitable not only for developing countries but also for developed nations and can help meet the Millennium Development Goal Target 7 on water and sanitation—halving the proportion of the 2.6 billion people in the world without access to improved sanitation by 2015.

21

SUCCESSFUL SANITATION PROJECTS IN THAILAND, MALAYSIA, AND SINGAPORE

JAY P. GRAHAM

21.1 RURAL SANITATION IN THAILAND: PERSISTENCE AND CREATIVITY PAYS OFF

21.1.1 Introduction

In 1960, less than 1% of Thailand's rural residents had access to basic sanitation, but by 2005, this figure had increased to 99.9% (1). This dramatic increase is the result of a concerted, decades-long campaign by the Thai government, which successfully employed a range of tools—health extension workers, revolving loans, competitions, local artisans, and new policies—to expand sanitation coverage. While Thailand's experiences are unique, and challenges still remain, the country's successful sanitation improvement efforts present potentially relevant lessons for other countries still struggling to improve basic sanitation coverage (Fig. 21.1).

21.2 BACKGROUND

The Kingdom of Thailand (Thailand) is located in Southeast Asia and has a population of nearly 66 million people—two-thirds of whom reside in rural areas. The country supports a free-market economy and has experienced relatively high economic growth over the past 50 years, with only 2 years of losses, both of which resulted from the 1997 Asian financial crisis. Thailand has effectively managed population growth, with growth rates dropping from 3.2% in 1960, to a rate of 0.6% today. The portion of the population living in rural areas has fallen relatively slowly—dropping roughly 13% since 1960 to 67% today. Rural development has been a

top priority of the Thai government since the 1960s, when the first 5-year National Economic and Social Development Plan began (2). A number of ministries have been involved in rural development and have substantially raised educational attainment, improved family planning, expanded access to safe water and adequate sanitation, and delivered basic health services and nutritional assistance to the poor (3). Since 1960, Thailand has experienced one of the fastest declines in infant mortality rate globally (Fig. 21.2).

21.2.1 Sanitation Progress Begins

Thailand's first significant steps toward universal rural sanitation came with the start of the Village Health and Sanitation Project in 1960, when basic sanitation coverage was less than 1% (see Fig. 21.3) (2). Directed by the Ministry of Public Health, the project aimed to reduce the incidence of diarrheal diseases associated with poor sanitation and hygiene. In 1961, the project was expanded and integrated into the National Economic and Social Development Plan, and subsequently renamed the Rural Environmental Sanitation Program (RESP) (1). Community health workers from the provincial health departments and village volunteers formed the foundation of the program. They promoted a variety of participatory activities and low-cost technologies converging around water, sanitation, hygiene, and solid waste disposal. RESP had its setbacks. Hardware for sanitation, for example, was initially given to households through government subsidies, with the assumption that recipient households would serve as models to the rest of the community. It became evident, however, that the free systems

Water and Sanitation-Related Diseases and the Environment: Challenges, Interventions, and Preventive Measures, First Edition.
Edited by Janine M. H. Selendy.

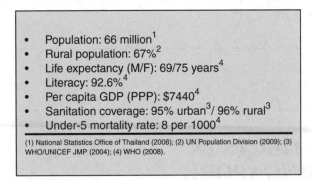

- Population: 66 million[1]
- Rural population: 67%[2]
- Life expectancy (M/F): 69/75 years[4]
- Literacy: 92.6%[4]
- Per capita GDP (PPP): $7440[4]
- Sanitation coverage: 95% urban[3]/ 96% rural[3]
- Under-5 mortality rate: 8 per 1000[4]

(1) National Statistics Office of Thailand (2008); (2) UN Population Division (2009); (3) WHO/UNICEF JMP (2004); (4) WHO (2008).

FIGURE 21.1   Kingdom of Thailand snapshot.

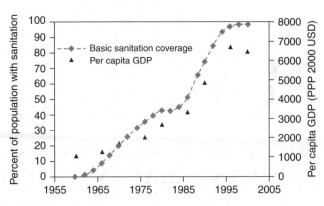

FIGURE 21.3   Events in Thailand's rural sanitation history.

1961: The Rural Environmental Sanitation Program begins.

1962–1963: Two successive prime ministers chair the National Health and Sanitation Development Conferences, raising awareness at high levels.

1980: Revolving loan program to help households finance sanitation improvements initiated.

1987: (i) Involvement of the private sector and local artisans began; (ii) an award scheme for provincial leaders and competition among provinces implemented.

1988: 100% Latrine Coverage Campaign Project initiated.

1989: The Ministry of Interior, with support from the Ministry of Public Health, adopts an ordinance requiring all new homes to have a sanitary toilet prior to registration.

were neither used nor maintained, and health officials and village committees consequently developed a more demand-driven process. In place of free systems, a revolving loan was created for each village so that households would finance their own sanitation improvements, helping to ensure user commitment and extend limited resources. Exceptions were made for poorest households who often would provide labor in exchange for participation in the program. At the same time, community health workers helped raise awareness and build demand for improved hygiene and sanitation. In many cases, health workers linked sanitation improvements to a household's desire for better access to water. The need to link sanitation to water supply improvements is generally an indicator of low demand for sanitation, and in some cases, sanitation systems subsequently failed or were unused. The Ministry of Public Health, however, persisted in their support and follow-up, ensuring continued use and maintenance. The national government committed financial resources to sanitation, dedicating approximately USD 2–6 million to sanitation annually (1990–1998 budget data) and committed health officers to each of the 75 provinces outside of Bangkok to raise awareness, train health workers, and monitor progress (1). Further, country- and province-level meetings, which involved volunteers, were conducted regularly to gauge progress, identify barriers, and develop solutions. The sanitation

technology of choice (selected by the government) was the pour-flush latrine, and although community participation and consumer choices were not part of the selection process, the technology fit well with Thai preferences. Additionally, the government helped develop the market for sanitation hardware by assisting the local private sector by training artisans to construct and sell products that included an array of low-cost, appropriate technologies (e.g., rainwater catchment systems) to meet the demands of rural consumers.

21.2.2   Recurring Commitments

In 1978, Thailand adopted the World Health Organization's strategy, Health for All by the Year 2000, using sanitation and water supply coverage as indicators of success. Nearly a decade later, toward the end of the International Drinking Water Supply and Sanitation Decade (1981–1990), efforts to increase sanitation coverage were accelerated by adoption of the "100% Latrine Coverage Campaign Project." In addition, a new policy by the Ministry of Interior (1989) bolstered sanitation efforts by requiring all households to have a sanitary toilet before a house could be registered. Through multiple strategies and projects, Thailand consistently renewed its efforts to increase access to basic sanitation (1).

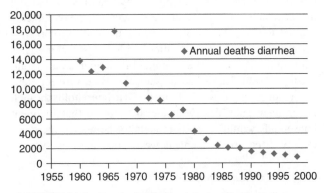

FIGURE 21.2   Deaths in Thailand due to diarrheal diseases.

21.2.3 Sanitation Today

In Thailand's 2005 national survey, 99.9% of the rural population reported having a sanitary toilet. Diarrheal morbidity, although relatively low, remains a focus of the Ministry of Health—annual incidence rate estimated at 1.8 cases per 100. Ongoing efforts by the government—including hygiene awareness campaigns, competitions, new sanitation and hygiene regulations, and a strong cadre of public health inspectors—demonstrate a continued commitment to universal sanitation (personal communication with the Ministry of Public Health, Thailand).

- Population: 28 million[1]
- Urban population: 68%[1]
- Life expectancy (M/F): 69/74 years[3]
- Literacy: 92%[5]
- Per capita GDP (PPP): $14,023[4]
- Sanitation coverage: 95% urban[2] / 93% rural[2]
- Under-5 mortality rate: 12 per 1000[3]

(1) Department of Statistics, Malaysia (2009); (2) WHO/UNICEF JMP (2006); (3) WHO, (2008); (4) IMF (2008); (5) UNICEF (2007).

FIGURE 21.4   Malaysia snapshot.

Lessons Learned 1

- *Persistence*: After 35 years and seven 5-year development plans—consistently including sanitation as part of their development criteria—Thailand has achieved near-universal sanitation.

- *High-Level Support*: The importance of basic sanitation was well recognized among government leadership and dedicated resources (financial and human) reflected a broad commitment to achieving universal sanitation.

- *Sanitation Policies*: Policies, such as the ordinance requiring households to have a sanitary toilet prior to registering a new house, helped accelerate gains in sanitation.

- *Reduced Hardware Subsidies*: The Ministry of Public Health quickly determined that hardware subsidies were undermining sustained use and maintenance of toilets and switched to a more demand-driven tactic that utilized revolving loans to finance hardware.

- *Private Producers*: To improve economic sustainability, the Ministry of Public Health provided technical assistance to train community artisans in building and marketing low-cost, appropriate technologies (e.g., pour-flush latrines and rainwater harvesting cisterns).

- *Competition and Awards*: Friendly competition among provinces and awards for provincial leaders created incentives to improve sanitation coverage.

- *Monitoring*: The government regularly monitored progress to identify barriers and develop solutions.

- *Integrated Programs*: Sanitation was made part of a larger "basic needs" package of goods aimed at improving the quality of life of rural communities.

21.3 SCALING UP SANITATION: THE CASE OF MALAYSIA

21.3.1 Background

The country of Malaysia has nearly 28 million inhabitants—up from nearly 6 million half a century ago. Following independence in 1957, Malaysia made rural development a fundamental part of the government's economic policies—generally focused on the twin objectives of achieving growth with equity. The success of rural development programs in Malaysia is reflected in the reduction of poverty in the rural sector from almost 60% in the 1970s to nearly 15% in 1995 (4). Further, between 1960 and 1990, income inequality fell as real GDP increased sevenfold—an annual growth rate of 6.8%. Under-5 mortality simultaneously dropped from 73 per 1000 live births in 1960 to 12 in 2008, as the government made great efforts to improve the health of "disadvantaged" communities. The improvement in the health status of the population has been significant, although disparities in health outcomes and wealth continue to exist between urban and rural populations (Fig. 21.4).

21.3.2 Rural Sanitation

Historically, diarrheal diseases accounted for a large portion of mortality in Malaysia, and major outbreaks of typhoid and cholera were not uncommon. In 1968, with technical support from the World Health Organization, the federal government initiated a number of pilot projects to improve household water supply, sanitation, and hygiene, as well as solid waste disposal in rural Malaysia (two-thirds of the population were rural in 1970) (Fig. 21.5). Following the pilot project, the government expanded its efforts and initiated the Rural Environment Sanitation Program. The program specifically focused on improving the health of rural communities and providing low-cost intermediate solutions (e.g., pour-flush latrines and protected wells) until piped water and sewage connections could be made. Community participation and

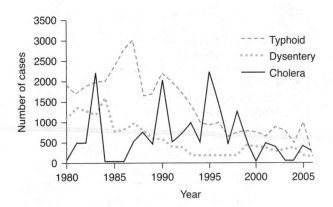

FIGURE 21.5 Gastrointestinal disease patterns in Malaysia, 1980–2006.

household-level hygiene education by community health workers were key elements of the program. Technical support, codes of practice, and construction materials were provided by the Ministry of Health, and communities supplied the labor. To ensure health improvements, the program linked sanitation improvements to household water supply projects—demand for water was generally much higher among households. Due to the fact that sanitation systems are often either not used or fall into a state of disrepair, the Ministry of Health established a monitoring program to ensure lasting changes. Currently, certain parts of the country still lag behind; the state of Sabah, one of Malaysia's poorest states, estimates coverage to be at approximately 93%. At the end of 2006, however, the ministry estimated that 98% of rural households (9.2 million people) had access to basic sanitation (estimate includes shared facilities) (5) (Box 21.1).

BOX 21.1 URBAN SANITATION

In the 1960s and 1970s, Malaysia's urban population grew at a pace of nearly 5%, and septic tanks at that time formed the majority of municipal sewerage systems (prior to this, manual collection of human excreta predominated). As population density increased and the economy grew, however, centralized systems became more common. The National Sewerage Development Program, launched in the 1970s, aimed to modernize sewerage systems in major urban centers and developed 19 sewerage master plans. Unable to gather the necessary resources, however, the government implemented only parts of the master plan in nine urban centers. Subsequently, sewerage management was decentralized to local authorities, which similarly struggled to gather the necessary resources to maintain and extend sewage

---

services. In an effort to slow the growth of the population without basic sanitation, new legislation was passed that required all new housing developments to have sewerage systems in place. The backlog of housing without sewerage services, however, remained and low-cost alternative wastewater treatment and disposal systems were increasingly sought. Many private companies attempted to fill this void—often marketing overly sophisticated systems unsuitable to local conditions. Further, local authorities often lacked the expertise to manage the systems. In response to the proliferation of a wide array of technologies, Malaysia created an inter-ministerial committee responsible for approving systems. Gains in sanitation coverage were made by the local authorities but rapid urbanization, compounded by the shortage of funds, manpower, and expertise limited their success (6). In 1991, the former prime minister of Malaysia, Dr. Mahathir Mohammad, initiated Wawasan 2020 (or Vision 2020). The Vision 2020 called for increased involvement of the country's private sector to provide services, while at the same time trying to ensure that the poor had access to basic services. Following Vision 2020, the federal government brought sewerage back under its purview (Sewerage Services Act of 1993). Under this act, the Sewerage Services Department was set up to regulate and monitor sewerage services, which were privatized shortly after passage of the new legislation. The concession, awarded to *Indah Water Konsortium* (IWK), was designed to upgrade rehabilitate and extend sewerage throughout the country (7). Initially, progress was slow, but over the last decade, the population served by IWK has gone up substantially (see Table 21.1). Further, in areas marked by high levels of poverty and high rates of diarrheal disease, the Ministry of Health assists households with connection fees. Resistance by the public to pay for sewage collection and treatment—something they are unaccustomed to paying for historically—remains high, and social marketing campaigns by IWK have been an important step to increase tariff collection (personal communication with Indah Water Konsortium Ir. Mohamed Haniffa Hj. Abdul Hamid and Shahrul Nizam Sulaiman). The country hopes to successfully reintroduce a combined tariff that would require consumers to pay for wastewater as well as potable water—a similar effort failed in the late 1990s. Currently, IWK manages the sewage from nearly 27 million Malaysians. Malaysia's National Water Services Commission is issuing guidelines to all water and sewerage asset owners and operators in the peninsula to have more transparent legislation, rules and regulations, tariff-setting principles and procedures, as well as standard key performance indicators and operating procedures.

TABLE 21.1   Change in the Population Served by Indah Water Konsortium (IWK) with Various Types of Sanitation Systems (2000–2008)

Type of Sanitation System	2000		2008	
	IWK Population Served	Percent of IWK population served	IWK Population Served	Percent of IWK population served
Centralized sewer system	8,564,722	51	17,481,291	65
Communal/individual septic tank	4,454,350	27	5,475,970	20
Pour-flush latrine	3,788,190	23	3,803,050	14
Total	16,807,262	100	26,760,311	100

Note: IWK operational area does not include the state of Kelantan, Sabah, and Sarawak; also excluded is the local authority of Johor Bahru.

Lessons Learned 2.

- *Participatory-Based Solutions*: Malaysia did not wait for a definitive sanitation solution in rural areas; it instead chose a community-based intervention to address immediate health concerns and quality of life issues.
- *Follow-up*: The government periodically monitors maintenance and use of household sanitation systems to ensure proper use and upkeep.
- *Pro-Poor Policies*: In both rural and urban areas, the Ministry of Health assists low-income households to gain access to basic sanitation.
- *Raising Awareness*: IWK is raising awareness of the importance of wastewater management and subsequently improving cost recovery through a social marketing campaign.
- *Appropriate Technologies*: Both IWK and the Ministry of Health have effectively selected sanitation-related technologies that do not go beyond the means of the community's ability to pay and maintain.
- *Integrated Programs*: Sanitation was made part of a larger package of goods aimed at improving the quality of life of rural communities (e.g., sanitary well with a household connection, rainwater collection system, and sullage and solid waste disposal systems).

21.4   SINGAPORE: FORWARD THINKING AND FORWARD MOVING

At the beginning of the twentieth century, a manual "two-pail" system was the prevailing method for collecting and removing human excreta from homes in Singapore (8). The two-pails allowed households to have one pail in use, while the other was transported to a disposal site where it was emptied and subsequently returned to the household,

replacing the pail in use. Around that same time, the annual infant mortality rate was estimated to be 354 per 1000. The development of modern sewage systems in Singapore started in 1910, and by the late 1930s, the population served by modern sanitation surpassed those using the manual disposal system. Singapore gained independence in August 1965, and that same month, the Public Utilities Board (PUB), which currently manages the entire water cycle of Singapore, came into being. In 1987, the manual system of excreta collection was completely phased out—at the same time, Singapore was gaining its reputation as a place of orderliness and effective administration. Health-related indicators had much improved—in 1986, the infant mortality rate was 9.1 per 1000, and life expectancy at birth was 71.4 years for males and 76.3 years for females (8). With many of the health problems associated with a contaminated water supply and lack of basic sanitation solved, environmentally driven efforts in Singapore increased. Efforts to clean up rivers and streams became a primary goal between 1977 and 1987. The Ministry of the Environment made great advances in extending the sewer system—reaching 100% of the population with modern sanitation by 1997. In 2001, authority to manage sewerage and drainage was shifted from the Ministry of the Environment to the PUB (9) (Fig. 21.6).

- Population: 4.8 million[1]
- Urban population: 100%[1]
- Life expectancy (M/F): 78/83 years[1]
- Literacy: 96%[1]
- Per capita GDP (PPP): $37,597[1]
- Sanitation coverage: 100%[2]
- Under-5 mortality rate: 3 per 1000[3]

(1) Singapore Government (2008); (2) WHO/UNICEF JMP (2004); (3) WHO (2008).

FIGURE 21.6   Singapore snapshot.

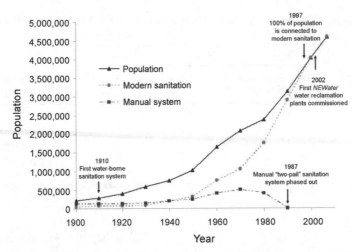

FIGURE 21.7 Graph based on census data and PUB presentation (2009).

The 263-square mile island city-state now has more than 36 square miles of parks, with 16 additional square miles reserved for park space (10). The whole population is served by modern sewage treatment that includes: 3200 km of sewers, 116 pumping stations, and 6 water reclamation plants (9). Singapore draws on a strategic development framework, the Concept Plan, which is updated every decade. The current plan, established in 2001, sets broad-based development plans for the next half century. It allows for an expected population of 5.5 million within 50 years. The Concept Plan provides details on land use needs (public, commercial, and industrial), transportation, and recreational facilities. Community participation is a core component of the planning process and neighborhoods are involved in local development planning (Fig. 21.7).

During the National Day Parade on August 9, 2002, Singapore celebrated the start of their NEWater program that effectively collects wastewater and processes it to a level surpassing international drinking water standards. At the launch of the program, the prime minister and the whole cabinet toasted a crowd of 60,000 parade goers with NEWater, which has grown in demand (11).

Singapore is currently developing a deep tunnel sewerage system that consists of two large tunnels that criss-cross the island at depths of 20–50 m. This effort will help eliminate the need for continual upgrading and expansion of the aging sewage system. The new deep tunnel system will allow Singapore to phase out 134 sewage pumping stations and smaller water reclamation plants—freeing up land and resources used for operations and maintenance (11).

REFERENCES

1. Aungkasuvapala N. Universal sanitation—Thailand. *East Asia Ministerial Conference on Sanitation and Hygiene*, 2007.

2. Luong TV, Chanacharnmongkol O, Thatsanatheb T. Universal sanitation—Thailand experiences. *Twenty-sixth WEDC Conference Proceedings*, Dhaka, Bangladesh, 2000.

3. Deolalikar AB. Poverty, growth, and inequality in Thailand. ERD Working Paper No. 8, April 2002.

4. The International Bank for Reconstruction and Development, Malaysia. *30 Years of Poverty Reduction, Growth and Racial Harmony*. The International Bank for Reconstruction and Development Malaysia, 2004.

5. Engku Azman Tuan Mat, Mohd Zaharon Mohd Talha. Rural environmental sanitation programme (RESP) Malaysia. *Proceedings from the East Asia Ministerial Conference on Sanitation and Hygiene*, 2007.

6. Pillay MS. Privatization of sewerage services in Malaysia. *WEDC Conference*. Ministry of Health, Malaysia, 1994.

7. Haarmeyer D, Mody D. Private capital in water and sanitation. *Finance Dev* 1997;March: 34–37.

8. Otaki Y, Otaki M, Sakura O. Water systems and urban sanitation: a historical comparison of Tokyo and Singapore. *J Water Health* 2007;5(2):259–265.

9. Tortajada C. Water management in Singapore. *Water Res Dev* 2006;22(2):227–240.

10. Hinrichsen D. Singapore: The Planned City. *Population Reports*, 2001.

11. Tan YS, Lee TJ, Karen T. *Clean, Green and Blue: Singapore's Journey Towards Environmental and Water Sustainability*. Singapore: ISEAS Publishing, 2009.

22

HOUSEHOLD-CENTERED ENVIRONMENTAL SANITATION SYSTEMS

BINDESHWAR PATHAK

22.1 INTRODUCTION

Environmental sanitation systems include the infrastructural arrangements required to protect and safeguard the environment of our habitats and promote the cleanliness and hygiene of the same. These could be organized at the community (city/towns/village) level or at the household level or as a combination of both. Disposal of solid and liquid wastes from the household and also for the community as a whole is the most critical component of the environmental sanitation systems and from the point view of public health, safe and sanitary disposal of human excreta is the single most important challenge for the environmental and public health managers. It is possibly in fairness of things that sanitation has been declared as the most important medical invention of the century, ahead of antibiotics and vaccines.

In most of the developed countries of the West, most urban and even rural communities are served by water carriage systems or sewerage systems for the disposal of human excreta and waste water from the households. Household toilets and bathrooms connected to sewerage systems is the symbol of Western civilization. However, the system is too capital intensive and requires huge amounts of water for flushing and maintaining the collection and conveyance systems. As such, a sewerage system is not affordable for most communities in the developing countries and, even where it has been built, it has run into problems of operation and maintenance and found to be unsustainable. The first sewerage system in the world was built in London in the year 1850, the second was in New York in 1860, and the third

sewerage system of the world was built in India in Calcutta in the year 1870. But, since then, during the past 140 years, out of 5,000 and odd towns in India sewerage systems have been provided in only about 200 and odd towns and of those too partially. It goes without saying that for the promotion of environmental health and sanitary disposal of wastes we will have to largely depend on household centered on-site sanitation systems, which are low-cost and which require much less water. These systems are user-friendly and sustainable. Even in the developed world, the concerns for long-term environmental sustainability demand a decentralized household-centered system requires less water.

In this chapter, we will discuss the various types of household-centered and decentralized human excreta and waste water disposal systems and also discuss the technologies developed and promoted by the Sulabh International Social Service Organization, on the same.

22.2 SANITATION SYSTEM COMPONENTS

A sanitation system includes all the components required for the management of human excreta and waste water. Although broadly speaking sanitation also includes safe disposal of solid waste, drainage, disposal of sullage and gray water, and so on, we will primarily focus on faecal waste water and human excreta disposal in this chapter. We should consider sanitation as a multistep process and not a single-point disposal of waste products and as such must be accounted far from the point of generation to the point of reuse or ultimate disposal.

Water and Sanitation-Related Diseases and the Environment: Challenges, Interventions, and Preventive Measures, First Edition.
Edited by Janine M. H. Selendy.
© 2011 Wiley-Blackwell. Published 2011 by John Wiley & Sons, Inc.

Basic components of and flow sheet of a sanitation system

```
┌──────────────┐   ┌──────────────────┐   ┌──────────────┐   ┌──────────────┐
│  Inputs or   │──▶│  User interface  │──▶│ Products or  │──▶│ Collection,  │
│   influent   │   │ (toilet hardwares)│   │  effluents   │   │ storage, and │
│              │   │                  │   │              │   │  treatment   │
└──────────────┘   └──────────────────┘   └──────────────┘   └──────────────┘

┌──────────────┐   ┌──────────────────┐
│ Products or  │──▶│   Conveyance/    │
│ effluents or │   │  disposal/reuse  │
│    sludge    │   │                  │
└──────────────┘   └──────────────────┘
```

To design a technically robust and sustainable system, which could be operated and maintained continually with minimum skill requirement, we must consider all of the influent and inputs that will be generated by the users and all of the products that might be created subsequently and then consider all the functions that these products would pass through before they can be appropriately used or disposed of without creating any environmental nuisance and public health hazards. Sanitation technologies are basically product-specific functions that will transform or transport products to another function or a final point of use or disposal. Inputs or influents to a sanitation system include the following.

22.2.1 Inputs and Products

The influent or inputs to a sanitation system include the following:

1. *Excreta* is the mixture of urine and faeces that is not mixed with any flushing water.
2. *Urine* is urine that is not mixed with faeces or water.
3. *Faeces* refers to semi-solid excrement without any urine or water.
4. *Anal cleansing water* is water that is collected after having been used to clean oneself after defecating (and/or urinating). It is generated by those who use water, rather than dry material for anal cleansing.
5. *Flush water* is the water that is used to move excreta, urine, or faeces from the user interface into the next technology.
6. *Storm water* is the general term for the rainfall that runs off of roofs, roads, and other surfaces before flowing toward low-lying land. It is the portion of rainfall that does not infiltrate into the soil.
7. *Gray water* the total of water generated from the washing of food, clothes, dishes, and people. It does not contain excreta, but it still contains pathogens and organics.
8. *Organics* refers to the organic material that must be added to some technologies in order to make them function properly (e.g., composting chambers).

9. *Dry cleansing materials* may be paper, corncobs, rocks, or other dry materials that are used for anal cleansing (instead of water). Depending on the system, the dry cleansing materials may be collected and disposed of separately.

Products or effluents from a sanitation system include the following:

1. *Black water* is the mixture of urine, faeces, and flushing water along with anal cleansing water (if anal cleansing is practiced) or dry cleansing material (e.g., toilet paper).
2. *Faecal sludge* is the general term for the undigested or partially digested slurry or solid that results from the storage or treatment of black water or excreta.
3. *Compost/humus* is the earth-like, brown/black material that is the result of decomposed organic matter. Generally it has been hygienized sufficiently that it can be used safely in agriculture.

22.3 CLASSIFICATION OF SANITATION TECHNOLOGIES

The technologies of human excreta and waste water disposal systems could be broadly classified into two groups, namely, on-site sanitation systems and off-site sanitation systems.

22.3.1 On-Site Sanitation Systems

1. Single-pit pour-flush systems
2. Double-pit pour-flush systems
3. Septic tank systems
4. Two-pit, single-pit VIP (ventilated improved pit latrine)
5. Urine separation and disposal system

22.3.2 Off-Site Sanitation Systems

1. Sewerage systems with centralized treatment plants and disposal

2. Simplified sewerage with semi-centralized system
3. Decentralized effluent and centralized faecal sludge treatment

The user interface, that is, the infrastructure of domestic toilet could be connected to an on-site sanitation system, which is generally managed by the user. Otherwise, it could be connected to a off-site sanitation system, which includes the collection, conveyance, and treatment disposal systems and is managed by community organizations like the municipalities or other local bodies.

22.4 DESCRIPTION OF VARIOUS HOUSEHOLD-CENTERED ON-SITE SANITATION SYSTEMS PRESENTLY BEING PRACTICED IN DEVELOPING COUNTRIES

22.4.1 Single-Pit, Pour-Flush Toilet System

Of the various on-site sanitation systems, previously listed, this is the simplest to construct at the household level. Here the toilets having a squatting plate with pan, trap, and water seal is placed directly on the top of a pit dug in the soil. All types of materials including anal cleansing water or solid cleansing material such as paper are generally discarded into the pit, though they may shorten the life of the pit and make pit emptying more difficult; whenever possible, solid cleansing materials should be disposed off separately. Anal cleansing water should be kept to a minimum to slow the rate of filling and to minimize the risk of groundwater pollution.

Although simple to construct and least costly, this system would require significant infrastructure for emptying the pit and treating the faecal sludge when the pit is full or, alternately, this will require sufficient space for constructing a new pit and then shifting the toilet infrastructure on top of the same. Because of this, sanitation programs undertaken in some of the developing countries using single-pit systems have often been found to be lacking in sustainability and continuity.

The success of the system depends on effective leaching of black water through the pit lining into the soil. While suitable for sandy soil with high absorption capacity under adverse hydrogeological conditions such as densely packed soil or fine clay or rocky or limestone formations, special precaution should be taken against groundwater pollution.

22.4.2 Two-Pit, Pour-Flush, Compost Toilet System

The two-pit, pour-flush, compost toilet system is an improvement on the single-pit system. In this system, the user-interface facilities include a squatting plate with pan and

trap with appropriate slope and water seal so that the hand flushing could be affected with minimum quantity of water (1–1.5 L). The toilet pan is connected through a distribution box to two-pits, dug in the soil. The pits are appropriately lined so that the black water could be easily leached into the surrounding soil. The system is entirely decentralized and household centered and all of the products can be treated on-site. However, the gray water could be transported off-site if there is a possibility to do so. The same would also be used for household horticulture and kitchen gardening or could be disposed of into a soakage pit or soakage trench.

This system is special in that, unlike septic tank, it does not produce faecal sludge, which has to be transported off-site. Here the black water is collected and processed into compost/humus *in situ*. After one of the pits is filled up, the flow is diverted to the other pit and it is left untouched for almost a year. During this period, the compost manure in the pit becomes totally pathogen-free and could be handled manually in the agricultural field as soil conditioner and manure. The system is particularly suitable for communities that are accustomed to using water for anal cleansing. It has a moderate level of capital investment toward construction and lining of the twin pits and the provision of a super structure of the toilet along with squatting pan with a water seal. The cost will vary with the quality of the super structure that the user will opt for. Since the pits are used in an alternating way, the system has an element of permanency and long-term sustainability.

The system operates very successfully in porous soil in high absorption capacity. But under adverse hydrogeological conditions such as fine clay, densely packed soil, rocky and limestone formations, special care should be taken against groundwater pollution. Under usual conditions, a distance of 10 m from the shallow groundwater sources should be practiced.

Sulabh International Social Service Organization (SISSO) has developed and promoted this type of household toilet system in India on a massive scale, which has started a nationwide sanitation movement. More than 1 million such toilets, *Sulabh-Shauchalayas*, as they are popularly known, have been constructed by Sulabh and almost 60% of the rural population has been covered by the central government under its Total Sanitation Campaign, which is using the same model. This has been further discussed in Section 22.6.

22.4.3 Alternating Double-Pit, Dry Toilet System (VIP)

In countries where people do not use water for anal cleansing, ventilated improved double-pit latrines are used successfully. This model was developed in sub-Saharan countries of Africa. In this system there is no need for transport or centralized treatment of the excreta since the same are decomposed on-site inside the pits. The process can take place in an alternating pit that is dug into the ground or in a constructed concrete chamber above the ground if required. Both the options will produce nutrient-rich safe humus-like product that can be used as a soil conditioner and manure.

The success of the system depends on an extended storage period. If a suitable and continuous source of organic matter (leaves, grass clipping, rice husks, etc.) is available, the decomposition process is enhanced and the storage period can be reduced. The digestion and storage period could be significantly reduced if the moisture content could be kept at the minimum level and the materials in the pit could be kept in an aerobic condition. It is therefore necessary that the gray water should be collected and treated separately. Dry anal cleansing materials could be discarded into the pit if they are organic in nature, but stone chips or other inert material used for the same should not be put into the pit.

VIP latrines are being successfully used in many African countries. As the name indicates, the smell and odor nuisance is prevented by the provision of a ventilation pipe, which is required in absence of the water seal and the flushing system.

22.5   ECOLOGICAL SANITATION: URINE SEPARATION, APPLICATION, AND DISPOSAL

This system requires a special user-interface, which will allow for separation of urine and faeces. In water-scarce areas, where there is a need and a desire to apply urine as a fertilizer to cropland, this system could be successfully used. Urine is collected on-site and either used in a household plot for horticulture or it could be transported to agricultural fields away from the point of generation. Due to the higher risk associated with the direct use, gray water and anal cleansing water should be treated separately.

The success of this system depends on intensive education of the community regarding separation of faeces and urine and keeping the faeces as dry as possible. Care should be taken to ensure that no water is introduced during cleaning. Also important is to supply ash, lime, or dry earth to cover the faeces to limit odor nuisance and provide a barrier between the faeces and potential vectors. The success of the system will also depend on available cropland that is accessible and appropriate for urine application. If there is no opportunity of appropriate application of urine, the same could be disposed of on-site. The solid materials used for anal cleansing should be discarded separately as far as possible. In case water is used for anal cleansing, it must be kept at the minimum level and must be separated from the faeces. The same could be mixed and treated along with gray water.

The technology is still in its infant stage and is yet to be applied in a large scale in countries like India, Bangladesh, and so on.

22.5.1   Decentralized Effluent and Semi-Centralized Faecal Sludge System (Septic Tank)

This system could be used if there is adequate space for a soakage pit or leach field and the soil has sufficient absorptive capacity. In this system the black water is collected, stored, and partially treated on-site. The effluent is disposed off-site in a soakage pit or a leach field, while the faecal sludge is transported to a centralized treatment facility. The grey water can be treated simultaneously or separately, depending on availability of space and maintenance requirement.

The capital investment of this system is considerable toward the construction of the on-site storage and treatment facility, popularly known as the septic tanks. This system is used in many urban homes in the developing countries and requires large volume of water and regular cleaning and transportation of the faecal sludge to a centralized treatment facility to be operated and maintained by the municipal authorities. In absence of regular desludging and appropriate effluent disposal systems, many septic tank systems are malfunctioning and causing health hazards by way of vector breeding and environmental pollution.

22.5.2   Simplified Sewerage with Semi-Centralized Treatment

This system is a combination of household-level technology to remove settleable solids from the black water and a simplified small diameter sewer system to transport the effluent to a semi-centralized treatment facility. A simple interceptor tank is required to remove the settleable solids at the household level. This system could be used as a means of upgrading underperforming on-site technologies like the septic tank by providing improved semi-centralized treatment. This system could be appropriately used in conjunction with semi-centralized biogas plant and effluent treatment systems for a cluster of households. Sulabh International Social Service Organization has developed and demonstrated the efficacy of such systems in a number of sites in India. The system and its operation and functioning are discussed in detail in Section 22.6.

22.5.3   Disposal of Effluents from Household-Centered, On-Site Treatment Systems

Some of the on-site treatment systems are described above like the two-pit, pour-flush, compost system, which comprehensively deals with the final disposal of the treated effluent. However, systems like septic tank, biogas digester, and so on require that the final effluent should be disposed off without creating any public health or environmental hazard. The following techniques are generally applied for the final disposal of the effluents.

1. Land irrigation
2. Aquaculture
3. Soakage pit
4. Leach field
5. Groundwater recharge
6. Disposal into water bodies

22.5.3.1 Land Irrigation

Waste water from household treatment system of varying qualities could be used for irrigating agricultural lands or household kitchen gardens. Although strictly from public health point of view, waste water should be used for crop irrigation or irrigating vegetable plants only after full secondary treatments, in the case of effluents from household treatment systems often partially treated effluents are also used. Waste water irrigation supplies water to plants through the use of channels or pipes to supplement the water supply as well as add valuable nutrients, such as nitrogen, phosphorus, and potassium (NPK).

The system, depending on the degree of treatment that the effluent has undergone, may be quite hazardous from the health point of view and vegetables, thus grown, should not be eaten raw.

22.5.3.2 Aquaculture

Under this system controlled cultivation of aquatic plants and animals are practiced in some countries. Cultivation of fish in a pond with waste water is widely practiced in countries like India, Bangladesh, Vietnam, and so on. Three kinds of systems exist.

1. Fertilization of fish ponds with faecal effluent/sludge
2. Fertilization of fish ponds with excreta
3. Fish grown directly in aerobic ponds

There is an element of health risk associated with this system and there is always a concern regarding the contamination of fish harvested from ponds fertilized with waste.

The aquaculture also includes the floating plant pond, which is essentially a modified maturation pond. The aerobic pond is populated with floating plants such as water hyacinth or duckweed that float on the surface and have roots that hang down into the water. The system can achieve high removal rates of both biochemical oxygen demand (BOD) and suspended solids, two key indicators of pollution potential of the waste water, although pathogen removal is not substantial. Duckweed ponds are used for fish fertilization and cultivation. It is, however, advisable that only cooked fish should be consumed, not raw.

22.5.3.3 Soakage Pit

A soakage pit, also known as soak away or leach pit, is a covered porous-walled chamber that allows water to slowly soak into the ground. Pre-settled effluent from a septic tank or biogas plant is discharged into soakage pit for dispersal in the soil. Soakage pits are best suited to soils with good absorptive properties. Fine clay or hard rocks are not appropriate for soak pits. Soakage pits can either be left empty or lined with appropriate material or it can be filled with burnt brick pieces or gravel.

22.5.3.4 Leach Field

The leach field is a network of trenches or slotted pipes that is used to disperse the effluent from the collection of storage point or on-site treatment systems. The pipes are generally laid at a depth of 15–20 cm from the surface to prevent the effluent from surfacing and to facilitate oxygen transfer into the leaching area. As in case of soakage pit, the application of leach field also requires porous soil with good absorptive capacity. A leach field should be at least 15–20 m away from a drinking water supply source like a dug well or a bore well.

22.5.3.5 Groundwater Recharge

The groundwater recharge is being considered as a ultimate disposal method for treated waste water in recent years. The groundwater recharge is gaining increasing popularity as the groundwater resources become increasingly depleted. However, before the final effluent from on-site treatment systems are recommended for recharging into the aquifer one has to be careful about the quality of the same. Although the soil is known to act as a filter for a variety of contaminants, groundwater recharge should not be viewed as a treatment alternative. The boundary conditions for groundwater recharge would be entirely guided by the local conditions and regulation.

22.5.3.6 Disposal into Water Bodies

The discharge of final effluent from on-site treatment systems like the septic tank or biogas plant directly into a receiving water body, like a river, lake, canal, is an acceptable disposal option provided the quality criteria regarding the treated effluent to be discharged into the water body is met, depending on the assimilative capacity of the water body. Water quality parameters, such as turbidity, temperature, suspended solids, biological oxygen demand, and nitrogen, potassium, phosphorus, should be carefully controlled and monitored before releasing any water into a natural body. The use of the water body whether it is used for industry, recreation, fish culture, or as a source of drinking water will determine the quality and quantity of waste water that could be introduced into the same without any adverse impact. Sulabh has experimented successfully with a decentralized biogas plant and effluent treatment, using ultraviolet ray and producing high-quality effluent with extremely low biological oxygen demand and faecal coliform count, which will have minimal impact on the water quality of the receiving water. This technology is being discussed in Section 22.6.

22.6 SULABH SANITATION TECHNOLOGIES

In India, due to lack of affordable sanitation technology, sanitation coverage is still far below the level of satisfaction. There was a major breakthrough in the field of sanitation when Sulabh developed and demonstrated the technology of two-pit, pour-flush toilets in 1970, for on-site disposal of household human wastes. For providing sanitation in slums, where people generally do not have space and resource to have their own toilets and at public places, Sulabh developed a novel concept of operation and maintenance of public toilets on a pay-and-use basis in 1974. Further, for the safe and hygienic disposal or use of human wastes, Sulabh developed a new technology for production and utilization of biogas from human wastes for different purposes and a convenient method of treatment of effluent of biogas plant to make it pathogen-free, odorless, and colorless, having biochemical oxygen demand value much lower than the permissible value for using it for different purposes or safe discharge in any water body. Duckweed-based waste water treatment having economic return, and composting of biodegradable wastes in much less time is another technical breakthrough of Sulabh in this field. It has developed good sanitation marketing system to make the system acceptable by mass. All these technologies are being described below in detail.

22.6.1 Sulabh Two-Pit, Pour-Flush, Compost Toilet

Sulabh two-pit, pour-flush, compost toilet is eco-friendly, technically appropriate, socio-culturally acceptable, and economically affordable. It is an indigenous technology; the toilet can easily be constructed by local labor and materials. It provides health benefits by on-site safe disposal of human excreta (Fig. 22.1).

There are two pits of varying size and capacity depending on the number of users. The capacity of each pit is normally designed for 3-years usage. Both pits are used alternately. When one pit is full, the incoming excreta is diverted to the second pit. In about 2 years, the sludge in the first pit not in use gets digested and is almost dry and pathogen-free, thus safe for handling as manure. Digested sludge is odorless and is a good manure and soil conditioner.

No vent pipe is needed: The gas gets absorbed in the soil facing the chamber because the brick lining inside is in lattice formation. The parameters change depending upon the coarseness of the soil and the type of terrain where the toilet is being constructed. Depending on the availability of space, shape of pits may be designed to suit the conditions. It may be rectangular, circular, or linear in shape. So far, Sulabh has implemented more than 1.2 million such household toilets in India.

The cost of Sulabh flush composting toilets varies widely to suit people of different economic stratum. The cost ranges from Rs 500 to Rs 50,000 (approximately USD 10–1,000) per unit. It depends upon material used for construction of pits and seat, as well as for superstructure.

An important aspect of the technology is that it requires only 1–1.5 L of water to flush excreta. That is the reason for

FIGURE 22.1 The Sulabh twin-pit, pour-flush, compost toilet—a simple solution to achieve the Millennium Development Goal on Sanitation.

s adoption in water scarcity areas in different countries. In he conventional system not less than 10 L of water is ⁻uired to flush; that is, the two-pit usage saves 8 L of water lush. Taking into account the number of units con- ⁻ed by Sulabh used by 5 persons per unit twice a day, 96 million L of water are saved per day.

This technology has been declared a Global Best Practice by United Nations HABITAT and Centre for Human Settlements, and is now recommended by the United Nations Development Programme (UNDP) for use by more than 2.6 billion people around the world.

22.6.2 Sulabh Public Toilet Complexes

Operation and maintenance of public toilets on a *pay-and-use* basis is a major breakthrough concept of Sulabh in the field of sanitation. This is more suitable for slums (where people generally do not have space and resources to have their own toilets), public places, market places, tourist places, religious places, and at other floating population sites. Sulabh has so far constructed more than 7,000 such public toilet complexes in different parts of the country, where maintenance is provided round the clock. User charge is Re. 1 (2 US cents) per use (Fig. 22.2).

The biggest public toilet of Sulabh has been constructed at Shirdi, in the district of Nasik, in the state of Maharashtra, having 120 WCs, 108 bathrooms, 28 special toilets (separate for ladies and gents), and 5,000 lockers for the convenience of the pilgrims.

The system of operation and maintenance of community toilets evolved by Sulabh has proved a boon for the local bodies in their endeavor to keep the towns clean and improve the environment. This is a unique example of partnership of local authorities, non-governmental organization, and the community.

These pay-per-use public facilities provide an economically sustainable, ecological, and culturally acceptable solution to hygiene problems in crowded slum communities and public places.

Both the Sulabh sanitation technologies together serve more than 10 million people daily.

22.6.3 Biogas from Public Toilet Complexes

Safe disposal of human wastes from public toilet complexes is a major challenge particularly in areas without sewage coverage. To overcome such a problem, Sulabh has developed a simple and affordable way to produce biogas from human wastes and on-site treatment of effluent from biogas digester for its safe reuse (Figs. 22.3–22.5).

Biogas from public toilets has multiple benefits. It improves sanitation, community health and hygiene, and the environment in addition to providing biogas for different purposes, such as cooking, lighting mantle lamps, warming oneself, and power generation for street lighting. For biogas generation no manual handling of excreta at any stage is required (Figs. 22.6 and 22.7).

Hydraulic retention time (HRT) of feed material is maintained for 30 days. Flow of human wastes into the digester is under gravity. One cubic foot of biogas is produced from the human excreta of one person per day. Human excreta-based biogas contains 65–66% methane,

FIGURE 22.2   Front and side view of the Sulabh Toilet Complex at Kothi compound, Rewa, Madhya Pradesh (constructed during 2006–2007).

FIGURE 22.3 Biogas digester, underground, below the flower bed, connected to Sulabh public toilet complex.

32–34% carbon oxide, about 1% hydrogen sulfide, and trace amounts of nitrogen and ammonia. A thousand cubic feet (30 m³) of biogas is equivalent to 600 cubic feet of natural gas, 6.4 gallons of butane, 5.2 gallons of gasoline, or 4.6 gallons of diesel oil.

Produced biogas is used for cooking, lighting through mantle lamps, warming oneself, and electricity generation. Cooking is the most convenient use of biogas. Initially diesel was required to run the engine—80% biogas and 20% diesel. Sulabh has modified the engine recently that does not require diesel at all; it runs on 100% biogas. So far Sulabh has constructed 200 public toilet linked biogas plants in India. It has also implemented five such plants in Kabul, Afghanistan, in collaboration with the Kabul Municipality, Afghanistan and the government of India. During the winter season when

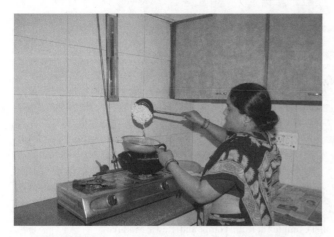

FIGURE 22.4 Food being cooked on a biogas burner.

FIGURE 22.5 A mantle lamp lit by biogas in a Sulabh complex.

FIGURE 22.6 Power generation from biogas with an engine operated by a battery for spark ignition.

the ambient temperature went down to subzero, biogas production remained more or less constant. It was due to suitable changes in the design of biogas plants and by providing adequate insulation to the plants.

22.6.4 Sulabh Biogas Plant Effluent Treatment (SET) System for Reuse of Effluent

Biogas plant effluent contains a good percentage of nitrogen, phosphorus, and potassium and other micronutrients for plants, but its bad odor, yellowish color, high biochemical oxygen demand, and pathogen contents limit its reuse for agriculture and horticulture; it is not safe for discharge in any water body. To make biogas plant effluent reusable, Sulabh has developed a simple and convenient technology named as Sulabh Effluent Treatment (SET) to further treat such effluent

FIGURE 22.7 Power generated from biogas is used for street lighting.

for its safe reuse. The technology is based on sedimentation and filtration of effluent through sand and activated charcoal followed by aeration in a tank and, finally, exposure of ultraviolet rays (Fig. 22.8).

(a)

FIGURE 22.8 (a) Treatment of waste water of human excreta-based biogas digester through Sulabh Effluent Treatment (SET) technology. (b) Diagrammatic representation of the SET technology.

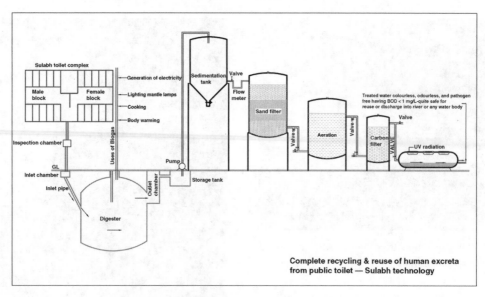

FIGURE 22.8   (*Continued*)

The treated effluent is colorless, odorless, and pathogen-free, having a biochemical oxygen demand less than 1 mg/L. It is quite safe for aquaculture, agriculture/horticulture purposes, or discharge in any water body without causing pollution and health risk. It can also be used for floor cleaning of public toilets in water-scarce areas.

22.6.5 Duckweed-Based Waste Water Treatment

Treatment of waste water is another major challenge to maintain sanitation in rural as well as urban areas due to lack of affordable technology. The available technologies are unaffordable due to high capital and maintenance costs. Due to noneconomic return, local authorities are generally not interested in taking up treatment of waste water causing severe health hazards and environmental pollution. Most of the untreated waste water is discharged into river or other water bodies. In rural areas, it is a common practice to discharge waste water/sullage without even collection. There is no question of treatment/recycle or even reuse of waste water/sullage as people are generally not aware of the technology (Figs. 22.9 and 22.10).

Waste water treatment through duckweed is a boon to overcome the problem. Duckweed—a small free-floating and fast-growth aquatic plant—has tremendous ability to reduce BOD, chemical oxygen demand (COD), suspended solids, and bacterial and other pathogens from waste water. It is a complete feed for fish and due to its high content of proteins and vitamins A and C; it is a highly nutritious feed for poultry and animals. The yield of fish increases by 2–3 times when fed with duckweed than that with conventional feeds in ponds. Reduction of BOD and COD of effluents varies from 80% to 90% at the retention time of 7–8 days. The first project funded by the Ministry of Environment &

FIGURE 22.9   Duckweed-based waste water treatment.

FIGURE 22.10   Duckweed-based pisciculture.

Forests, government of India, was completed in collaboration with the Central Pollution Control Board, New Delhi. The CPCB has made a guideline on the use of duckweed for the waste water treatment.

22.6.6 Sulabh Thermophilic Aerobic Composting (STAC) Technology

Sulabh has developed a new technology—Sulabh Thermophilic Aerobic Composter (STAC)—that requires only 10 days to make compost from any biodegradable wastes without any manual handling during composting. It is based on thermophilic aerobic method. The technology does not require recurring expenditure. The plant is a galvanized iron (GI) sheet made having a double wall filled with glass wool, partitioned with a perforated sheet into three chambers. After biodegradation, liquid is collected in the bottom chamber that can be easily taken out and used for agricultural/horticultural purposes. Manure that contains 30–35% moisture can be directly used for agriculture/landfill purposes, or it can be dried, granulated, and stored until further use (Fig. 22.11).

The practical utilities of this technology are (i) the organic solid waste can be efficiently converted into manure and soil conditioner, having a direct/indirect economic return; (ii) it will control diseases transmitted from wastes, because at high temperatures pathogens are eliminated from the waste; (iii) due to the reduction in volume, the carriage cost of wastes to a disposal site as well as an area needed for landfills will be drastically reduced; and (iv) the spread of weeds from wastes will also be controlled. The technology is more suited for rural areas because its by-products (compost) can be readily used for agricultural purposes, and it reduces health hazards.

FIGURE 22.11   Sulabh Thermophilic Aerobic Composting (STAC) Technology.

SECTION V

ENVIRONMENTAL FACTORS OF WATER POLLUTION

The broad extent, variety, and ramifications of environmental factors are covered in many chapters throughout the book, and particularly in Section V.

23

NATURALLY OCCURRING WATER POLLUTANTS

Lorraine C. Backer, Jonathan K. Kish, Helena M. Solo-Gabriele, and Lora E. Fleming

23.1 INTRODUCTION

Water quality is determined in part by the characteristics and concentrations of naturally occurring constituents. Many of these constituents are considered pollutants because they can harm the local ecology, aquatic organisms, or the people and other animals that use the water.

For the purpose of this chapter, a naturally occurring water pollutant (hereafter sometimes referred to as a "natural pollutant") is defined as an element, a molecule, a compound, or an organism that is pathogenic to humans and is found in groundwater, surface water, in aerosols generated from these waters, or in the food chain influenced by these waters. For a natural pollutant to be pathogenic (i.e., capable of producing disease in humans or other animals), there must be a route whereby potential victims can be exposed (e.g., ingestion, inhalation, or dermal contact). In addition, the individual must be susceptible to the specific disease and receive a dose of the pollutant that is sufficiently infectious or toxic to produce the disease.

Natural pollutants accumulate in waters as a result of physical and chemical processes (e.g., leaching from bedrock), biological processes (e.g., release of metabolic by-products or exuberant growth of harmful algae), or deposition from the air. Natural pollutants can also be present in waters as the result of human activities or man-made systems that expedite the transfer of contaminated water from one place to another (e.g., agricultural waterways). The contaminants listed in this chapter range from elements to complex compounds, from single-celled yeasts to multicellular helminthes, and they have been documented or are implicated in diseases in humans and aquatic flora and fauna.

In this chapter, we discuss direct exposures to these contaminants through drinking water and indirect exposures though the contaminants' accumulation in foods, particularly seafood. We also discuss the additional complexity of addressing environmental reality—exposure to mixtures. Finally, we discuss how physical forcers, such as changes in climate and land use, might mitigate or exacerbate the public health impacts of these contaminants.

23.2 NATURAL POLLUTANTS—CHEMICALS

In this section, we provide a brief overview of the impacts of selected naturally occurring chemical pollutants in water. Additional chapters in this volume provide detailed discussions about arsenic, mercury, dengue fever, soil-transmitted helminths, and malaria, as well as cyanobacteria and cyanobacterial toxins.

23.2.1 Inorganic Contaminants

A recent report by the United States Geological Survey provided data collected between 1991 and 2004 on contaminants in private drinking-water wells in the United States (1). The contaminants most often detected at concentrations higher than the human health benchmarks were the inorganic chemicals radon, strontium, arsenic, manganese, nitrate, uranium, boron, and fluoride (1). The health impacts associated with exposure to these contaminants are summarized in Table 23.1.

In addition to the risks associated with the heavy metals and radionuclides mentioned above, there are risks associated with other heavy metals that may be present in drinking-water

Water and Sanitation-Related Diseases and the Environment: Challenges, Interventions, and Preventive Measures, First Edition.
Edited by Janine M. H. Selendy.
© 2011 Wiley-Blackwell. Published 2011 by John Wiley & Sons, Inc.

TABLE 23.1 Inorganic Contaminants, Human Health Benchmark, and Possible Health Effects Associated with Exposure Above the Benchmark

Contaminant	Route of Exposure	Human Health Benchmark	Possible Health Effects or Perceptions of Poor Water Quality Associated with Exposure Above the Benchmark
Arsenic	Contaminated drinking water, food	10 µg/L (MCL[a])	Skin damage, problems with circulatory system, increased risk of some cancers
Boron	Contaminated drinking water	1000 µg/L (HBSL[b])	Gastrointestinal, reproductive, and developmental effects
Cadmium	Contaminated drinking water, food	5 µg/L (MCL)	Cadmium poisoning, cancer, Itai-Itai
Chromium	Contaminated drinking water	100 µg/L (MCL)	Chromium poisoning, cancer
Copper	Contaminated drinking water	1.0 mg/L (SMCL[c])	Perception of bad color, taste
Dissolved solids (inorganics)	Contaminated drinking water	500 mg/L (SMCL)	Perception of bad color, taste
Fluoride	Contaminated drinking water	4 mg/L (MCL)	Bone disease, mottled teeth in children, fluorosis
Iron	Contaminated drinking water	0.3 mg/L (SMCL)	Perception of bad color
Lead	Contaminated drinking water, food	15 µg/L (TT)	Lead poisoning
Manganese	Contaminated water inhalation, possibly food	300 µg/L (HBSL)	Neurologic effects
Mercury (methyl)	Contaminated fish	0.3 µg/KG (body weight) per day (ATSDR MRL[d])	Methylmercury poisoning
Nitrate	Contaminated drinking water	10 mg/L as N (MCL)	Methemoglobinemia in infants up to 6 months of age
Selenium (trace element—physiological importance, toxic high dose)	Contaminated drinking water	50 µg/L (MCL)	Selenium poisoning
Thallium	Contaminated drinking water	2 µg/L (MCL)	Thallium poisoning
Radon	Contaminated drinking water, inhalation	300 pCi/L (proposed MCL)	Increased risk of lung cancer
Strontium	Contaminated drinking water	4000 µg/L (HBSL)	Abnormal bone development
Uranium	Contaminated drinking water	30 µg/L (MCL)	Increased risk of cancer, kidney toxicity

[a] MCL (maximum contaminant level) = the highest level of a contaminant that is allowed in drinking water. MCLs are enforceable standards.
[b] HBSL (health-based screening level) = consistent with lifetime health advisories and risk-specific dose values. HBSLs are not enforceable.
[c] SMCL (secondary maximum contaminant level) = nonmandatory water quality standards; EPA does not enforce these SMCLs. They are established only as guidelines to assist public water systems in managing their drinking water for aesthetic considerations, such as taste, color, and odor. These contaminants are not considered to present a risk to human health at the SMCL.
[d] MRL (minimum risk level) = minimal risk level is an estimate of the level of daily human exposure to a hazardous substance likely to not cause appreciable risk of adverse noncancer health effects over a specified duration and route of exposure.
Source: Adapted from Ref. (1).

sources. Cadmium, mercury, and lead are widely distributed in the environment, and exposures are associated with renal, neurologic, and carcinogenic effects in humans and other animals (2, 3). Exposure to lead is primarily from contaminated air, food, and paint used in homes; however, drinking water may be a source of exposure, even in developed countries. For example, widespread lead exposure occurred in Washington, DC, between 2000 and 2004 when changes in drinking-water treatment resulted in the release of lead from distribution system pipes (4). Finally, exposure to arsenic from contaminated drinking water is common in many parts of the world. Chronic exposure to arsenic results in a number of adverse effects, including skin color changes, hard patches on the palms and soles of the feet, and cancer of the skin, bladder, kidney, and lung (5). The most well-known arsenicosis outbreak was associated with drinking water from shallow tube wells inadvertently drilled directly into highly contaminated aquifers in India and Bangladesh (5). More than 11,000 cases of arsenical skin lesions in Bangladesh and 29,000 cases in West Bengal, India, were attributed to drinking this contaminated water (6).

In addition to the direct impacts of inorganic drinking water contaminants, there is increasing evidence that exposure to low levels of these pollutants may increase the risk for chronic diseases as well as the long-term sequelae of these diseases. For example, using data from the Centers for Disease Control and Prevention's National Health and Nutrition Examination Survey (NHANES III), Navas-Acien et al. (7) reported that exposure to low levels of inorganic arsenic may play a role in the increasing prevalence of diabetes in the United States. Also using NHANES data, Schwartz et al. (8) reported that increased urinary cadmium levels were associated with both impaired fasting glucose and an increased risk for diabetes.

23.2.2 Algal Toxins

Marine and freshwater algae comprise an ancient microbial assemblage that includes dinoflagellates, diatoms, and cyanobacteria. Although ubiquitously present, periodically these organisms grow exuberantly to form blooms. The blooms are considered harmful (harmful algal blooms, or HABs) when they pose risks to the local ecology, such as from oxygen depletion, nutrient depletion, or light deprivation. HABs may also be associated with adverse health effects in wildlife, domestic animals, and people through the physical factors mentioned previously or because they produce potent toxins. Although these microalgae and bloom events have existed for millions of years, there is evidence that HABs are occurring more frequently and more extensively than in the past (9).

Most of the marine algae (such as the dinoflagellates) exert their adverse health effects indirectly through bioaccumulation in the food web. However, other algae (e.g.,

cyanobacteria) are able to live in freshwater environments and exert their known health effects through direct exposures such as contaminated drinking water or aerosols. Although many of these exposures and their subsequent health effects are acute (i.e., onset of illness occurs within hours to days), as with the inorganic pollutants discussed earlier, there is some evidence that chronic exposures and long-term health effects may also be important. For example, Falconer et al. (10) found evidence of liver toxicity (i.e., elevation of plasma levels of liver enzymes) in human populations during times when their drinking-water sources had a highly concentrated *Microcystis aeruginosa* bloom. This organism produces microcystins, which are highly potent liver toxins.

A number of reviews of the impacts of marine and freshwater algal toxins on human health have been published (11–14). A summary of the known toxins, responsible organisms, and health effects associated with direct waterborne exposures is presented in Table 23.2 (Fig. 23.1).

23.2.3 Organic Matter

Many drinking-water supplies accumulate organic compounds, such as humic and fulvic acids, from decaying vegetation. Typically, these compounds are not associated with adverse health effects. However, when drinking water is disinfected with chlorine, the chlorine reacts with the organic matter present to produce disinfection by-products (DBPs), some of which are biologically active. Trihalomethanes (THMs) and haloacetic acids (HAAs) are two classes of DBPs that are associated with adverse health effects, including cancers (18–23) and adverse reproductive outcomes (e.g., (24–27)). Removing organic matter from water before disinfection or using an alternative disinfection method, such as UV light, can reduce by-product formation. In countries where the more expensive methods of disinfection are not realistic options, the immediate benefits of reducing infectious disease risk by disinfecting drinking water with chlorine greatly outweigh the more long-term risks from disinfection by-products.

23.3 NATURAL POLLUTANTS—MICROBES

In this section, we provide a brief overview of the different types of naturally occurring microbes in water, and we include specific examples based upon morphological and transmission-based categories (Table 23.3). Other chapters in this volume of this text provide detailed discussions of specific microbial contaminants such as *Ascaris lumbricoides*.

Basic categories of microbes include bacteria, viruses, yeasts, fungi, and parasites. Bacteria are capable of self-replication and are thus the simplest of the "living" microbes. They are unicellular (usually in the $2\,\mu m$ size range), have a cell wall composed of peptidoglycan, and

TABLE 23.2 Algal Toxins, Responsible Organisms, and Health Effects from Water-Borne Exposures

Toxin[a]	Organism	Exposure Route or Transvector	Health Effects
Marine[a]			
Azaspiracid	Unspecified dinoflagellate	Contaminated water, mollusks	Azaspiracid shellfish poisoning (AZP)
Brevetoxin	Karenia brevis (formerly Gymnodinium breve)	Contaminated water, mollusks, possibly fish	Neurotoxic shellfish poisoning (NSP), possibly neurotoxic fish poisoning, respiratory irritation, skin irritation
Ciguatoxin	Gambierdiscus toxicus	Contaminated fish	Ciguatera fish poisoning (CFP)
Domoic acid	Pseudo-nitzschia spp.	Contaminated water, mollusks, possibly fish	Amnesiac shellfish poisoning (ASP)
Okadaic acid	Dinophysis spp., Prorocentrum lima	Contaminated water, mollusks	Diarrheic shellfish poisoning (DSP)
Saxitoxin	Gymnodinium catenatum, Pyrodinium bahamense var. compressum, Alexandrium spp.; some Cyanobacteria (e.g. Aphanizomenon)	Contaminated water, mollusks, pufferfish	Paralytic shellfish poisoning (PSP)
Tetrodotoxin	Possibly bacteria	Possibly contaminated water, pufferfish	Fugu or pufferfish poisoning
Freshwater[b]			
Anatoxin-a	Anabaena spp., Oscillatoria spp., Ahpanizomenon spp., Cylindrospermopsis raciborskii	Contaminated water, possibly food	Progression of muscle paralysis, cyanosis, convulsions, death (animals); also opisthotonos ("s"-shaped neck) in birds
Anatoxin-a (s)	Anabaena flos-aquae	Contaminated water, possibly food	Hypersalivation, tremors, fasciculationas, ataxia, diarrhea, recumbency (pigs); regurgitation, opishotonos, seizures (ducks); lacrimation, hypersalivation, urination, defecation, death from respiratory arrest (mice); red-pigmented tears (rats)
Cylindrospermopsin	Cylindrospermopsis raciborskii	Contaminated water, possibly food	Huddling, anorexia, slight diarrhea, gasping respiration (mice); enlarged liver, malaise, anorexia, vomiting (humans)
Microcystins	Microcystis spp., Anabaena spp.	Contaminated water, possibly food, possibly inhalation	Weakness, reluctance to move, anorexia, pallor of extremities and mucous membranes (animals); survivors may be photosensitized (animals); embryo lethality, teratogenicity (rats)
Nodularin	Nodularia spumigena	Contaminated water, possibly food	Skin and eye irritation (human)
Saxitoxin	Aphanizomenon flos-aquae, Anabaena circinalis	Contaminated water, possibly food	Incoordination, recumbency, death by respiratory failure (animals); paresthesia and numbness of lips and mouth extending to face, neck, and extremities; motor weakness; incoordination, respiratory and muscular paralysis (humans)

[a] Adapted from Refs (13, 15, 16).
[b] Adapted from Ref. (17).

(a)

(b)

FIGURE 23.1 (a) A mysterious sea foam washed up along the coast of Ocean Shores, Washington in the fall of 2009. The photos taken at Chance A La Mer State Park following a violent coastal storm (11/2009) show the foam responsible for the death of thousands of seabirds between September and November (b). Experts believe a toxic algae, *Akashiwo sanguinea*, is responsible for removing the natural oil and waterproofing from the feathers of the seabirds, rendering them unable to protect themselves from the elements. (Photographs courtesy of Julie Hollenbeck.) (*See insert for color representation of this figure.*)

lack a membrane-bound nucleus. Bacteria possess nuclear material, but this material is not organized within a membrane (i.e., they are prokaryotic or "pre-nuclear"). Viruses are considered "nonliving" because they require a host cell for reproduction. Viruses are a fraction of a micron in size and consist of genetic material (either DNA or RNA) surrounded by a protein coat. Fungi and yeasts (single-celled fungi) are eukaryotic, characterized by a membrane-bound nucleus. A distinguishing feature of fungi and yeasts is that their cell walls are composed of chitin.

The term *parasites* is a medical term that covers two categories of higher forms of life (i.e., protozoans and helminthes) capable of causing disease. Protozoans are eukaryotes. They can be unicellular (5–25 μm in size) or multicellular, whereas helminthes ("worms") are multicellular, relatively complex organisms that are part of the animal kingdom. Transmission of helminthes can occur by ingesting eggs (which are approximately 25 μm in diameter), by ingesting the organism, or by making dermal contact with the organism.

TABLE 23.3 **Identified and Probable Organisms (Bacteria, Viruses, Yeasts, Fungi, Parasites) Present in Water that can Cause Adverse Human or Ecologic Health Impacts**

Organism	Exposure Route or Transvector	Health Effects	Citation
Bacteria			
Genus species			
• Species			
Aeromonas • *A. hydrophila* • *A. veronii* • *A. sobria* • *A. caviae*	Contaminated water, shellfish, seafood	Gastroenteritis (at risk immunocompromised)	(28–30)
Bacillus cereus	Contaminated shellfish, water	Gastroenteritis	(31, 32)
Clostridium botulinum (Botulism Toxin E)	Contaminated water, fish	Botulism	(33–35)
Campylobacter	Contaminated water, mollusks	Gastroenteritis	(30, 36–39)
Edwardsiella tarda	Contaminated water, shellfish	Gastroenteritis	(40–42)
Escherichia coli • including enterotoxigenic	Contaminated irrigation and recreational water, shellfish, seafood	Gastroenteritis	(30, 43–45)
Helicobacter pylori	Possibly contaminated water, possibly food (transmission route not well understood)	Chronic gastritis, ulcers, gastric cancer	(46)
Legionella	Contaminated water (e.g., air-conditioning units)	Legionellosis (acute respiratory illness)	(30, 38, 47, 48)
Leptospira	Contaminated water	Leptospirosis	(49, 50)
Listeria monocytogenes	Contaminated water, seafood	Listeriosis	(51)
Non-tuberculosis *Mycobacterium* • *M. kansasii* • *M. marinum* • *M. xenopi* • *M. avium* (various other species have been identified in water)	Swimming pools and aquarium water, contaminated drinking water	Skin lesions particularly in immunocompromised, respiratory illness	(52, 53)
Pseudomonas aeruginosa	Drinking water	Infant mortality	(38, 47, 48);
Salmonella enterica • serotype typhi • serotype paratyphi	Contaminated irrigation water, mollusks	Septicemia, typhoid fever	(36, 54, 55)
Shigella	Contaminated water, mollusks	Gastroenteritis (bacillary dysentery)	(30, 38)
Staphylococcus aureus • Methicillin resistant (MRSA) • Nonmethicillin resistant (MSSA)	Marine water, sand reservoir	Skin infections, possibly respiratory infections	(34, 56, 57)
Vibrio sp. • *V. cholera* • *V. parahaemolyticus* • *V. mimicus* • *V. hollisae* • *V. fluvial* • *V. vulnificus*	Contaminated marine water, mollusks, crustaceans, fish	Gastroenteritis, septicemia (at risk immunocompromised, liver disease), gangrene (*vulnificus*)	(58–62)

TABLE 23.3   (*Continued*)

Organism	Exposure Route or Transvector	Health Effects	Citation
Viruses			
Family			
• *Genus*			
○ *species*			
Adenoviridae	Contaminated water, aerosolized droplets	Gastroenteritis	(63)
• *Adenovirus*			
Astroviridae	Contaminated water, shellfish	Gastroenteritis	(64, 65)
• *Astrovirus*			
Caliciviridae			(36, 64, 66);
○ Small round structured viruses (Norwalk, Cockle, Snow Mountain)	Contaminated water, shellfish	Gastroenteritis	(47, 48, 67–70);
○ Hepatitis E	Possibly contaminated water, possibly shellfish	Hepatitis	
Coronavirus			
○ SARS	Contaminated water (aerosol) (duck feces into pond water aerosolized droplets, fomites)	SARS (severe acute respiratory syndrome)	(71)
○ Influenza A	Contaminated water, shellfish	Gastroenteritis	(72)
Parvoviridae			
○ Parvovirus			(73)
Picornaviridae			
• *Enteroviruses*			
○ *echovirus*	Contaminated water	Gastroenteritis	(70, 74);
○ *poliovirus*	Contaminated water	Polio	(47, 48)
○ *cocksackie*	Contaminated water	Gastroenteritis	
• *Hepatovirus*			
○ Hepatitis A	Contaminated water, mollusks	Hepatitis	(65, 74)
Poliomaviridae	Contaminated water	Polyomavirus infection	(75, 76)
• *Polyomavirus*			
○ JC virus			
○ BK virus			
Reoviridae	Contaminated water, shellfish	Gastroenteritis	(64, 77)
○ Rotavirus			
Yeasts			
Candida tropicalis	Beach sand, ?marine water (limited epidemiologic evidence of transmission)	Possibly skin infections	(78)
Rhodotorula mucilaginosa	Beach sand, marine water (limited epidemiologic evidence of transmission)	Possibly skin infections	(78, 79)
Fungi			
Microsporidia (recently reclassified from parasite to fungi (see article))	Contaminated water	Microsporidiosis	(80)
Parasites			
Cryptosporidium parvum	Contaminated irrigation, marine water	Gastroenteritis	(81, 82)
Cyclospora cayetanensis	Contaminated water	Gastroenteritis	(82)
Entamoeba histolytica	Contaminated water, food	Amoebiasis, liver abscess	(47, 48, 82)
Fasciola hepatica	Contaminated food, water	Fascioliasis	(81)
Giardia lambia	Contaminated water, food	Giardiasis	(81)

(*Continued*)

TABLE 23.3   *(Continued)*

Organism	Exposure Route or Transvector	Health Effects	Citation
Naegleria fowleri	Inhalation of contaminated water	Primary amoebic meningoencephalitis (PAM or PAME) (fatal)	(83, 84)
Toxoplasma gondii	Contaminated water, food	Toxoplasmosis	(81, 85)
Nemathelminthes • Nematodes			
Anisakis spp.	Possibly contaminated water, shellfish, fish (raw)	Abdominal discomfort, eosinophilia, allergy	(86)
Ascaris lumbricoides (human roundworm)	Contaminated water, food	Ascariasis	(87)
Dracunculus medinensis (guinea worm)	Contaminated fresh water	Dracunculiasis	(87)
Enterobius vermicularis (hookworm)	Contaminated water, food	Enterobiasis	(88)
Eustrongylides sp.	Possibly contaminated water, shellfish, fish (raw)	Peritonitis	(89, 90)
Gnathostoma sp.	Possibly contaminated water, shellfish, fish	Abdominal discomfort, eosinophilia, allergy, eosinophilic meningitis	(81, 87)
Platyhelminthes • Trematoda (flukes)			
Fasciolopsis buski (giant intestinal fluke)	Contaminated water	Fasciolopsiasis	(87, 91)
Heterophyes heterophyes (intestinal fluke)	Contaminated water, shellfish, fish (raw)	Gastroenteritis	(81)
Nanophyetus salmincola	Contaminated water, shellfish, fish (raw)	Gastroenteritis, eosinophilia, "salmon-poisoning disease"	(92)
Schistosoma sp. (blood flukes)	Contaminated water	Schistosomiasis, bladder cancer	(87)
• Cestoda (tapeworms)			
Diphyllobothrium (broad fish form)	Possibly contaminated water, shellfish, fish (raw)	Gastroenteritis, anemia (B_{12}), eosinophilia	(93, 94)
Echinococcus granulosus (tapeworm)	Contaminated water	Echinococcosis (hydatid disease)	(87)
Hymenolepis nana (dwarf tape worm)	Contaminated water, food	Hymenolepiasis	(88)
Taenia spp. (tape worm)	Contaminated water	Taeniasis	(81)

Water-related routes of exposure to microbes include contacting or drinking contaminated water and inhaling contaminated aerosols. Transmission of disease through water contact includes situations where the microbe enters the body through an open wound (e.g., *Vibrio vulnificus* and *Staphylococcus aureus*) or where the developmental phase found in water penetrates into the skin (e.g., *Schistosoma* spp.). *Legionella* spp. can be found in aerosols released from air-conditioning cooling towers. The contaminated aerosols can subsequently enter nearby air-conditioning intake systems and can be dispersed throughout a building, such as a hospital or hotel, providing for an ideal means of transmitting the microbe to susceptible individuals.

One critically important water-related route of exposure is fecal–oral. Microbes transmitted this way are typically released into the environment in the feces of an infected person or animal. The microbes contaminate water, food, and fomites and are subsequently ingested by another individual. Classic examples of bacteria transmitted through fecal–oral routes include *Salmonella* spp. and *Shigella* spp. Viruses that can be transmitted via the fecal–oral route include hepatitis A and enterovirus. Common protozoa that can be transmitted by the fecal–oral route include *Cryptosporidium* spp. and *Giardia* spp. Both of these organisms produce highly resistant oocysts or cysts that survive for a long time in the environment. The oocyst and cyst forms have proven to be an exceptional challenge for drinking-water disinfection because they are not inactivated by chlorine at levels found in traditional drinking water treatment systems. Commonly used filtration technologies can effectively remove *Giardia* cysts from water; however, not all of these technologies can remove protozoa such as *Cryptosporidium*, and care must be exercised when one is selecting filter membranes (95).

The last group of microbes, helminthes, also poses a health risk through the fecal–oral route. Ingested helminth eggs are particularly dangerous because the subsequent larval-stage helminthes disseminate readily throughout the human host. By contrast, the ingestion of mature or larval helminthes (e.g., *Taenia* spp.) in contaminated meat results in disease localized to the gastrointestinal system.

In addition to the direct exposures to contaminated water mentioned previously, insects with water-bound life stages may be vectors for microbial diseases. Mosquitoes are the classic example of an insect vector whose development includes a water-based larval stage. However, water is not the means for transmitting the infectious agent; thus, discussion of mosquito-borne diseases such as malaria, dengue fever (both discussed in other chapters in this volume), and encephalitis is not included here.

Table 23.3 provides a summary of microbial water pollutants. It lists the organism, possible human exposure routes, health effects, and relevant citations.

23.3.1 Mixtures

Traditionally, exposures to naturally occurring contaminants in water and their impacts on health have been considered and evaluated one contaminant at a time. This method reflects the standard approaches of the associated scientific disciplines (e.g., toxicology and pharmacology) and the environmental contaminant regulatory process. However, in reality, naturally occurring contaminants are typically mixtures of chemicals and microbes in a highly variable environmental milieu. Components of these mixtures may interact antagonistically or synergistically, and an understanding at least that these interactions occur, if not their complete characterization, is critical to understanding, predicting, and preventing exposures and subsequent health effects.

One of the challenges to researchers, to public health, and to regulatory communities is to find tools and approaches to explore natural contaminant mixtures in water. An example of a contaminant mixture is leachate from a common wood preservative called CCA (chromate, copper, arsenate). Although CCA is released very slowly from the preserved wood, measureable amounts of chromium, copper, and arsenic can be detected in the environments where these materials are used (96, 97). Synergistic effects of the metals in CCA have been evaluated by dosing human nerve cells

with a mixture of chromium, copper, and arsenic and also dosing nerve cells with each of the metals independently. Results show that mixtures were twice as toxic as arsenic alone (Lynne Fieber, University of Miami, personal communication). In addition to the toxicity of the chemicals themselves, local environmental chemistry may also impact the ultimate health risk. For example, juvenile killifish (*Fundulus heteroclitus*) exposed to CCA leachates were found to be sensitive to copper, and the toxicity of a given concentration of copper was strongly related to the salinity of the water in which the fish were kept (Martin Grosell, University of Miami, personal communication). Thus, not only the synergistic effects of chemical mixtures but also the interactions with the local chemical environment should be considered when one is evaluating potential human and ecological health risks from environmental contaminants.

23.3.2 Physical Forcers

In this section, we provide a brief overview of how physical forcers—in particular the built environment, agricultural activities, and climate change—might affect public health consequence associated with exposure to naturally occurring water-borne contaminants.

23.3.2.1 The Built Environment Levels of many naturally occurring water pollutants increase as the environment becomes more developed. People, animals, and human activities may all contribute to environmental contaminant loads.

As discussed earlier, people and companion animals are an added source of some contaminants that naturally occur in the environment, such as infectious microorganisms. Thus, community sanitary infrastructure plays a critical role in minimizing the adverse effects associated with concentrating these pollutants in limited geographic areas.

To protect drinking-water sources and reduce contamination, most urban community infrastructure includes facilities to treat drinking water and to collect and treat wastewater before discharging it into the environment. However, over time, overburdening and deterioration of this infrastructure, coupled with urban runoff, have resulted in increased microbial contamination in all aquatic environments (98). This microbial pollution certainly consists of human and animal pathogens (e.g., bacteria, parasites, and viruses). It also represents a large nutrient load into aquatic environments, a load that may, in turn, encourage the growth and spread of other natural contaminants, such as the HAB organisms. This is another example of how complex mixtures of pollutants in the environment can generate different types and levels of health risks.

Intensive research on microbial pollution in aquatic environments has revealed that a huge number of previously unidentified organisms live naturally in aquatic environments (99). Some of these aquatic microbes are likely to be pathogenic to humans through direct skin contact or contamination of drinking water, food, and aerosols. For example, *Vibrio cholerae*, the organism responsible for cholera, is endemic in coastal waters around the globe (100). Its ability to survive may be enhanced by the simultaneous presence of concentrated algal blooms (101); thus, its public health impact may expand as HABs expand. Another example is the ability of pathogens to cross host species. For example, toxoplasmosis can be transmitted from cats to people through contact with contaminated feces. Recently, scientists have found that this organism can be transmitted through runoff to the oceans, subsequently infecting coastal sea lion populations (102).

In addition to the contributions to water pollution from people and their pets, human activities associated with producing the goods and services for dense urban communities exacerbate naturally occurring water pollution. Ore and gravel mining, fossil fuel extraction, and large-scale food production on land and water add to existing loads of heavy metals, petroleum-derived chemicals, nutrients, and infectious microorganisms.

In developed nations where wastewater is treated prior to discharge, rainwater or "urban" runoff is now recognized as an acute threat to the environment (4). Any contaminants (ranging from microbes to fertilizers and solvents) present in the soil or on impervious surfaces can be entrained in runoff and deposited in local water systems. Historically, communities have had only the capacity to collect and treat the "first flush" of runoff. More recently, communities have implemented systems to treat rainwater runoff through more elaborate technologies, including rain gardens, rain galleries, and disinfection (103, 104).

In addition to runoff, water collection and large-scale storage practices may affect concentrations of naturally occurring pollutants in drinking water. For example, in Hawaii, rainwater is harvested from household rooftops and stored for later use. Periodic volcanic eruptions produce airborne ash and aerosols called volcanic smog (VOG) that contain SO_2 and other contaminants (105). There is concern that the contaminants in ash and VOG can precipitate onto water-catchment surfaces, such as rooftops, and contaminated water could be collected for household use (106).

Dams built to control downstream water flow or to generate hydroelectric power may also concentrate naturally occurring pollutants. For example, there are two dams on the Klamath River in northern California. There is interest in removing the dams to restore historic water flow and to provide access to upstream waters to spawning salmon. However, recent testing has found high levels of pollutants, including arsenic, in the silt behind the dams, creating concerns that removing the dams will allow high levels of pollutants to impact downstream environments (107).

23.3.2.2 Agriculture

Agricultural practices can impact not only local water quality but also the environment and food products downstream from the pollution source. For example, contaminated irrigation is potentially a microbial contamination source for foods normally eaten raw, such as carrots and onions (108) and spinach (109). This potential may have been realized in an *Escherichia coli* O157:H7 outbreak associated with spinach in which 130 persons became ill, 31 persons developed hemolytic uremic syndrome, and 5 died, including 1 child (110). Jay et al. (111) identified the source of contamination as feral swine feces and suggested that *E. coli* O157:H7 could have been introduced into the spinach fields on harvesting equipment or in irrigation water.

Food animal production is another potential source of water pollution. Concentrated animal feeding operations (CAFOs) have the potential to release chemicals that can augment existing concentrations of naturally occurring environmental pollutants, including nitrogen and phosphorous as well as microorganisms (112, 113). Many CAFOs use liquid waste management systems, such as plastic-lined lagoons, which are susceptible to failure and subsequent offsite discharge, even during normal operating conditions. The public health threat can be greatly exacerbated during extreme flooding events. For example, in North Carolina in September, 1999, 39–50 cm of rain fell on eastern North Carolina in association with Hurricane Floyd (114). The satellite-based area of flooding contained 241 CAFO operations and impacted a population of approximately 170,000 persons. A large proportion of the affected population depended on groundwater for drinking water, and many of the wells were contaminated by flood waters. In addition to the direct impacts of CAFO manure pit releases to the environment, there are indirect impacts associated with runoff and subsequent environmental contamination by nutrients, veterinary pharmaceuticals, and pathogens when CAFO waste is used as fertilizer (115).

23.3.2.3 Climate Change

The predicted increases in coastal flooding associated with sea level rise and heavy rainfall events associated with regional climate changes will increase exposures to naturally occurring waterborne pollutants and the incidence of subsequent health impacts. As we learned during the flooding associated with Hurricane Katrina, deaths due to drowning and trauma comprise only the initial threats of a severe weather event. Subsequent flooding increases human exposures to the range of natural pollutants (including mixtures) described previously, resulting in health effects ranging from life-threatening *V. vulnificus* skin infections to increased risks for starvation, dehydration, and diarrheal diseases from ingesting contaminated water and food. As flood waters recede, natural pollutants remain not only in drinking water supplies, but also in the soil and plants, raising the prospect of future exposures from contaminated dusts and food chain contamination (116).

23.3.3 Response and Adaptation

In the previous sections, we briefly described some of the important naturally occurring water pollutants. Some pollutants are important as individual contaminants, and others may prove more dangerous when present in mixtures. We can expect ever-increasing amounts of naturally occurring pollutants to intermingle in our environment. The presence of pharmaceuticals (e.g., antibiotics) in ambient waters increases the risk of antimicrobial resistance in naturally occurring waterborne pathogens. Nutrients in runoff and weather changes that increase sea surface temperatures may support naturally occurring microbial communities that include pathogens and toxin-producing algae. Increasing interactions among aging or inadequate infrastructures, sea level rise, and discharges that create these complex mixtures may produce the "perfect storm" of public health threats. In the following sections, we outline some measures that could be taken to directly address naturally occurring water pollutants. These include reassessing our regulatory approach, protecting global water resources, viewing water resource protection with a global perspective, and creating public health capacity to champion these issues.

23.3.3.1 Regulating Water Quality

One important response to these threats involves reassessing how we view the environment. Currently, the environment is viewed and utilized as a repository for wastes generated by human activities, from basic sanitation to food production to nano particle manufacturing. Removing existing contaminants from water resources is an important component of protecting both public health and environmental health. This is demonstrated in the extensive regulations to limit the concentrations of contaminants in drinking and recreational waters, and it is supported by advances in engineering and chemical processes to improve drinking-water treatment.

As we have learned more about the acute and chronic health effects from exposures to contaminants that may occur naturally in water, the permissible concentrations of contaminants have decreased. While development of new technologies for contaminant removal may be keeping pace with enhanced regulations, many of these new technologies are not affordable by small systems or individuals not connected to public drinking-water systems. For example, in 2001, the U.S. Environmental Protection Agency (EPA) promulgated a rule to decrease the concentration of arsenic allowed in finished drinking water from 50 μg/L to 10 μg/L (http://www.epa.gov/safewater/arsenic/basicinformation.html#eight). The EPA estimated that the average annual cost per household to meet this new limit was from USD 38 to USD 327 (http://www.epa.gov/safewater/arsenic/funding.html). Because of the cost and the

difficulties associated with incorporating new arsenic removal technology into existing treatment processes, small systems serving up to 3300 people were allowed up to 14 years from the date of the ruling to comply (http://www.epa.gov/safewater/arsenic/funding.html).

Even aside from potentially burdensome regulatory requirements to protect drinking-water quality in the United States, the number of times public water systems incur violations of regulatory requirements appears to be increasing. Public water systems are systems that have at least 15 service connections or that regularly serve an average of at least 25 individuals daily for at least 60 days of the year (http://www.epa.gov/safewater/wsg/wsg_66a.pdf). For example, in 2000, there were 48,699 community drinking-water systems with reported violations, and nearly 68 million people were impacted. More than 10,000 of these violations involved exceeding maximum contaminant levels (MCLs). MCLs represent a concentration that incorporates existing information about health effects associated with exposure to a specific contaminant and the best practices needed to control concentrations of that contaminant in finished drinking water. In 2009, there were 49,472 community drinking-water systems with violations, 16,993 involving MCLs that impacted nearly 84 million people (http://www.epa.gov/safewater/databases/pivottables.html). Failure to enforce these standards adds to the underlying disease burden by putting large numbers of people at risk, probably without their knowledge, for exposure to unacceptable concentrations of drinking-water contaminants. The problems are exacerbated when drinking water is obtained from sources, such as private wells, that may not meet regulatory levels as specified by the Safe Drinking Water Act.

Several public health and environmental issues are contained within the current approach to regulate drinking-water pollutants, whether they are naturally occurring or created through anthropogenic activities. First, the single contaminant approach is unwieldy in both the level of scientific certainty needed to establish causality (and thus, the need to regulate) and the time needed to promulgate new legislation once the scientific evidence has been obtained and validated. Currently, over 100 chemicals and 12 microbes are on EPA's Contaminant Candidate List, a list of chemicals found in water that may be considered for future regulation (http://www.epa.gov/safewater/ccl/ccl3.html#chemical). It is clear that the current process simply cannot keep up with the numbers of new chemicals being released into the environment each year. Second, as discussed earlier, the single-chemical approach does not address the reality of chemical mixtures in aquatic environments. There is currently no agreement on alternative approaches, such as requiring proof that a new chemical does not cause biological harm before it is discharged into shared water resources, or the approach that uses statistical methods and chemical structure analysis to assess which chemicals are likely to have adverse biological effects and should thus not be released into the environment. One approach would be to regulate chemicals on the production end, rather than waiting until they are observed in the environment. This approach would require environmental impact assessments of new chemicals prior to their distribution to consumers.

Another public health issue, demonstrated by the earlier discussion of arsenic, is the disparity in access to safe drinking water even in developed nations. Urban centers with large populations can support sophisticated drinking-water treatment to remove at least the regulated contaminants. The chemical and physical processes used by large drinking-water utilities also remove additional contaminants, thus protecting their customers from exposures of unknown but potentially hazardous agents. By contrast, smaller communities or homeowners with private wells may not have the resources required to protect themselves from these health threats, whether regulated or not. One way to address this is by incorporating drinking-water source protection into land use planning and development, thereby encouraging or requiring existing communities to be responsible for providing safe drinking and, if appropriate, recreational waters. For example, land use has a large impact on water flow, runoff, and sediment deposition; protecting pristine landscapes with naturally occurring flora or even creating densely vegetated landscapes in urban areas can limit the amount of surface contaminants that find their way into our water resources. Sanitary surveys should be conducted periodically within the watersheds contributing to drinking-water sources, and regulations should be implemented to minimize the contamination of these watersheds from domestic, commercial, and industrial wastes. An economic analysis of the ability of a new community or project to sustainably provide safe resources for all of its members would be a useful first step in addressing disparities. Existing communities should be encouraged to use such resources as the EPA revolving fund (http://www.epa.gov/safewater/dwsrf/) to meet their obligations to provide safe water resources to their community members.

23.3.3.2 Water Resource Protection from a Global Perspective

We must put water pollution into the overall context of how global changes impact water quality and public health. Climate change cannot be discussed without considering drinking-water sources, physical and public health workforce infrastructure, sanitation, mixing of marine waters with fresh waters, and our seafood supply, all of which are affected by the naturally occurring pollutants we have discussed in this chapter. A realistic approach to assessing overall water resources must account for not only regional issues, but also global considerations.

One component of a global perspective is assessing water resources for coastal communities. Many coastal communities ironically face water shortages as groundwater extraction and sea level rise combine to increase the amount of salt water intrusion into local aquifers. Desalination is a costly process that is not likely to be an economically sustainable option for either large or small coastal communities. Alternative approaches to protecting existing freshwater resources must be incorporated into our daily lives, particularly conservation measures.

An indirect consequence of failure to adequately protect global water resources is contaminated food, even in the highly developed countries. Poor water quality in nations with lax environmental protection laws or lax enforcement of existing laws increases the risk that water-borne contaminants of all kinds will be incorporated into foods. For example, much of the seafood consumed in developed nations is harvested by developing nations. These countries increasingly produce seafood in aquaculture facilities that may be contaminated by locally generated water pollutants, including industrial organic compounds and metals, untreated or undertreated sewage (including enteric pathogens), and agricultural chemicals. In fact, this seafood has been found to be contaminated with a myriad of both naturally occurring and man-made pollutants, including those (e.g., dichlorodiphenyltrichloroethane [DDT], polychlorinated biphenyls [PCBs], etc.) no longer in use or heavily regulated in developed countries (117).

In addition to specific physical events, we should apply the global perspective to the relationship between health and the water environment by being receptive to cues from people and other living organisms. Disease surveillance activities should include enhancing our capacity to observe trends and interpret and respond to them to adequately protect public health. For example, a unique surveillance system, the Harmful Algal Bloom-related Illness Surveillance System (HABISS) (118) collects data on illnesses in people and animals, and also data on the environmental characteristics that describe the relevant algal blooms. Data from this surveillance activity will allow public health officials to predict the geographic locations of future blooms, thus allowing them to proactively conduct public health protection measures. Incorporating a response activity into other disease surveillance activities would improve our ability to respond rapidly to changing events.

Another cue for emerging environmental public health issues can come from other animals. For example, the One Health Initiative was created to forge collaborations between physicians, veterinarians, and other scientific/health-related disciplines to enhance all aspects of medicine and health, including public health (http://www.onehealthinitiative.com/) (119). Companion animals can serve as sentinels for diseases with an environmental etiology (14, 120–122). Wild animals, particularly aquatic animals and birds, can serve as sensitive indicators that the environment is out of balance. While we may not be able to directly relate disease in an animal population to the equivalent illness in people, we can certainly use the information to assess whether it makes sense to expect some adverse health effect in people living in the same environment. For example, high levels of heavy metals and other trace elements were found in dolphin populations residing along the eastern coast of the United States (123). Many of these elements have an anthropogenic source and accumulate in the fish comprising the dolphins' diet. People consuming fish from the same geographic area may also be at risk for accumulating unhealthy body burdens of these chemicals.

In addition to endorsing a cross-species approach to protecting public health from diseases associated with naturally occurring water pollutants, we should reconsider how we use our other global resources and how those activities might ultimately impact the sustainable availability of healthy water. For example, all forms of energy generation have an associated environmental and public health cost. The most obvious example is burning coal to produce electricity. Coal acquisition and burning adds many tons of naturally occurring heavy metals and metalloids, including mercury, into our terrestrial and aquatic environments each year, causing a reversal of some of the environmental improvements we have accomplished in the past. Designing buildings (including homes) to take advantage of regional conditions to capture sunlight and wind for heat and power would allow us to meet our energy needs while reducing the amount of mercury and other pollutants released into the air, and from there, into our water.

23.3.3.3 Environmental Public Health Infrastructure

In the United States, the environmental public health infrastructure has become dangerously understaffed and overcommitted (124), and the fact that environmental public health has a critical role in developing and maintaining quality water resources suggests an emerging crisis in public health protection. The number of environmental public health scientists and engineers graduating from the current public health and environmental engineering education system is insufficient to address current needs (125). We need to support training and workforce development to increase the number of public health professionals with the skills and knowledge to prevent and address public health emergencies.

The physical component of our environmental health infrastructure must also be addressed. For example, current drinking-water distribution systems and sanitary sewer systems in many U.S. cities were installed over 100 years ago. In some cases, particularly in older cities, sanitary sewers were intentionally "combined" with storm-water collection systems. During storm conditions, the combined sewer and storm-water flows often exceeded the capacity of wastewater treatment facilities, and the overflow was discharged directly into nearby waterways. While this design was acceptable at the time, the discharge of untreated sewage into the environment is no longer an acceptable practice. Damaged and leaking pipes, usually occurring in poorly maintained old distribution systems, result in the loss of an estimated 1 trillion gallons of treated drinking water each year, and they provide an opportunity for cross-contamination with sewage and runoff (http://www.epa.gov/awi/basic1.html). The public health community should support efforts to commit the estimated USD 388 billion in capital needs through 2019 for clean water in the United States alone (126).

23.4 CONCLUSION

There are many naturally occurring threats to the health of our water resources, and our response to these threats should include regulations, protecting water resources, initiating strong measures to protect global and local water resources, improvements to the sanitary infrastructure, and creating public health capacity to address these issues. Committing ourselves to the long view is critical to our success in adapting to our changing world. We need to improve our ability to predict the next environmental challenge, develop the necessary infrastructure to meet that challenge, and have our public health response plan in place before it happens.

REFERENCES

1. DeSimone LA, Hamilton PA, Gilliom RJ. The quality of our nation's waters—Quality of water from domestic wells in principal aquifers of the United States, 1991–2004—Overview of major findings. U.S. Geological Survey Circular 1332, 2009, 48 p.

2. De Burbure, C, Buchet, JP, Leryoer A, Misse C, Haguenoer, JM, Mutti A, Smerhovsky Z, Cikrt M, Trzcinka-Ochocka M, Razniewska G, Jakubowski M, Bernard A. Renal and neurologic effects of cadmium, lead, mercury and arsenic in children: evidence of early effects and multiple interactions at environmental exposure levels. *Environ Health Perspect* 2006;114:584–590.

3. Jarup L. Hazards of heavy metal contamination. *Br Med Bull* 2003;68:167–182.

4. U.S. Environmental Protection Agency (US EPA). Report of the Experts Scientific Workshop on Critical Research Needs for the Development of New or Revised Recreational Water Quality Criteria. Washington, DC, 2007. EPA 823-R-07-006.

5. WHO. WHO #6 Arsenicosis Case-Detection, Management and Surveillance, Report of a Regional Consultation New Delhi, India, November 5–9, 2002. 2008. Available at http://www.searo.who.int/LinkFiles/Arsenic_Mitigation_Arsenicosis_case-detection_management_and_surveillance.pdf. Accessed February 19, 2010.

6. Chowdhury UK, Biswas BK, Chowdhury TR. Groundwater arsenic contamination in Bangladesh and West Bengal, India. *Environ Health Perspect* 2000;108(4):393–397.

7. Navas-Acien A, Silbergeld EK, Pastor-Barriuso R, Guallar E. Arsenic exposure and prevalence of Type 2 diabetes in U.S. adults. *JAMA* 2009;300(7):814–822.

8. Schwartz GG, Il'yasova D, Ivanova A. Urinary cadmium, impaired fasting glucose, and diabetes in the NHANES III. *Diabetes Care* 2003;26(2):468–470.

9. Van Dolah F. Marine algal toxins: origins, health effects, and their increased occurrence. *Environ Health Perspect* 2000;108 (Suppl 1):133–141.

10. Falconer IR, Beresford AM, Runnegar MTC. Evidence of liver damage by toxin from a bloom of the blue-green algae, Microcystis aeruginosa. *Med J Aust* 1983;1:511–514.

11. Fleming LE, Backer L, Rowan A. The epidemiology of human illnesses associated with harmful algal blooms. In: Baden D, Adams D, editors. *Neurotoxicology Handbook*, Vol. 1. Totowa, NJ: Humana Press Inc, 2002, pp. 363–381.

12. Backer LC, Fleming LE, Rowan A, Cheng, Y-S, Benson J, Pierce RH, Zaias J, Bean J, Bossart GD, Johnson D, Quimbo R, Baden DG. Recreational exposure to aerosolized brevetoxins during Florida Red Tide events. *Harmful Algae* 2003;2:19–28.

13. Backer LC, Rogers HS, Fleming LE, Kirkpatrick B, Benson J. Phycotoxins in marine seafood. In: Dabrowski W, Sikorski ZE, editors. *Toxins in Food*. Boca Raton, FL: CRC Press, 2005, Chapter 7.

14. Zaias J, Backer LC, Fleming LE. Harmful algal blooms (HABs) In: Rabinowitz P, Conti L, editors. *Human–Animal Medicine: A Clinical Guide to Toxins, Zoonoses, and Other Shared Health Risks*. New York: Elsevier Science Publishers, 2010, pp. 91–104.

15. Backer L, Fleming LE, Rowan A, Baden D. Epidemiology and public health of human illnesses associated with harmful marine phytoplankton. In: Hallegraeff GM, Anderson DM, Cembella AD, editors. *UNESCO Manual on Harmful Marine Algae*. Geneva, Switzerland: UNESCO/WHO, 2003, pp. 725–750.

16. Fleming LE, Backer LC, Baden DG. Overview of aerosolized Florida Red Tide toxins: exposures and effects. *Environ Health Perspect* 2005;113(5):618–620.

17. Backer LC. Cyanobacterial harmful algal blooms (Cyano-HABs): developing a public health response. *Lake Reservoir Manage* 2002;18(1):20–31.

18. King WD, Marrett LD. Case-control study of bladder cancer and chlorination by-products in treated water (Ontario, Canada). *Cancer Causes Control* 1996;7:596–604.

19. Freedman DM, Cantor KP, Lee NL, Chen LS, Hei HH, Ruhl CE, et al. Bladder cancer and drinking water: a population-based case-control study in Washington County, Maryland (United States). *Cancer Causes Control* 1997;8:738–744.

20. Cantor KP, Lynch CF, Hildesheim ME, Dosemeci M, Lubin J, Alavanja M, et al. Drinking water source and chlorination byproducts I. Risk of bladder cancer. *Epidemiology* 1998;9:21–28.

21. Hildesheim ME, Cantor KP, Lynch CF, Dosemeci M, Lubin J, Alavanja M, et al. Drinking water source and chlorination by-products. II. Risk of colon and rectal cancers. *Epidemiology* 1998;9:29–35.

22. McGeehin M, Reif J, Becher J, Mangione E. A case-control study of bladder cancer and water disinfection methods in Colorado. *Am J Epidemiol* 1993;138:492–501.

23. Villanueva CM, Cantor KP, Grimalt JO, Malats N, Silverman D, Tardon A, et al. Bladder cancer and exposure to water disinfection by-products through ingestion, bathing, showering and swimming pool attendance. *Am J Epidemiol* 2006;165:148–156.

24. King WD, Dodds L, Allen AC. Relation between stillbirth and specific chlorination by-products in public water supplies. *Environ Health Perspect* 2000;108:883–886.

25. Savitz DA, Andrews KW, Pastore LM. Drinking water and pregnancy outcome in central North Carolina: source, amount, and trihalomethane levels. *Environ Health Perspect* 1995;103:592–596.

26. Savitz DA, Singer PC, Herring AH, Hartmann KE, Weinberg HS, Makarushka C. Exposure to drinking water disinfection by-products and pregnancy loss. *Am J Epidemiol* 2006;164:1043–1051.

27. Waller K, Swan SH, DeLorenze G, Hopkins B. Trihalomethanes in drinking water and spontaneous abortion. *Epidemiology* 1998;9:134–140.

28. Rusin PA, Rose JB, Haas CN, Gerba CP. Risk assessment of opportunistic bacterial pathogens in drinking water. *Rev Environ Contam Toxicol* 1997;152:57–83.

29. Albert MJ, Ansaruzzaman M, Talukder KA, Chopra AK, Kuhn I, et al. Prevalence of enterotoxin genes in Aeromonas spp. isolated from children with diarrhea, healthy controls, and the environment. *J Clin Microbiol* 2000;38: 3785–3790.

30. Ashbolt NJ. Microbial contamination of drinking water and disease outcomes in developing regions. *Toxicology* 2004;198:229–238.

31. Akharaiyi FC, Adebolu TT, Abiagom MC. A comparison of rainwater in Ondo State, Nigeria to FME approved drinking water quality standard. *Res J Microbiol* 2007;2:807–815.

32. Suthar S, Chhimpa V, Singh S. Bacterial contamination in drinking water: a case study in rural areas of northern Rajasthan, India. *Environ Monit Assess* 2009;159:43–50.

33. Eastaugh J, Shepherd S. Infectious and toxic syndromes from fish and shellfish consumption: a review. *Arch Intern Med* 1989;149:1735–1740.

34. Saavedra-Delgado AM, Metcalfe DD. Seafood toxins. *Clin Rev Allergy* 1993;11:241–260.

35. Long SC, Tauscher T. Watershed issues associated with *Clostridium botulinum*: a literature review. *J Water Health* 2006;4:277–288.

36. Rheinstein PH, Klontz KC. Shellfish-borne illnesses. *American Family Physician* 1993;47(8):1837–1840.

37. Savill MG, Hudson JA, Ball A, Klena JD, Scholes P, et al. Enumeration of *Campylobacter* in New Zealand recreational and drinking waters. *J Appl Microbiol* 2001;91:38–46.

38. Leclerc H, Schwartzbrod L, Dei-Cas E. Microbial agents associated with waterborne diseases. *Crit Rev Microbiol* 2002;28:371–409.

39. Devane ML, Nicol C, Ball A, Klena JD, Scholes P, et al. The occurrence of *Campylobacter* subtypes in environmental reservoirs and potential transmission routes. *J Appl Microbiol* 2005;98:980–990.

40. Van Damme LRV, Vandepitte J. Frequent isolation of *Edwardsiella tarda* and *Plesiomonas shigelloides* from healthy Zairese freshwater fish: a possible source of sporadic diarrhea in the tropics. *Appl Environ Microbiol* 1980;39: 475–479.

41. White FH, Simpson CF, Williams LE. Isolation of *Edwardsiella tarda* from aquatic animal species and surface waters in Florida. *J Wildlife Dis* 1973;9:204–208.

42. Hore C. Important unusual infections in Australia: a critical care perspective. *Crit Care Resusc* 2001;3:262–272.

43. Hughes J, Merson M, Gangarosa E. The safety of eating shellfish. JAMA: the journal of the American Medical Association 1977;237(18):1980.

44. Ayulo AM, Machado RA, Scussel VM. Enterotoxigenic *Escherichia coli* and *Staphylococcus aureus* in fish and seafood from the southern region of Brazil. *Int J Food Microbiol* 1994;24:171–178.

45. Rangel JM, Sparling PH, Crowe C, Griffin PM, Swerdlow DL. Epidemiology of *Escherichia coli* O157:H7 outbreaks, United States, 1982–2002. *Emerg Infect Dis* 2005;11:603–609.

46. Bellack NR, Koehoorn MW, MacNab YC, Morshed MG. A conceptual model of water's role as a reservoir in *Helicobacter pylori* transmission: a review of the evidence. *Epidemiol Infect* 2006;134:439–449.

47. Murray PR, Rosenthal KS, Kobayahsi GS, Pfaller MA. *Medical Microbiology*. St. Louis: Mosby Inc, 2002.

48. Levinson W, Jawestz E. *Medical Microbiology and Immunology*. New York: Lange Medical Books/McGraw-Hill Publishing Division, 2002.

49. Ganoza CA, Matthias MA, Collins-Richards D, Brouwer KC, Cunningham CB, et al. Determining risk for severe leptospirosis by molecular analysis of environmental surface waters for pathogenic *Leptospira*. *PLoS Med* 2006;3(8):e308. DOI: 10.1371/journal.pmed.0030308.

50. Monahan AM, Miller IS, Nally JE. Leptospirosis: risks during recreational activities. *J Appl Microbiol* 2009;107 (3):707–716.

51. Ivanek R, Gröhn YT, Wiedmann M. *Listeria monocytogenes* in multiple habitats and host populations: review of available data for mathematical modeling. *Foodborne Pathog Dis* 2006;3:319–336.

52. Dailloux M, Laurain C, Weber M, Hartemann PH. Water and nontuberculous mycobacteria. *Water Res* 1999;33:2219–2228.

53. Jernigan JA, Farr BM. Incubation period and sources of exposure for cutaneous *Mycobacterium marinum* infection: case report and review of the literature. *Clin Infect Dis* 2000;31:439–443.

54. D'Aoust JY. Salmonella and the international food trade. *Int J Food Microbiol* 1994;24:11–31.

55. Bhan MK, Bahl R, Bhatnagar S. Typhoid and paratyphoid fever. *Lancet* 2005;366:749–762.

56. Gabutti G, de Donno A, Bagordo F, Montagna MT. Comparative survival of faecal and human contaminants and use of *Staphylococcus aureus* as an effective indicator of human pollution. *Mar Pollut Bull* 2000;40:697–700.

57. Elmir SM, Wright ME, Abdelzaher A, Solo-Gabriele HM, Fleming LE, et al. Quantitative evaluation of bacteria released by bathers in a marine water. *Water Res* 2007;41:3–10.

58. Tauxe RV, Mintz ED, Quick RE. Epidemic cholera in the new world: translating field epidemiology into new prevention strategies. *Emerg Infect Dis* 1995;Oct–Dec;1(4):141–146.

59. Colwell RR. Global climate and infectious disease: the cholera paradigm. *Science* 1996;274:2025–2031.

60. Chakraborty S, Nair GB, Shinoda S. Pathogenic *Vibrios* in the natural aquatic environment. *Rev Environ Health* 1997;12:63–80.

61. Morris JG, Jr., Cholera and other types of vibriosis: a story of human pandemics and oysters on the half shell. *Clin Infect Dis* 2003;37:272–280.

62. Hsieh JL, Fries JS, Noble RT. Dynamics and predictive modelling of *Vibrio* spp. in the Neuse River Estuary, North Carolina, USA. *Environ Microbiol* 2008;10:57–64.

63. Jiang SC. Human adenoviruses in water; occurrence and health implications: a critical review. *Environ Sci Technol* 2006;40:7132–7140.

64. Metcalf TG, Melnick JL, Estes MK. Environmental virology: from detection of virus in sewage and water by isolation to identification by molecular biology—a trip of over 50 years. *Annu Rev Microbiol* 1995;49:461–487.

65. Pusch D, Oh DY, Wolf S, Dumke R, Schröter-Bobsin U, et al. Detection of enteric viruses and bacterial indicators in German environmental waters. *Arch Virol* 2005;150: 929–947.

66. Atmar RL, Neill FH, Romalde JL, Le Guyader F, Woodley CM, et al. Detection of Norwalk virus and hepatitis A virus in shellfish tissues with the PCR. *Appl Environ Microbiol* 1995;61:3014–3018.

67. Hopkins RS, Heber S, Hammond R, Rheinstein PH, Klontz KC. Shellfish-borne illnesses. *American Family Physician*, 1993;47(8),1837–1840.

68. Smith AW, Skilling DE, Cherry N, Mead JH, Matson DO. Calicivirus emergence from ocean reservoirs: zoonotic and interspecies movements. *Emerg Infect Dis* 1998;4:13–20.

69. Schaub SA, Oshiro RK. Public health concerns about caliciviruses as waterborne contaminants. *J Infect Dis* 2000;181 (Suppl 2):374–380.

70. Salvato JA, Nemerow NL, Agardy FJ. *Environmental Engineering*. Hoboken, NJ: John Wiley & Sons, Inc, 2003.

71. Casanova L, Rutala WA, Weber DJ, Sobsey MD. Survival of surrogate coronaviruses in water. *Water Res* 2009;43: 1893–1898.

72. Matsui S. Protecting human and ecological health under viral threats in Asia. *Water Sci Technol* 2005;51:91–97.

73. Wait DA, Sobsey MD. Comparative survival of enteric viruses and bacteria in Atlantic Ocean seawater. *Water Sci Technol* 2001;43:139–142.

74. Rajtar B, Majek M, Pola , ski , Polz-Dacewicz M. *Enteroviruses in water environment—a potential threat to public health. Ann Agric Environ Med* 2008;15:199–203.

75. Bofill-Mas S, Girones R. Role of the environment in the transmission of JC virus. *J Neurovirol* 2003;9 (Suppl 1): 54–58.

76. McQuaig SM, Scott TM, Lukasik JO, Paul JH, Harwood VJ. Quantification of human polyomaviruses JC virus and BK virus by TaqMan quantitative PCR and comparison to other water quality indicators in water and fecal samples. *Appl Environ Microbiol* 2009;75:3379–3388.

77. Hung T, Chen GM, Wang CG, Yao HL, Fang ZY, et al. Waterborne outbreak of rotavirus diarrhoea in adults in China caused by a novel rotavirus. *Lancet* 1984;1:1139–1142.

78. Vogel C, Rogerson A, Schatz S, Laubach H, Tallman A, Fell J. Prevalence of yeasts in beach sand at three bathing beaches in South Florida. *Water Res* 2007;41:1915–1920.

79. Gadanho M, Sampaio JP. Occurrence and diversity of yeasts in the mid-Atlantic ridge hydrothermal fields near the Azores Archipelago. *Microb Ecol* 2005;50:408–417.

80. Didier ES, Stovall ME, Green LC, Brindley PJ, Sestak K, Didier PJ. Epidemiology of microsporidiosis: sources and modes of transmission. *Vet Parasitol* 2004;126:145–166.

81. Macpherson CN. Human behaviour and the epidemiology of parasitic zoonoses. *Int J Parasitol* 2005;35:1319–1331.

82. Karanis P, Kourenti C, Smith H. Waterborne transmission of protozoan parasites: a worldwide review of outbreaks and lessons learnt. *J Water Health* 2007;5:1–238.

83. Schuster FL, Visvesvara GS. Free-living amoebae as opportunistic and non-opportunistic pathogens of humans and animals. *Int J Parasitol* 2004;34(9):1001–1027.

84. Yoder JS, Hlavsa MC, Craun GF, Hill V, Roberts V, Yu PA, Hicks LA, Alexander NT, Calderon RL, Roy SL, Beach MJ. Surveillance for waterborne disease and outbreaks associated with recreational water use and other aquatic facility-associated health events—United States, 2005–2006. *MMWR Surveill Summ* 2008;57(9):1–29.

85. Pozio E. Foodborne and waterborne parasites. *Acta Microbiol Pol* 2003;52 (Suppl):83–96.

86. Bouree P, Paugam A, Petithory JC. Anisakidosis: report of 25 cases and review of the literature. *Comp Immunol Microbiol Infect Dis* 1995;18:75–84.

87. Nithiuthai S, Anantaphruti MT, Waikagul J, Gajadhar A. Waterborne zoonotic helminthiases. *Vet Parasitol* 2004;126:167–193.

88. Mahvi AH, Kia EB. Helminth eggs in raw and treated wastewater in the Islamic Republic of Iran. *East Mediterr Health J* 2006;12:137–143.

89. Wittner M, Turner JW, Jacquette G, Ash LR, Salgo MP, Tanowitz HB, et al. Eustrongylidiasis: A parasitic infection acquired by eating sushi. *The New England Journal of Medicine* 1989;320(17):1124–1226.

90. Coyner DF, Spalding MG, Forrester DJ, Epizootiology of Eustrongylides ignotus in Florida: distribution, density, and natural infections in intermediate hosts. *Journal of Wildlife Diseases* 2002;38(3):483–499.

91. Dorny P, Praet N, Deckers N, Gabriel S. Emerging food-borne parasites. *Vet Parasitol* 2009;163:196–206.

92. Schlegel MW, Knapp SE, Millemann RE. "Salmon poisoning" disease. V. Definitive hosts of the trematode vector, Nanophyetus salmincola. *J Parasitol* 1968;54:770–774.

93. Dick TA, Nelson PA, Choudhury A. Diphyllobothriasis: update on human cases, foci, patterns and sources of human infections and future considerations. *SE Asian J Trop Med Public Health* 2001;32 (Suppl 2):59–76.

94. Scholz T, Garcia HH, Kuchta R, Wicht B. Update on the human broad tapeworm (genus *Diphyllobothrium*), including clinical relevance. *Clin Microbiol Rev* 2009;22:146–160.

95. United States Environmental Protection Agency (EPA). 2000 National Water Quality Inventory. http://water.epa.gov/lawsregs/guidance/cwa/305b/2000report_index.cfm. 2000. Accessed 3-14-11.

96. Townsend T, Solo-Gabriele H, Tolaymat T, Stook K, Hosein N. Chromium, copper, and arsenic concentrations in soil underneath CCA-treated wood structures. *Soil Sediment Contam* 2003;12:1–20.

97. Hasan AR, Hu L, Solo-Gabriele HM, Fieber L, Cai Y, Townsend TG. Field-scale leaching of arsenic, chromium, and copper from weathered treated wood. *Environ Pollut*, 2010.

98. Colford, JM, Wade TJ, Schiff KC, Wright CC, Griffith J, Sandhu SK, et al. Water quality indicators and the risk of illness at beaches with nonpoint sources of fecal contamination. *Epidemiology* 2007;18(1):27–35.

99. Boehm AB, Ashbolt NJ, Colford JM, Dunbar LE, Fleming LE, Gold MA, Hansel J, Hunter PR, Ichida AM, McGee CD, Soller JA, Weisberg SB. A sea change ahead for recreational water quality criteria. *J Water Health* 2009;7(1):9–20.

100. Faruque SM, Albert MJ, Mekalanos JJ. Epidemiology, genetics, and ecology of toxigenic *Vibrio cholerae*. *Microbiol Mol Biol Rev* 1998;62(4):1301–1314.

101. Belkin S, Colwell RR. *Oceans and Health: Pathogens in the Marine Environment*. New York: Springer, 2006.

102. Bossart GD. Case study; marine mammals as sentinel species for oceans and human health. *Oceanography* 2006;19(2): 134–137.

103. Moffa PE, editor. *The Control and Treatment of Industrial and Municipal Stormwater*. New York: Van Nostrand Reinhold, 1996.

104. Minton GR. *Stormwater Treatment: Biological, Chemical, and Engineering Principles*. Seattle, WA: RPA Press, 2005.

105. U.S. Geological Survey. *Frequently asked questions about air quality in Hawai'i*. 2008. Accessed December 15, 2009. Available at http://hvo.wr.usgs.gov/hazards/FAQ_SO2-Vog-Ash/P1.html.

106. U.S. Geological Survey (USGS). *Volcanic Air Pollution—A Hazard in Hawaii Survey Fact Sheet*, 169-97. Online Version 1.1, revised June 2000. Available at http://pubs.usgs.gov/fs/fs169-97/.

107. National Research Council of the National Academies. Hydrology, Ecology, and Fishes of the Klamath River Basin: Committee on Hydrology, Ecology, and Fishes of the Klamath

River, Board on Environmental Studies and Toxicology, Water Science and Technology Board, Division on Earth and Life Studies. Washington, DC: The National Academies Press, 2008.

108. Islam M, Doyle MP, Phatak SC, Millner P, Jiang X. Survival of *Escherichia coli* O157:H7 in soil and on carrots and onions grown in fields treated with contaminated manure composts or irrigation water. *Food Microbiol* 2004;22(1):63–70.

109. Cheong S, Lee C, Song SW, Cheio W, Lee CJ, Kim S-J. Enteric viruses in raw vegetables and groundwater used for irrigation in South Korea. *Appl Environ Microbiol* 2009;75(24): 7745–7751.

110. Ayers LT, Williams IT, Gray S, Griffin PM. Surveillance for foodborne disease outbreak—United States, 2006. *Morb Mortal Wkly Rep* 2006;58(22):609–615.

111. Jay MT, Cooley M, Carychao D, Wiscomb GW, Sweitzer RA, Crawford-Miksza L, Farrar JA, Lau DK, O'Collell J, Millington A, Asmundson RV, Atwill ER, Mandrell RE. *Escherichia coli* O157:H7 in feral swine near spinach fields and cattle, central California coast. *Emerg Infect Dis* 2007;13(12):1908–1911.

112. Voorburg JH. Pollution by animal production in the Netherlands: solutions. Review of science and technology. *Off Int. Epizootics* 1991;10(3):655–668.

113. Burkholder J, Libra B, Weyer P, Heathcote S, Koplin D, Thorne PS, Wichman M. Impacts of waste from concentrated animal feeding operations on water quality. *Environ Health Perspect* 2007;155(2):308–312.

114. Wing S, Freedman S, Band, L. The Potential Impact of Flooding on Confined Animal Feeding operations in Eastern North Carolina. *Environ Health Perspect* 2002;110:387–391.

115. Hatfield JL.Metrics for nitrate contamination of ground water at CAFO land application sites—Iowa swine study. U.S. EPA, 2009. EPA 600/R 09/045.

116. Sinigalliano CD, Gidley ML, Shibata T, Whitman D, Dixon TH, Laws E, et al. Impacts of Hurricanes Katrina and Rita on the microbial landscape of the New Orleans area. *PNAS* 2007;104(21):9029–9034.

117. Bowen R, Frankic A, Davis ME. The human development and resource use in the coastal zone: influences on human health. *Oceanography* 2006;19(2):62–71.

118. Glynn KM, Backer LC. Collecting public health surveillance data: creating a surveillance system. In: *Principles and Practices of Public Health Surveillance*, 3rd edition. New York, NY: Oxford University Press, in press.

119. One Health Initiative. Available at http://www.onehealthinitiative.com/. Accessed February 3, 2010.

120. Reif JS, Rhodes WH, Cohen D. Canine pulmonary disease and the urban environment. *Arch Environ Health* 1970;20: 145–150.

121. Backer LC, Grindem CB, Corbett WT, Cullins L, Hunter JL. Pet dogs as sentinels for environmental contamination. *Science Total Environ* 2001;274:161–169.

122. Backer L, Coss A, Wolkin A, Flanders D, Reif JS. Case-control study of lifetime exposure to drinking water disinfection by-products and bladder cancer in pet dogs. *J Am Vet Med Assoc* 2008;232(11):1663–1668.

123. Stavros H-CW, Bossard GD, Hulsey TC, Fair PA. Trace element concentrations in skin of free-ranging bottlenose dolphins (*Tursiops truncatus*) from the southeast Atlantic coast. *Sci Total Environ* 2007;388:300–315.

124. Centers for Disease Control and Prevention (CDC). *A strategy to revitalize environmental health services in the United States*. Atlanta, GA: CDC, 2002, 51 pp.

125. Association of Schools of Public Health, (ASPH). *Confronting the Public Health Workforce Crisis*. Washington, DC: ASPH, 2008, 9 pp.

126. U.S. Environmental Protection Agency (US EPA). *The Clean Water and Drinking Water Infrastructure Gap Analysis*. Washington, DC, 2002. EPA-816-R-02-020.

24

ANTHROPOGENIC SOURCES OF WATER POLLUTION: PARTS 1 AND 2

KERRY L. SHANNON, ROBERT S. LAWRENCE, AND M. DANIELLE MCDONALD

PART 1—INTRODUCTION

KERRY L. SHANNON AND ROBERT S. LAWRENCE

Anthropogenic sources of water pollution include any contamination of precipitation, surface water, wetlands, lakes, rivers, aquifers, estuaries, and oceans by human activity that releases pollutants into the environment. Human causes of water pollution can be grouped by municipal, agricultural, and industrial causes. They can also be categorized as point and nonpoint sources. Point sources include most municipal and industrial pollution. Nonpoint sources include most pollution from agriculture, mining operations, uncollected sewage, and urban storm-water runoff. All of these provide a number of different organic and inorganic contaminants that can affect human health (1).

24.1 MICROORGANISMS IN WATER

As described in Chapter 26 on wastewater microbiology, a number of pathogenic microorganisms in watersheds, including bacteria, viruses, protozoa, fungi, and algae, are capable of causing infection in humans and serving as zoonotic reservoirs of human disease. Globally over 10 million deaths and 250 million new cases of waterborne disease occur yearly, the majority of them in low- and lower-middle income countries where 2.5 billion people are still without access to improved sanitation and about 1 billion people lack access to potable water (1, 2). Bacteria such as *Vibrio cholera* and *Escherichia coli*; parasites such as

Giardia lamblia and *Entamoeba histolytica*; and viruses such as *norovirus* and *rotavirus* cause the majority of diarrheal disease. All of these are spread by fecal contamination of water used for drinking (1).

Many diarrheal diseases, as well as other conditions such as trachoma caused by *Chlamydia trachomatis*, can be prevented by access to clean water for hand washing to interrupt the fecal-oral transmission or the contamination of the conjunctivae with pathogens. Other water-based diseases such as schistosomiasis infect humans by skin contact with contaminated water rather than ingestion. Guinea worm disease (*Dracunculus medinensis*) is propagated by infected humans washing or bathing in water, allowing the guinea worm to release its larvae that are then consumed by tiny water fleas. The flea-infested water, when consumed by humans, introduces the guinea worm larvae into new hosts (1). Since 1986, the Carter Center has been leading the campaign to eradicate guinea worm disease. At the beginning of the campaign, 20 countries in Africa and Asia had an estimated 3.5 million cases; by 2010 fewer than 2000 cases were reported in the four countries still showing transmission—Sudan, Ghana, Mali, and Ethiopia (3).

In the United States, about 40% of estuaries do not meet ambient water quality standards because of pathogen contamination, most often detected and monitored by fecal coliform, *E. coli*, and *Enterococcus faecalis* (4). Higher concentrations of pathogens occur in warmer months. Urban streams have greater concentrations than forested streams (5). Studies in North Carolina found a correlation between fecal coliform densities and percent developed land as well as watershed population (6). Fecal coliform,

Water and Sanitation-Related Diseases and the Environment: Challenges, Interventions, and Preventive Measures, First Edition.
Edited by Janine M. H. Selendy.
© 2011 Wiley-Blackwell. Published 2011 by John Wiley & Sons, Inc.

E. coli, and enterococci levels have been correlated with housing density, population, development, and percent impervious surface covered area, as well as domestic animal density. Areas that had more domestic concentration and used sewers had more bacterial counts than areas that were less developed and used septic tanks. Microorganism die-off rates are affected by temperature, sunlight, salinity, and available nutrients (5).

24.2 URBAN RUNOFF

Urban runoff is of critical concern to water quality, principally in wet weather when the various pollutants that have collected on the ground are taken up by flowing water and carried to nearby streams, rivers, and lakes. Runoff of concern comes from numerous sources including streets, construction sites, parking lots, industrial storage yards, and lawns. The pollutants come from cars, trucks, leachate from asphalt paving, outdoor storage piles, construction sites, and pesticide spills (7).

Commercial and residential development of urban areas and increases in impervious surface coverage correlate with surface waters having an increase in biological oxygen demand or BOD, fecal coliform bacteria, orthophosphate, and surfactants (8). More impervious surfaces lead to a greater amount of water carrying pollutants that flows directly into drainage systems to streams, rivers, and lakes rather than soaking into soil and being filtered and/or broken down by soil bacteria.

Industries present threats to water quality both from effluents routinely released into water and air as part of the industrial process as well as in the risk for spills and accidents. Industries that are especially hazardous include manufacture of asbestos products, dyes, pesticide and mineral processing, petroleum refining, and chemical production. Many of these industries are moving to low- and lower-middle income countries where environmental regulations are not as stringent or weakly enforced and labor is cheaper (1).

In the Nairobi River Basin in Kenya, industries have been reported to release a large amount of chromium, nickel, copper, and lead in amounts that were much higher than the recommended concentrations (9).

Roads Runoff from roads is a serious source of pollution. A study of runoff from three highway sites in Los Angeles found a number of different heavy metals such as cadmium, copper, nickel, lead, and zinc as well as 16 different polycyclic aromatic hydrocarbons or PAHs. As the relative abundance of the different PAHs reflected the levels in gasoline engine exhaust, it is likely that incomplete combustion of residual fossil fuel in automobiles is the main source (10).

Thunqvist observed, "A road operation along with its traffic can pose a serious pollutant threat to groundwater and surface water in its vicinity. Examples of pollutants are metals from corrosion of vehicles, rails and poles and the wear of road surface and tyres; hydrocarbons from the wear of the road surface, tyres, exhaust, oil; and hazardous goods discharged in the case of an accident" (11). One study in Sweden noted that about 200,000-300,000 tons of sodium chloride are used for deicing each winter (11). This amount of road salt not only accounts for over half of the chloride load in the investigated river basin but also the chloride ion can be a tracer for deicing salt effects on groundwater, surface water, and the ecosystem.

Fecal and microbial contamination in urban areas comes from domestic animals and urban wildlife including pets, waterfowl, pigeons, rats, and raccoons (6). Some fecal contamination also comes from leaks in or illicit connections into the sewer system itself. This problem increases with heavy rain events since many urban storm water and sewage systems are connected, and high storm-water flows may exceed the capacity of intake by wastewater treatment plants (8). Bacteria in urban runoff almost always exceed public health standards for recreational swimming as determined by USEPA standards for recreational water allowances for fecal coliforms, *E. coli*, and enterococci. Fecal coliform levels in urban runoff are typically 20–40 times the amount allowed for safe swimming conditions (7). As with estuary systems, warmer weather has been correlated with higher concentrations of microbial contaminants in surface waters used for recreation (5).

Sediment from urban settings has a mix of metals from vehicles, particles from vehicle exhaust and industrial smoke-stacks, bits of tires, brake linings, and pavement (7). This sediment comes from roads, industrial sites, and commercial development. Areas that are under construction release the most significant amounts of sediment because there is little vegetation to stop erosion, and efficient drainage systems are often set up to sweep water, sediment, and other pollutants directly into the municipal drainage system and waterways. Areas under construction typically have 35–45 tons per acre per year of erosion compared to 1–10 tons per acre per year of cropland (7).

Nutrients such as phosphorous and nitrogen are at high levels in urban runoff, primarily as a result of gardening and turf management. Phosphorous loading is a major source of eutrophication in streams, rivers, ponds, and lakes. High levels of nitrates can cause methemoglobinemia by oxidizing iron from the ferrous state (Fe^{2+}) of the normal hemoglobin molecule to the ferric state (Fe^{3+}) to form methemoglobin, which is incapable of binding and transporting oxygen to the tissues. Children under 6 months of age exposed to nitrates in well water that exceed USEPA standards of 10 ppm of nitrate nitrogen are at risk for developing the blue-baby syndrome, a condition in which suffocation can result from elevated methemoglobin levels (1). Nitrates in the form of nitrosamines have also been implicated in causing cancers of the digestive tract (12).

Orthophosphates can stimulate bacteria and phytoplankton growth (8). Research has shown that residential applications to lawns and driveways add orthophosphate to the storm water load (13). This is worse in areas with high impervious surface coverage (14).

Surfactants are used in soaps and detergents and reduce surface tension at the air/water interface. They are common in wastewater as well as urban/suburban runoff in locations where soaps and detergents are used for floor cleaning, building cleaning, auto washing, and other similar purposes.

Toxic chemicals such as metals, pesticides, and persistent organic pollutants are also a concern for urban runoff. Metals such as lead, mercury, arsenic, cadmium, copper, zinc, and chromium may pose a risk to human health. *Vehicle traffic* is one of the largest contributors of exposure to metals such as zinc, cadmium, chromium, and lead. Concentrations of these metals correlate with the volume of traffic on streets with storm water drains (7). Lead levels are significantly lower now since the banning of leaded gasoline, but the levels still exceed water-quality standards. One study found that building walls make up 79% of the total sources of lead with brick walls and painted wood walls having especially high levels (15). *Roofs* can also be a source of metal contamination of water. Galvanized metal rooftops, gutters and spouts, and walls are a major source of zinc pollution in industrial areas, particularly as they often drain to pavement or storm water sewers instead of onto the ground (7). One study found that brick walls accounted for 58% of zinc runoff, followed by wear and tear of galvanized building components at 25% (15). The same study found that brake emissions from automobiles accounted for 47% of copper emissions, while building siding accounted for 21%. A study in Paris showed that roof runoff accounted for the largest release of trace elements such as heavy metals (16). *Outdoor storage* of scrap metal, coal, and salt can also contribute significantly to water pollution. Scrap metal piles contribute a large amount of mercury. Emissions from burning coal, oil, or waste may also release metals such as cadmium, copper, lead, and mercury that contaminate water sources. Chromium and lead from road salt piles and arsenic from scrap metal and coal piles add to the contamination of surface water (7).

Copper, silver, selenium, zinc, and chromium, although not toxic to humans at very low levels, can be very damaging to aquatic ecosystems (1). Metals affect aquatic life by forming particular complexes with the environment. Their ability to do this depends on their oxidation state and the pH, salinity, and temperature of the water (1). The sources of human exposure to metals are domestic wastewater effluents (As, Cr, Cu, Mn, and Ni), coal-burning power plants (As, Mn, and Se), nonferrous metal smelters (Cd, Ni, Pb, and Se), iron and steel plants (Cr, Mo, Sb, Zn), and dumping of sewage sludge or its application on croplands (As, Mn, and Pb) (1). A study by the USEPA showed that copper, lead, and

zinc are the most prevalent pollutants in urban runoff (1). Mining, particularly in low- and lower-middle income countries that lack adequate environmental regulations or enforcement, also contributes to metal contamination of surface waters. This is particularly the case in South America and Southeast Asia. The gold industry in the Amazon basin uses mercury to separate the fine gold particles from the other components of soil. Much of the mercury is lost to the environment (17).

24.3 CHEMICAL POLLUTANTS

Lead In addition to lead paint and lead in gas, industrial sources of lead include lead smelters, lead or silver ore mining, and lead refining (18). Lead can be damaging to the nervous system and kidneys, and can cause high blood pressure and digestive disorders in humans. It is particularly toxic to young children, affecting kidney function and the central and peripheral nervous system, depressing biosynthesis of protein, and disrupting normal nerve and red blood cell formation (19). Lead damages the proximal tubule of the kidney, leading to hyperuricemia and gout by inhibiting uric acid secretion and diminishing the glomerular filtration rate (GFR) (18). Lead has been shown to cause "neurotoxicity, developmental delays, hypertension, impaired hearing acuity, impaired hemoglobin synthesis, and male reproductive impairment" (20).

Lead has been classified as a "possible human carcinogen" because ample evidence exists in animals but not in humans at present. Oral exposure to lead in mice, rats, and hamsters is associated with an increase in renal tumors.

Mercury Mercury toxicity is magnified within aquatic food chains. Inorganic mercury can cause kidney damage and ulceration of skin, while the more toxic organic form, methyl mercury, often exists in surface waters. The organic form of mercury affects the central nervous system, kidneys, and the endocrine system, and may be fatal (1).

Arsenic Arsenic has a number of chronic effects if ingested, such as hyperkeratosis of palms and soles of feet and hyperpigmentation. The inorganic form can affect the GI tract and liver and can cause skin cancer (1). Chronic arsenic toxicity (CAT) or arsenicosis results from drinking water that is contaminated by arsenic and can cause a number of different health problems. Severe skin lesions with pigmentation and hyperkeratosis are characteristic of CAT. CAT also can result in chronic obstructive pulmonary disease and bronchiectasis, liver disease, polyneuropathy, peripheral vascular disease, hypertension, ischemic heart disease, diabetes mellitus, nonpitting edema of feet/hands, weakness, and anemia. Malignancies of the skin, lung, and urinary bladder are also associated with CAT. Although early skin cancer can be cured, many of the health problems associated with CAT have limited therapeutic options. The goal should be primary prevention by keeping the water from being contaminated with arsenic in the first place (21).

Arsenic occurs naturally and is an abundant element in the earth's crust. The inorganic forms of arsenite and arsenate are very toxic to human health. Modes of exposure include air, food, and water. Drinking water is contaminated through the use of pesticides, natural mineral deposits, improper disposal of chemicals with arsenic, and the use of organoarsenicals, such as Roxarsone, as growth promoters in industrial poultry production. The organoarsenicals are broken down by bacteria in the gut of the poultry or in the poultry litter to produce the toxic inorganic form. The Ganga–Brahmaputra–Meghna river basin in India and Bangladesh has a serious problem with arsenic contamination of groundwater. Over 25 million people in Bangladesh and 6 million people in West Bengal, India, have been exposed to water contaminated with arsenic at levels sufficient to pose a health risk (21).

Numerous studies have documented significant effects of arsenic *in utero* and early life exposures on child health and development. Arsenic exposure *in utero* has been associated with spontaneous abortion, stillbirths, reduced birth weight, and infant mortality, although there have been some mixed findings (22). With respect to cognitive function, a study in Thailand showed that arsenic exposure was associated with reduced visual perception, while a study in Taiwan reported reduced pattern memory and switching attention scores. A study in Bangladesh showed overall reduced intellectual function with arsenic exposure. Another study in India showed reduced vocabulary test scores and picture completion scores (23). Although an overall increase in childhood cancer has not yet been documented, one study shows a dramatic increase (RR = 10.6) in liver cancer with arsenic exposure in young children (23).

A study in western Bangladesh examined the levels of a variety of toxic metals in tube wells used for drinking water. Before conducting the study, researchers noticed that certain patients had symptoms that could not be entirely explained by arsenic exposure alone. The investigators suspected synergistic effects with other toxins, and they found that concentrations of a number of toxic metals exceeded drinking water guidelines of the World Health Organization (WHO). The percentages of contamination above WHO standards were very high: 78% for Mn, 48% for U, 33% for As, 1% for Ni, and 1% for Cr. Many wells had unsafe levels of both Mn and As or Mn and U, rarely both As and U. Many wells in the region are only tested for As. This study shows that more thorough and broader testing is needed to identify unacceptable health risks (20).

Manganese Manganese is required at low levels in the diet for adequate nutrition and good health. If accumulated in excess amounts, it may cause hepatic encephalopathy and neurologic damage in humans (20).

Cadmium Cadmium exposure is associated with renal disease. It has a long half-life for excretion from the body (greater than 30 years), so low-level exposure over time can lead to accumulation in certain tissues, particularly the kidney. Toxic levels cause the Fanconi syndrome, a reabsorptive defect in the proximal tubule of the kidney that inhibits ATP production and Na-K-ATPase activity. In addition to kidney problems, cadmium has also been associated with impaired lung function, a potential increase in risk of lung cancer, bone demineralization, yellow tooth discoloration, mild anemia, and disorders of calcium metabolism leading to osteomalacia (18, 19).

Cadmium is a by-product of zinc refining and is used for plating of steel, pigments, plastics, alloys, nickel–cadmium batteries, and in nuclear and electronic engineering (18, 24).

Lower levels of cadmium exposure have been related to renal tubular dysfunction. A study in Belgium showed a 10% chance of tubular dysfunction with 2 mg/day urinary cadmium levels (18). Studies suggest that the chronic low-level exposure to Cd that the general population has in many industrialized countries can be deleterious to the kidneys and bones of the population (19).

Aluminum Aluminum is the most abundant metal in the earth's crust. Normal levels of exposure are not harmful, and aluminum toxicity is usually associated with breathing aluminum dust in the workplace or eating foods that contain high levels of aluminum rather than being exposed to aluminum in water (1).

Chromium Chromium exposure can cause dermatitis, pulmonary congestion, and nephritis (1). One study in Karachi, Pakistan, showed significant contamination of chromium and lead in drinking water in almost all ground water sources with especially high levels in industrial areas. A positive correlation between chromium and lead levels suggested similar sources of contamination (25).

Organotin Complexes Organotin complexes are chemical compounds made by linking hydrocarbon compounds to tin. Toxicity varies with the compound, and some have been banned because of their biologic effects on marine life. Others are used as industrial biocides for their antifungal actions (1).

24.3.1 Radioactive Materials

Radionuclides such as radium-226, radium-228, radon-222, and uranium can contaminate drinking water and affect human health. Both natural and human caused sources of radionuclides occur in the form of gas or minerals. Mining sites or areas with previous mining activity are major sources of radioactive materials because rock layers with radioactive ore are exposed and left in the soils. Water runoff and leachate from the mines may contain toxic levels of radionuclides. Underground streams that go through these strata can also contribute to spreading radioactive materials to water used by humans. Smelters and coal-fired electrical-generating plants are also sources of natural radioactive materials. Uranium has been shown to cause proximal tubule dysfunction in the kidney; nephrotoxicity has been shown at very low concentrations of uranium. Toxicity may also affect bones (20).

Man-made radioactive materials can come from nuclear power plants, nuclear weapons facilities, radioactive materials disposal sites, and mooring sites for nuclear-powered ships. Hospitals contribute to exposure by dumping radioactive materials into sewers. Radon is not a problem if the water has been exposed to air in a reservoir before entering the house, since radon is a gas and will dissipate. If it enters directly through taps and showers, however, it can cause high levels of gas in a house. Radon has become the second leading cause of lung cancer in the United States (26).

24.3.2 Endocrine Disruptors

A number of synthetic endocrine-disruptors have been found in surface water in the United States and elsewhere. They are associated with discharge from wastewater treatment plants and agricultural runoff. In a study of different sites within the urban wastewater area in Oakland, California, at least one endocrine disruptor was found in 19 of 21 sites (27). Urban wastewater and receiving streams have been shown to contain high levels of endocrine disruptors.

There is growing concern that the way we are measuring the impacts of estrogenic, thyroid-disrupting, and anti-androgenic chemicals on humans is not sufficient, because most epidemiologic studies have focused on single chemicals. Humans are generally exposed to mixtures of toxins, so these single chemical studies do not fully capture the potential risks to health. It is important that mixtures are studied and that biomarkers that can capture cumulative exposures be developed (28).

Substantial evidence now exists that many endocrine disruptors combine in an additive effect on biologic systems, although some studies show results that are stronger than would be anticipated by addition alone, while others fail to show effects as expected. Kortenkamp believes that the additive model applies reasonably well for combinations of endocrine disruptors and can be used in mathematical modeling of expected combination effects. Combinations of different classes of endocrine disruptors such as estrogenic, anti-androgenic, and thyroid disrupting have not been well studied, however. The additive model tends to be the default mechanism when there are similar modes of action. Kortenkamp notes, however, that some studies have been conducted on thyroid-disrupting chemicals that have relatively different modes of action yet still resulted in effects when combined at doses well below the no-observed effect level (NOEL) (28).

24.3.2.1 *Health Effects of Endocrine Disruptors*   Strong evidence links environmental exposure to endocrine disruptors with reproductive, congenital, carcinogenic, and neurologic impacts on humans and the ecosystem in general.

Reproductive Effects   At a population level, fertility rates appear to be decreasing. In fact, more women are experiencing impaired fecundity than in the past, particularly those under 25. There is growing concern that decreased fertility and other reproductive problems result from exposure to the myriad of chemicals in the environment (29). A study of a population poisoned by ingestion of cooking oil contaminated with polychlorinated biphenyls (PCBs) and polychlorinated dibenzofurans (PCDFs) showed that prenatal exposure to these contaminants resulted in sperm with abnormal morphology, reduced motility, and reduced capacity to penetrate hamster oocytes (30). Similarly, a study of U.S. men living in the Midwest showed that those with high levels of exposure to the commercial pesticides alachlor, IMPY (2-isopropoxy-4-methylpyrimidinol), and atrazine (above the limit of detection), which are commercial herbicides and insecticides, had low sperm concentration, decreased motility, and abnormal morphology (OR: 30.0, 16.7, and 11.3 for alachlor, IMPY, and atrazine, respectively) (31). A study in frogs showed that a reduction of androgens and brain GnRH (gonadotropin-releasing hormone) levels was associated with gross limb malformations. The investigators propose that endocrine-disrupting chemicals play a role in these developmental abnormalities. "Atrazine is the most commonly used herbicide in the U.S. and probably the world. Atrazine contamination is widespread and can be present in excess of 1.0 ppb even in precipitation and in areas where it is not used" (32). Exposure to atrazine greater than or equal to 0.1 ppb resulted in retarded gonadal development and hermaphroditism (development of testicular oocytes) in leopard frogs. The investigators carried out environmental studies connecting atrazine exposure with developmental abnormalities in sites across the United States (33, 34). In one study, only one site had atrazine levels below the limit of detection, and this was the only site where no frogs had testicular oocytes (34). Interestingly, they found that 0.1 ppb atrazine had a greater association with abnormalities than the higher dose of 25 ppb (33).

Exposure to DDT (dichlorodiphenyltrichloroethane), a synthetic pesticide and persistent organic pollutant, has been associated with increased risk of breast cancer. A prospective cohort study found that women who were exposed to DDT before they were 14 years old and had high serum levels of *p,p'*-DDT (the most abundant metabolite of DDT) had a fivefold greater risk for developing breast cancer later in life compared with women not exposed to DDT as children (35). Another study, based on measuring PCB levels in the blood of women who were adults at the time of the study, did not find an overall significant increase in breast cancer rates (OR: 1.2,

95% CI: 0.9–1.6). They did, however, find that the cytochrome P-450 1A1 M2 variant genotype was associated with an increased risk of breast cancer in general (OR: 2.1, 95% CI: 1.1–3.9) but particularly for those with higher PCB levels (OR: 3.6, 95% CI: 1.5–8.2). This was especially true for postmenopausal women (OR: 4.3, 95% CI: 1.6–12.0) (36).

Neurotoxicity has also been associated with exposure to numerous industrial chemicals. Colborn suggests a link between the increase in attention deficit hyperactivity disorder, autism, and other associated neurodevelopmental and behavioral problems with exposure of the fetus to thyroid disturbance from a large number of synthetic chemicals including a number of pesticides and herbicides, PCBs, thiocyanates, military and aerospace chemicals, perchlorates, dioxin, hexachlorobenzene, and the insecticide fenvalerate (37). The large range of neurodevelopmental effects, ranging from very mild to severe, and the large numbers of complicated mixes that people are often exposed to makes it very challenging to pinpoint the specific chemicals responsible. Increasing evidence, however, suggests a strong association between a number of different pesticides, herbicides, and other chemicals and neurodevelopmental problems (38). One study showed a dose–response association between response inhibition, size of the splenium or the corpus callosum, and PCB exposure (39). Grandjean and Landrigan note that the developing brain is much more vulnerable to injury from toxic agents than the adult brain. During early neurodevelopment the map of neurons is put in place with neurons moving on precise pathways to create a highly interconnected system. The placenta can provide protection against some toxins, but many others, such as mercury, move across freely into the fetal circulation. The blood–brain barrier is not fully formed until 6 months after birth (40). In the first few years of life the growth of glial cells and myelination continue, so during this period the developing brain continues to be highly vulnerable to chemicals. Breast milk can also be a source of halogenated industrial compounds such as PCBs, pesticides, and other persistent lipophilic substances stored in maternal fat. These exposures can be 100-fold greater than the mother's own exposure when normalized for body weight (40). Many neurotoxins can cause developmental delays, and evidence in human studies confirms a large number of metals, organic solvents, and pesticides that can injure the nervous system (40).

Endocrine disrupting hormones can affect the hypothalamic-pituitary-thyroid axis in the brain. One study showed that polybrominated diphenyl ether (PBDE), a flame retardant, affects the thyroid by "depressing peripheral levels of T4 ... and can alter TH signaling at multiple levels of the hypothalamic-pituitary-thyroid axis" (41). Exposure to PBDE can result in reduced thyroxin in plasma, accompanied by elevated TSHβ in the pituitary. The levels of the transcripts for the different TH receptors were modified in the brain but not in the liver. Because these receptors are important in regulating neurodevelopment, this effect is of particular concern.

Levels of PBDEs have been increasing in breast milk, suggesting that it may be important to look carefully at exposures children are getting through this mechanism and *in utero* and how to reduce the mother's exposure. PBDEs are similar structurally to PCBs and PBBs; thus, exposures to these and other persistent organic pollutants or POPs also raises concerns about neurotoxicity mediated through this mechanism (42).

Triclosan is an antibacterial agent in hand soap, dish soap, toothpaste, cutting boards, countertops, and other consumer products, and is used in pesticides. It likely affects thyroid, androgen, and estrogen hormone systems (27). Triclosan was found in all clinic and hospital waste sites measured in Oakland (27).

24.3.3 Phthalates

Phthalates (1,2-benzenedicarboxylic acid) are a group of chemicals used in a number of industrial applications. High molecular weight phthalates such as DEHP (bis (2-ethylhexyl) phthalate) are used for plasticizers in manufacturing polyvinyl chloride (PVC), which is used in flooring, wall coverings, food contact applications, and medical devices (43, 44). Low molecular weight phthalates such as DBP (dibutyl phthalate) are used as solvents in perfumes, lotions, and cosmetics, as well as lacquers, varnishes, coatings, and some pharmaceuticals (43). One study showed that levels of phthalates in the sewage lagoon as well as treatment stream were significantly higher than the water criteria set by the USEPA, and the investigators expressed concern of impacts on the ecosystem and the people who lived downstream (45). This is not a new problem. A study conducted on the Charles and Merrimack Rivers published in 1973 documented contamination with phthalates (46).

When looking at the many different chemicals and plastic waste, Thompson observed that our current use of plastics not only affects human health and ecosystems but also is not sustainable at current production levels (47). He noted that plastics constitute about 10% of the municipal waste stream by weight (47).

Exposure routes are varied and include ingestion, inhalation, and dermal contact. Phthalates may have endocrine function and reproductive or developmental effects, although more research is necessary to confirm this (43). Phthalates inhibit estrogen receptors in fish (27) and bioaccumulate in aquatic plants. In humans, particular phthalates have been shown to have a number of impacts related to the male reproductive system including reduced penile size, incomplete testicular descent, and decrease in anogenital distance (31, 44). Other studies show increased damage to sperm DNA and decreased motility, reduced concentration, and altered morphology of sperm related to exposure to phthalates (44). Both males and females are shown to have shorter gestational age at birth with exposure to phthalates (44, 48). Low birth weight has also been associated with phthalates such as DBP and DEHP (49). Females have premature thelarche, or development of breast buds, associated with DEHP. Respiratory problems, allergy, and asthma have been associated with DEHP and BzBP exposure in both male and female children (44). Decreased pulmonary function, increased waist circumference, insulin resistance, and thyroid effects have also been associated with phthalates (44).

As the different phthalates act by similar mechanisms, it is not surprising that studies have shown that a mixture of different phthalates acts in a dose-additive manner (50).

24.3.4 Organic Pollutants

Organic pollutants are also of serious concern, especially as many are very long lasting, and once they are in the environment, it is hard to get rid of them, hence, the term *persistent organic pollutants* or POPs. Most are lipophilic and hydrophobic; they will attach strongly to particulate matter in water and will often be found in sediment. Thus, any disturbance of sediment can provoke release of POPs. Most organic pollutants in the water come from industrial activities, including petroleum refining, coal mining, manufacturing of synthetic chemicals, steel production, textile production, and wood pulp processing. Domestic use of petroleum and heating oils, atomizers, pesticides, and fertilizers also contribute to POPs burden, and urban runoff contains a number of different organic pollutants (1).

Polychlorinated biphenyls (PCBs) are a group of 209 compounds that do not easily degrade, burn, dissolve in water, or conduct electricity (51). They are used for insulation, coolants, dielectric fluids, hydraulic fluids, and lubricants in transformers, capacitors, and a variety of other electrical equipment (52). PCBs are normally released into the atmosphere and then deposited onto land and water. They are of concern to human health because they can cause skin sores and liver problems, and in the long term are carcinogenic and cause problems with reproduction, fetal development, liver function, and compromise of the immune system (7). Both PCBs and DDTs were found at higher levels near urban areas in a study done in the Mekong delta, indicating urban areas as a source for these chemicals (53). Dioxins and furans are other POPs that create health risks. These chemicals are released during waste incineration whenever a chlorine source and organic material are exposed to high temperature (52).

Polycyclic aromatic hydrocarbons (PAHs) consist of about 10,000 different compounds and are by-products of incomplete combustion of coal, oil, wood, tobacco, charbroiled meats, garbage, and other materials (54). They are often released as point sources from refineries and production plants in the petroleum, power plant, coal-tar, coking, asphalt and bitumen, paper, wood, aluminum, and industrial machinery industries (54). PAHs are also in gasoline and exhaust from vehicles, asphalt, tar, and wood preservatives. They are likely carcinogenic and can be irritating to the skin and lungs and cause liver and kidney damage (54).

Organic solvents are often found in groundwater as a result of industrial processes such as plating and electronics manufacturing, and leaching from landfills. These solvents are highly volatile. They are common in household products such as cleaners and paint thinners. Particular solvents have been associated with cancer, birth defects, and cardiovascular diseases (1).

Disinfection by-products (DBPs) are formed when chlorine or other chemical disinfectants are added to water to kill pathogens before distributing the water for drinking, washing, and cooking. These chemicals react with organic and inorganic matter in the water to form noninfectious but potentially hazardous compounds. Trihalomethanes, halo acetic acids, and chlorinated hydroxyfuranones are among the common DBPs, a number of which are carcinogenic (1, 55).

In the Zambezi River Basin in South Africa it is estimated that 93,000 metric tons of industrial waste are released into the Zambezi River annually, affecting surface and groundwater quality. Sources include sewage-treatment facilities, pulp and paper mills, fertilizer factories, abattoirs, textile and cloth manufacturing entities, mining activities, agriculture, and chemical industries (9).

24.3.5 Chemical Mixtures

In regulating and evaluating toxicity of waste, the general practice is to assess individual toxicities for particular chemicals. This approach, however, often underestimates the impact of the whole effluent toxicity because there may be interactions among the different pollutants in the wastewater (56).

24.4 LOSS OF ECOSYSTEM SERVICES

Ecosystem services are the resilience factors in nature that, among other things like providing food and clean water, also help detoxify pollutants and waste through natural flocculation, action of soil microorganisms, filtering, and biotic recycling through root systems (57). Outwater observed in 1996, "This country's waterways have been transformed by omission. Without beavers, water makes its way too quickly to the sea; without prairie dogs, water runs over the surface instead of sinking into the aquifer; without bison, there are no groundwater-recharge ponds in the grasslands and the riparian zone is trampled; without ..." (58). Rather than focusing on pollution, Outwater speaks to the omission of crucial components of the ecosystem by human activity that have diminished resilience and compromised ecosystem services.

The processes of sedimentation, aeration, mixing, and bacterial processing of various water bodies help to break down and filter wastes (1). But our destruction of ecosystems and our increasing pollution are decreasing the services provided and overburdening those that are left. In areas that have undergone rapid industrialization, waterways have often become very polluted. In Czechoslovakia, for example, 70% of source waters were heavily polluted, and 30% were not capable of sustaining fish (1).

The ecosystem provides numerous services that make life on this planet possible. In 1997 Robert Costanza and his colleagues calculated the economic value or natural capital of different ecosystem areas based on 17 different ecosystem services they provided. For the entire biosphere he concluded that the value was between USD 16 and 54 trillion per year, with an average of USD 33 trillion per year. He examined particular types of environments and calculated their contribution to different ecosystem services. Arboreal forests have a value of USD 302 per hectare per year, with the principal services being climate regulation, waste treatment, and food production. For wetlands, the value is USD 14,785 per hectare per year, and swamps and floodplains generate as much as USD 19,580 per hectare per year in natural capital. The largest ecosystem services of wetlands are disturbance regulation, water supply, and waste treatment. Coastal estuaries are also valued particularly high at USD 22,832 per hectare per year; sea grass/algae beds generate USD 19,004 per hectare per year, mainly for their role in nutrient cycling. Rivers and lakes are valued at USD 8,498 per hectare per year, predominantly for water regulation, water supply, waste treatment, and recreation (59).

24.5 DAMS AND OTHER STRUCTURAL DISTURBANCES TO ECOSYSTEMS

Many governments, lending agencies, and other proponents of dams note that the benefits that dams provide are imperative to our modern infrastructure. According to the World Commission on Dams, the irrigation from dams has helped increase world food production 12–16% (60). Proponents argue that although dams are not perfect, they provide a better source of electricity than fossil fuels or other less renewable resources. In many areas, dams have also been able to control floods around large population centers and use the excess water to supply many people with water for general use and irrigation.

In recent years, however, it has come to light that these benefits may not be as straightforward or as beneficial as originally projected. Many communities, activist groups,

and researchers have challenged the virtues of dams, focusing on the negative impacts dams have on the environment and local communities, as well as their inefficiency in power generation. It is often suggested that the project proposals for dams de-emphasize their effects on the environment and the local community, and exaggerate the benefits. They argue that if the proposals had been accurate, most of the dams would not have been viewed as economically viable, and when the full impact of dams on society is considered, including the environmental degradation and the social costs of relocation, it is apparent that the benefits simply do not outweigh the costs.

Given the immense resources that the ecosystem provides, tampering with it can have economic implications. Although dams have in some cases provided energy and water for irrigation, industry, and other uses, large dam projects have many negative effects on the environment and the communities around them. Large reservoir dams can result in rapid change to the ecosystems from the disruptions of river flow and changes to preexisting lakes. Certain species are unable to live in the new environment, while others thrive. Such dams often flood wetlands, which, when left intact, help to clean water, serve as a place for fish spawning, and support great biodiversity. Other projects have flooded large forests and other local resources. Dams block migration of fisheries. Although some projects have tried to address the problem with fish ladders and other engineering solutions, they are often unsuccessful in sustaining the local fish population (61). Although fisheries are often greatly diminished, other species much prefer the new reservoir environment. The slow flowing environment has been found to be a breeding ground for a number of infectious agents that can have deleterious impacts on human health. Common examples include snail-vector diseases such as schistosomiasis and arthropod-borne diseases such as malaria. Following the creation of the Volta Lake in Ghana, the incidence of onchocerciasis or river blindness decreased, as the blackfly (*Simulium damnosum*), which transmits this illness, is dependent on free-flowing water. The incidence of schistosomiasis, however, dramatically increased (62).

Although hydropower is often touted as clean and sustainable and a much better alternative to the use of fossil fuels, mounting evidence suggests that reservoirs created by large dams emit significant amounts of greenhouse gases, in some cases exceeding the amount that would be created by generating equivalent amounts of electricity by burning fossil fuels. This effect results from the release of gases from previous sinks, the decay of flooded biomass and methane-producing bacteria that feed on the decaying vegetation both from the original flooding, and from biomass carried into the reservoir from upstream (61). The concept of sustainability is also a bit of a falsehood because although water may not disappear, dams have finite lifetimes. In India, large deposits of rich topsoil have been flooded or swept into the rivers and deposited in the reservoirs, dramatically reducing their storage capacity and shortening their useful life (63). Furthermore, river valleys and estuaries have been permanently altered and degraded as a result of dams, such that future generations will not be able to receive their ecosystem benefits.

Another concern is that there is new evidence that shows that dams can vastly increase the chances of earthquakes and trigger seismicity. Over 70 dams have been linked with earth tremors (61). There are also concerns about the environmental and social implications of dam failures, particularly in the case of older dams.

As far as the local communities are concerned, dams often destroy much of their livelihoods not only by flooding out wetlands where food and wood are collected and reducing fisheries dramatically, but also by flooding out farmlands and even villages, forcing people to relocate to cities or other locations to find work. This process can be very disruptive to families and communities, particularly for those already struggling with poverty. Tilt and his colleagues note that the most common social implications of large dam projects are "the migration and resettlement of people near the dam sites; changes in rural economy employment structure; effects on infrastructure and housing; impacts on non-material or cultural aspects of life; and impacts on community health and gender relations" (64).

24.6 GROUNDWATER CONTAMINATION

In *The Science of Water: Concepts & Applications,* Spellman provides a long list of sources of pollution of groundwater (and the specific contaminants of concern with each source) including: subsurface percolation, injection wells, land application, landfills, open dumps, residential disposal, surface impoundments, waste tailings, waste piles, materials stockpiles, graveyards, animal burial, above ground storage tanks, underground storage tanks, containers, open burning and detonating sites, radioactive disposal sites, pipelines, material transportation and transfer operations, irrigation practices, pesticide application, animal feeding operations, deicing salts applications, urban runoff, percolation of atmospheric pollutants, mining and mine drainage, production wells, and construction excavation (55).

The main components of groundwater pollution are "organic chemicals, such as trichloroethylene, 1,1,1-trichloroethane, benzene, perchlorate, gasoline (and gasoline additives such as MTBE), pesticides and soil fumigants, disease-causing organisms, and nitrates" (65). About 150 million people in the United States, or half of the current population, depend on groundwater for drinking, which makes control of pollution sources a public health priority (65). Ninety-eight percent of rural populations and 32% of those served by municipal water systems rely on groundwater (65). Groundwater moves on the order of less

than a foot to a few feet per day, a much slower pace than surface water for which flow is measured in feet/second. Because of this slow rate of flow, contamination may remain for a long period and may not be detected (65).

Groundwater provides about 25% of freshwater used in the United States. In many areas, particularly cities, water is being withdrawn from aquifers at a rate greater than it can be replenished. As water levels decline, pumping costs increase, underground pressure and flow dynamics change, and seawater intrusion or saline water being drawn up from lower areas may occur. Poor-quality water can also leak down into deeper areas of the aquifer.

Anthropogenic contamination of groundwater is often overlooked as it is not visible, making it difficult to generate political will for environmental protection and stronger regulation. Major contributors to the problem of groundwater contamination are improper disposal of wastes and spills of hazardous substances onto the ground.

24.6.1 Underground Storage Tanks

Underground storage tanks are used all over the United States. Leaks in these tanks or other structural failures are a significant source of contamination in groundwater. Leaks of petroleum from underground storage tanks have become a large problem despite regulations enacted in 1987 to address the issue (55).

24.6.2 MTBE

Methyl tertiary-butyl ether (MTBE) is a volatile organic compound used as an octane enhancer in gasoline that helps to reduce carbon monoxide emission and ozone levels. In areas with poor air quality, the Clean Air Act of 1990 mandated that reformulated gasoline (RFG) be used to help reduce the ozone and smog problems. RFG currently makes up 30% of gasoline sold in the United States. Ethanol and MTBE are most commonly used to meet this requirement (55). Despite its benefits in reducing toxic air pollutants, however, MTBE poses a threat to drinking water because it can leak into the ground from petroleum storage tanks, or from motors used in boats or other recreational vehicles on the water. Air contamination by emissions from cars and trucks can also contribute to surface water pollution. MTBE can move easily into groundwater. Although no current tests exist for human health effects of MTBE, animal studies show a number of cancer and other effects at high exposure levels over prolonged periods (55).

24.6.3 Septic Tanks

Septic tanks are a concern as they age and leak fecal bacteria and viruses into the groundwater. Detergents, nitrates, chlorides, and solvents are also often disposed of in septic tanks or used to treat the septic tanks (trichloroethylene). These contaminants can be a risk to human health (55).

24.6.4 Landfills

In the United States landfills are falling out of favor since the deleterious environmental effects have been observed. Nonetheless, some newer and many of the older landfills continue to pollute groundwater and aquifers, especially those located in lower lying areas close to the water table. "Leachate, high BOD [biochemical oxygen demand], chloride, organics, heavy metals, nitrate, and other contaminants" leach into water from these sites (55).

24.7 LEGISLATION AND REGULATION OF WATER SUPPLIES

After World War II growing awareness of the increase in pollution of U.S. waterways moved the U.S. Congress to pass the Federal Water Pollution Control Act or Clean Water Act in 1948. The original statute (Ch. 758; P.L. 845) was a public health act, authorizing the Surgeon General of the U.S. Public Health Service to develop comprehensive programs for the elimination or reduction of pollution in interstate waters and the improvement of the "sanitary condition of surface and underground waters" (66). Although initiated as a public health act, the statute also called for steps necessary to protect and conserve waters for recreation, agriculture, industry, healthy environments for fish and other aquatic life, in addition to assuring safe water for public consumption. Subsequent amendments to the original Clean Water Act added authorization for stricter standards, additional water quality projects, and refinement of allowable discharges. Then in June 1969 the Cuyahoga River in Cleveland, OH, caught fire. It was not the first time that oil slicks and debris had caught fire on the Cuyahoga or other rivers, but this event served as a rallying cry for protection of the environment, contributing to the political will necessary for the creation of the USEPA, declaration of the first Earth Day, and passage of the 1970 Clean Water Act (66), which authorized USD 50 billion for cities and states to build wastewater treatment facilities. The Federal Water Pollution Control Act Amendments of 1972 granted states the authority to set standards for surface waters sufficient to be able to support aquatic life and recreation. But regulations without adequate enforcement are insufficient to protect surface and groundwater from continued anthropogenic pollution. As water resources become more limited and population pressure increases demand, new and stronger regulatory methods must be implemented to assure the right to water as a major component of the right to health.

PART 2—PHARMACEUTICALS AND PERSONAL CARE PRODUCTS IN THE ENVIRONMENT

M. Danielle McDonald

Pharmaceuticals and Personal Care Products in the Environment

With advances in environmental residue analysis resulting in higher detection efficiencies, scientists have now become aware that pharmaceuticals and the active ingredients in personal care products are present in the world's rivers, lakes, and marine coastal ecosystems (67, 68); many are found in high enough concentrations to cause harm to aquatic organisms. Compounds such as nonsteroidal anti-inflammatory drugs (pain relievers), beta-blockers (blood pressure modulators), blood lipid lowering agents (cholesterol reducers), neuroactive compounds (e.g., antidepressants), steroidal hormones (e.g., contraceptives), antibiotics, surfactants (detergents), musks (perfumes), and UV filters (sunscreens) can enter the environment through the agricultural runoff from veterinary-treated livestock (69, 70), during manufacturing, by irrigation with reclaimed wastewater and by landfill leachates (71–75). However, the main route of entry is through human consumption, excretion and the subsequent treatment, or lack thereof, and release by sewage treatment plants (STPs). Pharmaceuticals and personal care products (PPCPs) also enter STPs through bathing and by the improper disposal of unused and expired pharmaceuticals either down the toilet or sink (76). STP elimination efficiencies span a large range from completely ineffective (0% elimination) to relatively effective (99% elimination) in removing these compounds, depending on the STP treatment technology, hydraulic retention time, and season (77, 78). Furthermore, while adsorption to suspended solids (sewage sludge), one of the two main elimination processes important in wastewater treatment (the other being biodegradation) removes PPCPs from wastewater, degradation of these compounds in sludge does not always occur and many compounds, such as steroid hormones, end up being present in sludge in measurable amounts. This becomes an environmental problem when the contaminated sludge is then used as land fertilizer (79, 80).

In terms of the toxicity of environmental PPCP contamination, aquatic organisms are continuously exposed to low levels of these chemicals over long periods of time or even over their entire life cycle. Chronic toxicity is usually not lethal, but it is this long-term exposure that can result in changes in the reproductive success, behavior, and general fitness of aquatic biota, all which ultimately affect the ecosystem. Moreover, chronic, sublethal exposure of aquatic organisms to low concentrations of PPCPs, or to any toxicant for that matter, can have implications on human health, since it allows for the organism to accumulate the toxicant in its tissues, sometimes to very high levels, which could then be transferred to other aquatic organisms and finally to humans through the food chain. Because of their lipophilicity, many PPCPs are readily taken up by aquatic organisms and bioaccumulated. While the direct risk to human health through exposure to or consumption of polluted water or contaminated aquatic organisms is not yet cause for concern in most cases; that is, we would have to eat hundreds of thousands of kilograms of fish or drink an equivalent volume of water per day to get a daily dose of any given pharmaceutical, the synergistic effects of low amounts of multiple compounds if bioaccumulated in fish and then consumed by humans is, at this time, not predictable.

Unlike many other environmental toxicants, pharmaceuticals are designed to target specific metabolic and molecular pathways in humans. When introduced into the environment, these compounds will likely affect the same pathways and have similar side-effects in aquatic organisms (i.e., aquatic plants, invertebrates, and vertebrates) as they do in mammals since the target organs or receptors are very similar. However, this is not always the case and the lack of a similar mechanism seldom results in a "noneffect" of the compound; instead, an unpredicted response is usually measured. Since the specific modes of action in aquatic organisms are not well known for many drugs and personal care products, toxicity analysis of PPCPs, namely, the identification of PPCP-sensitive endpoints, in aquatic organisms has been an ongoing challenge.

In addition to their predicted effect, research has shown that many of the PPCPs listed above also exert negative effects on the endocrine system of aquatic organisms, even when the endocrine system is not their primary target in humans. Compounds that interfere with the endocrine system of organisms are referred to as endocrine disrupting compounds (EDCs) and steroid hormones, such as those used in birth control pills, are unique in this sense in that their target in humans is in fact the endocrine system. For this reason, this class of compounds has had a substantial impact to our aquatic environment. In general, most of the estrogenic metabolites tend to result in a similar ecological effect, which is a significant reduction in reproduction, whether it be due to a decrease in reproductive behavior and fecundity (81, 82), an increase in feminization, often resulting in "intersex" species (83–85) or a decrease in fertilization rate (86). While the changes in reproductive abilities of aquatic biota are devastating ecologically and will ultimately have a global impact if left unchecked, evidence also suggests that environmental EDCs also pose a clear risk to human health (87, 88), that may include a role in the decline in the human sex ratio (89–92), several female reproductive disorders (reviewed by Crain et al. (93)) and obesity (reviewed by Newbold et al. (94)).

In addition to environmental contamination with EDCs, another potentially serious problem to the aquatic ecosystem and human health is the indiscriminant use and misuse of

antibiotics, which is the leading proposed cause of the increased and spreading resistance among bacteria (95). Drug resistance in bacteria is not believed to develop in surface waters, since environmental concentrations of antibiotics are not high enough to be toxic and promote the selection of resistant bacteria. However, if our current habits do not change, antibiotics will reach levels in the environment that are toxic, the surviving bacteria will multiply, and very quickly a resistant strain will be developed.

Essentially, the most straightforward way to reduce the quantity of pharmaceuticals in the environment is to limit their consumption (96). This can be done in many ways, the first being to educate health care practitioners to ensure they fully understand the importance and environmental implications of selecting the right medication and therapy for each patient. They could also minimize the volume of pharmaceutical waste by identifying the lowest effective dosage on a per person basis and this would only be feasible if pharmaceuticals were packaged in small enough quantities. Repackaging is especially important for PPCPs that are prone to being thrown out because they are purchased in quantities too great to be used before they expire. Lastly, educating patients of the importance of completing treatments and following their physician's directions precisely will ensure the proper dosing of pharmaceuticals and prevent the recurrence of illness due to improper pharmaceutical usage.

Currently, unused medications are commonly disposed of in the trash, are flushed down the toilet or sink or they are shared with other individuals (76, 97). These methods not only lead to detrimental effects on environmental health but can have harmful effects on human health directly. Australia, Canada, and many nations within the European Union have been the global leaders in terms of their proactive efforts in reducing the harmful effects of improper PPCP disposal on human and environmental health by providing the legal framework and resources required to allow health care facilities, patients, and the public to return unwanted and expired pharmaceuticals to be reused or disposed of by incineration. These programs not only reduce the amount of pharmaceutical waste introduced to the environment by keeping these compounds out of landfills and the water supply but they help to avoid dosing errors, drug abuse, and accidental poisonings that result in the accumulation of unused pharmaceuticals in the home. Here in the United States, pharmaceutical take-back programs are becoming more common, depending on, household trash and recycling service of individual local governments. The U.S. Office of National Drug Control Policy recently released federal guidelines on the disposal of prescription drugs (www.whitehousedrugpolicy.gov). These guidelines inform the public not to flush prescription drugs down the toilet or drain; instead, if pharmaceutical take-back programs are not available in your community, place the unused pharmaceutical, mixed with an undesirable substance such as cat litter or used coffee grounds, in the trash. The involvement of the U.S. federal government is an important first step in getting on top of this problem in this country; now it is up to us as individuals to hold ourselves and our community leaders accountable.

REFERENCES

1. Nash L. Water quality and health. In: Gleick, PH, Pacific Institute for Studies in Development, Environment, and Security & Stockholm Environment Institute, editors. *Water in Crisis: A Guide to the World's Fresh Water Resources*. New York: Oxford University Press, 1993, pp. 25–39.

2. UNICEF. *Water, Sanitation and Hygiene*, 2008. Retrieved 30 March, 2010, from http://www.unicef.org/wash/index_statistics.html.

3. The Carter Center. *Guinea Worm Disease Eradication*: The Road to Eradication:Countdown to Zero. Retrieved 17 March 2011, from http://www.cartercenter.org/health/guinea_worm/mini_site/index.html.

4. USEPA. *The Quality of Our Nation's Waters—A Summary of the National Water Quality Inventory: 1998 Report to Congress*. No. EPA841-S-00-001. Washington, DC: USEPA, 2000.

5. Smith JEJ, Perdek JM. Assessment and management of watershed microbial contaminants. *Crit Rev Environ Sci Technol* 2004;34(2):109–139.

6. Mallin MA, Williams KE, Esham EC, Lowe RP. Effect of human development on bacteriological water quality in coastal watersheds. *Ecol Appl* 2000;10:1047–1056.

7. Johnson CD, Juengst D. *Polluted Urban Runoff: A Source of Concern*. No. I-02-97-5M-20-S. Madison, WI: University of Wisconsin Extension, 1997.

8. Mallin MA, Johnson VL, Ensign SH. Comparative impacts of stormwater runoff on water quality of an urban, a suburban, and a rural stream. *Environ Monit Assess*, 2008. DOI 10.1007/s10661-008-0644-4.

9. Nweke OC, Sanders WHI. Modern environmental health hazards: a public health issue of increasing significance in Africa. *Environ Health Perspect* 2009;117(6):863–870.

10. Lau S, Han Y, Kang J, Kayhanian M, Stenstrom MK. Characteristics of highway stormwater runoff in Los Angeles: metals and polycyclic aromatic hydrocarbons. *Water Environ Res* 2009;81(3):308–318.

11. Thunqvist E. Regional increase of mean chloride concentration in water due to the application of deicing salt. *Sci Total Environ* 2004;325:29–37.

12. Larsson SC, Bergkvist L, Wolk A. Processed meat consumption, dietary nitrosamines and stomach cancer risk in a cohort of Swedish women. *Int J Cancer* 2006;119(4):915–919.

13. Bannerman RT, Owens DW, Dodds RB, Hornewer NJ. Sources of pollutants in Wisconsin stormwater. *Water Sci Technol* 1993;28(3–5):241–259.

14. Holland AF, Sanger DMG, Lerberg SB, Santiago MS, Riekerk GHM, et al. Linkages between tidal creek ecosystems and the landscape and demographic attributes of their watersheds. *J Exp Mar Biol Ecol* 2004;298(2):151–178.

15. Davis AP, Shokouhian M, Ni S. Loading estimates of lead, copper, cadmium, and zinc in urban runoff from specific sources. *Chemosphere* 2001;44:997–1009.

16. Gromaire-Mertz MC, Garnaud S, Gonzalez A, Chebbo G. Characterisation of urban runoff pollution in Paris. *Water Sci Technol* 1999;39(2):1–8.

17. Malm O, Pfeiffer WC, Souza CMM, Reuther R. Mercury pollution due to gold mining in the Madeira river basin, Brazil. *AMBIO AMBOCX* 1989;19(1):11–15.

18. Bernard BP, Becker C. Environmental lead exposure and the kidney. *Clin Toxicol* 1988;26:1–34.

19. Gonick HC. Nephrotoxicity of cadmium and lead. *Indian J Med Res* 2008;128:335–352.

20. Frisbie SH, Mitchell EJ, Mastera LJ, Maynard DM, Yusuf AZ, Siddiq MY, et al. Public health strategies for western Bangladesh that address arsenic, manganese, uranium, and other toxic elements in drinking water. *Environ Health Perspect* 2009; 117(3):410–416.

21. Mazumder DNG. Chronic arsenic toxicity & human health. *Indian J Med Res* 2008;128:436–447.

22. Raqib R, Ahmed S, Sultana R, Wagatsuma Y, Mondal D, Hoque AM, et al. Effects of *in utero* arsenic exposure on child immunity and morbidity in rural Bangladesh. *Toxicol Lett* 2009;185(3):197–202.

23. Smith AH, Steinmaus CM. Health effects of arsenic and chromium in drinking water: recent human findings. *Ann Rev Public Health* 2009;30:107–122.

24. Nikic D, Stojanovic D, Stankovic A. Cadmium in urine of children and adults from industrial areas. *Cent Eur J Pub Health* 2005;13(3):149–152.

25. Nadeem-ul-Haq Arain MA, Haque Z, Badar N, Mughal N. Drinking water contamination by chromium and lead in industrial lands of Karachi. *J Pakistan Med Assoc* 2009;59(5): 270–274.

26. Field RW, Steck DJ, Smith BJ, et al. Residential radon gas exposure and lung cancer: the Iowa radon lung cancer study. *Am J Epidemiol* 2000;151(11):1091–1102.

27. Jackson J, Sutton R. Sources of endocrine-disrupting chemicals in urban wastewater, Oakland, CA. *Sci Total Environ* 2008;405:153–160.

28. Kortenkamp A. Low dose mixture effects of endocrine disrupters: implications for risk assessment and epidemiology. *Int J Androl* 2008;31:233–240.

29. Barrett JR. Fertile grounds for inquiry: environmental effects on human reproduction. *Environ Health Perspect* 2006;114(11): A644–A649.

30. Guo YL, Hsu P, Hsu C, Lambert GH. Semen quality after prenatal exposure to polychlorinated biphenyls and dibenzofurans. *Lancet* 2000;356:1240–1241.

31. Swan SH, Kruse RL, Liu F, Barr DB, Drobnis EZ, Redmon JB, et al. Semen quality in relation to biomarkers of pesticide exposure. *Environ Health Perspect* 2003;111(12): 1478–1484.

32. Sower SA, Reed KL, Babbitt KJ. Limb malformations and abnormal sex hormone concentration in frogs. *Environ Health Perspect* 2000;108(11):1085–1090.

33. Hayes T, Haston K, Tsui M, Hoang A, Haeffele C, Vonk A. Atrazine-induced hermaphroditism at 0.1 ppb in American leopard frogs (*Rana pipiens*): laboratory and field evidence. *Environ Health Perspect* 2003;111(4):568–575.

34. Hayes TB, Haston K, Tsui M, Hoang A, Haeffele C, Vonk A. Feminization of male frogs in the wild. *Nature* 2002;419:895–896.

35. Cohn BA, Wolff MS, Cirillo PM, Sholtz RI. DDT and breast cancer in young women: new data on the significance of age at exposure. *Environ Health Perspect* 2007;115(10):1406–1414.

36. Zhang Y, Wise JP, Holford TR, Xie H, Boyle P, Zahm SH, et al. Serum polychlorinated biphenyls, cytochrome P-450 1A1 polymorphisms, and risk of breast cancer in Connecticut women. *Am J Epidemiol* 2004;160:1177–1183.

37. Colborn T. Neurodevelopment and endocrine disruption. *Environ Health Perspect* 2004;112(9):944–949.

38. Colborn T. A case for revisiting the safety of pesticides: a closer look at neurodevelopment. *Environ Health Perspect* 2006;114(1):10–17.

39. Stewart P, Fitzgerald S, Reihman J, Gump B, Lonky E, Darvill T, et al. Prenatal PCB exposure, the corpus callosum, and response inhibition. *Environ Health Perspect* 2003;111(13):1670–1677.

40. Grandjean P, Landrigan PJ. Developmental neurotoxicity of industrial chemicals. *Lancet* 2006;368:2167–2178.

41. Lema SC, Dickey JT, Schultz IR, Swanson P. Dietary exposure to 2,2′,4,4′-tetrabromodiphenyl ether (PBDE-47) alters thyroid status and thyroid hormone-regulated gene transcription. *Environ Health Perspect* 2008;116(12):1694–1699.

42. Hooper K, McDonald TA. The PBDEs: an emerging environmental challenge and another reason for breast-milk monitoring programs. *Environ Health Perspect* 2000;108(5):387–392.

43. Meeker JD, Sathyanarayana S, Swan SH. Phthalates and other additives in plastics: human exposure and associated health outcomes. *Philos Trans R Soc* 2009;364:2097–2113.

44. Swan SH. Environmental phthalate exposure in relation to reproductive outcomes and other health endpoints in humans. *Environ Res* 2008;108:177–184.

45. Ogunfowokan AO, Torto N, Adenuga AA, Okoh EK. Survey of levels of phthalate ester plasticizers in a sewage lagoon effluent and a receiving stream. *Environ Monit Assess* 2006;118:457–480.

46. Hites RA. Phthalates in the Charles and the Merrimack rivers. *Environ Health Perspect* 1973;17–21.

47. Thompson RC, Moore CJ, vom Saal FS, Swan SH. Plastics, the environment and human health: current consensus and future trends. *Philos Trans R Soc* 2009;364:2153–2166.

48. Latini G, de Felice C, Presta G, del Vecchio A, Paris I, Ruggieri F, et al. *In utero* exposure to di-(2-ethylhexyl)phthalate and duration of human pregnancy. *Environ Health Perspect* 2003;111(14):1783–1785.

49. Zhang Y, Lin L, Cao Y, Chen B, Zheng L, Ge R. Phthalate levels and low birth weight: a nested case-control study of Chinese newborns. *J Pediatr* 2009;1–5.

50. Howdeshell KL, Wilson VS, Furr J, Lambright CR, Rider CV, Blystone CR, et al. A mixture of five phthalate esters inhibit fetal testicular testosterone production in the Sprague-Dawley rat in a cumulative, dose-additive manner. *Toxicol Sci* 2008;105(1):153–165.

51. Jones KC, de Voogt P. Persistent organic pollutants (POPs): state of the science. *Environ Pollut* 1999;100:209–221.

52. Keita-Ouane F. Persistent organic pollutants: a global issue, 1997. Retrieved 30 March 2010, from http://www.chem.unep.ch/pops/POPs_Inc/proceedings/lusaka/OUANE.html.

53. Minh NH, Minh TB, Kajiwara N, Kunisue T, Iwata H, Viet PH, et al. Pollution sources and occurrences of selected persistent organic pollutants (POPs) in sediments of the Mekong river delta, South Vietnam. *Chemosphere* 2007;67:1794–1801.

54. ATSDR. *Public health statement for polycyclic aromatic hydrocarbons (PAHs)*, 2008. Retrieved 30 March 2010, from http://www.atsdr.cdc.gov/toxprofiles/phs69.html.

55. Spellman FR. *The Science of Water: Concepts and Applications*, 2nd edition. Boca Raton, FL: CRC Press, 2008.

56. Tisler T, Zagorc-Koncan J. The "whole-effluent" toxicity approach. *Int J Environ Pollut* 2007;31(1–2):3–12.

57. Raudsepp-Hearne C, Reid W. *Millennium Ecosystem Assessment*, 2005. Retrieved 30 March 2010, from http://www.millenniumassessment.org/en/index.aspx.

58. Outwater A. *Water: A Natural History*. New York: Basic Books, 1996.

59. Costanza R, d'Arge R, de Groot R, Farber S, Grasso M, Hannon B, et al. The value of the world's ecosystem services and natural capital. *Nature*, 1997;387:253–260.

60. World Commission on Dams. *Dams and Development: A New Framework for Decision-Making. The Report of the World Commission on Dam*. London: Earthscan, 2000.

61. McCully, P. *Silenced Rivers: The Ecology and Politics of Large Dams*. London: Zed Books, 2001.

62. Fobil JN. *Remediation of the environmental impacts of the Akosombo and Kpong Dams in Ghana*, 2003. Retrieved 30 March 2010, from http://www.solutions-site.org/artman/publish/article_53.shtml.

63. Allvares C, Billorey R. *Damming the Narmada*. Penang: Jutaprint, 1988.

64. Tilt B, Braun Y, He D. Social impacts of large dam projects: a comparison of international case studies and implications for best practice. *J Environ Manage* 2009;90:S249–S257.

65. Nemerow NL, Salvato JA, Agardy FJ. *Environmental Engineering: Water, Wastewater, Soil and Groundwater Treatment and Remediation*, 6th edition. Hoboken, NJ: John Wiley & Sons, 2009.

66. Federal Water Pollution Control Act (Clean Water Act). Retrieved 30 March 2010, from http://www.fws.gov/laws/lawsdigest/FWATRPO.HTML.

67. Kolpin DW, Furlong ET, Meyer MT, Thurman EM, Zaugg SD, Barber LB, Buxton HT. Pharmaceuticals, hormones, and other organic wastewater contaminants in US streams, 1999–2000: a national reconnaissance. *Environ Sci Technol* 2002; 36:1202–1211.

68. Vassog T, Anderssen T, Pedersen-Bjergaard S, Kallenborn R, Jensen E. Occurrence of selective serotonin reuptake inhibitors in sewage and receiving waters at Spitsbergen and in Norway. *J Chromatogr* 2008;1185A:194–205.

69. Boxall ABA, Blackwell P, Cavallo R, Kay P, Tolls J. The sorption and transport of a sulphonamide antibiotic in soil systems. *Toxicol Lett* 2002;131:19–28.

70. Boxall AB, Kolpin DW, Halling-Sorensen B, Tolls J. Are veterinary medicines causing environmental risks? *Environ Sci Technol* 2003;37:286A–294A.

71. Daughton CG, Ternes TA. Pharmaceuticals and personal care products in the environment: agents of subtle change? *Environ Health Perspect* 1999;107:907–938.

72. Balcioglu IA, Ötker M. Treatment of pharmaceutical wastewater containing antibiotics by O-3 and O-3/H_2O_2 processes. *Chemosphere* 2003;50:85–95.

73. Cordy GE, Duran NL, Bouwer H, Rice RC, Furlong ET, Zaugg SD, Meyer MT, Barber LB, Kolpin DW. Do pharmaceuticals, pathogens, and other organic waste water compounds persist when waste water is used for recharge? *Ground Water Monitor Remed* 2004;24:58–69.

74. Kinney CA, Furlong ET, Werner SL, Cahill JD. Presence and distribution of wastewater-derived pharmaceuticals in soil irrigated with reclaimed water. *Environ Toxicol Chem* 2006;25:317–326.

75. Kinney CA, Furlong ET, Zaugg SD, Burkhardt MR, Werner SL, Cahill JD, Jorgensen GR. Survey of organic wastewater contaminants in biosolids destined for land application. *Environ Sci Technol* 2006;40:7207–7215.

76. Seehusen DA, Edwards J. Patient practises and beliefs concerning disposal of medications. *J Am Board Fam Med* 2006;19:542–547.

77. Ternes TA. Occurrence of drugs in German sewage treatment plants and rivers. *Water Res* 1998;32:3245–3260.

78. Fent K, Weston AA, Caminada D. Ecotoxicology of human pharmaceuticals. *Aquat Toxicol* 2006: 76:122–159.

79. Dizer H, Fischer B, Sepulveda I, Loffredo E, Senesi N, Santana F, Hansen PD. Estrogenic effect of leachates and soil extracts from lysimeters spiked with sewage sludge and reference endocrine disrupters. *Environ Toxicol* 2002;17:105–112.

80. Golet EM, Xifra I, Siegrist H, Alder AC, Giger W. Environmental exposure assessment of fluoroquinolone antibacterial agents from sewage to soil. *Environ Sci Technol* 2003; 37:3243–3249.

81. Nash JP, Kime DE, Van der Ven LT, Wester PW, Brion F, Maack G, Stahlschmidt-Allner P, Tyler CR. Long-term exposure to environmental concentrations of the pharmaceutical ethinylestradiol causes reproductive failure in fish. *Environ Health Perspect* 2004;112:1725–1733.

82. Brown KH, Schultz IR, Cloud JG, Nagler JJ. Ahneuploid sperm formation in rainbow trout exposed to the environmental estrogen 17α-ethynylestradiol. *Proc Natl Acad Sci USA* 2008;105:19786–19791.

83. Metcalfe CD, Metcalfe TL, Kiparissis Y, Koenig BG, Khan C, Hughes RJ, Croley TR, March RE, Potter T. Estrogenic potency of chemicals detected in sewage treatment plant effluents as determined by *in vivo* assays with Japanese medaka (*Oryzias latipes*). *Environ Toxicol Chem* 2001;20:97–308.

84. Hirai N, Nanba A, Koshio M, Kondo T, Morita M, Tatarazako N. Feminization of Japanese medaka (*Oryzias latipes*) exposed to 17β-estradiol: formation of testis-ova and sex-transformation during early ontogeny. *Aquat Toxicol* 2006;77:78–86.

85. Salierno JD, Kane AS. 17 Alpha-ethinylestradiol alters reproductive behaviors, circulating hormones, and sexual morphology in male fathead minnows (*Pimephales piromelas*). *Environ Toxicol Chem* 2009;28:953–961.

86. Van den Belt K, Verheyen R, Witters H. Reproductive effects of ethynylestradiol and 4t-octylphenol on the zebrafish (*Danio rerio*). *Arch Environ Contam Toxicol* 2001;41:458–467.

87. Mclachlan JA. Environmental signaling: what embryos and evolution teach us about endocrine disrupting chemicals. *Endocrine Rev* 2001;22:319–341.

88. Mclachlan JA, Newbold RR, Burow ME, Li SF. From malformations to molecular mechanisms in the male: three decades of research on endocrine disrupters. *Apmis* 2001;109:263–272.

89. WHO/IPCS. Global assessment of the state-of-the-science of endocrine disruptors. World Health Organization/International Program on Chemical Safety, 2002. WHO/PCS/EDE/0. 02. Available at www.who.int/ppcs/emerg_site/edc/global_edc_ch5.pdf.

90. Jongbloet PH, Zielhuis GA, Groenewoud HMM, Pasker-de Jong PCM. The secular trends in male:female ratio at birth in postwar industrialised countries. *Environ Health Perspect* 2001;109:749–752.

91. Hood E. Are EDCs blurring issues of gender? *Environ Health Perspect* 2005;113:A670–A677.

92. Mackenzie CA, Lockridge A, Keith M. Declining sex ration in a first nation community. *Environ Health Perspect* 2005;113:1295–1298.

93. Crain D, Janssen S, Edwards T, Heindel J, Ho S, Hunt P, Iguchi T, Juul A, McLachlan J, Schwatz J, Skakkebaek N, Soto AM, Swan S, Walker C, Woodruff TK, Woodruff TJ, Guidice MLC, Guillette LJ Jr. Female reproductive disorders: the roles of endocrine-disrupting compounds and developmental timing. *Fertil Steril* 2008;90:911–940.

94. Newbold RR, Padilla-Banks E, Jefferson WN. Environmental estrogens and obesity. *Mol Cell Endocrinol* 2009;304:84–89.

95. Gould IM. Antibiotic resistance: the perfect storm. *Int J Antimicrob Agents* 2009;34:S2–S5.

96. Daughton LG. Cradle-to-cradle stewardship of drugs minimizing their environmental disposition while promoting human health II. Drug disposal, waste reduction and future directions. *Environ Health Perspect* 2003;111:775–785.

97. Kuspis DA, Krenzelok EP. What happens to expired medications? *Vet Human Toxicol* 1996;38:48–49.

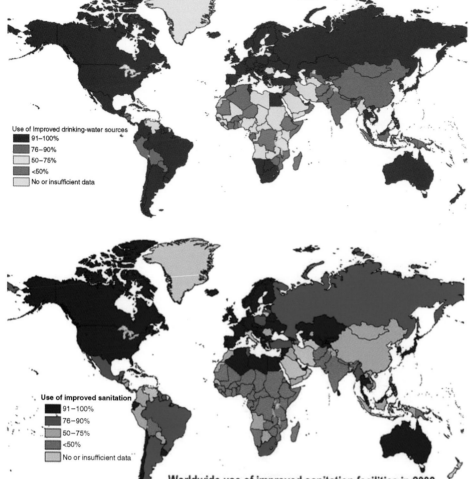

FIGURE 1.7 Countries represented by the percentage of population using improved drinking water supplies. (*See page 11 for text discussion.*)

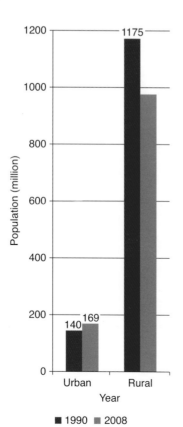

Worldwide use of improved sanitation facilities in 2008

FIGURE 2.5 Use of improved sanitation facilities by country (2008). (*See page 20 for text discussion.*)

FIGURE 2.7 Number of people engaging in open defecation in urban and rural sectors worldwide (1990–2008). (*See page 21 for text discussion.*)

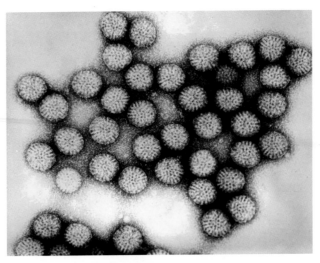

Colorized transmission electron micrograph of **ROTAVIRUS VIRIONS**. (Content provider: CDC/ Dr. Erskine Palmer. *Source*: CDC, Public Health Image Library, Image ID # 178.) (*See page 51 for text discussion.*)

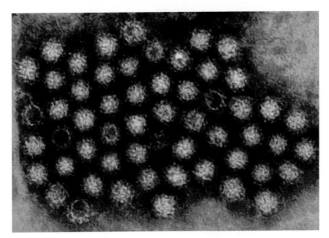

Colorized transmission electron micrograph of **NOROVIRUS VIRIONS**. (Content provider: CDC/ Charles D. Humphrey. *Source*: CDC, Public Health Image Library, Image ID # 10708.) (*See page 52 for text discussion.*)

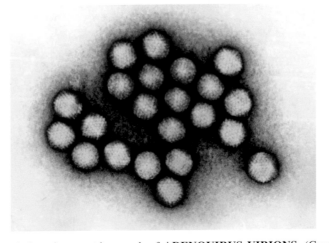

Colorized transmission electron micrograph of **ADENOVIRUS VIRIONS**. (Content provider: CDC/Dr. G. William Gary, Jr. *Source*: CDC, Public Health Image Library, Image ID # 10010.) (*See page 53 for text discussion.*)

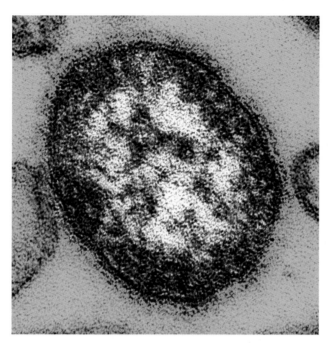

Colorized thin-section electron micrograph transmission of a single **MEASLES VIRION**. (Content provider: CDC/Cynthia S Goldsmith; William Bellini, Ph.D. *Source*: CDC, Public Health Image Library, Image ID # 10707.) (*See page 53 for text discussion*.)

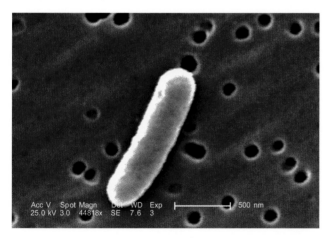

Colorized scanning electron micrograph a single enterotoxigenic *E. COLI* **BACTERIUM**. (Content provider: CDC/Janice Haney Carr. *Source*: CDC, Public Health Image Library, Image ID # 10577.) (*See page 54 for text discussion*.)

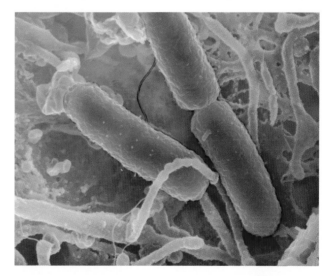

Colorized scanning electron micrograph of a *SHIGELLA* **sp. BACTERIA**. (Copyright Institut Pasteur.) (*See page 55 for text discussion*.)

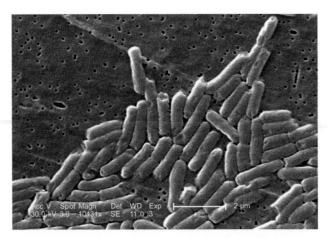

Colorized scanning electron micrograph of a *SALMONELLA* **sp. COLONY**. (Content provider: CDC/Janice Haney Carr. *Source*: CDC, Public Health Image Library, Image ID # 10896.) (*See page 55 for text discussion.*)

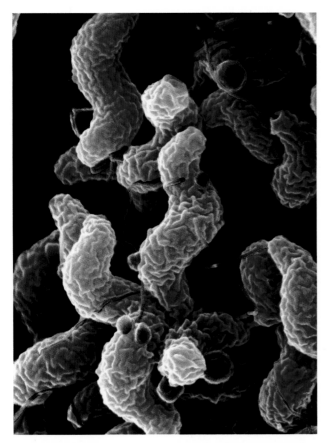

Colorized scanning electron micrograph of *CAMPYLOBACTER JEJUNI* **BACTERIA**. (*Source*: Agricultural Research Service (ARS) is the U.S. Department of Agriculture's Chief Scientific Research Agency.) (*See page 56 for text discussion.*)

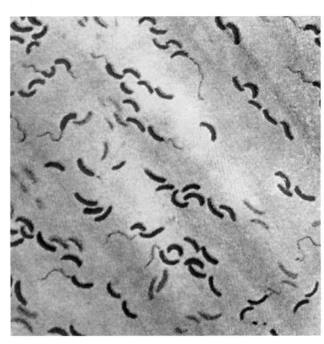

Gram-stained light micrograph of *V. CHOLERAE*. (Content provider: CDC. *Source*: CDC, Public Health Image Library, Image ID # 5324.) (*See page 56 for text discussion.*)

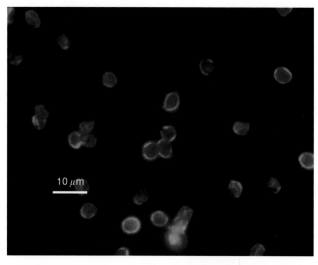

Immunofluorescence light micrograph of purified ***CRYPTOSPORIDIUM PARVUM* OOCYSTS**. (Photo credit: H.D.A Lindquist, U.S. EPA. *Source*: http://www.epa.gov/nerlcwww/cpt_seq1.html.) (*See page 57 for text discussion*.)

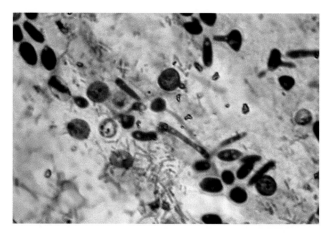

Light micrograph of ***CRYPTOSPORIDIUM* sp. OOCYSTS** in direct fecal smear, acid-fast stained red. (Content provider: CDC/*J Infect Dis* 1983 May;147(5):824–828. *Source*: CDC, Public Health Image Library, Image ID # 5242.) (*See page 57 for text discussion*.)

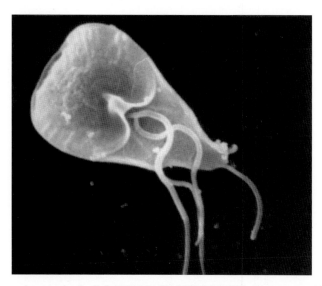

Scanning electron micrograph of a ***G. LAMBLIA* TROPHOZOITE**. (Content provider: CDC/Janice Carr. *Source*: CDC, Public Health Image Library, Image ID # 8698.) (*See page 58 for text discussion*.)

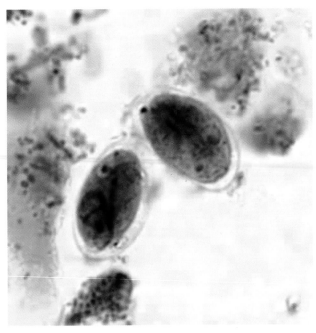

Light micrograph of **G. LAMBLIA** stained with trichrome. (Content provider: CDC/Division of Parasitic Diseases. *Source*: http://www.dpd.cdc.gov/dpdx/HTML/ImageLibrary/Giardiasis_il.htm.) (*See page 58 for text discussion.*)

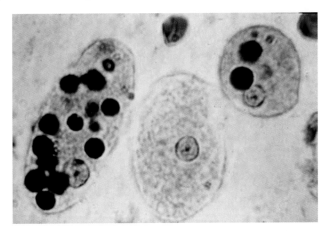

Light micrograph of **E. HISTOLYTICA TROPHOZOITES** with ingested erythrocytes stained with trichrome. (Content provider: CDC/Division of Parasitic Diseases. *Source*: http://www.dpd.cdc.gov/dpdx/ HTML/ImageLibrary/A-F/Amebiasis/body_Amebiasis_il6.htm.) (*See pages 58–59 for text discussion.*)

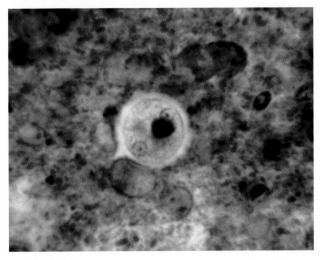

Light micrograph of **E. HISTOLYTICA CYST** stained with chlorazol black. (Content provider: CDC/Dr. George Healy. *Source*: CDC, Public Health Image Library, Image ID # 1474.) (*See pages 58–59 for text discussion.*)

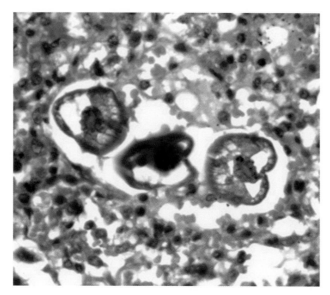

FIGURE 7.7 *A. lumbricoides* larvae in lung tissue. (Courtesy Division of Parasitic Diseases/CDC.) (*See page 85 for text discussion.*)

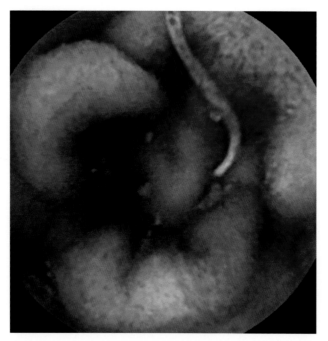

FIGURE 7.9 Hookworm in intestine (video capsule endoscopy image). (Courtesy Dr. Lauren Gerson, Stanford University School of Medicine.) (*See page 86 for text discussion.*)

FIGURE 8.1 Mixed bloom of *Anabaena circinalis*, *Microcystis flos-aquae*, and *Cylindrospermopsis raciborskii*; Murray River upstream of Torrumbarry Weir, New South Wales, Australia. Image courtesy NSW Office of Water. (*See page 101 for text discussion.*)

FIGURE 8.2 *Nodularia spumigena*; recreational lake, Logan Shire, Queensland, Australia. (*See page 103 for text discussion.*)

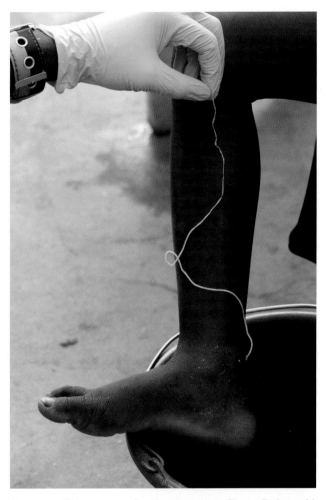

FIGURE 10.1 Emerging Guinea worm. Savelugu, northern Ghana: Patient with a Guinea worm emerging, at the Savelugu Case Containment Center. (Photo credit: The Carter Center/Louise Gubb.) (*See page 126 for text discussion.*)

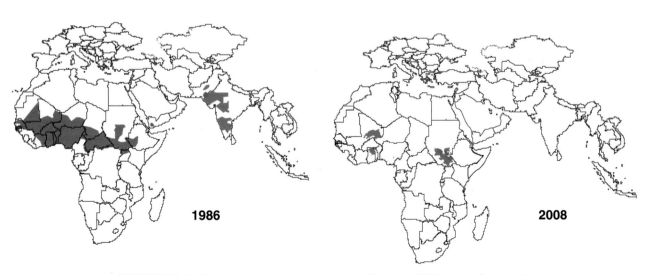

1986

2008

FIGURE 10.3 Guinea worm reduction over time. (*See page 128 for text discussion.*)

(a)

(b)

FIGURE 10.2   (a) Cloth filter. Savelugu, northern Ghana: Technical assistant Eugene Yeng and a young community-based surveillance volunteer, Abdulhai Idiris. (Photo credit: The Carter Center/Louise Gubb.) (b) Pipe filters . (Photo credit: The Carter Center/Emily Staub.) (*See page 127 for text discussion*.)

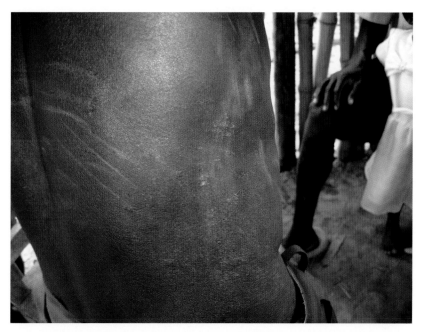

FIGURE 11.2 The typical changes of early onchocerciasis infection. (*See page 136 for text discussion.*)

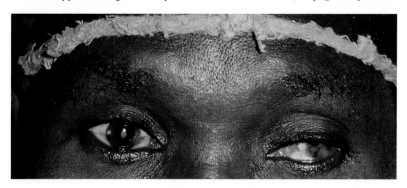

FIGURE 11.5 A patient from Sudan showing typical chronic uveitis with secondary cataract in her right eye and sclerosing keratits in her left eye. (*See page 138 for text discussion.*)

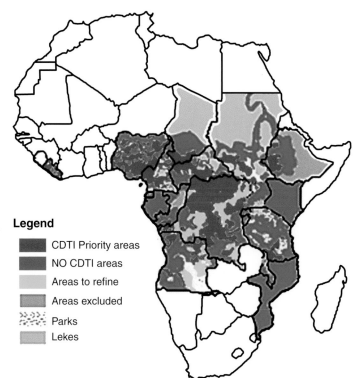

Legend

- CDTI Priority areas
- NO CDTI areas
- Areas to refine
- Areas excluded
- Parks
- Lekes

FIGURE 11.7 The current REMO map seen on the APOC website. http://www.who.int/blindness/partner-ships/APOC/en/ (*See page 141 for text discussion.*)

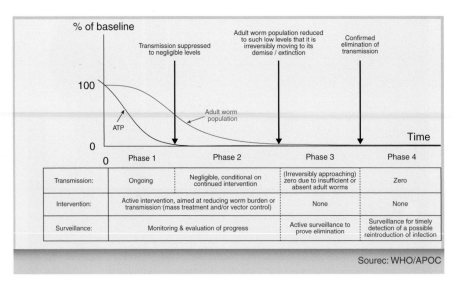

FIGURE 11.14 A schematic representation of the phases in programs for elimination of onchocerciasis transmission, in relation to the theoretical fall-off of the adult worm population and ATP. http://www.who.int/blindness/partnerships/APOC/en/ (*See page 145 for text discussion.*)

Schistosomiasis

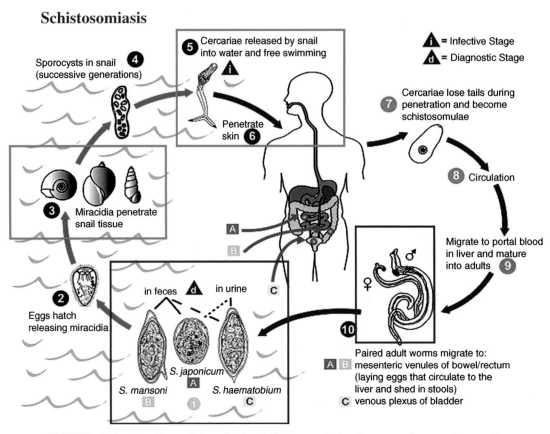

FIGURE 13.1 The Transmission Cycle of Schistosomiasis. Courtesy of www.cdc.gov. The numbers inserted in the text correspond to numbers indicated on the drawing. (*See pages 167–168 for text discussion.*)

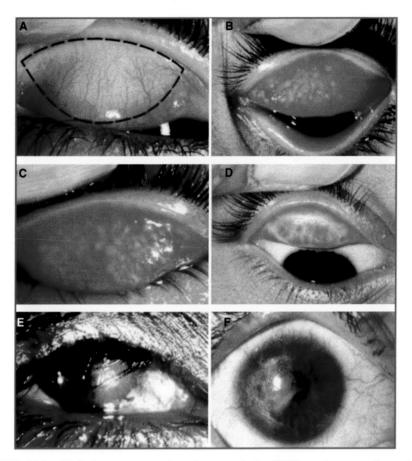

FIGURE 14.1 Clinical appearance of trachoma and the WHO trachoma grading scheme. (a) Normal everted upper tarsal conjunctiva; pink, smooth, thin, and transparent. Large deep-lying blood vessels that run vertically are present over the tarsal conjunctiva. The dotted line shows the area to be examined. (b) Trachomatous inflammation follicular (TF). Five or more follicles of >0.5 mm. (c) Trachomatous inflammation-follicular and intense (TF + TI). Inflammatory thickening obscuring >1/2 the normal deep tarsal vessels. (d) Trachomatous scarring (TS). (e) Trachomatous trichiasis (TT). (f) Corneal opacity (CO). These photographs are reproduced with permission from the WHO Programme for the Prevention of Blindness and Deafness. (*See page 176 for text discussion.*)

FIGURE 15.2 Characteristic dysenteric stool of a patient with shigellosis, containing blood, mucus, and small amounts of fecal matter. (*See page 190 for text discussion.*)

(a)

(b)

FIGURE 23.1   (a) A mysterious sea foam washed up along the coast of Ocean Shores, Washington in the fall of 2009. The photos taken at Chance A La Mer State Park following a violent coastal storm (11/2009) show the foam responsible for the death of thousands of seabirds between September and November (b). Experts believe a toxic algae, *Akashiwo sanguinea*, is responsible for removing the natural oil and waterproofing from the feathers of the seabirds, rendering them unable to protect themselves from the elements. (Photographs courtesy of Julie Hollenbeck.) (*See pages 275–276 for text discussion.*)

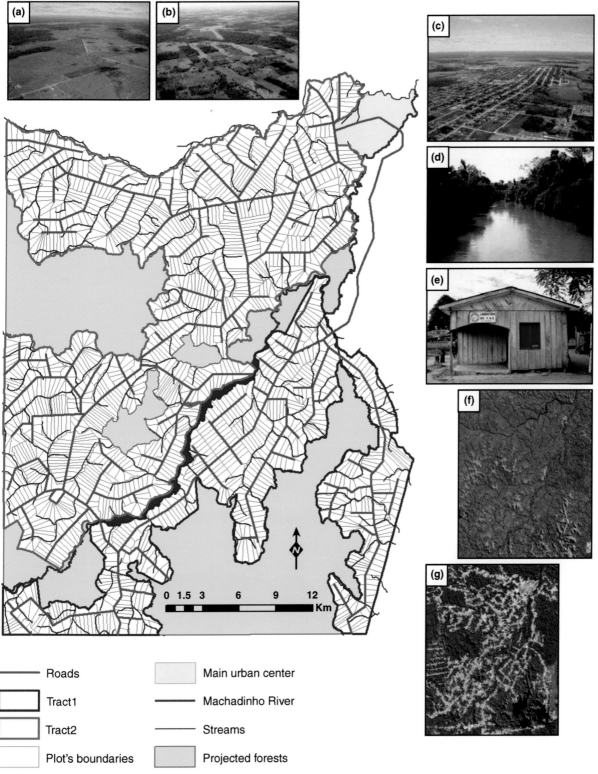

A. Area with pasture. **B.** Area with diversified crop production and patches of original forest. **C.** Aerial view of the main urban center. **D.** Machadinho River. **E.** Laboratory where blood slides

FIGURE 32.4 Location and design of the Machadinho settlement project. (a) Area with pasture. (b) Area with diversified crop production and patches of original forest. (c) Aerial view of the main urban center. (d) Machadinho River. (e) Laboratory where blood slides are prepared and analyzed. (f, g) Satellite images taken in 1985 and 1995, respectively (Landsat TM, 30 m of resolution). (*See pages 408–409 for text discussion.*)

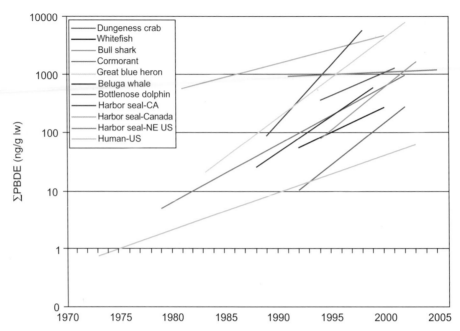

FIGURE 34.6 PBDE concentrations have been increasing in North American wildlife and human tissues over the past 30 years. Trend lines show average PBDE concentrations reported for the first and last year of the studies. References: Dungeness crab (Ref. 107); whitefish (Ref. 108); bull shark (Ref. 109); cormorant (Ref. 110); great blue heron (Ref. 110); beluga whale (Ref. 111); bottlenose dolphin (Ref. 109); harbor seal—San Francisco Bay, CA (Ref. 73); harbor seal—British Columbia, Canada (Ref. 112); harbor seal—NW Atlantic United States (Ref. 104); humans—United States (Ref. 80). (*See page 446 for text discussion.*)

Proportion of the population using improved drinking water sources, 2008
Percent Total

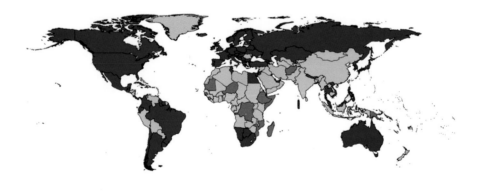

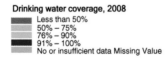

Drinking water coverage, 2008
- Less than 50%
- 50% – 75%
- 76% – 90%
- 91% – 100%
- No or insufficient data Missing Value

Note: The boundaries and the names shown and the designations used on these maps do not imply official endorsement or acceptance by the United Nations.

FIGURE 35.3 Global map of drinking-water coverage in 2008. *Source*: WHO/UNICEF JMP, 2010. (*See page 466 for text discussion.*)

25

THE HUDSON RIVER: A CASE STUDY OF PCB CONTAMINATION

David O. Carpenter and Gretchen Welfinger-Smith

25.1 INTRODUCTION

Originating in Lake Tear of the Clouds on Mt. Marcy, situated in the Adirondack Mountains in northeastern New York State, the Hudson River runs south to the Battery on Manhattan Island, eventually empting into the New York–New Jersey Harbor and the Atlantic Ocean. The Hudson is 315 river miles (RM) long with a 35,000 km^2 watershed located in four states, and serves as part of the border between New York (NY) and New Jersey (NJ). Named after Henry Hudson, the first to explore the river in 1609, the Hudson has served as a route of trade and has played a major role throughout the history of the United States. It was a British bribe for control of the Hudson that led to the treachery of Benedict Arnold during the Revolutionary War, and is where Newburgh, NY, lies, home to George Washington's headquarters from 1782 through the end of the Revolutionary War. The Hudson River was a major trade route from the mid-1800s to early 1900s with its connection to the Erie Canal, linking the Atlantic Ocean to Lake Erie. The Hudson also connects to the Champlain Canal, leading to Lake Champlain, which was another key transportation route during that time period, and continues to be in use today. For these reasons and many others, the Hudson River was federally designated by President Clinton as an American Heritage River for its role in American history and culture in 1998.

While the Hudson River grew as a trade route, industry and commerce along its banks also grew, continuing through to present day. First, in the eighteenth and nineteenth centuries, countless mills were erected along the banks of the Hudson. Many mills were constructed in the presence of waterfalls, such as the Hadley's Falls, to harness waterpower. This led to the development of power companies, such as the Palmer Water Power Company in 1866 (http://www.hudsonrivermillproject.org/pages/theme02a.htm). With the creation of power companies came the creation of reservoirs intended to stabilize the river's water levels in order to eliminate seasonal power supply fluctuations. This stable power supply led to increased mill development in the surrounding areas. The geological characteristics of the Hudson have been permanently altered by the creation of these dams and reservoirs. The changes affected the Hudson's shape and water levels, influencing the cycling and transport of modern day pollutants. The environmental impact resulting from industrial expansion paralleled the rate of industrial growth and development.

Today's greatest toxic legacy of this industrial era is the polychlorinated biphenyl (PCB) contamination of the Hudson. General Electric (GE) built two electrical capacitor manufacturing plants on the banks of the Hudson at the villages of Hudson Falls and Fort Edward, and the capacitors were filled with PCBs, primarily Aroclor 1242. Purchase records show that GE bought some 133 million pounds of PCBs, and it is assumed that about 1% of this amount escaped (1). Beginning sometime in 1947, and continuing until 1977, approximately 1.3 million pounds (589,670.08 kg) of PCBs leaked and were washed into the Hudson. Beginning on January 31, 1975, GE was permitted by the EPA to discharge 30 lb/d of chlorinated hydrocarbons (2). GE ceased the direct discharging PCBs into the river in 1977 (2). However, PCBs continue to leak from fractured bedrock into the Hudson to the present day. In

Water and Sanitation-Related Diseases and the Environment: Challenges, Interventions, and Preventive Measures, First Edition.
Edited by Janine M. H. Selendy.
© 2011 Wiley-Blackwell. Published 2011 by John Wiley & Sons, Inc.

addition PCBs from the two plant sites enter the Hudson via surface runoff, soil infiltration, erosion of contaminated sediment, and the redeposition of some volatized PCB.

The Hudson River is considered nontidal freshwater until about river mile (RM) 153, where it becomes tidal freshwater. At about RM 60, near Newburgh, through Manhattan it becomes estuarine, and the NY–NJ Harbor is a marine ecosystem (3). The upper part of the river is lined with forests, agricultural land, emergent marshes, nontidal wetlands, cities, towns, and industry. As the river flows south, there also are agricultural lands with some deciduous forests, with industry and cities becoming increasingly dominant. The southern part contains a larger variety of ecosystems than the northern, with both fresh and brackish tidal marshes and swamps, aquatic beds, and intertidal shores and mudflats. The Hudson tidal estuary has a gradient from freshwater (salinity <0.5%) to polyhaline (salinity 18–30%) with the divide between fresh and saline waters generally ranging from RM 15 to RM 100, varying primarily as a result of tidal action and freshwater runoff (2).

Salinity levels, especially in estuarine systems, can result in gradients and the stratification of water. As the freshwater and saltwater layers stratify, so do the suspended particulate matter and organic carbon. The federal dam (RM 153.9), at Troy, NY, separates the Upper Hudson River from the Lower. The Mohawk River merges with the Hudson at RM 156. Between Fort Edward (RM 195) and the federal dam the Hudson passes through a series of locks and dams associated with the Champlain Canal system. These structures have created several pools that currently hold upstream sediment deposits.

25.2 PCB CONTAMINATION SOURCES OF THE UPPER HUDSON

Based on levels of contamination, EPA has divided the 200 miles of the contaminated portion of the Hudson into three parts. The upper Hudson is the river from Hudson Falls down to the federal dam at Troy. The middle Hudson is that stretch from the Troy dam to Poughkeepsie, while the lower Hudson is from Poughkeepsie down to the Atlantic Ocean. The upper Hudson is nontidal freshwater, while the middle Hudson is tidal freshwater, and the lower Hudson is tidal saltwater.

The EPA considers that there are four continued sources of PCBs to the upper Hudson River, but only one is considered to be a primary source, the GE Hudson Falls plant site. Extensive evidence shows the facility continues to leak PCBs into the Hudson through the underlain layer of fractured bedrock beneath the plant site. The largest documented leakage event occurred from 1991 to 1993 when a gate structure within the Allen Mill, near Bakers Falls, failed in September 1991 releasing 260 kg of PCBs (4). Sediment/soil

samples taken from the mill in the 1990s had a PCB range of <1–73,000 ppb, water samples ranged from 0.1 to 390,000 ppb, and oil samples had a PCB range of 94,000–940,000 ppm (5). Water column data from 1998 to 2000 shows the leakage from the Hudson Falls plant contributes to the vast majority of the 3–8 kg/month load which travels past Rogers Island (RM 194), located just to the south, based on the fact that the congener patterns in the PCBs loads at Rogers Island resemble unweathered Aroclor 1242, consistent with the nonaqueous phase PCB bearing oils from bedrock beneath the GE faculties at the Hudson Falls plant site (4).

The second source location listed by the EPA as a continued potential source is the GE Fort Edward plant, which is underlain by a layer of silt and clay. Since the elimination of PCB usage and an upgrade of the site's wastewater treatment plant, there has been a marked reduction of leakages and discharges into the Hudson. Thus, it is considered that the bulk of the post-1977 contamination comes from bank erosion of contaminated soils and sediments near the former discharge pipe (4).

The third source is the Remnant Deposit 1 site, which is near to Rogers Island. It is the only Remnant Deposit site that was not remediated by the 1989–1991 remedial efforts of GE because the majority of this Remnant Deposit has washed downstream, and capping the island proved impractical. Therefore, this sediment is available for resuspension and transport during high velocity flow conditions and diffusive exchange during low velocity flow. This source is expected to emit a weathered congener pattern, unlike the Hudson Falls plant site. The sampling evidence suggests that resuspension and transport, and diffusive exchange processes are currently not major contributors to the annual PCBs load at Rogers Island under normal flow conditions (4).

Remnant Deposits 2 through 5 are considered to be the fourth potential sources of PCBs to the Hudson north of Rogers Island. In 1991 these sites were capped by GE as part of their remedial efforts, raising the elevation of these deposits to a level above the 100-year flood level, significantly limiting the movement of sediment into the Hudson. Although the potential exists for PCBs to enter the river via groundwater, the EPA sampling results from 1997 to 1999 do not suggest significant transport of PCBs into the river from these sites based on their congener patterns.

Secondary sources of PCBs to the upper Hudson River also exist. Paper mills introduced PCBs throughout New York State by recycling carbonless copy paper containing Aroclor 1242, with an estimated maximum total discharge of PCBs during 1977 and 1978 at <2.3 kg/year into the Hudson from Bakers Falls to Troy (6, 7). Other sources include electric utilities and manufacturers using PCB-containing equipment, transportation, and landfills containing PCB contamination. Atmospheric deposition and urban runoff also contribute to the PCB load of the Hudson.

25.2.1 PCB Contamination Sources of the Lower Hudson/Hudson Estuary and NY–NJ Harbor

The primary source of PCBs to the lower Hudson, or Hudson Estuary, is the upper Hudson (Fig. 25.1). In the early 1990s data showed the upper Hudson discharging 5–10 lbs of PCB daily, which was reduced to approximately 0.5 kg daily by the late 1990s (8). In addition to the upper Hudson the Mohawk River, the Hudson's largest tributary, is also considered to be a secondary source of PCB to the lower Hudson (9). Additional secondary sources include other tributary rivers originating in both New York and New Jersey,

industrial runoff, wastewater treatment plant discharges, combine sewer overflows, atmospheric deposition (urban and industrial released), and the decomposition of contaminated biota (10, 11).

The Upper Hudson River has been the primary source of PCBs to the NY–NJ Harbor since the 1960s. Studies have shown that from the early 1970s to 1989 nearly two-thirds of PCB deposits in the harbor sediments along the main stem of the Hudson were derived from PCB inputs of the upper Hudson, which were transported downstream (12). As the strength of this source decreased with time, other more local

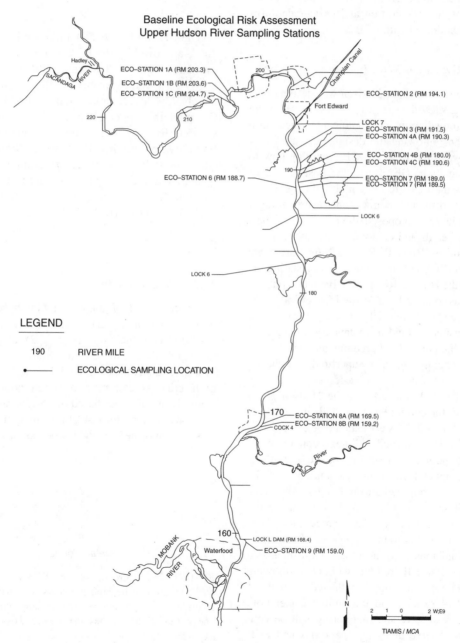

FIGURE 25.1 Map of the Upper Hudson River.

sources began to dominate the PCB loads into the harbor such as municipal wastewater discharge (13). Water pollution control plants (WPCPs), for example, have been shown to contribute to NY–NJ Harbor PCB levels by up to 3% (14). Evidence also exists for additional PCB sources in the western side of the harbor, with observed PCB levels higher there than those observed along the stem of the Hudson River (12). One possible source is the Passaic River in NJ. Studies have found the lower Passaic to be contaminated with PCBs and have identified several sources including industrial, municipal, and urban runoff (15, 16). Possible additional sources of PCB loading include other tributaries, wastewater treatment plant discharges, combined sewer overflows, storm water outflows, surface runoff, atmospheric deposition, and the degradation of biotic material contaminated with PCBs.

25.2.2 Hudson River PCB Remediation

While the 200 mile stretch of the Hudson was designated a National Priorities List (NPL) site in 1984, legally requiring its clean up and remediation, no significant remediation occurred until 2009. Following the designation, a lengthy debate ensued between the EPA and GE regarding how to handle the contamination. GE argued the best course of action was to allow the PCBs to remain buried in sediment rather than risk reexposing the sediment and resuspending it into the river, while EPA proposed dredging as the best remedy. Based on health and ecological risk assessments EPA issued the Hudson River PCBs Site New York State Record of Decision (ROD) (4) in February 2002, stating the decision to dredge the Hudson River of PCBs in two phases.

This decision was contested and debated for several years, as was the method and costs of dredging and the party to be held legally responsible to fund it. In January 2009 a final agreement was reached on Phase I dredging and on May 15, 2009, 25 years after being listed as a Superfund site, Phase I dredging of the Hudson River commenced.

According to the ROD the goal of the Hudson River dredging is to reduce the current levels of PCBs in the Upper Hudson River by 65%. The EPA estimated that 2.65 million cubic yards of PCB-contaminated sediment, containing an estimated 146,166 pounds (66,300 kg) of PCB, will be removed from the upper Hudson upon completion of the dredging. The ROD also required the dredging be conducted in two phases, Phase I and Phase II. Phase I dredging, scheduled for the first year, is required to proceed slowly in order to determine the ability to meet preestablished performance standards and make any necessary adjustments to dredging operations. Phase II dredging will occur following an evaluation of Phase I dredging, conducted by both an independent panel of experts and the public. Under both stages of dredging, extensive site monitoring will be conducted in order to observe dredging operations and PCB levels, while protecting public health.

TABLE 25.1 Comparison of ROD Estimates and Current Design Targets by River Section

	EPA ROD Estimates		GE Design Estimates	
	Volume of Sediment (cy)	Mass of PCB (kg)	Volume of Sediment (cy)	Mass of PCB (kg)
River Section 1	1,492,000	36,000	939,800	60,600
River Section 2	565,000	23,600	364,000	28,500
River Section 3	393,000	6,700	491,000	24,000
Total	2,450,000	66,300	1,794,800	113,100

The GE Dredge Area Delineation Plan of 17 December 2007, approved by the EPA in January 2008, exceeds the estimated required PCB removal listed in the EPA ROD by nearly 59% and estimates it will need to dredge approximately 73% less sediment to do so (Table 25.1). Under the GE plan the upper Hudson will be divided into three sections: River Section 1 running from the site of the former Fort Edward Dam (RM 194.8) to the Thompson Island Dam (RM 188.5), River Section 2 extending from the Thompson Island Dam to the Northumberland Dam (RM 183.4), and River Section 3 stretching from the Northumberland Dam to the Federal Dam (RM 153.9). Phase I dredging will be conducted in River Section 1 and Phase II includes River Sections 1–3.

25.2.3 What Are PCBs?

PCBs are a group of nonpolar, hydrophobic chemicals composed of two connected benzene rings with a variable number (1–10) of chlorine atoms (Fig. 25.2). PCBs were manufactured globally, and in the United States were manufactured by Monsanto under the commercial name, "Aroclor." A number of different Aroclor mixtures were produced, most of which were given a number based on the average degree of chlorination (17). Aroclor 1242, the primary mixture used by GE, was 42% chlorine by weight, whereas Aroclor 1260 was 60%

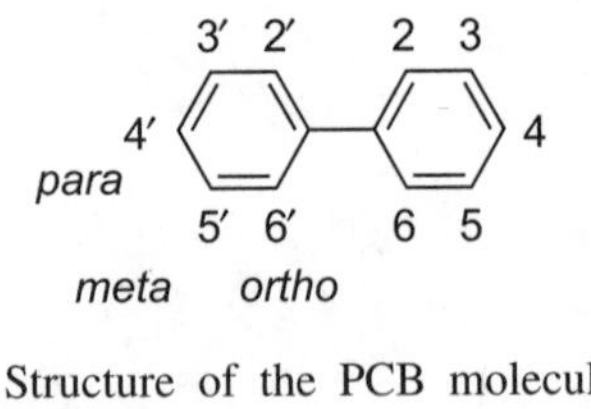

FIGURE 25.2 Structure of the PCB molecule. There are 10 positions around the biphenyl rings that may have hydrogen or chlorides. These positions are traditionally numbered from 2 to 6 on one ring and 2' to 6' on the other. The 2 and 6 positions are identified as "*ortho*," the 3 and 5 positions as "*meta*," and the 4 position as "*para*."

chlorine by weight. There are theoretically 209 different PCB congeners, depending upon the number and positions of the chlorine atoms around the biphenyl ring, although only about half that number are present in significant amounts. PCBs are stable, nonflammable oils with high boiling points. They have good electrical insulating properties and were used as insulators and cooling fluids for transformers and capacitors. They were also widely used as hydraulic fluids, in carbonless reproduction paper, in caulking, in early fluorescent light fittings, as plasticizers in paints and cements, and in PVC coatings for electrical wiring and wiring components. Because of the chlorination, PCBs are very persistent both in the environment and in biologic systems.

While all PCB congeners share some common properties, individual congeners differ with regard to water solubility, octanol–water partition coefficient, Henry's Law constants, enthalpy, partitioning within ecosystems, ability to undergo microbial transformations, and degradation (18–25). By tradition the positions are numbered, as shown in Fig. 25.2, with the 1 and 6 positions described as *ortho,* the 3 and 5 positions as *meta,* and the 4 position as *para.* Lower chlorinated congeners are more volatile and water soluble, while higher chlorinated congeners are more persistent. PCBs can be grouped based on extent of chlorination (mono, di, trichloro, etc.) and by the presence or absence of *ortho* chlorines. The variations in chemical properties among the congeners determine how each interacts in terms of accumulation and distribution, and also their metabolism and elimination by biota and humans (26–30). Because of their lipophilic nature and resistance to degradation, PCBs accumulate in body fat of mammals, and in the waxy cuticle and stem surfaces of plants (31–33).

Removal of PCBs from the body is quite inefficient and is primarily due to the activity of various cytochromes P450 in the liver and other tissues. Congeners with no *ortho* chlorines exist in a planar configuration, and those with four or more chlorines are able to bind to the aryl hydrocarbon (Ah) receptor, leading to induction of cytochromes P450 1A1, 1A2, 2A1, and 1B1 (32). The complex binds to DNA and results in up- and downregulation of many different genes that mediate dioxin-like toxicity (34). Interestingly it is now clear that the same PCB congener (PCB 126) will induce different genes in different tissues (35), which adds yet another level of complexity. Some mono-*ortho* congeners also have dioxin-like activity, and while they are not as potent as some of the dioxin congeners, they are often present at sufficiently higher concentrations so as to contribute significantly to the total toxic equivalents (TEQs) (36). The di- and higher *ortho*-substituted congeners have little or no dioxin-like activity, but activate cytochrome P450s 2B1 and 2B2. These congeners induce different patterns of genes (37). While many of the health effects of exposure to PCBs in both animals and humans result from gene induction, individual PCB congeners also may alter physiological function

by changing the metabolism of endogenous hormones, by binding to transport proteins and by direct actions on other cell processes (38).

25.2.4 PCBs Health Risks and Exposure Pathways—Human

Exposure to PCBs has been shown to result in a wide array of adverse health affects in humans. PCBs are proven carcinogens in animals and probable carcinogens in humans. PCBs cause immune suppression, exposure results in reduced IQ in children and poorer memory function in adults, with shortened attention span and reduced ability to deal with frustration. In addition, PCBs disrupt both estrogenic and androgenic sex hormone systems, and increase risk of liver and cardiovascular disease and diabetes. Details of the evidence for these health risks will not be presented here beyond those studies focused specifically on the Hudson River, but information on these risks can be obtained from recent reviews (39–42).

Exposure of humans to PCBs can be via ingestion, inhalation, or absorption through the skin. Ingestion of contaminated fish is a major route of exposure. The USEPA developed a Human Health Risk Assessment (HHRA) for the Upper and Middle Hudson River sections. This HHRA examined potential cancer and noncancer risks for known area receptors including anglers, recreators (adult, adolescent and child), and local residents using dose–response relationships for carcinogenicity and systemic toxicity through ingestion of fish, incidental ingestion of sediment, dermal contact with sediment and river water, and the inhalation of volatilized PCBs in air (6). The HHRA was based on the assumption that fish consumption is the primary pathway for human PCB exposure and for potential adverse noncancer risks (43). The fish recommendation was based on weighted PCBs concentration distributions in brown bullhead (*Ameiurus nebulosus*), largemouth bass (*Micropterus salmoides*), and yellow perch (*Perca flavescens*), while area-weighted sediment and area-weighted water concentrations were derived from the USEPA's Baseline Modeling (6).

The HHRA baseline results developed by EPA includes both the central tendency risk (CTR), which is used to describe more fully the health effects associated with average exposure, and the reasonably maximally exposed (RME) risk, which estimates the increased risk for an individual to develop cancer over a lifetime (44). EPA uses a Hazard Index (HI) to correlate REM values with exposure levels. The reference HI level is 1. EPA results for noncancer health effects for fish consumption in the upper Hudson for a young child show the $HI = 104$ (over 100 times higher than the HI reference levels of one), adolescents $HI = 71$, and for adults the $HI = 65$ (43). Study results found that the CTR for both cancer risks and noncancer health hazards from fish consumption in the upper Hudson, which are based on approximately one meal every 2

TABLE 25.2   U.S. EPA Upper 2000 Point Estimate Cancer Risk Summary—Upper Hudson (43)

Pathway	Central Tendency Risk	RME Risk
Ingestion of fish		
Total[a]	3×10^{-5} (3 in 100,000)	1×10^{-3} (1 in 1,000)
Adult	1×10^{-5} (1 in 100,000)	6×10^{-4} (6 in 10,000)
Adolescent	7×10^{-6} (7 in 1,000,000)	4×10^{-4} (4 in 10,000)
Child	1×10^{-5} (1 in 100,000)	4×10^{-4} (4 in 10,000)
Exposure to sediment		
Baseline recreator	2×10^{-7} (2 in 10,000,000)	2×10^{-6} (2 in 1,000,000)
Avid recreator	1×10^{-6} (1 in 1,000,000)	9×10^{-6} (9 in 1,000,000)
Exposure to water		
Baseline recreator	3×10^{-8} (3 in 100,000,000)	2×10^{-7} (2 in 10,000,000)
Avid recreator	1×10^{-7} (1 in 10,000,000)	1×10^{-6} (1 in 1,000,000)
Inhalation of air	2×10^{-8} (2 in 100,000,000)	1×10^{-6} (1 in 1,000,000)

[a] Total risk for young child (aged 1–6), adolescent (aged 7–18), and adult (over 18).

months, are also above the EPA's level of concern. Cancer risks and noncancer health hazards from PCB exposure in the Hudson River through dermal contact with sediments and water, incidental ingestion of sediments, and the inhalation of volatized PCB are generally within or below the EPA's level of concern (43) (Tables 25.2–25.5).

A risk assessment was also determined for an avid recreator's scenario, where the CTR was evaluated on exposure estimates being twice per week for 6 months a year (52 d/y) and RME estimated as four times per week for 6 months a year (104 d/y) for adults, adolescents, and young children (43).

There have been relatively few studies that have examined human health effects directly of individuals who live near the Hudson. However, we have performed studies specifically studying the rates of hospitalization for various diseases (cardiovascular disease, acute myocardial infarction, stroke, diabetes, respiratory infections) among individuals living in zip codes abutting the contaminated portions of the Hudson River, and have demonstrated that hospitalization rates are elevated even after adjusting for other risk factors (45–49). In a study of acute myocardial infarction, Sergeev and Carpenter (46) found an odds ratio of 1.39 (95% CI: 1.18–1.63) for

individuals living adjacent to the Hudson as compared to those living elsewhere in upstate New York in a zip code that did not contain a hazardous waste site. This is in spite of the fact that residents living near the Hudson had higher income, smoked less, and exercised more than other New Yorkers. It is unlikely that fish consumption is responsible for this relationship, which suggests that the exposure to vapor-phase PCBs was via inhalation. This may indicate that EPA's risk assumptions for inhalation are not sufficiently stringent to protect human health.

Fitzgerald et al. (50) evaluated neuropsychological status in 253 persons between the ages of 55 and 74 years who lived along the contaminated portions of the upper Hudson. They found that an increase in serum PCB concentration from 250 to 500 ppb (lipid adjusted) was associated with a 6.2% decrease in verbal learning and a 19.2% increase in depressive symptoms. These effects on memory and depression are occurring at very low PCB concentrations. Similar findings have been obtained in other studies on adults (51).

Clearly PCB exposure poses serious threats to human health, whether that exposure comes from the Hudson or elsewhere. While the carcinogenic and cognitive actions of PCBs have been known for a long time, it is becoming

TABLE 25.3   U.S. EPA Upper 2000 Point Estimate Cancer Risk Summary—Mid Hudson (43)

Pathway	Central Tendency Risk	RME Risk
Ingestion of fish		
Total[a]	1×10^{-5} (1 in 100,000)	7×10^{-4} (7 in 10,000)
Adult	6×10^{-6} (6 in 1,000,000)	3×10^{-4} (3 in 10,000)
Adolescent	3×10^{-6} (3 in 1,000,000)	2×10^{-4} (2 in 10,000)
Child	5×10^{-6} (1 in 1,000,000)	2×10^{-4} (2 in 10,000)
Swimming/wading exposure to sediment[a]	2×10^{-8} (2 in 100,000,000)	2×10^{-7} (2 in 10,000,000)
Swimming/wading exposure to water[a]	9×10^{-9} (9 in 1,000,000,000)	6×10^{-8} (6 in 100,000,000)
Consumption of drinking water[a]	3×10^{-8} (3 in 100,000,000)	1×10^{-7} (1 in 10,000,000)

[a] Total risk for child (aged 1–6), adolescent (aged 7–18), and adult (over 18).

TABLE 25.4 U.S. EPA Upper 2000 Point Estimate Noncancer Hazard Summary—Upper Hudson (43)

Pathway	Central Tendency Risk Hazard Index	RME Risk Hazard Index
Ingestion of fish		
Adult	7	65
Adolescent	8	71
Child	12	104
Exposure to sediment[a]		
Baseline recreator	0.03	0.04
Avid recreator	0.20	0.30
Exposure to water[a]		
Baseline recreator	0.01	0.02
Avid recreator	0.06	0.10
Inhalation of air[b]	Not calculated	Not calculated

[a] Values for young child or adolescent, which are higher than adult for these pathways.
[b] Noncancer hazards were not calculated for the inhalation pathway due to a lack of noncancer toxicity values for this pathway.

increasingly clear that PCBs increase the risk of a number of chronic diseases such as diabetes (52, 53) and cardiovascular disease (54, 55). It is remarkable that these adverse effects are occurring at PCB levels found commonly in the general population. It is also clear that the fetus is the most vulnerable of all, as PCB exposure during development not only causes decrements in cognitive function that last for life, but also alters endocrine systems (both sex hormones and thyroid) and predisposes to other diseases later in life, including cancer. Because of the long half-life of PCBs, it is especially important that even young girls limit their exposure to PCBs, because they will pass them on to their fetus even many years after exposure. These findings indicate how important it is to dredge contaminated sites such as the

TABLE 25.5 U.S. EPA Upper 2000 Point Estimate Noncancer Hazard Summary—Mid Hudson (43)[b]

Pathway	Central Tendency Risk Hazard Index	RME Risk Hazard Index
Ingestion of fish		
Adult	3	34
Adolescent	4	37
Child	62	53
Exposure to sediment[a]	0.002	0.0004
Exposure to water[a]	0.005	0.007
Consumption of drinking water[a]	0.01	0.02

[a] Values for young child or adolescent, which are higher than adult for these pathways.
[b] Noncancer hazards were not calculated for the inhalation pathway due to a lack of noncancer toxicity values for this pathway.

Hudson River, since the PCBs are not going to go away without active removal.

25.2.5 Exposure Pathways and Risks—Ecological

The ecosystems and habitats are very different in the upper and lower Hudson. The upper is a freshwater, nontidal environment, and the lower is tidal turning increasingly brackish to the south. The diversity in environments results in a large array of ecosystems exposed to PCB.

The bioaccumulation of PCBs in the aquatic invertebrates, amphibians, birds, fish, eels, and mammals is of great concern. The Hudson River supports 206 species of fish and 143 species of birds, resident and migrating, with 64 of these fish and bird species listed as Threatened, Endangered, Rare, or of Special Concern by federal and New York State authorities. Thirty-nine areas of significant habitat have been identified in the Lower Hudson River (4).

Areas of focus when determining ecological risk regarding wildlife include effects on reproduction, growth, metabolism, and survival. Commonly observed effects from PCB exposure among a variety of species include the following: hepatotoxicity, immunotoxicity, neurotoxicity, weight loss, increased abortion, low birth weight, embryolethality, teratogenicity, feminization of males, impaired development, reduced growth, histological changes, alterations in biochemical processes, gastrointestinal ulceration and necrosis, bronchitis, impaired growth, and dermal toxicity/chloracne, increases in cytochrome P-450 levels, and cancer (8, 56, 57).

Several PCB exposure pathways exist for species living along the Hudson. Semiaquatic and aquatic species are exposed via direct uptake from water through egg membranes, gills, skin, mouth, and gastrointestinal linings. Aquatic and semiaquatic species are also exposed to PCBs through ingestion of contaminated sediment and the food chain. Terrestrial species' exposure pathways include surface water ingestion, inhalation of air, food uptake, incidental sediment ingestion, and contact with floodplain sediments and soils. Diet has been found to be a major PCB uptake route for many fish species. However, fish can also concentrate significant levels of PCBs directly from the water column (58, 59).

The USEPA developed a list of measurement endpoints for wildlife PCB exposure along the Hudson River in order to evaluate risk. Measurement endpoints were selected to represent toxicity mechanisms and exposure pathways. This analysis concluded that PCB levels in the Hudson exceed effect-level thresholds for adverse effects on fish reproduction. In addition, results showed when comparing sediment guidelines established for the protection of benthic life to the observed and modeled sediment concentrations of the Hudson, the threshold effect concentration (TEC), mid-range effect concentration (MEC), and the extreme effect concentration (EEC) were exceeded at the upper Hudson sampling locations and the TEC was also exceeded at all lower Hudson sampling locations. The MEC was exceeded at most lower Hudson sampling locations (43). Water column samples were compared to NYSDEC Water Quality Criteria (WQC) for PCBs. Chronic freshwater WQC for PCB is 0.014 μg/L, significantly less than the average levels of 0.071 μg/L found in the upper Hudson water column. All collected water column samples from RM194.6–RM156.5 exceeded the WQC for the surface water standard for protection of wildlife of 1.2×10^{-4} g/L, indicating that some aquatic and wildlife species may be adversely affected (3).

Studies have been conducted of PCB exposure and adverse health effects for species native to the river. McCarty and Secord conducted a study in 1999 comparing the quality of tree swallow (*Tachycineta bicolor*) nests, a native bird species, at a known PCB contaminated site at Remnant Deposit 4 in the Hudson to tree swallow nests in other areas further downstream from the primary PCB contamination sources and also at a site in western New York (60). Nest mass and the number of feathers lining the nest cup were used as measurements of nest quality. The mass was measured on the first day an egg was laid and again on the day they hatched. The feathers lining the cup were also counted on the day the first egg was laid, again after the clutch of eggs was completed, and on the day the eggs hatched. Results showed nest-building behavior for both males and females was impaired in regions of PCB contamination. Decreases in nest quality can lead to decreases in clutch size and overall population numbers.

Deer mice (*Peromyscus maniculatus*) are another species native to the Hudson. Johnson et al. published a study in 2009 relating point source PCB exposure to bone density loss in deer mice and the associated increased susceptibility to bone fracture (61). Another study regarding PCB-exposed deer mice showed serious damage to the seminiferous tubules, which can impair reproduction and in turn population numbers (62). Mink (*Mustela vison*) are also commonly found along the Hudson. Compared to other mammals, mink have been demonstrated to be very sensitive to PCB exposure. Mink have been relied upon as an indicator species for ecological exposure assessment to PCB. Signs of PCB contamination in mink include anorexia, bloody stools, fatty liver, kidney degeneration, and hemorrhagic gastric ulcers (63).

Hudson River fish have very high levels of PCBs. Decades after the direct discharge of PCBs into the Hudson River ended levels in fish remain significantly elevated. In 1998 NYSDEC data showed average tissue concentrations for carp (*Cyprinus carpio*), brown bullhead (*Ameiurus nebulosus*), and largemouth bass (*Micropterus salmoides*) at the Thompson Island Pool (RM 188–195) to be 32.7, 16.2, and 18.2 μg/g, respectively (43). As referenced in Baker et al. (64), recent PCB levels in the fillets of Hudson fish have been reported as high as 480 ppm in common carp, 290 ppm in white sucker (*Catostomus commersonii*), 160 ppm in American eel (*Anguilla rostrata*), 150 ppm in largemouth bass, 50 ppm in red-breasted sunfish (*Lepomis auritus*), and 27 ppm in black crappie (*Pomoxis nigromaculatus*); lower Hudson species showed recent maximum concentrations of 77 ppm in shortnose sturgeon (*Acipenser brevirostrum*) liver, 42 ppm in Atlantic sturgeon (*Acipenser oxyrinchus oxyrinchus*) gonad, and 31 ppm in striped bass (*Morone saxatilis*) fillet. These levels are to be compared to the EPA Guidance for Assessing Chemical Contaminant Data for Use in Fish Advisories (6), which recommends no consumption of any fish that has a PCB concentration greater than 0.094 ppm. This is yet more evidence that fish in the Hudson will never be safe to eat unless the PCBs are removed.

The USEPA issued a series of ecological risk assessment reports regarding the exposure of PCBs to the ecosystems and wildlife of the Hudson River. The toxicity quotient (TQ) was used to determine PCB toxicity to the various wildlife populations. TQs>1 are typically considered as a potential risk to ecological receptors. The TQ is the direct numerical comparison of a measured or modeled exposure concentration or dose to a benchmark dose or concentration calculated as:

$$TQ = \text{Modeled Dose or Concentration}/$$
$$\text{Benchmark Dose or Concentration}$$

TQs consider the potential for general effects of exposure upon individual animals within a local population operating under the premise that if effects are not deemed to occur at the average individual level, they are probably insignificant at the population level. Risk characterization and assessment studies conducted by the USEPA for Hudson River PCBs used the following endpoints: sustainability of a benthic invertebrate community, sustainability of local forage, omnivorous, and piscivorous fish populations, and the sustainability of local wildlife including waterfowl, insectivorous, semipiscivorous and piscivorous birds and mammals, and omnivorous mammals. The major findings of the reports, based on TQ, risk characterization, collected data and computer modeling, concluded that PCBs in the Hudson River result in the following:

- Fragile populations of threatened and endangered species, such as the bald eagle (*Haliaeetus leucocephalus*), are particularly susceptible to the adverse effects from PCB exposure.
- Birds and mammals eating PCB-contaminated fish from the Hudson River, including the bald eagle, belted kingfisher (*Megaceryle alcyon*), great blue heron (*Ardea herodias*), raccoons (*Procyon lotor*), river otter (*Lontra canadensis*), and mink are at risk at the population level because PCBs may adversely affect their survival, growth, and reproduction with the piscivorous

mammals being at the greatest risk due to their feeding patterns.

- Measured and modeled water concentrations of PCBs are high enough to be of concern to piscivorous wildlife for the duration of the modeling period (through 2018) throughout the Hudson.
- Piscivorous fish, including the yellow perch (*Perca flavescens*), striped bass and largemouth bass, and omnivorous fish, such as the brown bullhead and short nose sturgeon, may experience adverse reproductive effects, particularly in the upper river. Forage fish, like the pumpkinseed (*Lepomis gibbosus*) and spottail shiner (*Notropis hudsonius*), are unlikely to experience adverse reproductive effects based on the exposure levels during the study time (late 1990s–early 2000s).
- Insectivorous mammal populations, including the little brown bat (*Myotis lucifugus*), may experience adverse reproductive effects from exposure (3, 43).

In April 1986, the United States Fish and Wildlife Service released a study (58) pertaining to the health effects experienced by animals exposed to PCBs. This review examined the existing information regarding PCB toxicity to aquatic organisms and fish, birds, reptiles, and mammals. Review findings regarding aquatic organisms showed that, in general, toxicity increases with increasing exposure to PCBs, and crustaceans and organisms in the younger developmental stages are the most sensitive groups. In addition to this, Eisler surmised that lower chlorinated PCBs were more toxic to aquatic species than the higher chlorinated congeners. Birds were found to be more resistant than mammals to acute toxic effects of PCBs. Signs of PCB toxicity in birds include morbidity, tremors (which may become continuous), upward pointed beaks, and muscular incoordination. Birds were shown to have hemorrhagic areas in the liver and gastrointestinal tracts filled with blackish fluid upon necropsy. Overall conclusions regarding the toxicity of PCBs exposure to any species is dependent on PCB concentration and duration, and on the Aroclor mixture, as different Aroclor mixtures yield different ratios of lower to higher chlorinated congeners (58, 65).

25.2.6 PCB Bioaccumulation Pathways

Bioaccumulation is the process in which a chemical, or group of chemicals, increases in concentration within organisms each step up the food chain. Given that PCBs are very lipid soluble and do not quickly or readily depurate, they readily bioaccumulate. Not all congeners bioaccumulate equally, nor do they bioaccumulate evenly amid different species. PCB bioaccumulation is partly controlled by exposure, which is affected by the species proximity to the source and ecological and physiological factors including diet, feeding

and assimilation rate, growth rate, and lipid loss through metabolism or reproduction (66). Congener composition (chlorine number and ring position) also directly affects their ability to bioaccumulate in organisms. For example, of the four hexachlorobiphenyls, the $2,2',3,3',6,6'$ was eliminated and metabolized more rapidly than the others because of having the 4,5 unsubstituted carbons (65). A laboratory study where several animal species were exposed to six planar PCB congeners (non-*ortho*, mono-*ortho*, and di-*ortho* congeners) found marked differences in retention times among the congeners after 29 days. Differences were dependent on congener structure and species, likely a result of varying fat levels and different patterns of degradation. PCB levels varied by species with the highest found in the rat followed by the rabbit, guinea pig, trout, and the lowest levels in Japanese quail. The quail was found to only retain non-*ortho* congeners, even at low levels, while rabbits retained the highest levels of the di-*ortho* and mono-*ortho* congeners. Fish were recorded to retain all congeners used in the study fairly evenly and at lower concentrations than the other organisms (65).

The USEPA uses a food-chain multiplier (FM), a factor applied to the bioconcentration factor, to measure biomagnification of hydrophobic chemicals in higher trophic level organisms with the equation BAF = FM × BCF, where BAF = bioaccumulation factor and BCF = bioconcentration factor (67). BAFs are dependent on environmental conditions like temperature and organic carbon (OC) content, chemical properties including octanol–water partition coefficients, and species characteristics such as lipid content and weight.

Bioaccumulation of PCB is directly associated with the equilibrium partitioning of PCBs with freely dissolved chemicals in overlying water for plankton (phyto- and zoo-) and sediment pore water for benthic organisms (68). Factors affecting dietary uptake of PCBs include feeding rate, dietary uptake efficiency, the concentration in diet, and dietary feeding preferences (proportion of each prey species in the diet) (68). One point to be taken into consideration regarding bioaccumulation is growth dilution. Growth dilution at the microalgae level may in principle propagate up the food chain, ultimately causing diminished contaminant concentrations in game fish and other top predators. It has been shown that in aquatic ecosystems transient growth dilution (the prompt response to perturbation) occurs at all levels of the food chain; however, the new steady-state concentrations can fluctuate depending on functional relationships such as ratio of prey biomass to the ratio of predator biomass (69).

The introduction of invasive zebra mussels (*Dreissena polymorpha*) to the Hudson has affected the food chain, and in turn, PCB concentration and distribution (70). Zebra mussels decrease phytoplankton quantity and has demonstrated the ability to control the number of phytoplankton available in the pelagic food web (71). Zebra mussels alter phytoplankton community composition directly and indirectly. Indirect effects include altering nutrient or light regimes in a manner that favors certain phytoplankton groups, removing phytoplankton from the water column at a higher rate, resulting in increased abundance of faster growing species and removal of nonbuoyant algal forms, since zebra mussels are bottom filter feeders. This leads to the dominance of buoyant algal forms (71). Direct effects include selective removal, selective ingestion, and differential digestion of phytoplankton (71). Zebra mussels influence PCB dynamics through direct partitioning of PCBs. PCB levels are affected by ingestion of contaminated particles and the transfer of PCB-contaminated particulates from the water column to sediments through the biodeposition of feces and pseudofeces (72). Since zebra mussels are able to filter a wide range of particles, they have the ability to increase residence time of PCBs in the Hudson River via redeposition to sediments and also through transmission of PCBs to food chain organisms via benthic exposure and significant bioconcentration in mussel tissues (72).

25.2.7   PCB Loss from Hudson

PCB concentrations in the Hudson River are dynamic. In addition to seasonal fluxes and inputs of additional loads, there is PCB loss. PCB loss is primarily associated with degradation, volatilization, biotic uptake and transfer, and sediment and water column transport out of the system. Different congeners react differently in identical conditions, and individual congener behavior is also condition dependent.

25.2.8   Degradation and Transformation

PCB degradation and transformations in the environment occur by microbial, photochemical, and pathways. All involve the dechlorination of the benzene rings, but undergo different reactions with different end products.

Microbial PCB degradation and transformation has been researched in-depth. The microbial degradation of PCB can occur under aerobic and anaerobic conditions, in marine and freshwater environments (21, 73–75). While GE argued for years that microbial PCB degradation would relatively rapidly solve the problem of PCBs contamination of the Hudson (76–78), this has not proven to be the case. Several species of bacteria found in aquatic environments, both gram negative and positive, have been demonstrated to partially dechlorinate PCBs. Examples of such bacteria belong to the genres of *Rhodococcus, Burkholderia, Acinetobacter, Ralstonia,* and *Pseudomonas* and include species such as *Alcaligenes eutrophus* H850, *Pseudomonas putida* LB400, *Pseudomonas* sp. P2, *Rhodococcus globerulus* P6, and *Rhodococcus erythropolis* TA421, which have all proved to degrade PCBs (21, 73, 79–84). The comparison of congener patterns for lost congeners and product congeners has led to

the identification of at least eight distinct microbial dechlorination processes (74, 85–87). Half-lifetimes of PCBs in water and sediment generally exceed the half-lifetimes of PCBs in plankton and benthic invertebrates, consistent with the assumption that changes in concentrations in the plankton and benthos mirror those occurring in water and sediments (68).

Rates and degrees of microbial dechlorination and dechlorination routes are determined by the type and concentrations of microbial populations, the concentration of PCBs present and environmental factors including pH, temperature, and organic carbon, available nutrients and the available electron donors, such as H_2, and electron acceptors (88, 89). Several studies have demonstrated that sediment carbon enrichment enhanced microbial dechlorination rates and/or reduced lag times (80, 90–92). One result of microbial degradation of PCBs and other compounds is isotopic fractionation. Generally it is carbon fractionation, but evidence exists showing the isotopic fractionation of other elements, including hydrogen (93, 94). These fractionation products can be used to measure degradation rates (95, 96).

Aerobic dehalogenation and transformations of PCBs can result in the creation of hydroxylated PCBs, which have been shown to exhibit estrogenic or antiestrogenic activities in humans (89, 97). Under aerobic conditions, mono-, di-, and trichlorinated biphenyls biodegrade relatively rapidly compared to the slower rate observed for tetrachlorinated biphenyls. Higher chlorinated congeners do not biodegrade well in aerobic conditions (65).

Anaerobic dechlorination is the primary method of microbial action on highly chlorinated congeners, possibly via chlororespiration (the initial use of the PCB as an electron acceptor). However, this process is not effective in destroying the PCB molecule. Thus, dechlorination converts higher chlorinated into lower chlorinated congeners and from more hydrophobic congeners to less (75). Typical anaerobic-mediated dechlorination of PCB removes the *meta* chlorines, leading to the generation of primarily *para-* and *ortho-* substituted congeners (88, 98). However, the process is very dependent on the initial PCB concentration and stops relatively quickly after only a relatively small reduction in average number of chlorines (99, 100).

Excluding *ortho*-dechlorination, dechlorination reactions for anaerobic microbial dechlorination tend to be those that release the greatest amounts of reaction heat/energy $(\Delta H^{\circ}_{\tau})$ (101). Anaerobic-mediated reductive dechlorination removes chlorines from higher chlorinated PCBs. The enantiomeric compositions of chiral compounds are vulnerable to the biological degradation processes. PCB congeners with 3–4 *ortho* chlorines, which restrict the rotation along the central C–C axis, making them atropisomers, meet these chiral requirements. This characteristic also makes them excellent resources in the monitoring of microbial dechlorination (102, 103).

Organic substrates have also been demonstrated to affect anaerobic microbial dechlorination in sediment. Nies and Vogel (90) demonstrated that microbe containing anaerobic sediment treated with nutrients, and then separately with the substrates acetate, acetone, methanol, and glucose, showed similar dechlorination patterns. Sediment treated with methanol, glucose, and acetone had greater rates and extents of dechlorination than sediment treated with acetate. Batches treated with nutrients but not substrate showed no significant dechlorination and no dechlorination was observed in autoclaved controls that received both substrate and nutrients. Dechlorination occurred primarily on the *meta-* and *para*-positions of the highly chlorinated congeners resulting in the accumulation of the less-chlorinated products, primarily the *ortho*-substituted. Similar affects have been observed in aerobic microbial degradation (84). Specific PCB congeners have been demonstrated to stimulate microbial PCB-degrading enzymes, increasing the amount of microbial dechlorination of other congeners within an Aroclor. It was found that halogenated compounds, including halogenated aromatic compounds, can be used to stimulate (prime) microbial dechlorination of PCBs and increase PCB-degrading microbes (86, 104). Priming with 2,6-dibromobiphenyl has been shown to stimulate the growth of microorganisms that dehalogenate PCBs, resulting in *meta*-dechlorination (105). A field study by Harkness et al. (106) showed that the addition of nonanalogous compounds, such as inorganic nutrients, biphenyl, and oxygen, enhanced PCB biodegradation. Another study examined the use of bioaugmentation of PCB-contaminated sediment with a PCB-dechlorinating enrichment, but without a primer as a method of PCB dechlorination. It was concluded that bioaugmentation alone was able to decrease the concentration of hexa-nona chlorinated benzenes (CBs), but a primer was necessary for the most effective dechlorination (107). The growth of PCB dechlorinators requires the presence of PCBs (89). For example, it has been found that the bacterium Double—Flanked-1 (DF-1), which reductively dechlorinates congeners with doubly flanked chlorines, grows in response to the dechlorination of specific PCBs (108).

Ming Chen et al. (101) compared reductive dechlorination occurrences of PCBs in anaerobic microorganisms collected from contaminated sediments at Er-Jen River (Tainan, Taiwan), Hudson River (Ft. Edward, NY), Puget Sound (WA), and Silver Lake (Pittsfield, MA). They found that the microorganisms collected from sediment exposed to a greater amount of PCB over decades (Hudson River and Silver Lake) were capable of dechlorinating PCB through reactions with smaller ΔH°_{τ} values and smaller relative retention times. With this they concluded that areas subjected to high PCB levels over long periods of time provide suitable environments for the growth and development of PCB utilizers and dechlorinators, and acclimated microorganisms at these locations have developed diversified dechlorination systems, affecting a wide range of PCB congeners.

The effects of chlorine substitution patterns on oxidative breakdown from microbes fall into the following general patterns:

1. Hydroxylation is favored at the *para* position in the least-chlorinated phenyl ring, unless this site is sterically hindered (i.e., 3,5-dichloro substitution).

2. In the less-chlorinated biphenyls, the *para* position of both biphenyl rings and the carbon that is *para* to the chlorine substituent are all readily hydroxylated.

3. The availability of two vicinal unsubstituted carbon atoms (particularly C-5 and C-4 in the biphenyl nucleus) also facilitates oxidative metabolism of the PCB substrate, but it is not a necessary requirement for metabolism.

4. As the degree of chlorination increases on both phenyl rings, the rate of oxidative metabolism decreases.

5. The metabolism of specific PCB isomers by different species can result in considerable variations in metabolite distribution (109).

Several studies have demonstrated photodegradation as a valuable method for PCB degradation, with variations in reaction pathways, products, and product distribution, depending on reaction conditions (110–116). Due to steric effects, congeners with chlorine atoms at the *ortho*-positions are more likely to undergo photodegradation than those without (110, 115). Some of these degradation products include the dioxin-like non-*ortho* coplanar PCBs and lower chlorinated congeners (110, 111, 114, 117). Several studies have demonstrated the symmetrical and coplanar PCB congeners to be less photoreactive than other congeners (110, 117). Photodegradation is not as effective means of degradation as other types. The half-life of PCB degradation by photodechlorination was determined to be shorter than that by anaerobic biological dechlorination (117).

The degradation of PCBs via hydroxyl radicals in the atmosphere is probably the most important global loss mechanism for PCBs (118). Atmospheric OH radicals form when ozone is photolyzed (119). The OH radical behaves as an electrophile in PCB atmospheric reactions, with each electronegative chlorine substituent withdrawing electron density from the aromatic ring (120). Reactions with OH molecules occur primarily in the gas phase (119). Dechlorination has been determined as the major mechanism of photodegradation, compared to hydroxylation and arylation (115).

The loss of PCB congeners in the atmosphere is dominated by the reaction of the less chlorinated congeners (121). Air pollution, including automotive exhaust, increases OH levels in the atmosphere, thereby increasing PCB OH degradation in urban environments. Lighter congeners have been demonstrated to react readily with OH radicals and are less likely to sorb to particles, making OH degradation the significant pathway for PCB loss. The higher chlorinated congeners are less likely to undergo OH degradation and are more likely to strongly sorb to suspended particles, making deep-sea transfer a significant pathway for atmospheric loss of PCB (118, 121). A study conducted by Totten et al. (119) examined atmospheric PCB degradation and found that the particle-bound fraction of PCBs is rarely more than 10% of total airborne PCB burden for PCBs containing 1–6 chlorines under sampling conditions (northern U.S. cities in summer).

A study by Matykiewiczova et al. (116) demonstrated that PCBs can undergo photolysis in snow. Results showed reductive dechlorination as the major photochemical process with dehalogenated congeners of the parent PCB and biphenyl to be the primary photoproducts along with chloroquaterphenyls and hydroxychlorobiphenyls. Snow quality and the matrix location of the PCBs molecules were concluded to be the primary factors in determining whether the PCBs will remain in the snow long enough to undergo photodegradation. In order for photolysis to take place the PCBs need to be situated at the surface of the snowpack, exposed to sunlight. PCBs located deeper in the snowpack will not undergo photolysis.

OH radicals have been shown to play a key role in the degradation of volatilized PCBs, reducing their half-lives, especially in urban environments where are abundant. OH reactions appear to be the major permanent loss process of PCBs from the atmosphere (122, 123). Urban areas generate higher concentrations of OH radicals, and during daylight hours they react with lighter congeners, removing them from the atmosphere. This reaction occurs, to a lesser degree, in rural areas due to smaller atmospheric OH concentrations (119). The degree to which PCBs react with OH is dependent on OH concentrations, gaseous PCB concentrations, and the rate constant (K_{OH}) for the OH/PCB reaction (119, 121). PCB/OH reactions are at their maximum during the daytime in summer months. Globally, the reaction occurs to the greatest degree over the tropical/subtropical regions (118). Although atmospheric PCBs can react with O_3 and NO_3, the reactions are negligibly slow compared to those with OH, as are reactions with NaCl occurring in marine environments (119). Lower chlorinated congeners tend to react more readily with OH molecules than higher chlorinated (118, 119). Products of atmospheric OH-biphenyl reactions include hydroxylated biphenyls, benzoic acid/methyl benzoate, benzaldehyde, and salicylic acid (122, 123).

25.2.9 PCB Sediment Sorption, Desorption, and Transport

PCBs are transported within the Hudson River system in sediment, water, and biota. Due to their strong hydropho-

bicity, when PCBs are first introduced into the Hudson, they are either deposited onto river sediment or sorb to suspended particulate matter in the water column. Sorbed PCBs will remain sorbed, solubilize, bioaccumulate, volatilize, or move among these phases (19, 124–126). Sediments can behave as a reservoir for PCBs, storing them until proper conditions exist to release them back into the environment via scouring, desorption, volatilization, and ingestion. Sediments in rivers are transported by sedimentation (the tendency for particles in suspensions to settle out), coagulation, resuspension, and burial. Sediment associated PCBs may be buried, settle out of suspension and/or be resuspended, or chemically diffuse between sediment and water. The chemical properties of PCBs, in conjunction with environmental conditions, determine how they migrate and partition within the Hudson. The particulate PCB contribution increases with increased suspended sediment, while dissolved PCBs decrease relative to total PCBs transported. There are three operationally defined phases of contaminant transport in fluvial systems: the dissolved phase ($<0.45\,\mu m$), the suspended (particulate) phase, and the bed load phase (127).

Particle size is a primary determining factor for transport potential and, therefore, has a significant impact on contaminant loadings (127). Larger particle size sediment ($>63\,\mu m$) will be less contaminated as compared to smaller particle size sediment, due to less surface area. However, it is still possible to have a highly contaminated load resulting from a large concentration of large particle sediment (128).

Due to the *in situ* conditions (organic carbon, temperature, precipitation, salinity, river velocity, etc.) of the Hudson, PCBs frequently exchange between the sorbed and dissolved phases. A multitude of factors determine whether PCB contaminated sediments will be resuspended and relocated, diffuse into the water column, or volatilize. Parameters governing the sorptive properties of PCBs include partition coefficients like Henry's constant, Gibbs free energy, and vapor pressure. For equilibrium adsorption below saturation in either phase the partition coefficient can be thought of as the equilibrium constant of a reaction, proportional to the Gibbs free energy change (129). These parameters also determine the fate of water column PCBs.

Freshwater sorbate is a three-component system: the particulate fraction, water fraction, and dissolved (129). Suspended matter–water partitioning is a major process in determining the composition of PCBs in Hudson River sediment (13). PCBs are nonpolar, hydrophobic compounds with high *n*-octanol/water partition coefficient (K_{ow}) values. K_{ow} values are the driving force behind how PCBs interact with the various types of sediment, with K_{ow} increasing with congener chlorination (23). The higher the K_{ow} value, the more likely they are to sorb to sediment. The K_{ow} values and organic carbon fraction of the sediment (f_{oc}) are directly related to the partitioning of PCB between the freely dissolved and sediment phases K_d, where as $K_d = f_{oc} K_{ow}$ (130).

Sediment properties, such as organic matter content, mineralogic composition, and especially grain size, influence the adsorption and retention of PCBs to sediments (131, 132). For example, small particles with large surface areas have strong binding properties for pollutants, due to an increase in the number of chemical reaction sites (127). Lower chlorinated congeners have a stronger affinity for coarser sediment (grain size>63 μm), while the heavier congeners prefer finer sediment (<63 μm) (128). Congener shape and chlorine number factor into its ability to sorb and desorb with sediment; congeners in the planar form are able to maximize their contact with planar adsorbent surfaces (133).

Lower chlorinated congeners partition more strongly into the water phase or volatilize more readily from suspended particles whereas higher chlorinated congeners sorb more strongly to sediment and particles (13, 121). Partitioning to colloids appears to be a significant component of total concentration for mono- and dichlorobiphenyls in the Hudson, but not for the higher chlorinated congeners (129).

Organic carbon content plays a key role in PCB distribution within aquatic environments. Sorption to soil is directly related to soil OC content (134, 135). In the aquatic environment organic carbon is present in three phases: dissolved (DOC), colloidal (COC), and particulate (POC). POC is the primary sorbent of nonionic organic pollutants in the aquatic environment, including PCBs (136, 137). PCB partitioning from water to POC varies based on the congener, temperature, particulate concentration, nonlinear adsorption, and changes in the nature of the sorbate, weather particulates or OC, and water saliently (9). In aquatic systems PCBs prefer to sorb to particulate organic matter, particularly the vegetal fragment fraction (128, 138). It has been demonstrated that colloidal material does not appear to have a significant effect on water column partitioning for PCBs in the Hudson estuary, except for the most hydrophobic compounds; however, it proved to have a significant effect in marine sediment pore water, with the DOC-sorbed component accounting for 80% or more of the PCBs (139, 140).

Sorption to DOC was found not to have a major effect on the particle-sorbed concentrations for most congeners within the limited range of DOC observed in the upper Hudson (129). Planar PCBs sorb to carbonaceous geosorbents, like black carbon, more than nonplanar (19). Some lower chlorinated congeners have comparable values for K_{POC} and K_{DOC} in the Hudson, with the DOC-sorbed fraction accounting for approximately one third to one half of the total congener mass in the water column. As chlorination increases K_{POC} increases and K_{DOC} decreases. Partitioning to DOC accounts for <10% of the total mass in the water column for most congeners, with lower chlorinated congeners comprising a larger fraction of total PCB mass. This also accounts for a significant proportion of the dissolved concentration of the higher chlorinated congeners. Whether DOC has a significant effect on overall PCB phase distribution depends on DOC concentration, the nature of the DOC, and the hydrophobicity of the congener (129).

Carbonaceous material (CM) refers to all noncarbonated carbon-containing matter, including soil (primarily humus), sediment matter, organic carbon, black carbon, and soot. Humic substances play an important role in PCBs sorption to soils. Soot is a highly carbonaceous material containing mostly carbon, with values measured as high as 85% (141). Mono-*ortho*-PCBs, opposed to multiple-*ortho*-PCBs, appear to sorb preferentially to sediments containing soot and soot-like material, possibly because planar molecules may have a high affinity for soot, based on its pore size (133, 142).

25.2.10 Diurnal and Seasonal Fluxes Affecting PCB Transport

PCB transport in the Hudson varies with the seasons because of changes in temperature, precipitation, organic carbon availability, and runoff. Although there is no direct evidence of this occurring in the Hudson River, it is important to note that PCB contaminated fish whose migration patterns cause them to spawn and die in the same location at the same time, such as salmon, have been found to add to PCBs loads, which can result in some seasonal variations in PCBs concentration (143, 144).

The river flow varies throughout the year due to seasonal fluctuations in precipitation and temperature, which has been demonstrated to have a direct effect on PCB concentrations at various locations throughout the Hudson River. The high flow season for the Hudson River is in the spring (March–April) and the low flow period is from summer through winter (May–February) (145). Measured average PCBs loads in the Hudson River at the Thompson Island Dam (RM 194.6) and Waterford (RM~156), each had an average measured flow of 1.16 kg/d during the low flow periods to 18.1 kg/d during the high flow, and loads measured at Rogers Island (RM 194) ranging from 0.49 kg/d during low flow periods and 17.9 kg/d during high flow periods (9).

Jaward et al. (146) observed that diurnal cycling for some PCB congeners over the open ocean—indicating environmental processes like atmospheric OH radical concentrations, the atmospheric boundary layer height, wind speed/direction, air and surface water temperatures, light penetration, stratification, and biogeochemical cycles—can affect ambient PCB concentrations. Diurnal cycling has been observed in costal zones where vegetated surfaces raise daytime temperatures and OH radicals are present from urban sources, as well as land–sea breezes and emissions (119, 147). Concentrations of OH radicals are highest during the daytime (118). Lee et al. (147) found diurnal PCB cycling decreased as day/night temperature differences decreased, indicating that temperature plays a key role in diurnal cycling. Zooplankton also play a role in the diurnal cycling of PCBs. Zooplankton migrate daily, resulting in the transport of OC and the potential for the sorption and removal of PCBs from the dissolved phase, in turn lowering gas phase concentrations (146, 148).

25.2.11   Temperature

Temperature is a major factor influencing the partitioning and fate of PCBs. It has a direct relationship with the solubility of PCBs with solubility increasing with temperature. An increase in temperature raises the potential for PCBs to volatilize from both water and sediment (121, 126, 149). Volatilization is more prominent in warm seasons, like spring and summer (150, 151). In a study of vapor-phase PCBs near a contaminated site on the St. Lawrence River, Chiarenzelli et al. (152) reported the highest air levels in May. While May was not the warmest month, the air levels reflected the drying sediments deposited on the shoreline as the spring floods receded. They concluded that drying sediments are a major source of vapor-phase PCBs, perhaps even more important than those coming directly from the water column. Temperature affects PCB reactions other than volatilization. Desorption, diffusion, solubilization, and biological process-es increase with an increase in sediment temperature during the late spring and summer months (22). Lower chlorinated congeners are more likely to volatilize than the higher chlorinated, but all PCB congeners undergo volatilization cycles with changing air temperature (147, 153). Partition coefficients between sorbed and dissolved phases decrease with temperature at equilibrium conditions (129).

Concentrations of the less volatile, heavier PCBs have shown less temperature dependence in terms of plant PCB concentrations than the more volatile, less chlorinated PCB congeners (154). Temperature also influences the sorption/desorption processes of PCBs onto sediment. The more hydrophobic the PCB congener is, the greater the influence temperature has on sediment desorption, with a direct rela-tionship existing between temperature and PCB accumula-tion (153). Temperature is an important factor in PCB parti-tioning between OC and water in the upper Hudson (129). Slow desorption processes, like those existing inside clays, require greater temperatures to desorb PCBs. This is because in clay-dominated sediment PCBs become trapped within the intercrystalline water layers existing inside the clay and are retained more strongly than PCBs sorbed to other sediment materials such as organic matter (155).

The saturation liquid-phase vapor pressure (ρ°_L, Pa) of each congener as a function of temperature predicts the degree of association of PCBs with particles (156). *Ortho*-chlorine substituted congeners have been found to preferentially sorb to aerosol particles, with a degree of adsorption within a homolog following the order: $0>1>2>3>4$ (156). Mean atmospheric lifetimes for PCBs, when taking into account OH concentrations, are indirectly related to temperature, with atmospheric lifetimes increasing with chlorine numbers (121).

25.2.12   Precipitation

Precipitation is a driving force behind localized PCB levels in the Hudson River and for the transport of PCBs. Particle scavenging by rain and snow dominates wet deposition; therefore, the influence of temperature on gas/particle parti-tioning and the predominance of snow during the winter in high latitudes enhances the removal of organic chemicals from the atmosphere (157). Because of its porosity, snow is more effective at scavenging atmospheric particles than rain (157). When precipitation is low, water levels tend to remain more stagnant, and PCBs remain more localized. However, the slow-moving water can increase the solubility of the PCBs by providing the necessary time and conditions for them to diffuse from sediments into the water column. As precipitation levels increase, PCBs tend to stay sorbed more to the sediments. Sorption to sediment can increase the possibility for transport of PCBs due to the increased levels of sediment scouring which occur at higher river velocities.

During the spring, water levels in the Hudson are signif-icantly higher than during other periods due to snowmelt and increased rainfall. Snow and ice cover prevent the exchange of organic vapors between the atmosphere, the hydrosphere, and the terrestrial environment (158). In addition to this, snow cover can deposit PCBs to an area because PCBs are stored in the particulate matter associated with snow in a process known as snow scavenging (157). Upon snow melt, a flux of PCBs are released into surrounding soils, water, and vegetation through degradation, volatilization, meltwater runoff, and sorption to soils and freshwater (158). These losses to the environment can lead to seasonal increases in PCB concentrations in surface water (159, 160). The parti-cles, along with their contaminants, can also be deposited to snowpack in the absence of precipitation in a process known as dry particle deposition (158). PCB photodegradation for the same congeners has been observed to be much faster in liquid water than in frozen (116). PCB molecules trapped within closed ice cavities, rather than on the ice crystals surface which can diffuse out of the snow, have the greatest chance to undergo long-term photodegradation (116). The total surface area of the snow pack, based on density, specific surface area, and snow depth, is a controlling factor of the sorbed quantity of PCBs (161). Therefore, a reduction in snow surface will result in the desorption of chemicals from the ice surface into the interstitial pore space (161).

Rain has been demonstrated to scavenge both gas phase and particulate phase PCBs (162). Rainfall also inhibits PCB air–surface exchange (119). High concentrations of OC in rain and fog are thought to cause an enhancement in the equilibrium gas scavenging of hydrophobic organic compounds due to the strong partitioning of contaminants to POC and DOC (163). Partition coefficients for PCBs in rainwater are generally much higher than those measured in surface waters, suggesting the atmospheric particles are more sorptive than solids in surface waters (164).

25.2.13 Volatilization

PCBs are semivolatile compounds and global transport of vapor-phase PCBs constitutes one important mechanism whereby they spread. PCBs are volatilized from dump sites (165), and during drying of contaminated sediments, such as that which occurs every spring when the river level first rises, then recedes, depositing sediment on land (166), and they can volatilize directly from the water column. However, there is two-way traffic here, with vapor-phase PCBs, especially those of higher molecular weight, also being deposited back into the water (167, 168).

Duinker and Bouchertall (169) collected samples from an urban area and analyzed them for PCBs. They found that the filtered air was dominated by low chlorinated congeners and the aerosols in the rain were dominated by highly chlorinated congeners, with the vapor phase representing up to 99% of total atmospheric concentrations for the most volatile congeners. The greater volatility of lower chlorinated congeners is shown clearly in Fig. 25.3. In this study air was passed over pure Aroclor 1242, the major PCB mixture used at the General Electric plants. As is apparent, the lower chlorinated congeners selectively appear in air. There is some volatilization of higher chlorinated congeners but the congener pattern in air is very different from that of the pure parent Aroclor. The higher chlorinated congeners

also volatilize, but the congener pattern in the air is very different from that in the pure Aroclor. Several studies have demonstrated the ability for PCBs to exchange across the air–water interface (146, 170). PCB air–water exchange is dependent on wind speed, and since wind speed tends to be faster during the day, the PCB flux might be expected to be greater during the day, resulting in higher PCB concentrations (119).

PCBs volatilize in warmer temperatures and are transported to colder areas where they condense onto the earth's surface (171). PCB concentrations are higher at low altitudes and latitudes where atmospheric lifetimes (τ_A) are shorter, rather than at higher altitudes and latitudes where τ_A is longer (121). Lighter congeners are found predominantly in the gas phase, as opposed to being sorbed to airborne particles (119). Some congeners, such as CB 153, have shown to be lost equally from the atmosphere via atmospheric OH degradation and deep-sea transfer. The average atmospheric lifetime of PCBs due to reactions with OH radicals range from less than 1 month for the less chlorinated congeners, to over a year for heavy congeners. The average annual global loss of PCBs due to reactions with OH radicals for the period of the early 1970s to the early 2000s was in the range of hundreds to thousands of tons of PCBs annually, consisting mostly of the lighter congeners (121, 122).

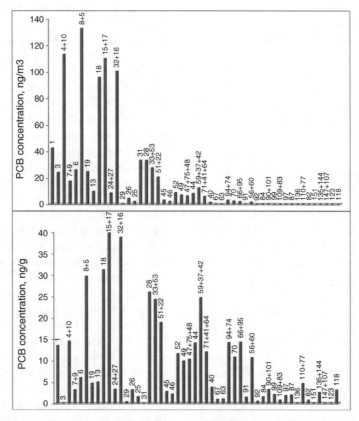

FIGURE 25.3 Congener patterns in pure Aroclor 1242 (top, in ng/g) and in the vapor phase after air was blown over the parent mixture (bottom, in ng/m³). The degree of chlorination increases from left to right. Congener numbers are given above the bars. Unpublished data from Ann Casey, with thanks to Dr. Zafar Aminov.

The major source of PCB to vegetation is the transfer of vapor-phase PCB from air to the aerial above ground portions of plants (172). Tolls and McLachlan (173) and Hung et al. (174) have demonstrated that there are two compartments for dry gaseous deposition of PCBs in plants, a large storage compartment on the interior of the leaf, with a smaller surface compartment. Root uptake is negligible, and the PCBs migrate from the surface to the interior compartment. Congeners with fewer *ortho*-chlorines have a lower liquid-phase vapor pressure; therefore, *mono*- and non-*ortho* PCBs will be associated with aerosols to a greater degree than multi-*ortho* PCBs of the same homolog group, thereby increasing their chances for wet and dry deposition (156). Vapor pressure is a strong predictor as to whether or not PCBs will diffuse into the atmosphere, and to their long-term fates. Long-range transport of individual PCB congeners is heavily dependent on congener vapor pressures, which decrease with increasing weight (119).

The spread of vapor-phase PCBs from the Hudson River has been reported to have migrated into the drinking water basins for New York City in studies by Commoner et al. (175). In this study they measured vapor-phase PCB concentrations, and then modeled the spread to the city reservoirs. They found high levels of PCBs (up to 1,740 pg/m^3) in the Neversink watershed at Frost Valley, reflective of the long-term accumulation and revolatilization processes occurring along the Hudson River and found modeled concentration estimates of up to 1,050 pg/L of water in the 13 surrounding watersheds. In a study from Anniston, AL, site of one of the PCB manufacturing plants, Hermanson and Johnson (176) found high levels of PCBs in tree bark near to the plant, which fell off over a distance of about 3 miles. Tree bark has sufficient lipid to accumulate lipophilic substances like PCBs, and their study provides at least some indication of the distance that is impacted by this process.

25.2.14 PCBs in Drinking Water

Several communities along the Hudson River derive their drinking water from the river. The degree to which this poses a health hazard has not been well evaluated. Most water treatment plants will remove particulate-bound PCBs, but those dissolved in water are more difficult to remove unless there is an excellent activated charcoal system in place. This is the same issue that applies to pharmaceuticals that get into the drinking water after having been released into wastewater. The EPA standard for PCBs in drinking water is 500 ppt, but the evidence that this is a safe level is lacking. We took a drinking-water sample from the Waterford, NY, city hall; Waterford is the only town in the upper Hudson to derive drinking water directly from the Hudson. The total PCB concentration was 110 ppt. This sample was taken prior to dredging. During the dredging, river-water concentrations have frequently exceeded the 500 ppt standard, causing the dredging to be temporarily stopped.

The problems with PCBs in drinking water is very similar to that of pharmaceuticals such as birth control drugs, antibiotics, chemotherapeutic drugs, and antipsychotic medications that get into the drinking water after having been released into wastewater, an issue which is getting increasing attention. The concentrations of pharmaceuticals in finished drinking water is usually very low, but measurable (177). Estrogenic substances have been convincingly demonstrated to feminize male fish (178). To date no one has convincingly demonstrated adverse human health effects from pharmaceuticals in drinking water, but such studies are difficult and few have been done. However the presence of antibiotics, which can add to the problem of antibiotic resistance, and estrogenic chemicals, which can cause endocrine disruption, is of major concern.

25.2.15 PCB Global Fate

Environmental conditions are dynamic and PCB congeners vary in their characteristics. Therefore, there is no single path or long-term fate for the PCBs in the Hudson. PCBs can be transported out of the Hudson through biological uptake and transfer vaporization, and through the water column into the Atlantic. Long-term fates for PCBs include: removal by degradation, loss to the stratosphere, leaching into the ground with irreversible sorption to the terrestrial environment, burial in glaciers, icecaps and peat bogs, and transfer to the deep sea adsorbing to sediments, plastic debris and phytoplankton eventually settling and down welling, and burial and irreversible sorption in freshwater and marine shelf sediments (121). Polar transport and redeposition is the principal fate for most PCBs. Although PCBs were never produced in the Arctic and rarely used in this region, they have been measured in surface and seawater, air, sediment, terrestrial and marine mammals, fish, birds, and humans, often at high concentrations (179–182).

Goldberg (183) first used the term "global distillation" to describe the ability of semivolatile compounds like PCBs to spread from local sites of contamination. PCBs, along with other contaminants, have a strong affinity for polar migration, a process that involves evaporation, wet/dry deposition, and atmospheric transfer of heat (184). They are considered to be "hoppers," contaminants who undergo repeated cycles of deposition and reevaporation as they migrate, in this case to the Arctic (185, 186). PCBs have been grouped into four categories based on their relative mobilities to move away from sources and toward the poles. The groupings were as follows: 0–1 chlorine atom PCBs—highly mobile worldwide with no deposition, 1–4 chlorine atoms—relatively high mobility with deposition in the polar altitudes, 4–8 chlorine atoms—relatively low mobility with deposition in the middle latitudes, and 8–9 chlorine atoms—low mobility with deposition close to the source (187). As a greater percentage of PCB production and use was in the northern hemisphere, the Arctic region has received a significant amount of the polar migrated PCBs (188). PCBs also reach the Arctic through biotransport. As PCBs accumulate up the food chain from phytoplankton to the top predator of the chain (human or animal), they also migrate with the species, thereby being transferred globally. Many marine and bird species migrate from the lower latitudes to the polar latitudes, transferring PCBs with them (179).

Another global impact is the contamination of the Atlantic Ocean from the outflow of the Hudson and other contaminated rivers and streams. Sediment material dredged from the NY–NJ Harbor has been deemed too contaminated to be dumped into the Atlantic Ocean (189). Despite the ban on the disposal of the dredged sediment into the Atlantic, contaminated sediment, water, and biotic material continue to flow from the harbor into the ocean. We have been accustomed to thinking of the oceans as sites so large that they cannot be polluted, but this is clearly not the case. Levels of PCBs and other persistent contaminants are significant in wild ocean fish, and are even higher in farmed ocean fish that are fed fish oil/fish meal obtained from wild fish (190, 191). The PCBs reach the ocean by all of the routes discussed previously— through being dissolved in water, bound to particulates that wash out into the ocean, and volatilizing and redepositing into the ocean. Unfortunately most of the loss of PCBs from the Hudson has had the effect of spreading the PCBs around the earth.

The PCB contamination sites of the Hudson River have global implications. These "hot spots" threaten local populations and ecosystems, including those as far away as the Arctic. While some of the Hudson River PCBs will undergo degradation and permanent sorption and settling, most will not. The majority of PCBs will travel regionally and eventually globally, exposing humans and wildlife.

Fish consumption is a major source of PCBs exposure for humans worldwide. Salmon, for example, is a popular fish in many nations and cultures and has been proven to contain PCBs regardless of origin. Hites et al. (190) found farmed salmon to have higher concentrations of PCBs than wild salmon, yet all the salmon tested showed elevated PCBs levels. Larger fish species, like sharks and swordfish (*Xiphias gladius*), have also been demonstrated to be a dietary source of PCBs to humans (192, 193) as well as bottom feeders such as catfish. While fish consumption is a primary source of dietary PCB exposure, other dietary exposures exist. In areas located near industrial sites that once produced or used PCBs, and in areas that experienced PCB contamination from leaks, disposal, and other sources, local agricultural livestock, dairy products, and produce have been found to have elevated levels of PCBs (194, 195). PCBs have also been observed in meat, dairy, and vegetable products originating in areas not considered to be elevated in PCBs (196, 197).

Some populations demonstrate immediate adverse health effects from PCB exposure, primarily at high levels, while other populations may show little to no adverse health effects

from being exposed at lower levels. Regardless of whether or not adverse health effects appear in the exposed population, offspring of exposed populations generally demonstrate adverse health effects due to the transfer of PCBs from parent to offspring (179). Due to the persistence and toxicity of PCBs, it is critical to remediate PCB-contaminated sites in the Hudson and globally. While it is likely impossible to remove all PCBs from the environment, reductions in levels can reduce exposure and toxicity rates for all species, including humans.

While the Hudson River makes an excellent case study of the PCB problem, it is but one of many contaminated bodies of water found in every country. Even the oceans are contaminated, serving as both a source and the ultimate sink for PCBs (198). There are also huge concentrations of PCBs in landfills and soils, from which PCBs spread by volatilization and migration (199). Sewage sludge has been found to contain PCBs (200), and it is spread on farm fields as a fertilizer. In addition some PCB-containing equipment and products, such as transformers, paint, window caulking (201), joint sealants (202), and even wood floor finish (203) are still in use in old buildings, and as they are discarded, they continue to contribute to the spread of PCBs. These household products lead to elevated PCBs in house dust (204). PCBs contaminate much of our food supply (205). While freshwater fish from many contaminated bodies of water are most contaminated, almost all animal fat contains levels of PCBs that have the potential to accumulate and cause harm. The perception that the problem of PCBs has passed when their manufacture ended is not correct, and action at many different levels, from government agencies to individual behaviors is needed to reduce the possibility of further spread of PCBs.

Unfortunately our present lack of any cost-effective method of destroying PCBs remains a serious problem. Current technology is usually either placing contaminated soils or sediments in a landfill, where hopefully they will not spread but unfortunately will not be destroyed, or be incinerated in a process that is both extremely expensive and destroys all organic material. While there are promising methods for destruction of PCBs under development, ranging from use of bioreactors (206), photocatalytic degradation using TiO_2 (207), electrochemical peroxidation (208), use of products found in molds and fungi (209, 210) to use of supercritical extraction and wet oxidation (211), none of these methods have reached widespread application. Further development of the technology for PCB destruction is urgently needed.

25.3 CONCLUSIONS

PCBs are dangerous chemicals and unfortunately they are very persistent. Hudson River PCBs have harmed wildlife and humans that live near the river. Most of the PCBs released into the river were still there until the most expensive dredging operation ever began in May 2009. Even with the planned removal of 249,343 pounds of PCBs, there will be a significant amount of PCBs remaining in the river. Furthermore, because the dredged sediments are being land-filled in Texas, even those removed from the river are not going to be destroyed. Even if the landfill does not leak, there remains the possibility that 100 years from now everyone will have forgotten what was in the landfill and will dig a basement or disturb it in some other way that results in human exposure. An even more serious consideration is that Hudson River PCBs continue to spread globally, along with other persistent chemicals. Because they are both very persistent and very toxic, they directly impact the health of both animals and people living at distant sites. These considerations indicate how important it is that we find cost-effective ways to destroy PCBs and other persistent organics.

REFERENCES

1. Sanders JE. PCB pollution in the upper Hudson River. In: *Contaminated Marine Sediments Assessment and Remediation, National Research Council.* Washington, DC: National Academy Press, 1989, pp. 365–400.

2. McCarty JP. The Hudson River—PCB Case Study. In: Hoffman DJ, Rattner BA, Burton GA Jr., Cairns J Jr.editors. *Handbook of Ecotoxicology*, 2nd edition. CRC Press LLC, 2003, pp. 813–831.

3. USEPA. Phase 2 Report—Review Copy Further Site Characterization and Analysis. Baseline Ecological Risk Assessment Hudson River PCBs Reassessment RI/FS. Vol. 2E. TAMS Consultants, Inc., Menzie-Cura & Associates, Inc., for U.S. Environmental Protection Agency Region II and U.S. Army Corps of Engineers Kansas City District, 1999.

4. USEPA. Hudson River PCBs Site New York: Record of Decision, 2002.

5. Rhea J, Connolly J, Haggard J. Hudson River PCBs: a 1990s perspective. *Clearwaters* 1997;27:24–28.

6. USEPA. Hudson River PCBs Reassessment RI/FS Responsiveness Summary for Human Health Risk Assessment. Vol. 2F. TAMS Consultants, Inc., Menzie-Cura & Associates, Inc., for U.S. Environmental Protection Agency Region II and U.S. Army Corps of Engineers Kansas City District, 2000.

7. NYSDEC (New York State Department of Environmental Conservation). A survey of PCB in wastewater from paper recycling operations, 1978.

8. Levinton JS, WaldmanJr., *The Hudson River Estuary.* Cambridge University Press, 2006.

9. USEPA. Phase 2 Report—Further Site Characterization and Analysis. Data Evaluation and Interpretation Report, Hudson River Reassessment RI/FS. Vol. 2C. TAMS Consultants, Inc., Menzie-Cura & Associates, Inc., for U.S. Environmental Protection Agency Region II and U.S. Army Corps of Engineers Kansas City District, 1997.

10. Connolly JP, Zahakos HA, Benaman J, Ziegler CK, Rhea JR, Russell KA. Model of PCB fate in the Upper Hudson River. *Environ Sci Technol* 2000;34:4076–4087.

11. Asher BJ, Wong CS, Rodenburg LA. Chiral source apportionment of polychlorinated biphenyls to the Hudson River estuary atmosphere and food web. *Environ Sci Technol* 2007;41:6163–6169.

12. Bopp RF, Chillrud SN, Shuster EL, Simpson HJ, Estabrooks FD. Trends in chlorinated hydrocarbon levels in Hudson River basin sediments. *Environ Health Perspect* 1998;106:1075–1081.

13. Bopp RF, Simpson HJ, Olsen CR, Kostyk N. Polychlorinated biphenyls in sediments of the tidal Hudson River, New York. *Environ Sci Technol* 1981;15:210–216.

14. Durell GS, Lizotte RD Jr. PCB levels at 26 New York City and New Jersey WPCPs that discharge to the New York/New Jersey Harbor Estuary. *Environ Sci Technol* 1998;32:1022–1031.

15. Huntley SL, Iannuzzi TJ, Avantaggio JD, Carlson-Lynch H, Schmidt CW, Finely BI. Combined sewer overflows (CSOs) as sources of sediment contamination in the lower Passaic River, New Jersey. II. Polychlorinated dibenzo-*p*-dioxins, polychlorinated dibenzofurans, and polychlorinated biphenyls. *Chemosphere* 1997;34:223–250.

16. Walker WJ, McNutt RP, Maslanka CK. The potential contribution of urban runoff to surface sediments of the Passaic River: sources and chemical characteristics. *Chemosphere* 1999;38:363–377.

17. Johnson GW, Chiarenzelli J, Quensen JF III, Hamilton MC. Chapter 10: Polychlorinated biphenyls. In: Morrison R, Murphy B, editors. *Environmental Forensics: A Contaminant Specific Guide.* Amsterdam: Elsevier, 2006. pp. 187–225.

18. Brodsky J, Ballschmiter K. Reversed phase liquid chromatography of PCBs as a basis for calculation of water solubility and log K_{ow} for polychlorobiphenyls. *Fresenius Z Anal Chem* 1988;331:295–301.

19. Cornelissen G, Gustafsson O, Bucheli TD, Jonker MT, Koelmans AA, van Noort PC. Extensive sorption of organic compounds to black carbon, coal, and kerogen in sediments and soils: mechanisms and consequences for distribution, bioaccumulation, and biodegradation. *Environ Sci Technol* 2005;39:6881–6895.

20. Dunnivant FM, Coates JT, Eizerman AW. Experimentally determined Henry's Law constants for 17 polychlorobiphenyl congeners. *Environ Sci Technol* 1988;22:448–453.

21. Kolar B, Hrsak D, Fingler S, Cetkovic H, Petri I, Kolic NU. PCB-degrading potential of aerobic bacteria enriched from marine sediments. *Int Biodeter Biodegr* 2007;60:16–24.

22. McDonough KM, Dzombak DA. Microcosm experiments to assess the effects of temperature and microbial activity on polychlorinated biphenyl transport in anaerobic sediment. *Environ Sci Technol* 2005;39:9517–9522.

23. Moret I, Gambaro A, Piazza R, Ferrari S, Mandori L. Determination of polychlorobiphenyl congeners (PCBs) in the surface water of the Venice lagoon. *Mar Pollut Bull* 2005;50:167–174.

24. Rusling JF, Miaw CL. Kinetic estimation of standard reduction potentials of polyhalogentated biphenyls. *Environ Sci Technol* 1989;23:476–479.

25. Van Drooge BL, Grimalt JO. Atmospheric semivolatile organochlorine compounds in European high-mountain areas (Central Pyrenees and High Tatras). *Environ Sci Technol* 2004;38:3525–3532.

26. Safe S. Toxicology, structure–function relationship, and human and environmental health impacts of polychlorinated biphenyls: progress and problems. *Environ Health Perspect* 1992;100:259–269.

27. LeRoy KD, Thomas P, Khan IA. Thyroid hormone status of Atlantic croaker exposed to Aroclor 1254 and selected PCB congeners. *Comp Biochem Physiol* 2006;144:263–271.

28. Newman J, Gallo MV, Schell LM, DeCaprio AP, Denham M, Deane GD, the Akwesasne Task Force on the Environment. Analysis of PCB congeners related to cognitive functioning in adolescents. *Neurotoxicology* 2009;30:686–696.

29. Bazzanti M. Distribution of PCB congeners in aquatic ecosystems: a case study. *Environ Int* 1997;23:799–813.

30. Antunes P, Amado J, Vale C, Gil O. Influence of the chemical structure on mobility of PCB congeners in female and male sardine (*Sardina pilchardus*) from Portuguese coast. *Chemosphere* 2007;69:395–402.

31. Safe S, Brown KW, Donnelly KC, Anderson CS, Markiewicz KV, McLachlan MS, Reischl A, Hutzinger O. Polychlorinated dibenzo-*p*-dioxins and dibenzofurans associated with wood-preserving chemical sites: biomonitoring with pine needles. *Environ Sci Technol* 1992;26:394–396.

32. Safe S. Polychlorinated biphenyls (PCBs): environmental impact, biochemical and toxic responses, and implications for risk assessment. *Crit Rev Toxicol* 1994;24:87–149.

33. Dushenko WT, Grundy SL, Reimer KJ. Vascular plants as sensitive indicators of lead and PCB transport from local sources in the Canadian Arctic. *Sci Total Environ* 1996; 188:29–38.

34. Johnson CD, Balagurunathan Y, Tadesse MG, Falahatpisheh MH, Brun M, Walker MK, Dougherty ER, Ramos KS. Unraveling gene-gene interactions regulated by ligands of the aryl hydrocarbon receptor. *Environ Health Perspect* 2004;112:403–412.

35. Maier MS, Legare ME, Hanneman WH. The aryl hydrocarbon receptor agonist 3,3′,4,4′, 5-pentachlorobiphenyl induces distinct patterns of gene expression between hepatoma and glioma cells: chromatin remodeling as a mechanism for selective effects. *Neurotoxicology* 2007;28: 594–612.

36. Van den Berg M, Birnbaum LS, Denison M, De Vito M, Farland W, Feeley M, Fiedler H, Hakansson H. The 2005 World Health Organization reevaluation of human and mammalian toxic equivalency factors for dioxins and dioxin-like compounds. *Toxicol Sci* 2006;93:223–241.

37. Vezina CM, Walker NJ, Olson Jr., Subchronic exposure to TCDD, PeCDF, PCB126, and PCB153: effect on hepatic gene expression. *Environ Health Perspect* 2004;112: 1636–1644.

38. Carpenter DO, Arcaro K, Spink DC. Understanding the human health effects of chemical mixtures. *Environ Health Perspect* 2002;110:25–42.

39. Agency for Toxic Substances and Disease Registry, (ATSDR). *Toxicological Profile for Polychlorinated Biphenyls (PCBs).* Atlanta, GA: U.S. Department of Health and Human Services, Public Health Service, 2000.

40. Carpenter DO. Polychlorinated biphenyls (PCBs): routes of exposure and effects on human health. *Rev Environ Health* 2006;21:1–23.

41. Carpenter DO. Environmental contaminants as risk factors for developing diabetes. *Rev Environ Health* 2008;23:59–74.

42. Carpenter DO, Nevin R. Environmental causes of violence. *Physiol Behav.* 2009;99:260–268.

43. USEPA. Phase 2 Report—Further Site Characterization and Analysis. Revised Baseline Ecological Risk Assessment Hudson River PCBs Reassessment RI/FS. Vol. 2E. TAMS Consultants, Inc., Menzie-Cura & Associates, Inc., for U.S. Environmental Protection Agency Region II and U.S. Army Corps of Engineers Kansas City District, 2000.

44. USEPA. Phase 2 Report—Revised Copy Further Site Characterization and Analysis. Human Health Risk Assessment Hudson River PCBs Reassessment RI/FS. Vol. 2F. TAMS Consultants, Inc., Gradient Corp., for U.S. Environmental Protection Agency Region II and U.S. Army Corps of Engineers Kansas City District, 1999.

45. Kudyakov R, Baibergenova A, Zdeb M, Carpenter DO. Respiratory disease in relation to patient residence near to hazardous waste sites. *Environ Toxicol Pharmacol* 2004;18:249–257.

46. Sergeev AV, Carpenter DO. Hospitalization rates for coronary heart disease in relation to residence near areas contaminated with POPs and other pollutants. *Environ Health Perspect* 2005;113:756–761.

47. Shcherbatykh I, Huang X, Lessner L, Carpenter DO. Hazardous waste sites and stroke in New York State. *Environ Health* 2005;4:18.

48. Huang X, Lessner L, Carpenter, DO. Exposure to persistent organic pollutants and hypertensive disease. *Environ Res* 2006;102:101–106.

49. Kouznetsova M, Huang X, Ma J, Lessner L, Carpenter DO. Increased rate of hospitalization for diabetes and residential proximity of hazardous waste sites. *Environ Health Perspect* 2007;115:75–79.

50. Fitzgerald EF, Belanger EE, Gomez MI, Cayo M, McCaffrey RJ, Seegal RF, Jansing RL, Hwang SA, Hicks HE. Polychlorinated biphenyl exposure and neuropsychological status among older residents of upper Hudson River communities. *Environ Health Perspect* 2008;116:209–215.

51. Haase RF, McCaffrey RJ, Santiago-Rivera AL, Morse GS, Tarbell A. Evidence of an age-related threshold effect of polychlorinated biphenyls (PCBs) on neuropsychological functioning in a Native American population. *Environ Res* 2009;109:73–85.

52. Codru N, Schymura MJ, Negoita S, the Akwesasne Task Force on the Environment, Rej R, Carpenter DO. Diabetes in relation

to serum levels of polychlorinated biphenyls (PCBs) and chlorinated pesticides in adult Native Americans. *Environ Health Perspect* 2007;115:1442–1447.

53. Lee DH, Lee IK, Song K, Steffes M, Toscano W, Baker BA, Jacobs DR Jr., A strong dose–response relation between serum concentrations of persistent organic pollutants and diabetes: results from the National Health and Examination Survey 1999–2002. *Diabetes Care* 2006;29:1638–1644.

54. Ha M-H, Lee D-H, Jacobs DR. Association between serum concentrations of persistent organic pollutants and self-reported cardiovascular disease prevalence: results from the national health and nutrition examination survey. *Environ Health Perspect* 2007;115:1204–1209.

55. Goncharov A, Haase RF, Santiago-Rivera A, Morse G, Akwesasne Task Force on the Environment, McCaffrey RJ, Rej R, Carpenter DO. High serum PCBs are associated with elevation of serum lipids and cardiovascular disease in a Native American population. *Environ Res* 2008;106: 226–239.

56. Eisler R. *Handbook of Chemical Risk Assessment: Health Hazards to Humans, Plants, and Animals*, Vol. 2 Boca Raton, FL: Lewis Publishers, 2000.

57. Lang, V. Review: polychlorinated biphenyls in the environment. *J Chromatogr* 1992;595:1–43.

58. Eisler R. Polychlorinated biphenyl hazards to fish, wildlife, and invertebrates: a synoptic review. *USFWS* 1986.

59. Bush B, Kadlec MJ. Dynamics of PCBs in the aquatic environment. *Great Lakes Res Rev* 1995;1:24–30.

60. McCarty JP, Secord AL. Nest-building behavior in PCB-contaminated tree swallows. *Auk* 1999;116:55–63.

61. Johnson KE, Knopper LD, Schneider DC, Olson CA, Reimer KJ. Effects of local point source polychlorinated biphenyl (PCB) contamination on bone mineral density in deer mice (*Peromyscus maniculatus*). *Sci Total Environ* 2009;407:5050–5055.

62. Alston DA, Tandler B, Gentles B, Smith EE. Testicular histopathology in deer mice (*Peromyscus maniculatus*) following exposure to polychlorinated biphenyl. *Chemosphere* 2003;52:283–285.

63. Aulerich RJ, Ringer RK. Current status of PCB toxicity to mink, and effect on their reproduction. *Arch Environ Contam Toxicol* 1977;6:279–292.

64. Baker JE, Bohlen WF, Bopp R, Brownawell B, Collier TK, Farley KJ, Geyer WR, Nairn R. *PCBs in the Upper Hudson River: The Science Behind the Dredging Controversy*. White paper prepared for the Hudson River Foundation, 2001, 47 pp.

65. Rice CP, O'Keefe PW, Kubiak TJ. Sources, pathways, and effects of PCBs dioxins, and dibenzofurans. In: Hoffman DJ, Rattner BA, Burton GA Jr., Cairns J Jr., editors. *Handbook of Ecotoxicology*, 2nd edition. CRC Press LLC, 2003, pp. 813–831.

66. Ashley JT, Horwitz R, Steinbacher JC, Ruppel B. A comparison of congeneric PCB patterns in American eels and striped bass from the Hudson and Delaware River estuaries. *Mar Pollut Bull* 2003;46:1294–1308.

67. Gobas APC. Assessing bioaccumulation factors of persistent organic pollutants in aquatic food chains. In: Harrad S, editor. *Persistent Pollutants: Environmental Behavior and Pathways of Human Exposure*. Norwell, MA: Kluwer Academic Publishers, 2001.

68. Gobas FAPC, Z'Graggen MN, Zhang X. Time response of the Lake Ontario ecosystem to virtual elimination of PCBs. *Environ Sci Technol* 1995;29:2038–2046.

69. Herendeen RA, Hill WR. Growth dilution in multilevel food chains. *Ecol Model* 2004;178:349–356.

70. Smith TE, Stevenson RJ, Caraco NF, Cole JJ. Changes in phytoplankton community structure during the zebra mussel (*Dreissena polymorpha*) invasion of the Hudson River (New York). *J Plankton Res* 1998;20:1567–1579.

71. Bastviken DTE, Caraco NF, Cole JJ. Experimental measurements of zebra mussel (*Dreissena polymorpha*) impacts on phytoplankton community compositions. *Freshwater Biol* 1998;39:375–386.

72. Cho, Y-C, Frohnhoefer RC, Rhee G-Y. Bioconcentration and redeposition of polychlorinated biphenyls by zebra mussels (*Dreissena polymorpha*) in the Hudson River. *Water Res* 2004;38:769–777.

73. Bedard DL, Unterman R, Bopp LH, Brennan MJ, Haberl ML, Johnson C. Rapid assay for screening and characterizing microorganisims for the ability to degrade polychlorinated biphenyls. *Appl Environ Microbiol* 1986;51:761–768.

74. Bedard DL, May RJ. Characterization of the polychlorinated biphenyls in the sediments of woods pond: evidence for microbial dechlorination of Aroclor 1260 *in situ*. *Environ Sci Technol* 1996;30:237–245.

75. Abraham WR, Nogales B, Golyshin PN, Pieper DH, Timmis KN. Polychlorinated biphenyl-degrading microbial communities in soil and sediments. *Curr Opin Microbiol* 2002;5:246–253.

76. Brown J.F. Jr., Wagner RE, Bedard DL, Brennan MJ, Carnahan JC, May RJ. PCB transformations in upper Hudson sediments. *NE Environ Sci* 1984;3:167–179.

77. Brown J.F. Jr., Wagner RE, Feng H, Bedard DL, Brennan MJ, Carnahan JC, May RJ. Environmental dechlorination of PCBs. *Environ Toxicol Chem* 1987;6:549–593.

78. River, Watch. *A GE Report on the Hudson River*. Fall 2000, Which future for the Hudson?.

79. Gibson DT, Cruden DL, Haddock JD, Zylstra GJ, Brand JM. Oxidation of polychlorinated biphenyls by *Pseudomonas* sp. Strain LB400 and *Pseudomonas pseudoalcaligenes* KF707. *J Bacteriol* 1993;175:4561–4564.

80. Novakova H, Vosahlikova M, Pazlarova J, Mackova M, Burkhard J, Demnerova K. PCB metabolism by *Pseudomonas* sp. P2. *Int Biodeter Biodegr* 2002;50:47–54.

81. Asturias JA, Timmis KN. Three different 2,3-dihydroxybiphenyl-1, 2-dioxygenase genes in the gram-positive polychlorobiphenyl-degrading bacterium *Rhodococcus globerulus* P6. *J Bacteriol* 1993;175:4631–4640.

82. Chung S-Y, Maeda M, Song E, Horikoshi K, Kudo T. A gram-positive polychlorinated biphenyl-degrading bacterium, *Rhodococcus erythropolis* strain TA421, isolated from

termite ecosystem. *Biosci Biotechnol Biochem* 1994;58: 2111–2113.

83. Seto M, Kimbara K, Shimura M, Hatta T, Fukuda M. A novel transformation of polychlorinated biphenyls by *Rhodococcus* sp. strain RHA1. *Appl Environ Microbiol* 1995;61:3353–3358.

84. Luo W, D'Angelo EM, Coyne MS. Carbon effects on aerobic polychlorinated biphenyl removal and bacterial community composition in soils and sediments. *Chemosphere* 2008;70:364–373.

85. Wu Q, Bedard DL, Wiegel J. Temperature determines the pattern of anaerobic dechlorination of Aroclor 1260 primed by 2,3,4, 6-tetrachlorobiphenyl in Woods Pond sediment. *Appl Environ Microbiol* 1997;63:4818–4825.

86. Van Dort HM, Smullen LA, May RJ, Bedard DL. Priming microbial meta-dechlorination of polychlorinated biphenyls that have persisted in Housatonic River sediments for decades. *Environ Sci Technol* 1997;31:3300–3307.

87. Quensen J. F. III, Boyd SA, Tiedje JM. Dechlorination of four commercial polychlorinated biphenyl mixtures (Aroclors) by anaerobic microorganisms from sediments. *Appl Environ Microbiol* 1990;56:2360–2369.

88. Wiegel J, Wu Q. Microbial reductive dehalogenation of polychlorinated biphenyls. *FEMS Microbiol Ecol* 2000;32: 1–15.

89. Kim J, Rhee G. Population dynamics of polychlorinated biphenyl-dechlorinating microorganisms in contaminated sediments. *Appl Environ Microb* 1997;63:1771–1776.

90. Nies L, Vogel TM. Effects of organic substrates on dechlorination of Aroclor 1242 in anaerobic sediments. *Appl Environ Microb* 1990;56:2612–2617.

91. Alder AC, Haggblom MM, Oppenheimer SR, Young LY. Reductive dechlorination of polychlorinated biphenyls in anaerobic sediments. *Environ Sci Technol* 1993;27:530–538.

92. Morris PJ, Mohn WW, Quensen J.F. III, Tiedje JM, Boyd SA. Establishment of a polychlorinated biphenyl-degrading enrichment culture with predominantly meta dechlorination. *Appl Environ Microb* 1992;58:3088–3094.

93. Ward JAM, Ahad JME, Lacrampe-Couloume G, Slate GF, Edwards EA, Sherwood Lollar B. Hydrogen isotope fractionation during methanogenic degradation of toluene: potential for direct verification of bioremediation. *Environ Sci Technol* 2000;34:4577–4581.

94. Abraham WR, Hesse C, Pelz O. Ratios of carbon isotopes in microbial lipids as an indicator of substrate usage. *Appl Environ Microbiol* 1998;64:4202–4209.

95. Reddy CM, Heraty LJ, Holt BD, Sturchio NC, Eglinton TL, Drenzek NJ, Xu L, Lake JL, Maruya KA. Stable chlorine isotopic compositions of Aroclors and Aroclor-contaminated sediments. *Environ Sci Technol* 2000;34:2866–2870.

96. Meckenstock RU, Morasch B, Griebler C, Richnow HH. Stable isotope fractionation analysis as a tool to monitor biodegradation in contaminated aquifers. *J Contam Hydrol* 2004;75:215–255.

97. Moore M, Mustain M, Daniel K, Chen I, Safe S, Zacharewski T, Gillesby B, Joyeux A, Balaguer P. Antiestrogenic activity of hydroxylated polychlorinated biphenyl congeners identified in human serum. *Toxicol Appl Pharmacol* 1997;142: 160–168.

98. Liu X, Sokol RC, Kwon OS, Bethoney CM, Rhee G-Y. An investigation of factors limiting the reductive dechlorination of polychlorinated biphenyls. *Environ Toxicol Chem* 1996;15:1738–1744.

99. Sokol CS, Bethoney CM, Rhee GY. Effect of Aroclor 1248 concentration on the rate and extent of polychlorinated biphenyl dechlorination. *Environ Toxicol Chem* 1998; 17: 1922–1926.

100. Sokol CS, Bethoney CM, Rhee GY. Reductive dechlorination of preexisting sediment polychlorinated biphenyls with long-term laboratory incubation. *Environ Toxicol Chem* 1998;17:982–987.

101. Ming Chen I, Chang F-C, Hus M-F, Wang Y-S. Comparisons of PCBs dechlorination occurrences in various contaminated sediments. *Chemosphere* 2001;43:649–654.

102. Wong PW, Garcia EF, Pessah IN. Ortho-substituted PCB95 alters intracellular calcium signaling and causes cellular acidification in PC12 cells by an immunophilin-dependent mechanism. *J Neurochem* 2001;76:450–463.

103. Brändli RC, Bucheli TD, Kupper T, Mayer J, Stadelmann FX, Tarradellas J. PCBs, PAHs and their source characteristic ratios during composting and digestion of source-separated organic waste in full-scale plants. *Environ Pollut* 2007;148: 520–528.

104. Deweerd KA, Bedard DL. Use of halogenated benzoates and other halogenated aromatic compounds to stimulate the microbial dechlorination of PCBs. *Environ Sci Technol* 1999;33:2057–2063.

105. Wu Q, Bedard DL, Wiegel J. 2, 6-Dibromobiphenyl primes extensive dechlorination of Aroclor 1260 in contaminated sediment at 8–30oC by stimulating growth of PCB-dehalogenating microorganisms. *Environ Sci Technol* 1999;33: 595–602.

106. Harkness MR, McDermott JB, Abramowicz DA, Salvo JJ, Flanagan WP, Stephens ML, Mondello FJ, May RJ, Lobos JH, Carroll KM, et al. *In situ* stimulation of aerobic PCB biodegradation in Hudson River sediments. *Science* 1993;259: 503–507.

107. Bedard DL, Van Dort HM, May RJ, Smullen LA. Enrichment of microorganisms that sequentially *meta, para*-dechlorinate the residue of Aroclor 1260 in Housatonic River sediment. *Environ Sci Technol* 1997;31:3308–3313.

108. Wu Q, Watts JEM, Sowers KR, May HD. Identification of bacterium that specifically catalyzes the reductive dechlorination of polychlorinated biphenyls with doubly flanked chlorines. *Appl Environ Microbiol* 2002;68:807–812.

109. Safe S. Polyhalogentated aromatics: uptake, disposition and metabolism. In: Kimbrough RD, Jensen AA, editors. *Halogenated Biphenyls, Terphenyls, Napthalenes, Dibenzodioxins and Related Products*. New York: Elsevier, 1989.

110. Miao XS, Chu SG, Xu XB. Degradation pathways of PCBs upon UV irradiation in hexane. *Chemosphere* 1999;39: 1639–1650.

111. Lepine F, Masse R. Degradation pathways of PCB upon gamma irradiation. *Environ Health Perspect* 1991;89: 183–187.

112. Oida T, Barr JR, Kiata K, McClure PC, Lapeza C.R. Jr., Hosoya K, Ikegami T, Smith CJ, Patterson D.G. Jr., Tanka N. Photolysis of polychlorinated biphenyls on octadecylsilylated silica particles. *Chemosphere* 1990;39:1795–1807.

113. Izadifard M, Achari G, Langford CH. The pathway of dechlorination of PCB congener by a photochemical chain process in 2-propanol: the role of medium and quenching. *Chemosphere* 2008;73:1328–1334.

114. Lin YJ, Teng LS, Lee A, Chen YL. Effect of photosensitizer diethylamine on the photodegradation of polychlorinated biphenyls. *Chemosphere* 2004;55:879–884.

115. Bunce JJ, Landers JP, Langshaw J, Nakal JS. An assessment of the importance of direct solar degradation of some simple chlorinated benzenes and biphenyls in the vapor phase. *Environ Sci Technol* 1989;23:213–218.

116. Matykiewiczova N, Klanova J, Klan P. Photochemical degradation of PCBs in snow. *Environ Sci Technol* 2007;41: 8308–8314.

117. Chang FC, Chiu TC, Yen JH, Wang YS. Dechlorination pathways of *ortho*-substituted PCBs by UV irradiation in *n*-hexane and their correlation to the charge distribution on carbon atom. *Chemosphere* 2003;51:775–784.

118. Mandalakis M, Berresheim H, Stephanou EG. Direct evidence for destruction of polychlorinated biphenyls by OH radicals in the subtropical troposphere. *Environ Sci Technol* 2003;37: 542–547.

119. Totten LA, Eisenreich SJ, Brunciak PA. Evidence for destruction of PCBs by the OH radical in urban atmospheres. *Chemosphere* 2002;47:735–746.

120. Kwok ES, Atkins R, Arey, J. Rate constants for the gas-phase reactions of the OH radical with dichlorobiphenyls, 1-chlorodibenzo-p-dioxin, 1, 2-diemnthoxbenzene, and diphenyl ether: estimation of OH radical reaction rate constants for PCBs, PCDDs, and PCDFs. *Environ Sci Technol* 1995;29:1591–1598.

121. Wania F, Daly GL. Estimating the contribution of degradation in air and deposition to the deep sea to the global loss of PCBs. *Atmos Environ* 2002;36:5581–5593.

122. Anderson PN, Hites RA. OH radical reactions: the major removal pathway for polychlorinated biphenyls from the atmosphere. *Environ Sci Technol* 1996;30:1756–1763.

123. Brubaker WW, Hites RA. Gas-phase oxidation products of biphenyl and polychlorinated biphenyls. *Environ Sci Technol* 1998;32:3913–3918.

124. Bremle G, Larsson P. PCB in the air during landfilling of a contaminated lake sediment. *Atmos Environ* 1998;32: 1011–1019.

125. Allen-King RM, Grathwohol P, Ball WP. New modeling paradigms for the sorption of hydrophobic organic chemicals to heterogeneous carbonaceous matter in soils, sediments, and rocks. *Adv Water Resour* 2002;25:985–1016.

126. Vilanova RM, Fernandez P, Grimalt JO. Polychlorinated biphenyl partitioning in the waters of a remote mountain lake. *Sci Total Environ* 2001;279:51–62.

127. Droppo IG, Jaskot C. Impact of river transport characteristics on contamination sampling error and design. *Environ Sci Technol* 1995;29:161–170.

128. Pierad C, Budzinski H, Garrigues P. Grain-size distribution of polychlorobiphenyls in coastal sediments. *Environ Sci Technol* 1996;30:2776–2783.

129. Butcher JB, Garvery EA, Bierman VJ Jr., Equilibrium partitioning of PCB congeners in the water column: field measurements from the Hudson River. *Chemosphere* 1998;36: 3149–3166.

130. Di' Toro DM, Zarba CS, Hansen DJ, Berry WJ, Swartz RC, Cowan CE, Pavlou, SP. Technical basis for establishing sediment quality criteria for nonionic organic chemicals using equilibrium partitioning. *Environ Toxicol Chem* 1991;10: 1541–1583.

131. Klamer JC, Hegeman WJM, Smedes F. Comparison of grain size correction procedures for organic micropollutants and heavy metals in marine sediments. *Hydrobiologia* 1990; 208:213–220.

132. Ollivon D, Garban B, Blanchard M, Teil MJ, Carru AM, Cesterikoff C, Cevreuil M. Vertical distribution and fate of trace metals and persistent organic pollutants in sediments of the Sein and Marne rivers (France). *Water Air Soil Pollut* 2002;134:57–79.

133. Jonker MTO, Smedes F. Preferential sorption of planar contaminants in sediments from Lake Ketelmeer, the Netherlands. *Environ Sci Technol* 2000;34:1620–1626.

134. Burgess RM, Ryba SA, Cantwell MG, Gundersen JL. Exploratory analysis of the effects of particulate characteristics on the variation in partitioning of nonpolar organic contaminants to marine sediments. *Water Res* 2001;35: 4390–4404.

135. Ayris S, Harrad S. The fate and persistence of polychlorinated biphenyls in soil. *J Environ Monit* 1999;1:395–401.

136. Means JC, Wood SG, Hassett JJ, Banwart WL. Sorption of polynuclear aromatic hydrocarbons by sediments and soils. *Environ Sci Technol* 1980;14:1524–1528.

137. Schwarzenbach RP, Westall J. Transport of nonpolar organic compounds from surface water to groundwater. Laboratory sorption studies. *Environ Sci Technol* 1981;15:1360–1367.

138. Karickhoff SW. Organic pollutant sorption in aquatic systems. *J Hydraul Eng* 1984;110:707–735.

139. Brownawell BJ, Achman DR.Partitioning of PCBs in the water column of the Hudson River estuary. Report to Hudson River Foundation, New York, 1994.

140. Burgess RM, McKinney RA, Brown WA. Enrichment of marine sediment colloids with polychlorinated biphenyls: trends resulting from PCB solubility and chlorination. *Environ Sci Technol* 1996;30:2556–2566.

141. Nejar NM, Makkee M, Illan-Gomez MJ. Catalytic removal of NOx and soot from diesel exhaust: oxidation behaviour of carbon materials used as model soot. *Appl Catal B Environ* 2007;75:11–16.

142. Jonker MTO, Koelmans AA. Sorption of polycyclic aromatic hydrocarbons and polychlorinated biphenyls to soot and soot-like materials in the aqueous environment: mechanistic considerations. *Environ Sci Technol* 2002;36:6725–6734.

143. Krümmel EM, Gregory-Eaves I, Macdonald RW, Kimpe LE, Demers MJ, Smol JP, et al. Concentrations and fluxes of salmon derived polychlorinated biphenyls (PCBs) in lake sediments. *Environ Sci Technol* 2005;39:7020–7026.

144. MDEQ (Michigan Department of Environmental, Quality). Total maximum daily load for polychlorinated biphenyls for the Kawkawlin River Bay County, Michigan, 2002.

145. Farley KJ, Thoman RV, Cooney TF, III., Damiani DR, Wands, Jr. *An Integrated Model of Organic Chemical Fate and Bioaccumulation in the Hudson River Estuary.* Environmental Engineering Department Manhattan College, March 1999.

146. Jaward FM, Barber JL, Booij K, Dachs J, Lohman R, Jones, KC. Evidence for dynamic air–water coupling and cycling of persistent organic pollutants over the Atlantic Ocean. *Environ Sci Technol* 2004;38:2617–2625.

147. Lee RG, Hung H, Mackay D, Jones, KC. Measurement and modeling of the diurnal cycling of atmospheric PCBs and PAHs. *Environ Sci Technol* 1998;32:2172–2179.

148. Steinberg DK, Carlson CA, Bates NR, Goldthwait SA, Madin LP, Michales, AF. Plankton vertical migration and the active transport of dissolved organic and inorganic carbon in the Sargasso Sea. *Deep Sea Res PT I* 2000;47:137–158.

149. Huang Q, Hong, CS. Solubilities of non-ortho and mono-ortho PCBs at four temperatures. *Water Res* 2002;36:33543–33552.

150. Halsall CJ, Howsam GM, Lee RGM, Ockenden WA, Jones, KC. Temperature dependence of PCBs in the UK atmosphere. *Atmos Environ* 1999;33:541–552.

151. Miskewitz RJ, Hires RI, Korfiatis GP, Sidhoum M, Douglas WS, Su, TL. Laboratory measurements of the volatilization of PCBs from amended dredged material. *Environ Res* 2008;106:319–325.

152. Chiarenzelli J, Bush B, Casey A, Barnard E, Smith B, O'Keefe P, Gilligan E, Johnson, G. Defining the sources of airborne polychlorinated biphenyls: evidence for the influence of microbially dechlorinated congeners from river sediment. *Can J Fish Aquat Sci* 2000;57:86–94.

153. Borghini F, Grimalt JO, Sanchez-Hernandez JC, Barra R, García CJ, Focardi, S. Organochlorine compounds in soils and sediments of the mountain Andean Lakes. *Environ Pollut* 2005;136:253–266.

154. Grimalt JO, Borghini F, Sanchez-Hernandez JC, Barra R, Torres Garcia CJ, Focardi, S. Temperature dependence of the distribution of organochlorine compounds in the mosses of the Andean Mountains. *Environ Sci Technol* 2004;38: 5386–5392.

155. Gdaniec-Pietryka M, Wolska L, Namiesnik, J. Physical speciation of polychlorinated biphenyls in the aquatic environment. *Trends Anal Chem* 2007;26:1005–1012.

156. Falconer RL, Bidleman, TF. Vapor pressures and predicted particle/gas distributions of polychlorinated biphenyl congeners as functions of temperature and ortho-chlorine substitution. *Atmos Environ* 1994;28:547–554.

157. Franz TP, Eisenreich, SJ. Snow scavenging of polychlorinated biphenyls and polycyclic aromatic hydrocarbons in Minnesota. *Environ Sci Technol* 1998;32:1771–1778.

158. Daly GL, Wania, F. Simulating the influence of snow on the fate of organic compounds. *Environ Sci Technol* 2004;38: 4176–4186.

159. Quemarais B, Lemieux C, Lum, KR. Temporal variation of PCB concentrations in the St. Lawrence River (Canada) and four of its tributaries. *Chemosphere* 1994;28:947–959.

160. Pham T, Lum K, Lemieux, C. Seasonal variation of DDT and its metabolites in the St. Lawrence River (Canada) and four of its tributaries. *Sci Total Environ* 1996;179:17–26.

161. Hanot L, Domine, F. Evolution of the surface area of a snow layer. *Environ Sci Technol* 1999;33:4250–4255.

162. Poster DL, Baker, JE. Influence of submicron particles on hydrophobic organic contaminants in precipitation. 1. Concentrations and distributions of polycyclic aromatic hydrocarbons and polychlorinated biphenyls in rainwater. *Environ Sci Technol* 1996;30:341–348.

163. Capel PD, Leuenberger C, Giger, W. Hydrophobic organic chemicals in urban fog. *Atmos Environ* 1991;25A:1335–1346.

164. Ko F-C, Baker, JE. Partitioning of hydrophobic organic contaminants to resuspended sediments and plankton in the mesohaline Chesapeake Bay. *Mar Chem* 1995;49:171–188.

165. Hermanson MH, Hites, RA. Long-term measurements of atmospheric polychlorinated biphenyls in the vicinity of superfund dumps. *Environ Sci Technol* 1989;23: 1253–1258.

166. Chiarenzelli J, Scrudato R, Arnold G, Wunderlich M, Rafferty, D. Violation of polychlorinated biphenyls from sediment during drying at ambient conditions. *Chemosphere* 1996;33:899–911.

167. Tsai P, Hoenicke R, Yee D, Bamford HA, Baker, JE. Atmospheric concentrations and fluxes of organic compounds in the northern San Francisco Estuary. *Environ Sci Technol* 2002;36:4741–4747.

168. Rowe AA, Totten LA, Cavallo GI, Yagecic, Jr. Watershed processing of atmospheric polychlorinated biphenyl inputs. *Environ Sci Technol* 2007;41:2331–2337.

169. Duinker JC, Bouchertall, F. On the distribution of atmospheric polychlorinated biphenyl congeners between vapor phase, aerosols, and rain. *Environ Sci Technol* 1989;23:57–62.

170. Achman DR, Hornbuckle KC, Eisenreich, SJ. Volatilization of polychlorinated biphenyls from Green Bay, Lake Michigan. *Environ Sci Technol* 1993;27:75–87.

171. Wania F, Mackay, D. Global fractionation and cold condensation of low volatility organochlorine compounds in polar regions. *Ambio* 1993;22:10–18.

172. O'Connor GA, Kiehl, D. Plant uptake of sludge-borne PCBs. *J Environ Qual* 1990;19:113–118.

173. Tolls J, McLachlan, MS. Partitioning of semivolatile organic compounds between air and *Lollum multiflorum* (Welsh Ray Grass). *Environ Sci Technol* 1994;28:159–166.

174. Hung H, Thomas GO, Jones KC, Mackay, D. Grass-air exchange of polychlorinated biphenyls. *Environ Sci Technol* 2001;35:4066–4073.

175. Commoner B, Bush B, Bartlett PW, Couchot K, Eisl, H. The exposure of the New York City Watershed to PCBs emitted

from the Hudson River. CBNS, Queens College, CUNY, Flushing, NY, 1999.

176. Hermanson MH, Johnson, GW. Polychlorinated biphenyls in tree bark near a former manufacturing plant in Anniston, Alabama. *Chemosphere* 2007;68:191–198.

177. Schwab BW, Hayes EP, Fiori JM, Mastrocco FJ, Roden NM, Cragin D, Meyerhoff RD, D'Aco VJ, Anderson, PD. Human pharmaceuticals in US surface waters: a human health risk assessment. *Regul Toxicol Pharm* 2005;42:296–312.

178. Milnes MR, Bermudez DS, Bryan TA, Edwards TM, Gunderson MP, Larkin IL, Moore BC, Guillette LJ, Jr. Contaminant-induced feminization and demasculinization of nonmammalian vertebrate males in aquatic environments. *Environ Res* 2006;100:3–17.

179. Cone, M., *Silent Snow*. New York: Grove Press, 2005.

180. AMAP. *Assessment Report: Arctic Pollution Issues*. Oslo, Norway: AMAP, 1998.

181. Borgå K, Di Guardo, A. Comparing measured and predicted PCB concentrations in Arctic seawater and marine biota. *Sci Total Environ* 2005;342:281–300.

182. Fisk AT, de Wit CA, Wayland M, Kuzyk ZZ, Burgess N, Letcher R, Braune B, Norstrom R, Blum SP, Sandu C, Lie E, Larsen HJS, Skaare JU, Muir, DCG. An assessment of the toxicological significance of anthropogenic contaminants in Canadian arctic wildlife. *Sci Total Environ* 2005;*351–352*: 57–93.

183. Goldberg, ED. Synthetic organohalides in the sea. *Proc R Soc Lond B* 1975;189:277–289.

184. Hargrave BT, Harding GC, Vass WP, Erickson PE, Fowler BR, Scott, V. Organochlorine pesticides and polychlorinated biphenyls in the Arctic Ocean. *Arch Environ Contam Toxicol* 1992;22:41–54.

185. Wania, F. Assessing the potential of persistent organic chemicals for long-range transport and accumulation in polar regions. *Environ Sci Technol* 2003;37:1344–1351.

186. Wania, F. Potential of degradable organic chemicals for absolute and relative enrichment in the Arctic. *Environ Sci Technol* 2006;40:569–577.

187. Wania F, Mackay, D. Tracking the persistence of persistent organic compounds. *Environ Sci Technol* 1996;30:390A.

188. Lohmann R, Breivik K, Dachs J, Muir, D. Global fate of POPs: current and future research directions. *Environ Pollut* 2007;150:150–165.

189. Wakeman TH, Themelis NJ. A basin-wide approach to dredged material management in New York/New Jersey Harbor. *J Hazard Mater* 2001;85:1–13.

190. Hites RA, Foran JA, Carpenter DO, Hamilton MC, Knuth BA, Schwager, SJ. Global assessment of organic contaminants in farmed salmon. *Science* 2004;303:226–229.

191. Huang X, Hites RA, Foran JA, Hamilton C, Knuth BA, Schwager SJ, Carpenter DO. Consumption advisories for salmon based on risk of cancer and noncancer health effects. *Environ Res* 2006;101:263–274.

192. de Azevedo e Silva CE, Azeredo A, Lailson-Brito J, Torres JPM, Malm, O. Polychlorinated biphenyls and DDT in sword-fish (*Xiphias gladius*) and blue sharks (*Prionace glauca*) from Brazilian coast. *Chemosphere* 2006;67:S48–S53.

193. Storelli MM, Marcotrigiano, GO. Persistent organochlorine residues and toxic evaluation of polychlorinated biphenyls in sharks from the Mediterranean Sea (Italy). *Mar Pollut Bull* 2001;42:1323–1329.

194. Turrio-Baldassari L, Alivernini S, Carasi S, Casella M, Fuselli S, Iamiceli AL, LaRocca C, Scarcella C, Battistelli, CL. PCB, PCDD and PCDF contamination of food of animal origin as the effect of soil pollution and the cause of human exposure in Brescia. *Chemosphere* 2009;76:278–285.

195. Bobovnikova TsI, Alekseeva LB, Dibtseva AV, Chernik GV, Orlinsky DB, Priputina IV, Pleskachevskaya, GA. The influence of a capacitor plant in Serpukhov on vegetable contamination by polychlorinated biphenyls. *Sci Total Environ* 2000;246:51–60.

196. Baars AJ, Bakker MI, Baumann RA, Boon PE, Freijer JI, Hoogenboom LAP, Hoogerbrugge R, van Klaveren, JD, Liem AKD, Traag WA, deVries J. Dioxins, dioxin-like PCBs and non-dioxin-like PCBs in foodstuffs: occurrence and dietary intake in The Netherlands. *Toxicol Lett* 2004;151:51–61.

197. Zuccato E, Grassi P, avoli E, Valdicelli L, Wood D, Reitano G, Fanelli, R. PCB concentrations in some foods from four European countries. *Food Chem Toxicol* 2008;46:1062–1067.

198. Jurado E, Lohmann R, Meijer S, Jones KC, Dachs J. Latitudinal and seasonal capacity of the surface oceans as a reservoir of polychlorinated biphenyls. *Environ Pollut* 204; 128:149–162.

199. Meijer SN, Ockenden WA, Sweetman A, Breivik K, Grimalt JO, Jones, KC. Global distribution and budget of PCBs and HCB in background surface soils: implications for sources and environmental processes. *Environ Sci Technol* 2003;37: 667–672.

200. Baker EL, Landrigan PJ, Glueck CJ, Zack MM, Liddle JA, Burse VW, Housworth WJ, Needham, LL. Metabolic consequences of exposure to polychlorinated biphenyls (PCB) in sewage sludge. *Am J Epidemiol* 1980;112:553–563.

201. Herrick RF, McClean MD, Meeker JD, Baxter LK, Weymouth, GA., (2004) An unrecognized source of PCB contamination in schools and other buildings. *Environ Health Perspect* 2004;112:1051–1053.

202. Kohler M, Tremp J, Zennegg M, Seiler C, Kohler SM, Beck M, Lienemann P, Wegmann L, Schmid, P. Joint sealants: an overlooked diffuse source of polychlorinated biphenyls in buildings. *Environ Sci Technol* 2005;39:1967–1973.

203. Rudel RA, Seryak JM, Brody, JG. PCB-containing wood floor finish is a likely source of elevated PCBs in residents' blood, household air and dust: a case study of exposure. *Environ Health* 2008;7:2 doi: 10.1186/1476-069X-7-2

204. Franzblau A, Zwica L, Knutson K, Chen Q, Lee SY, Hong B, Adriaems P, Demond A, Garabrant D, Gillespie B, Lepkowski J, Luksemburg W, Maier M, Towey, T. An investigation of homes with high concentrations of PCDDs, PCDFs, and/or dioxin-like PCBs in house dust. *J Occup Environ Hyg* 2009;6:188–199.

205. IOM (Institute of, Medicine). *Dioxins and Dioxin-like Compounds in the Food Supply. Strategies to Decrease Exposure.* Washington, DC: The National Academies Press, 2003, 318 pp.

206. Pagano JJ, Scrudato RJ, Roberts R, Bemis, JC. Reductive dechlorination of PCB-contaminated sediments in an anaerobic bioreactor system. *Environ Sci Technol* 1995;29: 2584–2589.

207. Huang Q, Hong, CS. TiO$_2$ photocatalytic degradation of PCBs in soil-water systems containing fluoro surfactant. *Chemosphere* 2000;41:871–879.

208. Arienzo M, Chiarenzelli J, Scrudato R, Pagano J, Falanga L, Connor, B. Iron-mediated reactions of polychlorinated biphenyls in electrochemical peroxidation process (ECP). *Chemosphere* 2001;44:1339–1346.

209. Seto M, Nishibori K, Masai E, Fukuda M, Ohdaira, Y. Degradation of polychlorinated biphenyls by a "Maitake" mushroom, *Grifola frondosa. Biotechnol Lett* 1999;21: 27–31.

210. Aust SD, Swaner PR, Stahl JD. Detoxification and metabolism of chemicals by white-rot fungi. In: Gan JJ, Zhu PC, Aust SD, Lemley AT,editors. *Pesticide Decontamination and Detoxification.* ACS Symposium Series 2003;863: 3–14.

211. Anitescu G, Tavlarides LL. Supercritical extraction of contaminants from soils and sediments. *J Supercrit Fluids* 2006;38:167–180.

26

OTHER WATER POLLUTANTS: ANTIBIOTIC-RESISTANT BACTERIA

Amy R. Sapkota

26.1 INTRODUCTION

Beyond fungi, protozoa, viruses, algae, and susceptible bacteria, wastewater can serve as an important reservoir of antibiotic-resistant bacteria and their associated resistance genes. Studies conducted in the United States, Europe, Asia, and other countries have documented elevated levels of antibiotic-resistant commensal and pathogenic bacteria in both untreated and treated wastewater present at various stages within wastewater treatment plants (WWTP) (1–7). These microorganisms and their antibiotic resistance genes enter WWTPs in wastewater originating from a variety of point and nonpoint sources including private homes, hospitals, pharmaceutical manufacturers, agricultural runoff, and urban runoff (8) (Fig. 26.1). Resistant bacteria and resistance genes (9) can then be disseminated back to humans, animals, and the environment through exposures to untreated and treated wastewater, completing a continual cycle where bacterial antibiotic resistance is spread and possibly amplified among different compartments of the biosphere.

The consequences of increasing rates of antibiotic resistance equate to a crisis in the treatment of bacterial infections in both human and veterinary medicine. Antibiotic-resistant infections are increasingly more difficult to treat (if not, untreatable), particularly when the responsible microorganism is resistant to multiple classes of antibiotics—a scenario that is observed more often than not in clinical settings. Because of this, antibiotic-resistant infections are generally more severe than those caused by susceptible bacteria, and they are associated with higher

mortality rates. These factors contribute to considerable medical and societal costs. In a recent study by Roberts et al. (10), it was estimated that antibiotic-resistant infections among a cohort of patients in one Chicago hospital were characterized by a 6.5% mortality rate—twice the rate associated with susceptible infections. In addition, it was estimated that the societal costs associated with resistant infections in this cohort of only 1391 patients were between USD 10 million and USD 15 million (10).

26.2 DEVELOPMENT OF ANTIBIOTIC-RESISTANT BACTERIA

But, how do bacteria become resistant to antibiotics in the first place? And how do antibiotic-resistant bacteria and resistance genes get into wastewater? Any and all uses of antibiotics can contribute to increases in antibiotic-resistant bacteria. Typically, many people tend to assume that the use (and sometimes abuse) of antibiotics by humans accounts for the greatest extent of antibiotic usage. However, antibiotics are also widely used (i) at nontherapeutic levels in animal feed to promote growth, improve feed efficiency (11, 12), and combat coccidiosis in combination with ionophores (2); (ii) at therapeutic levels in animal agriculture to treat disease (11); and (iii) at nontherapeutic levels in aquaculture feed to promote growth, prevent disease, and control parasites (12, 13). In addition, while it is often forgotten or left out of review articles and textbooks, antibiotics have been sprayed on food crops

Water and Sanitation-Related Diseases and the Environment: Challenges, Interventions, and Preventive Measures, First Edition.
Edited by Janine M. H. Selendy.
© 2011 Wiley-Blackwell. Published 2011 by John Wiley & Sons, Inc.

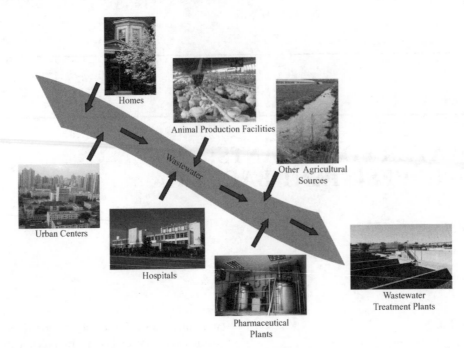

FIGURE 26.1 Antibiotic-resistant bacteria and their resistance genes enter raw wastewater from both point and nonpoint sources.

(i.e., apples, pears, tomatoes, peppers, and potatoes) and ornamental plants since the development of antibiotics for the purposes of controlling bacterial plant diseases (14).

In the United States, the Food and Drug Administration regulates the use of antibiotics in food animal production and the Environmental Protection Agency (EPA), Office of Pesticide Programs (OPP), regulates the use of antibiotics on food crops. It is important to note that even in the certified organic farming of pears and apples, two antibiotics (streptomycin and oxytetracycline) are approved for use by the EPA OPP to control fire blight (15). In contrast, the European Union has banned the use of human antibiotics as growth promoters in animal feed, and does not allow the use of antibiotics to control plant diseases on organic crops (16).

So, who uses the most antibiotics and is, therefore, the largest contributor to increases in antibiotic-resistant bacteria that may be observed in wastewater and other environmental media? This is a hot potato question that numerous researchers, industry groups, and nonprofits have attempted to answer for many years, and to date we still do not have clear answers. The Union of Concerned Scientists (17) and the Animal Health Institute (18) have reported vastly different estimates of antibiotic use among humans and animals (Fig. 26.2) and incomplete data exists concerning the use of antibiotics in aquaculture (13) and crop production (19). But regardless of who is "using" antibiotics, acquired resistance among exposed bacterial populations can develop quickly due to antibiotic selective pressures (20, 21).

Here is how this selective pressure works. At any given time, a bacterial population—whether it is present in a human, an animal, wastewater, a pile of manure, or a fish pond—is characterized by a mixed population of antibiotic-susceptible microorganisms and a few antibiotic-resistant microorganisms (Fig. 26.3). When an antibiotic is added to the mix, the drug knocks out the susceptible microorganisms generally by attacking (i) bacterial cell wall synthesis, (ii) protein synthesis, (iii) DNA synthesis, or (iv) folic acid synthesis (21). Then, the resistant bacteria that remain, carrying antibiotic resistance genes or mutations, can predominate and flourish (Fig. 26.3). These resistance genes and/or gene mutations encode for proteins that combat the effects of the antibiotics, and are, therefore, grouped under four main categories of antibiotic resistance mechanisms. These categories include (i) limiting uptake of antibiotics, (ii) active efflux of antibiotics, (iii) enzymatic inactivation of antibiotics, and (iv) modification or protection of cellular antibiotic targets (21). A fifth, lesser understood, antibiotic resistance mechanism involves the failure of an antibiotic to be activated by the bacterium (21). Table 26.1 provides specific examples of each of these antibiotic resistance mechanisms.

If an antibiotic resistance mechanism is mediated by a mutation, this mutation can be passed on to future bacterial generations; hence, this is one way that antibiotic resistance can spread. However, becoming resistant to antibiotics through genetic mutations is viewed as a metabolically expensive and inefficient process for a bacterium (21). A quicker and more efficient way for a bacterium to gain

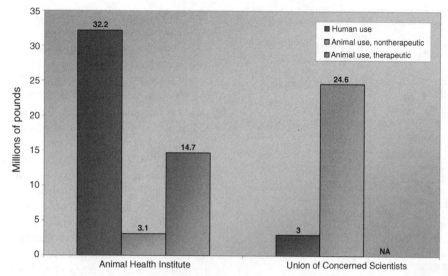

FIGURE 26.2   Estimates of antibiotic use in the United States by the Animal Health Institute (AHI) and the Union of Concerned Scientists (UCS) (UCS did not estimate therapeutic animal use).

an antibiotic resistance trait is to acquire a resistance gene (or set of genes) from another bacterium through horizontal gene transfer (HGT) events. Many researchers believe that the most common HGT mechanism is conjugation (21), where genes are directly transferred from one bacterium to another through a conjugative pilus (a filamentous appendage extending from the surface of a bacterium). Although it is difficult to document, it is thought that HGT events among and between bacterial species occur often in environmental media such as wastewater and animal manure, which tend to be rich reservoirs for antibiotic-resistant bacteria and resistance genes.

26.3   TYPES OF ANTIBIOTIC-RESISTANT BACTERIA IDENTIFIED IN WASTEWATER

A variety of antibiotic-resistant bacteria, expressing various resistance genes, have been detected in both untreated and treated wastewater (Table 26.2). When we are talking about wastewater, we are talking about environmental media that is chock-full of fecal and urinary material. Because of its fecal nature, many researchers have focused their investigations of antibiotic-resistant bacteria in wastewater on fecal indicator microorganisms (Box 26.1), including *Escherichia coli* and *Enterococcus*. *E. coli* expressing resistance to amoxicillin, ciprofloxacin, cephalothin, gentamicin, trimethoprim–sulfamethoxazole, and tetracycline have been isolated from untreated (raw) wastewater influent, as well as treated wastewater effluent (3, 4, 22, 23). In particular, trimethoprim–sulfamethoxazole-resistant *E. coli* belonging to clonal group A (CGA) was recently recovered from four WWTPs located across

BOX 26.1   THE FECAL INDICATOR CONCEPT

Analyzing environmental samples, such as wastewater, for the presence of enteric pathogens on a routine basis can be costly, time-consuming, and difficult (1). Therefore, instead of examining samples for the pathogens themselves, researchers often look for fecal indicator microorganisms whose presence signifies that (i) the sample may be contaminated with fecal matter; and (ii) pathogens originating from this fecal matter may be present as well. Ideally, a good fecal indicator microorganism should meet the following criteria (1):

- Is useful for all water types
- Is present when enteric pathogens are present and absent when enteric pathogens are absent
- Has a longer survival time than the most hardy enteric pathogen
- Does not grow in water
- Is easy to detect using standard methods
- Is directly correlated with the amount of fecal contamination in a given sample
- Is a nonpathogenic member of the normal intestinal flora of warm-blooded animals

While it is recognized that no microorganism meets all of these criteria, the microorganisms that are most commonly used as fecal indicators, include the fecal coliforms, *E. coli*, and *Enterococcus*, all of which can be easily isolated and quantified using standard microbiological methods.

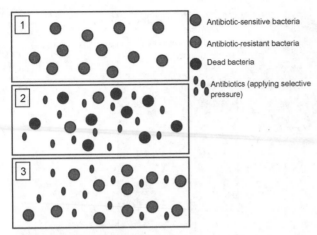

- ● Antibiotic-sensitive bacteria
- ● Antibiotic-resistant bacteria
- ● Dead bacteria
- ❜ Antibiotics (applying selective pressure)

These are photographs showing actual antibiotic-resistant bacteria

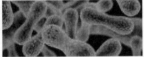

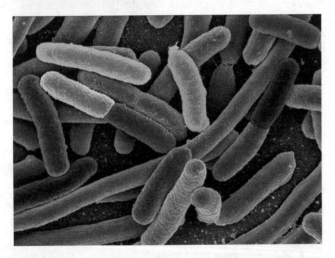

FIGURE 26.3 Antibiotic selective pressures acting on a bacterial population. ⬚1 Bacterial populations are naturally comprised of antibiotic-sensitive bacteria, as well as a few antibiotic-resistant microorganisms that carry either a mutation or a resistance gene. ⬚2 When an antibiotic selective pressure is applied, the majority of the sensitive bacteria are killed or inhibited and the resistant bacteria are selected. ⬚3 These resistant microorganisms then flourish and reproduce, creating future generations of antibiotic-resistant bacterial populations.

the United States (3). It has been reported that this microorganism could account for nearly 50% of trimethoprim–sulfamethoxazole-resistant *E. coli* recovered from women with urinary tract infections and pyelonephritis (an infection of the ureters and kidneys) (3, 24). The presence of *E. coli* CGA in wastewater could contribute to the dissemination of this microorganism in the environment, as well as additional human and animal populations (3).

Enterococcus resistant to vancomycin, ampicillin, rifampicin, tetracycline, erythromycin, nitrofurantoin, and ciprofloxacin also have been recovered from untreated and treated wastewater (5, 25–27). Of particular concern is the finding of vancomycin-resistant *Enterococcus* (VRE) in wastewater (25, 26, 28). Vancomycin is an important antibiotic in human clinical settings and is oftentimes viewed as a drug of last resort against gram-positive bacterial infections, including those caused by enterococci. Because of this, there are high mortality rates associated with VRE infections, particularly during nosocomial (hospital-acquired) outbreaks. A factor that contributes to these observed mortality rates is that VRE isolates are typically resistant to other antibiotics in the clinical arsenal as well. These multidrug resistant enterococci have also been recovered from wastewater. For example, Beier et al. (25) isolated VRE from community wastewater in Texas that were not only resistant to vancomycin but also resistant to 8 fluoroquinolone antibiotics and 10 out of 17 additional antibiotics that are utilized to combat gram-positive infections.

In addition to *Enterococcus,* researchers have isolated other antibiotic-resistant, gram-positive, bacterial pathogens from wastewater, including methicillin-resistant *Staphylococcus aureus* (MRSA) and coagulase-negative *Staphylococcus* spp. (CoNS) (29–31). MRSA can cause severe, and oftentimes fatal, skin, soft tissue, bloodstream, and lung infections depending on where an individual is infected. Previously, it was documented that the majority of MRSA cases appeared in hospital settings (hospital-acquired MRSA or HA-MRSA). However, recently there have been increases in community-acquired MRSA (CA-MRSA) infections in otherwise healthy individuals. The finding of MRSA in wastewater (29, 30) is troubling and additional research is necessary to evaluate whether the presence of MRSA in wastewater could be contributing to increases in CA-MRSA infections.

Beyond gram-positive microorganisms, multiple gram-negative, antibiotic-resistant bacteria have also been isolated from wastewater. These microorganisms include *Acinetobacter* spp., *Pseudomonas* spp., *Shigella* spp., *Klebsiella* spp., and *Salmonella enterica* serovars (4, 32–34). These microorganisms can cause a range of human illnesses from severe nosocomial infections (*Acinetobacter* spp.) to foodborne gastrointestinal illnesses (*Shigella* spp. and *Salmonella* spp.). It is important to note, however, that in many cases, these microorganisms have been isolated from raw, influent

TABLE 26.1   Examples of Antibiotic Resistance Mechanisms

Category	Example
1. Limiting uptake of antibiotics	Porins in the outer membranes of gram-negative bacteria that can exclude antibiotics such as vancomycin
2. Active efflux of antibiotics	Efflux pumps in gram-negative and gram-positive bacteria that can pump tetracycline out of the cell's cytoplasm
3. Enzymatic inactivation of antibiotics	β-Lactamase, an enzyme in gram-negative bacteria that cleaves the β-lactam ring of antibiotics like penicillin and cephalosporins, inactivating the antibiotic
4. Modification or protection of cellular antibiotic targets	Presence of an encoded protein in the cytoplasm protects the ribosome from antibiotics such as tetracycline

Source: Ref. 21.

wastewater. Future work is necessary to understand the prevalence of these gram-negative, antibiotic-resistant bacteria in treated effluent that is discharged into the environment.

26.4   IMPACT OF WASTEWATER TREATMENT PROCESSES ON ANTIBIOTIC RESISTANCE

Some studies have been conducted to determine the effects of wastewater treatment processes on levels of specific antibiotic-resistant bacterial species in treated wastewater. And the findings of these studies have been mixed. Ferreira da Silva et al. (27) observed that in a comparison of raw and

secondary-treated wastewater, the prevalence of amoxicillin-, ciprofloxacin-, tetracycline-, and cephalothin-resistant *E. coli* increased by 5–10% in treated wastewater (4). In the same study, these researchers found that wastewater treatment processes were accompanied by increases in resistance to the following antibiotics among other gram-negative microorganisms including *Klebsiella* spp. and *Shigella* spp.: ciprofloxacin, sulfamethoxazole–trimethoprim, tetracycline, and gentamicin. Similarly, Kim et al. (35) observed that increases in organic loading and bacterial growth rates in activated sludge at a WWTP resulted in increases in tetracycline-resistant bacteria during the treatment process. Martins da Costa et al. (5) and Ferreira da

TABLE 26.2   Examples of Antibiotic-Resistant Bacteria Recovered from Untreated (Raw) and Treated Wastewater

Microorganism	Recovered from Untreated (Raw) Wastewater	Recovered from Treated Wastewater	References
Gram-negative			
Ciprofloxacin-resistant *E. coli*	Yes	Yes	4
Sulfamethoxazole–trimethoprim-resistant *E. coli*	Yes	Yes	(3, 4)
Gentamicin-resistant *E. coli*	Yes	Yes	4
Ciprofloxacin-resistant *Shigella* spp.	No	Yes	4
Sulfamethoxazole–trimethoprim-resistant *Shigella* spp.	No	Yes	4
Tetracycline-resistant *Shigella* spp.	No	Yes	4
Gentamicin-resistant *Shigella* spp.	No	Yes	4
Ciprofloxacin-resistant *Klebsiella* spp.	Yes	Yes	4, 27
Sulfamethoxazole–trimethoprim-resistant *Klebsiella* spp.	Yes	Yes	4, 27
Ciprofloxacin-resistant *Salmonella enterica* serovars	Yes	No	29
Sulfamethoxazole–trimethoprim-resistant *Salmonella enterica* serovars	Yes	No	29
Gentamicin-resistant *Acinetobacter* spp.	Yes	Yes	33
Tetracycline-resistant *Acinetobacter* spp.	Yes	Yes	33
Gram-positive			
Gentamicin-resistant *Enterococcus* spp.	Yes	Yes	20, 22
Erythromycin-resistant *Enterococcus* spp.	Yes	Yes	20, 22
Ciprofloxacin-resistant *Enterococcus* spp.	Yes	Yes	20, 22
Vancomycin-resistant *Enterococcus* spp.	Yes	No	5, 20, 21
Methicillin-resistant *Staphylococcus aureus* (MRSA)	Yes	No	24, 25
Erythromycin-resistant *Staphylococcus saprophyticus*	Yes	Yes	26
Gentamicin-resistant *Staphylococcus saprophyticus*	Yes	Yes	26

Silva et al. (27) also showed that wastewater treatment processes resulted in either increases or no changes in antibiotic-resistant *Enterococcus* spp. For example, the prevalence of ciprofloxacin-resistant enterococci increased in treated wastewater compared to raw wastewater (27). Additional research also has shown that the conditions present within WWTP may favor HGT events among bacteria (2, 36, 37), which may explain why the amplification of antibiotic resistance has been observed in treated wastewater in some studies.

On the other hand, other researchers have observed decreases in levels of antibiotic-resistant bacteria after wastewater treatment. In a study of two wastewater treatment facilities (one that received only community wastewater and one that received hospital wastewater), Garcia et al. (22) observed overall trends of decreases in antibiotic-resistant enterococci and antibiotic-resistant *E. coli*. In particular, levels of multidrug-resistant *E. coli* were significantly reduced in treated wastewater (22). Guardabassi et al. (38) observed significant reductions in the number of total and antibiotic-resistant bacteria present in raw versus treated wastewater at two tertiary WWTP. Specifically, levels of *Acinetobacter* spp. and presumptive coliforms resistant to tetracycline, gentamicin, and ampicillin were 10–1000 lower in treated versus raw wastewater (38).

A potential explanation for the discrepancies observed in the effects of wastewater treatment on reductions of antibiotic-resistant bacteria is that different wastewater treatment processes were investigated in the various studies. For instance, in many of the studies that have observed increases in antibiotic resistance after wastewater treatment, it seems as though the wastewater underwent only primary and secondary treatment (4, 27), while the few studies that have observed decreases in resistance in treated wastewater have included analyses of wastewater that underwent primary, secondary, and tertiary treatment (38). However, an important note to make is that oftentimes researchers do not specify the exact type of wastewater treatment processes that were investigated in their studies. Therefore, it is difficult to know whether apples are being compared to apples or oranges when one attempts to compare the effects of wastewater treatment on the prevalence of antibiotic-resistant microorganisms. Important future research that is necessary should include studies that compare the effects of different specific types of wastewater treatment processes on levels of resistance observed in gram-negative and gram-positive pathogens, as well as commensal bacterial microorganisms.

26.5   SUMMARY

Whether present in untreated or treated wastewater, antibiotic-resistant bacteria can pose significant risks among exposed populations, particularly when the identified microorganism is a true or opportunistic pathogen. Therefore, in studies that continue to assess potential human exposures and health outcomes that may be associated with direct and/or indirect contact with wastewater, it is prudent for researchers to evaluate levels of resistant bacteria along with susceptible bacterial populations and other microbes that may be present in the waste stream.

REFERENCES

1. Maier RM, Pepper IL, Gerba CP. *Environmental Microbiology*, 2nd edition. Burlington, MA: Academic Press, 2009.
2. Kim S, Aga DS. Potential ecological and human health impacts of antibiotics and antibiotic-resistant bacteria from wastewater treatment plants. *J Toxicol Environ Health B Crit Rev* 2007; 10(8):559–573.
3. Boczek LA, Rice EW, Johnston B, Johnson JR. Occurrence of antibiotic-resistant uropathogenic *Escherichia coli* clonal group A in wastewater effluents. *Appl Environ Microbiol* 2007;73(13):4180–4184.
4. Ferreira da SM, Vaz-Moreira I, Gonzalez-Pajuelo M, Nunes OC, Manaia CM. Antimicrobial resistance patterns in *Enterobacteriaceae* isolated from an urban wastewater treatment plant. *FEMS Microbiol Ecol* 2007;60(1):166–176.
5. Martins da Costa P, Vaz-Pires P, Bernardo F. Antimicrobial resistance in *Enterococcus* spp. isolated in inflow, effluent and sludge from municipal sewage water treatment plants. *Water Res* 2006;40(8):1735–1740.
6. Schluter A, Szczepanowski R, Kurz N, Schneiker S, Krahn I, Puhler A. Erythromycin resistance-conferring plasmid pRSB105, isolated from a sewage treatment plant, harbors a new macrolide resistance determinant, an integron-containing Tn402-like element, and a large region of unknown function. *Appl Environ Microbiol* 2007;73(6):1952–1960.
7. bu-Ghazaleh BM. Fecal coliforms of wastewater treatment plants: antibiotic resistance, survival on surfaces and inhibition by sodium chloride and ascorbic acid. *New Microbiol* 2001; 24(4):379–387.
8. Rosenblatt-Farrell N. The landscape of antibiotic resistance. *Environ Health Perspect* 2009;117(6):A244–A250.
9. Pruden A, Pei R, Storteboom H, Carlson KH. Antibiotic resistance genes as emerging contaminants: studies in northern Colorado. *Environ Sci Technol* 2006;40(23):7445–7450.
10. Roberts RR, Hota B, Ahmad I, et al. Hospital and societal costs of antimicrobial-resistant infections in a Chicago teaching hospital: implications for antibiotic stewardship. *Clin Infect Dis* 2009;49(8):1175–1184.
11. Sapkota AR, Lefferts LY, McKenzie S, Walker P. What do we feed to food-production animals? A review of animal feed ingredients and their potential impacts on human health. *Environ Health Perspect* 2007;115(5):663–670.
12. Sapkota A, Sapkota AR, Kucharski M, et al. Aquaculture practices and potential human health risks: current knowledge and future priorities. *Environ Int* 2008;34(8):1215–1226.

13. Hernández Serrano P. Responsible use of antibiotics in aquaculture. Rome: FAO, 2005.

14. Khachatourians GG. Agricultural use of antibiotics and the evolution and transfer of antibiotic-resistant bacteria. *CMAJ* 1998;159(9):1129–1136.

15. U.S. Department of Agriculture AMS. National Organic Program: National List of Allowed and Prohibited Substances, 2008.

16. U.S. Department of Agriculture AMS. National Organic Standards Board (NOSB) Item for Public Comment: Draft Recommendation on US/EU Equivalency, NOSB International Committee, 2002.

17. Mellon M, Benbrook C, Benbrook KL. *Hogging It: Estimates of Antimicrobial Abuse in Livestock*. Cambridge, MA: Union of Concerned Scientists, 2001.

18. Carnevale R. Industry Perspective on Antibiotic Resistance in Food Producing Animals. Pork Industry Conference on Addressing Issues of Antibiotic Use in Livestock Production. University of Illinois, Urbana, IL, 2000.

19. USDA NASS. Agricultural Chemical Usage: 2007 Field Crops Summary. Report No.: Ag Ch 1 (08), 2008.

20. Rosenblatt-Farrell N. The landscape of antibiotic resistance. *Environ Health Perspect* 2009;117(6):A244–A250.

21. Salyers AA, Whitt DD. *Bacterial Pathogenesis: A Molecular Approach*, 2nd edition. Washington, DC: ASM Press, 2002.

22. Garcia S, Wade B, Bauer C, Craig C, Nakaoka K, Lorowitz W. The effect of wastewater treatment on antibiotic resistance in *Escherichia coli* and *Enterococcus* sp. *Water Environ Res* 2007;79(12):2387–2395.

23. Watkinson AJ, Micalizzi GB, Graham GM, Bates JB, Costanzo SD. Antibiotic-resistant *Escherichia coli* in wastewaters, surface waters, and oysters from an urban riverine system. *Appl Environ Microbiol* 2007;73(17):5667–5670.

24. Johnson JR, Murray AC, Kuskowski MA, et al. Distribution and characteristics of *Escherichia coli* clonal group A. *Emerg Infect Dis* 2005;11(1):141–145.

25. Beier RC, Duke SE, Ziprin RL, et al. Antibiotic and disinfectant susceptibility profiles of vancomycin-resistant *Enterococcus faecium* (VRE) isolated from community wastewater in Texas. *Bull Environ Contam Toxicol* 2008;80(3):188–194.

26. Harwood VJ, Brownell M, Perusek W, Whitlock JE. Vancomycin-resistant *Enterococcus* spp. isolated from wastewater and chicken feces in the United States. *Appl Environ Microbiol* 2001;67(10):4930–4933.

27. Ferreira daSilva M, Tiago I, Verissimo A, Boaventura RA, Nunes OC, Manaia CM. Antibiotic resistance of enterococci

28. Nagulapally SR, Ahmad A, Henry A, Marchin GL, Zurek L, Bhandari A. Occurrence of ciprofloxacin-, trimethoprim-sulfamethoxazole-, and vancomycin-resistant bacteria in a municipal wastewater treatment plant. *Water Environ Res* 2009; 81(1):82–90.

29. Börjesson S, Matussek A, Melin S, Löfgren S, Lindgren PE. *J Appl Microbiol*. 2010. Methicillin-resistant Staphylococcus aureus (MRSA) in municipal wastewater: an uncharted threat? Apr;108(4):1244–51. Epub 2009 Aug 6.

30. Borjesson S, Melin S, Matussek A, Lindgren PE. A seasonal study of the mecA gene and *Staphylococcus aureus* including methicillin-resistant *S. aureus* in a municipal wastewater treatment plant. *Water Res* 2009;43(4):925–932.

31. Faria C, Vaz-Moreira I, Serapicos E, Nunes OC, Manaia CM. Antibiotic resistance in coagulase negative *staphylococci* isolated from wastewater and drinking water. *Sci Total Environ* 2009;407(12):3876–3882.

32. Prado T, Pereira WC, Silva DM, Seki LM, Carvalho AP, Asensi MD. Detection of extended-spectrum beta-lactamase-producing *Klebsiella pneumoniae* in effluents and sludge of a hospital sewage treatment plant. *Lett Appl Microbiol* 2008; 46(1):136–141.

33. Tumeo E, Gbaguidi-Haore H, Patry I, Bertrand X, Thouverez M, Talon D. Are antibiotic-resistant *Pseudomonas aeruginosa* isolated from hospitalized patients recovered in the hospital effluents? *Int J Hyg Environ Health* 2008;211 (1–2):200–204.

34. Berge AC, Dueger EL, Sischo WM. Comparison of *Salmonella enterica* serovar distribution and antibiotic resistance patterns in wastewater at municipal water treatment plants in two California cities. *J Appl Microbiol* 2006;101(6):1309–1316.

35. Kim S, Jensen JN, Aga DS, Weber AS. Fate of tetracycline resistant bacteria as a function of activated sludge process organic loading and growth rate. *Water Sci Technol* 2007;55 (1–2):291–297.

36. Mach PA, Grimes DJ. R-plasmid transfer in a wastewater treatment plant. *Appl Environ Microbiol* 1982;44(6): 1395–1403.

37. Kruse H, Sorum H. Transfer of multiple drug resistance plasmids between bacteria of diverse origins in natural microenvironments. *Appl Environ Microbiol* 1994;60(11):4015–4021.

38. Guardabassi L, Lo Fo Wong DM, Dalsgaard A. The effects of tertiary wastewater treatment on the prevalence of antimicrobial resistant bacteria. *Water Res* 2002;36(8):1955–1964.

SECTION VI

CURRENT AND FUTURE TRENDS IN PREVENTING WATER POLLUTION

Ameliorating conditions and taking preventive measures to reduce pollutants found in ground and surface waters are covered from toxic chemical pollutants resulting from natural sources such as arsenic and lead to anthropogenic sources. Preventive measures are discussed.

27

CONTROLLING WATER POLLUTANTS

Jeffery A. Foran

27.1 INTRODUCTION

At Owls Head (Water Pollution Control Plant, Brooklyn, New York), a swimming pool's worth of sewage and wastewater was rushing in every second. A little after 1 am . . . Owls Head reached its capacity, workers started shutting intake gates . . . and untreated feces and industrial waste started spilling into the Upper New York Bay (from "Sewers at Capacity, Waste Poisons Waterways," New York Times, 23 November 2009).

More than 20% of the nation's [drinking] water treatment systems have violated key provisions of the Safe Drinking Water Act over the last five years. (from "Millions in U.S. Drink Dirty Water," New York Times, 8 December 2009)

The *New York Times* paints a grim picture of water pollution and the nation's water quality—water used for swimming, fishing, drinking, irrigation, power generation, transportation, and many others. They're not alone. The U.S. Environmental Protection Agency (1) reports that, of the rivers and streams assessed by states in the United States, the water quality of only half is characterized as good, while the remaining half have water quality that is threatened or impaired. A full 66% of assessed lakes, reservoirs, and ponds have water quality that is threatened or impaired.

The causes of impairments include contamination by pathogens such as *Escherichia coli* (*E. coli*) and coliforms (from fecal waste), chemical contaminants such as mercury and polychlorinated biphenyls (PCBs), nutrients including phosphorus and nitrogen, and others such as exotic species,

turbidity, and oxygen depletion. According to the U.S. Environmental Protection Agency (EPA) (2), agriculture, atmospheric deposition, municipal sewage discharges, storm water runoff, industrial discharges, and legacy/historical (in-place) pollution are all sources of pollutants that impair surface waters.

The year 2012 will mark the 40th anniversary of the passage of the 1972 Federal Water Pollution Control Act Amendments—popularly known as the U.S. Clean Water Act (CWA) (3). Two years later we will celebrate the 40th anniversary of the U.S. Safe Drinking Water Act (SDWA) (4). A significant component of the nation's water quality problems has been resolved over the past 40 years. No longer do rivers such as the Cuyahoga near Cleveland catch fire, rarely are masses of fish and waterfowl found dead on shorelines, Lake Erie has itself risen from the dead, and rarely do the United States and other developed nations experience disease epidemics associated with contaminated drinking water. Less visible but no less important problems, such as bioaccumulation of persistent organic pollutants including dichlorodiphenyltrichloroethane (DDT), which lead to reproductive failure in osprey, eagles, and other species, have also declined although not disappeared. But as the events described by the *New York Times* and the U.S. EPA indicate, passage of the U.S. Clean Water and Safe Drinking Water Acts, and nearly 40 years of subsequent enforcement, have not resolved the nation's surface and drinking water quality problems. States, the U.S. Food and Drug Administration (FDA), and the U.S. EPA continue to issue warnings to human consumers to either limit or avoid consumption of some commercially sold marine fish species and many freshwater, sport-caught fish because of

Water and Sanitation-Related Diseases and the Environment: Challenges, Interventions, and Preventive Measures, First Edition.
Edited by Janine M. H. Selendy.
© 2011 Wiley-Blackwell. Published 2011 by John Wiley & Sons, Inc.

elevated levels of mercury and other contaminants in fish tissues. And as the *New York Times* indicated, violations of the U.S. CWA and SDWA are rampant, suggesting that these statutes have not achieved their potential to eliminate water pollution.

This chapter examines what is being, and what can be, done about water pollutants. Efforts under the U.S. CWA and SDWA, as well as efforts to control water pollutants in Europe, are generally treatment oriented; thus, this chapter presents the treatment-based approach to pollutant management. The chapter also explores less common prevention-based approaches to pollutant management, where they are used presently and how they could be used to achieve statutory and health-based goals of ensuring that freshwater and marine resources are protected at levels commensurate with their value to humans and as part of natural ecosystems.

27.2 WHAT IS BEING DONE ABOUT WATER POLLUTANTS?

Pollutants of anthropogenic origin enter surface and groundwater systems from two major sources: point sources, which are typically discharges from industrial or municipal pipes; and nonpoint sources, discharges from diffuse sources such as contaminated sediments, agricultural runoff, air deposition of pollutants (which may derive from both point and nonpoint discharges to air), soil erosion, mine drainage, and storm water runoff from urban areas, among others. Pollutants from point sources are most often controlled using a treatment-based approach. The control of pollutants from nonpoint sources remains a significant challenge. While nonpoint sources are largely unregulated, they are often managed via prevention-based approaches to pollutant control.

27.2.1 Treatment

Prior to the 1900s, surface waters were viewed as convenient receptacles of pollutants from shipping, industry, municipalities, and from individuals. The U.S. Congress passed the Rivers and Harbors Act of 1899 (5) in an attempt to address what was a growing problem of lakes, harbors, and other water bodies clogged with floating debris. The occurrence of disease epidemics associated with ingestion of pathogen-contaminated drinking water drawn from polluted surface water stimulated actions to move pollutant discharges away from drinking-water intakes, and to disinfect with chlorine and other halogens drinking water that could not be physically separated from pollutant sources. However, these and many subsequent steps between 1900 and 1972 did not stop the discharge of

asphalt, grease, bleaches, detergents, paints and thinners, wastes from mining, coal and ash, coal tar, metals, chlorine-based wastes including pesticides such as DDT, PCBs, human sewage, and myriad other pollutants from being dumped into surface waters (for a detailed discussion of the evolution of regulatory approaches to pollutant control in the early to mid-1900s, see Refs 6 and 7).

Passage of the U.S. Rivers and Harbors Act in 1899, the U.S. Federal Water Pollution Control Act (FWPCA) in 1956 (8), the U.S. Water Quality Act amendments to the FWPCA in 1965 (9), the U.S. Clean Water Restoration Act in 1966 (10), and the U.S. Water Quality Improvement Act in 1970 (11), as well as the U.S. National Environmental Policy Act (NEPA) (12) in 1969 did little to stem the tide of pollutant discharges to the nation's lakes and streams. But as important as what the statutes did not do was what they did—they sanctioned the "assimilative capacity" approach to pollutant control that had been used since the first pollutant discharge; that is, the statues tacitly approved the concept that surface waters could absorb pollutants without being adversely affected. Stated another (and perhaps more cynical) way, these statutes endorsed what polluters had assumed for a hundred years—that, surface waters provided a convenient, cheap, out-of-sight/out-of-mind place to put garbage, sewage, and wastes from residential, municipal, industrial, agricultural, and other sources.

The assimilative capacity of the nation's lakes and streams was not unlimited, of course, and floating debris, fish kills, odors, and river fires of the mid-1900s burst the out-of-sight/out-of-mind bubble. So strong was the wake-up call of the 1950s and 1960s that the U.S. Congress declared in its work leading up to the Clean Water Act that the use of any river, lake, stream, or ocean as a waste treatment system was unacceptable (4). The U.S. Clean Water Act of 1972 (3), passed over President Nixon's veto, called for the restoration and maintenance of the chemical, physical, and biological integrity of the nation's waters and a *national goal of elimination of the discharge of pollutants to navigable waters* (by 1984).

As a matter of practicality, or perhaps tradition, during its writing of the CWA, the U.S. Congress also established a national policy of prohibition of the discharge of toxic substances *in toxic amounts*, and an *interim* goal, to be achieved by 1983, of attaining water quality that provided protection of fish, shellfish, wildlife, and water-based recreation. Thus, was born the modern approach to pollutant management for surface waters, an approach that continues today to sanction the use of the perceived assimilative capacity of surface waters for industrial, municipal, and other wastes, that continues a long history of using lake, streams, and oceans as part of the waste treatment process, and that has not achieved, and indeed conflicts with the CWA goals of restoring the integrity of the nation's waters and eliminating pollutant discharges. Surface waters remain today approved receptacles of industrial, agricultural, municipal, and other wastes and their pollutants. Thus, modern approaches to pollutant control are based on the assumption that surface waters provide treatment or assimilative capacity for pollutants and, therefore, some discharge of pollutants to surface waters is acceptable.

With the concept of surface water pollutant assimilative capacity comes the question—how much pollution is acceptable; that is, what quantity of pollutants can be discharged to surface waters without harming human health, wildlife, or freshwater and marine ecosystems? Efforts by Congress to write, and by the U.S. Environmental Protection Agency to implement, the Clean Water Act have focused heavily on this question. The result is a two-tiered approach to control pollutant discharges. The technology-based component of the two-tiered approach requires specific treatment methods (technologies) for discrete industrial categories and for wastewater treatment plants. A water quality component, which determines the amount (mass) and concentration of pollutants that can be discharged from industries and wastewater treatment plants without posing unacceptable risks to humans and aquatic life, overlies the technology-based component.

The technology-based approach (Section 301 of the CWA) requires that industries and Waste Water Treatment Plants (WWTP) discharging pollutants to surface waters install and use the best available technology (BAT) economically achievable to treat pollutants prior to their discharge. Industries that discharge to WWTPs rather than directly to surface water must meet pretreatment guidelines, although the ultimate discharge of the pollutants to surface waters is, in this case, managed by the WWTP. BAT requirements, which apply to existing dischargers in a variety of industrial categories, are developed by the U.S. Environmental Protection Agency. New Source Performance Standards (NSPS), also developed by the U.S. EPA, are applied to new categories or new sources within an industrial category, and are usually more restrictive than BAT requirements. Both BAT and NSPS are not based on a determination of whether treatment technologies achieve zero discharge or whether they actually protect water quality; rather, they are based on whether they meet the definition of "best" and "economically achievable." The zero discharge goal of the U.S. CWA was to be achieved (by 1985) via incremental improvements in treatment technology, established by EPA through national effluent guidelines that would become more stringent with advances in treatment technologies and implementation of pollution prevention practices (7). However, EPA has failed to continuously update BAT standards and to force the use of the newest, most stringent treatment technologies, which has rendered the approach ineffective in achieving zero discharge (7). It has also, at times, provided an incentive to industries to keep older facilities online, with their less

protective treatment technologies, to avoid more costly NSPS. BAT and NSPS have also never been applied to nonpoint sources such as pesticide runoff from farmlands. As a result, the use of BAT and NSPS have not achieved the zero discharge goal of the U.S. CWA, nor have they achieved the act's objectives—to restore and maintain the chemical, physical, and biological integrity of the nation's waters, to eliminate the discharge of pollutants to surface waters, or to attain water quality that protects fish, wildlife, and recreation.

To address the shortcomings of the technology-based approach to pollutant control, Section 303 of the U.S. CWA requires states and/or the federal government to develop and adopt enforceable water quality standards, including numeric criteria that define the maximum acceptable concentrations of toxic and other pollutants that can occur in surface waters. Water quality standards (WQS) and water quality criteria (WQC) define "acceptable" as a concentration of an individual pollutant that will not harm humans, fish, shellfish, wildlife, recreational activities (e.g., swimming or fishing), or other uses of a surface water system. Acceptable concentrations, defined by numeric water quality criteria, are typically risk-based pollutant concentrations derived from toxicity tests conducted with laboratory animals, from epidemiologic studies of exposed human populations, tests of whole effluent, and other tests. WQC are intended to protect aquatic life from the effects of acute and chronic exposure to pollutants, and humans from cancer and noncancer effects of pollutant exposure. Exposure routes for both humans and aquatic animals include direct water contact as well as exposure through consumption of contaminated drinking water and food (e.g., consumption of contaminated fish by humans and fish-eating birds). Water quality criteria are to be achieved in all areas of surface water systems except in the immediate area where a point source (a pollutant discharge from an industry or WWTP) mixes with a receiving water body. This area, called a *mixing zone*, typically has less stringent criteria to protect aquatic life, terrestrial animals, and humans and is an important feature of the process to apply water quality standards and criteria to the control of pollutant discharges from industries and WWTPs.

An industry or WWTP in the United States that discharges pollutants to surface waters must obtain a permit from its state or the federal government to do so. The permit is issued under the National Pollutant Discharge Elimination System (NPDES) and typically incorporates both the technology-based and water quality–based approaches to pollutant control. The technology-based approach requires BAT to be used by the discharger to treat its wastes. The water quality–based approach requires further treatment, if BAT does not result in pollutant concentrations that will protect the surface water system adequately; that is, will not protect aquatic and terrestrial biota and human health.

Water quality–based effluent limits (WQBEL) are determined by a wasteload allocation for the receiving water body. The wasteload allocation is effectively determined by the receiving water's *assimilative capacity*; that is, its ability to dilute, absorb, and treat (through natural degradation, for example) the pollutant to levels that do not exceed water quality standards and criteria. However, the WQBEL does not apply at the immediate point of discharge; rather, it applies after the pollutant discharge has had an opportunity to mix with the receiving water (outside the mixing zone). Less stringent criteria apply, usually providing a minimal level of protection for aquatic life, within the mixing zone up to the point of discharge. There are many other components of the NPDES system that require detailed analysis of the pollutant, the receiving water, and the organisms that are to be protected, as well as issues such as pollutant levels of detection, background concentrations of pollutants, and issues associated with the discharge and toxicity of combinations of pollutants (i.e., additive or synergistic toxicity) that must be, but are not always, addressed as an effluent limit is developed for an industrial discharger or WWTP under the National Pollution Discharge Elimination System (6). However, despite the aggressive name and complex nature of the system, none of its components are designed to eliminate discharges of pollutants; rather, the system is designed to reduce or limit pollutant discharges to levels that pose an acceptable level of risk to a water body and its uses. Equally important, and discussed further later, the NPDES system was designed to control pollutant discharges from point sources (industries and WWTP), not from nonpoint sources such as polluted storm water runoff, residential and agricultural use of pesticides, animal feeding lots, and many others.

Water pollutant control in Europe is very similar to the approach used in the United States. The European Union (EU) *Water Framework Directive* (13), adopted in October 2000, in conjunction with EU legislation on the Registration, Evaluation, Authorization, and Restriction of Chemicals (REACH) (14) and other Directives, establishes a "legal framework to protect and restore clean water across Europe and ensure its long-term, sustainable use." The EU Water Framework Directive bases water management on natural geographical and hydrologic units (e.g., river basins), promotes sustainable water use based on long-term protection of available water resources, and encourages measures that result in the reduction, cessation, or phasing-out of discharges of priority hazardous substances to marine and freshwater surface and ground water systems. The ultimate aim of the Water Framework Directive is to achieve concentrations of naturally occurring substances near background, and to achieve concentrations of anthropogenic, synthetic substances close to zero. However, while the Framework calls for cessation or phase-out of some pollutant discharges, the implementation of the Framework, similar to implementation of the U.S. CWA, falls back on the concept of acceptable levels of pollutants in water bodies and the development of pollutant discharge controls from industries and WWTPs that draw on receiving water assimilative capacity.

The EU strategy to control pollution of surface waters is described in Article 16 of the Framework, which requires establishment of a list of priority substances, a procedure for the identification of priority substances/priority hazardous substances, and adoption of specific measures against pollution with these substances (Table 27.1). Article 16 states:

> The European Parliament and the Council shall adopt specific measures against pollution of water by individual pollutants or groups of pollutants presenting a significant risk to or via the aquatic environment, including such risks to waters used for the abstraction of drinking water. For those pollutants, measures shall be aimed at the progressive reduction and, for priority hazardous substances, as defined in Article 2(30), at the cessation or phasing-out of discharges, emissions and losses.

While the goals of the EU Framework are laudable, its implementation falls back on an approach to water quality management that is built upon the concept of acceptable pollutant concentrations in surface waters and an inherent pollutant assimilative capacity of lakes and streams. Directive 2008/105/EC (December 2008) (15) establishes Environmental Quality Standards (EQS) for priority substances and certain other pollutants; that is, it defines acceptable concentrations of pollutants in surface waters, all of which are above zero. The directive implies that EQS also define acceptable concentrations of pollutants in discharges to surface waters from industries and treatment plants, after the discharge is mixed with and diluted by the receiving water; that is, it endorses the use of mixing zones and receiving water dilution, which are virtually identical to the use of those concepts under the U.S. CWA National Pollutant Discharge Elimination System.

Directive 2008/1/EC (16) provides more specific guidance to control pollutant discharges from industries and treatment plants. This directive mirrors the approach to pollutant discharge control under the U.S. CWA by requiring a permit-based discharge program, the use of best available treatment technology and, where BAT does not adequately protect surface waters, the use of Environmental Quality Standards (EQS) for individual pollutants to limit point source discharges. Where EQS are used, both available dilution capacity of the receiving water and mixing zones are allowed and incorporated in the calculation of emission limit values for individual toxic or hazardous substances. Progress reports required by the EU Framework have been developed to address the status of monitoring programs, the state of administrative programs

TABLE 27.1 Substances Classified as Hazardous (Candidates for Ban or Phase Out) Under the EU Water Framework Directive and the Stockholm Convention on Persistent Organic Pollutants

Compound	EU Priority Hazardous Substances[a]	Persistent Organic Pollutants[b]
Aldrin		x
Anthracene	x	
Brominated diphenylethers	x	
Cadmium and its compounds	x	
Chlordane		x
Chlordecone		x
C_{10-13}-chloroalkanes	x	
Dieldrin		x
DDT		x
Endosulfan	x	
Endrin		x
Heptachlor		x
Hexabromobiphenyl		x
Hexa/heptabromodiphenyl ether		x
Hexachlorobenzene	x	x
Hexachlorobutadiene	x	
Hexachlorocyclohexane	x	x
Lindane		x
Mercury and its compounds	x	
Mirex		x
Nonylphenols	x	
Polychlorinated biphenyls		x
Polychlorinated dibenzo-p-dioxins		x
Polychlorinated dibenzofurans		x
Perfluorooctane sulfonic acid		x
Pentachlorobenzene	x	x
Polyaromatic hydrocarbons	x	
Tributyltin compounds	x	
Penta/tetrabromodiphenyl ether		x
Toxaphene		x

[a] Annex X to Directive 2008/105/EC—List of priority substances in the field of water policy.
[b] Stockholm Convention on persistent organic pollutants (POPs).

to implement the Framework, and levels of Framework-related program implementation. However, the status of pollutant emission reductions from point sources and the status of compliance with EQS in specific water bodies have not been assessed.

Water treatment to remove pollutants is also a fundamental approach to protect drinking water, whether it is derived from surface or groundwater. The U.S. Safe Drinking Water Act of 1974 (4) is a treatment-based statute designed to manage toxic and pathogenic pollutants in drinking water. Under the SDWA, the U.S. Environmental Protection Agency sets health-based standards for naturally occurring drinking-water contaminants, such as bacteria and viruses, and for contaminants of anthropogenic origin such as metals and pesticides. Drinking-water treatment plants are required either by individual states or the U.S. EPA to meet SDWA standards before water leaves the treatment plant.

Much of the U.S. SDWA focuses on treatment to reduce contaminants, although recent revisions to the act have focused more heavily on protecting ground and surface water that serves as a source of drinking water (discussed further later). The Safe Drinking Water Act also requires water suppliers to notify the public when there is a contamination problem in a drinking-water system and it requires treatment plants to provide annual reports to their users on the quality of their tap water.

The SDWA has been successful in reducing the occurrence of water-borne diseases in the United States. However, it has, on occasion, had unintended adverse consequences. For example, Washington, DC, like most metropolitan areas, disinfects its drinking water with chlorine, as required by the SDWA. Not uncommonly, the use of chlorine for disinfection creates carcinogenic by-products and, as a result, the city added corrosive chloramines to reduce the by-products as required by the U.S. EPA. An unintended consequence of this action, however, was corrosion of aging metal pipes and mobilization of lead with resultant concentrations up to 20 times allowable levels in drinking water serving over 6000 homes in the DC area. To reduce lead exposures, the city is in the process of replacing 23,000 lead-containing water service pipes. This will take several years and, in the interim, the city is notifying individuals served by the pipes and providing recommendations to reduce lead exposure. It is also adding orthophosphate during the treatment process to reduce pipe corrosion and associated release of lead, as well as other metals such as copper, into the water supply.

Despite the goal of pollutant discharge elimination called for in both the United States and European Union, neither has achieved the goal and, in many cases, treatment-based approaches to water pollutant control continue to allow significant concentrations and loads of pollutants to enter lakes, streams, and coastal and marine systems. Lack of controls on nonpoint sources has also contributed to the degradation of surface and ground water. As a result of both historic and more recent pollutant discharges, significant contamination of some freshwater and marine systems has occurred, requiring massive cleanup efforts as yet another approach to controlling water pollutants.

27.2.2 Cleaning Up Pollutants

The legacy of 150 years of pollutant discharges to lakes and streams became apparent gradually during the early to mid-1900s. A burning Cuyahoga River, a dead Lake Erie, widespread kills of fish and wildlife, and scum-fouled lakes and rivers brought new attention to the plight of the nation's surface waters during the second half of the century and nowhere was attention more focused than in the U.S. Great Lakes ecosystem. Efforts to curb inputs of nutrients, particularly phosphorus, that had caused gross eutrophication in the Great Lakes were implemented in the 1970s, and resulted rather quickly in water quality improvements (17, 18). However, the extent and impacts of toxic pollutant discharges were only beginning to be understood.

During the 1980s, the U.S./Canadian International Joint Commission (IJC), guided by the U.S.–Canada Great Lakes Water Quality Agreement (available at www.epa.gov/glnpo/glwqa/1978/index.htm), identified 43 Great Lakes Areas of Concern (AOC) (19); areas within the basin that are so severely degraded that they could not support aquatic life and wildlife, and other "beneficial uses" such as swimming, fishing (and fish consumption), and aesthetic pursuits. The Great Lakes Water Quality Agreement directed the United States and Canada to develop and implement Remedial Action Plans (RAPs) for each of the 43 AOCs (listed at http://www.great-lakes.net/envt/pollution/aoc.html) and to take steps toward implementation of the plans (more information about Great Lakes AOCs and RAPs can be found at http://www.epa.gov/glnpo/aoc/). Today, 40 of the 43 AOCs remain substantially impaired while 3 have taken adequate steps to improve water quality to a point where they have been removed from the list.

The most significant challenge at many of the 43 Great Lakes AOCs was addressing in-place pollutants—persistent, toxic, and often bioaccumulative pollutants that had been discharged from industries, wastewater treatment plants, and agriculture and that had accumulated in ecosystem components including sediments and biota. Nowhere is the problem more challenging than in the lower Green Bay and Fox River in Wisconsin. This region has been home to paper mills and other industries since the early 1900s, and the river and lower Green Bay have been the recipients of industrial discharges since that time. During the first half of the century, the river and lower bay were plagued by discharges of raw sewage, oil slicks, fish kills, and beach closures. While some of these problems were addressed during the middle of the century, over 300,000 kg of PCBs were discharged to the Fox River between 1950 and 1970 by P.H. Glatfelter Co, Appleton Papers, Wisconsin Tissue, Fort James (formerly Fort Howard); and the Appleton and Neenah/Menasha wastewater treatment plants (which received and treated discharges from these and other facilities). It has been estimated by the Wisconsin Department of Natural Resources (DNR) (20) that the lower Fox River contains nearly 30,000 kg of PCBs and 11 million cubic yards of contaminated sediment, while the lower Green Bay contains over 70,000 kg of PCBs in sediments. The contamination has caused impairment and degradation of fish and wildlife populations, deformities and reproductive problems in birds and other animals, and degradation of benthic (sediment dwelling) populations. Human health has also been threatened. There are upwards of 40,000 highly exposed recreational and subsistence anglers and their families, including Hmong anglers and members of the Oneida band of Chippewa Indians, in the Fox River/Green Bay area (21). Estimated cancer risks from consuming fish from the Fox River and lower Green Bay among these individuals are as high as 1×10^{-3}, equivalent to the cancer risk of smoking 1–3 packs of cigarettes each day.

A remedial action plan for the AOC was developed during the late 1980s with the goals of protecting humans who consume fish from the river and bay, protecting ecological components and systems, and reducing transport of PCBs from the Fox River to Green Bay and to Lake Michigan. In 2003, agreement was reached to remove some of the PCB-contaminated sediment from the Fox River. The agreement stated that approximately 30,000 kg of PCBs in about 7 million cubic yards of sediment would be removed by dredging at a cost of several hundred million dollars (the most common cost estimate for cleanup is about $700 million). The remaining 70,000 kg still in the river and lower Green Bay are to be left in place. Of the original 300,000 kg discharged to the river and bay, 200,000 kg remain unaccounted for, perhaps transported to and distributed throughout Lake Michigan, lost via volatilization, or simply missed during accounting studies. What is clear is that, despite massive removal of contaminated sediments at a cost approaching $1 billion, the legacy of toxic pollutant discharges to the Fox River and Green Bay—impaired fish and wildlife and threats to human health from fish and wildlife consumption—will not be resolved in our lifetime.

27.2.3 Reducing Exposure

Concentrations of contaminants in sediments may be reduced through dredging or neutralized by capping; contaminants in drinking water may be neutralized by disinfection or their concentrations reduced via filtration; and pollutants discharged from Waste Water Treatment Plants and industries, and storm water and agriculture runoff to surface waters may be reduced through process and treatment controls. However, not all contaminants can be removed from all media, nor can some contaminants be removed completely from specific media; thus, additional efforts may be required to reduce exposure to pollutants in water. Two such efforts are described below.

27.2.3.1 Fish Consumption Advisories Many persistent toxic pollutants discharged to surface waters, such as DDT, PCBs, and mercury, accumulate and remain in freshwater and marine ecosystems. These substances may accumulate in sediments, such as in the Fox River and Green Bay, and in aquatic biota. Of particular concern are persistent, highly lipophilic toxic pollutants that bioconcentrate or biomagnify to high levels in tissues of fish and fish-eating animals.

The U.S. Food and Drug Administration (FDA) is charged with regulating contaminants in commercially sold foods, including fish, under the Federal Food Drug and Cosmetic Act (FFDCA). FDA develops tolerance and action levels for contaminants in fish (and for all other foods), and when a toxicant concentration exceeds a tolerance or action level in a particular species, the FDA may remove it from interstate commerce. FDA may also issue consumption warnings for contaminated fish sold commercially without removing a species from commerce, such as the agency has done in conjunction with the U.S. EPA for shark, swordfish, tile fish, mackerel, and some tuna contaminated with methyl mercury (22).

Fish that are caught and consumed by sport anglers (rather than being purchased from fish stores and supermarkets) are not regulated under the FFDCA. Yet, these fish may contain the highest concentrations and the greatest variety of contaminants. For example, fish caught by sport anglers from inland lakes in the upper Midwest are so widely contaminated with mercury that states in the region have issued generic advice to reduce consumption without regard to species or the lakes and rivers from which they are caught (23).

Since the U.S. FDA has chosen not to regulate contaminated sport-caught fish, the issue has largely been left to the states, which develop and issue consumption advice for contaminated fish species or, in some cases, for groups of species from certain water bodies. Consumption advice for sport-caught fish was first issued by Great Lakes states in the 1970s. Contaminants, including PCBs, DDT, polybrominated diphenyl ethers (PBDEs), dioxins, aldrin, dieldrin, methyl mercury, and many others have accumulated in tissues of sport-caught fish both within and outside of the Great Lakes region, prompting 48 states in the United States to develop and issue consumption advice for local, sport-caught fish species.

Most states draw on the U.S. FDA tolerance or action levels to develop consumption advice for sport-caught fish. Consumption advice in the form of do-not-eat warnings or for reduced consumption (e.g., do not eat more than three meals of species × per week or month) may be issued for an entire species, for various size-classes of species, or for species or size-classes from specific water bodies. However, the U.S. FDA tolerance or actions levels for many contaminants are outdated and incorporate consumption rate and other assumptions that may not be appropriate for sport-caught fish. As a result, the U.S. EPA has issued guidance for the development of risk-based consumption advice (24) that can be used by any state, and the states in the Great Lakes basin have developed uniform consumption advice for shared waters of the Great Lakes that is a hybrid of the risk-based advice recommended by EPA and the chemical-specific tolerance and action levels issued by the FDA (for a detailed discussion of fish consumption advice development, see Ref. 6).

27.2.3.2 Swimming Advisories Human exposure to pathogens occurs typically through consumption of contaminated food or drinking water. However, ingestion of pathogen-contaminated water while swimming is also a public health concern. Studies indicate that beachgoers can become ill after swimming in open waters with diseases such as ear, nose, and throat infections, respiratory infections, and diarrhea (25, 26).

One approach to reducing human exposure to pathogen-contaminated surface waters is closing a beach to swimming and other activities. The Natural Resources Defense Council (27) reported that, during 2008, there were more than 20,000 days of swimming beach closings and advisories in the United States. These closings were due primarily to untreated storm water and sewage spills and overflows that enter surface waters. State and local public health agencies monitor water quality at some swimming beaches, focusing on *E. coli* or fecal coliforms, which occur in the gastrointestinal tracts of higher animals and, as a result, are common in animal feces and serve as indicators of the presence of human pathogens such as *Giardia* and *Cryptosporidium*. When concentrations of indicator organisms exceed regulatory thresholds, agencies may close a beach or warn individuals that contact with water may be hazardous to health.

The Beaches Environmental Assessment and Coastal Health (BEACH) Act of 2000 (28) provided funding to states to test water at swimming beaches for fecal contamination. However, monitoring activities are sporadic and uneven, as are efforts to identify and remediate sources of pathogen contamination at swimming beaches. Beach Protection Act bills now pending in the U.S. Congress would reauthorize and strengthen the BEACH Act by providing funding to identify and clean up contaminant sources and would also require states to use rapid testing methods for pathogen indicators so that timely warnings of contamination could be issued.

27.2.4 Preventing Pollution

Section 319 of the U.S. CWA amendments of 1987 (29) created the State Nonpoint Source Management Program, and the U.S. Congress passed the Coastal Zone Nonpoint Pollution Control Program of 1990 to address nonpoint sources of pollutants to freshwater and marine systems. Similarly, the U.S. Safe Drinking Water Act amendments of 1996 (30) required states to develop and implement source water assessment programs. Primary among the goals of the 1996 SDWA amendments is identification and analysis of threats to public drinking water, whether drawn from surface or ground water, and identification and implementation of protection measures to address threats to those sources. What is noteworthy about the U.S. CWA and SDWA programs to address nonpoint sources, however, is their emphasis on pollution prevention rather than pollutant treatment.

While there is no comprehensive federal statute or regulation in the United States to develop and implement programs to address nonpoint sources or to implement source water protection plans and initiatives, activities at the local and state levels provide examples of prevention-based approaches to water protection.

For example, the City of Chatham, Massachusetts (MA), has enacted an ordinance to manage, and protect groundwater from fuel storage systems (available at http://www.epa.gov/safewater/sourcewater/pubs/techguide_ma_chatham_fuelstoragesyst.pdf). Generally, underground fuel storage systems are not allowed and, if they are to be allowed, require a permit and a variance from the Board of Health. Above ground storage systems must have clearance of at least 12 in. between the bottom of the system and the earth's surface. The area beneath the system must be kept free of grass and weeds and should be covered with a bed of white or light pea stone. Tanks within 100 feet of a public or private well, a storm drain, or any part of a water resource area must be registered with the Board of Health and any leaking tank, or any tank that shows signs of aging, rust, wear, must be removed by a licensed hazardous waste removal company. Similarly, Falmouth, MA, has created Water Resource Protection Districts (WRPD) to prevent degradation of surface water and groundwater utilized for public water supplies (available at www.epa.gov/safewater/sourcewater/pubs/techguide_ord_ma_falmouth_aquifer.pdf). Land in a WRPD may be not be used for junkyards, solid waste disposal, public sewage treatment facilities with on-site disposal of effluent, car washes, coin-operated or commercial laundries, trucking or bus terminals, or airports or commercial accommodations. Subsurface hazardous chemical, gasoline, and oil storage in corrodible containers is also prohibited. Activities involving large-scale use, production, or storage of chemicals, pesticides, herbicides, fertilizers, industrial wastes, or other potentially hazardous wastes or materials will only be allowed with a special permit from the Board of Appeals and requires a demonstration that any proposed use will not result in the degradation of surface water or groundwater currently used for public water supply or proposed for a future water supply.

These and many other approaches are often simply common sense. Gas stations or other businesses that use hazardous substances, septic systems, or the application of pesticides pose significant risks to source waters; thus, their control through prohibitions or use restrictions is appropriate. Similarly, standards that mandate the use of particular construction and operating devices, operating and maintenance practices, and product and waste disposal procedures are effective approaches to protect source waters. Alternatively, land can be purchased or conservation easements constructed to protect groundwater recharge zones, drinking-water wellheads, or sensitive surface water systems. All of these initiatives to restrict or prohibit activities or land uses can be enhanced through education programs, which increase awareness of threats to drinking-water sources, encourage voluntary source water protection, and build support for regulatory initiatives (for more information about these initiatives, visit http://cfpub.epa.gov/safewater/sourcewater/).

Standard operating procedures, use and control of human-made systems and devices, and other initiatives that focus on activity management to reduce threats to water supplies are often collectively known as best management practices (BMPs), similar although not identical to BAT under the Clean Water Act. Where BAT is a well-defined and enforceable mandate for the control of pollutant discharges from specific industrial categories or processes, BMPs are often voluntary or loosely implemented through regulation at the regional or local level, without national standards or mandates, and are often not used in a permit-based program to manage nonpoint pollutant sources. However, despite the lack of a national regulatory mandate to control nonpoint sources of pollutants, their focus on pollution prevention rather than pollutant treatment provides an important alternative to the treatment-based approach to pollutant management for point sources.

27.3 WHAT CAN BE DONE ABOUT WATER POLLUTANTS?

The United States and the European Union have established zero discharge as a goal for pollutant control in surface waters, although both rely on a treatment-based approach to manage pollutant discharges from point sources that has not, and will not achieve the goal. Further, nonpoint sources contribute greater quantities of some pollutants to surface waters than point sources; yet, in contrast to point source control programs, a strict regulatory structure for control of nonpoint sources does not exist.

The treatment-based approach to pollutant control has a number of weaknesses. First, success of the treatment-based approach has been defined not by zero discharge but by the assumption that there are acceptable concentrations of pollutants in surface waters and, where those concentrations are achieved (i.e., by meeting water quality criteria or standards), human health, wildlife, and aquatic ecosystems will be protected. However, the basis for this assumption is flawed. Concentrations of individual pollutants in surface waters are deemed acceptable if they occur at or below an effect or risk threshold, as determined typically in studies of laboratory animals. However, Suter (31) examined over 90 water pollutants and found that each caused significant reductions in aquatic organism survival and fecundity at or below concentrations defined by water quality criteria. Similarly, concentrations of single cancer-causing pollutants are typically deemed acceptable if they do not raise cancer risk above 1×10^{-5} (one additional cancer death in 100,000 exposed people). Yet, the adoption of a 1×10^{-5} cancer risk level as acceptable has no scientific basis; rather, it has been based, in part, on convenience and economic achievability (32).

Determination of acceptable pollutant concentrations for both humans and nonhuman organisms is also limited by a failure to consider the adverse effects associated with concurrent exposure to multiple contaminants, such as those that typically occur in a discharge from an industrial or waste treatment plant. Further, there is ample evidence that non-chemical stressors (e.g., disease, thermal stress, habitat alteration, and many others) influence an organism's response to chemical stressors (and vice versa) in ways that may render them significantly more susceptible to chemical stressors than predicted from chemical-specific, single-stressor assessments (33); yet, chemical-specific water quality criteria do not account for the concurrent impacts of multiple chemical and nonchemical stressors.

Second, while a certain level of treatment technology (e.g., BAT) is required for all point sources (industries and WWTPs), regulatory programs typically require in-plant treatment only to a point where water quality criteria or standards are achieved. However, in-plant treatment technology alone often does not result in meeting water quality criteria; rather, "treatment" that occurs through pollutant degradation, absorption, sediment deposition, accumulation in biota, volatilization, and transport, among others, is also provided by the receiving water itself. This pollutant assimilative capacity of the receiving water is incorporated into treatment-based point source control programs through, for example, calculations of end-of-pipe pollutant concentrations that allow zones of pollutant mixing and dilution prior to meeting water quality criteria, and by considering the biological availability of pollutants before setting and requiring compliance with ambient water quality criteria or standards in surface water. Use of a receiving water's assimilative capacity in this fashion serves as an opportunity or excuse not to require more aggressive efforts to control pollutant discharges from point sources.

Finally, while scientific knowledge as well as in-plant treatment technology could conceivably advance to a point where end-of-pipe pollutant concentrations meet water quality criteria that fully protect human health, wildlife, and all components of aquatic ecosystems, the political and economic costs of treatment to achieve these levels would be prohibitive. Therefore, an alternative approach to pollutant control is necessary.

Adler (34) has argued that the mandatory and enforceable permitting and treatment obligations for municipal and industrial point sources under the U.S. Clean Water Act, as well as under the EU Water Framework Directive, have created a significant gap between contemporary approaches to regulate point and nonpoint sources of pollutants. He also suggests that the best way to close the gap between point and nonpoint source pollutant control is to adopt a mandatory system of enforceable "best practice" standards for nonpoint source pollution, similar to the mandatory technology-based controls for point sources under the U.S. CWA (35). While differences between the two systems may be significant because of the comparative diversity of nonpoint sources associated with the nature of the sources themselves as well as differences in climate, soils, topography, land uses, and others, a mandatory system of enforceable, prevention-based best practice standards for nonpoint sources would make major strides toward eliminating these sources of pollutants to surface as well as ground waters. Concurrently, the existing enforceable approach to pollutant control for point sources to surface waters must remain enforceable but shift from a treatment-based to a prevention-based approach.

The 40th anniversary of contemporary efforts to control water pollutants, as well as recent reports by the *New York Times* and the U.S. EPA (at the beginning of this chapter) of blatant, continuing water pollution problems, should provide an impetus to pursue water pollutant management holistically. Pollutant control programs for all waters—surface, ground, fresh, brackish, salt, inland, coastal, and marine—should incorporate a mandatory, prevention-based approach that requires measurable progress toward and, ultimately, achievement of zero discharge for every source, whether direct or indirect, point or nonpoint, within a strictly defined time frame. The approach should be prevention based, as called for by the U.S. Government Accountability Office (36), and result in a net reduction of toxic pollutants discharged to surface and ground waters as well as other media that may interact, directly or indirectly, with surface water or groundwater. It should require use reductions of hazardous chemicals, implemented through process changes, chemical substitutions, recycling, and other approaches including chemical bans and phase-outs, product changes or bans, and behavior changes that affect product consumption and use.

A system of penalties and/or rewards should encourage technology development to achieve zero discharge, accompanied by requirements that incremental advancement toward zero be achieved consistently over the time frame. The approach should not require, nor should it prohibit, government participation in technology development. It may, however, require some aggressive governmental actions to control pollutant discharges, such as restricting, phasing out, or banning particularly hazardous chemicals. For example, the U.S./Canadian International Joint Commission (37) proposed during the 1980s and early 1990s to phase out the use of chlorine in manufacturing and other industrial processes. The impetus for the proposal was the recognition that many of the persistent, bioaccumulative toxicants plaguing the Great Lakes were chlorine based (dieldrin, DDT, PCBs, chlorinated dioxins, etc.), and that the most effective method to ensure that these and many other chlorine-based compounds did not contaminate the Great Lakes ecosystem was to prohibit chlorine use. Of course, this would have required major process changes in many industries as well as in agriculture, at very significant cost. The proposal was not formally adopted, perhaps because of the perceived costs among other reasons.

The European Union has begun to take more aggressive steps to control chemicals through its Water Framework Directive, as described above, as well as through REACH (14), which strives to ensure that substances of very high concern are progressively substituted by safer substances or technologies. Where substances of very high concern are identified, European industries must seek authorization for the continued use of the substances, and/or face restrictions on their manufacture, use, or marketing. Finally, efforts to control chemicals through phase out and ban have also occurred internationally. The Stockholm Convention on Persistent Organic Pollutants (POPs), held in 2001, addressed chemicals that "remain intact in the environment for long periods, become widely distributed geographically, and accumulate in the fatty tissue of humans and wildlife" (Table 27.1). The convention requires parties (ratifying countries) to eliminate or reduce the release of POPs into the environment. Over 150 countries, including most in Europe, have ratified the convention as of December 2009, although the United States, Russia, and a few others have not.

Aggressive steps such as phasing out chemicals, altering manufacturing processes and agricultural activities to eliminate use of and reliance on hazardous chemicals, and other initiatives such as those under the U.S. SDWA source protection program are critical components of pollution prevention efforts designed to protect surface and groundwater. Of course, any system to prevent pollution should also prohibit shifting of pollutant discharges to other media, for example, the system should not allow shifting discharges from water to air; rather, the emphasis of these programs should be on technological changes and other prevention activities that reduce and ultimately eliminate pollutant discharges completely.

The elimination of pollutant discharges to surface and ground water through a mandatory, prevention-based approach to pollutant control has the potential to result in achievement of the United States and EU goals of zero discharge as well as full protection of human health, wildlife, and aquatic ecosystems. The statutory language to achieve discharge elimination is already in place in the United States and in Europe. What is required is a rejection of the concepts that treatment alone is an appropriate approach to manage pollutant discharges, that surface and groundwater provide pollutant treatment through an inherent assimilative capacity, and that there are acceptable levels of pollutants that can occur in surface and ground water.

REFERENCES

1. U.S. Environmental Protection Agency. 2004 National Water Quality Assessment Database, 2010. Available at www.epa.gov/waters/305b/index_2004.html. Accessed 28 January 2010

2. U.S. Environmental Protection Agency. Water Quality Assessment and Total Maximum Daily Load Information, 2010. Available at www.epa.gov/waters/ir/. Accessed 28 January 2010.

3. U.S. Federal Water Pollution Control Act. Pub. L. 92-500, 86 Stat. 816 (codified generally as 33 U.S.C. § 1251 et seq.), 1972.

4. U.S. Safe Drinking Water Act. Pub. L. 93-523, 88 Stat. 1660 (codified as 42 U.S.C. §300f et seq.), 1974.

5. U.S. Rivers and Harbors Act. Pub. L. 30 Stat. 1151 (codified at 2 U.S.C. §401 et seq.), 1899.

6. Foran JA. *Regulating Toxic Substances in Surface Waters.* Boca Raton, FL: Lewis Publishers, CRC Press, 1993.

7. Adler RW, Landman JC, Cameron DM. *The Clean Water Act: 20 Years Later.* Washington, DC: Island Press, 1993.

8. Federal Water Pollution Control Act Amendments. Pub. L. 84-660 (codified generally as 33 U.S.C. § 1251 et seq.), 1956.

9. Water Quality Act. Pub. L. 89-234 (codified generally as 33 U.S.C. § 1251 et seq.), 1965.

10. Clean Water Restoration Act. Pub. L. 89-753 (codified generally as 33 U.S.C. § 1251 et seq.), 1966.

11. Water Quality Improvement Act. Pub. L. 91-224 (codified generally as 33 U.S.C. § 1251 et seq.), 1970.

12. National Environmental Protection Act. Pub. L. 91-190 (codified as 42 U.S.C. § 4321 et seq.), 1969.

13. Water Framework Directive. European Union Directive 2000/60/EC (published at OJ L 327, 22 December 2000), 2000.

14. Registration, Evaluation, Authorization, and Restriction of Chemicals (REACH). (published at OJ L 1907, 18 December 2006), 2006.

15. European Union Directive 2008/105/EC (Published at OJ L 348/84, 24 December 2008).

16. European Union Directive 2008/1/EC (published at OJ L 24/8, 20 January 2008).

17. Chapra SC, Robertson A. Great Lakes eutrophication: the effect of point sources control of total phosphorus. *Science* 1977;196:1448–1450.

18. Cullen P, Forsberg C. Experiences with reducing point sources of phosphorus to lakes. *Hydrobiologia* 1988;170:321–336.

19. U.S./Canadian International Joint Commission. *Seventh Biennial Report on Great Lakes Water Quality*. Washington, DC; Ottawa, Ontario: IJC, 1994.

20. Wisconsin Department of Natural Resources (DNR). The Fox River/Green Bay Cleanup Project, 2010. Available at http://dnr.wi.gov/org/water/wm/foxriver/index.html. Accessed 28 January 2010.

21. Wisconsin Department of Natural Resources, (DNR). PCBs and Health: Fish Consumption Advisories, 2010. Available at http://dnr.wi.gov/org/water/wm/foxriver/health.html. Accessed 28 January 2010.

22. U.S. Environmental Protection Agency. What You Need to Know about Mercury in Fish and Shellfish, 2004. Available at http://www.epa.gov/fishadvisories/advice/. Accessed 28 January 2010.

23. Wisconsin Department of Natural, Resources. Choose Wisely: A Health Guide for Eating Fish in Wisconsin, 2009. Available at http://dnr.wi.gov/fish/consumption/. Accessed 28 January 2010.

24. U.S. Environmental Protection Agency. Guidance for Assessing Chemical Contaminant Data for Use in Fish Consumption Advisories, 2009. Available at http://www.epa.gov/waterscience/fish/technical/guidance.html. Accessed 28 January 2010.

25. Colford JM, Wade TJ, Schiff KC, Wright CC, Griffith JF, Sandhu SK, Burns S, Sobsey M, Lovelace G, Weisberg SB. Water quality indicators and the risk of illness at beaches with nonpoint sources of fecal contamination. *Epidemiology* 2007;18(1):27–35.

26. Wade TJ, Calderon FL, Brenner KP, Sams E, Beach M, Haugland R, Wymer L, Dufour AP. High sensitivity of children to swimming-associated gastrointestinal illness: results using a rapid assay of recreational water quality. *Epidemiology* 2008;19(3):375–383.

27. Natural Resources Defense Council. Testing the Waters: A Guide to Water Quality at Vacation Beaches, 2009. Available at http://www.nrdc.org/water/oceans/ttw/titinx.asp. Accessed 28 January 2010.

28. Beaches Environmental Assessment and Coastal Health (BEACH) Act. Pub. L. 106-284 (codified at 33 U.S.C. 1313), 2000.

29. Water Quality Act. Pub. L. 100-4 (codified generally as 33 U.S.C. § 1251 et seq.), 1987.

30. Safe Drinking Water Act Amendments. Pub. L. 104-182 (codified as 42 U.S.C. §300f et seq.), 1996.

31. Suter GW, Rosen AE, Linder E, Parkhurst DF. Endpoints for responses of fish to chronic toxic exposures. *Environ Toxicol Chem* 1987;6:793–810.

32. Travis CC, Richter SA, Crouch EAC, Wilson R, Klema ED. Cancer risk management: a review of 132 Federal regulatory decisions. *Environ Sci Technol* 1987;21:415–420.

33. Foran JA, Ferenc. *Multiple Stressors in Ecological Risk and Impact Assessment*. Pensacola, FL: SETAC Press, 1999.

34. Adler RW, Resilience, restoration, and sustainability: revisiting the fundamental principles of the Clean Water Act. Washington Univ. Journal of Law and Policy 2010;32:139–174.

35. Adler RW. Freshwater: sustaining use by protecting ecosystems. In: Dernbach J, editor. *Agenda for a Sustainable America*. Washington, DC: Environmental Law Institute. Reprinted in *Environ Law Reporter* 2009;39:10,309–10,315.

36. U.S. Government Accounting Office. Water Pollution: Stronger Efforts Needed by EPA to Control Toxic Water Pollution. Report to the Chairman, Environment, Energy, and Natural Resources Subcommittee, Committee on Government Operations, U.S. House of Representatives, Report # GAO/RCED-91-154, 1991.

37. U.S./Canadian International Joint Commission (IJC). *Sixth Biennial Report on Great Lakes Water Quality*. Washington, DC; Ottawa, Ontario: IJC, 1992, 59 pp.

28

GLOBAL SUBSTITUTION OF MERCURY-BASED MEDICAL DEVICES IN THE HEALTH SECTOR

JOSHUA KARLINER AND PETER ORRIS

28.1 INTRODUCTION

Mercury, one of the world's most ubiquitous heavy metal neurotoxicants, has been extensively used in health care since antiquity. It has been an integral part of many medical devices, most prominently thermometers and blood pressure devices (sphygmomanometers). In recent decades this has led to a paradox where institutions and professionals, whose mission is to heal and promote health, have been contributing to a significant global environmental health problem—mercury contamination—through the use of their health care instruments.

In response, in the first decade of this century, the health sector in the United States largely phased out mercury thermometers. Most U.S. hospitals no longer purchase mercury sphygmomanometers, and as of 2008, more than one third of the U.S. population lived in states that have banned or severely restricted their sale and/or use. Similarly, the European Union has banned mercury thermometers in 2007 and is considering similar strictures for sphygmomanometers. Unfortunately, despite some outstanding efforts in a growing list of countries, the health sector in much of the rest of the world still uses mercury-based medical devices.

Recently several developing countries have initiated a shift toward mercury-free health care. These countries have responded to policy guidelines issued by the World Health Organization. They have also been convinced of the technical and economic viability of the alternatives. Most notably, the ministries of health in Argentina, Costa Rica, Mongolia and the Philippines have issued national policies phasing out mercury-based medical devices. Health care systems in countries ranging from India, Nepal, South Africa, Tanzania, Mexico, Chile, and Brazil also have programs and policies to demonstrate and expand mercury-free health care.

In addition to eliminating this important source of mercury contamination, these efforts to substitute mercury-based medical devices educate and activate health care leaders around the world. Some have become advocates not only for cleaning up health care's mercury pollution but also for more comprehensive control of this global environmental contaminant, at the local, national, and global level where an international treaty to control the trade and use of mercury is being negotiated.

This article briefly explores first the toxicity and environmental health impacts of mercury, the use of mercury in the health sector, several examples of health systems moving away from mercury-based medical devices, obstacles encountered, and perspectives for the future.

28.2 THE PROBLEM: MERCURY IN THE ENVIRONMENT

Mercury is a naturally occurring heavy metal. At ambient temperature and pressure, mercury is a silvery-white liquid that readily vaporizes. When released into the air, mercury may stay in the atmosphere for up to a year, and is transported and deposited globally. It is within this environment that inorganic and organic compounds of mercury are formed.

Since the start of the industrial era, the total amount of mercury circulating in the world's atmosphere, soils, lakes, streams, and oceans has increased by a factor of between two

Water and Sanitation-Related Diseases and the Environment: Challenges, Interventions, and Preventive Measures, First Edition.
Edited by Janine M. H. Selendy.
© 2011 Wiley-Blackwell. Published 2011 by John Wiley & Sons, Inc.

and four (1). This increase has been affected by human endeavors, which include the removal of mercury from its subterranean home through mining and the extraction of fossil fuels. Human exposure to mercury can result from a variety of sources, including, but not limited to, consumption of fish rich in methylmercury, and due to spills or leaks of the metallic element itself.

Mercury causes a variety of significant adverse impacts on human health and the global environment. High levels of metallic mercury vapor may produce pneumonitis and pulmonary edema if inhaled, and though skin absorption is 50-fold less than lung, toxic levels have been reported due to handling of the liquid metal. This exposure is especially significant if the epithelial barrier has been broken due to cuts or abrasions. Target organs other than the lungs include the kidneys, the nervous system, and the GI tract. Anecdotal reports from hospitals utilizing mercury thermometers report breakage ranging from several to several hundred a month. These reports are paralleled by those noting leakages of mercury-containing sphygmomanometers as well with the potential for significantly larger amounts released.

Yet, of even more concern is potential for developmental neurotoxicity produced by low-dose methylmercury exposure through food. Elemental mercury accumulates in lake, river, stream, and ocean sediments, where it is transformed into methylmercury, which then accumulates in fish tissue. This contamination of fish stock is ubiquitously present in oceans and lakes throughout the world, concentrating several hundred thousand times as it moves up the aquatic food chain (2).

Methylmercury is of special concern for fetuses, infants, and children because it impairs neurological development. When a woman eats seafood that contains mercury, it accumulates in her body, requiring months to excrete. If she becomes pregnant within this time, her fetus is exposed to methylmercury in the womb, which can adversely affect the fetus's central nervous system. Impacts on cognitive thinking, memory, attention, language, and fine motor and visual spatial skills have been documented in children with exposure *in utero* to levels of methylmercury found in women of child-bearing age in the United States (3).

In 2005, Transande et al. using national blood mercury prevalence data from the U.S. Centers for Disease Control estimated that in this century between 316,588 and 637,233 U.S. children each year have cord blood mercury levels >5.8 μg/L, a level associated with loss of IQ. They estimated that lost productivity due to this amounts to $8.7 billion annually (range, $2.2–43.8 billion) (4).

The United Nations Environment Programme (UNEP) and World Health Organization have identified the adverse effects of mercury pollution as a serious global environmental and human health problem (5). The UNEP Governing Council has targeted reducing methylmercury accumulation in the global environment as a major global priority (6).

28.3   MERCURY IN HEALTH CARE

Mercury thermometers and blood pressure devices add to the global burden of mercury removed from its below-ground repository and spread about on the surface to form highly neurotoxic organomercury compounds. Further, these devices break or leak with regularity, exposing health care workers and patients to the acute effects of the inhalation of the metal itself. In view of this, as part of a global initiative to reduce the use and spread of mercury in all aspects of society, health care providers and institutions have begun to replace mercury-based medical devices with affordable, accurate, and safer alternatives.

The health care sector is far from the greatest source of organic mercury compounds in the environment. Rather, coal-fired power plant emissions and mercury cell chloralkali plants, along with artisanal gold mining and battery disposal are all far more significant polluters. However, the health care sector does play an important role as a source of global emissions, and can serve as a leader in efforts to phase out mercury in daily commerce.

Mercury can be found in many health care devices and is present in fluorescent bulbs as well as dental amalgams. Mercury is also found in many chemicals and measurement devices used in health care laboratories (7). Medical waste incinerators, as well as municipal waste incinerators, emit mercury into the atmosphere when they burn wastes that contain mercury. According to the U.S. Environmental Protection Agency (EPA), in 1996, prior to the mercury phaseout in U.S. health care, medical waste incinerators were the fourth largest source of mercury emissions to the environment. Hospitals were also known to contribute 4–5% of the total wastewater mercury load. And mercury fever thermometers alone contributed about 15 metric tons of mercury to solid waste landfills annually (8).

While no comprehensive figures are available, anecdotal evidence suggests that in most of Asia, Africa, and Latin America, mercury spills are not properly cleaned, nor is the waste segregated and managed properly. Rather, it is either incinerated, flushed down the drain, or sent, via solid waste, to a landfill.

Thermometer breakages on a case-by-case basis pose some harm to patients, nurses, and other health care providers when mercury is absorbed through the skin or mercury vapor is inhaled. Only a relatively small amount of mercury— roughly—1 g[1]—is released when each thermometer breaks. However, when taken cumulatively on a hospital ward, in an entire hospital, nationally and globally, the situation takes on more serious dimensions.

---

[1] Mercury mass in thermometers may vary from 0.50 to 1 g per thermometer. For the purposes of this article we will use the widely reported estimate of 1 g per thermometer.

In Buenos Aires, for instance, the city government, which runs 33 hospitals and more than 38 clinics, was purchasing nearly 40,000 new thermometers a year, until it began to switch over to alternatives in 2006 (9). Given that nurses and other health care professionals often buy their own thermometers to supplement the city's procurement, the city's health system was using well over 40,000 thermometers a year, most of which would break, and some of which would be taken home (where most would ultimately break as well). The system was ultimately emitting in excess of 40 kg of mercury into the local hospital environment and into the global ecosystem every year.

If one were to use this figure and extrapolate for the entire country, one can estimate that until recently thermometers broken in Argentina's health care system were spilling 826 kg, or nearly 1 metric ton of mercury, into the global environment every year.

In Mexico City, the 250-bed "Federico Gomez" Children's Hospital is a medical service, teaching, and research hospital affiliated with the National Autonomous University of Mexico. This prestigious children's hospital documented a thermometer breakage rate of 385 per month, or well over 4,000 per year (see Table 28.1). The total number of estimated broken thermometers in this one hospital between 2002 and early 2007 is nearly 22,000—the equivalent of 22 kg of mercury (10).

TABLE 28.1 Monthly Mercury Thermometer Breakage at Federico Gomez Children's Hospital, Mexico City, Prior to Substitution with Alternatives

Services	Number of Thermometers Broken per Month
Intensive care unit	20
Post-operatory recovery	20
Emergency room	30
Out-patient studies recovery	6
Surgery	15
Pediatric ICU	15
Surgery ICU	15
Nephrology	30
External consultation	20
General consultation	30
Out-patient surgery	2
Pediatrics III, IV	15
Pediatrics I, II	30
Immunosuppressive illnesses	30
Chemotherapy	2
Urological surgery	45
Special care	30
Orthopedics	30
Total	385
Approximate yearly total	4620

Source: Ref. 10.

While the Federico Gomez hospital has substituted its mercury devices with alternatives, when it undertook its initial assessment there was no clean-up protocol for mercury spills. Rather, mercury waste was deposited with both infectious and biological hazardous wastes, or with municipal wastes. Broken fluorescent lamps were also treated as municipal waste. Mercury-containing equipment was not repaired if broken, and the procedure followed was to merely register the loss and replace it with new equipment (10).

The regular and ongoing breakage of thermometers and the lack of mercury waste management protocols and practices found at the Federico Gomez hospital is not an exception, but more generally the rule in hospitals throughout much of the Global South, where patients and health care workers are regularly and unknowingly exposed to this toxin.

This is the case, for instance, in India, where far fewer thermometers are employed in many hospitals (11). In a study of New Delhi hospitals, the NGO Toxics Link found dangerously high levels of mercury in a series of indoor air samples. They found the "substantial presence of mercury in ambient air of both the hospitals" studied. These levels, which ranged from 1.12 $\mu g/m^3$ to 3.78 $\mu g/m^3$, were all higher than numerous international standards (12).

One of the biggest mercury hot spots that Toxics Link found in its study was the room used to calibrate blood pressure devices (sphygmomanometers), which contain 80–110 g of mercury or roughly 100 times the amount found in a single fever thermometer (13).

Mercury release and contamination from sphygmomanometer calibration is a common problem throughout the world. Louis Havinga, manager of Health Technology Services for the KwaZulu Natal Province Department of Health in South Africa explained:

This is the most important point why the Health Technology Services has moved away from the use of mercury products. The technicians were exposed to mercury when they repaired mercury column sphygmomanometers. Special precautions and equipment is needed if working with mercury products like a dedicated fume/vapour extraction unit within the maintenance department. The mercury is extracted from the device and placed in a special marked container. The container must be able to seal and should remain inside the fume/vapour extraction unit. Once the container is full, the container must be disposed of in a well documented and controlled manner by making use of a recognized hazardous waste disposal company which is very costly. (14)

And while sphygmomanometers break less frequently than thermometers, the spillage is significant and therefore problematic from an environmental health perspective. At the Mayo Clinic in the United States, between 1993 and 1995, 50 spills were documented relating to leakage and spills from sphygmomanometers (15).

28.4 RESPONSES

28.4.1 The United States

In 1998, the NGO Health Care Without Harm began working with EPA, American Hospital Association, the American Nurses Association, and our health care partners to reduce and eliminate mercury use in health care in the United States.

As a result, we have witnessed the progressive phase-out of the use of mercury-based medical devices from the U.S. health care community both through voluntary initiatives and legislative mandates.

Over the course of 10 years, with support from Health Care Without Harm, Hospitals for a Healthy Environment (now Practice Greenhealth), and environmental health advocates across the country, the U.S. health care sector has made significant progress in addressing mercury in health care. Noteworthy results include

- All the top pharmacy chains in the nation have stopped selling mercury thermometers, representing approximately 31,844 retail stores, making it next to impossible to purchase a mercury thermometer in the United States (16).
- Mercury thermometer bans have been passed in 28 states and bans or severe restrictions on sphygmomanometers cover one third of the U.S. population.
- More than 1200 hospitals have signed a pledge to eliminate the use of mercury through Hospitals for a Healthy Environment, and more than 400 have become virtually mercury-free.
- Most large Group Purchasing Organizations have taken mercury sphygmomanometers "off contract," including Consorta (buys for 480 hospitals), Premier, and Novation (buy for 3,100 hospitals).

Over 97% of 554 hospitals surveyed by the American Hospital Association are aware of and have taken steps to address the mercury issue. These steps include the following

Mercury in Clinical Devices
- Over 80% have completely eliminated mercury thermometers from their facilities, and 18.7% have replaced some or most with a plan in place for eliminating the remainder.
- Over 73% have completely eliminated mercury sphygmomanometers, with 25% having replaced some or most with a plan in place for eliminating the remainder.
- About 75% have completely eliminated other clinical items (cantor tubes, bougies, etc.) with about 10% having replaced some or most with a plan in place for eliminating the remainder.

Mercury in Facilities
- Over 72% have inventoried all devices and labeled them as mercury-containing, where appropriate.
- About 75% are recycling fluorescent bulbs.

Other Environmental Improvements Made in the Health Care Sector
- 80% report that they have a waste reduction policy.
- 90% have a regulated medical waste minimization program (17).

28.4.2 The European Union

Several European countries have taken the lead in substituting mercury-based medical devices. Evidence from accuracy studies, serious concerns about the hazards of mercury, and ready availability of alternatives has led to several EU countries and a number of health care facilities and associations to completely prohibit mercury in most of its applications. For instance, Sweden, Netherlands, and Denmark have all banned the use of mercury thermometers, blood pressure devices, and a variety of other equipment (18).

In July 2007, the European Union banned the sale of mercury thermometers for use in health care. The ban went into effect in 2008 (19).

Under this recently passed Europe-wide legislation, mercury sphygmomanometers and other measuring devices are also banned for sale to the general public. The European Union is now considering banning sphygmomanometers for clinical use.

A recent report by Committee on Emerging and Newly Identified Health Risks, under the Directorate General for Health and Consumers of the European Commission found that "mercury-free blood pressure measuring devices (when clinically validated) are generally reliable substitutes for mercury-containing sphygmomanometers in routine clinical practice." It also concluded that "there is no evidence of adverse effects on patients' health in clinical settings due to the replacement of mercury containing sphygmomanometers by validated mercury-free alternatives." The report further finds that mercury-based sphygmomanometers are "not essential" for calibration. Therefore, due to the acute toxic hazard to health care workers and chronic hazard to society, these devices have no place in clinical care (20). This has set the stage for a phase-out of mercury-sphygmomanometers in the European Union.

The European Union has also already banned mercury in a number of products including batteries and electronic and electrical equipment. The European Union further encourages member-states to advise citizens about the risks to pregnant women and children of mercury exposure from frequent consumption of predatory fish.

28.4.3 Developing Countries

As it becomes clear to health care leaders in developing countries that accurate, cost-effective alternatives are available, more and more hospitals, health care systems, and entire nations are beginning to make the switch. The World Health Organization (WHO) policy on mercury in health care has provided a framework for this transition.

In 2005 the World Health Organization advised, in its eloquently crafted policy paper on the topic, a global transition of the health care sector toward the use of mercury-free care alternatives. Specifically, the paper projected a 3-step approach:

Short Term: Develop and implement plans to reduce the use of mercury equipment and replace it with mercury-free alternatives. Address clean-up, storage and disposal.

Medium Term: Increase efforts to reduce the use of unnecessary mercury equipment in hospitals.

Long Term: Support a ban of mercury-containing devices and promote alternatives. (21)

In December 2008 the World Health Organization and Health Care Without Harm (HCWH) launched a Global Initiative to achieve virtual elimination of mercury-based thermometers and sphygmomanometers over the next decade and their substitution with accurate, economically viable alternatives. The initiative is a component of the UN Environment Programme's Mercury Products Partnership. Its goal is:

By 2017, to phase out the demand for mercury-containing fever thermometers and sphygmomanometers by at least 70% and to shift the production of all mercury-containing fever thermometers and sphygmomanometers to accurate, affordable, and safer non-mercury alternatives.(22)

The UNEP Products Partnership is in turn part of a larger global effort to address the toxic environmental health impacts of mercury accumulation in the global environment. This effort consists of a series of other voluntary partnerships in areas of major mercury emissions such as chlor-alkali production, artisanal gold mining, coal-fired power plants, and mercury waste management (23).

Working under this framework, or in parallel to it, a number of developing countries are taking significant strides toward mercury elimination in the health care sector. The following are some examples:

Argentina: In February 2009 Argentina's Minister of Health issued a declaration banning the health sector from purchasing new mercury-based medical devices (24).

Philippines: The Department of Health (DOH) of the Philippines issued an Administrative Order in August 2008 to phase out all mercury-based medical devices in the country over a 2-year period (25).

In both countries these policies effectively establish the framework required for the phase out of mercury thermometers and blood pressure devices out of the more than 1,700 hospitals in the health care sector in each nation.

Taiwan: The country has banned mercury thermometers (26).

Uruguay: The country is phasing out mercury thermometers.

Brazil: The City of São Paulo is the first in Brazil to eliminate the use of devices containing mercury in its public hospitals. To date 34 public hospitals/emergency rooms and 85 primary health care centers have been recognized for their elimination of mercury devices. This adds to the more than 100 private hospitals in São Paulo that have already made the switch.

Mexico: In late 2009, the Health Secretariat of Mexico City—which oversees a health care system of more than 28 major hospitals and hundreds of health centers—joined the HCWH–WHO Global Initiative to substitute mercury-based medical devices with safer, economically viable alternatives. This adds to more than a dozen other major hospitals in Mexico that are already making the switch (27).

India: The city of Delhi has issued a policy and is phasing out mercury-based medical devices in all of its hospitals, while the Central Government has issued similar guidelines for all federally controlled facilities (28).

South Africa: The province of KwaZulu Natal has phased out mercury-based medical devices (29).

Costa Rica: Since 2007, the Costa Rican Social Security administration has been moving toward mercury elimination in the country's public hospitals that it controls. Several hospitals have already made the switch (30).

Chile: More than 23 hospitals have switched over from mercury-based medical devices to the alternatives.

28.4.4 The Importance of China

To date only a handful of hospitals in China have substituted mercury-based medical devices for alternatives. At the same time, China is by far the single largest global producer and supplier of mercury thermometers *and* the alternatives.

A report by China's environmental protection agency, SEPA, shows that the country is producing an increasing number of mercury-based medical devices for both domestic consumption and export.

Eight factories in China produced 150 million thermometers in 2004—a 20% increase from the year 2000. In the process, these factories ran through nearly 200 metric

TABLE 28.2   Total Consumption of Mercury in Thermometer and Sphygmomanometer Production in China 2003–2005 and Production and Export of Mercury-Containing Thermometers from China 2003–2005

Year	2003	2004	2005
(a) Total consumption of mercury			
Sphygmoman-ometers (kg)	51,736	94,872.6	81,484.4
Thermometers (kg)	169,609	185,325	200,907.9
(b) Production and export			
Total produc-tion (ten thousands)	14,309	15,820.5	17,363
Export (ten thousands)	5,500	6,325	7,000
Domestic sales (ten thousands)	8,809	9,495.5	10,363
Percentage by export (%)	38.4	39.8	40.3

Source: Ref. 31.

tons of mercury. Meanwhile, mercury sphygmomanometer production in three factories increased nearly 50% in the same time period, to 1.5 million in 2004 (see Table 28.2a). Overall, this production made up about 10% of China's total mercury consumption (31).

Roughly 40% of all thermometers—or 60 million devices—are exported from China, mostly to Southeast Asia, Latin America, and Africa, with a small amount entering the U.S. and European markets (see Table 28.2b) (31).

By comparison, Hicks, an Indian company, produces about 570,000 mercury thermometers annually for domestic consumption. This makes up roughly 50% of the Indian market (32).

In addition to spillage in a health care or home setting, the production of mercury thermometers and sphygmomanometers themselves pose serious occupational health and safety hazards to factory workers, as well as local and global pollution problems. The best-documented case in this regard is that of Kodaikanal, India.

In 1983, Cheesborough Ponds Corporation relocated a factory from New York in the United States to the lakeside tourist destination of Kodaikanal in Southern India's Palani Hills. The factory produced mercury thermometers primarily for the U.S., European, South American, and Australian markets. In 1998, Hindustan Lever, a subsidiary of the Anglo-Dutch corporation Unilever, bought the factory and ran it until it was closed down in 2001.

Data from when the plant was operational report levels of airborne mercury well above internationally recognized safety limits. Not surprisingly, a series of serious mercury-related occupational health problems have been reported by former workers at the plant, including fatigue, headaches,

nausea, skin complaints, respiratory dysfunction, kidney dysfunction, and more. Once it was closed, the plant and its surrounding areas, including the local lake, were found to be contaminated with high levels of mercury. Meanwhile, offsite, mercury-laden waste was sold to other businesses, including 7.4 tons to a local scrap dealer—thereby spreading the pollution (33). Ultimately, the company was compelled to ship back about 285 tons of mercury waste to the original suppliers of mercury in the United States. More than 5,000 tons of contaminated soil remains at the factory site, while offsite contamination in some places is so serious that it triggers national hazardous waste cleanup criteria (34).

What is clear from this example is that a great deal of mercury employed in the thermometer production process never makes it to the hospital, but rather ends up as hazardous local and global pollution. As the SEPA report observes, a significant amount of the mercury "consumed by the thermometer manufacturing industry in China is lost to the environment" before the thermometers even leave the factory (32).

Assuming 1 g of mercury in each thermometer, based on Table 28.2, thermometer production facilities in China are spilling more than 27 metric tons of mercury into the environment every year.[2]

28.5   CONCLUSION

While significant progress has been made in only one decade in shifting global consumption of mercury-based medical devices toward safer, accurate, and affordable alternatives, much work remains to be done.

An essential step in achieving the goal of substituting 70% of mercury-based medical devices with safer alternatives will be the phasing out of China's mercury thermometer and blood pressure device industry. In order for this to be achieved, China's health sector must help lead the way by shifting its consumption toward alternatives. This shift is eminently achievable from a technical and economic standpoint, as has been shown in a variety of health care settings in a diversity of countries ranging from Argentina, to the Philippines, Sweden, and the United States.

Mercury-free health care is not only possible, but if the right forces converge, the day is not far off when most health care institutions will be virtually mercury-free.

Challenges, of course, remain. The quality of the non-mercury devices range from high to very low. Health care systems must develop standards by which to base their procurement of this equipment. To date, no international standards exist to facilitate the shift. Furthermore, the myth

---

[2] There are 200,907 kg of mercury consumed to produce 176 million thermometers with 1 g of mercury each; this leaves 27,227 kg of mercury unaccounted for or 27.23 metric tons.

that mercury blood pressure devices are necessary to maintain a "gold standard" for blood pressure measurements, while negated in several studies, continues to persist.

Similarly, it is proven in many settings that for every 10 mercury thermometers that break on a ward, health care professionals can employ 1 digital thermometer, and that this ends up saving hospitals and national health care systems money. However, this evidence must be developed through pilot experiences on a country-by-country basis. This takes time, energy, and resources.

Finally, disposal of mercury waste is a significant issue. Currently there is no clear solution for mercury waste management. And while stopping the throughput of mercury thermometers reverses mercury contamination patterns that have existed for years, hospitals often end up with toxic mercury waste—old thermometers and blood pressure devices—tucked away in a storage room. Finding a solution at the national, regional, and global levels to manage mercury waste, not just from health care but also from other important sources is imperative.

In 2009 the world's governments agreed to negotiate a global treaty to restrict the use of mercury in products and processes around the world, while also addressing the waste issue. Ultimately, a legally binding treaty that should be concluded by the mid-2010s will help consolidate the gains made by the health sector in substituting mercury, while projecting those gains into other economic sectors so as to protect the world's environment and human health from mercury contamination.

REFERENCES

1. Health Canada. Available at http://www.hc-sc.gc.ca/ewh-semt/pubs/contaminants/mercur/q1-q6_e.html.

2. Health Canada. Available at http://www.hc-sc.gc.ca/ewh-semt/pubs/contaminants/mercur/q47-q56_e.html.

3. United States Environmental Protection Agency. Available at http://www.epa.gov/mercury/effects.htm.

4. Trasande L, Landrigan PJ, Schechter C. Mount Sinai School of Medicine, New York, New York. *USA Environ Health Perspect* 2005;113:590–596.

5. UNEP. *Global Mercury Assessment.* UNEP Chemicals, Geneva, December 2002.

6. http://www.chem.unep.ch/mercury/Decision%2024-3.pdf.

7. Making medicine mercury-free, a resource guide for mercury free medicine. In: *Going Green: A Resource Kit for Pollution Prevention in Health Care.* Health Care Without Harm, Arlington, June 6, 2007. Available at http://www.noharm.org/goinggreen.

8. U.S. Environmental Protection Agency. Mercury Study Report to Congress, Science Advisory Board Review Draft, Vol. 2, 1996, pp. 4–19, ES-3.

9. Presentation by Dr. Dra. Adriana Grebnicoff, Ministerio de Salud del GCBA, Coordinación de Salud Ambiental, Primera Conferencia Latinoamericana sobre la Eliminación del Mercurio en el Cuidado de la Salud, Buenos Aires, August 5, 2006.

10. HCWH/CAATA. Progress Report to the Commission on Environmental Cooperation: Partnership Project to Reduce Use, Discharges and Emissions of Toxic Substances in the Healthcare Sector in Mexico, With an Emphasis on Mercury Reductions, Mexico City/Buenos Aires, July 2007.

11. A 2004 study revealed that on average, 70 thermometer breakages occur per month in a typical 300–500 bed hospital. *Lurking Menace: Mercury in the Health Care Setting,* Toxics Link, New Delhi, June 2004.

12. Pastore P, Singh R, Jain N. *Mercury in Hospital Indoor Air: Staff and Patients at Risk.* New Delhi: Toxics Link, January 2007.

13. http://www.sustainablehospitals.org/HTMLSrc/IP_mercury_amounts.html.

14. Personal communication between Rico Euripidou, GroundWork, South Africa and Louis Havinga, Manager, Health Technology Services, KwaZulu Natal Department of Health, Durban, June 20, 2007.

15. Cited in A new era: the elimination of mercury sphygmomanometers. In: *Going Green, A Resource Kit for Pollution Prevention in Health Care.* Arlington: Health Care Without Harm. Available at http://www.noharm.org/library/docs/Going_Green_2-9_A_New_Era_The_Elimination_of_M.pdf.

16. Personal communication with Ken Scott Walden, Asset Management, 2005.

17. Making Medicine Mercury Free: A 2005 Report on the Status of Virtual Mercury Elimination in the Health Care Sector. Hospitals for a Healthy Environment, Arlington, 2005. Available at http://www.h2e-online.org/docs/h2e2005MercuryReport.pdf. Also see, http://www.noharm.org/us/mercury/issue.

18. KEMI. Mercury Free Blood Pressure Measurement Equipment: Experiences in the Swedish Health Care Sector. Swedish Chemicals Inspectorate, Stockholm, 2005. Available at http://www.kemi.se/upload/Trycksaker/Pdf/PM/PM_7_05 web.pdf. Mercury In Health Care. Stay Healthy, Stop Mercury Campaign, EEN/Health Care Without Harm Europe, June 2006.

19. Ban on sale of mercury measuring instruments. European Parliament, July 7, 2007. Available at http://www.europarl.europa.eu/news/expert/infopress_page/064-8949-190-07-28-911-20070706IPR08897-09-07-2007-2007-false/default_en.htm. EU bans sale of thermometers for use in health care. Health Care Without Harm news release, July 12, 2007. Available at http://www.noharm.org/details.cfm?ID=1655&type=document.

20. Committee on Emerging and Newly Identified Health Risks, under the Directorate General for Health and Consumers of the European Commission on *Mercury Sphygmomanometers in Healthcare and the Feasibility of Alternatives* (September 23,

2009). Available at http://ec.europa.eu/health/ph_risk/commit-tees/04_scenihr/docs/scenihr_o_025.pdf.

21. WHO. Mercury in Health Care Policy. WHO/SDE/WSH/05.08, 2005. Available at http://www.who.int/water_sanitation_health/medicalwaste/mercurypolpaper.pdf.

22. www.mercuryfreehealthcare.org.

23. http://www.chem.unep.ch/MERCURY/.

24. http://www.noharm.org/lib/downloads/mercury/Argentina_Plan_Min_Exposure_Mercury.pdf.

25. http://www.noharm.org/lib/downloads/mercury/Philippines_Gradual_Phaseout_Mercury.pdf.

26. http://www.noharm.org/lib/downloads/mercury/Taiwan_Ban_Mercury_Thermometers.pdf.

27. http://www.mercuryfreehealthcare.org/SSDF.pdf.

28. http://www.toxicslink.org/pub-view.php?pubnum=244.

29. Memorandum, Re: Phasing out of mercury sphygmoman-ometers. Kwazulu-Natal Department of Health, Central Medico Technical Division, January 30, 2003; Memorandum, Re: The phasing out of products that contain mercury in health care institutions, Kwazulu-Natal Department of Health, Health Technology Unit, March 15, 2006.

30. http://www.ccss.sa.cr/html/comunicacion/noticias/2009/12/n_1100.html.

31. Research Analysis Report on Mercury Use in China 2003–2005—The Measuring Devices Industry of China, Chemical Registration Center of State Environmental Protection Administration of China (SEPA), Natural Resources Defense Council, May 2007.

32. Personal communication with Prashant Pastore, Toxics Link, India, September 3, 2007.

33. Bridgen K, Stringer R. *Atmospheric Dispersal of Mercury from the Hindustan Lever Limited Thermometer Factory, Kodai-kanal, Tamil Nadu, India, Using Lichen as a Barometer*. Greenpeace Research Laboratories, Exeter, November 2003; Karunasagar D, et al. Studies of mercury pollution in a lake due to a thermometer factory situated in a tourist resort. *Environ Pollut* 2006;143:153–158.

34. Personal communication with Nityanand Jayaraman, environmental researcher, August 25, 2007.

29

ADDRESSING SOURCES OF PCBs AND OTHER CHEMICAL POLLUTANTS IN WATER

Gretchen Welfinger-Smith and David O. Carpenter

29.1 INTRODUCTION

Chemical pollutants such as persistent organic pollutants (POPs), other organic compounds, personal care and pharmaceutical products, nitrates, heavy metals, and radioactive materials are a serious threat to water health. Water is essential for life, and a clean and uncontaminated water supply is essential for a healthy life. Dangerous water contaminants can come from natural sources, but many are a result of human activities, including industry, agriculture, military activities, urban development, and waste disposal methods.

Chemical pollutants in water pose a threat to both human and ecological health. For centuries waste of all kinds was either dumped directly into water bodies and/or was deposited in unlined pits or surface sites, where contaminants were susceptible to leaching into groundwater or streams. Contamination of water with chemicals has become so frequent and severe that action had to be taken to control, monitor, and reduce the concentrations of contaminants in water.

In the 1970s the United States and other developed nations launched a series of legislative actions acknowledging the need to prevent contamination of water through new pollution prevention regulations as well as new methods to treat and remove contaminants from both drinking and wastewater. These regulations fostered the development of new methods of treating drinking water and sanitation and wastewater treatment plant methodologies.

The need to identify water contaminants and their sources, find methods to minimize or eliminate their presence, and enact policy to implement this continues to the present.

Today there are new threats to water from chemicals, industry, and agriculture that must be further identified, studied, and regulated.

The early part of the twentieth century was a period where most developed countries made significant progress in reducing microbial contamination of water. Raw sewage was no longer allowed to be dumped directly into lakes and streams, and chlorination of drinking water became a standard treatment. These actions had immediate positive effects in reducing human morbidity and mortality for infectious disease, especially diarrhea, and resulted in great improvement in the appearance of lakes and streams. Basic water and sanitation issues persist, especially in developing, overpopulated, and conflict-stricken regions. However, the positive steps taken to rid water of infectious organisms are often ineffective in removing chemical contamination. Here we review both traditional and emerging threats to clean water, their treatment, remediation, and reduction, policies and regulations pertaining to water contamination, and society's role in maintaining a fresh water supply. While contaminants in water bodies have the ability to seriously impact ecosystems, much of our focus in this chapter will be on drinking water.

29.2 PERSISTENT ORGANIC POLLUTANTS (POPs)

POPs are organic chemicals that are persistent in both the environment and in the biological systems including the human body. POPs bioaccumulate in human and animal

Water and Sanitation-Related Diseases and the Environment: Challenges, Interventions, and Preventive Measures, First Edition.
Edited by Janine M. H. Selendy.
© 2011 Wiley-Blackwell. Published 2011 by John Wiley & Sons, Inc.

tissues, and biomagnify through the food web. Many, but not all, POPs are lipophilic and are thus found in animal fats. Others bind to proteins. Most POPs are semivolatile, and because of their volatility (even though it is limited), they undergo long-range transport and are redistributed globally. POPs may vary in water solubility, which is usually relatively low for those that are lipophilic but much higher for those that are not. POPs is a term used specifically for persistent organic compounds, but there also are other kinds of persistent chemicals. Therefore, some prefer use of the term "persistent toxic substances," which then will include things such as toxic metals. In this chapter, we will distinguish POPs from metals, and deal with each separately.

The manufacture and use of many POPs was discontinued in the late 1970s and 1980s in most developed countries. The Stockholm Convention on Persistent Organic Pollutants is a global treaty of the United Nations, which went into effect in 2004 after signature by 160 countries (still not including the United States!). The original treaty identified 12 substances (the "dirty dozen") for elimination or drastic restriction. These included 12 chlorinated pesticides (aldrin, endrin, mirex, chlordane, heptachlor, toxaphene, dieldrin, hexachlorbenzene, dichlorodiphenyltrichloroethane (DDT), polychlorinated biphenyls (PCBs), dioxins, and furans (PCDD/PCDF)). The persistence and toxicity of these compounds is in great part a consequence of the chlorines they contain. Of the 12, dioxins and furans were never intentionally produced, but are rather by-products either of chemical manufacture or products of combustion. Only DDT was listed with a specific allowance for restricted use for prevention of malaria, and that only in countries where malaria is a major public health problem.

In 2009, nine additional POPs were added to the Stockholm Convention (1). These include chlordecone, lindane, α-hexachlorocyclohexane, β-hexachlorocyclohexane (all chlorinated pesticides), and pentachlorobenzene (with many industrial uses), hexabromobiphenyl, tetra- and pentabromodiphenylether, hexa- and heptabromodiphenylether (all brominated flame retardants), and perfluorooctane sulfonic acid, its salts and perfluorooctane sulfonyl fluoride (used in polymers and for water-repellant purposes). All of these are lipophilic, except for the perfluorinated compounds, which bind to proteins. Exemptions were made for use of lindane for treatment of head lice and scabies, for the brominated biphenyl ethers for purposes of recycling, and for use of the perfluorinated compounds for some critical production purposes.

Restriction of manufacture and use of individual POPs was most often a consequence of demonstration of persistence even before there was clear evidence of adverse health effects. Further understanding of how dangerous these chemicals are continues, especially for the newer POPs that have not been studied for so long. But in addition to sharing the properties of being persistent and bioaccumulative,

individual POPs have many common health effects. While only dioxin is rated as a proven human carcinogen by both the United States Environmental Protection Agency (2) and World Health Organization (WHO) (3), chlordane, DDT, heptachlor, hexachlorobenzene, lindane, mirex, toxaphene, and PCBs and most other POPs are all rated as probable human carcinogens, based on proof that they are carcinogenic to animals and evidence in humans is consistent with this conclusion. In addition, many of these POPs have adverse effects on other organ systems. Table 29.1 presents a summary of organ systems affected by the best-studied POPs, derived from information from IRIS (4), USEPA (1), IARC (3), and ATSDR (5–7).

In addition to sharing properties of persistence and toxicity, and because these substances have become global pollutants, they are often found together. This is particularly true in large bodies of water that are repositories of global as well as local pollution sources. Figure 29.1 shows the correlation among 11 POP concentrations obtained in a large sample of wild and farmed salmon from around the world (8). Each point in this figure represents levels in a single composite of fish. The figure and the analysis that accompanied this figure show clearly that there was a statistically significant relationship among levels of each of the contaminants in salmon. That is to say, a fish high in one POP was also high in each of the other POPs that were monitored. The source of the contaminants came from the oceans, either from the natural diet of wild fish or from the fish meal/fish oil fed to the farmed salmon. The analysis shows that fish from northern European waters were the most contaminated, and this is the area that has been industrialized the longest. Farmed fish from North America are less contaminated, and those from Chile the least contaminated, again reflecting the history of industrialization. Wild fish from the North Pacific are less contaminated than those that are farmed, probably because the wild fish feed lower on the food chain, but the pattern of contaminants was the same even though the levels were lower. This study shows how important water is in being a reservoir for POPs and how this pollution of water directly impacts human health, in this case in the form of POPs contamination of an important human food. The fish were also found to have high levels of the polybrominated diphenylethers (PBDEs) (9), persistent substances for which there is less information currently available regarding their human health hazards (http://www.who.int/entity/ipcs/publications/hsg/en/index.html).

Dioxin and PCBs deserve some special consideration, both because they have been better studied than most of the other POPs and because of their high toxicity. Dioxins are sometimes considered to be the most toxic substance known to man. They are persistent unwanted by-products of combustion, and also by-products of manufacture of other chemicals, such as the herbicide mixture Agent Orange, used in Vietnam. Dioxins and furans form photochemically from

TABLE 29.1 **Health Effects of the Major POPs**

Contaminant RfD (mg/(kg d))	Immune Suppression	IQ/CNS Toxicity	Fetotoxicity	Reproductive Effects	Musculo-Skeletal Toxicity	Liver Toxicity	Kidney Toxicity	CV/Blood Toxicity	Endocrine Dysfunction
Chlordane (5×10^{-4})	X	X				X		X	X
DDT/DDE (5×10^{-4})	X	X	X	X		X		X	X
Dieldrin (5×10^{-5})	X	X	X	X	X	X	X	X	X
Dioxin ($-$)	X	X	X	X	X	X		X	X
Endrin (3×10^{-4})		X		X	X				X
HCB (8×10^{-4})	X	X	X		X	X	X	X	X
Heptachlor (1.3×10^{-5})		X		X	X	X		X	
Lindane (3×10^{-4})	X	X	X	X		X	X		
Mirex (2×10^{-4})			X	X	X	X	X	X	X
PCBs (2×10^{-5})	X	X	X	X		X	X	X	X
Toxaphene (2.5×10^{-4})	X	X				X	X		X

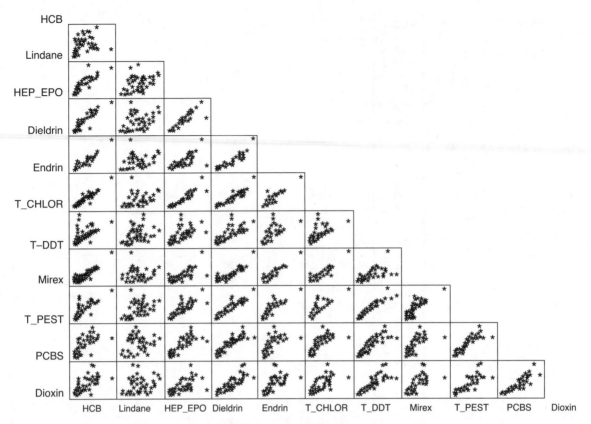

FIGURE 29.1   Scatter-plot matrix for the content of pairs of 10 contaminants and total pesticides measured in salmon samples from various locations where the salmon were produced or purchased. HEP_EPO, heptachlor; T_CHLOR, total chlordane; T_DDT, total DDT; T_PEST, total pesticide. From Ref. 8. Reproduced with permission.

precursors at room temperature by ultraviolet irradiation (10) and enzymatically (11). They are also formed during the degradation of unburned carbon species in the presence of chlorine (12) and during the combustion of other POPs (13). Dioxins form in the incineration of municipal solid waste, hospital waste, hazardous waste, and sewage sludge and through combustion processes in cement kilns, wood burning, diesel vehicles, crematoria, and coal-fired utilities (14). Table 29.2 shows sources of dioxins in the United States at three time points, and indicates how effective regulations through the Clean Air Act have been in promoting reductions in dioxin release, especially into the air. The current major source is backyard burn barrels. But since the majority of dioxin emissions are into air, this is a clear pathway for deposition into water. Applying filters, scrubbers, and activated carbon to treat incineration and combustion emissions are all effective mechanisms in collecting dioxins and preventing release into air. Dioxins have many adverse health effects both to fish and wildlife and to humans. Dioxin and dioxin-like chemicals were responsible for the almost total disappearance of lake trout from the U.S. Great Lakes in the past because of their great early life-stage toxicity (15). A number of human diseases are associated with dioxin

exposure, including cancer (16), cardiovascular disease (17), hypertension, diabetes, and chronic respiratory disease (18), birth defects (19), suppression of the immune system (20), and decrements of cognitive function (21, 22).

Unlike dioxins, PCBs were manufactured throughout the world from the late 1920s until relatively recently, and had a large number of useful purposes ranging from electrical insulators to hydraulic fluids to solvents for paints and window caulking. PCBs are a mixture of up to 209 different congeners consisting of various degrees of chlorine substitution around the biphenyl rings. PCBs have high boiling points and are relatively nonflammable, making them ideal for use as insulators and cooling fluids for transformers and capacitors. PCBs are transported globally in water, sediment, air, and biota.

Some PCBs have dioxin-like activity through binding to a cytosolic receptor that is a member of the steroid family of hormone receptors, known as the aryl-hydrocarbon receptor (AhR). Activation of the AhR results in gene induction, with up- and downregulation of many different genes (23). This gene induction is almost certainly the basis of the major health effects of dioxins. Since these dioxin-like PCBs act by the same mechanism as dioxin, everything that dioxins do

TABLE 29.2 Inventory of Sources of Dioxin in the United States, May 2000 (from EPA)

Inventory of Sources of Dioxin in the United States-May, 2000	1987 Emissions (g TEQdf-WHO98fyr)	1995 Emissions (g TEQdf-WHO98/yr)	2002/4 Emissions (g TEQdf-WHO98/yr)
Municipal Solid Waste Incineration, air	8877.0	1250.0	12.0
Backyard Barrel Burning, air	604.0	628.0	628.0
Medical Waste Incineration, air	2590.0	488.0	7.0
Secondary Copper Smelting, air	983.0	271.0	5.0
Cement Kilns (haz. waste), air	117.8	156.1	7.7
Sewage Sludge/land applied, land	76.6	76.6	76.6
Residential Wood Burning, air	89.6	62.8	62.8
Coal-fired Utilities, air	50.8	60.1	60.1
Diesel Trucks, air	27.8	35.5	35.5
Secondary Aluminum Smelting, air	16.3	29.1	29.1
2,4-D, land	33.4	28.9	28.9
Iron Ore Sintering, air	32.7	28.0	28.0
Industrial Wood Burning, air	26.4	27.6	27.6
Bleached Pulp and Paper Mills, water	356.0	19.5	12.0
Cement Kilns (nonhaz. waste), air	13.7	17.8	17.8
Sewage Sludge Incineration, air	6.1	14.8	14.8
EDC/Vinyl chloride, air	NA	11.2	11.2
Oil-fired Utilities, air	17.8	10.7	10.7
Crematoria, air	5.5	9.1	9.1
Unleaded Gasoline, air	3.6	5.9	5.9
Hazardous Waste Incineration, air	5.0	5.8	3.5
Lightweight ag. kilns, haz. waste, air	2.4	3.3	0.4
Kraft Black Liquor Boilers, air	2.0	2.3	2.3
Petrol Refine Catalyst Reg., air	2.2	2.2	2.2
Leaded Gasoline, air	37.5	2.0	2.0
Secondary Lead Smelting, air	1.2	1.7	1.7
Paper Mill Sludge, land	14.1	1.4	1.4
Cigarette Smoke, air	1.0	0.8	0.8
EDC/Vinyl chloride, land	NA	0.7	0.7
EDC/Vinyl chloride, water	NA	0.4	0.4
Boilers/industrial furnaces, air	0.8	0.4	0.4
Tire Combustion, air	0.1	0.1	0.1
Drum Reclamation, air	0.1	0.1	0.1
Totals	13,995	3,252	1,106
Percent Reduction from 1987		77%	92%

PCBs also do. However, the majority of PCB congeners are not dioxin-like, and act via a variety of other mechanisms, some of which involve gene induction through different pathways, while other congeners act in other ways. There is some evidence that more of the carcinogenic activity of PCBs resides with the non-dioxin-like than dioxin-like congeners (24). In addition to being carcinogens, PCBs suppress immune system function, alter thyroid and sex steroid endocrine systems, increase risk of diabetes and cardiovascular disease, and cause reductions in IQ and altered behavior (25).

The primary source of human POPs exposure is through ingestion, although inhalation and dermal absorption are also important. Because of the global nature of the spread of POPs, almost all animal fats contain some of the lipophilic POPs. Fish bioaccumulate POPs from the aquatic environ-

ment. Domestic animals contain POPs because of being fed waste animal fats and fish oils (26), as well as from contamination of animal foods from air transport and other uses in agriculture. Inuits and Native Americans whose diet includes high fish and marine mammal consumption also have high serum and breast milk POPs levels that have been attributed primarily to their diets (27–30). This is a concern for anyone consuming a high fish diet since POPs are distributed in aquatic systems globally (31, 32).

29.2.1 Other Contaminants in Water that Are of Major Public Health Concern

While POPs are particularly dangerous because of both their persistence and toxicity, there are a number of other chemical contaminants of concern in water. Some, such as metals, are also persistent. Contamination of waterways with metals is a particular problem in areas near to leather production, which uses many toxic metals in the process of tanning. Most of the metal released (chromium, arsenic) remain primarily in the sediments.

Organo-metals, such as methylmercury and organotin compounds, are of particular concern because they bioaccumulate in fish and are particularly toxic. Organotin compounds are widely used as anti-fouling treatments in marine paints, and leach into the water, and bioaccumulate in the food chain. They have adverse effects both on aquatic species and mammals, causing immune system dysfunction (33) and a variety of toxic actions on the nervous system (34, 35). Exposure of humans to organotin compounds comes almost exclusively through consumption of seafood (36). Farmed and wild fish from all over the world have been found to contain organotin compounds (37, 38), but little study has been done to determine health effects in humans.

Mercury is a particular problem related to water because of the fact that both elemental mercury, which is atmospherically transported, and inorganic mercury, which often leaches from rock into bodies of water and is also sometimes atmospherically transported bound to particulate matter, can be changed into methylmercury by aquatic microorganisms. Mercury in the past was also frequently discharged directly into streams (39), leading to significant amounts in sediments in major bodies of water. The only long-term sink for mercury in the biosphere is deep ocean sediments (40). Once methylmercury is formed it, like POPs, bioaccumulates in both freshwater and marine fish and poses significant adverse health effects to wildlife and humans who eat the fish.

Methylmercury moves up the aquatic food web to cause mortality, reproductive failure and other health effects in animals that eat fish (41). The bioaccumulation potential is significant, resulting in fish tissues having over 10^6 times higher methylmercury concentrations than ambient water (42). Consumption of fish containing methylmercury poses significant risk to human health, especially to the developing fetus. USEPA (43) advises no consumption of swordfish, shark, tilefish, and king mackerel because of high levels of methylmercury. The U.S. National Academy of Sciences (44) issued a report on methylmercury and concluded that at present in the United States 10% of women of reproductive age exceed the reference dose for methylmercury, and that this number is 50% of women who eat fish on any given day. The problem is serious for both fresh water and ocean fish, with most of the mercury in inland lakes and watersheds coming from combustion point sources by atmospheric transport. A child's IQ is reduced by prenatal or early life exposure to methylmercury (45). Methylmercury exposure in adults also causes serious disease, including an elevated risk of cardiovascular disease (46) and a series of nervous system and behavioral abnormalities (47). The presence of both methylmercury and POPs in fish

raises serious questions about the safety of consuming many species of fish in spite of the presence of healthy omega-3 fatty acids in seafood (48).

Many other organics do not have the same persistence as POPs or metals, but lack of persistence does not necessarily mean lack of toxicity or even lower toxicity. Contaminants of this sort include other kinds of pesticides and agricultural chemicals, solvents, polyaromatic hydrocarbons (PAHs), disinfection by-products, and pharmaceuticals. The toxicities of these chemicals will be discussed in the context of the sources of the substances to water bodies.

29.2.2 Sources of Contamination of Water Bodies

29.2.2.1 Wastewater   While wastewater treatment plants are effective in removing pathogens, sediment, and some contaminants, they are ineffective in removing many pharmaceuticals, personal care products, and household chemicals. Flushing of unused and out-of-date prescriptions is a source of antibiotics in water systems (49), and is pretty standard practice. The largest source of pharmaceuticals and their metabolites in wastewater is from human excretion in urine or feces. But personal care products used on the skin also get washed down the drain, as do detergents and chemicals used in the home and in industry, such as solvents and degreasers. Many water treatment plants were never designed to remove anthropogenic chemicals (50). While the waste resulting from the production of personal care products is regulated by the U.S. EPA, the disposal of only a few of these chemicals, themselves, is regulated (51). Therefore, all of these chemicals are regularly detected in stream and raw-water supplies. Fragrance compounds, bisphenol A, caffeine, carbamazepine (an anticonvulsant), cotinine (a metabolite of nicotine), flame retardants, plasticizers, and trichloroethylene and tetrachloroethylene (solvents used for cleaning metals and dry cleaning) are also frequently found in drinking water that has undergone drinking-water treatment consisting of treatment with either bituminous or lignite granular activated carbon or anthracite to retain fine solids and bacteria. These facts demonstrate that conventional water treatment methods are ineffective in removing all pharmaceuticals and chemicals, especially the hydrophobic compounds (52).

The U.S. Geological Survey investigated 139 streams in the United States, and found over 80 chemicals, with one or more present in 80% of streams sampled (53). The chemicals included antibiotics, various prescription and nonprescription drugs, benzo-a-pyrene, phthalates, pesticides, detergents and their metabolites, and caffeine. Most did not exceed drinking water standards. Similar results have been seen in Taiwan (54) and in the surface water used for drinking water for New York City (55). In a study examining the presence of 113 organic compounds in the source water of a drinking water treatment facility, over 75% of the 72 samples analyzed contained polycyclic musk fragrances, pharmaceuticals and their degradation products, insect repellants, flame retardants, plasticizers, and PAHs (56). Finished drinking-water samples from this study all contained up to 21 compounds at detectable concentrations, with all of the samples containing carbamazepine and N,N-diethyl-*meta*-toluamide (DEET). Ghosh et al. (57) have even reported the presence of an active Tamiflu metabolite in river water, suggesting the possibility of the development of viral as well as bacterial resistance as a consequence of antibiotic discharge into wastewater.

In the United States and other countries hospital effluent has been cited as being a significant source of antibiotics in wastewaters. Even iopromide, the iodinated X-ray contrast agent, has been attributed to hospital effluent (58). Brown et al. (59) found hospitals to likely be the primary contributors of antibiotics to municipal wastewaters in New Mexico. The same study showed that wastewater treatment was able to remove 8 of 11 antibiotics that were analyzed, with removal efficiencies of the remaining 3 (sulfamethoxazole, trimethoprim, and ofloxacin) ranging between 20% and 77%. The Rio Grande has tested positive for antibiotic-resistant bacteria (60), likely as a consequence of the presence of antibiotics. In addition to being associated with the development of antibiotic-resistant bacteria, antibiotics in wastewaters have been shown to destroy the microbial populations employed in water treatment plants (61).

Pesticides are another contaminant commonly found in wastewater. In addition to agricultural runoff (discussed later), pesticides originate from an array of sources including applications to lawns and wooded areas, treated building materials, and indoor applications eventually ending up in the wastewater. Conventional municipal treatment plants are not effective at totally removing pesticides (62, 63). Additional treatments like activated carbon have been demonstrated to effectively remove degradation products of the herbicide, chloroacetamide, while others, like ozonation, remove herbicidal degradation products but create other contaminants (63).

Most POPs are not very water soluble, but in contaminated bodies of water the levels of PCBs in the soluble phase can be significant. The solubility of PCB congeners decreases with the degree of chlorination, is greater for *ortho*-substituted congeners than non-*ortho* congeners with the same number of chlorines, and increases exponentially with temperature (64). While most PCBs in water will be bound to particulates that will be removed in the filtration steps, these soluble PCBs will not be removed by water treatment processes other than by activated charcoal or other more expensive methods. Even use of these processes may not remove them completely. We, for example, analyzed for total PCBs in Hudson River water prior to the onset of dredging, and found a concentration of 340 ppt. However, the drinking water taken from a water fountain in the town hall of a village

that derived drinking water from the Hudson had a level of 110 ppt (Welfinger-Smith and Carpenter, unpublished). This level is in spite of the report that the village uses activated charcoal filtration, which is obviously not very effective in removing PCBs. This is to be compared with the USEPA action level of 500 ppt (http://www.epa.gov/ogwdw000/pdfs/factsheets/soc/pcbs.pdf). While this concentration is not terribly high, if one drinks this water every day, it does pose a risk of significant exposure. There is also a risk of inhalation exposure when bathing and cleaning with the contaminated water.

A significant body of evidence indicates adverse effects of wastewater effluents on fish and wildlife. Most of the biological effects are caused by estrogenic substances, which cause feminization and demasculinization of fish, amphibians, and reptiles (65). These actions result in altered morphology of reproductive organs, reduced fertility, especially of males and altered levels of sex hormones. In addition, sex-specific behaviors are altered. In addition to estrogenic chemicals that cause reproductive failure in fish (66), many chemicals in wastewater are also anti-androgenic (67), which further adds to the overall feminization. Sewage effluent has been found to cause synthesis of vitellogenin, the egg protein, in male fish (68), and this is accompanied by reduced reproductive success. In certain situations, such as effluent from a pulp mill, even the sex ratio of fish embryos became male-biased due to releases of androgenic substances. Synergistic effects of estrogenic chemicals has been observed in other species (69, 70), and it has also been demonstrated that chemicals behave differently in mixtures than they do individually (71), indicating the potential for both unexpected and synergetic effects in humans when exposed to these mixtures.

Most of the alterations seen in fish and aquatic wild life involve alteration of sex hormonal status, as these are gross anatomic changes that are easily detected. However, fish exposed to environmental estrogens also show genotoxic damage, modulated immune function, and altered metabolism (72). A meta-analysis of the effects of the herbicide, atrazine, on freshwater fish and amphibians documented no direct mortality, but reduced growth, altered metamorphosis, altered immunity, and increased infections and gonadal abnormalities (73).

While the evidence for ecologic alterations secondary to chemicals in wastewater is convincing, there is less certainty about the degree of hazard to humans. Exposure to humans will come from drinking water, not sewage effluent. However, when the drinking-water supply comes from surface water, there will be some level of the same chemicals that are not removed either by dilution or by the treatment processes. Groundwater also may contain these chemicals, although usually at lower concentrations.

Several categories of pharmaceuticals and chemicals in wastewater pose particular concern for humans. This is in spite of the fact that to date there has not been clear proof that exposure to these substances through drinking water has caused human disease. Antibiotics are a problem because of the important issue of antibiotic resistance. Overuse of antibiotics by humans and in agriculture results in the release of antibiotics into the environment, and this promotes the development of antibiotic resistance in bacteria. While the pathway is not direct, the end result will almost certainly be that strains of infectious organisms develop that are capable of causing human disease but are not sensitive to antibiotics. The potential adverse health effects of antibiotic resistance are enormous.

The estrogenic, anti-estrogenic, androgenic, and anti-androgenic compounds are also of great concern to human health. There is strong evidence that endocrine disruptive chemicals alter human physiology, but less evidence that the route of exposure is via drinking water. Anti-androgenic pesticides given to rats cause reproductive malformations (74). Prenatal phthalate exposure causes both anatomic and behavioral feminization of baby boys (75), and in adults has been associated with obesity and insulin resistance (76). In rats perinatal exposure to bisphenol A causes an increase in prostate size (77) and increased risk of prostate cancer development (78). Bisphenol A is one chemical that supports use of municipal drinking water, because the major source to humans is bottled water in polycarbonate bottles (79). Because human sexual structure and function is very dependent upon the ratio of estrogenic to androgenic hormonal balances, there is every reason to be concerned about chemicals in drinking water that may disrupt this balance, especially during fetal development, even if they are present at low concentrations (80).

Another developing area of concern is the increasing use of nanoparticles. While nanotechnology is an exciting new field, these very small particles will easily pass through water treatment facilities (81), and the health effects are presently unknown (82–84). Many of the active agents in suntan lotions, cosmetics, hair care products, and pet care supplies used for protecting against ultraviolet radiation are nanoparticles. Although these and the other chemicals found in such lotions are frequently found in wastewater, they are not regulated, nor has the human and ecological toxicity been determined for many. As the field of nanoscience continues to develop it is likely that this problem will only grow.

29.2.2.2   Combined Sewer Overflow and Urban Runoff
Urban and storm water runoff originates from a variety of sources including pavement, automobiles, airports, industry, buildings and rooftops, and at times agriculture. While much of urban and storm water runoff is captured by the wastewater treatment process, an increasing problem is combined sewer overflow. Sewage system overflows are the most frequent violators of the Clean Water Act (84). These events have the

potential to release large concentrations of contaminants into surface waters. Combined sewer overflows occur when the capacity of a treatment plant is exceeded by storm runoff from heavy rainfall events. Thus, the subject of urban and storm water runoff deserves attention.

Urban runoff contains sediment, debris, chemicals, heavy metals, pathogens, road salts, oils and grease, and nutrients. The combination of a decrease in vegetative areas and a rise in impermeable surfaces significantly increases runoff in urban environments compared to rural environments. On average, a typical city block generates over five times the runoff as a woodland area of identical size (85). Runoff is a significant source of heavy metals to waterways (86) with the heavy metals in urban storm water runoff resulting from atmospheric deposition, vehicle emissions, and from the dissolution of building materials like concrete, galvanized materials, brick, and tile (87). Urban and storm water runoff accounted for the impairment of 22,559 miles of U.S. streams and rivers and 701,024 acres of lakes (88).

Rooftop runoff is a major component of total urban runoff. It consists of the compounds contained in roofing materials and the airborne pollutants and organic substances added to roofs via interception and deposition including chemicals, organics, and biological contaminants and debris like leaves, dead insects, and bird wastes (89, 90). In addition, because of direct solar radiation and lack of shade, roofs have higher temperatures that can result in accelerated chemical reactions and the organic decomposition of materials on the rooftops (89, 91). Rooftops may be significant sources of heavy metals such as cadmium, lead, and zinc (92) and PAHs (93).

Vegetated roofs are intended to increase greenspace, mitigate the urban heat island effect, and reduce urban runoff. Under the right conditions they can behave as a sink for some atmospherically deposited pollutants. The roofs range from simple lawns to complex gardens with bushes and small trees. Despite their environmentally friendly qualities, these roofs have also been demonstrated to be sources of runoff pollution. Phosphorus, dissolved organic carbon, and nutrients and metals like potassium, calcium, manganese, zinc, and copper were all found in runoff from vegetative roofs (94). The contaminants varied on a case-by-case basis and can likely be attributed to the gardening practices and roof setups, which can be modified to reduce runoff. When properly constructed to minimize runoff, however, vegetative rooftops are very beneficial to the environment.

Contaminants from construction site runoff have impaired approximately 5% of U.S. surface waters (95). Erosion control techniques for construction include exposing minimal land for minimal time, especially during peak runoff periods, using mulches and fast-growing cover vegetation, retaining existing trees and shrubs when possible, roughening the soil surface or constructing berms to reduce runoff velocity, and the construction of retention basins to encourage the settling out of suspended sediment (95).

Methods to reduce pollution from urban runoff include frequent street-sweeping, proper disposal of pet wastes, litter control, the application of organic mulches to reduce erosion on bare ground, incorporating more green space within urban areas along waterways, the use of surface ponds, holding tanks, rooftop catchments, subsurface tunnels or similar structures to hold, retain, and gradually release storm water, using judicious choices in building materials, and the careful and limited application of pesticides, fertilizers, road salt, and sand (92, 95). Sustainable drainage systems are another option for minimizing the contamination of water from urban runoff. They employ a variety of techniques including swales, filter strips, wetlands, and ponds to harvest, store, treat, and recycle runoff following a premise of acknowledging the use of vegetative structures as an important part of the storm water system and the surface management of as much storm water as possible (96).

Permeable pavement systems are a type of sustainable drainage system that has proven successful in reducing urban runoff. The general principle is to collect, treat, and infiltrate surface runoff to support groundwater recharge, while reducing runoff and saving water through recycling and pollution prevention (97, 98). Permeable pavement has been demonstrated to reduce or remove metals such as lead and copper, diesel fuel, and motor oil (99). It has even been proven successful in reducing pollutants on fine-grained soils, like clays, during small storm events (100). However, because they are not as effective on fine-grained soils during large storm events, it is sensible to use permeable pavement systems in conjunction with other approaches.

Typical methods used to reduce storm water pollutant loads include gross pollutant traps, sedimentation ponds, infiltration, artificial wetlands, and source controls, which are employed based on the best management practice approach. While these meet regulatory requirements for prescribed pollutants on a site-specific basis, the site-specific basis approach is inadequate for achieving real improvements in downstream water quality on a catchment-wide basis (101). Reducing pollution from runoff on a catchment-wide basis is the best method for decreasing contamination, and the most cost effective. This methodology should be further researched, funded, and implemented as a management practice in order to successfully mitigate water contamination from urban and storm water runoff. Urban governments can also enact city-specific legislation and requirements to reduce runoff. For example, New York City has instituted zoning laws requiring new parking lots to include landscaped areas to absorb rainwater, has established a tax credit for roofs with absorbent vegetation, and is spending millions of dollars on environmentally friendly infrastructure projects, while Philadelphia plans to spend $1.6 billion over 20 years to build rain gardens and sidewalks of porous pavement and plant thousands of trees (84). With such an array of options available to alleviate pollution from urban runoff, there is little excuse for local governments not implementing at least some form of runoff reduction methodology.

29.2.2.3 Groundwater

While groundwater is usually less contaminated with chemicals and rarely is very contaminated with microbiological agents, groundwater is very vulnerable to contamination with metals and other naturally occurring toxic substances found in rock formations. Perhaps the most serious human health problem coming from groundwater is contamination with arsenic. Natural arsenic in groundwater above levels of $10\,\mu g/L$ (the current U.S. standard for municipal drinking water) is common, and is a particular problem in many parts of the world, especially in Bangladesh, India, Vietnam, and parts of South America (102). In New Hampshire, where wells are the primary water source for 40% of the population, 10% of the wells contain arsenic levels greater than $10\,\mu g/L$ (103). While municipal water supplies are now required to reduce arsenic levels below $10\,\mu g/L$, it is obvious that people using well water are going to be exposed. Arsenic is a very dangerous metal, causing cancer (104), cardiovascular disease (105), and neurotoxicity (106).

Groundwater is also frequently contaminated with pesticides and other organic chemicals such as chlorinated solvents that have leached through the soil. In less urban areas where septic systems are common, the groundwater can also be contaminated with pharmaceuticals (107). This may result in significant human exposure, because areas with septic systems often are also areas where drinking water comes from wells without treatment (108, 109). Underground storage tanks are another common source for groundwater contamination. Underground storage tanks are used to store gasoline at gas stations, oil for heating (residential and industrial), petroleum at petroleum distribution points, jet fuel at airports and military facilities, and for industrial storage. These tanks are subject to corrosion and fracture during geologic events, resulting in the release of their contents into subsurface soils, frequently leaching into groundwater. Perchlorate, used as an oxidant in explosives and aerospace fuels, is very water soluble, and frequently contaminates groundwater. Because perchlorate blocks iodine uptake into the thyroid gland, its presence in drinking water is of concern (110).

29.2.2.4 Atmospheric Transport

Contaminants can be transported into surface water both directly from the vapor phase and through rainwater. Rainwater has been found to contain a variety of contaminants including pesticides, herbicides, metals, radioactive compounds, and polar compounds photochemically formed in the atmosphere like acetic acids and nitrophenols (111). Atmospheric deposition affects every body of water.

Atmospheric transport is an important route by which POPs enter water bodies. POPs, other pesticides, and PAHs are semivolatile compounds that are deposited into water bodies from the air (112, 113), and also volatize from water into the air (114). Ackerman et al. (115) found that concentrations of PCBs, PBDEs, pesticides, and some PAHs in fish from remote lakes in U.S. National Parks were in some cases higher than those found in non-park lakes and even Pacific Ocean salmon. The source of these chemicals is atmospheric transport. For the Great Lakes studies have found that while the net fluxes of α and γ hexachlorocyclohexane are into the lakes, the net flux of PCBs is out of the lakes (116). These observations are important indicators of why POPs are global pollutants, constantly moving from land to air to water and back again. This pattern also explains why the polar regions are so contaminated, because contaminants in air come out of the vapor phase in the cold polar environment but do not move back into the vapor phase as readily as in warmer climates. We have found very elevated serum levels of PCBs in Siberian Yupiks from St. Lawrence Island, Alaska (27). This is a remote island with no local sources of contamination beyond some former military sites, but the diet of these people consists of marine mammals and fish contaminated primarily by atmospheric transport of POPs. This also suggests that the problem of global transport will only get worse with global warming that will increase the volatilization of POPs.

Atmospheric mercury is a global pollutant and is transported in a vapor phase by air currents over very long distances (117). It is estimated in the United States that 97% of total anthropogenic mercury emissions originate from combustion and industrial sources (42). Coal combustion is a significant contributor to the total mercury emitted in the combustion of fossil fuels. In China over 50 million tons of coals were combusted annually for industrial and domestic purposes, most without cleaning or flue gas controls (118). Discussion of the health effects resulting from seafood contamination with methylmercury was presented earlier. Mercury remains one of the contaminants of greatest concern from atmospheric transport.

The atmospheric deposition of nitrogen is also a concern. Large-scale animal operations contribute to atmospheric nitrogen through the release of ammonia, which is in turn redeposited into water bodies acting as a source of nitrogen pollution. For example, in North Carolina, a state with a significant percentage of the U.S. poultry and swine concentrated animal feeding operations, 50% of the state's atmospheric nitrogen deposition comes from the associated waste lagoons (119). Aquatic nitrogen is an environmental concern as well as threat to human health, which is later discussed in this chapter.

29.2.2.5 Agriculture Runoff
The growth and development of society in recent decades cannot be separated from the improvements in agriculture and modern food production technologies. But these improvements have been made by extensive use of herbicides, pesticides, fungicides, and fertilizers. Unfortunately these chemicals, as well as animal wastes, have not stayed on the farm, but have leached into water bodies causing both chemical contamination and eutrophication because of the excessive phosphates and nitrogen. Agricultural runoff, containing chemicals, metals, excess nutrients, sediment, antibiotics, and fecal matter, is the leading nonpoint source of water pollution of U.S. rivers and lakes, the second largest source of impairments to wetlands, and is a major contributor to contamination of surveyed estuaries and groundwater (120).

Runoff in agricultural areas contains a variety of pesticides that are then transferred to groundwater without any significant decrease in concentrations, especially in areas with more permeable soils, which allow for rapid infiltration (121). These pesticides also may contaminate surface water. Pesticides are highly mobile and persistent, and may leach to subsurface drains, which are important contributors to surface water pesticide levels (122). Agricultural irrigation has played an important role in agricultural water pollution. Since the 1950s global irrigated agriculture has grown 174%, accounting for ~90% of global water consumption (123). Irrigation has contributed to lowering water tables resulting in water salinization and the leaching of fertilizer into aquifers (124).

Large agricultural operations that congregate and confine animals and generally carry out all phases of the production process (feeding, manuring, slaughtering, processing) on a small land area are known as concentrated animal feeding operations (CAFOs). The estimated 376,000 CAFOs in the United States generate about 28 billion pounds of manure each year (2). The manure is often stored in lagoons that are required to be lined to prevent or minimize leakage of wastes into groundwater. The manure is often eventually applied to agricultural fields, and may leach off the fields during heavy rains.

Antibiotic use in agriculture is a major problem, especially because of its use in CAFOs. Veterinary pharmaceuticals are sometimes used to treat and prevent infectious diseases, manage reproductive processes and production, and control noninfectious diseases (125). However, antibiotics are routinely administered to promote growth rather than to fight infectious disease. Drugs are administered to animals through feed or water, injection, implant, drench, paste, orally, topically, or as a bolus (49). Following treatment, livestock will excrete 50–90% of the administered dose, the parent drug accounting for 9–30%, and the rest being metabolites (126). Sulfonamides are one of the most widely used antibiotics in both humans and animals and have low chelating abilities, and a low tendency to sorb to soils making them mobile in the environment, and capable of contaminating groundwater (126). They are only partially removed in wastewater treatment plants at a rate of about 60% (127).

Manure from animals administered antibiotics has been found to contain antibiotic compounds, which can be transported to surface, subsurface, and groundwater when the manure is applied to agricultural crops (49). This intensive use of antibiotics promotes the development of antibiotic resistance both on the farm and in the aquatic environments into which the antibiotics move.

Excessive fertilizer use results in the mobility of excess nutrients, like nitrates and phosphates, into aquatic environments, including drinking-water sources. In addition to causing eutrophication, nitrates in drinking water have been linked to the disease; methemoglobinemia (blue baby syndrome) (128); reproductive and developmental effects (129); bladder, ovarian, uterine, and rectal cancers (130); and thyroid dysfunction (131). Because nitrates are highly water soluble, they have great potential to leach into groundwater (126). Nitrates can also enter surface waters through runoff and erosion.

A variety of steps can be taken to reduce water pollution from agricultural activities. Best management practices are a combination of methods intended to reduce the transport and loss of sediment, nutrients, pesticides, and other pollutants from agricultural lands to water bodies, while reducing the need for anthropogenic chemicals and fertilizers in farming. Management practices includes techniques to catch and filter runoff, often using noncrop vegetation like vegetated field boarder strips, grassed waterways, riparian buffers, and constructed wetlands. No-till agriculture, drainage pipes, and cover crops are other examples of management practices designed to reduce the occurrence of runoff. Often a combination of practices proves to be the most successful means of pollutant reduction. For example, the use of winter weeds with control pipes (used to channel runoff, reducing erosion) was proved to be a cost-effective supplemental best management practice for no-till systems in the Mississippi Delta (132).

It is possible for farmers and ranchers to reduce erosion and sedimentation from 20% to 90% by applying management practices that control the volume and flow rate of runoff water, keep the soil in place, and reduce soil transport (120). Reducing sedimentation also reduces the transport of soil-bonding contaminants, especially those that bond readily to organic matter like many POPs and metals.

Pesticides applied in agriculture frequently mobilize and can leach into groundwater, run off into surface water, or volatilize into the atmosphere to be redeposited globally. One possible prevention method is to adsorb the pesticides prior to mobilization into water by building a barrier of soil mixed with small amounts of natural organic substrates like peanut shells and avocado stones, which will in turn retain the pesticides. For example, 0.5 g of date stones was able to remove over 90% of a variety of the pesticides, atrazine, simazine, and dieldrin, from 50 mL of a pesticide solution (133). Spray drift from pesticide application can be reduced by decreasing exposure to wind with vegetative or artificial windbreaks and the application of drift-reducing technologies like drift-reducing nozzles, spray additives to coarsen the droplet size distribution, and shielded and band sprayers (134). Incorporating the pesticide into the soil can reduce pesticide volatilization (135). Fungicide concentrations in municipal wastewater treatment plants have been found to be elevated during pesticide application periods, which can often be attributed to the improper application of pesticides (62). The use of pesticides should be limited and replaced with nontoxic alternatives whenever possible.

Manure is a significant source of agricultural pollution, but simple best management practices like storing manure in lagoons, spreading manure on fields in the spring rather than the winter, and alterations in livestock diets have proved successful at reducing contaminant loads. The dairy industry generates as much as a million gallons of waste daily (136), which is equivalent of approximately 120 lbs of wet manure per day per cow, the same amount of waste excrement generated by 20–40 people (http://www.epa.gov/region09/animalwaste/problem.html). Simply reducing the amount of phosphorous in diets of dairy cows has been proven to reduce the concentration of manure phosphorous, reducing the land area required to disperse the manure to meet phosphorous output regulations and lessening the need for storage (137). The more management practices are employed the less the risk is of water contamination. Organic farming practices, small-scale animal and agricultural operations, and good land management will also reduce the pollution resulting from agriculture.

29.2.2.6 Hazardous Waste and Military Sites

Hazardous waste sites pose a threat to water quality, especially older sites that were not constructed to adequately contain wastes that then leak into groundwater. In the past many solid waste sites also contained hazardous wastes, although they lacked the necessary structural designs to properly store these materials. Many landfill sites were not lined, did not have leachate collection systems, and therefore became a source of groundwater contamination. This poses particular problems in areas where nearby residents obtain drinking water from wells, which is common in rural and out-of-town areas where urban landfills are often located. The leachate can also contaminate surface waters. Numerous studies document elevations in rates of cancer (138), congenital malformations (139, 140), and neurobehavioral effects (141) among individuals living near hazardous waste sites. However, for most of these studies it is unclear whether the major route of exposure is via water or air. The agricultural use of leachate-contaminated water is also a concern. Plant uptake of heavy metals and other contaminants has been well documented (142–144).

Military sites are a special form of hazardous waste sites and often pose a significant threat to the environment. They frequently contain many different contaminants including fuels, oils, multiple solvents, herbicides, pesticides, metals, cyanide, explosives and their metabolites, dyes, perchlorates, dioxins, furans, and polychlorinated napthalenes, which can all potentially migrate into bodies of water (145). In areas of former U.S. military sites in Vietnam, dioxins from Agent Orange are still highly prevalent in soil, water sediment, biota, and humans. There were active policies that often included the disposal of herbicides directly into water bodies or created situations for the gradual seepage of them through the soil into groundwater (146). Nuclear storage sites like the Hanford nuclear waste storage facility have also been determined to be a source of water pollution by leaking radioactive contaminants like ^{137}Cs (cesium) into soils (147, 148). This site is located on the edge of the Columbia River, the major irrigation source for the highly agricultural food-producing zone of the Columbia Basin in central Washington State. Contamination of this river has the potential to expose many people through food consumption, drinking water, and recreation.

29.2.2.7 Metal Contamination from Mining and Coal Power Plants

Mining is a major source of heavy metal contamination of surface water and groundwater, and both leads to human exposure through consumption of well water and significant additional costs to municipalities who must apply technology appropriate for removal of metals from municipal drinking water. Most metals are toxic at relatively low concentrations, and mining waste and mine tailings are often sources of contamination of water. Some mining operations pose greater problems than others, especially those containing radioactive metals such as uranium. Uranium mine tailings pose very severe health problems in Central Asian countries such as Kyrgyzstan (149), as well as in the Four Corners region of the American Southwest (150), with surface and well water contamination greatly in excess of standards with resultant direct metal and radiation toxic effects. The remediation and proper containment of abandoned mines in conjunction with pollution management practices can reduce the toxic outputs of mines.

The amalgamation method of gold mining can result in increases in aquatic mercury, because it remains in the processing water and will be discharged into the environment if the water is not treated properly (151). Mercury mines that release mine drainage water can generate very high concentrations of iron-rich mercury-bearing precipitates that can be a significant source of bioavailable mercury to surface waters, leading to formation of methylmercury (152). Calcines in mine drainage react with water, releasing and transporting mercury from mine sites into streams (153).

Abandoned mines have also been determined as significant sources of mercury to watersheds (154).

In 2009 the United States announced plans to improve regulations of coal plant emissions of toxic metals into waterways, with the new rules scheduled to be unveiled in 2012 (155). Regulations currently exist requiring the reduction of atmospheric emissions from coal plants, but there are no regulations concerning the disposal of the removed contaminants. The recent implementation of scrubbers on the smoke stacks of a coal-fired power plant in Pennsylvania has become a new source of water pollution through the discharge of contaminated wastewater into the Monongahela River, the drinking-water source for 350,000 people, because of a lack of federal regulations governing the disposal of power plant discharges into waterways or landfills (156). When what regulations that do exist are broken or ignored, the existing fines and penalties are often not implemented. This has resulted in 90% of the 313 coal-fired power plants that have violated the Clean Water Act since 2004 not being fined or otherwise sanctioned by federal or state regulators (157). Accidents from coal mines can result in serious occurrences of water contamination. For example, in December 2008 approximately 300 million gallons of sludge and water was released from the Kingston Fossil Plant in Kingston, Tennessee, after a storage pond wall broke damaging homes and flowed into the Emory and Clinch Rivers, tributaries of the Tennessee River (157).

Poor coal-mining practices also result in water contamination. The dumping of coal ash into a sand and gravel pit by the Baltimore Gas and Electric Company in Anne Arundel County, Maryland, during the mid-1990s lead to the contamination of residential wells with heavy metals including thallium, cadmium, and arsenic via leaching through the pit (157). Underground coal gasification has been determined as a source of groundwater pollution resulting in the introduction of contaminants such as PAHs, metals, radioactive materials, and ionic species (158).

29.2.2.8 Radioactive Substances

Radionuclide's presence in drinking water is usually due to erosion of naturally occurring radioactive material (159). The most serious problem is radon, which equilibrates with groundwater as it flows through radioactive rock. The radon can be ingested, but the more important route of exposure is when the radon comes out as a gas in hot showers and is inhaled. In the United States radon has been found at a higher than average concentration in New England, some Southwestern states, Appalachian and Rocky Mountain states, and some areas in the Great Plains (160). Uranium mine tailings are an important source of radioactive contamination of surface water in places such as the Navaho Reservation in the U.S. Southwest and in many parts of Central Asia. In these and other sites the mine tailings have been simply left as small mountains of material, which then erode following rain and the radioactive elements migrate into surface water and groundwater. The situation is made more serious because of the fact that people in many of these regions obtaining drinking water from either surface water or wells without any treatment.

Drilling for natural gas may result in significant levels of uranium in drilling wastewater, which is generally treated through municipal and industrial treatment plants that are not equipped to remove radioactive contaminants (161). This can lead to contamination of municipal drinking-water sources as well as private wells.

While a major contributor to anthropogenic marine radioactivity is global fallout from nuclear testing in the 1950s and 1960s, some regions like the Irish and North Seas receive significant concentrations of radionuclides from discharges from processing plants, and some regions like the Baltic and Black Seas have been affected by accidental discharges like the 1986 Chernobyl accident (162). Surface waters are important transporters of radioactivity from contaminated zones (163) in cases of leaks from nuclear storage locations and facilities and power plants and in nuclear accidents. Release of tritium from nuclear power plants is a concern, since nuclear power plants routinely and accidentally release significant amounts of tritium into both air and water, and there are no available methods for removing it. Tritium has a half-life of more than 12 years.

Standard drinking-water treatment facilities are effective in removing several radionuclides, but were ineffective in removing ^{90}Sr (strontium) (164). There is still a need to develop better methods of removing radionuclides from water sources, as well as regulating exposure levels. Action also needs to be taken to minimize the erosion of radioactive materials into water and to remediate locations contaminated with radioactive soil and sediment.

29.2.2.9 Disinfection By-Products

Another source of contaminants to water is the creation of disinfection by-products as a result of drinking and/or wastewater treatment. Disinfection by-products result from reactions between organic materials and contaminants in the wastewater with the disinfectant products used in the treatment processes like chlorine and ozone. Disinfection by-products are a series of chlorinated and brominated organic compounds, of which chloroform and bromoform are particularly common products. They have been linked with health effects including cancer and reproductive and developmental issues (159, 165). According to the U.S. EPA over 500 disinfection by-products have been identified (166), while other estimates have stated the number to exceed 600 (165). Currently only 11 disinfection by-products are regulated in the United States. There is a belief among U.S. EPA regulators that if the 11 regulated by-products are reduced to levels at or below the maximum contaminant levels allowed this will result in an associated reduction in the entire suite of disinfection by-products, in turn reducing the risk of exposure and adverse health effects (165). One problem with this assumption is that some disinfection by-products increase in levels when regulated substances decreased (167). Some unregulated substances with potential toxicity include formaldehyde, acetaldehyde, N-nitrosodimethylamine (165), bromoacetonitrile, chloroacetonitrile, and trichloroacetonitrile (168). Disinfection by-products are proven to be human carcinogens both when orally consumed and through dermal exposure as when showering or bathing (169). Reducing the input of organic matter in water prior to treatment can effectively reduce the production of disinfection by-products.

29.3 WASTEWATER AND DRINKING-WATER TREATMENT

Since the source of much of the exposure to chemicals that people experience from water comes from their municipal drinking-water supply, and at least some of the concern of the ecologic effects of man-made chemicals is related to wastewater discharge, it is important to understand how these processes work and what they remove or do not remove. Many contaminants are present in very low concentrations, often making their detection challenging. Frequently contaminants are toxic at trace and undetectable amounts, making them difficult to treat and leaving uncertainties as to their adequate removal.

Wastewater contains human feces and urine, detergents, prescription medications, illegal drugs, suspended solids, fats, oils and greases, and solvents. One of the major goals of wastewater treatment is to remove solids and greatly reduce the microorganisms found there. This is most often done by a series of screens and sedimentation tanks, usually followed by chlorination to kill bacteria and viruses. This process is called primary treatment, and in some cases the wastewater is then discharged into a body of water. Primary treatment alone is not considered to be adequate, although it does result in removal of solids and the sludge is digested to a great degree. Most wastewater treatment facilities will also use secondary treatment in which the wastewater is oxygenated so as to allow microorganisms to break down organic material. Secondary treatment will reduce the levels of total suspended solids but will not remove viruses, metals, dissolved minerals, or dissolved chemicals. Tertiary treatment, if used, involves adding a coagulant to form a gelatinous floc that can be removed by filtration. Tertiary treatment will greatly reduce phosphates, nitrates, ammonia, and organics that adhere to the floc. The wastewater is then chlorinated prior to being released. Additional treatment steps are frequently used for industrial wastes, but because of cost are not commonly used in municipal wastewater treatment. Thus, the waste water can be passed through activated carbon beds to absorb inorganic and organic material, and ion-exchange resins to remove certain chemicals or oxidants (chlorine, ozone, hydrogen peroxide, UV) may be used to oxidize and/or convert contaminants into inorganic compounds.

Drinking water comes either from groundwater (about 50% in the United States) or surface water derived from lakes, rivers, or reservoirs. In developed countries like the United States, more than 90% of drinking water comes from municipal water supplies. In general, groundwater is safer than surface water, containing fewer microorganisms and much less organic material, but groundwater can contain dangerous metals depending upon the rock formations through which the water flows and can also contain toxic contaminants as a result of surface leaching in recharge areas. When drinking water is obtained from surface water, it is often from the same body of water into which the wastewater has been released. Surface water will have more organic material, more microorganisms, and more chemicals coming from the variety of sources discussed earlier. Drinking water treatment has some of the same steps used in wastewater treatment. There is first a preliminary screen to remove debris and sometimes a microstrainer to remove fine material such as algae and protozoa. Then a coagulation process is used, most commonly by adding hydrated aluminum sulfate, where the aluminum ions react with negatively charged colloidal suspended matter and hydroxyl anions to form a gelatinous mass called floc. The water is then slowly stirred to promote coagulation onto the floc, which then precipitates in the process of sedimentation. The water is then filtered through a layer of sand or gravel, sometimes containing charcoal, which removes the floc.

Drinking water treatment has a primary goal of removing microbiological agents that might cause disease, including bacteria and viruses. This is primarily accomplished by chlorination. Protozoa are also a major problem, but these are not very sensitive to chlorination and must be removed through the fine screen used in the preliminary treatment step.

Occurrence and type of wastewater and drinking water contaminants varies regionally and are based on population number and demographics like education, income, industry, and religion, along with cultural traditions and local environmental conditions. In both the western Balkan region and in the United States, surfactants were determined to be the most abundant chemical class in examined wastewaters (53, 170), while wastewater in Denmark was found to frequently contain analgesics like salicylic acid and ibuprofen, and its metabolites, fragrances, and the sunscreen compound hydrocinnamic acid (171). In a Turkish drinking water study some of the most frequently detected organic contaminants were chloroform, bromoform, benzene, toluene, and naphthalene (172). The major surface water bodies in China contain significant amounts of hexachlorobenzene, with pesticide usage and industry being an important source (173). In Tanzania, by 2002, only 50% of the rural population and 70% of the urban population had access to a reliable, safe water supply (174). Water quality deterioration in Tanzania has been attributed primarily to discharge of untreated wastewater from towns, industries and mines, agricultural and livestock discharge, increased sediment loads from deforestation, and inappropriate cultivation practices (175). For example, the Wami River basin, Tanzania, which empties into the Indian Ocean, is rapidly being degraded by charcoal making, the disposal of excess molasses from the sugar industry, and the use of poisonous substances and herbs in fishing. In addition, the increase in unplanned urbanization of the Wami basin has increased the disposal of untreated domestic wastes and sewage into the river (174).

The standard processes used for treatment of both waste and drinking water are very effective at removal of particulates and substances that bind to the floc. They are relatively ineffective without additional steps at the removal of dissolved chemicals and metals. The only effective process for removal of dissolved chemicals is filtration through activated charcoal (powdered activated carbon or granular activated carbon), if that step is taken. This is used most often to improve odor, but also does reduce levels of both natural and synthetic organic compounds. Natural organic compounds will compete with synthetic ones, and a saturation of sites can occur rapidly, depending on the levels in water. The Freundlich Adsorption Isotherm parameters vary greatly with the chemical, and factors such as contact time, bed depth, and hydraulic loading rate are important variables (176). POPs are in general relatively well removed by activated charcoal, but the trihalomethanes and volatile organic compounds are removed less efficiently. Synthetic resins can also be used to remove organic compounds, but they are more selective than activated charcoal.

Metals can be removed by ion exchange, activated alumina adsorption, special resins (177), reverse osmosis, membrane nanofiltration, electrodialysis, ultrafiltration or microfiltration (178), but these procedures add significantly to the cost of water treatment and therefore are commonly not used unless water quality standards are exceeded.

There are a number of experimental and natural methods for destruction of chemicals in water, but at present these have limited use on a large scale. Advanced oxidation processes, where an oxidative species like hydroxyl radicals are generated, have been shown to be effective at destroying endocrine disrupting compounds like bisphenol-A, nonylphenol, and phthalates in water (179). Non-isothermal bioreactors have also proven successful at remediating this type of contamination (180). Aerobic biofilm is another treatment processes that can remove 99% or more of BTEX (benzene, toluene, ethylbenzene, xylenes) from the liquid phase (181). Some compounds, like organolead compounds, are more effectively removed when a combination of physical and chemical treatment methods are employed (182). Electrochemical peroxidation has been found valuable in removing metals from aqueous systems (183). Point-of-use treatment methods can also be employed to reduce and eliminate chemical contaminants and pharmaceuticals in drinking water. Carbon filters, distillation, or reverse osmosis devices are all possible methods (184).

Personal care products and pharmaceuticals have been proven to be reduced over time, in some cases in excess of 80%, by several methods, which include vertical flow constructed wetlands, horizontal flow constructed wetlands, biofilters, and sandfilters, with vertical flow constructed wetlands proving to be the most reliable and robust of the methods (171). Constructed wetlands can be used to treat pollution from an array of sources including agricultural, industrial, and urban runoff. Horizontal and vertical constructed wetlands can be constructed separately or together in a staged fashion—a combined wetland. Combined wetlands, known as "hybrid constructed wetlands," have been shown to be more effective in the removal of contaminants, such as ammonia, than individual wetlands (185, 186). Naturally existing wetlands can also be used to treat runoff. The current trend in new construction is not to destroy wetlands, but rather to incorporate them into the construction design in an environmentally friendly and useful fashion.

Groundwater remediation can be accomplished both *in situ* and *ex situ*, depending on contaminant type and remediation method. Permeable reactive barriers are a method of *in situ* treatment that intercepts and decontaminates plumes in the subsurface for contaminants such as POPs, nitrate ammonia, and hydrocarbons. They have several advantages over *ex situ*, pump-and-treat methods. Permeable reactive barriers utilize the natural gradient of groundwater flow to carry contaminants, reducing the need for a continuous energy input. Also, degradation of the contaminants is achieved rather than just a phase change, thus avoiding the technical and regulatory issues related to the ultimate discharge requirements of effluent from pump-and-treat systems (187). One disadvantage, however, is that they are restricted to shallow plumes, up to 50 ft below the ground surface (188).

There are always technical issues in monitoring levels of contaminants in water. Electron capture detectors (ECDs), in conjunction with a gas chromatogram, measures halogenated compounds at lower concentrations (189). Liquid chromatography-tandem mass spectrometry (LC-MS/MS) with solid-phase extraction (SPE) has proven effective at detecting organic pollutants in water at the part per trillion (ppt) level (190). Inductively coupled plasma mass spectroscopy (ICP-MS) is the common instrumentation method used for the detection of most metals. One exception is mercury. Cold vapor atomic fluorescence spectrometry can measure mercury at levels as low as 0.5 ppt in water. There is a continuing movement to improve techniques allowing for contaminant detections at lower levels and also for methods of rapid detection.

29.4 POLICY AND REGULATION

Legislation is the backbone to water protection. Environmental regulations exist at the international, national, and local levels. The United Nations Environment Programme (UNEP) coordinates activities for the United Nations, although other components also deal with environmental concerns. International treaties such as the Stockholm and Rotterdam Conventions are activities of the United Nations and are examples of global efforts working towards ridding the world of the toxic contaminants plaguing the environment. The European Environment Agency is the lead organization within the European Union, and is designed to aid member nations in making responsible environmental decisions and to coordinate environmental information in Europe. The U.S. EPA works to implement and enforce U.S. environmental regulations, and other countries have comparable organizations.

The Stockholm Convention, as discussed previously, outlaws nine of the "Dirty Dozen" chemicals, limits the use of DDT for only malaria control, and works to reduce dioxin and furan production, while adapting restrictions on the creation and use of future listed substances, including the more recently added "Dirty Nine." The Stockholm Convention also requires the development and implementation of strategies for identifying stockpiles, products, and articles that contain or are contaminated with POPs, as well as managing and destroying the POPs, and successfully remediating the area.

As of the 2009 Stockholm Convention effectiveness evaluation, 44 parties reported their progress, of which 82% indicated they had adopted legal or administrative measures to eliminate the production and use of chemicals of Annex A (listed for elimination), 91% reported to have restricted the production or use of DDT in accordance to Annex B (restricted POPs), and 39 reported on regulatory measures taken for PCBs. All together 88 implementation plans received included total quantities of POPs (1).

The Rotterdam Convention shares similar objectives as the Stockholm Convention. The convention objectives are to promote shared global responsibility and cooperative efforts in the international trade of hazardous chemicals to protect human and ecological health and to contribute to the environmentally sound use of the chemicals with the use of a national decision-making process (191).

Programs like the Stockholm and Rotterdam Conventions have led to the development of other programs like the Strategic Approach to International Chemicals Management (SAICM). In 2006 the International POPs Elimination

Network (IPEN) developed SAICM as a management tool for chemicals adopted by over 100 international governments and organizations in the public health and environmental communities. The goal of SAICM is to "achieve the sound management of chemicals throughout their life-cycle so that, by 2020, chemicals are used and produced in ways that lead to the minimization of significant adverse effects on human health and the environment" (192).

The International Code of Conduct on the Distribution and Use of Pesticides was first adopted in 1985 by the United Nations Food and Agriculture Organization (FAO) and established international standards regarding the distribution and use of pesticides. This was later revised in 2002. The Pesticide Action Network (PAN) has worked largely with the FAO in developing these standards (192). PAN is a global network of over 600 nongovernmental organizations, institutions, and individuals with the goal to replace the use of hazardous pesticides with ecologically sound alternatives. Recently PAN has worked to have the Rotterdam Convention and Stockholm Conventions ratified (in both cases unsuccessfully with the United States). PAN also created the PAN Pesticide Database, which provides toxicity and regulatory information for over 6,500 pesticides (193). They also provide a variety of resources for local and regional activism, local information regarding pesticide usage, and general consumer information.

The Vector Management Global Program (Global DSSA Program) was established to protect human and environmental health by reducing the use of DDT by the use of sustainable alternatives. A group of organizations have been involved in implementing this plan, including the UNEP, WHO, and Global Environmental Facility. Multiple projects are currently underway globally to find sustainable alternatives to DDT. One project that was completed in Mexico and Central America, a 5-year long pilot study, showed success with DDT alternatives cutting malaria cases by 63% (1). The North American Regional Action Plan on DDT (NARAP) was designed to reduce exposure to DDT in North America. Mexico focused on three objectives to eliminate DDT. These were eliminating parasites in people, improving personal and household hygiene, and using environmental management practices to eliminate breeding sites. The NARAP initiative resulted in the elimination of use of DDT for malaria control in Mexico in 2000 (194).

The European Community Legislation Registration, Evaluation, Authorization, and Restriction of Chemical substances (REACH) entered into law in June 2007 with the purpose of protecting human health through improved and earlier identification of the properties of chemical substances while innovating capability and competitiveness of the European Union chemical industry, and at the same time giving greater responsibility to the industry to manage the risks from chemicals and provide safety information

regarding them. In addition to this, REACH calls for the progressive substitution of the most dangerous chemicals when suitable alternatives have been identified (195). REACH uses the "Substitution Principle," the concept of substituting hazardous substances with safer alternatives whenever possible (196). It forces the study and knowledge of the toxicology of chemicals. A problem with this legislation is that it could potentially drive producers to develop newly designed less lipophilic/bioaccumulative chemicals that will be inherently more difficult to remove by traditional water treatment techniques (197). REACH is not yet perfect, and several recommendations for improvement have been offered (198), but it is vastly superior to programs in other countries, specifically in the United States.

The European Union enacted the Nitrates Directive (91/676/EEC) with the intent of reducing water pollution resulting from agriculture, and designated a maximum contaminant level of 50 mg/L in drinking water. This directive also requires the identification of waters where nitrate levels exceed, or are likely to exceed, directive nitrate levels, which led to the designation of nitrate vulnerable zones. In the United Kingdom designations of these zones impose strict limits on the timing and application of organic and inorganic nitrogen-based materials on agricultural lands, affecting ~14% of all land in Scotland and ~12,000 farmers. The Action Program for nitrate vulnerable zones in Scotland 2003 (SI 2003, no. 51) established rules regarding: the quantities, timing, and application conditions for applying nitrogen fertilizer to land; storage capacity for livestock manures; and the planning and recording of fertilizer use (199). However Barnes et al. found that only 62% of farmers surveyed prepared the required nutrient management plan, the majority did keep fertilizer and manure application records for individual fields, but only about 40% invested in obtaining farming or agronomic advice for improved recordkeeping. Most did not change to practices that would improve water quality, and many were skeptical of the science designating nitrate vulnerable zones. It was concluded that policymakers need to provide more clear information regarding the purpose of the nitrate policy in order to encourage farmers to implement better nitrate reduction practices.

In the United States there have been a number of national laws implemented that are for the purpose of establishing national standards for drinking and wastewater, as well as use and release of dangerous substances. Most of these federal mandates are updated on an irregular basis as knowledge and technology grows.

One of the oldest U.S. mandates affecting water was the 1948 Water Pollution Control Act, which funded research on water quality, and was amended in 1956 to provide funds for municipal sewage treatment plants. It was amended again in 1972 as the Clean Water Act, with the noble goal of elimination of all pollutant discharges into navigable waters and

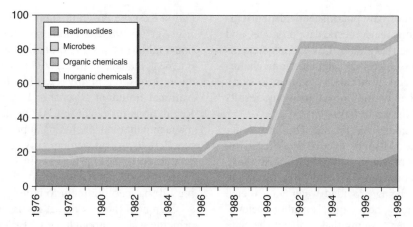

FIGURE 29.2   Number of contaminants regulated under the Safe Drinking Water Act, by contaminant type (from EPA).

to make water bodies swimmable and fishable. It published limits for priority pollutants in water, created a system for identifying new point sources, and established pretreatment standards for industrial waste prior to discharge. In the United States the number of organic chemicals regulated by the Clean Water Act increased by more than threefold in the early 1990s (Fig. 29.2). But many chemicals are still not specifically regulated, and the potential health effects of many emerging contaminants present in water have not been evaluated (197).

The Clean Water Act was the first U.S. act to distinguish point sources from nonpoint sources, and to pay specific attention to the nonpoint sources. A point source originates from an identifiable source at an exact place such as a factory, municipal sewage treatment plant, discharge from a drain pipe, an oil spill, the direct dumping of materials and chemicals into water bodies, and leaching and runoff from military facilities, industrial plants, and underground storage tanks. Nonpoint sources include diffuse sources that cannot be traced back or isolated to a direct source and are difficult to control like storm water runoff, atmospheric deposition, and leaching and runoff from urban areas, agriculture, mining, and logging.

The Clean Water Act was violated over 506,000 times from 2004 to 2007 by over 23,000 companies and facilities with less than 3% of the violations resulting in fines or other significant punishments by state officials (200). Although the foundation for this act does protect water quality, it will not work unless it is enforced properly.

The Safe Drinking Water Act of 1974 was the first to establish national standards for contaminants in drinking water. This act authorizes the U.S. EPA to set national health-based standards for drinking water for both naturally occurring and man-made contaminants known as the National Primary Drinking Water Standards (NPDWS). The NPDWS are divided into two subgroups. The first group is

Maximal Contaminant Levels (MCLs), which are the highest levels of a regulated contaminant allowed in finished drinking water. Six classifications for contaminants exist: inorganic chemicals, organic chemicals, microorganisms, radionuclides, disinfectants, and disinfectant by-products. The MCLs are set as close to maximum contaminant level goals as possible (the level of a contaminant in drinking water below which there is no known or expected health risks) (201). The second group is treatment techniques, which are used when there is no reliable method to measure contaminants at low concentrations due to cost and/or available analytical techniques and set a required procedure or level of technological performance. As of December 2009, 87 contaminants, 7 microorganisms, 4 disinfectant by-products, 3 disinfectants, 16 inorganic chemicals, 53 organic chemicals, and 4 radionuclides were regulated (202). A set of guidelines has been established under the Safe Drinking Water Act regarding how and when to sample water and methods of analysis to use in order to ensure accurate, consistent readings. The results are required to be reported both to government agencies and to all water customers annually.

Secondary drinking regulations have also been developed, but are nonenforceable and intended to be used as guidelines regarding contaminants that may cause cosmetic or aesthetic effects in drinking water. Although these guidelines are not federally enforceable, some states choose to enforce them through incorporating them into their state drinking-water regulations.

Currently over 90 contaminants are listed under either the primary or secondary standards of the NPDWR (203). Some regulated contaminants have levels set so high that the risk of exposure is still great. As of January 2010 over 51,770,000 organic and inorganic contaminants, excluding biosequences, had been indexed by the American Chemical Society's Chemical Abstracts Service in their CAS Registry (204). This number will continue to grow with the discovery and invention of new pharmaceuticals, agricultural and industrial chemicals, domestic care products, and weapons. It is imperative that studies be conducted examining the health effects of these new and emerging chemicals, both individually and in mixtures, and also to study their environmental behavior and impact, using the results to update the NPDWR list accordingly.

In addition to the lax guidelines, the Safe Drinking Water Act is not being enforced. Although over 20% of the 54,700 water treatment systems regulated in the United States have violated key provisions between 2004 and 2009, leading to over 49 million people being provided water with illegal concentrations of chemicals, heavy metals, and/or bacteria, less than 6% of the water systems that violated the Safe Drinking Water Act were ever fined or punished by state or federal officials, including the EPA. The majority of the violations occurred at water systems serving less than 20,000 residents, where resources and experience are often insufficient (205). It is clear that in order for the Safe Drinking Water Act to be effective, it is necessary to increase funding for research, monitoring, and enforcement. One way to increase funding would be through proper enforcement of the Safe Drinking Water Act, holding polluters accountable. Proper enforcement would provide a source of funding while discouraging bad practices. This regulation has the potential to greatly reduce the contamination of the water supply in the United States, if it is provided the necessary resources and enforced effectively.

In the 1987 Water Quality Act Congress announced regulation of storm water runoff with the Federal Storm Water Rule. This was implemented in two phases. Phase I began in 1990 requiring National Pollutant Discharge Elimination System (NPDES) permits for storm water discharges from medium and large municipal separate storm sewer systems and Phase II commenced in 2003, requiring NPDES permits for storm water discharges from certain regulated small municipal storm sewer systems and on construction activity disturbing one to five acres of land. Regulation calls for the identification of sources and types of pollutants that could be present in storm water runoff from their premises and implement controls to prevent potential contaminants from washing into storm sewers, preventing and controlling spills, and adopting street deicing methods less environmental damaging than the use of road salt. All potential contributors to storm water contamination, municipalities and industries, are required to obtain discharge permits stipulating exactly how they intend to minimize polluted runoff by developing and implementing comprehensive management programs.

The Environmental Quality Incentives Program, reauthorized in the Farm Security and Rural Investment Act of 2002, provides a voluntary conservation program with financial and technical aids for implementing management practices on eligible agricultural land (206). The Conser-

vation Reserve Program, authorized under the Food Security Act of 1985, is another voluntary program offering financial incentives and cost-share assistance for establishing best management practices like resource-conserving cover crops on highly erodible land (207). Under the National Pollutant Discharge Elimination System Permit Regulations for CAFOs, CAFOs must implement nutrient management plans including appropriate best management practices to protect water quality (208). Much of the current regulation regarding CAFOs in the United States allows farms to self-audit and certify that they are not polluting (209), enabling them to largely escape regulation.

Currently, there is no set MCL for radon levels in U.S. drinking water, although the USEPA was mandated to do so under the 1996 amendments (210). Instead, the USEPA has established an alternative standard based on the contribution of radon from drinking water to radon levels in indoor air equivalent to the national average concentration of radon in outdoor air, known as the Alternative Maximum Contaminant Level (211). Some states have opted to set their own standards for radon in drinking water. Other radionuclides are regulated under the Safe Drinking Water Act. The EU has set standards for radon and other radionuclides in drinking water under the European Drinking Water Directive 98/83/EC (212).

Modern water management approaches have been shifting away from the pollutant and source approach to more of an integrated approach with a crux on the watershed, or a "place-based" management. For example, a study of the San Francisco Estuary by Oros et al. (213) examined the total pollutants to the estuary in order to determine its health and safety rather than looking at a single pollutant or pollutant types and sources. Results revealed the presence of a variety of chemical contaminants including fire retardants, atrazine, pesticides, and herbicides like metolachlor, molinate and terbuthylazine, piperonyl butoxide, an insecticide synergist used in pesticide formulations, the insect repellant N,N-diethyltoluamide, pharmaceuticals like acetaminophen, compounds found in personal care products such as fragrances, octyl methoxycinnamate, a ultraviolet beta-blocker used in sunscreens and cosmetics, 4-nonylphenol that is used to produce non-ionic surfactants, triallyl isocyanurate employed as an accelerator agent in the vulcanization process for rubber and high-temperature thermoplastic production, and also plasticizers. Many of these chemicals are highly persistent in the environment and originate from many sources including atmospheric deposition, sewage treatment plant and industrial wastewater effluent discharge, boating activities, and agricultural and urban runoff (213). This study demonstrates the presence of a variety of water contaminants and sources on a single water body and the need to take into account the synergistic effect of all sources and types of water pollution for the security and welfare of the earth's water.

29.5   BEHAVIORAL CHANGE AS A WAY TO REDUCE WATER CONTAMINATION

While treatment methods and management practices are effective in reducing concentrations and/or removing contaminants from water bodies, the most effective way of doing this is by reducing and eliminating the production and use of the contaminants whenever possible. Everyday lifestyle changes that reduce the usage of potential water pollutants like toxic chemicals are among the most important. Such changes include simple modifications in diet and exercise, which would reduce the need for pharmaceuticals and prescription drugs, minimizing the use of antimicrobials, and substituting naturally occurring personal care products, pharmaceuticals, and cleaning agents for chemical ones. Reducing the consumption of meat will both improve individual health and will also greatly reduce the generation of agricultural pollutants, along with the need to manage and treat them. Purchasing and promoting the use of organic, locally grown, and seasonally available foods will also reduce agricultural sources of water pollution. Reducing the use of consumables will in turn reduce the chemical pollutants resulting from their production and transport, as well as their potential to become a pollution source when disposed of, properly or improperly. The occurrence of pharmaceuticals in water can be reduced by enacting "take-back" programs for prescription drugs, preventing their disposal into wastewater. Such programs have been implemented in many nations but are only beginning to be used in the United States at the state level (214). Citizens can also encourage their governments to adopt environmental policies that reduce the use and production of contaminants while calling for their remediation and destruction, like the Stockholm Convention.

Industry can also take steps to reduce their use of potential water contaminants. One step is to employ less toxic chemicals, like regents, as suggested for domestic situations. When applicable, industries can also switch to new technologies like microchannel technology. Microchannel technology turns raw materials into products without by-products or pollution, requiring less energy, and eliminating the need for downstream abatement and treatment equipment (215). The use of Analytical Green Chemistry will also allow industry to reduce its generation and emitting of pollutants by developing analytical methodologies, which minimize or eliminate the use of toxic substances and the generation of wastes. Examples include the use of screening methodology to reduce the need to process large sample sets, replacing toxic reagents with nontoxic reagents, replacing mercury electrodes with less toxic ones like bismuth film electrodes, recovery of reagents, online decontamination of wastes, and the development of reagent-free methodologies like FT-Ramen spectrometry (216).

These programs would manage all materials as perpetual resources rather than as wastes, having the potential to manage the life-cycle flow of materials while reducing their impact on the environment, and simultaneously improving conditions for people and industry (214).

29.6   CONCLUSIONS

Water is contaminated both by natural elements and man-made chemicals and pharmaceuticals. Contaminated surface water has adverse effects on fish and wildlife, and contaminated drinking water may cause adverse human health effects. Developed countries have made major progress in getting infectious organisms out of drinking water, and developing methods of treating wastewater so as to remove solids and reduce infectious agents. Access to clean drinking water remains a major public health issue in many developing countries, however. The technology for removal of most chemical and pharmaceutical contaminants from drinking water exists but can be expensive. In practice in the United States and other developed countries, we weigh the cost of removal of substances from drinking water against the cost to ecologic and human health of not doing so. Providing the public with clean drinking water will always remain one of the most important priorities of society.

REFERENCES

1. UNEP (United Nations Environment Programme). A big step forward towards a DDT-free world. Fourth Meeting of the Conference of the Parties to the UNEP-Linked Stockholm Convention on Persistent Organic Pollutants, 2009. (Online) Available at http://www.unep.org/Documents.Multilingual/Default.Print.asp?DocumentID=585&ArticleID=6148&l=en. Accessed December 12, 2009.

2. USEPA (United States Environmental Protection Agency). Office of Water, December 2000. Proposed regulations to address water pollution from concentrated animal feeding operations. EPA 883-F-00-016. (online) Available at http://www.p2pays.org/ref/01/00830.pdf. Accessed January 15, 2010.

3. IARC (International Agency for Research on Cancer). Monographs Programme on the Evaluation of Carcinogenic Risks to Humans, 2004. Available at www.iarc.fr.

4. IRIS (Integrated Risk Information System). All searches conducted online through Toxnet in 1999 unless specifically noted with another year. Database developed and maintained by the USEPA, Office of Health and Environmental Assessment, Environmental Criteria and Assessment Office, Cincinnati, OH. Available at www.epa.gov/iris/subst/index.html.

5. ATSDR (Agency for Toxic Substances and Disease Registry). Toxicological Profile for Chlordane. U.S. Department of

Health and Human Services, Public Health Service, Atlanta, GA, 1994.

6. ATSDR (Agency for Toxic Substances and Disease Registry). Toxicological Profile for Hexachlorobenzene. U.S. Department of Health and Human Services, Public Health Service, Atlanta, GA, 1996.

7. ATSDR (Agency for Toxic Substances and Disease Registry). Toxicological Profile for Polychlorinated Biphenyls (PCBs). U.S. Department of Health and Human Services, Public Health Services, Atlanta, GA, 2000.

8. Huang X, Hites RA, Foran JA, Hamilton C, Knuth BA, Schwager SJ, Carpenter DO. Consumption advisories for salmon based on risk of cancer and non-cancer health effects. *Environ Res* 2006;101:263–274.

9. Hites RA, Foran JA, Schwager SJ, Knuth BA, Hamilton MC, Carpenter DO. Global assessment of polybrominated diphenyl ethers in farmed and wild salmon. *Environ Sci Technol* 2004;38:4945–4949.

10. Choudry GG, Webster GRB. Environmental photochemistry of polychlorinated dibenzofurans (PCDFs) and dibenzo-*p*-dioxins (PCDDs): a review. *Toxicol Environ Chem* 1987;14:43–61.

11. Oberg L, Rappe C. Biochemical formation of PCDDs/Fs from chlorophenols. *Chemosphere* 1992;25:49–52.

12. Addink R, Olie K. Mechanisms of formation and destruction of polychlorinated dibenzo-*p*-dioxins and dibenzofurans in heterogeneous systems. *Environ Sci Technol* 1995;29:1425–1435.

13. Weber R. Relevance of PCDD/PCDF formation for the evaluation of POPs destruction technologies—review on current status and assessment gaps. *Chemosphere* 2007;67:S109–S117.

14. Kulkarni PS, Crespo JG, Afonso CAM. Dioxins sources and current remediation technologies: a review. *Environ Int* 2008;34:139–153.

15. Cook PM, Robbins JA, Endicott DD, Lodge KB, Guiney PD, Walker MK, et al. Effects of aryl hydrocarbon receptor-mediated early life stage toxicity on lake trout populations in Lake Ontario during the 20th century. *Environ Sci Technol* 2003;37:3864–3877.

16. Crump KS, Canady R, Kogevinas M. Meta-analysis of dioxin cancer dose response for three occupational cohorts. *Environ Health Perspect* 2003;111:681–687.

17. Humblet O, Birnbaum L, Rimm E, Mittleman MA, Hauser R. Dioxins and cardiovascular disease mortality. *Environ Health Perspect* 2008;116:1443–1448.

18. Kang HK, Dalager NA, Needham LL, Patterson DG Jr., Lees PS, Yates K, Matanoski GM. Health status of Army Chemical Corps Vietnam veterans who sprayed defoliant in Vietnam. *Am J Ind Med* 2006;49:875–884.

19. Ngo AD, Taylor R, Roberts CL, Nguyen TV. Association between Agent Orange and birth defects: systematic review and meta-analysis. *Int J Epidemiol* 2006;35:1220–1230.

20. ten Tusscher GW, Steerenberg PA, van Loveren H, Vos JG, von dem Borne AE, Westra M, et al. Persistent hematologic and immunologic disturbances in 8-year-old Dutch children associated with perinatal dioxin exposure. *Environ Health Perspect* 2003;111:1519–1523.

21. Schantz SL, Bowman RE. Learning in monkeys exposed perinatally to 2,3,7, 8-tetrachlorodibenzo-*p*-dioxin (TCDD). *Neurotoxicol Teratol* 1989;11:13–19.

22. Barrett DH, Morris RD, Akhtar FZ, Michalek JE. Serum dioxin and cognitive functioning among veterans of Operation Ranch Hand. *Neurotoxicology* 2001;22:491–502.

23. Vezina CM, Walker NJ, Olson JR. Subchronic exposure to TCDD, PeCDF, PCB126, and PCB153: effect on hepatic gene expression. *Environ Health Perspect* 2004;112:1636–1644.

24. van der Plas SA, Sundberg H, van den Berg H, Scheu G, Wester P, Jensen S, Bergman A, de Boer J, Koeman JH, Brouwer A. Contribution of planar (0-1 *ortho*) and nonplanar (2-4 *ortho*) fractions of Aroclor 1260 to the induction of altered hepatic foci in female Sprague-Dawley rats. *Toxicol Appl Pharmacol* 2000;169:255–268.

25. Carpenter DO. Polychlorinated biphenyls (PCBs): routes of exposure and effects on human health. *Rev Environ Health* 2006;21:1–23.

26. IOM (Institute of Medicine). *Dioxins and Dioxin-like Compounds in the Food Supply. Strategies to Decrease Exposure.* Washington, DC: The National Academies Press, 2003.

27. Carpenter DO, DeCaprio AP, O'Hehir D, Akhtar F, Johnson G, Scrudato RJ, et al. Polychlorinated biphenyls in serum of the Siberian Yupik people from St. Lawrence Island, Alaska. *Int J Circumpolar Health* 2005;64:322–335.

28. DeCaprio AP, Johnson GW, Tarbell AM, Carpenter DO, Chiarenzelli JR, Morse GS, et al. Polychlorinated biphenyl (PCB) exposure assessment by multivariate statistical analysis of serum congener profiles in an adult Native American population. Environ Res 2005;98:284–302.

29. Cone M. *Silent Snow: The Slow Poisoning of the Arctic.* Grove Press, 2005.

30. Hansen JC. Environmental contaminants and human health in the Arctic. *Toxicol Lett* 2000;112–113:119–125.

31. Corsolini S, Ademollo N, Romeo T, Greco S, Focardi S. Persistent organic pollutants in edible fish: a human environmental health problem. *Microchem J* 2005;79:115–123.

32. Berger U, Glynn A, Holmstrom KE, Berglund M, Ankarberg EH, Tornkvist A. Fish consumption as a source of human exposure to perfluorinated alkyl substances in Sweden - analysis of edible fish from Lake Vattern and the Baltic Sea. Chemosphere 2009;76:799–804.

33. Smialowicz RJ, Riddle MM, Rogers RR, Luebke RW, Copeland CB, Ernst GG. Immune alterations in rats following subacute exposure to tributyltin oxide. *Toxicology* 1990;64:169–178.

34. Oyama Y. Modification of voltage-dependent Na^+ current by triphenyltin, an environmental pollutant, in isolated mammalian brain neurons. *Brain Res* 1992;26:93–99.

35. Oortgiesen M, Visser E, Vijverberg HP, Seinen W. Differential effects of organotin compounds on voltage-gated potassium currents in lymphocytes and neuroblastoma cells. *Naunyn Schmiedebergs Arch Pharmacol* 1996;353:136–143.

36. Appel KE. Organotin compounds: toxicokinetic aspects. *Drug Metab Rev* 2004;36:763–786.

37. Guerin T, Sirot V, Volatier JL, Leblanc JC. Organotin levels in seafood and its implications for health risk in high-seafood consumers. *Sci Total Environ* 2007;388:66–77.

38. Harino H, Fukushima M, Kawai S. Accumulation of butyltin and phenyltin compounds in various fish species. *Arch Environ Contam Toxicol* 2000;39:13–19.

39. Hanisch C. Where is mercury deposition coming from? *Environ Sci Technol* 1998;32:176–179.

40. Goldman LR, Shannon MW, The Committee on Environmental Health. Technical report: mercury in the environment: implications for pediatricians. *Pediatrics* 2001;108:197–205.

41. Cristol DA, Brasso RL, Condon AM, Fovargue RE, Friedman SL, Hallinger KK, et al. The movement of aquatic mercury through terrestrial food webs. *Science* 2008;320:335.

42. USEPA (United States Environmental Protection Agency). Mercury Study Report to Congress, 1997. EPA-452/R-97-004.

43. USEPA (United States Environmental Protection Agency). What you need to know about mercury in fish and shellfish, March 2004. EPA 823-R-04-055. Available at http://www.epa. gov/nps/agmm/. Accessed January 16, 2010.

44. NAS (National Academy of Sciences). *Toxicological Effects of Methylmercury*. Washington DC: National Academy Press, 2000.

45. Axelrad DA, Bellinger DC, Ryan LM, Woodruff TJ. Dose–response relationship of prenatal mercury exposure and IQ: an integrative analysis of epidemiologic data. *Environ Health Perspect* 2007;115:609–615.

46. Guallar E, Sanz-Gallardo MI, van't Veer P, Bode P, Aro A, Gómez-Aracena J, et al. Mercury, fish oils, and the risk of myocardial infarction. *N Engl J Med* 2002;347:1747–1754.

47. Hightower JM, Moore D. Mercury levels in high-end consumers of fish. *Environ Health Perspect* 2003;111:604–608.

48. Bushkin-Bedient S, Carpenter DO. Benefits versus risk of consumption of fish and other seafood. *Rev Environ Health* 2010;25(3):161–191.

49. Sarmah AK, Meyer MT, Boxall AB. A global perspective on the use, sales, exposure pathways, occurrence, fate and effects of veterinary antibiotics (VAs) in the environment. *Chemosphere* 2006;65:725–759.

50. Daughton C. Non-regulated water contaminants: emerging research. *Environ Impact Assess* 2004;24:711–732.

51. Daughton CG. Environmental stewardship of pharmaceuticals: the green pharmacy. USEPA NGWA, 2003. Available at http://www.epa.gov/esd/bios/daughton/ngwa2003.pdf. Accessed November 30, 2009.

52. Stackelberg PE, Furlong ET, Meyer, MT, Zaugg SD, Henderson AK, Reissman DB. Persistence of pharmaceutical compounds and other organic wastewater contaminants in a conventional drinking-water-treatment plant. *Sci Total Environ* 2004;329:99–113.

53. Kolpin DW, Furlong ET, Meyer MT, Thurman EM, Zaugg, SD, Barber LB, Buxton HT. Pharmaceuticals, hormones, and other organic wastewater contaminants in U.S. streams, 1999-2000: a national reconnaissance. *Environ Sci Technol* 2002;36: 1202–1211.

54. Lin AY, Yu TH, Lin CF. Pharmaceutical contamination in residential, industrial, and agricultural waste streams: risk to aqueous environments in Taiwan. *Chemosphere* 2008;74: 131–141.

55. Palmer PM, Wilson LR, O'Keefe P, Sheridan R, King T, Chen CY. Sources of pharmaceutical pollution in the New York City Watershed. *Sci Total Environ* 2008;394:90–102.

56. Stackelberg PE, Gibs J, Furlong ET, Meyer MT, Zaugg SD, Lippincott RL. Efficiency of conventional drinking-water-treatment processes in removal of pharmaceuticals and other organic compounds. *Sci Total Environ* 2007;377:255–272.

57. Ghosh GC, Nakada N, Yamashita N, Tanaka H. Oseltamivir carboxylate, the active metabolite of oseltamivir phosphate (tamiflu), detected in sewage discharge and river water in Japan. *Environ Health Perspect* 2010;118:103–107.

58. Grunheid S, Amy G, Jekel M. Removal of bulk dissolved organic carbon (DOC) and trace organic compounds by bank filtration and artificial recharge. *Water Res* 2005;39:3219–3228.

59. Brown KD, Kulis J, Thomson B, Chapman TH, Mawhinny DB. Occurrence of antibiotics in hospital, residential, and dairy effluent, municipal wastewater, and the Rio Grande in New Mexico. *Sci Total Environ* 2006;366:772–783.

60. Raloff J. Waterways carry antibiotic resistance. Science News Online: The Weekly Newsmagazine of Science, 1999;155. Available at http://www.sciencenews.org/sn_arc99/6_5_99/fob1.htm. Accessed January 17, 2010.

61. Kummerer K. Antibiotics in the aquatic environment—a review—part I. *Chemosphere* 2009;75:417–434.

62. Stamatis N, Hela D, Konstantinou I. Occurrence and removal of fungicides in municipal sewage treatment plants. *J Haz Mater* 2010;175:829–835.

63. Hladik ML, Bouwer EJ, Roberts AL. Neutral degradates of chloroacetamide herbicides: occurrence in drinking water and removal during conventional water treatment. *Water Res* 2008;42:4905–4914.

64. Huang Q, Hong CS. Aqueous solubilities of non-ortho and mono-ortho PCBs at four temperatures. *Water Res* 2002;36:3543–3552.

65. Milnes MR, Bermudez DS, Bryan TA, Edwards TM, Gunderson MP, Larkin IL, Moore BC, Guillette LJ Jr. Contaminant-induced feminization and demasculinization of nonmammalian vertebrate males in aquatic environments. *Environ Res* 2006;100:3–17.

66. Nash JP, Kime DE, Van der Ven LT, Wester PW, Brion F, Maack G, Stahlschmidt-Allner P, Tyler CR. Long-term exposure to environmental concentrations of the pharmaceutical ethinyl estradiol causes reproductive failure in fish. *Environ Health Perspect* 2004;112:1725–1733.

67. Jobling S, Burn RW, Thorpe K, Williams R, Tyler C. Statistical modeling suggests that anti-androgens in effluents from wastewater treatment works contribute to widespread sexual disruption in fish living in English rivers. *Environ Health Perspect* 2009;117:797–802.

68. Robinson CD, Brown E, Craft JA, Davies IM, Moffat CF, Pirie D, Robertson F, Stagg RM, Struthers S. Effects of sewage

effluent and ethynyl oestradiol upon molecular markers of oestrogenic exposure, maturation and reproductive success in the sand goby (*Pomatoschistus minutus,* Pallas). *Aquat Toxicol* 2003;62:119–134.

69. Thorpe KL, Hutchinson TH, Hetheridge MJ, Scholze M, Sumpter JP, Tyler CR. Assessing the biological potency of binary mixtures of environmental estrogens using vitellogenin induction in juvenile rainbow trout (*Oncorhynchs mykiss*). *Environ Sci Tech* 2001;35:2476–2481.

70. Lin LL, Janz DM. Effects of binary mixtures of xenoestrogens on gonadal development and reproduction in zebrafish. *Aquat Toxicol* 2006;80:382–395.

71. McMurry CS, Dickerson RL. Effects of binary mixtures of six xenobiotics on hormone concentrations and morphometric endpoints of northern bobwhite quail (*Colinus virginianus*). *Chemosphere* 2001;43:829–837.

72. Filby AL, Neuparth T, Thorpe KL, Owen R, Galloway TS, Tyler CR. Health impacts of estrogens in the environment, considering complex mixture effects. *Environ Health Perspect* 2007;115:1704–1710.

73. Rohr JR, McCoy KA. A qualitative meta-analysis reveals consistent effects of atrazine on freshwater fish and amphibians. *Environ Health Perspect* 2010;118:20–32.

74. Gray L.E. Jr., Wolf C, Lambright C, Mann P, Price M, Cooper RL, Ostby J. Administration of potentially antiandrogenic pesticides (procymidone, linuron, iprodione, chlozolinate, *p,p′*-DDE, and ketoconazole) and toxic substances (dibutyl- and diethylhexyl phthalate, PCB 169, and ethane dimethane sulphonate) during sexual differentiation produces diverse profiles of reproductive malformations in the male rat. *Toxicol Ind Health* 1999;15:94–118.

75. Swan SH, Liu F, Hines M, Kruse RL, Wang C, Redmon JB, Sparks A, Weiss B. Prenatal phthalate exposure and reduced masculine play in boys. *Int J Androl* 2009;32:1–9.

76. Stahlhut RW, van Wijngaarden E, Dye TD, Cook S, Swan SH. Concentrations of urinary phthalate metabolites are associated with increased waist circumference and insulin resistance in adult U. S. males. *Environ Health Perspect* 2007; 115:876–882.

77. Timms BG, Howdeshell KL, Barton L, Bradley S, Richter CA, vom Saal FS. Estrogenic chemicals in plastic and oral contraceptives disrupt development of the fetal mouse prostate and urethra. *Proc Natl Acad Sci USA* 2005;102: 7014–7019.

78. Prins GS, Tang WY, Belmonte J, Ho SM. Perinatal exposure to oestradiol and bisphenol A alters the prostate epigenome and increases susceptibility to carcinogenesis. *Basic Clin Pharmacol Toxicol* 2008;102:134–138.

79. Carwile JL, Luu HT, Bassett LS, Driscoll DA, Yuan C, Chang JY, Ye X, Calafat AM, Michels KB. Polycarbonate bottle use and urinary bisphenol A concentrations. *Environ Health Perspect* 2009;117:1368–1372.

80. Weiss B. Endocrine disruptors and sexually dimorphic behaviors: a question of heads and tails. *Neurotoxicology* 1997;18:581–586.

81. Bystrzejewska-Piotrowska G, Golimowski J, Urban PL. Nanoparticles: their potential toxicity, waste and environmental management. *Waste Manage* 2009;29:2587–2595.

82. Xia T, Li N, Nel AE. Potential health impact of nanoparticles. *Annu Rev Public Health* 2009;30:137–150.

83. Shvedova AA, Kagan VE, Fadeel B. Close encounters of the small kind: adverse effects of man-made materials interfacing with the nano-cosmos of biological systems. *Annu Rev Pharmacol Toxicol* 2010;50:63–88.

84. Duhigg C. As sewers fill, waste poisons waterways. *The New York Times,* November 23, 2009.

85. USEPA (United States Environmental Protection Agency). Protecting water quality from urban runoff, February 2003. EPA 841-F. Available at http://www.epa.gov/nps/toolbox/other/epa_nps_urban_facts.pdf. Accessed December 26, 2009.

86. Walker WJ, Mc Nutt RP, Maslanka CK. The potential contribution of urban runoff to surface sediments of the Passaic River: sources and chemical characteristics. *Chemosphere* 1999;38:363–377.

87. Davis AP, Shokouhian M, Ni S. Loading estimates of lead, copper, cadmium, and zinc in urban runoff from specific sources. *Chemosphere* 2001;44:997–1009.

88. USEPA (United States Environmental Protection Agency). National Water Quality Inventory Report to Congress: 2004 Reporting Cycle. EPA 841-R-08-001, January 2009.

89. Chang M, McBroom MW, Beasley RS. Roofing as a source of nonpoint water pollution. *J Environ Manage* 2004; 73:307–315.

90. Abdulla FA, Al-Shareef AW. Roof rainwater harvesting systems for household water supply in Jordan. *Desalination* 2009;243:195–207.

91. Chang M, Crowley CM. Preliminary observations on water quality of storm runoff from four selected residential roofs. *Water Resour Bull* 1993;29:777–783.

92. Gromaire MC, Garnaud S, Saad M, Chebbo G. Contribution of different sources to the pollution of wet weather flows in combined sewers. *Water Res* 2001;35:521–533.

93. Tsakovski S, Tobiszewski M, Simeonov V, Polkowska Z, Namiesnik J. Chemical composition of water from roofs in Gdansk, Poland. *Environ Pollut* 2010;158:84–91.

94. Berndtsson JC, Bengtsson L, Jinno K. Runoff water quality from intensive and extensive vegetated roofs. *Ecol Eng* 2009;35:369–380.

95. Nadakavukaren A. *Our Global Environment: A Health Perspective,* 6th edition. Long Grove, IL: Waveland Press Inc., 2006.

96. Astebol SO, Hvitved-Jacobsen T, Simonsen O. Sustainable storm water management at Fornebu—from an airport to an industrial and residential area of the city of Oslo, Norway. *Sci Total Environ* 2004;334-335:239–249.

97. Scholz M, Grabowiecki P. Review of permeable pavement systems. *Build Environ* 2007;42:3830–3836.

98. Pratt CJ, Newman AP, Bond PC. Mineral oil biodegradation within a permeable pavement: long-term observations. *Water Sci Technol* 1999;39:109–130.

99. Brattebo BO, Booth DB. Long-term stormwater quantity and quality performance of permeable pavement systems. *Water Res* 2003;37:436–4376.

100. Dreelin EA, Fowler L, Carroll CR. A test of porous pavement effectiveness on clay soils during natural storm events. *Water Res* 2006;40:799–805.

101. Davis BS, Birch GF. Catchment-wide assessment of the cost-effectiveness of stormwater remediation measures in urban areas. *Environ Sci Policy* 2009;12:84–91.

102. Nordstrom DK. Public health. Worldwide occurrences of arsenic in ground water. *Science* 2002;296:2143–2145.

103. Karagas MR, Stukel TA, Tosteson TD. Assessment of cancer risk and environmental levels of arsenic in New Hampshire. *Int J Hyg Environ Health* 2002;205:85–94.

104. Moore LE, Lu M, Smith AH. Childhood cancer incidence and arsenic exposure in drinking water in Nevada. *Arch Environ Health* 2002;57:201–206.

105. Navas-Acien A, Sharrett AR, Silbergeld EK, Schwartz BS, Nachman KE, Burke TA, Guallar E. Arsenic exposure and cardiovascular disease: a systematic review of the epidemiologic evidence. *Am J Epidemiol* 2005;162: 1037–1049.

106. Wasserman GA, Liu X, Parvez F, Ahsan H, Factor-Litvak P, Kline J, van Geen A, Slavkovich V, Loiacono NJ, Levy D, Cheng Z, Graziano JH. Water arsenic exposure and intellectual function in 6-year-old children in Araihazar, Bangladesh. *Environ Health Perspect* 2007;115:285–289.

107. Swartz CH, Reddy S, Benotti MJ, Yin H, Barber LB, Brownawell BJ, Rudel RA. Steroid estrogens, nonylphenol ethoxylate metabolites, and other wastewater contaminants in groundwater affected by a residential septic system on Cape Cod, MA. *Environ Sci Technol* 2006;40:4894–4902.

108. Barnes KK, Kolpin DW, Furlong ET, Zaugg SD, Meyer MT, Barber LB. A national reconnaissance of pharmaceuticals and other organic wastewater contaminants in the United States—(I) groundwater. *Sci Total Environ* 2008; 402:192–200.

109. Focazio MJ, Kolpin DW, Barnes KK, Furlong ET, Meyer MT, Zaugg SD, et al. A national reconnaissance for pharmaceuticals and other organic wastewater contaminants in the United States—(II) untreated drinking water sources. *Sci Total Environ* 2008;402:201–216.

110. Zewdie T, Smith CM, Hutcheson M, West CR. Basis of the Massachusetts reference dose and drinking water standard for perchlorate. *Environ Health Perspect* 2010;118:42–48.

111. Díaz-Cruz MS, Barceló D. Trace organic chemicals contamination in ground water recharge. *Chemosphere* 2008;72:333–342.

112. Rowe AA, Totten LA, Xie M, Fikslin TJ, Eisenreich SJ. Air–water exchange of polychlorinated biphenyls in the Delaware River. *Environ Sci Technol* 2007;41:1152–1158.

113. Sun P, Basu I, Blanchard P, Brice KA, Hites RA. Temporal and spatial trends of atmospheric polychlorinated biphenyl concentrations near the Great Lakes. *Environ Sci Technol* 2007;41:1131–1136.

114. Hafner WD, Hites RA. Potential sources of pesticides, PCBs, and PAHs to the atmosphere of the Great Lakes. *Environ Sci Technol* 2003;37:3764–3773.

115. Ackerman LK, Schwindt AR, Simonich SL, Koch DC, Blett TF, Schreck CB, et al. Atmospherically deposited PBDEs, pesticides, PCBs, and PAHs in western U.S. National Park fish: concentrations and consumption guidelines. *Environ Sci Technol* 2008;42:2334–2341.

116. Buehler SS, Hites RA. The Great Lakes' integrated atmospheric deposition network. *Environ Sci Technol* 2002;36:354A–359A.

117. Ryaboshapko A, Bullock OR Jr., Christensen J, Cohen M, Dastoor A, Ilyin I, Petersen G, Syrakov D, Artz RS, Davignon D, Draxler RR, Munthe J. Intercomparison study of atmospheric mercury models: 1. comparison of models with short-term measurements. *Sci Total Environ* 2007;376:228–240.

118. Tang S, Feng X, Qiu J, Yin G, Yang Z. Mercury speciation and emissions from coal combustion in Guiyang, southwest China. *Environ Res* 2007;105:175–182.

119. Costanza JK, Marcinko SE, Goewert AE, Mitchell CE. Potential geographic distribution of atmospheric nitrogen deposition from intensive livestock production in North Carolina, USA. *Sci Total Environ* 2008;398:76–86.

120. USEPA (United States Environmental Protection Agency). Protecting water quality from agricultural runoff, March 2005. EPA 841-F-05-001. Available at: http://www.epa.gov/owow/nps/Ag_Runoff_Fact_Sheet.pdf. Accessed December 31, 2009.

121. Verstraeten IM, Thurman EM, Lindsey ME, Lee EC, Smith RD. Changes in concentrations of triazine and acetamide herbicides by bank filtration, ozonation, and chlorination in a public water supply. *J Hydrol* 2002;266:190–208.

122. Brown CD, van Beinum W. Pesticide transport via sub-surface drains in Europe. *Environ Pollut* 2009;157:3314–3324.

123. Shiklomanov IA. Appraisal and assessment of world water resources. *Water Int* 2000;25:11–32.

124. Scanlon BR, Jolly I, Sophocleous M, Zhang L. Global impacts of conversions from natural to agricultural ecosystems on water resources: quantity versus quality. *Water Resour Res* 2007;43:W03437, doi:10.1029/2006WR005486.

125. Rice DN, Straw BE. G92-1093 Use of Animal Drugs in Livestock Management, 1992. Extension: Historical Materials from University of Nebraska–Lincoln Extension. Available at http://digitalcommons.unl.edu/cgi/viewcontent.cgi?article=1223&context=extensionhist. Accessed January 17, 2010.

126. García-Galán MJ, Garrido T, Fraile J, Ginebreda A, Díaz-Cruz MS, Barceló D. Simultaneous occurrence of nitrates and sulfonamide antibiotics in two groundwaters bodies of Catalonia (Spain). *J Hydrol* 2009;383:93–101.

127. Carballa M, Omil F, Lema JM, Llompart M, García-Jares C, Rodríguez I, et al. Behavior of pharmaceuticals, cosmetics and hormones in a sewage treatment plant. *Water Res* 2004;38:2918–2926.

128. Knobeloch L, Salna B, Hogan A, Postle J, Anderson H. Blue babies and nitrate-contaminated well water. *Environ Health Perspect* 2000;108:675–678.

129. Fan AM, Steinberg VE. Health implications of nitrate and nitrite in drinking water: an update on methemoglobinemia occurrence and reproductive and developmental toxicity. *Regul Toxicol Pharmacol* 1996;23:35–43.

130. Weyer PJ, Cerhan JR, Kross BC, Hallberg GR, Katamnei J, Breuer G, et al. Municipal drinking water nitrate level and cancer risk in older women: the Iowa women's health study. *Epidemiology* 2001;12:327–338.

131. Gatseva PD, Argirova MD. High-nitrate levels in drinking water may be a risk factor for thyroid dysfunction in children and pregnant women living in rural Bulgarian areas. *Int J Hyg Environ Health* 2008;211:555–559.

132. Yuan Y, Dabney SM, Bingner RL. Cost effectiveness of agricultural BMPs for sediment reduction in the Mississippi Delta. *J Soil Water Conserv* 2002;57:259–267.

133. El Bakouri H, Morillo J, Usero J, Ouassini A. Natural attenuation of pesticide water contamination by using ecological adsorbents: application for chlorinated pesticides included in European framework directive. *J Hydrol* 2009;364:175–181.

134. Brown C, Alix A, Alonso-Prados JL, Auteri D, Gril JJ, Hiederer R, et al. Landscape and mitigating factors in aquatic risk assessment. Volume 1: Extended summary and recommendations. Report of the FOCUS Working Group on landscape and mitigation factors in ecological risk assessment, 2004. Available at http://eusoils.jrc.ec.europa.eu/ESDB_Archive/eusoils_docs/other/FOCUS_Vol1.pdf. Accessed December 12, 2009.

135. Reichenberger S, Bqach M, Skitshak A, and Frede HG. Mitigation strategies to reduce pesticide inputs into ground and surface water and their effectiveness: a review. *Sci Total Environ* 2007;384:1–35.

136. Duhigg C. Health ills abound as farm runoff fouls wells. *The New York Times*, September 18, 2009.

137. Toor GS, Sims JT, Dou Z. Reducing phosphorus in dairy diests improves farm nutrient balances and decreases the risk of nonpoint pollution of surface and ground waters. *Agr Ecosyst Environ* 2005;105:401–411.

138. Budnick LD, Logue JN, Sokal DC, Fox JM, Falk H. Cancer and birth defects near the Drake Superfund site, Pennsylvania. *Arch Environ Health* 1984;39:409–413.

139. Shaw GM, Schulman J, Frisch JD, Cummins SK, Harris JA. Congenital malformations and birthweight in areas with potential environmental contamination. *Arch Environ Health* 1992;47:147–154.

140. Geschwind SA, Stolwijk JA, Bracken M, Fitzgerald E, Stark A, Olsen C, Melius J. Risk of congenital malformations associated with proximity to hazardous waste sites. *Am J Epidemiol* 1992;135:1197–1207.

141. White RF, Feldman RG, Eviator II, Jabre JF, Niles CA. Hazardous waste and neurobehavioral effects: a developmental perspective. *Environ Res* 1997;73:113–124.

142. Wilson D. *Fateful Harvest*. New York: HarperCollins Publishers Inc., 2010.

143. Kabata-Pendias A. *Trace Elements in Soils and Plants*, 3rd edition, Boca Raton, FL: CRC Press LLC, 2001.

144. McLaughlin MJ, Parker DR, Clarke JM. Metals and micronutrients—food safety issues. *Field Crops Res* 1999;60: 143–163.

145. Clausen J, Robb J, Curry D, Korte N. A case study of contaminants on military ranges: Camp Edwards, Massachusetts, USA. *Environ Pollut* 2004;129:13–21.

146. Dwernychuk LW, Cau HD, Hatfield CT, Boivin TG, Hung TM, Dung PT, Thai ND. Dioxin reservoirs in southern Viet Nam- a legacy of Agent Orange. *Chemosphere* 47:117–137.

147. Mon J, Youjun D, Flury M, Harsh J. Cesium incorporation and diffusion in cancrinite, sodalite, zeolite, and allophone. *Micropor Mesopor Mat* 2005;56:277–286.

148. Flury M, Czigany S, Chen G, Harsh JB. Cesium migration in saturated silica sand and Hanford sediments as impacted by ionic strength. *J Contam Hydrol* 2003;71:111–126.

149. Sevcik M. Uranium tailings in Kyrgyzstan: catalyst for cooperation and confidence building? *Nonproliferation Rev* 2003;10:147–154.

150. Raymond-Whish S, Mayer LP, O'Neal T, Martinez A, Sellers MA, Christian PJ, et al. Drinking Water with Uranium below the U.S. EPA Water Standard Causes Estrogen Receptor– Dependent.

151. Wang Q, Kim D, Dionysiou DD, Sorial GA, Timberlake D. Sources and remediation for mercury contamination in aquatic systems—a literature review. *Environ Pollut* 2004;131:323–336.

152. Rytuba JJ. Mercury mine drainage and processes that control its environmental impact. *Sci Total Environ* 2000;260:57–71.

153. Feng X, Qiu G. Mercury pollution in Guizhou, southwestern China—an overview. *Sci Total Environ* 2008;400:227–237.

154. Qui G, Feng X, Wang S, Shang L. Environmental contamination of mercury from Hg mining areas in Wuchuan northeastern Guizhou, China. *Environ Poll* 2006;142:549–558.

155. Cappiello D.U.S. to place limits on power plant water pollution. *U.S. News and World Report*, September 15, 2009. Available at http://www.usnews.com/science/articles/2009/09/15/us-to-place-limits-on-power-plant-water-pollution.html. Accessed December 15, 2009.

156. Duhigg C.Cleansing the air at the expense of waterways. *The New York Times*, October 13, 2009.

157. Dewan S.Coal ash spill revives issue of its hazards. *The New York Times*, 2008.

158. Shu-qin L, Jing-gang LI, Mei M, Dong-lin D. Groundwater pollution from underground coal gasification. *J China Univ Mining Technol* 2007;17:0467–0472.

159. Afzal BM. Drinking water and women's health. *J Midwifery Women's Health* 2006;51:12–18.

160. The National Academies. Radon in drinking water constitutes small health risk, 1998. Available at http://www8.nationalacademies.org/onpinews/newsitem.aspx?RecordID=6287. Accessed December 29, 2009.

161. Sapien J, Shankman S.Unworkable options for drill wastewater. *Times Union*, December 29, 2009. Available at http://www.timesunion.com/AspStories/story.asp?storyID=882817&TextPage=1. Accessed December 29, 2009.

162. Livingston HD, Povinec PP. Anthropogenic marine radioactivity. *Ocean Coast Manage* 2000;43:689–712.

163. Smith JT, Voitsekhovitch OV, Håkanson L, Hilton J. A critical review of measures to reduce radioactive doses from drinking water and consumption of freshwater foodstuffs. *J Environ Radioact* 2001;56:11–23.

164. Jimenez A, De La Montaña Rufo M. Effect of water purification on its radioactive content. *Water Res* 2002;36:1715–1724.

165. Richardson SD, Plewa MJ, Wagner ED, Schoeny R, DeMarini DM. Occurrence, genotoxicity, and carcinogenicity of regulated and emerging disinfection by-products in drinking water: a review and roadmap for research. *Mutat Res* 2007;636:178–242.

166. Weinberg HS, Krasner SW, Richardson D, Thurston D Jr. The Occurrence of Disinfection By-products (DBPs) of Health Concern in Drinking Water: Results of a Nationwide DBP Occurrence Study. EPA-600-R02-068. USEPA, National Exposure Research Laboratory, Athens, GA, 2002.

167. Krasner SW, Weinberg HS, Richardson SD, Pastor SJ, Chinn, R, Sclimenti MJ, et al. The occurrence of a new generation of disinfection by-products. *Environ Sci Technol* 2006; 40:7175–7185.

168. Mullner MG, Wagner ED, McCalla K, Richardson SD, Woo YT, Plewa MJ. Haloacetonitriles vs. regulated haloacetic acids: are nitrogen-containing DBPs more toxic? *Environ Sci Technol* 2007;41:645–651.

169. Uyak V. Multi-pathway risk assessment of trihalomethanes exposure in Istanbul drinking water supplies. *Environ Int* 2006;32:12–21.

170. Terzić S, Senta I, Ahel M, Gros M, Petrović M, Barcelo D, et al. Occurrence and fate of emerging wastewater contaminants in Western Balkan Region. *Sci Total Environ* 2008;399:66–77.

171. Matamoros V, Arias C, Brix H, Bayona JM. Preliminary screening of small-scale domestic wastewater treatment. *Water Res* 2009;43:55–62.

172. Kavcar P, Odabasi M, Kitis M, Inal F, Sofuoglu SC. Occurrence, oral exposure and risk assessment of volatile organic compounds in drinking water for Izmir. *Water Res* 2006;40:3219–3230.

173. Wang G, Lu Y, Han J, Luo W, Shi Y, Wang T, Sun Y. Hexachlorobenzene sources, levels and human exposure in the environment of China. *Environ Int* 2010;36:122–130.

174. Madulu NF. Environment, poverty, and health linkages in the Wami River basin: a search for sustainable water resource management. *Phys Chem Earth* 2005;30:950–960.

175. Mwaka, I, Sayi CN, Lupimo S, Odhiamo S, Kashililah H, Sosola E, et al. Water law, water rights, and water supply (Africa) Tanzania—study country report, Tanzania, DFID, 1999. Available at http://allafrica.com/download/resource/main/main/id/00010043.pdf. Accessed January 17, 2010.

176. Snoeyink VL, Summers RS. Adsorption of organic compounds. In: Letterman RD, editor. *Water Quality and Treatment*, 5th edition. New York: McGraw-Hill, Inc., 1999.

177. Clifford DA. Ion exchange and inorganic adsorption. In: Letterman RD, editor. *Water Quality and Treatment*, 5th edition. New York: McGraw-Hill, Inc., 1999.

178. Taylor JS, Wiesner M. Membranes. In: Letterman RD, editor. *Water Quality and Treatment*, 5th edition. New York: McGraw-Hill, Inc., 1999.

179. Gultekin I, Ince NH. Synthetic endocrine disruptors in the environment and water remediation by advanced oxidation processes. *J Environ Manage* 2007;85:816–832.

180. Diano N, Grano V, Fraconte L, Caputo P, Ricupito A, Attansio A, et al. Non-isothermal bioreactors in enzymatic remediation of waters polluted by endocrine disruptors: BPA as a model of pollutant. *Appl Catal B Environ* 2007; 69:252–261.

181. Farhadian M, Duchez D, Vachelard C, Larroche C. Mono-aromatics removal from polluted water through bioreactors—a review. *Water Res* 2008;42:1325–1341.

182. Andreottola G, Dallago L, Ferrarese E. Feasibility study for the remediation of groundwater contaminated by organolead compounds. *J Haz Mater* 2008;156:488–498.

183. Arienzo M, Chiarenzelli J, Scrudato R. Remediation of metal-contaminated aqueous systems by electrochemical peroxidation: an experimental investigation. *J Hazard Mater* 2001;87:187–198.

184. McKinney M, Schoch RM, Yonavjak L. *Environmental Science: Systems and Solutions*, 4th edition. Sudbury, MA: Jones and Bartlett Publishers, 2007.

185. Vymazal J. Horizontal sub-surface flow and hybrid constructed wetlands. *Ecol Eng* 2005;25:478–490.

186. Vymazal J. The use constructed wetlands with horizontal sub-surface flow for various types of wastewater. *Ecol Eng* 2009;35:1–17.

187. Thiruvenkatachari R, Vigneswaran S, Naidu R. Permeable reactive barrier for groundwater remediation. *J Ind Eng Chem* 2008;14:145–156.

188. Xenidis A, Moirou A, Paspaliaris I. Reactive materials and attenuation processes for permeable reactive barriers. *Mineral Wealth* 2002;123:35–48.

189. Sankararamakrishnan N, Sharma AK, Sanghi R. Organochlorine and organophosphorous pesticide residues in groundwater and surface waters of Kanpur, Uttar Pradesh, India. *Environ Int* 2005;31:113–120.

190. Rodil R, Quintana JB, López-Mahía P, Muniategui-Lorenzo S, Prada-Rodriguez D. Multi-residue analytical method for the determination of emerging pollutants in water by solid-phase extraction and liquid chromatography-tandem mass spectrometry. *J Chromatogr A* 2009; 1216:2958–2969.

191. Rotterdam Convention: Share Responsibility. Available at http://www.pic.int/home.php?type=t&id=5&sid=16. Accessed January 17, 2010.

192. Weinberg J, IPEN (International POPs Elimination Network). An NGO guide to SAICM: The strategic approach to international chemicals management, 2008. Available at http://www.ipen.org/campaign/documents/education/saicm%20introduction%20english.pdf.

193. PAN Pesticide Database. Available at http://www.pesticideinfo.org/. Accessed January 17, 2010.

194. Chanon KE, Méndez-Galván JF, Galindo-Jaramillo JM, Olguín-Bernal H, Borja-Aburto VH. Cooperative actions to achieve malaria control without the use of DDT. *Int J Hyg Environ Health* 2003;206:387–394.

195. EUROPA. REACH. Available at http://ec.europa.eu/environment/chemicals/reach/reach_intro.htm. Accessed January 11, 2010.

196. Garcia-Serna J, Pérez-Barrigón L, Cocero MJ. New trends for design towards sustainability in chemical engineering: green engineering. *Chem Eng J* 2007;133:7–30.

197. Schriks M, Heringa MB, van der Kooi MM, de Voogt P, van Wezel AP. Toxicological relevance of emerging contaminants for drinking water quality. *Water Res* 2010;44 (2):461–476.

198. Ruden C, Hansson SO. Registration, Evaluation, and Authorization of Chemicals (REACH) is but the first step—how far will it take us? Six further steps to improve the European chemicals legislation. *Environ Health Perspect* 2010;118:6–10.

199. Barnes AP, Willock J, Hall C, Toma L. Farmer perspectives and practices regarding water pollution control programmes in Scotland. *Agric Water Manage* 2009;96:1715–1722.

200. Duhigg C. Clean water laws are neglected, at a cost in suffering. *The New York Times*, September 13, 2009.

201. USEPA (United States Environmental Protection Agency). Drinking water standards and health effects, June 2004. EPA 816-F-04-034. Available at http://www.epa.gov/nps/agmm/. Accessed January 16, 2010.

202. USEPA (United States Environmental Protection Agency). Drinking water contaminants, May 2009. EPA 816-F-09-0004. Available at http://www.epa.gov/safewater/contaminants/index.html. Accessed November 30, 2009.

203. USEPA (United States Environmental Protection Agency). National primary drinking water regulations, May 2009. EPA 816-F-09-004. Available at http://www.epa.gov/ogwdw000/consumer/pdf/mcl.pdf. Accessed January 12, 2009.

204. CAS. The Latest CAS Registry Number and Substance Count, American Chemical Society, 2010. Available at http://www.cas.org/cgi-bin/cas/regreport.pl. Accessed January 12, 2010.

205. Duhigg C. Millions in U.S. drink dirty water, records show. *The New York Times*, December 8, 2009.

206. USDA (United States Department of Agriculture). Environmental Quality Incentives Program, 2003. Available at http://www.nrcs.usda.gov/programs/eqip/. Accessed January 8, 2010.

207. USEPA (United States Environmental Protection Agency). Management measures to control nonpoint source pollution from agriculture, July 2003. EPA 841-B-03-004. Available at http://www.epa.gov/nps/agmm/. Accessed January 16, 2010.

208. USEPA (United States Environmental Protection Agency). Fact Sheet: NPDES permit regulation and effluent limitations guidelines for concentrated animal feeding operations, January 2003. EPA 821-F-03-003. Available at http://www.p2pays.org/ref/32/31164.pdf. Accessed January 8, 2010.

209. USEPA (United States Environmental Protection Agency). Federal Register, Vol. 73, No. 225, November 2008. Available at http://www.epa.gov/npdes/regulations/cato_final_rule_preamble2008.pdf.

210. USEPA (United States Environmental Protection Agency). Office of Ground Water and Drinking Water. Safe drinking water act amendments of 1996, 1996. Available at http://www.epa.gov/safewater/sdwa/summ.html#3c. Accessed December 26, 2009.

211. MDEH (Maine Department of Environmental Health). Environmental and Occupational Health Program, 2006. Maximum exposure guideline for radon in drinking water. Registry Number: 10043-92-2. Available at http://www.maine.gov/dhhs/eohp/wells/documents/radonMEG.pdf. Accessed January 13, 2010.

212. EC (European Commission) (1998) European Commission (EC) Council Directive 98/83/EC of November 3, 1998 on the quality of water intended for human consumption, L330/32.

213. Oros DR, Jarman WM, Lowe T, David N, Lowe S, Davis JA. Surveillance for previously unmonitored organic contaminants in the San Francisco Estuary. *Mar Pollut Bull* 2003;46:1102–1110.

214. Daughton CG. Non-regulated contaminants: emerging issues. Presented at: From Source Water to Drinking Water: Emerging Challenges for Public Health, National Academies, Institute of Medicine, Roundtable on Environmental Health Sciences, Research, and Medicine (EHSRT), Workshop #5, October 16, 2003, Washington DC. Available at http://epa.gov/nerlesd1/chemistry.

215. Lerou JJ, Tonkovich AL, Silva L, Perry S, McDaniel J. Microchannel reactor architecture enables greener processes. *Chem Eng Sci* 2010;65:380–385.

216. Armenta S, Garrigues S, de la Guardia M. Green analytical chemistry. *Trends Anal Chem* 2008;27:497–510.

30

ADDITIONAL MEASURES TO PREVENT, AMELIORATE, AND REDUCE WATER POLLUTION AND RELATED WATER DISEASES: GLOBAL WATER GOVERNANCE

Nikhil Chandavarkar

30.1 INTRODUCTION

On March 22, 2010, the United Nations celebrated World Water Day with the theme "Water Quality" (1). Ensuring clean drinking water is perhaps the single most important measure that can prevent disease across the world. Efforts need to be undertaken at the national, regional, and international levels in both water supply and sanitation for such benefits to be derived.

Water is essential for life. Unfortunately, some 1 billion people around the world lack access to clean drinking water and some 2.5 billion people do not have access to basic sanitation (2). Consequently, water-borne diseases are spreading worldwide and know no borders.

Ensuring health through clean water is a global challenge and requires a global solution, through effective global water governance. Accordingly, ensuring clean water forms an integral part of the water and sanitation agenda of the most representative global governance institution, the United Nations. The UN system's water and sanitation policy and operational work in support of member states is coordinated by UN-Water, an interagency mechanism that brings together 26 UN entities that deal with water and sanitation (3). Among the many accomplishments of UN-Water are annual assessments of water and sanitation situation in the world, three editions of the *World Water Development Report*, which provide a strategic overview to policymakers and policy shapers on global water challenges, and a series of policy briefs for decision makers on such subjects as water and

climate change, gender and water, and transboundary water cooperation.

Water governance refers to political, social, economic, and administrative systems that regulate the development and management of water resources and provision of water services. The means by which water resources and services are governed depends on the particular socioeconomic, political, and historical situations of individual countries (4).

Most nations are beset by comparable water challenges, such as ensuring water infrastructure, including pipe systems, pumping stations, filtering plants, and sewage systems. Additionally, countries must address political and social issues of improving service coverage, effectiveness, and affordability, raising quality standards, ensuring transparency and accountability of water operations, and resolving international water conflicts. Nations need to meet public health needs as well as ecosystem needs of water.

The massive capital costs and economies of scale that characterize water resources management make it a natural monopoly that needs government regulation, no matter what type of governance is put in place. All forms of water governance need to meet two basic practical and ethical goals: improving the water supply and defending the public interest, recognizing that water is a public good essential for life.

Possible governance structures range from a fully public option, to public–private partnerships, to a completely private option. Water governance arrangements vary greatly across the world. However, globally, who are in both developing

Water and Sanitation-Related Diseases and the Environment: Challenges, Interventions, and Preventive Measures, First Edition.
Edited by Janine M. H. Selendy.
© 2011 Wiley-Blackwell. Published 2011 by John Wiley & Sons, Inc.

and developed countries, some 85% of drinking water is publicly provided.

Fully public governance of water is usually the responsibility of national or municipal authorities. Public officials make decisions, and funding is provided from public coffers, loans, or user charges. Public authorities oversee water operations, set quality and performance standards, and promote public information and participation.

Public–private water partnerships imply privatizing some of the assets or operations of a public water system (5). Such partnerships involve service contracts, management contracts, leases, and concessions. Additionally, ownership of water systems can be split among private and public shareholders in a corporate utility. Majority ownership, however, is usually maintained within the public sector, while private ownership is often legally restricted.

At the other extreme, private governance, which is highly exceptional in today's world, implies that the government transfers the water operations to the private sector. As such, infrastructure, capital investment, commercial risk, and operations and management become the responsibility of the private provider.

Fully private businesses and entrepreneurs tend to occur in areas where public water service is poor or nonexistent. These private businesses range from large multinational water companies to small-scale private water vendors at local level.

Good national governance of water resources, however, is only part of the story. Climate change is making the whole world aware of the interdependence of water resources across boundaries. Ensuring safe water supplies for current and future generations will depend on effective supranational governance of water.

Supranational governance of water involves nonbinding, hortatory resolutions of international governing bodies, particularly the United Nations General Assembly, which covers 192 countries, as well as binding instruments under international law, such as the Protocol on Water and Health under the Convention on the Protection and Use of Transboundary Watercourses and International Lakes, which require ratifying countries to enact legislation in line with the provisions of the convention, once they enter into force (6).

The Convention on the Protection and Use of Transboundary Watercourses and International Lakes, adopted by the UN General Assembly in May 1997, ratified by 36 countries and the European Union, seeks to promote local, national, and regional measures to protect and ensure the ecologically sustainable use of transboundary surface waters and groundwater, preventing transboundary impacts, protecting ecosystems, including from hazardous substances, and preserving ecosystems for future generations as well as promoting public participation in decision making.

The Convention is intended to strengthen national measures for the protection and ecologically sound management of transboundary surface waters and groundwater. The Convention obliges Parties to prevent, control, and reduce water pollution from point and nonpoint sources. The Convention also includes provisions for monitoring, research and development, consultations, warning and alarm systems, mutual assistance, institutional arrangements, and the exchange and protection of information, as well as public access to information (7).

Moreover, several bilateral or multilateral agreements between European countries are based on the principles and provisions of this Convention. A first example was the Danube River Protection Convention in 1994, which develops the Convention's provisions in a more specific subregional context. Other examples are the agreements on the rivers Bug, Meuse, Rhine and Scheldt, on Lake Peipsi, as well as on Kazakh–Russian and Russian–Ukrainian transboundary waters. The most recent examples include the 1999 Rhine Convention and the European Union Water Framework Directive.

The Protocol on Water and Health, under the above water convention, tackles the issue of ensuring access to safe water and adequate sanitation. The Protocol goes beyond the Millennium Development Goal of halving the percentage of those without access to water and sanitation and calls for access to affordable and potable drinking water for all people across the world.

Water is widely recognized in civil society as a human right. The official status of safe drinking water with regard to human rights under international law will be clarified in the course of the coming 2 years by the Independent Expert appointed by the United Nations Secretary General on the issue of human rights obligations related to access to safe drinking water and sanitation (8).

Ultimately ensuring human health through clean water will involve both national and supranational governance, whether at the regional or global levels. International legislative outcomes such as resolutions and conventions on water, ranging from the nonbinding to the legally binding, will be essential to ensure that water as a public good is provided worldwide to all humanity.

REFERENCES

1. World Water Day website. Available at http://www.worldwaterday.org/

2. UN—Water Joint Monitoring Programme report, 2010. Available at http://www.unwater.org/activities_JMP2010.html.

3. UN—Water website. Available at http://www.unwater.org/flashindex.html.

4. UNDP, Water Governance for Poverty Eradication, Available at http://www.undp.org/water/pdfs/241456_UNDP_Guide_Pages.pdf.

5. WaterWiki, Public Private partnerships in water sector. Available at http://waterwiki.net/index.php/Public-Private_Partnerships_in_the_Water_Sector.

6. UN—Water, Institutional Capacity Development in Transboundary Water Management. Available at http://www.unwater.org/downloads/05_Institutional_Capacity_Development_in_Transboundary_Water_Management.pdf.

7. UNECE. The Convention on the Protection and Use of Transboundary Watercourses and International Lakes. Available at http://www.unece.org/env/water/text/text.htm.

8. UN Independent Expert on the issue of human rights obligations related to access to safe drinking water and sanitation. Available at http://www2.ohchr.org/english/issues/water/iexpert/Ind_expert_DeAlbuquerque.htm.

SECTION VII

EMERGING ISSUES IN ECOLOGY, ENVIRONMENT, AND DISEASE

An important fraction of the burden of water-related diseases—in particular water-related vector-borne diseases—is attributable to the way water resources are developed and managed. In many parts of the world the adverse health impacts of water pollution, dam construction, irrigation development, and flood control cause significant preventable disease.

31

CHANGING GEOGRAPHIC DISTRIBUTION OF DISEASE VECTORS

Mary E. Wilson

31.1 INTRODUCTION

Vectors of human disease are responsive to environmental conditions. The survival, distribution, abundance, and behavior of vectors, such as mosquitoes and ticks, are intimately linked to the landscape, the local ecology, and the physicochemical environment. The built environment can also affect vector distribution and behavior, sometimes providing protected niches where vectors can survive. Human activities, especially via travel and trade, move vectors around the world, providing abundant opportunities for vectors to become established in new geographic areas (1). Ticks and perhaps other vectors can be carried long distances by migratory birds.

This chapter will examine how vectors change in geographic distribution and the factors that influence whether presence of a vector will be associated with transmission of human disease, and if so, the intensity of transmission (2). The most important vectors of human disease globally, mosquitoes, are inextricably linked to water, which is essential for their propagation. Other vectors, such as ticks, sandflies, other flies, mites, lice, and fleas, can also transmit human diseases, but the primary focus in this chapter will be on mosquitoes, with a few examples given of other vectors. Vector-borne pathogens span many types of pathogenic organisms, with the most important being the malaria parasite, viruses (e.g., dengue, West Nile, chikungunya, Japanese encephalitis, yellow fever viruses), bacteria (e.g., Lyme disease spirochetes, rickettsiae) (3), and helminths (e.g., filarial worms, onchocerciasis) and nonmalaria protozoa (e.g., those causing Chagas disease, African trypanosomia-

sis) (see Table 31.1). Although this chapter will focus on vector-borne infections of humans, it is important to note that vector-borne infections also affect animals and plants (4). Infections in these populations can have profound implications for food security and access to materials derived from plants and animals. Disruption of food sources can affect socioeconomic and political stability of countries and regions. Among emerging infections, vector-borne infections are relatively common (5). In an analysis of emerging infectious diseases (EID) events between 1940 and 2004, Jones and colleagues found that vector-borne diseases were responsible for 22.8% of EID events and 28.8% in the last decade (6).

31.2 HOW VECTORS MOVE

31.2.1 Mosquitoes

31.2.1.1 Aedes Aegypti Although some movement of vectors occurs through natural movement, such as flying, or passive carriage on winds or with water, extensive transport of vectors is a result of human activity (7). The flight distance of mosquitoes can be measured in meters and infrequently has been demonstrated to be as long as 1–2 km (8, 9). Adult mosquitoes, eggs, and immature forms, in contrast, can easily be carried around the world by air, sea, or land. In past centuries, mosquitoes were transported in ships and introduced into new areas, where they could become established if eco-climatic conditions were appropriate (10). *Aedes aegypti*, the most important vector for dengue viruses, is thought to have originated in West Africa but was introduced into the

Water and Sanitation-Related Diseases and the Environment: Challenges, Interventions, and Preventive Measures, First Edition.
Edited by Janine M. H. Selendy.
© 2011 Wiley-Blackwell. Published 2011 by John Wiley & Sons, Inc.

TABLE 31.1 Most Common Human Vector-Borne Infections by Vector

Vector	Pathogens/Infections
Mosquito	
Aedes	Dengue, yellow fever, chikungunya viruses, filarial pathogens
Anopheles	Malaria (*Plasmodium falciparum, P. vivax, P. malariae, P. ovale, P. knowlesi,* filarial pathogens
Culex	West Nile virus, Japanese encephalitis, filarial pathogens
Flies	
Sandflies	Leishmaniasis, Mayaro virus, sandfly fever virus
Tsetse flies	African trypanosomiasis
Blackflies (*Simulium*)	Onchocerciasis
Ticks	
Argasidae (soft ticks)	*Borrelia* (relapsing fever)
Ixodidae (hard ticks)	*Borrelia burgdorferi* (Lyme), rickettsiae, ehrlichiae, babesia
Bugs (Hemiptera)	
Reduviidae (genus *Triatoma*)	Chagas disease (*Trypanosoma cruzi*)

Americas, probably aboard slave ships in the fifteenth through the seventeenth centuries. Although the voyage from Africa to the Americas could take 4–6 weeks, the water storage casks on the ships were suitable breeding sites, and blood meals were available from the humans on board. Outbreaks of yellow fever occurred on board ships. From the coastal areas where *Ae. aegypti* was introduced and became established, it subsequently spread into the interior. Its presence allowed the introduction and establishment of the yellow fever virus, which caused massive outbreaks in the Americas as far north as Boston (10). The yellow fever virus also became established in nonhuman primates in the Americas, where the virus is maintained today. Although an effective yellow fever vaccine has been available since the 1930s, human cases still occur when unvaccinated humans come into contact with mosquitoes carrying the virus. Mosquitoes can also transmit the yellow fever virus from one human to another in urban settings. Today an estimated 200,000 cases of yellow fever and 30,000 deaths occur worldwide each year, the majority in Africa (11). The number of human cases has increased in the last two decades.

Because of its importance in the transmission of the deadly yellow fever virus, *Ae. aegypti* was targeted by the Pan American Health Organization (PAHO), which adopted a resolution in 1947 calling for the hemisphere-wide eradication of *Ae. aegypti* (12). Twenty-two countries in the Americas were successful (Fig. 31.1; map of the Americas showing distribution of *Ae. aegypti* during 3 different years) and were declared free of *Ae. aegypti*. The largest country, Brazil, eradicated the mosquito by 1958, but the mosquito

reinfested the country in 1967 (12). Attempts at reeradication followed and were initially successful, but the country was reinfested in 1976 and has been infested since then. Today all countries of Latin America are infested with *Ae. aegypti* with the exception of mainland Chile. *Ae. aegypti* is also present in the southern part of the United States. *Ae. aegypti* is now also widely distributed widely in Asia and is also found in parts of Australia (10). It was once established in large areas of Europe. Figure 31.2 shows the contour lines of the January and July isotherms, which are the potential geographical limits in the northern and southern hemispheres for the year-round survival of *Ae. aegypti*.

Ae. aegypti is extremely well adapted to the contemporary urban environment. It enters houses and prefers humans (to animals) as a source of blood meals (13). It is often found in dark places, such as closets, and may rest among clothing. It usually stays within 100 m of human habitats and breeds in artificial containers holding water, such as discarded plastic cups, flowerpots, buckets, and discarded tires (9). For many residents of urban areas in tropics and subtropics piped water is unavailable or supplies are erratic, leading them to store water in containers in the home. In an urban area in Rio de Janeiro state, Brazil, investigators studied the relative contribution of different containers and sites to the population of *Ae. aegypti* mosquitoes (14). They inspected 747 containers in 300 dwellings for the presence of *Ae. aegypti* larvae and pupae. The containers with the highest percentage positive for larvae were tires, jugs, barrels, cisterns, and water tanks—primarily containers that are necessary because of inadequate or unreliable piped water supplies. Overall they found that 90.2% of the larvae and 89.9% of pupae were in containers related to the water supply or trash. Water storage containers also hold a large volume of water, hence, can support larger populations of larvae and pupae than small volume containers can (Fig. 31.3, see photo of Salvador and water tanks in slum areas, by Felix Lam). Although programs to cover the water storage tanks have been started in many areas, many tanks remain uncovered. The habit of *Ae. aegypti* of distributing its eggs among several oviposition sites, makes control of these mosquito populations especially difficult (13).

In an analysis of projected impact of climate change on the range of *Ae. aegypti* in Australia, where drought is already a serious problem and where climate change is expected to exacerbate water shortages, investigators concluded that changes in water storage practices (shifting to storage in and around the home) likely would have a greater effect on mosquito populations and risk for vector-borne infections, such as dengue, than climate change would (16). Many other investigators have assessed or tried to model the potential impact of climate change on the distribution and activity of *Ae. aegypti* (17–20).

31.2.1.2 Aedes Albopictus *Aedes albopictus*, the so-called Asian tiger mosquito, continues to invade new areas;

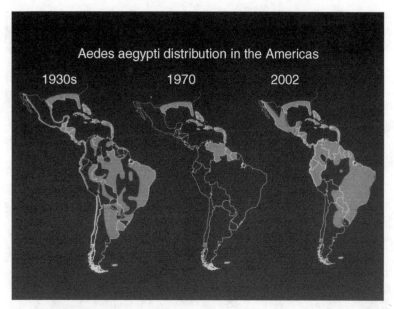

FIGURE 31.1   Map of the Americas showing distribution of *Ae. aegypti* during three different years.

in some regions it has displaced populations of *Ae. aegypti* (7). It is less discriminating than *Ae. aegypti* in its selection of hosts for blood meals and will feed on many vertebrates, including birds and reptiles, though it prefers mammals and will feed on humans if they are available. In a study of *Ae. albopictus* in the Chinese Territory of Macao (situated in the Pearl River delta), for example, humans were the most common host, accounting for 44% of blood meals in the mosquitoes examined (21). *Ae. albopictus* can breed in artificial containers, including used tires, but also breeds in tree holes, other plants, including the decorative bromeliads, and other sites in the natural environment. It is more abundant in areas with vegetation, and is found in suburban and rural areas as well as in cities.

Although it originated in Asia, *Ae. albopictus* has been transported to most continents and has become established widely in the Americas, Europe, and Africa. It was identified in the United States in Memphis, Tennessee, in 1983 (22). When it was found near Houston, Texas, in 1985, local surveillance studies determined that it was already widespread and common in that area (23). A survey of 12 states in 1985 found that it was already present in 84% of the 57 counties surveyed. It has subsequently spread broadly in the continental United States (24). It is able to survive in colder environments than *Ae. aegypti* and has been found as far north as Illinois and Nebraska. It has become established in Mexico and all countries of Central and South America with the exception of continental Chile. It is present in at least

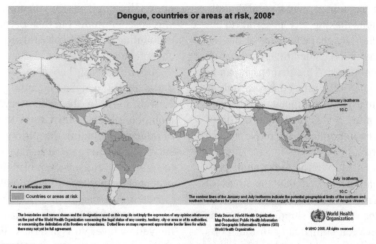

FIGURE 31.2   Map showing countries with dengue and limits for year-round survival of *Ae. aegypti* (WHO).

FIGURE 31.3   Photo of Salvador and water tanks in slum areas, by Felix Lam.

16 European countries with Italy being the most heavily infested. It was first reported to be present in Italy in 1990. In many European countries, the presence of *Ae. albopictus* was first documented after 2000. The mosquito is also established in parts of Africa (at least Nigeria, Gabon, Equatorial Guinea, and Cameroon) (25).

Much of the global dispersal of *Ae. albopictus* can be attributed to international trade in used tires, which provide an ideal habitat for mosquito eggs or immature forms. One examination of 22,000 used tires arriving in Brazil from Japan found that 25% of them held water (26). Investigators found *Ae. aegypti* and three other species of mosquitoes that were not native to that area. Tires are typically shipped in large containers stacked on transport ships. Millions are shipped each year with some ships large enough to carry more than 14,500 units. The containers may be sent directly from dockside to destination, making any meaningful inspection virtually impossible. The characteristics of the *Ae. albopictus* mosquitoes found in the United States suggested that they had originated in northern Asia, perhaps Japan or South Korea (23). At least one of the infestations in Italy is thought to have originated in used tires shipped to Italy from Atlanta, Georgia, in the United States. Interstate traffic of used tires is also thought to be important in the movement of the mosquito within countries.

An analysis of international ship traffic movements coupled with climatic information showed that the areas of vector dispersal were remarkably predictable (27). Tatem and colleagues found that the volume of shipping traffic was more than twice as high between climatically similar ports where *Ae. albopictus* had become established in comparison with those where it had not yet invaded.

Among the reasons for the intense concern about the expanding distribution of *Ae. albopictus* is that this mosquito is competent to transmit a number of viruses that are pathogenic for humans. In the wild, *Ae. albopictus* mosquitoes infected with Eastern equine encephalitis and LaCrosse viruses have been found. It has been the primary vector for several of the outbreaks of chikungunya fever, a viral infection that has caused explosive outbreaks in parts of Africa, the Indian Ocean islands, in many parts of Asia, and even caused a summertime outbreak in northern Italy in 2007 (28–31). A mutation has been identified in the envelope protein gene of the chikungunya virus that allows it to replicate more efficiently in *Ae. albopictus* mosquitoes. Presence of the mutation is associated with increased midgut infectivity of the mosquito and more rapid viral dissemination into the mosquito salivary glands (32). In one outbreak, the mutated strain was absent initially but then became the predominant strain, accounting for >90% of viral sequences, suggesting a fitness advantage of this virus (33).

Ae. albopictus has also been the vector for several outbreaks of dengue, including ones in Macao (21) and Hawaii (34). In the laboratory, it can be infected with a number of other arboviruses that can cause severe disease in humans, including the yellow fever virus. Because it can survive in forests as well as in urban areas it has the potential to serve as a bridge vector, bringing viruses from nonhuman primates and perhaps other animals, into urban areas (35, 36). Concern continues that yellow fever virus could be introduced into urban areas by humans infected and viremic with the yellow fever virus and thus able to spark epidemic urban yellow fever in unvaccinated populations in areas heavily infested with *Ae. aegypti* (37). Although *Ae. albopictus* mosquitoes naturally infected with yellow fever virus have not yet been identified, this vector could enter forested areas inhabited by primates, which can be infected with yellow fever virus in some tropical areas. Areas at highest risk would be those lying in the tropics or subtropics (25, 37, 38).

Laboratory studies of *Ae. aegypti* and *Ae. albopictus* strains of mosquitoes from Florida (USA) have shown that both are susceptible to a chikungunya virus from recent outbreaks and capable of transmitting the virus (39). In the summer of 2009, a cluster of dengue fever cases occurred in the Florida Keys (40). Imported cases of chikungunya fever have been confirmed in the United States in travelers who have visited areas with active transmission, but to date no local transmission has been documented. Although mosquitoes are changing in distribution, it is the human travelers who are infected with the chikungunya or dengue viruses who introduce it into new areas. If competent mosquito vectors are present and climatic conditions are appropriate, local transmission can occur (41). In temperate areas, local transmission is possible only during the hot months of the year. The movement of dengue viruses globally is almost exclusively attributable to the movement of people (42–44).

31.2.1.3 Anopheles gambiae and Brazil

An historical example is relevant in showing the importance of the specific type of vector in risk of transmission of an infection. In 1930, in Natal, Brazil, an entomologist saw, to his surprise, the mosquito *Anopheles gambiae*. Furthermore, he counted approximately 2000 larvae (45). This was unexpected because this mosquito was an African mosquito, not known to inhabit South America. Because dikes had been built along the local river, the saltwater flats had been transformed into freshwater hay fields that could support the breeding of mosquitoes. It was later postulated that rapid French boats (avisos) that made mail runs between Dakar, Senegal and Natal, Brazil were the most likely source. The boats were able to cover the 3300 km in <100 h and anchored within 1 km of the site where the larvae were found. The possibility of introduction via an aircraft could not be excluded. From this small colony that *An. gambiae* established in the marsh near the coast, the mosquitoes spread along the coastal region and moved 200 miles inland into the Jaguaribe River valley. Factors that favored its dissemination from Natal were the confluence of traffic linking the port to other areas. Natal was an ocean port and received steamships that traveled along the coast as well as international steamships. Small sailing vessels that traveled to fishing villages stopped in Natal, and the city served as the terminus of two railroad lines and was the center of a truck transportation network. Malaria was already present in this region of Brazil, so the parasite was present, but the local mosquitoes were not efficient vectors for the malaria parasite. *An. gambiae*, in contrast, entered houses, bred in open pools outside of the forest, and was

highly efficient in transmitting malaria parasites. In 1938 and 1939, almost a decade after the mosquito had first been identified in this region, outbreaks of malaria exploded, causing more than 100,000 cases and about 20,000 deaths. At this point a massive and intense campaign was launched to eradicate this invading vector. With a staff of >3000 persons and at a cost of >USD 2 million (1940 USD), the mosquito was eradicated from Brazil. The last Brazilian-bred *An. gambiae* was collected in November 1940 (45). This eradication effort was completed without DDT, which was not yet available. Today, increasing resistance of vectors to insecticides complicates control efforts (46).

Tatem and colleagues (27) also looked at *An. gambiae*, which in contrast to *Ae. albopictus*, has rarely spread from its geographic origin in Africa. They postulated that the low volume of sea traffic from African ports to climatically similar destinations may partly explain this. The increase in the volume of air traffic from Africa that is currently underway could increase the risk of introduction of this mosquito into climatically similar areas.

31.2.2 Air Transport of Vectors

Mosquitoes and other insects can survive airplane flights, and for this reason disinsection (fumigation of airplane), usually with pyrethroid aerosol sprays, is sometimes carried out (47). After the disastrous consequences of the establishment of *An. gambiae* in Natal in the 1930, the Brazilian government required inspections of aircraft arriving from Africa. In one 9-month period in 1941–1942, *An. gambiae* was found seven times on aircraft. Inspectors also found 132 other mosquitoes and 2 live tsetse flies (vector for African trypanosomiasis, also known as sleeping sickness) (48). The U.S. Public Health service published results of insects found in aircraft during a 13-year period. They reported finding 20,000 insects, including 92 species of mosquitoes, 51 of which were not known to live in mainland United States, Hawaii, or Puerto Rico (49). Mosquitoes have been found in the airplane cabins, in luggage storage compartments, in storage areas, and in airfreight containers. It is clear that mosquitoes infected with malaria parasites can complete a long international flight and survive long enough to infect individuals in areas that are not endemic for malaria, such as Europe. As of August 1999, 89 cases of airport malaria had been reported, the overwhelming majority from Europe (France, Belgium, and United Kingdom had the highest numbers), reflecting the large volume of nonstop flights from Africa (50). In 1994 during a three-week period when six cases of airport malaria occurred in or near Charles de Gaulle Airport, an estimated 2000–5000 anopheline mosquitoes were imported in the 250–300 aircraft that arrived from malaria-endemic areas of Africa (51). Mosquitoes can even travel, rarely, in suitcases (baggage malaria). At least two cases of malaria acquired from such mosquitoes have been reported (48).

Although mosquitoes can survive air travel, most arrivals by air travel do not lead to the establishment in the new region. On aircraft, typically only a few adults arrive, in contrast to boats with tires and water containers, in which large numbers of immature forms can be imported. In many instances, the eco-climatic conditions in the new environment are not favorable for survival and spread.

As West Nile virus spread throughout continental United States, Kilpatrick and colleagues (52) assessed the pathways by which the West Nile virus could potentially reach Hawaii, an event that could be devastating for the local bird populations, as well as causing human disease and disrupting tourism to the islands. They concluded that the virus could reach Hawaii from North America in humans, wind-transported mosquitoes, human-transported mosquitoes, human-transported birds or other vertebrates, and migratory birds. After doing a quantitative estimate for each, they concluded that the highest risk was posed by mosquitoes transported by airplanes and human-transported birds that are exempt from quarantine (52).

31.2.3 Japanese Encephalitis

Transmission of the mosquito-transmitted Japanese encephalitis (JE) virus has expanded into new territories in Asia, in addition to Australia, as described later. JE virus is transmitted primarily by the mosquito *Culex tritaeniorhynchus*, which often breeds in irrigated rice fields. Water birds serve as reservoir hosts, but pigs are amplifying hosts; other animals, as well as humans, can be infected. In contrast to dengue, movement of infected humans is unimportant to JE epidemiology because the level of viremia in humans is usually low grade and of short duration, even though infection can be fatal or leave those infected with severe neurologic impairment. Migrating birds may help to disperse the virus. Although an effective vaccine is available, it is not used routinely in the poorer countries in the region. Reasons for the expansion of JE human infections in the regions affected in Asia include marked increase in the pig rearing and expansion of irrigated rice farming in size and intensity. Because of population growth and the changes in rice irrigation and pig rearing, the number of people living in high-risk areas has increased. In Myanmar, for example, between 1990 and 2005, the rice paddy area increased 47% and pork production increased almost fourfold (53). The largest number of people living in close proximity to rice irrigation live in Bangladesh (more than 116 million), a country where JE vaccine is not routinely used (53). The epidemiology of JE could shift in the future if climate changes lead to alternations in bird migratory patterns or agricultural practices. Overall, the number of cases may decrease in the future because of large-scale vaccination programs in China, but JE human infections are likely to continue and potentially worsen in countries that cannot afford the vaccine.

31.2.4 Shift in Distribution Related to Climate Change

Because vectors are responsive to eco-climatic conditions, changes in temperature and rainfall and increased occurrence of extreme events could influence geographic range, abundance, and activity of some vectors of human infections (54). Cases of tick-borne encephalitis, for example, have occurred at higher altitudes as well as latitudes, in Europe (55). This is thought to be attributable, at least in part, to rising temperatures. The main tick vector, *Ixodes ricinus* in Europe, has extended its range north and westwards (56). For many vector-borne infections, it is difficult to determine the relative contribution of climate change when many other ecological and environmental factors and human activities are changing simultaneously (17). Warmer or drier conditions in some areas can potentially disrupt or reduce transmission of some infections. Under warmer conditions, insects may be able to overwinter in areas where they currently cannot (57).

31.2.5 Dispersal without Human Assistance

Although mosquitoes are weak flyers, unusual weather patterns may carry mosquitoes long distances. Japanese encephalitis, a mosquito-borne viral infection, first appeared in Australia in 1995 in the Torres Strait. Investigators used a computer simulation to assess whether weather conditions before the 1995 and 1998 JE outbreaks generated winds that were sufficient to have carried mosquitoes from Papua New Guinea (PNG) to the Torres Strait and Cape York of Australia (58). They found that during both years low pressure systems west of the Cape York Peninsula provided conditions that could have carried mosquitoes to Australia. For this to have occurred, large numbers of JE-infected mosquitoes must have been present in PNG and the mosquito species involved must be able to reach high altitudes. The authors note that female *Culex annulirostris* mosquitoes, the implicated species, have been collected at altitudes of 310 m. Transport of infected vectors (Culicoides midges) by wind has been thought to be responsible for incursions of blue-tongue virus, an animal pathogen, into northern Australia from Indonesia (59).

In contrast, some species of black flies, for example, the *Simulium damnosum*—the vector for onchocerciasis, have been reported to travel 150–200 km without human assistance. Movement may be aided by wind with the passage of the InterTropical Convergence Zone (60).

31.2.6 Role of Migratory Birds

Migrating birds that carry ticks may introduce these potential vectors and the pathogens they carry into new areas. In an examination of a potential pathway for the introduction of vectors and pathogens into a new region, in Sweden, Bjoers-dorff and colleagues (61) examined migratory passerine birds for ticks at a stopover site on the east coast of Oland in 1996. Of the 3054 birds of 56 species examined, *Ixodes ricinus* ticks were collected from 2.4% of birds of 18 species. Ehrlichia DNA was identified in 8% of the 112 nymph-stage ticks. Based on the estimated 100 million birds that migrate through Sweden annually, the authors projected that almost 600,000 Ehrlichia-infected ticks were imported into Sweden that year. The birds that they examined arrived from continental Europe, North Africa, and sub-Saharan Africa. The tick species found is also the main vector for *Borrelia burgdorferi* (the cause of Lyme disease) and tick-borne encephalitis virus in Europe. Unanswered by this study is whether this movement has increased and whether it has led to the establishment of the vector and pathogen in a new geographic area.

Migrating birds can also play a role in transporting human pathogens into areas where competent vectors already exist (62). In the fall of 1998, West Nile virus was isolated from migrating white storks (*Ciconia ciconia*) that had landed in a town in southern Israel (63). The flock of 1200 had migrated from southern Europe, where West Nile virus had been causing outbreaks since 1996. See also the earlier discussion about JE virus infections.

31.2.7 Vector without Disease

Presence of disease vectors in a region does not mean that a vector-borne infection will occur. Although multiple species of mosquitoes that were competent to transmit West Nile (WN) virus infested the Americas, the disease, West Nile fever, did not exist (except for occasional instances of imported cases) until the virus was introduced into the eastern United States in 1999. It then rapidly spread across the North America. By the end of 2007, more than 11,000 cases of neurologic infection from WN virus and >1,000 deaths had been reported in the United States (64). Although the means of introduction was never documented, it was most likely imported in an infected mosquito or bird (64). Genetic sequences of the WN isolates found in New York were similar to those in isolates from Israel, suggesting the likely origin was the Middle East or eastern Europe (65, 66). Over a period of several years WN virus also spread into some parts of the Caribbean and Central and South America. Already in place were abundant potential vectors, amplifying hosts (birds), and favorable environmental conditions to allow the virus to flourish and expand its distribution. Because birds, the most important reservoir host, could fly and some migrated, the virus spread easily. It is now endemic in parts of the Americas and is expected to continue to cause cases of human infection. Although human infections have a distinct seasonal pattern, being more common in the warmer months

of the year, the virus can survive the winter in hibernating (diapause) adult female mosquitoes, and probably in some birds, and perhaps also in rodents (64).

As already noted, mosquitoes that are competent to transmit dengue and yellow fever viruses—as well as JE virus and malaria parasites—infest parts of the United States (as well as parts of Europe and many other regions).

Populations can be naturally resistant or can be made resistant to some vector-borne infections. One common approach is the use of vaccines. Available vaccines, for example, are highly effective in protecting against JE and yellow fever infections, but they are often not available to those populations at highest risk for infection.

31.2.8 Presence of Vectors and Pathogens in the Absence of Recognized Disease

Vectors and pathogens can be present in an area where human disease is not recognized. In a recent study investigators collected triatomine bugs, the vector for Chagas disease (American trypanosomiasis), in and around homes in the Tucson, Arizona (southwestern United States) areas (67). They examined the triatomines with polymerase chain reaction (PCR) for the presence of *Trypanosoma cruzi*. They found that 42% of the 164 insects collected were infected. The authors speculated that the absence of reports of human disease in this area might be attributed to good housing conditions (absence of crevices where insects can reside), availability of animal hosts for blood meals, misdiagnosis of human cases, and characteristics of the vector or the local parasite associated with reduced infectivity or virulence.

Studies of urban populations along the United States (Texas)–Mexico border on either side of the Rio Grande River found that residents of Mexico were more likely to have serologic evidence of dengue infection than were residents in Texas. Although competent vectors (*Ae. aegypti*) infested urban areas on both sides, air conditioning was more common in Texas and was associated with a reduced likelihood of being seropositive for dengue infection (68). Housing conditions, such as presence of screens and availability of running water (so that water is not stored in containers in the household), could reduce exposure to some vector-borne infections.

31.2.9 Tools

Many new tools have been useful in tracing the origin of vectors and in anticipating where they might become established. Analysis of genetic characteristics of vectors, isozyme analyses, examination of DNA sequences in a mitochondrial subunit of nicotinamide adenine dinucleotide (NAD) dehydrogenase, and other approaches have been used to determine the relatedness and origin of vectors. Surveillance networks, such as GeoSentinel, and organizations, such

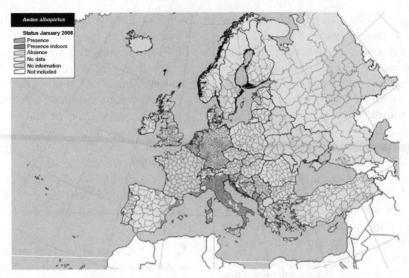

FIGURE 31.4 Map: current distribution of *Ae. albopictus* in Europe [map 1 in annex of ECDC technical report 2009] Available at: http://ecdc.europa.eu/en/activities/pages/programme_on_emerging_and_vector-borne_diseases_maps.aspx.

as World Health Organization (WHO), Centers for Disease Control and Prevention (CDC), European Centre for Disease Control and Prevention (ECDC), and ProMED, can help to track the spread of vector-borne infections. Other tools via the Internet provide access to essential data, such as the Disease Vector Database, a global, free, public, Web-accessible resource with data on the geographical distribution of vectors of infectious diseases and reservoirs (69). It can be accessed at http://www.diseasevectors.org. Groups have also convened, drawing on many kinds of expertise including disease modeling, climate sciences, and entomology to develop risk maps, so that public health experts can better prepare for the potential introduction and spread of vector-borne infections. An excellent recent example is the ECDC Technical Report that includes risk maps for *Ae. albopictus* in Europe. The report, freely accessible online, includes maps of the current distribution of *Ae. albopictus* in Europe, the potential weeks of vector activity, and projected distribution and activity under different climate change scenarios (Fig. 31.4).

31.3 SUMMARY

Global changes today will lead to continued changes in the geographic distribution of vectors and the diseases that they spread among humans, animals, and plants. Among the most potent forces changing the distribution of vectors are travel and trade, though changes in the natural environment through human activities (e.g., deforestation), alterations in the built environment (e.g., dams, urbanization), climate change, and human demography

and activities all contribute as well. A theme of water is a common element throughout—too little, too much, wrong place.

REFERENCES

1. Wilson ME. Global travel and emerging infections. In: *Infectious Disease Movement in a Borderless World*. Microbial Threats Forum, Institute of Medicine. Workshop Summary. Washington, DC: The National Academies Press, 2010, pp. 90–104, 126–129.

2. Wilson ME. Geography of infectious diseases. In: Cohen J, Powderly WG,editors. *Infectious Diseases*, 2nd edition. New York: Mosby, 2004, pp. 1419–1427.

3. Walker DH, Paddock CD, Dumler JS. Emerging and re-emerging tick-transmitted rickettsial and ehrlichial infections. *Med Clin N Am* 2008;92(6):1345–1362.

4. Bram RA, George JE, Reichard FE, Tabachnick WJ. Threat of foreign arthropod-borne pathogens to livestock in the United States. *J Med Entomol* 2002;39(3):405–416.

5. Gubler DJ. Resurgent vector-borne diseases as a global health problem. *Emerg Infect Dis* 1998;4:442–450.

6. Jones KE, Patel NG, Levy MA, Storeygard A, Balk D, Gittleman JL, Daszak P. Global trends in emerging infectious diseases. *Nature* 2008;451:990–994.

7. Lounibos LP. Invasions by insect vectors of human disease. *Annu Rev Entomol* 2002;47:233–266.

8. Shannon R, Davis N. The flight of *Stegomyia aegypti* (L.). *Am J Trop Med Hyg* 1930;10:145–150.

9. Reiter P. Oviposition, dispersal, and survival in *Aedes aegypti*: implications for the efficacy of control strategies. *Vector-borne Zoonotic Dis* 2007;7(2):261–273.

10. Reiter P. Yellow fever and dengue: a threat to Europe? *Euro-surveillance*. Published on March 11, 2010. Available at www.eurosurveillance.org

11. World Health Organization. Yellow fever. Fact sheet, December 2009. Accessed March 30, 2010. Available at http://www.who.int/mediacentre/factsheets/fs100/en/index.html.

12. PAHO. A timeline for dengue in the Americas to December 31, 2000 and noted first occurrences. Prepared by J. Schneider, D. Droll, June, 2001. Available at www.paho.org.

13. Suwonkerd W, Mongkalangoon P, Parbaripai A, Grieco J, Achee N, Roberts D, Chareonviriyaphap T. The effect of host type on movement patterns of *Aedes aegypti* (Diptera: Culicidae) into and out of experimental huts in Thailand. *J Vector Ecol* 2006;31(2):311–318.

14. Medronho RA, Macrini L, Novellion DM, Lagotta MTF, Camara VM, Pedreia CE. *Aedes aegypti* immature forms distribution according to type of breeding site. *Am J Trop Med Hyg* 2009;80(3):401–404.

15. Reiter P, Amador MA, Anderson RA, Clark GG. Short report: dispersal of *Aedes aegypti* in an urban area after blood feeding as demonstrated by rubidium-marked eggs. *Am J Trop Med Hyg* 1995;52(2):177–179.

16. Kearney M, Porter WP, Williams C, Ritchie S, Hoffmann AA. Integrating biophysical models and evolutionary theory to predict climatic impacts on species' ranges: the dengue mosquito *Aedes aegypti* in Australia. *Functional Ecol* 2009;23:528–538.

17. Gubler DJ, Reiter P, Ebi KL, Yap W, Nasci R, Patz JA. Climate variability and change in the United States: potential impacts on vector- and rodent-borne diseases. *Environ Health Perspect* 2001;109(Suppl 2):223–233.

18. Jetten TH, Focks DA. Potential changes in the distribution of dengue transmission under climate warming. *Am J Trop Med Hyg* 1997;57(3):285–297.

19. Hales S, de Wet N, Maindonald J, Woodward A. Potential effect of population and climate changes on global distribution of dengue fever: an empirical model. *Lancet* 2002;360:830–834.

20. Johansson MA, Dominica F, Glass GE. Local and global effects of climate on dengue transmission in Puerto Rico. *PLoS NTD* 2009;3(2):e382.

21. Almeida APG, Baptista SSSG, Sousa CAGCC, Novo MTLM, Ramos HC, Panella NA, Godsey M, Simoes MJ, Ansetmo AL, Komar N, Mitchell CJ, Ribeiro H. Bio-ecology and vectorial capacity of *Aedes albopictus* (Diptera: Culicidae) in Macao, China, in relation to dengue virus transmission. *J Med Entomol* 2005;42(3):419–428.

22. Reiter P, Darsie R. *Aedes albopictus* in Memphis, Tennessee (USA): an achievement of modern transportation? *Mosquito News* 1984;44:396–399.

23. Hawley WA, Reiter P, Copeland RS, Pumpini CB, Craig GB. *Aedes albopictus* in North America: probable introduction in used tires from northern Asia. *Science* 1987;1114–1116.

24. Moore DG, Mitchell CJ. *Aedes albopictus* in the United States: ten-year presence and public health implications. *Emerg Infect Dis* 1997;3(3):329–334.

25. Reiter P. A mollusc on the leg of a beetle: human activities and the global dispersal of vectors and vector-borne pathogens. In: *Infectious Disease Movement in a Borderless World*. Microbial Threats Forum, Institute of Medicine. Workshop Summary. Washington, DC: The National Academies Press, 2010, pp. 150–165, 175–178.

26. Craven RB, Eliason DA, Francy DB, Reiter P, Campos EG, Jakob WL, Smith GC, Bozzi DJ, Moore CG, Maupin GO, Monath TP. Importation of *Aedes albopictus* and other exotic mosquito species into the United States in used tires from Asia. *J Am Mosquito Control Assoc* 1988;4(2):138–142.

27. Tatem AJ, Hay SI, Rogers DJ. Global traffic and disease vector dispersal. *Proc Natl Acad Sci* 2006;103(16):6242–6247.

28. Rezza G, Nicolietti L, Angelini R, Romi R, Finarelli AC, et al. Infection with chikungunya virus in Italy: an outbreak in a temperate regions. *Lancet* 2007;370:1840–1846.

29. Reiter P, Fontenille D, Paupy C. *Aedes albopictus* as an epidemic vector of chikungunya virus: another emerging problem? *Lancet Infect Dis* 2003;3:463.

30. Simon F, Savini H, Parola P. Chikungunya: a paradigm of emergence and globalization of vector-borne diseases. *Med Clin North Am* 2008;92(6):1323–1434.

31. Charrel RN, de Lamballiere X, Raoult D. Chikungunya outbreaks—the globalization of vector-borne diseases. *N Engl J Med* 2007;356(8):769–771.

32. Tsetsarkin KA, Vanlandingham DL, McGee CE, Higgs S. A single mutation in chikungunya virus affects vector specificity and epidemic potential. *PLoS Pathog* 2007;3(12):e201.

33. Schuffenecker I, Iteman I, Michault A, et al. Genome micro-evolution of chikungunya viruses causing the Indian Ocean outbreak. *PLoS Med* 2006;3:e263.

34. Effler PVL, Pang L, Kitsutani P, Vorndam V, Nakata M, Ayers T, Elm J, Tom T, Reiter P, Rigau-Perez JG, Hayes JM, Mills K, Napier M, Clark GG, Gubler DJ, Hawaii Dengue Outbreak Investigation Team. Dengue fever, Hawaii, 2001–2002. *Emerg Infect Dis* 2005;11(5); 742–749.

35. Lourenco-de-Oliveira R, Castro MG, Braks MAH, Lounibos LP. The invasion of urban forest by dengue vectors in Rio de Janeiro. *J Vector Ecol* 2004;29(1):94–100.

36. Maciel-de-Freitas R, Neto RB, Goncalves JM, Codeco CT, Lorenco-de-Olivera R. Movement of dengue vectors between the human modified environment and an urban forest in Rio de Janeiro. *J Med Entomol* 2006;43(6):1112–1120.

37. Monath T. Facing up to the re-emergence of urban yellow fever. *Lancet* 1999;353(9164):1541.

38. Massad E, Burattini MN, Coutinho FAB, Lopez LF. Dengue and the risk of urban yellow fever reintroduction in Sao Paulo State, Brazil. *Rev Saude Publica* 2003;37(4):477–484.

39. Reiskind MH, Pesko K, Westbrook CJ, Mores CN. Susceptibility of Florida mosquitoes to infection with chikungunya virus. *Am J Trop Med Hyg* 2008;78:422.

40. Centers for Disease Control and Prevention. Locally acquired dengue – Key West, Florida, 2009–2010. MMWR. 2010;59:577-81, http://www.cdc.gov/dengue/epidemiology/local_dengue.html

41. Charrel RN, de Lamballerie X, Raoult D. Seasonality of mosquitoes and chikungunya in Italy. *Lancet* 2008;8:5–6.

42. Wilder-Smith A, Gubler DJ. Geographic expansion of dengue: the impact of international travel. *Med Clin North Am* 2008;92:1377–1390.

43. Stoddard ST, Morrison AC, Vazquez-Prokopic GM, Soldan VP, Kochel TJ, Kitron U, Elder JP, Scott TW. The role of human movement in the transmission of vector-borne pathogens. *PLoS NTD* 2009;3(7):e481.

44. Harrington LC, Scott TW, Lerdthusnee K, Coleman RC, Costero A, Clark GG, Jones JJ, Kitthawee S, Kittayapong P, Sithiprasasna R, Edman JD. Dispersal of the dengue vector *Aedes aegypti* within and between rural communities. *Am J Trop Med Hyg* 2005;72(2):209–220.

45. Soper FL, Wilson DB. *Anopheles gambiae in Brazil, 1930–1940.* New York: The Rockefeller Foundation, 1943, 262 pp.

46. Hemingway J, Ranson H. Insecticide resistance in insect vectors of human disease. *Annu Rev Entomol* 2000;45: 371–391.

47. Russell RC. Survival of insects in the wheel bays of a Boeing 747B aircraft on flights between tropical and temperate airports. *Bull WHO* 1987;65:659–662.

48. Gratz NG, Steffen R, Cocksedge W. Why aircraft disinsection? *Bull WHO* 2000;78:995–1004.

49. Hughes JH. Mosquito interceptions and related problems in aerial traffic arriving in the United States. *Mosquito News* 1961;21:93–100.

50. Isaaccson M. Airport malaria: a review. *Bull World Health Organ* 1989;67:737–743.

51. Giacomini T, et al. Study on six cases of malaria contracted near Roissy-Charles de Gaulle in 1994. Preventive measures necessary in airports. *Bull de l'Acad Natl Med* (France) 1995;179(2):335–353 [in French].

52. Kilpatrick AM, Gluzberg Y, Burgett J, Daszak P. Quantitative risk assessment of the pathways by which West Nile virus could reach Hawaii. *EcoHealth* 2004;1:205–209.

53. Erlanger TE, Weiss S, Keiser J, Utzinger J, Wiedenmayer K. Past, present, and future of Japanese encephalitis. *Emerg Infect Dis* 2009;15(1):1–7.

54. Dohm DJ, O'Guinn ML, Turell MJ. Effect of environmental temperature on the ability of *Culex pipiens* (Diptera: Culicidae) to transmit West Nile virus. *J Med Entomol* 2002;39 (1):221–225.

55. Lukan M, Bullova E, Petko B. Climate warming and tick-borne encephalitis, Slovakia. *Emerg Infect Dis* 2010;16(3):524–526.

56. Mansfield KL, Johnson N, Phipps LP, Stephenson JR, Fooks AR, Solomon T. Tick-borne encephalitis virus—a review of an emerging zoonosis. *J Gen Virol* 2009;90:1781–1794.

57. Bale JS, Hayward SAL. Insect overwintering in a changing climate. *J Exp Biol* 2010;213:980–994.

58. Richie SA, Rochester W. Wind-blown mosquitoes and introduction of Japanese encephalitis into Australia. *Emerg Infect Dis* 2001;7(5):900–903.

59. Melville LF, Prichard LI, Hunt NT, Daniels PW, Eaton B. Genotypic evidence of incursions of new strains of bluetongue viruses in the Northern Territory. *Arbovirus Res Aust* 1997;7:181–186.

60. Garms R, Walsh JF. The migration and dispersal of black flies: *Simulium damnosum* S.L., the main vector of human onchocerciasis. In: Kim KC, Merritt RW,editors. *Black Flies: Ecology, Population Management, and Annotated World List.* University Park, PA: Pennsylvania University Press, 1987, pp. 201–214.

61. Bjoersdorff A, Bergstrm S, Massung RF, Haemig PD, Olsen B. Ehrlichia-infected ticks on migrating birds. *Emerg Infect Dis* 2001;7(5):877–879.

62. Rapole JH, Derrickson SR, Hubalek Z. Migratory birds and spread of West Nile virus in the Western Hemisphere. *Emerg Infect Dis* 2000;6(4):319–328.

63. Malkinson M, Banet C, Weisman Y, Pokamonski S, King R, Duebel V. Intercontinental transmission of West Nile virus by migrating white storks. *Emerg Infect Dis* 2001;7(3):540.

64. Peterson L, Hayes E. West Nile virus in the Americas. *Med Clin North Am* 2008;92(6):1307–1322.

65. Deubel V, Gubler DJ, Layton M, Malkinson M. West Nile virus: a newly emergent disease. *Emerg Infect Dis* 2001;7(3):536.

66. Gubler DJ. The continued spread of West Nile virus in the western hemisphere. *Clin Infect Dis* 2007;45(8):1039–1046.

67. Reisenman CE, Lawrence G, Guerenstein PG, Gregory T, Dotson E, Hildebrand JG. Infection of kissing bugs with *Trypanosoma cruzi*, Tucson, Arizona, USA. *Emerg Infect Dis* 2010;16(3):400–405.

68. Reiter P, Lathrop S, Bunning M, Biggerstaff B, Singer D, Tiwari T, Baber L, Amador M, Thirion J, Hayes J, Seca C, Mendez J, Ramirez B, Robinson J, Fawlings J, Vorndam V, Waterman S, Gubler D, Clark G, Hayes E. Texas lifestyle limits transmission of dengue virus. *Emerg Infect Dis* 2003;9:86–89.

69. Moffett A, Strutz S, Guda N, Ganzalez C, Ferro MC, Sanchez-Cordero V, Sarkar, S. A global public database of disease vector and reservoir distributions. *PLoS NTD* 2009;3(3):e378.

70. Barrett ADT, Higgs S. Yellow fever: a disease that has yet to be conquered. *Annu Rev Entomol* 2007;52:209–229.

71. Beebe JW, Cooper RD, Mottram P, Sweeney AW. Australia's dengue risk driven by human adaptation to climate change. *PLoS NTD* 2009;3(5):e429.

72. European Centre for Disease Prevention and, Control., ECDC Technical Report. Development of *Aedes albopictus* risk maps. Stockholm, May 2009. Available at http://ecdc.europa.eu/en/activities/pages/programme_on_emerging_and_vector-borne_diseases_maps.aspx.

32

MALARIA IN THE BRAZILIAN AMAZON

Marcia C. Castro and Burton H. Singer

32.1 INTRODUCTION

Malaria transmission in Brazil is restricted to the Amazon region (1), and accounts for more than 50% of all cases in Latin America. It is characterized by a dynamic interplay of social, environmental, economic, political, and behavioral factors (2, 3), operating at multiple spatial and temporal scales (4). Transmission patterns are likely to be impacted in the future by climate change, recently considered as the "biggest global health threat of the 21st century" (5–7). In addition, continued changes to the local environment alter the nature and intensity of the human–vector contact (8). Planned and future infrastructure projects, either aimed at promoting development of the region or to better respond and adapt to climate change, may exacerbate the burden of malaria if their negative impacts are not anticipated and mitigated (9–11). Moreover, intense human mobility in the region contributes to the spread of malaria across the region (12–14).

Mitigation of malaria in the Amazon is no easy task. The Amazon Basin covers 7% of the Earth's surface and roughly half of South America. Extending over nine countries—Bolivia, Brazil, Colombia, Ecuador, French Guiana, Guyana, Peru, Suriname, and Venezuela, it has significant importance for the regional climate (15). Brazil holds the largest fraction of the forest, approximately 70% (16). Brazil itself is divided into five major administrative regions (North, Northeast, Center-West, Southeast, and South), and the legal Brazilian Amazon occupies 61% of the Brazilian territory. The Brazilian Amazon is comprised of the states of the North region, one state from the Center-West region (Mato Grosso), and most of Maranhão state (west of longitude 44°), located in the Northeast region (Fig. 32.1). According to the last Brazilian Census (2010), 24.9 million people lived in the Amazon region, or 13.4% of the national population. In 2008, approximately 59% and 63% of the population in the Amazon had access to sanitation and clean water (17).

The Brazilian Amazon experienced its most significant population growth during the past four decades. Between 1970 and 2007, the region grew from 7.8 million people to 23.6 million, an average growth rate of 3% per year against 1.8% observed during the same period for the entire country (18). This was mostly a result of in-migration in response to governmental incentives in support of agriculture, mineral extraction, and wide-ranging human settlement (3, 19–23). During the same period, the Amazon also observed significant environmental change. In 1978, approximately 169.9 thousand km^2 of the forested area had been removed, and by 2003 the cleared area amounted to 648.5 thousand km^2 (16.2% of the initially forested area of the Amazon) (24). Much of the disturbances to the forest contributed to an increase in the number of water habitats suitable for *Anopheles* mosquitoes (the malaria vectors) breeding (25, 26). Migratory movements in response to governmental incentives brought to the region a largely malaria naive population (27, 28). In such a dynamically changing environment, malaria took its toll. A steady increase in malaria cases was observed during the 1970s and 1980s, and the levels have fluctuated since the 1990s, most likely in response to new control strategies and, possibly, to the occurrence of extreme climatic events (Fig. 32.2).

Since 1993, the Amazon region accounted for more than 99% of all malaria cases in Brazil. Cases are heterogeneously distributed in the region, mostly concentrated in agricultural settlement areas, mining camps, and areas of land invasion. A new concept of malaria (frontier malaria) was introduced

Water and Sanitation-Related Diseases and the Environment: Challenges, Interventions, and Preventive Measures, First Edition.
Edited by Janine M. H. Selendy.

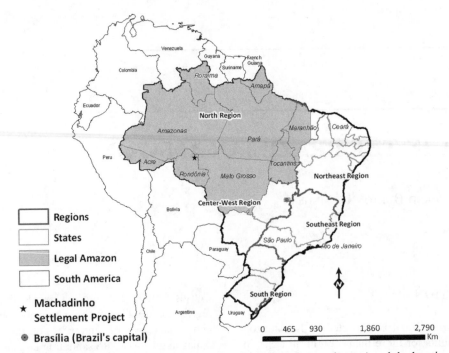

FIGURE 32.1   Major administrative divisions of Brazil (regions and states) and the location of the Brazilian Amazon.

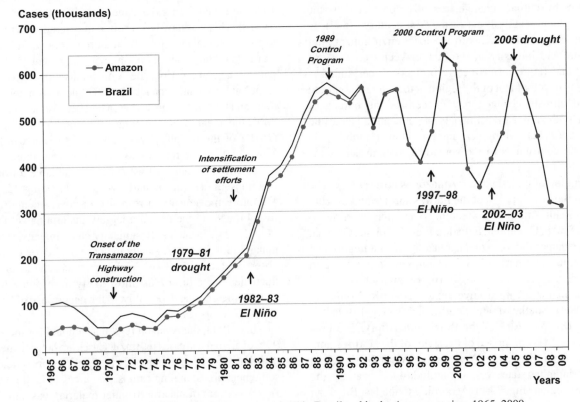

FIGURE 32.2   Number of malaria cases reported in Brazil and in the Amazon region, 1965–2009. Source: Brazilian Ministry of Health. *Note*: Historical events connected with the expansion of the Amazon frontier, and extreme climatic conditions of significance are indicated in the graph. The 1989 Control Program was closed in 1996 (136). Data for 2009 are subjected to change.

to describe the complex factors, operating at multiple scales in space and time, which are associated with malaria risk in the Amazon. The formulation is multidisciplinary in nature, and brings in the need to incorporate social sciences in the conceptual framework of transmission (2, 4, 29). It is particularly suitable to describe the dynamics of malaria transmission in the rapidly transforming ecosystems of settlement areas in the Amazon, and has no parallel in other endemic areas.

In this chapter we discuss the process of malaria transmission in the Brazilian Amazon, characterizing its idiosyncrasies, framed within an historical context, and highlighting current and future challenges for controlling the disease. Lessons taken from the Amazon case improve our understanding of the potential for malaria outbreaks consequential to environmental disturbance accompanied by inadequate planning for development of new settlement areas. Ultimately, the Amazon example sheds light on the choice of strategies that should be adopted in the future in order to minimize disease transmission, to prevent transmission following the construction of new infrastructure projects, and to mitigate potential changes in the pattern of malaria transmission due to extreme climatic events.

The remainder of this chapter is organized in six sections. We start with a brief review (Section 32.2) of important historical facts regarding malaria in Brazil generally and in the Amazon in particular. The objective is to facilitate understanding of current patterns of transmission. Section 32.3 contains a summary of evolutionary and behavioral aspects of *Anopheles* malaria vectors of importance in the Brazilian Amazon. Section 32.4 specifically addresses the drivers and characteristics of frontier malaria, and the following section provides an example of frontier malaria, based on a spatiotemporal analysis of a settlement project in the western Brazilian Amazon. Section 32.6 emphasizes some of the current and future challenges for malaria control in the Brazilian Amazon, highlighting areas where further research is urgently needed. Finally, we conclude with some remarks on lessons learned and on alternative strategies for mitigation of the current and likely future malaria transmission in the region.

32.2 HISTORICAL PERSPECTIVE

Available evidence based on historical documents, studies in genetics and evolutionary biology, epidemiological issues, and archaeological discoveries suggests that human malaria was not present in the New World prior to the European conquest (30). The earliest references to "fevers" in Brazil are letters written by Jesuits in the sixteenth century (31–33). One of the most detailed references dates back to 1587 (34), which provided evidence that at certain times and places, indigenous people suffered from tertian and quartan fevers.

Usual treatment among the Portuguese colonizers was based on blood-letting and purging, and later on quinine (31). The latter, however, was often overused (35), and the first cases of *Plasmodium falciparum* resistance to quinine were observed in Brazil as early as 1910 (36), including cases in the Amazon region (37).

The earliest map of the geographical distribution of malaria in Brazil was published in 1885 (38). It indicated that the disease was present and endemic in areas along the major rivers of the country, particularly in the Amazon region, and less pronounced in the most southern portion of the country. This spatial pattern was corroborated by field observations collected along river valleys (39), and by observations of enlarged spleens among individuals living along the margins of major rivers from São Paulo state to the Amazon region (40). In the early 1900s, Rio de Janeiro (the capital of Brazil at the time) and the states of the Amazon region were considered as areas where malaria transmission was severe (39).

In the case of the Amazon, a series of reports released in the early 1900s revealed the poor sanitary conditions of the region, and portrayed malaria as very severe in rubber plantations and proximal to rivers (37, 41–44). Health assistance was precarious or totally absent, and malaria was considered to be the most severe disease prevailing in the region. A survey conducted among children from 6 months to 10 years of age in Manaus, the capital of Amazonas state, between 1905 and 1909, indicated that living in proximity of swamps considerably increased the risk of malaria. Blood examination through microscopy showed that 51.6% and 40.1% of children living near swamps in the suburbs and in the city, respectively, were infected with malaria parasites (44).

The sheer size of the Amazon region, local climate and hydrologic characteristics, and vector behavior are important challenges for controlling malaria. During the rainy season the water level of rivers increases dramatically, flooding the areas immediately proximal to the margins. As the rainy season ends, the water level decreases, and pools of water become widespread. Many of these offer ideal conditions for mosquito breeding. Larval control is not feasible given the size of the region and the extremely difficult task of identifying and reaching all, or even a large majority of breeding habitats. Hydraulic construction (ditches) near the margins of rivers, which would allow the accumulated water to flow back to the river, was proposed in 1916, but never implemented on a large scale (41). Finally, an outdoor biting pattern has been observed for some Amazon malaria vectors since the mid-1980s and is likely to vary in intensity by location (45–49). Although it is unclear which factor(s) triggered this exophilic behavior, some studies suggested a possible link with indoor residual spraying (50).

Nationwide antimalaria campaigns accompanied by engineering works (e.g., drainage and filling of marshes) were prominent in the first half of the 1900s, and facilitated the control of malaria in most regions of the country (51–53). The Amazon, however, was an exception, and virtually all cases of malaria in Brazil are currently recorded in the region. In the remainder of this section we discuss multiple aspects that potentially led to this outcome.

32.2.1 The Construction of the Madeira-Mamoré Railway

The construction of the Madeira-Mamoré Railway, located in the state of Rondônia (Fig. 32.1), is a good example of the diversity and magnitude of health challenges that forest disturbance can bring about (54). The railway was planned to facilitate the flow of Brazilian and Bolivian products for export (mainly rubber). The closest city to the construction area was initially founded as a Jesuit mission in 1737, but later abandoned due to the prevalence of fevers. In 1910, the area had no infrastructure and during the rainy season was often transformed into a swamp (55).

After four failed attempts to construct the railway by different companies, mainly due to loss of workforce consequential to malaria and logistic difficulties (55–57), the final construction phase took place between 1907 and 1912. During that time, approximately 21,000 employees worked on the railway, and probably more than 3,000 died. Malaria was the major cause of incapacity for work in 87.5% of the cases. However, these numbers are likely to be underreported, since only workers in critical condition would go to a hospital (55, 58, 59). In 1910, a detailed set of recommendations for malaria control and prevention were proposed for the construction area, and this represented the first antimalaria campaign in the state of Rondônia (60, 61). Among the recommendations were: the mandatory distribution of quinine, sanitary improvement, construction of screened houses for workers, avoidance of outdoors at dusk, and mandatory treatment of all sick people (60). The Madeira-Mamoré Railway operated for 60 years, and the main cause of its deactivation was the decline in rubber export activity.

32.2.2 Control Efforts

The first official antimalaria campaign in Brazil[1] was implemented in the state of São Paulo in 1905, under the direction of the Oswaldo Cruz Institute,[2] in order to improve

the health of employees of the port of Santos (64). Although the campaign started with the use of quinine, the concept of malaria as a "household infection" (a disease acquired inside the houses and rarely outside) was proposed for the first time,[3] leading to novel approaches of malaria control (60, 65). Mosquito screening inside the houses was done every 80 days, nets installed in all windows, and in-house fumigation with sulfur, pyrethrum, or tobacco adopted (62, 64). It was the first time that a measure aimed at killing the adult mosquito was used (66), and the concept of malaria as a household infection became the basis for campaigns relying on indoor residual spraying. In 1910, Brazilian physicians introduced the concept of malaria as a local disease, suggesting that the success of interventions depended on local characteristics, and therefore antimalaria campaigns should be developed *in situ* (67, 68). Also, the idea of malaria as a disease confined to the tropics was refuted by Brazilian scientists, who claimed that the health status of countries across the globe were not dependent on latitude but on knowledge and local efforts (54).

In 1930, the highly effective African malaria vector, *Anopheles gambiae*, was discovered in Ceará state (Northeast region of Brazil—Fig. 32.1), most likely brought to the area by a French ship coming from Dakar, Senegal, in 1929 (69). Precise identification of the invading mosquito was not possible at that time, since *An. gambiae* was recognized as a complex of seven species only in the 1960s. Recent DNA analysis of preserved museum species collected between 1932 and 1940 in Ceará suggest that *An. arabiensis* (a mosquito of the *An. gambiae* complex) is likely to have been the invading species (70).

In 1938–1939, as a direct consequence of the introduction of *An. gambiae*, there were 150,000 malaria cases, and 14,000 deaths in only 8 months (69, 71). With the support of the Rockefeller Foundation, control efforts started at the end of 1938. Approximately 4,000 workers were employed during 19 months, and measures of control included monthly house fumigation with pyrethrum, early case detection and rapid treatment, spraying of cars and trucks leaving or entering the endemic area, elimination of breeding sites, spreading Paris green on mosquito breeding habitats, and house capture. In 1940, *An. gambiae* had been eradicated from Brazil (72–74). A similar campaign was repeated in Egypt in 1943–1944, inspiring the eradication strategy that was the priority in WHO's agenda regarding malaria between 1955 and 1969 (75, 76).

In the early 1940s, a successful environmental management strategy eradicated malaria transmitted by a bromeliad-breeding mosquito from the southern states of São Paulo, Paraná, Santa Catarina, and Rio Grande do Sul. Bromeliads

---

[1] In 1898, the same year of Ronald Ross's discovery that *Anopheles* transmitted malaria, Adolfo Lutz (1855–1940), a Brazilian scientist and naturalist, associated the anopheline to malaria transmission during the construction of the São Paulo-Santos railway, and recommended that houses should be built far from the forest, where mosquito breeding sites were located (62).

[2] The Oswaldo Cruz Institute, founded in 1900, was the first research institute in Brazil's history. Research results from the institute led to Brazil's international reputation in science (63).

[3] The concept was proposed by Carlos Chagas (1878–1934), a Brazilian physician who coordinated many successful antimalaria campaigns in Brazil, and who is internationally known for the discovery of Chagas disease.

were gradually removed from the urban areas, and eucalyptus trees (on which bromeliads do not grow) were planted in forested areas (72, 77). This intervention was based on the findings of Adolpho Lutz, who first discovered and described forest malaria and its vectors, as reported in 1903 (78, 79).

Nationwide antimalaria campaigns started in 1940, when the number of malaria cases was on the order of 4 to 5 million (out of a population slightly larger than 41 million) (72). In 1945, DDT (dichlorodiphenyltrichloroethane) was first utilized as an adulticide (80), and in 1947 it was used nationwide to spray the interior of houses every 6 months (69, 72, 81).[4] In the Amazon, DDT spraying started in 1945 in one municipality of Pará state. It was expanded to 27 and 91 municipalities by 1947 and 1948, respectively, located in the states of Amazonas, Pará, Amapá, and Rondônia (former Guaporé) (83, 84). Efforts were initially mostly concentrated in capitals and major cities, progressively targeted to areas along streams (a common breeding ground for the anopheline) (84). The DDT program continued to expand, favored by an adequate number of well-paid personnel (85), and in 1959 it formally became a component of the national control effort (86).

In 1953, 6 years after the onset of DDT use, the number of confirmed malaria cases in 11 localities in the South, Southeast, and Northeast regions of Brazil was reduced by 99% (87). In an effort to tackle the malaria problem in the Amazon more efficiently, an unconventional method was introduced in 1953: distribution of chloroquinated salt to the population (72, 88). The method was incorporated by the WHO as part of the strategies for malaria eradication, and 13 countries in South America, Africa, and Asia used medicated salt between 1952 and 1978 (89). However, difficulties in the distribution and storage of the salt, as well as evidence of *P. falciparum* resistance to chloroquine both in Brazil and other countries, lead to an interruption of the intervention (85, 89–91).

The total number of malaria cases in the country continued to decrease after the introduction of DDT, reaching its minimum level in 1970 when 52,371 cases were registered, 31,733 of them in the Amazon. That time, however, coincided with the commencement of important changes in the geopolitical strategy adopted by the Brazilian government (92, 93), which would bring about some of the most significant transformations of the Amazon region, both socially and environmentally. Migration to the Amazon, and most important, the human mobility inside the Amazon region, became intense. Financial and human resources were scarce, and the type of housing (e.g., with partial walls or walls and roof made of tree leaves) in some areas imposed serious constraints on indoor spraying (69, 85).

----

[4] DDT was prohibited for agricultural use in 1985, but continued to be used in antimalaria campaigns until 1997 (82).

32.2.3 Opening of the Amazon Frontier

Historical occupation of the Amazon rarely promoted settlement in the area (94). Instead, economic booms (e.g., rubber, gold, minerals) often brought massive numbers of temporary migrants, which invariably resulted in malaria epidemics (21, 58, 95, 96). Although previous attempts to integrate the Amazon with the rest of country took place before 1960, it was not until that year, when the federal capital was moved from Rio de Janeiro to the interior of Brazil (Fig. 32.1), and the first road into the Amazon was opened, that the modern Amazon frontier expansion was commenced (21, 97). The major goal was to build roads linking the coast with the interior of Brazil, promoting national development, and increasing the industrial power of the country (98).

After 1964, during the military dictatorial government, geopolitical strategies for integration, and ultimately for the strengthening of national security, included a phase of intense occupation of the Amazon (92, 99). The construction of highways continued in the late 1960s and early 1970s (20, 100, 101). Following the opening of roads, large numbers of migrants started to arrive in search of land and employment, putting pressure on services, infrastructure, and the environment. The human population of the Amazon grew from 7.2 million in 1970 to 11 million in 1980 and then to 18.7 million by 1996. A major attempt to organize the occupation of the Amazon was the Northwest Region Integrated Development Program—POLONOROESTE (102).

Focusing on the area occupied by Rondônia and part of Mato Grosso states, POLONOROESTE was established in 1981, partly financed by the World Bank. It was a complex and ambitious project, designed to avoid problems faced by previous development efforts in the Amazon. However, a combination of factors contributed to early failures, including massive migration, lack of infrastructure, economic recession and fiscal crisis in the country, inability of government agencies to coordinate and execute the different phases of the project, lack of adequate information regarding soil quality of planned settlement areas, and absence of a detailed study regarding the sustainable carrying capacity of the project (102–104). Among the most serious consequences of POLONOROESTE were very high rates of deforestation, conflicts with indigenous populations, and severe outbreaks of malaria (105). These outbreaks resulted from a combination of factors that included human-made transformations to the environment, lack of acquired immunity among most of the settlers, precarious habitat conditions that offered no protection against mosquitoes, and modified vector behavior that included outdoor biting and peak biting times at dawn and dusk (3, 26).

After malaria reached its lowest level in Brazil in 1970, an increasing trend was observed following the rise of malaria in the Amazon region (Fig. 32.2). In 1986, when approximately

443,000 cases of malaria were reported in Brazil, Rondônia state was considered the "capital" of malaria in Brazil, containing 43% of all cases registered in the nation, and 46% of all cases in the Amazon. In 1989, more than 577,000 cases were observed in the Amazon, and a new control strategy adopted (as detailed later). Since then, malaria cases have been fluctuating, with a peak of more than 635,000 registered in 1999.

32.3   *ANOPHELES* VECTORS IN THE AMAZON

There is considerable variation among current estimates of the origin of *An. darlingi*, the primary malaria vector in the Amazon basin of Brazil. Recent work places the expansion out of the central area at 304,300 years BP, and out of the southeast and south areas at 121,492 years BP (106). This places expansion out of their ancestral groups in the late Pleistocene. In contrast, Mirabello & Conn (107) estimated the expansion of *An. darlingi* throughout South America at 25,311 years BP. These estimates for the vector have corresponding widely varying counterparts for expansions of *plasmodium* in South America (108). The vector and *plasmodium* timings for expansions are both estimated to be in the thousands of years, in contrast to the limits of actual documentation of human malaria in the region to hundreds of years (30). The direct human evidence, however, does not include DNA-based studies, and this gap is likely responsible for the substantial, qualitative disparity between vector, *plasmodium*, and human estimates. This point notwithstanding, we are still left with great uncertainty about the evolution of the components of the malaria transmission system in the Brazilian Amazon. Much more fine-grained geographic and molecular-level studies than heretofore will be required to reduce the extant levels of uncertainty.

Given the continental scale distribution of *An. darlingi*, there is also significant additional variability due to biting preferences or resistance to control measures. Geographical barriers such as rivers and mountain ranges may have contributed to substantially diversify *An. darlingi* in different areas of its range (106). A recent evaluation of the population genetics and historical expansion of *An. darlingi* in South America identified six population groups: Colombia, central Amazonia (geographically located around the Amazon River, covering most of the central and western Amazon basin, extending to Iquitos in Peru), southern Brazil (that actually spans from Pará to Paraná states), southeastern Brazil (Rio de Janeiro and Espírito Santo states), and two distinct groups in the northeast of Brazil. Based on orderings (in time) of *An. darlingi* haplotypes, the two oldest haplotypes are only distributed in Amazonian populations on either bank of the Amazon River (central group) (106). Thus, the likely ancestral distribution of *An. darlingi* in South America is in this region, as has also been previously

suggested from analyses of mitochondrial and nuclear loci (107, 109, 110).

32.3.1   Classification, Physiology, and Behavior

Systematic description and classification of *Anopheles* mosquitoes in Brazil began with the 1901 publication by Oswaldo Cruz (111), where, as mentioned previously, he also referred to the work of Adolpho Lutz starting in 1898, and incriminating what is now known as *An. cruzi* as the vector responsible for a malaria epidemic consequential to the construction of the Santos–São Paulo railroad. Lutz also made the striking observation that *An. cruzi* bred in bromeliad plants, as previously discussed in Section 32.2.2. In a review of research on vectors of malaria in Brazil, Deanne (112), listed more than 50 species of anophelines (Fig. 32.3) that had been described and classified, mostly on the basis of morphological features, from 1900 until the early 1980s.

Of the more than 50 species of anophelines found in Brazil, all known vectors of human malaria belong to the subgenera *Nyssorhynchus* and *Kerteszia*. These are *An. darlingi*, *An. aquasalis*, *An. albatarsis*, *An. triannulatus*, *An. nuneztovari*, *An. oswaldi*, *An. strodei*, and *An. evansae* from the *Nyssorhynchus* subgenus; and *An. cruzi*, *An. bellator*, and *An. homunculus* from the *Kerteszia* subgenus. As indicated previously, the most important primary vector is *An. darlingi*, found throughout the inland malarious areas of the legal Amazon and southward to the northern part of Paraná state (South of Brazil).

Molecular biological techniques have facilitated refinement of the earlier morphologically based taxonomies to identify some previously cryptic subspecies (113). In addition, as indicated in the previous subsection, considerable progress has been made in tracing the spatial expansion and population structure of a few of the *Anopheles* species in Brazil (106, 107). However, this is a topic in need of much further development, as outlined by Donnelly (114).

An. darlingi can maintain endemicity even at low densities (45), and is the only vector that can transmit all three malaria parasites found in the Amazon (115). Surveys conducted in the Amazon between July 1942 and June 1946, before DDT was introduced in the region, indicated that *An. darlingi* represented 21% of all anophelines captured, and 66% of all anophelines captured inside the houses. In addition, among all *An. darlingi* captured during the survey period, 88% were found indoors (116). Therefore, before the large-scale indoor residual spraying with DDT (as described in Section 32.2.2), *An. darlingi* was mainly an endophilic vector, in contrast to the prevalent outdoor pattern observed since the mid-1980s (45–49). Although it is unclear which factor(s) triggered this exophilic behavior, a few studies suggested a possible link with the use of DDT (50). This could be a result of mosquito behavior change or of

Genus _Anopheles_	Subgenus _Nyssorhynchus_	_An. darlingi, An. argyritarsis argyritarsis, An. argyritarsis sawyeri, An. albitarsis, An. braziliensis, An. lanei, An. aquasalis, An. oswaldoi, An. ininii, An. dunhami, An. evansae, An. rangeli, An. nuneztovari, An. benarrochi, An. galvāoi, An. strodei, An. triannulatus triannulatus, An. triannulatus davisi, An. rondoni, An. lutzi, An. parvus, An. antunesi, An. nigritarsis_
	Subgenus _Kerteszia_	_An. cruzi cruzi, An. cruzi laneanus, An. homunculus, An. bellator, An. bambusicolus, An. neivai_
	Subgenus _Anopheles_	_An. eiseni, An. peryassui, An. mattogrossensis, An. tibiamaculatus, An. maculipes, An. mediopunctatus, An. punctimacula, An. intermedius, An. fluminensis, An. minor, An shannoni, An. neomaculipalpus, An. evandroi, An. bustamantei, An. rachoui, An. anchietai_
	Subgenus _Lophopodomyia_	_An. squamifemur, An. gilesi, An. pseudotibiamaculatus_
	Subgenus _Stethomyia_	_An. nimbus, An. kompi, An. thomasi_
Genus _Chagasia_		_Ch. fajardoi, Ch. bonneae, Ch. rozeboomi_

FIGURE 32.3 Anophelines found in Brazil. Source: Deane (112).

a selection process favoring mosquitoes that bite and rest outdoors. On the one hand, _An. darlingi_ has been considered as a truly anthropophilic and endophilic mosquito, based on observations from several areas in South America. On the other hand, when analyzed in interior regions, where human population density is low and relatively mobile, the mosquito reveals a zoophilic and exophilic behavior (46, 85, 117). Whether or not these two groups of mosquitoes are genetically different remains unanswered (118).

32.4 AGRICULTURAL SETTLEMENT AND ENVIRONMENTAL CHANGE

As of 2007, almost 2,500 agricultural settlements had been opened in the Brazilian Amazon, reaching approximately 378,000 families, 77% of the total estimated targeted population (119). The settlements varied in terms of available resources, average size of land, and soil quality. Impacts were equally varied, such as rate of turnover, health outcomes, and patterns of deforestation and land use driven by a multitude of factors operating at varied temporal and spatial scales (120–122).

In any settlement project, the first environmental transformation is the process of clearing a small area for house construction and preparation for cultivation. The most common agriculture practice is slash-and-burn, which increases levels of soil pH, phosphorus, calcium, and magnesium, and decreases levels of toxic aluminum ions. A poor clearing and burning compromises the crop yields, can cause the obstruction of steams, and is likely to leave taller trees standing, providing the necessary partial shade for _An. darlingi_

breeding (123). This brings up the notion of forest fringe, a frontier between the forest and the property, where anopheline density is very high, and the risk of malaria transmission is augmented (26).

In this rapidly transforming Amazon ecosystem, characterizing malaria risk requires a new concept of malaria that considers biological and ecological phenomena acting at multiple spatial scales, juxtaposed with behavioral and economic conditions. This concept is defined as frontier malaria (2, 124), and operates at three spatial scales (4). First, at a micro/individual level, vector densities are high, as a consequence of ecosystem transformations that promote _An. darlingi_ larval habitats (partial shade near the forest fringe and along river edges, clear standing water of high pH) (49, 125, 126). Human exposure is intense, reflecting limited knowledge of transmission among settlers. _An. darlingi_ has a bimodal biting pattern (47)—at dawn and dusk—just when settlers are going to and returning from their fields. _P. falciparum_ is the primary parasite augmented by limited abundance of _P. vivax_. Morbidity is high and mortality is low (reflecting an unusual evolution of virulence of _P. falciparum_ in the Amazon), and immunity is low among new settlers (they mostly come from malaria-free areas). Housing quality is poor, thereby rendering indoor residual spraying ineffective. Curative health services are sparsely available, thus limiting antimalarial drug distribution.

Second, at a community level, frontier malaria is characterized by weak institutions, minimal community cohesion, political marginality of the settlers, and high rates of both in- and out-migration. This combination of conditions severely limits organized attempts at ecosystem management to minimize malaria risk and development of health clinics.

The human mobility ensures proliferation of parasites. Third, at a state and national level, frontier malaria is characterized by unplanned development of new settlement areas, stimulated by agricultural failures at previous settlement localities and by a desire of people to avoid further malaria episodes. This process, however, only serves to promote further transmission.

Frontier malaria was also conjectured to follow a distinctive time path (29). At the opening of a settlement area, malaria rates rise rapidly, and the first two levels of the above spatial characterization are fully operative. After 6–8 years, the unstable human migration (both in and out) and the highly variable ecological transformations (driven by variation in land clearance practices and local ecology) is replaced by a more organized process of urbanization and development of community cohesion. Frontier malaria is gradually replaced by more stable low levels of transmission and lower malaria rates. The process of urbanization itself (especially the introduction of impervious surfaces and drains) is an important intervention, as it creates environments that are inhospitable to *An. darlingi* larvae, and that are increasingly remote from forest fringes, thereby substantially reducing human exposure. These patterns have been empirically supported at the settlement and state levels (4).

32.5 FRONTIER MALARIA: AN EXAMPLE

To illustrate the context in which frontier malaria emerges, as well as its spatial and temporal characteristics, we use the Machadinho settlement project (Fig. 32.1) as a case study. Started in late 1984, Machadinho was one of the settlement areas promoted by POLONOROESTE (102). The area was primarily jungle before the settlement started, but sparsely populated by rubber tappers. It was the first settlement project that incorporated a plan of action to prevent the most harmful consequences of frontier expansion observed previously in the Amazon (27). Instead of adopting the traditional fishbone pattern frequently used in most settlement projects in the Amazon, Machadinho had an original and carefully planned plot design (Fig. 32.4). The shape of the plots was irregular, following the course of rivers and streams, so that every plot would have a natural source of water in its rear. Roads were planned in such a way that during the rainy season the increased water volume of rivers and streams would not leave them impassable. Areas with very irregular elevation were assigned as protected forest, and therefore not subjected to settlement or forest clearance. In total, the project had an area of 209,960 ha, of which approximately 68,000 ha were assigned as forest reserves. Such a design can only be implemented in areas of jungle, where there is no previous construction or agricultural practice in place (103).

In addition, the project included the construction of auxiliary roads, schools, health units, governmental agencies involved in rural development, forestry control posts, commercial and recreational areas; land use planning; execution of complementary soil analyses; provision of financial support for the acquisition of seeds and initial equipment; building of crop drying and storage in accordance with the local climatic and agricultural conditions; protection of natural parks and reserves, water sources, and consideration of endangered species; acquisition of all necessary equipment for schools; and establishment of community organization. Finally, the plans included a health project, which had three main goals: strengthen malaria control, set up a network of health care facilities, and stimulate research (104).

However, a combination of flaws in the planning and implementation processes, and adverse economic conditions resulted in major social, environmental, and health problems in Machadinho (13). By the time POLONOROESTE was approved (December 1981), Brazil was experiencing serious economic constraints: rising inflation, a critical balance of payments, and trade balance deficits. In 1982, Brazil started an economic stabilization program resulting in resource cutbacks. As a consequence, the federal government could not provide its share of financial resources for POLONOROESTE. Technical and financial support to new settlers were compromised (104). Assistance with housing construction was limited due to scarce resources, and only 30% of the families benefited from the distribution of wood to build a house (127). The majority of settlers who did not receive assistance relied on plastic, palm thatch, cardboard, and other precarious material to build a temporary house, which had extremely poor protection against mosquitoes, if any at all (Fig. 32.5). Construction of schools had serious delays in the early years of the project, and there was a shortage of teachers and furniture, all resulting in constraints on the provision of education for all children (128).

Regarding health care services, in 1985 there was only one health post and a mobile unit that was not sufficient to cover the entire area (129). The activities of health agents were also compromised: In 1986 only one car and five bicycles were available (128). Training was deficient and health professionals were scarce. Finally, soil quality was much lower than initially anticipated: 95% of plots assigned to settlers lacked adequate soil fertility at some level, and 58% of the area had good suitability for agricultural practice only if mechanization and intensive investment to improve the land were implemented (103, 130), inputs that poor settlers could not afford.

These problems were augmented by the local context. On average, settlers moving to Machadinho were low-income migrants (mainly from the southern portion of Brazil), with some agricultural experience (but little knowledge of agricultural potential or the techniques necessary for farming in a tropical rain forest area), and with low education. Most had no previous exposure to malaria, and consequently no acquired immunity against it, and very little knowledge on

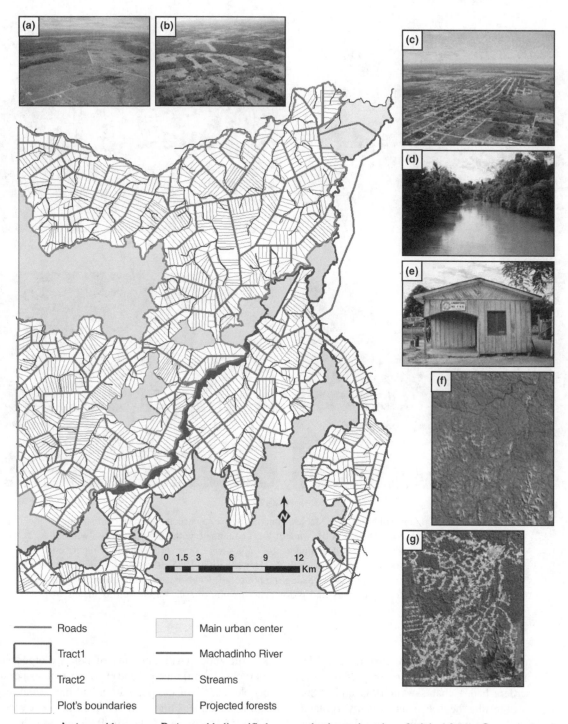

Roads

Tract1

Tract2

Plot's boundaries

Main urban center

Machadinho River

Streams

Projected forests

A. Area with pasture. **B.** Area with diversified crop production and patches of original forest. **C.**

Aerial view of the main urban center. **D.** Machadinho River. **E.** Laboratory where blood slides

FIGURE 32.4 Location and design of the Machadinho settlement project. (a) Area with pasture. (b) Area with diversified crop production and patches of original forest. (c) Aerial view of the main urban center. (d) Machadinho River. (e) Laboratory where blood slides are prepared and analyzed. (f, g) Satellite images taken in 1985 and 1995, respectively (Landsat TM, 30 m of resolution). (*See insert for color representation of this figure.*)

FIGURE 32.5 Examples of housing in Machadinho. (a) House constructed with wood distributed by INCRA in 1984, photographed in 2001, when it was abandoned, without windows and without the original door. However, the overall sealing of the walls was of superior quality in comparison with houses that had plastic or thatch on the walls. (b) House built in 1985, all made of palm thatch. (c) House built in 1985; there is no door closing the house. (d) Rubber tappers' housing, located in the border between forest reserves and plots, near streams (the canoe is used as means of transportation).

malaria transmission, prevention, and treatment (103). Regarding the use of inputs in agriculture, only 28.6% and 33.5% of the settlers had a chainsaw in 1985 and 1986, respectively. As a result, the speed and quality of land clearance near the house was far from ideal. The number of partially shaded water bodies suitable for *Anopheles* breeding proliferated, and settlers became highly exposed to malaria transmission due to proximity to the forest fringe (Fig. 32.6). Rubber tappers living in the protected forest reserves of Machadinho were most likely asymptomatic to malaria, and may have served as reservoirs of malaria parasites (103).

An outbreak of malaria was soon observed, following the arrival of the first settlers in late 1984. In 1985, the Annual Parasite Index (API—number of positive blood slides per total population) reached 3,400 positive slides per thousand people, 65.7% of the population had malaria at least once, and this number jumped to 90.1% in the next year. Also in 1986, 55.9% of people had malaria episodes in more than 5 months of the year (27, 128). By 1995, however, the level of transmission had been reduced, following the consolidation of the settlement project, and corroborating the transition pattern of frontier malaria previously proposed (29).

Four household surveys conducted in 1985–1987 and 1995 appraised levels of malaria infection through self-report information and gathered data that allowed the identification of spatio-temporal malaria risk profiles based on environmental, behavioral, and economic characteristics (4).

(a)

(b)

FIGURE 32.6 Forest clearance and potential breeding habitats. (a) Forest fringe in 1986, the limit between the forest and the cleared area, located in very short distance from the house. (b) Water body ideal for *Anopheles* breeding, common after forest clearance, which leaves small streams exposed to partial shade and partially clogged by fallen trees.

Important findings included: (a) in the initial years of settlement, environmental transformations and conditions dominated the high malaria risk profiles; (b) by 1995, risky conditions essentially reflected personal behavior and economic circumstances, as the substantial land-clearance process left much of Machadinho relatively inhospitable to *An. darlingi* larval development; (c) low-risk conditions were initially dominated by behavioral and economic measures, particularly reflecting ownership of capital equipment (chain saws) and agricultural expertise that lead to rapid clearance of land and construction of good-quality housing; and (d) by 1995, there was substantial urban development in Machadinho city, a

decided cattle-ranching activity, and good-quality housing, conditions that dominated low malaria risk profiles.

Most importantly, malaria risk profiles changed over time and across space, revealing idiosyncrasies of settlement efforts that that need to be considered in the selection of combinations of control strategies (4). These findings suggest some interventions that could be adopted in new settlement areas in the Amazon including rapid initial clearance of land for agriculture, and house construction that is protective and facilitated by government support. Land clearance should be of good quality, and extend through a buffer area around the house that offers protection from the high density of adult mosquitoes and larval habitats located at the forest fringe. Indeed, field observations in the early twentieth century revealed that anophelines captured far from houses were not infected with the malaria parasite, and therefore efforts to reduce malaria needed to incorporate recommendations for house construction, including clearance of forest within a 100 m radius of the houses, and avoidance of their construction at a distance less than 500 m from rivers, streams, lakes, and swamps (131). Studies conducted during the 1980s in the Amazon suggested that the distance between the house and the forest should ideally be between 1,000 and 1,500 m (45). Recommendations along these lines, however, were never implemented in past settlement projects. It is notable that the average plot size in some settlement programs, such as Machadinho (frontage size along the road averages 400–500 m, and the depth ranges between 700 and 900 m (27)), would make it unfeasible to achieve recommendations of large distances.

32.6 MALARIA PREVENTION AND MITIGATION: CURRENT AND FUTURE CHALLENGES

Malaria transmission in Brazil is, at present, restricted to the Amazon region. However, its spatial distribution is very irregular, with most high-risk locations concentrated in colonization projects and mining areas. This spatial pattern highlights the need to target malaria control interventions spatially (132, 133). In fact, after witnessing the number of malaria cases surpassing the mark of 600,000 in 1999, the Brazilian government implemented the Intensification Plan of Malaria Control Activities in the Legal Amazon (PIACM) in July 2000, which was expected to be operational until December 2002 (134). The PIACM targeted 254 municipalities in the Amazon (32.1% of the total number of municipalities), which accounted for 93.6% of the malaria cases. The criteria for selection of those municipalities were: (i) those that had an API equal or greater than 50 cases per 1,000 people; (ii) those where *P. falciparum* malaria was responsible for 20% or more of the total number of cases; (iii) the capitals of the nine states that comprise the Amazon region; (iv) the set of municipalities that accounted for at least 80% of all malaria cases in each state; and (v) those where urban malaria was observed (135). A reduction of 45% in the number of cases was registered between 1999 and 2002, but this gain was not homogeneous across the Amazon region (134). The reduction ranged from 78% to 35% in different states, and Rondônia was the only state that registered an increase of 12% in the number of cases.

Spatially targeted interventions were a significant improvement in the use of financial resources for control policies. As an example, in 1986, 60% of all malaria cases in the Amazon were concentrated in 458 municipalities, but 70% of the budget for malaria control was being spent in municipalities with only 3% of cases (136). Targeted interventions demand good quality health data, and important improvements have been implemented by the Brazilian government in the past two decades. A new epidemiological surveillance system is in place and drug resistance is constantly monitored in sentinel sites in the Amazon (137).

In 2003, the government launched the National Malaria Control Program (PNCM), which focused on improvement of local health services, surveillance, diagnosis, and treatment; building local capacity; targeted vector control; research; and community sensitization (134). Its impacts, however, were not reflected in declines of malaria cases until 2006. After more than 607,000 cases were recorded in 2005, the number dropped to 315,000 in 2008, and to 307,000 in 2009. Further success in controlling malaria in the Amazon, however, is likely to face three additional challenges, each operating at different scales, as detailed next.

32.6.1 Asymptomatic Infections

At a local scale, natives and long-term residents, who have been continually exposed to malaria, and most likely developed acquired immunity,[5] often do not present symptoms of the disease, but carry *plasmodium* in the blood (140). These asymptomatic individuals are potentially able to infect mosquitoes after a blood meal (141), and therefore can act as reservoirs of malaria, facilitating the spread of the disease to non-immune populations (140). Since asymptomatic individuals are unlikely to search for health care, they remain "silently" infectious. When vector control is not effective, either because it has low coverage or because the local conditions impose barriers to its successful implementation, the presence of asymptomatic infections can potentially contribute to sustained levels of transmission (142). Given the importance of outdoor biting in the Amazon (47, 48), the occurrence of asymptomatic infections is likely to produce local epidemics and/or maintain local levels of transmission. Indeed, asymptomatic infections have been reported in the Amazon (142–149).

An alternative to address asymptomatic infections is to implement an aggressive active case detection (AACD) strategy, in which members of a locality are tested monthly and treated for malaria (142). While AACD did show success in smaller and more isolated areas, its widespread use in the Amazon region is likely to face significant logistic and financial difficulties (142, 149). Accurate spatio-temporal patterns of the prevalence of asymptomatic infections across the Amazon region, and a characterization of the profile of asymptomatic individuals, are required. Such information would have two important and immediate applications. First, it would allow decision makers to establish the feasibility of the targeted implementation of AACD in combination with other control strategies. Second, it could guide planning activities of new infrastructure projects (e.g., dams and roads), which often result in large immigration of malaria naive individuals (150, 151). Finally, it would be desirable to have an analysis of the potential role and feasibility of using AACD as part of any elimination/eradication plan, as discussed in much of the contemporary discourse (152). Indeed, asymptomatic carriers can represent a limiting factor in malaria elimination plans unless an AACD program is implemented and maintained.

32.6.2 Planned Large-Scale Infrastructure Projects

At a regional scale, development and integration plans (e.g., dams, roads, hydroways) that include or specifically target the Amazon region will bring about further changes to the social and environmental landscape. Large-scale infrastructure projects, although often justified by their perceived contribution to economic and regional development, commonly result in population displacements (voluntary and involuntary), increase in the incidence of vector- and water-borne diseases, and significant environmental disturbances (153, 154). These and other negative impacts can potentially be avoided or minimized if comprehensive impact assessments are conducted, and mitigation plans proposed, implemented, and carefully monitored (9, 10, 155). Yet, the implementation of these plans demands financial and human resources beyond the capacity of local health sectors, and therefore should not solely rest under their responsibility.

The Decadal Plan for Energy Expansion (PDEE), issued in 2006, put forth a plan to construct 83 new large hydropower plants in Brazil, 22 located in the Amazon (156). Specifically, two of those dams are located in Rondônia state, along the Madeira River (the longest tributary of the Amazon River), in the same area where the Madeira–Mamoré railway was initially constructed. Recent epidemiological surveys conducted in the area to be impacted by the dams (prior to the onset of construction activities) indicated a prevalence of malaria infection of approximately 24%, many asymptomatic (157). Therefore, unless a comprehensive set of mitigation

---

[5] Immunity is acquired faster against *P. vivax* than against *P. falciparum*, and is lost when exposure to malaria stops (76, 138, 139).

strategies are implemented in the area, the onset of construction is likely to mimic past experiences (158, 159), and bring about a surge of malaria epidemics among non-immune migrants attracted by job and economic opportunities.

On a larger scale, the Initiative for the Integration of Regional Infrastructure in South America (IIRSA), a continent-wide development effort aiming at regional integration, is expected to promote the construction of more than 400 infrastructure projects (160). Overall, the projects in the IIRSA portfolio focus on transportation (roads, ports, airports, waterways, bridges, and railroads) and energy (hydropower, gas pipelines, and transmission lines). It has been estimated that the IIRSA will directly impact approximately 2.5 million km^2 in South America, including, just in Brazil, 137 conservation units, 107 indigenous areas, and 484 areas considered of high priority for conservation due to biodiversity (161). The extent to which these projects will impose further challenges to control malaria in the Amazon depend on how their negative impacts will be properly anticipated and mitigated in a timely fashion through targeted interventions.

32.6.3 Extreme Climatic Events

The Amazon's fragile ecosystem can be significantly impacted by climate change. The effects can occur directly, through extreme climate events (e.g., drought, flooding, excessive heat). Also, the region can be indirectly impacted by climate-driven population displacement, reduced crop production, and threatened water and food security, all of which impact individuals' vulnerability to infections. Future scenarios put forth by the Fourth Assessment Report of the Intergovernmental Panel on Climate Change (IPCC) suggested that the Amazon region is likely to experience reduced rainfall due to climate change, followed by decreases in malaria transmission (162). This scenario, however, carries much uncertainty. The IPCC report indicates a trend of drier dry seasons in the eastern portion, and wetter wet seasons in the western side (15). That variability could play a major role in the future spatial distribution and intensity of malaria transmission in the Amazon.

Although droughts are believed to reduce malaria incidence (162), the relationship between them is not straightforward, and is modified by several confounding factors. At least six dry periods have been observed in the Amazon in the past 50 years. Three were a result of the El Niño Southern Oscillation phenomenon, and tended to have a short duration. They have been connected to increased malaria during the second half of the year and during the following year in other parts of South America (163, 164); no evaluation has been done for the Brazilian Amazon. A recent and prolonged drought was observed in 2005, not a result of El Niño, but related to a circulation pattern powered by warm seas in the Atlantic (165–167). While there were more than 600,000

malaria cases reported that year, a there has been no assessment of a connection with climate.

Even though these extreme climate events are expected to become more numerous, their impacts on malaria, adjusted for other driving forces of transmission, have not been assessed. Moreover, it is unknown the extent to which coping strategies are adopted by local residents in response to these events. Improved understanding of malaria-climate change links is needed to inform policymakers and facilitate the proposal of National Adaptation Programs of Action (NAPAs), as proposed by the Nairobi work program under the United Nations Framework Convention on Climate Change (http://unfccc.int/adaptation/sbsta_agenda_item_adaptation/items/3991.php).

32.7 CONCLUSIONS

Malaria transmission in Brazil is currently restricted to the Amazon, and this pattern is not expected to change in the near future. It is unclear, however, if the decline in the number of cases observed since 2005 will continue, level off, or reverse. This uncertainty is a consequence of several factors, including the coverage and effectiveness of the national malaria control program, the nature of development plans for the Amazon, the eventual occurrence of extreme climatic events, and the future demographic and environmental dynamics of the region, to name a few.

In the initial section of this chapter we reviewed important facts related to malaria endemicity and attempts at control over the past 100 + years in Brazil. These experiences hold lessons for contemporary efforts at malaria control in the Amazon region, including the planning of large-scale development projects. For example, the Madeira River hydroelectric project, which partially overlaps with the site of the Madeira–Mamore Railroad, could (and should) incorporate malaria prevention schemes and impose an ongoing mitigation strategy for impacted communities if negative experiences from a century ago are to be avoided. Whether or not this will happen remains to be seen. However, the increasing international pressure from nongovernmental organizations and human rights activist groups to include health impact assessments (HIA) in advance of project construction, and follow-up mitigation well into the operating phase will hopefully influence the Madeira River dams. The recent International Finance Corporation (IFC) guidelines for HIAs (168) represent an important step forward toward prevention and mitigation of malaria problems associated with large-scale projects in the Amazon region and elsewhere in the tropics.

A common theme of agricultural settlement initiatives (planned and unplanned), mineral extraction projects, and large dams is that there is substantial ecosystem transformation associated with them (122, 169). An immediate corollary of these transformations has frequently been a considerable expansion of desirable breeding and larval habitats for the primary malaria vector in the Amazon region, *An. darlingi*. In particular, cutting into primary forest for agriculture and mining with limited early availability of mechanized equipment, and precarious establishment of cohesive communities has led to the phenomenon of frontier malaria previously described. This process continues to the present time in the formation of new settlements, some created as a way to "legalize" land invasions (and therefore allow these areas to receive basic infrastructure). This process of invasion, and later legalization, highlights problems in the current Brazilian legislation regarding land ownership, which contribute to increased deforestation, land conflicts, and, to a large extent, malaria outbreaks (170, 171). Lessons from the past—for example, the Machadinho settlement project discussed in Section 32.5—are not always taken into account, and the frontier malaria phenomenon is a direct consequence. In other western Amazon localities—for example, Acre state—a comprehensive and innovative plan of forest management (172) is currently being implemented, which is a clear sign of improvement in a situation that has been troublesome in the Amazon region since the early 1970s.

In Section 32.3, we gave a brief overview of evolutionary and changing behavioral features of *Anopheles* vectors in the Amazon, emphasizing *An. darlingi*. Considerable uncertainty prevails about the timing and geographical details of *Anopheles* expansions in the region. Regarding the origin and timing of expansions of malaria vectors throughout Brazil and other parts of South America, much more refined DNA sequencing and intensive geographical sampling of *Anopheles* species than heretofore will be necessary to narrow the range of uncertainty—currently tens of thousands of years—about these population dynamics. Of necessity, such vector studies should be linked with much more refined evolutionary studies of human and animal plasmodia in the Amazon, so that a defensible historical picture of malaria in the Amazon can be put forward. This should be augmented by shorter-term studies to assess the root causes of behavioral changes in circulating *Anopheles*, which are consequential to the major human migrations and ecosystem transformations initiated in the 1960s.

Finally, Brazil has been cited in recent years (173) as a country where malaria eradication should be feasible. Such commentary is based on coarse-level data focused on children and weighted on malaria rates from health posts/clinics. We view this as particularly problematic, since the substantial number of asymptomatic carriers in the Amazon region represents a limiting factor for any eradication effort. Without aggressive active surveillance and appropriate treatment of asymptomatic cases—assuming that this is even a practical option—more nuanced mitigation strategies will, of necessity, have to be part of the future story of malaria in the Brazilian Amazon.

REFERENCES

1. PNCM. Casos confirmados de Malária, segundo mês de notificação. Brasil, Grandes Regiões e Unidades Federadas. 2008. Programa Nacional de Controle da Malária—PNCM, Ministério da Saúde, Sistema de Informação da Vigilância Epidemiológica—Malária (Sivep-Malária). Available at http://portal.saude.gov.br/portal/arquivos/pdf/casos_conf_malaria_mes_notificacao_2008.pdf. Accessed May 2009.

2. Sawyer DR. Frontier malaria in the Amazon region of Brazil: types of malaria situations and some implications for control. Brasília: PAHO/WHO/TDR. Technical Consultation on Research in Support of Malaria Control in the Amazon, 1988, 19 pp.

3. Sawyer DR. Malaria on the Amazon frontier: economic and social aspects of transmission and control. *SE Asian J Trop Med Public Health* 1986;17:342–345.

4. Castro MC, Monte-Mór RL, Sawyer DO, Singer BH. Malaria risk on the Amazon frontier. *Proc Natl Acad Sci* 2006;103:2452–2457.

5. Costello A, Abbas M, Allen A, Ball S, Bell S, et al. Managing the health effects of climate change. *Lancet* 2009;373:1693–1733.

6. WHO. *Protecting Health from Climate Change: Global Research Priorities*. Geneva: World Health Organization, 2009, 32 pp.

7. Githeko AK, Lindsay SW, Confalonieri UE, Patz JA. Climate change and vector-borne diseases: a regional analysis. *Bull WHO* 2000;78:1136–1147.

8. Deane LM, Ribeiro CD, Oliveira RL, Oliveira-Ferreira J, Guimarães AE. Study on the natural history of malaria in areas of the Rondônia state—Brazil and problems related to its control. *Revista do Instituto de Medicina Tropical de São Paulo* 1988;30:153–156.

9. Krieger N, Northridge M, Gruskin S, Quinn M, Kriebel D, et al. Assessing health impact assessment: multidisciplinary and international perspectives. *J Epidemiol Commun Health* 2003;57:659–662.

10. Wood C. Environmental impact assessment in developing countries: an overview. *Conference on New Directions in Impact Assessment for Development: Methods and Practice*, Manchester, UK, 2003.

11. World Health Organization. *Human Health and Dams*. Geneva: The World Health Organization's submission to the World Commission on Dams (WCD), 2000.

12. Monte-Mór RL. *Modernities in the Jungle: Extended Urbanization in the Brazilian Amazonia* [PhD]. Los Angeles: University of California–Los Angeles, 2004, 378 pp.

13. Monte-Mór RL. Urban and rural planning: impact on health and the environment. In: Shahi GS, Levy BS, Binger A, Kjellstrom T, Lawrence R, editors. *International Perspectives on Environment, Development, and Health: Toward a Sustainable World*. New York: Springer, 1997, pp. 554–566.

14. Sawyer DO, Monte-Mór RL. Malaria risk factors assessment in Brazil. Brasília. Interregional Meeting on Malaria, 1992, 23 pp.

15. Malhi Y, Roberts JT, Betts RA, Killeen TJ, Li W, et al. Climate change, deforestation, and the fate of the Amazon. *Science* 2008;319:169–172.

16. IDB. Amazonia without myths. Washington, DC: Inter-American Development Bank and Commission on Development and Environment for Amazonia, 1992, 99 pp.

17. IBGE. Pesquisa Nacional por Amostra de Domicílios, Síntese de Indicadores 2008. Available at http://www.ibge.gov.br/home/estatistica/populacao/trabalhoerendimento/pnad2008/default.shtm. Accessed March 2010.

18. IBGE. (2009) SIDRA—Sistema IBGE de Recuperação Automática, 2009. Available at http://www.sidra.ibge.gov.br/. Accessed December 2009.

19. Benchimol S. Population changes in the Brazilian Amazon. In: Hemming J, editor. *Change in the Amazon Basin.* Manchester: Manchester University Press, 1985, pp. 37–50.

20. Moran EF. An assessment of a decade of colonization in the Amazon Basin. In: Hemming J, editor. *Change in the Amazon Basin: The Frontier after a Decade of Colonization.* Manchester: Manchester University Press, 1985, pp. 91–102.

21. Browder JO, Godfrey BJ. *Rainforest Cities: Urbanization, Development, and Globalization of the Brazilian Amazon.* New York: Columbia University Press, 1997, 429 pp.

22. Schmink M, Wood CH. *Contested Frontiers in Amazonia.* New York: Columbia University Press, 1992, 387 pp.

23. Schmink M, Wood CH. *Frontier Expansion in Amazonia.* Gainesville: University of Florida Press, 1984, 502 pp.

24. Fearnside PM. Deforestation in Brazilian Amazonia: history, rates, and consequences. *Conserv Biol* 2005;19:680–688.

25. Coimbra CEA Jr. Human factors in the epidemiology of malaria in the Brazilian Amazon. *Hum Organ* 1988;47:254–260.

26. Sawyer DR. *Malaria and the Environment.* Brasília: Instituto SPN, 1992, 37 pp.

27. Sawyer DR, Sawyer DO. Malaria on the Amazon frontier: economic and social aspects of transmission and control. Belo Horizonte: CEDEPLAR, 1987.

28. Singer BH, Castro MC. Agricultural colonization and malaria on the Amazon frontier. In: *Population Health and Aging: Strengthening the Dialogue between Epidemiology and Demography.* New York: Annals of the New York Academy of Sciences, 2001, pp. 184–222.

29. Sawyer DR, Sawyer DO. The malaria transition and the role of social science research.In: Chen LC, editor. *Advancing the Health in Developing Countries: The Role of Social Research.* Westport: Auburn House, 1992, pp. 105–122.

30. Castro MC, Singer BH. Was malaria present in the Amazon before the European conquest? Available evidence and future research agenda. *J Archaeol Sci* 2005;32:337–340.

31. Santos Filho LC. *História Geral da Medicina Brasileira.* São Paulo: Editora Humanismo Ciência e Tecnologia, 1977, 397 pp.

32. Azpilcueta Navarro J. *Cartas Avulsas.* Belo Horizonte: Editora da Universidade de São Paulo, 1988, 529 pp.

33. Leite S. *Cartas dos Primeiros Jesuítas do Brasil.* São Paulo: Comissão do IV Centenário da cidade de São Paulo, 1954, 1094 pp.

34. Soares de Sousa G, Varnhagen FA. *Tratado Descritivo do Brasil em 1587.* São Paulo: Companhia Editora Nacional, 1971, 389 pp.

35. Bomtempo JM. *Memória Sobre Algumas Enfermidades do Rio de Janeiro, e mui Particularmente Sobre o Abuso Geral, e Pernicioso Efeito da Aplicação da Preciosa Casca Peruviana, ou Quina.* Rio de Janeiro: Typographia Nacional, 1825, 74 pp.

36. Neiva A. Formação de raça do hematozoario do impaludismo resistente a quinina. *Memórias do Instituto Oswaldo Cruz* 1910;2:131–140.

37. Cruz O. *Relatório Sobre as Condições Médico-Sanitárias do Valle do Amazonas.* Rio de Janeiro, 1913, 111 pp.

38. Costa DAM. A malária e suas diversas modalidades clínicas. Rio de Janeiro: Imprensa a Vapor Lombaerts & Comp, 1885.

39. Penna B. *Saneamento do Brasil.* Sanear o Brasil é povoal-o; é enriquecel-o; é moralisal-o. Rio de Janeiro: Typ. Revista dos Tribunais, 1918, 114 pp.

40. von Spix JB, von Martius KFP, Lloyd HE. *Travels in Brazil, in the Years 1817–1820: Undertaken by Command of His Majesty the King of Bavaria.* London: Longman, 1824, 298 pp.

41. Peixoto A. *O Problema Sanitário da Amazônia.* Rio de Janeiro, 1917, 28 pp.

42. Chagas C. *Notas Sobre a Epidemiologia do Amazonas.* Rio de Janeiro, 1913, 19 pp.

43. Chagas C. *Estudos Hematológicos no Impaludismo.* Rio de Janeiro, 1903, 57 pp.

44. Thomas HW. *The Sanitary Conditions and Diseases Prevailing in Manaos, North Brazil, 1905–1909, with Plan of Manaos and Chart. Expedition to the Amazon, 1905–1909.* Rockefeller Archive Center, 1910. Casa de Oswaldo Cruz (FIOCRUZ) collection. DOC. 007.

45. Tadei WP, Santos JMM, Costa WLS, Scarpassa VM. Biologia de Anofelinos Amazônicos. XII. Ocorrência de espécies de *Anopheles,* dinâmica da transmissão e controle da malária na zona urbana de Ariquemes (Rondônia). *Revista do Instituto de Medicina Tropical de São Paulo* 1988;30:221–251.

46. Bustamante FM. Efeito das aplicações intradomiciliárias de DDT sobre a densidade do *Anopheles darlingi* em várias regiões do Brasil. *Revista Brasileira de Malariologia* 1951; III:571–590.

47. Klein TA, Lima JBP. Seasonal distribution and biting patterns of *Anopheles* mosquitoes in Costa Marques, Rondônia, Brazil. *J Am Mosquito Control Assoc* 1990;6:700–707.

48. Lourenço-de-Oliveira R, Guimarães AEG, Arlé M, Silva TF, Castro MG, et al. Anopheline species, some of their habits and relation to malaria in endemic areas of Rondônia state, Amazon region of Brazil. *Memórias do Instituto Oswaldo Cruz* 1989;84:501–514.

49. Tadei WP, Thatcher BD, Santos JMM, Scarpassa VM, Rodrigues IB, et al. Ecologic observations on anopheline vectors of malaria in the Brazilian Amazon. *Am J Trop Med Hyg* 1998;59:325–335.

50. Bustamante FM, Pinto OS, Freitas Jr. Observações sobre o comportamento do *Anopheles darlingi* em casas experimentais tratadas com DDT e BHC na área de Engenheiro Dolabela, Estado de Minas Gerais. *Revista Brasileira de Malariologia e Doenças Tropicais* 1952;IV:347–360.

51. Hochman G, Mello MTB, Santos PRE. A malária em foto: imagens de campanhas e ações no Brasil da primeira metade do século XX. *História, Ciências, Saúde* 2002;9:233–273.

52. Neiva A. Profilaxia da malária e trabalhos de engenharia: notas, comentários, recordações. *Revista do Clube de Engenharia* 1940;VI:60–75.

53. Mello MTVB. Imagens da memória: uma história visual da malária (1910–1960). Niterói, RJ: Universidade Federal Fluminense, Departamento de História, 2007, 287 pp.

54. Peixoto A. Clima e saúde; introdução bio-geográfica à civilização Brasileira. Rio de Janeiro: Companhia Editora Nacional, 1938, 295 pp.

55. Cruz O. Madeira–Mamoré Railway Company—Considerações gerais sobre as condições sanitárias do Rio Madeira. Rio de Janeiro, 1910, 43 pp.

56. Craig NB, Vasconcelos MN. Estrada de ferro Madeira–Momoré, história trágica de uma expedição. São Paulo: Companhia Editora Nacional, 1947, 449 pp.

57. Ferreira MR. A ferrovia do diabo: história de uma estrada de ferro na Amazônia. São Paulo: Melhoramentos: Secretaria de Estado da Cultura, 1981, 400 pp.

58. Perdigão F, Bassegio L. Migrantes Amazônicos—Rondônia: a trajetória da ilusão. São Paulo, SP: Edições Loyola, 1992, 221 pp.

59. Thomas HW. Malaria in the Amazon region, and the protection of ships. In: Ross R, editor. *The Prevention of Malaria*. New York: E.P. Dutton & Company, 1910, pp. 382–389.

60. Cruz O. Prophylaxis of malaria in central and southern Brazil. In: Ross R, editor. *The Prevention of Malaria*. New York: E.P. Dutton & Company, 1910, pp. 390–399.

61. Coimbra MELS. SUCAM and malaria control. Belo Horizonte: CEDEPLAR, 1985, 26 pp.

62. Silveira AC, Rezende DF. Avaliação da estratégia global de controle integrado da malária no Brasil. Brasília: Organização Pan-Americana da Saúde, 2001, 120 pp.

63. Stepan N. *Beginnings of Brazilian Science: Oswaldo Cruz, Medical Research and Policy, 1890–1920*. New York: Science History Publications, 1976, 225 pp.

64. Chagas C. *Luta Contra a Malária*. Rio de Janeiro, 1934, 24 pp.

65. Chagas C. *Prophylaxia do Impaludismo*. Rio de Janeiro: Typ. Besnard Frères. (Trabalho do Instituto de Manguinhos), 1905, 48 pp.

66. Deane LM. A história da evolução dos conhecimentos sobre a malária. Rio de Janeiro, 1986, 51 pp.

67. Cruz O, Chagas C, Peixoto A. Sobre o saneamento da Amazônia. Manaus: P. Daou, 1972, 205 pp.

68. Stepan NL. The only serious terror in these regions: malaria control in the Brazilian Amazon. In: Armus D, editor. *Disease in the History of Modern Latin America: From Malaria to AIDS*. Durham, NC: Duke University Press, 2003, pp. 25–50.

69. Tadei WP. Considerações sobre as espécies de *Anopheles* e a transmissão da malária na Amazônia. Manaus: INPA—Instituto Nacional de Pesquisas da Amazônia, 1991, 16 pp.

70. Parmakelis A, Russello MA, Caccone A, Marcondes CB, Costa J, et al. Short report: historical analysis of a near disaster: *Anopheles gambiae* in Brazil. *Am J Trop Med Hyg* 2008;78:176–178.

71. Deane LM. A história do *Anopheles gambiae* no Brasil. Rio de Janeiro. III Seminário de Vetores Urbanos e Animais Sinatrópicos—4a. Reunião Brasileira sobre Simulídeos, 1990.

72. Deane LM. Malaria studies and control in Brazil. *Am J Trop Med Hyg* 1998;38:223–230.

73. Soper FL, Wilson DB. *Anopheles gambiae* in Brazil, 1930 to 1940. New York City: The Rockefeller Foundation, 1943, 262 pp.

74. Killeen GF, Fillinger U, Kiche I, Gouagna LC, Knols BG. Eradication of *Anopheles gambiae* from Brazil: lessons for malaria control in Africa? *Lancet Infect Dis* 2002;2:618–627.

75. Litsios S. *The Tomorrow of Malaria*. Karori, NZ: Pacific Press, 1997, 183 pp.

76. Bailey NTJ. *The Biomathematics of Malaria*. London: C. Griffin, 1982, 210 pp.

77. Pinotti M. The biological basis for the campaign against the malaria vectors in Brazil. *Trans Roy Soc Trop Med Hyg* 1951;44:663–682.

78. Lutz A. Waldmosquitos und Waldmalaria. Centralblatt für Bakteriologie, Parasitenkunde und Infektionskrankheiten 1903;33:282–292.

79. Benchimol JL, Sá MR. Adolpho Lutz: Obra completa. Febre amarela, malária e protozoologia. Rio de Janeiro, RJ: Editora Fiocruz, 2005, 956 pp.

80. Deane LM. Observações sobre a malária na Amazônia Brasileira. *Revista do Serviço Especial de Saúde Pública* 1947;1:3–60.

81. Loiola CCP. The use of DDT in malaria control programs in Brazil, 1998. Puerto Iguazú, Argentina.

82. Mendes RA, Jesus IM, Santos ECO, Faial KF, Lima MO, et al. Níveis séricos de DDT total de trabalhadores expostos no programa de controle da malária no estado do Pará, Brasil. *Cadernos de Saúde Coletiva* 2007;15:559–568.

83. Deane LM, Freire EPS, Tabosa W, Ledo J. A aplicação domiciliar de DDT no contrôle da malária em localidades da Amazônia. *Revista do Serviço Especial de Saúde Pública* 1948;1:1121–1161.

84. Deane LM, Ledo JF, Freire EPS, Cotrim J, Sutter VA, et al. Contrôle da Malária na Amazônia pela aplicação domiciliar de DDT e sua avaliação pela determinação do índice de transmissão. *Revista do Serviço Especial de Saúde Pública* 1948;2:545–560.

85. Tauil P, Deane L, Sabroza P, Ribeiro C. A malária no Brasil. *Cadernos de Saúde Pública* 1985;1:71–111.

86. Roberts DR, Alecrim WD. Behavioral response of *Anopheles darlingi* to DDT-sprayed house walls in Amazonia. *Bull PAHO* 1991;25:210–217.

87. SNM. Casos de malária confirmados por exame de laboratório, 1945 e 1953. SNM—Serviço Nacional de Malária, 1953. Ministério da Saúde. Casa de Oswaldo Cruz (FIOCRUZ) collection. CC/DT/19252040.

88. Pinotti M, Lôbo AGS, Damasceno G, Soares R. Experiências de campo com o sal cloroquinado. *Revista Brasileira de Malariologia e Doenças Tropicais* 1955;7:5–23.

89. Payne D. Did medicated salt hasten the spread of chloroquine resistance in *Plasmodium falciparum*? *Parasitol Today* 1988;4:112–115.

90. Coimbra MELS. Política pública e saúde: o caso da malária no Brasil. *Análise e Conjuntura* 1987;2:72–90.

91. Wernsdorfer WH. Drug resistance of malaria parasites. Geneva: 20th Expert Committee on Malaria, 1998. Working Paper MAL/ECM/20/98/15.

92. Silva GC. *Geopolítica do Brasil*. Rio de Janeiro: José Olympio, 1967, 275 pp.

93. Mattos CM. *Uma Geopolítica Pan-Amazônica*. Rio de Janeiro, RJ: Biblioteca do Exército Editora, 1980, 215 pp.

94. Becker BK. Geopolítica da Amazônia. *Estudos Avançados* 2005;19:71–86.

95. Singer BH, Sawyer DO. Perceived malaria illness reports in mobile populations. *Health Policy Plann* 1992;7:40–45.

96. Aron JL, Patz J. *Ecosystem Change and Public Health: A Global Perspective*. Baltimore: Johns Hopkins University Press, 2001, 480 pp.

97. Wood CH, Wilson J. The magnitude of migration to the Brazilian frontier. In: Schmink M, Wood CH, editors. *Frontier Expansion in Amazonia*. Gainesville: University of Florida Press, 1984, pp. 142–152.

98. Diegues ACS, Millikan B. A Dinâmica social do desmatamento na Amazônia: populações e modos de vida em Rondônia e sudeste do Pará. São Paulo: United Nations Research Institute for Social Development: Núcleo de Apoio à Pesquisa sobre Populações Humanas e Áreas Úmidas Brasileiras, 1993, 155 pp.

99. Silva GC. *Aspectos Geopolíticos do Brasil*. Rio de Janeiro: Biblioteca do Exército, 1957, 81 pp.

100. Mahar DJ. *Frontier Development Policy in Brazil: A Study of Amazonia*. New York: Praeger, 1979, 182 pp.

101. Moran EF. *Developing the Amazon*. Bloomington: Indiana University Press, 1981, 292 pp.

102. World Bank. *Brazil: Integrated Development of the Northwest Frontier*. Washington: The World Bank, 1981, 101 pp.

103. Castro MC. *Spatial Configuration of Malaria Risk on the Amazon Frontier: The Hidden Reality behind Global Analysis* [PhD]. Princeton: Princeton University, 2002, 293 pp.

104. World Bank. *World Bank Approaches to the Environment in Brazil: A Review of Selected Projects. Vol. 5: The POLO-NOROESTE Program*. Washington: The World Bank, 1992, 344 pp.

105. BENFAM. Colonização: um projeto que pode fracassar. *População & Desenvolvimento* 1987;21:47–48.

106. Pedro PM, Sallum MAM. Spatial expansion and population structure of the neotropical malaria vector, *Anopheles darlingi* (Diptera: Culicidae). *Biol J Linn Soc* 2009;97:854–866.

107. Mirabello L, Conn JE. Molecular population genetics of the malaria vector *Anopheles darlingi* in Central and South America. *Heredity* 2006;96:311–321.

108. Escalante AA, Barris E, Ayala FJ. Evolutionary origin of human and primate malarias: evidence from the circumsporozoite protein gene. *Mol Biol Evol* 1995;12:616–626.

109. Manguin SWJ, Conn JE, Rubui-Palis Y, Danoff-Burg JA, Roverts DR. Population structure of the primary malaria vector in South America, *Anopheles darlingi*, using isozyme, random amplified polymorphic DNA, internal transcribed spacer 2, and morphologic markers. *Am J Trop Med Hyg* 1999;60:364–376.

110. Conn JE, Mirabello L. The biogeography and population genetics of neotropical vector species. *Heredity* 2007;99:245–256.

111. Cruz OG. Contribuição ao estudo dos culicídeos do Estado do Rio de Janeiro. *Brasil Médico* 1901;15:423–426.

112. Deane LM. Malaria vectors in Brazil. *Memórias do Instituto Oswaldo Cruz* 1986;81:5–14.

113. Mirabello L, Vineis JH, Yanoviak SP, Scarpassa VM, Povoa MM, et al. Microsatellite data suggest significant population structure and differentiation within the malaria vector *Anopheles darlingi* in Central and South America. *BMC Ecol* 2008;8:3, doi:10.1186/1472-6785-8-3.

114. Donnelly M. Bloodthirsty amazons? *Heredity* 2006;96:421. doi:410.1038/sj.hdy.6800812.

115. Consoli RAGB, Oliveira RL. Principais mosquitos de importância sanitária no Brasil. Rio de Janeiro: Editora Fiocruz, 1998, 228 pp.

116. SESP. Adultos de anophelinos examinados pelo Laboratório Central e Divisão de Malária. Julho de 1942–Junho de 1946. SESP—Programa da Amazônia. Casa de Oswaldo Cruz (FIOCRUZ) collection. LD/TP/19420740.

117. Giglioli G. Biological variations in *Anopheles darlingi and Anopheles gambiae:* their effect on practical malaria control in the neotropical region. *Bull WHO* 1956;15:461–471.

118. Chadee DD, Kitron U. Spatial and temporal patterns of imported malaria cases and local transmission in Trinidad. *Am J Trop Med Hyg* 1999;61:513–517.

119. MDA/INCRA/DTI. Projetos de Reforma Agrária Conforme Fases de Implementação—Período da Criação do Projeto: 01/01/1900 até 05/10/2007. Brasília: Ministério do Desenvolvimento Agrário—MDA; Instituto Nacional de Colonização e Reforma Agrária—INCRA; Diretoria de Obtenção de Terras e Implantação de Projetos de Assentamento—DT; Coordenação–Geral de Implantação—DTI–Sipra, 2007.

120. Brondízio ES, McCracken SD, Moran EF, Siqueira AD, Nelson DR, et al. The colonist footprint: toward a conceptual framework of land use and deforestation trajectories among small farmers in the Amazonian frontier. In: Wood CH, Porro R,editors. *Deforestation and Land Use in the Amazon*. Gainesville: University Press of Florida, 2002, pp. 133–161.

121. Browder JO. Reading colonist landscapes: social factors influencing land use decisions by small farmers in the Brazilian Amazon. In: Wood CH, Porro R, editors. *Deforestation*

and Land Use in the Amazon. Gainesville: University Press of Florida, 2002, pp. 218–240.

122. Wood CH, Porro R. *Deforestation and Land Use in the Amazon.* Gainesville: University Press of Florida, 2002, 385 pp.

123. Fearnside PM. *Human Carrying Capacity of the Brazilian Rainforest.* New York: Columbia University Press, 1986, 293 pp.

124. Sawyer DR. Deforestation and malaria on the Amazon frontier. Campinas, Brazil. Seminar on Population and Deforestation in the Humid Tropics, International Union for the Scientific Study of Population, 1992.

125. Charlwood JD. Biological variation in *Anopheles darlingi* root. *Memórias do Instituto Oswaldo Cruz* 1996;91:391–398.

126. Charlwood JD. Observations on the bionomics of *Anopheles darlingi* root (Diptera: Culicidae) from Brazil. *Bull Entomol Res* 1980;70:685–692.

127. INCRA. Exposição de motivos. Porto Velho: INCRA, 1991. INCRA/SE-17/G/No 02/91 INCRA/SE-17/G/No. 02/91.

128. Sydenstricker JM. Parceleiros de Machadinho: história migratória e as interações entre a dinâmica demográfica e o ciclo agrícola em Rondônia [Master degree dissertation] Campinas, SP: Universidade de Campinas, 1992, 190 pp.

129. Mahar DJ. Government policies and deforestation in Brazil's Amazon Region. Washington, DC: World Bank, 1989, pp. v, 56.

130. Fearnside PM. Settlement in Rondônia and the token role of science and technology in Brazil's Amazonian development planning. *Interciencia* 1986;11:229–236.

131. Leão FA. Em torno do saneamento do Brasil. Rio de Janeiro: Typographia Leuzinger, 1918.

132. Carter R, Mendis KN, Roberts D. Spatial targeting of interventions against malaria. *Bull WHO* 2000;78:1401–1411.

133. Castro MC, Sawyer DO, Singer BH. Spatial patterns of malaria in the Amazon: implications for surveillance and targeted interventions. *Health Place* 2006;13:368–380.

134. Brasil. Programa Nacional de Prevenção e Controle da Malária PNCM. Brasília: Ministério da Saúde. Secretaria de Vigilância em Saúde, 2003, 132 pp.

135. FUNASA. Plano de Intensificação das Ações de Controle da Malária na Amazônia Legal. 2000. Available at http://www.funasa.gov.br/epi/malaria/pdfs/plano_malaria.PDF. Accessed January 2009.

136. Akhavan D, Musgrove P, Abrantes A, Gusmão RA. Cost-effective malaria control in Brazil: cost-effectiveness of a malaria control program in the Amazon basin of Brazil, 1988–1996. *Social Sci Med* 1999;49:1385–1399.

137. MS. *Situação Epidemiológica da Malária no Brasil 2005.* Brasília: Ministério da Saúde, 2005.

138. McGregor IA, Wilson RJM. Specific immunity: acquired in man. In: Wernsdorfer WH, McGregor I, editors. *Malaria: Principles and Practice of Malariology.* Edinburgh; New York: Churchill Livingstone, 1988, pp. 559–619.

139. Molineaux L. The epidemiology of human malaria as an explanation of its distribution, including some implications for its control. In: Wernsdorfer WH, McGregor I, editors.

Malaria: Principles and Practice of Malariology. Edinburgh; New York: Churchill Livingstone, 1988, pp. 913–998.

140. Coura JR, Suárez-Mutis M, Ladeia-Andrade S. A new challenge for malaria control in Brazil: asymptomatic *Plasmodium* infection—a review. *Memórias do Instituto Oswaldo Cruz* 2006;101:229–237.

141. Alves FP, Gil LHS, Marrelli MT, Ribolla PEM, Camargo EP, et al. Asymptomatic carriers of *Plasmodium* spp. as infection source for malaria vector mosquitoes in the Brazilian Amazon. *J Med Entomol* 2005;42:777–779.

142. Macauley C. Aggressive active case detection: a malaria control strategy based on the Brazilian model. *Social Sci Med* 2005;60:563–573.

143. Alves FP, Durlacher RR, Menezes MJ, Krieger H, Silva LHP, et al. High prevalence of asymptomatic *Plasmodium vivax* and *Plasmodium falciparum* infections in native Amazonian populations. *Am J Trop Med Hyg* 2002;66:641–648.

144. Camargo EP, Alves F, Silva LHP. Symptomless *Plasmodium vivax* infections in native Amazonians. *Lancet* 1999;353: 1415–1416.

145. Ladeia-Andrade S, Ferreira MU, Scopel KKG, Braga EM, Bastos MS, et al. Naturally acquired antibodies to merozoite surface protein (MSP)-1$_{19}$ and cumulative exposure to *Plasmodium falciparum* and *Plasmodium vivax* in remote populations of the Amazon Basin of Brazil. *Memórias do Instituto Oswaldo Cruz* 2007;102:943–951.

146. Tada MS, Marques RP, Mesquita E, Martha RCD, Rodrigues JA, et al. Urban malaria in the Brazilian Western Amazon Region. I. High prevalence of asymptomatic carriers in an urban riverside district is associated with a high level of clinical malaria. *Memórias do Instituto Oswaldo Cruz* 2007;102:263–269.

147. Suárez-Mutis M, Cuervo P, Leoratti FMS, Moraes-Avila SL, Ferreira AW, et al. Cross sectional study reveals a high percentage of asymptomatic *Plasmodium vivax* infection in the Amazon Rio Negro area, Brazil. *Revista do Instituto de Medicina Tropical de São Paulo* 2007;49: 159–164.

148. Katsuragawa TH, Gil LHS, Tada MS, Silva LHP. Endemias e epidemias na Amazônia. Malária e doenças emergentes em áreas ribeirinhas do Rio Madeira. Um caso de escola. *Estudos Avançados* 2008;22:111–141.

149. Ladeia-Andrade S, Ferreira MU, Carvalho ME, Curado I, CouraJr., Age-dependent acquisition of protective immunity to malaria in riverine populations of the Amazon Basin of Brazil. *Am J Trop Med Hyg* 2009;80:452–459.

150. Keiser J, Castro MC, Maltese MF, Bos R, Tanner M, et al. Effect of irrigation and large dams on the burden of malaria on global and regional scale. *Am J Trop Med Hyg* 2005;72:392–406.

151. Fearnside PM. Environmental impacts of Brazil's Tucuruí dam: unlearned lessons for hydroelectric development in Amazonia. *Environ Manage* 2001;27:377–396.

152. Smith DL, Hay SI. Endemicity response timelines for Plasmodium falciparum elimination. *Malaria J* 2009;8:87. doi:10.1186/1475-2875-8-87.

153. Couldrey M, Morris T, editors. *Dilemmas of Development-Induced Displacement (Forced Migration Review,* Vol. 12) Oxford, UK: Refugee Studies Center, 2002.

154. Lerer LB, Scudder T. Health impacts of large dams. *Environ Imp Assess* 1999;19:113–123.

155. Winkler MS, Divall MJ, Krieger GR, Balge MZ, Singer BH, et al. Assessing health impacts in complex eco-epidemiological settings in the humid tropics: advancing tools and methods. *Environ Imp Assess Rev* 2009;30:52–61.

156. Brasil. Plano Decenal de Expansão de Energia Elétrica—2005–2015. Secretaria de Planejamento e Desenvolvimento Energético, Ministério de Minas e Energia—MME; Empresa de Pesquisa Energética—EPE, 2006.

157. Katsuragawa TH, Cunha RPA, Souza DCA, Gil LHS, Cruz RB, et al. Malária e aspectos hematológicos em moradores da área de influência dos futuros reservatórios das hidrelétricas de Santo Antônio e Jirau, Rondônia, Brasil. *Cadernos de Saúde Pública* 2009;25:1486–1492.

158. Fearnside PM. Social impacts of Tucuruí dam. *Environ Manage* 1999;24:483–495.

159. Scudder T. *The Future of Large Dams: Dealing with Social, Environmental, Institutional and Political Costs.* London; Sterling, VA: Earthscan, 2006, 389 pp.

160. IDB. *Building a New Continent: A Regional Approach to Strengthening South American Infrastructure.* Initiative for the Integration of Regional Infrastructure in South America (IIRSA) Washington, DC: Inter-American Development Bank—IDB. Report 10039, 2006, 50 pp.

161. Wanderley IF, Fonseca RL, Pereira PGP, Prado ACA, Ribeiro AP, et, al., Implicações da Iniciativa de Integração da Infra-estrutura Regional Sul-Americana e projetos correlacionados na política de conservação no Brasil. *Política Ambiental* 2007;3:3–42.

162. Confalonieri U, Menne B, Akhtar R, Ebi KL, Hauengue M, et al., Human health. In: Parry M, Canziani O, Palutikof J, van der Linden P, Hanson C, editors. *Climate Change 2007: Impacts, Adaptation and Vulnerability Contribution of Working Group II to the Fourth Assessment Report of the Intergovernmental Panel on Climate Change.* Cambridge, UK: Cambridge University Press, 2007, pp. 391–431.

163. Gagnon A, Smoyer-Tomic K, Bush A. The El Niño Southern Oscillation and malaria epidemics in South America. *Int J Biometeorol* 2002;46:81–89. DOI: 10.1007/s00484-00001-00119-00486.

164. Poveda G, Rojas W, Quinones M, Velez I, Mantilla R, et al. Coupling between annual and ENSO timescales in the malaria-climate association in Colombia. *Environ Health Perspect* 2001;109:489–493. DOI: 410.2307/3454707.

165. Cox PM, Harris PP, Huntingford C, Betts RA, Collins M, et al. Increasing risk of Amazonian drought due to decreasing aerosol pollution. *Nature* 2008;453:212–215.

166. Marengo JA, Nobre CA, Tomasella J, Oyama MD, Oliveira GS, et al. The drought of Amazonia in 2005. *Am Meteorol Soc* 2008;21:495–516.

167. Zeng N, Yoon J-H, Marengo JA, Subramaniam A, Nobre CA, et al. Causes and impacts of the 2005 Amazon drought. *Environ Res Lett* 2008;3:9 pp. doi:10.1088/1748-9326/1083/1081/014002.

168. IFC. *Introduction to Health Impact Assessment.* Washington, DC: Environment and Social Development Department, International Finance Corporation (IFC), 2009, 67 pp.

169. Moran EF. Deforestation in the Brazilian Amazon. In: Sponsel LE, Headland TN, Bailey RC, editors. *Tropical Deforestation: The Human Dimension.* New York: Columbia University Press, 1996, pp. 149–164.

170. Alston LJ, Libecap GD, Mueller B. Land reform policies, the sources of violent conflict, and implications for deforestation in the Brazilian Amazon. *J Environ Econ Manage* 2000;39:162–188.

171. Kirby KR, Laurance WF, Albernaz AK, Schroth G, Fearnside PM, et al. The future of deforestation in the Brazilian Amazon. *Futures* 2006;38:432–453.

172. Souza JJV. O Programa de Desenvolvimento Sustentável do Acre: uma análise à luz do desenvolvimento sustentável e da cooperação internacional: Universidade Federal de Santa Catarina, Centro de Ciências Jurídicas, Curso de Pós-Graduação em Direito—CPGD, 2008, 188 pp.

173. Guerra CA, Gikandi PW, Tatem AJ, Noor AM, Smith DL, et al. The limits and intensity of *Plasmodium falciparum* transmission: implications for malaria control and elimination worldwide. *PLoS Med* 2008;5(2):e38. DOI: 10.1371/journal.pmed.0050038.

33

ECOSYSTEM SERVICES, WATER RESOURCE DEVELOPMENT, AND HUMAN INFECTIOUS DISEASE

Uriel N. Safriel

33.1 INTRODUCTION

That human health depends on water, not just for drinking but also for sanitation, is common knowledge. However, water is also just as essential for the production of food and other commodities like fiber and biofuel, without which human health is severely compromised. Therefore, maintaining the well-being of a persistently growing global human population requires an intensification of irrigated agriculture, for which, especially in drylands, water resource development is indispensable. However, though the ultimate goal is human well-being, of which health is a subset, water resource development affects the environment in ways that often have direct and indirect negative effects on human health, as well as on other attributes of human well-being. The direct effects are waterborne infectious diseases, and the indirect ones are non-waterborne diseases that emanate from the effect of water resource development on the nonwater environment as well as other environmental changes that impinge on human well-being. Understanding these linkages to the environment, of water resource development, water-related diseases and human health and well-being, requires a new perception of what "environment" is.

This chapter argues (a) that the "environment" provides water to people, as well as other "services" on which human life and its well-being depend, including the regulation of infectious diseases; (b) that the intensification of the environmental water provision through water resource development often impairs the ability of the environment to support humanity. The first section of this chapter presents the concept of ecosystems, their biodiversity, and services and

elaborates on the drivers and impacts of water resource development on ecosystems and their services. The second section zooms in on the ecosystem service of disease regulation and elaborates on the impact of water resource development on regulation, eruption, and emergence of human infectious diseases. The last section concludes with water resource management that optimizes the provision of ecosystem services, including that of disease regulation.

33.2 ECOSYSTEMS SERVICES AND WATER RESOURCE DEVELOPMENT

33.2.1 Environment, Ecosystems, Biodiversity, and Services

33.2.1.1 Environment and Ecosystems The term "environment" is mostly used in association with human actions, and always in a nonneutral context—what people do is either "good for the environment" or "bad for the environment." To circumvent this value-laden concept of "environment" practitioners of life sciences coined the term "ecosystem," used for addressing and operationalizing the environment's functionality. The term applies to any area unit on earth, implying that this area is a dynamic system by virtue of the interactions among all the living organisms within it, and between them and its physical and chemical attributes. This approach is not new, but the Millennium Ecosystem Assessment (MA) project (1) was innovative in determining that not only lakes, rivers, forests, and deserts but also the "nonnatural" areas, cities, and farmlands, are

Water and Sanitation-Related Diseases and the Environment: Challenges, Interventions, and Preventive Measures, First Edition.
Edited by Janine M. H. Selendy.
© 2011 Wiley-Blackwell. Published 2011 by John Wiley & Sons, Inc.

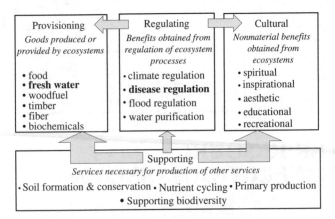

FIGURE 33.1 Ecosystem services. Bolded services are specifically addressed in this chapter. (Adapted from Ref. 1.)

"ecosystems." This approach reflects the recognition that most ecosystems on earth are affected and managed by humans to a certain extent, actively or passively, either deliberately or unintentionally, and humankind is an integral and interactive component in all ecosystems, that jointly comprise the global "environment."

33.2.1.2 Ecosystems and Ecosystem Services

It has been customary by economists to divide the economic output into physical "goods" and intangible "services." These terms have also been applied to outputs of ecosystems, such as marketable "goods" and invaluable but priceless "services." However, economists recognized that the economic output actually constitutes a continuum rather than divided between two discrete entities. This prompted the MA to incorporate "goods" into "services" and define "ecosystem services" as "benefits people obtain from ecosystems," irrespective of whether they are "natural" or intensively managed by people, yet can be classified into four functional groups (Fig. 33.1). The provisioning services refer to "goods" produced by ecosystems, like food that is mostly provided by cultivated ecosystems (i.e., agricultural fields and orchards) but also by "natural" ones (e.g., fisheries derived). Similarly, freshwater, produced in the atmosphere but provided to people by ecosystems (i.e., river, stream, and lake ecosystems) is a provision service. The cultural services constitute the nonmaterial benefits obtained from ecosystems, such as the recreation options, often highly valued when provided by freshwater ecosystems, as well as the spiritual and aesthetic values provided by many ecosystems. The regulating services are benefits obtained from regulation of ecosystem processes such as surface runoff or climate, thus buffering the flow of provisioning and cultural services against environmental changes. The supporting services, which are critical for the provision of all other services, derive from basic ecosystem functions such as primary production that generate the material basis for all

life on earth, nutrient cycling, which is tightly linked to primary production, and soil formation and conservation, instrumental in maintaining the infrastructure for biological productivity. Finally, there is the service of supporting the biodiversity that is involved in the provision of all ecosystem services.

33.2.1.3 Ecosystems Services and Biodiversity

The seminal component of all ecosystems, the one that makes them function, that is, provide services, is their biota—the totality of animals, plants, and microorganisms that inhabit ecosystems. Each of these is involved directly and indirectly in ecosystem functions that serve people. What is critical to the quality and sustainability of this provision is not just the dimensions of the biota (i.e., the number of species and the abundance of each), but mainly the diversity within it (i.e., the degrees by which these species differ from each other) (2). Thus, whereas "biota" just means the living component of the ecosystem, "biodiversity" pertains to the functionality of that component in the provision of services (3), secured by the quality, range, and extent of differences between the living entities dwelling in a given ecosystem (4).

33.2.1.4 Biodiversity and Service Provision

Though ecosystems differ in the composition and dimension of their biodiversity (number of species, differences between them in population sizes, structure, and function), all have a large number of species and the specific role of each in provision of a specific service is often hard to determine. Each species may be directly involved in the provision of a single provisioning or a single cultural service, while many more species jointly collaborate in providing each of the regulating and supporting services. Nevertheless, at any given time or site many species may seem to be redundant, though this need not be the case. Either the role of a single species in service provision is subtle yet critical, or it is expressed when local conditions change, hence over decades or centuries, ecological redundancy in species composition of an ecosystem is unlikely (1). Therefore, biodiversity conservation is required not just "for the sake of the environment," but for securing the provision of the whole suite of ecosystem services critical to human existence and its well-being. Furthermore, even if the specific role in service provision of a certain endangered species is unknown, applying the "precautionary principle" in molding and implementing policies pertaining to ecosystems and their biodiversity is advisable.

33.2.2 Water-Related Ecosystem Services

People intuitively associate the provision of freshwater with freshwater ecosystems (e.g., lakes, rivers, springs, small streams, ponds, as well as groundwater storages). However, the ability of these ecosystems to provide water, the major nonliving component of these ecosystems, depends on adjacent or even distant terrestrial ecosystems, since these determine both the quantity and quality of the water collected and stored by freshwater ecosystems. Thus, terrestrial and freshwater ecosystems are jointly involved in the water-provisioning service. Moreover, they also provide other services, some indirectly linked to water provision, but all contribute to human well-being.

33.2.2.1 Water-Related Services of Terrestrial Ecosystems

Water provision by freshwater ecosystems depends on the water-regulating services of terrestrial ecosystems. This service is provided by a specific biodiversity component, the soil vegetation cover, made of myriads of plant species highly diverse in size, structure, and function. On the global scale, this plant biodiversity component regulates the single largest flux from the biosphere to the atmosphere through the ecosystem function of evapotranspiration (5). The physiological process of plant transpiration drives part of this flux, while evaporation from the soil surface is regulated by the degree of protection from solar radiation that penetrates through the plant cover.

Plant biodiversity also regulates the rainfall flux, when raindrops encounter the canopy of the soil plant cover. The structural diversity of this cover, made of the physical architecture of each of its species, determines the proportion of rainfall, which

(a) is directed to storage in the soil, through reducing the raindrop impact on the soil surface, thus providing for the services of terrestrial primary productivity, forage and crop production;

(b) generates surface runoff, thus redistributing the rainwater to create source–sink spatial pattern (6) and promote forage production (5);

(c) is directed to freshwater ecosystems, thus maintains their water provision service, or joins a flashflood leading to the ocean;

(d) finds its way through soil and nonsaturated rocky media to groundwater storages, thus feeding springs and rivers; and

(e) is used by the plants, partly as raw material for primary production and mostly lost to the atmosphere through transpiration.

Thus, depending on human needs and the variations in the physical properties of the ecosystem and the biological properties of the vegetation cover, the benefits derived from the water regulation service of terrestrial ecosystems are diverse and mutually substitutable. For example, an intense surface flow promotes the water provision service

of a distant freshwater ecosystem, but this is at the expense of the same water used locally for forage provision, if it is locally stored as soil moisture. Similarly, the amount of rainwater transpired by the vegetation cover may constitute only a proportion of the rainfall input (e.g., in Israel) (7), or it may exceed it (e.g., in China) (8, 9). Finally, the same vegetation cover that is instrumental in the provision of the water regulation service of terrestrial ecosystems is also instrumental in the provision of soil conservation service, protecting it from water and wind erosion.

33.2.2.2 Water-Related Services Provided by Freshwater Ecosystems

Wetlands (e.g., swamps and marshes) provide the service of water purification through breaking down, detoxifying, and removing chemicals harmful to the biodiversity of the freshwater ecosystem itself, and for the people to whom the water is provisioned (10). The relatively slow water movement in wetlands allows suspended material to be deposited, and provides time for the mineralization of organic compounds and the biodegradation of toxic chemicals (11). Wetlands also support a submerged vegetation, which further slows water movement, thus accelerating deposition of suspended matter, which increases water purity. The deposition reduces the wetland's depth, which increases its spatial expansion, thus augmenting the flood regulation and water provision services by providing for storage during floods and promoting slow downstream release of water. The flood regulation service lowers flood peaks, thus reducing flood damages, including soil erosion and the clogging of water reservoirs (11, 12).

Rivers and streams provide the service of water purification, including at least partial treatment of wastewater. The instrumental biodiversity here is that of freshwater microorganisms, whose high oxygen demand for processing the wastewater's organic load is satisfied by the oxidizing properties of the stream current (11). Freshwater herbivores and predators maintain the diversity of wastewater-treating species, through matching it with the diversity of compounds to be degraded or recycled. The vegetation of riverbanks contributes to their physical stability—when inundated from rising waters the banks' vegetation moderates the velocity of water flow. The services of soil conservation and water regulation of the riparian ecosystem mold the physical features of the stream, such as channel width and channel depth, which affect the water purification service. When the pollution load exceeds the capacity of the ecosystem for providing the water purification service, the ecosystem becomes "polluted," with most of its other services degraded.

Finally, lakes regulate the flow of outgoing rivers and provide water storage, a capacity often managed to enhance the water provision service. Lakes also provide wastewater treatment, mostly by allowing suspended solids to settle and decompose.

33.2.2.3 Nonwater-Related Services of Freshwater Ecosystems

The primary productivity service of most freshwater ecosystems, including man-made ones, supports provisioning services such as fibers (reeds of wetlands, papyrus of marshes, trees of riparian ecosystems) and food (fish and crustaceans, aquatic birds, edible unicellular algae). Nearly all species cultivated in artificial freshwater ecosystems have wild progenitors and close relatives in "natural" freshwater ecosystems. While the declining resistance to emerging environmental stresses endanger the cultivated species, their progenitors and wild relatives not only maintain the genetic variability required for resistance but also continue to evolve in "natural" ecosystems in response to the changing environment. Hence, freshwater ecosystems that support progenitors of cultivated species constitute a repository of transferable genetic variability, which is instrumental in developing resistant varieties. Finally, all freshwater ecosystems provide the freshwater biodiversity-supporting service. Even constructed wetlands and other artificial freshwater ecosystems often attract wildlife, especially migrating or wintering birds (13). Depending on local cultures, the same biodiversity components of a freshwater ecosystem can be consumed by people, or they provide a cultural service, mainly where such ecosystems are scarce (e.g., in drylands) (14).

33.2.3 Valuating Ecosystems and Their Conservation

33.2.3.1 The Benefit—Valuation of Ecosystem Services

Comparing available market prices, estimating the "willingness to pay" and the cost of restoring or synthetically replacing services, values of 17 services across 16 major ecosystems (cultivated and urban excluded) were estimated by Costanza et al. (15). Though attracting criticism (16, 17), this pioneering work provides an

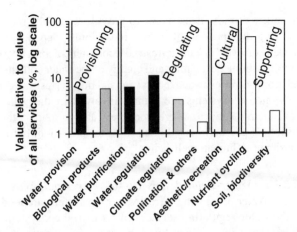

FIGURE 33.2 Estimated global value of ecosystem services expressed as percentage of the aggregated value of all global ecosystems (annual mean of USD33 trillion). Columns: Black indicates water-related services; Grey indicates services that are only partly water-related; Blank columns indicate services not related to water. (Data from Ref. 15.)

insight by comparing its valuation of water-related services to that of other services. It is then evident that water-related regulating services are more valuable than other regulating services (Fig. 33.2). Furthermore, despite their small global area cover, freshwater ecosystems are of high value (Table 33.1). Whereas the combined area of all freshwater ecosystems is only 2.4% of all terrestrial ecosystems, the value of their services is 40% of the value of terrestrial ecosystems—the average annual value of services per hectare of a freshwater ecosystem is 16.8 times higher than that of a hectare of a terrestrial ecosystem. Thus, despite its shortcomings this valuation exercise draws attention to the disproportionate value of water-related ecosystem services, and the freshwater ecosystems that generate them.

33.2.3.2 The Cost—Valuation of Ecosystem Conservation

To avoid decline in the provision of ecosystem services, ecosystems need protection. James et al. (18)

TABLE 33.1 Valuation of Freshwater Ecosystems in Relation to Valuation of All Terrestrial, Nonmarine Global Ecosystems (Including the Freshwater Ecosystems)

	Global Area (million ha)	Value per Area Unit (USD/(ha year))	Total Global Value (USD/year × 10⁹)
Terrestrial ecosystems (freshwater ecosystems included)	15,323	804	12,319
Freshwater ecosystems			
Freshwater wetlands	165	19,580	3,231
Lakes and rivers	200	8,498	1,700
Total freshwater ecosystems	365	13,508	4,931
Freshwater relative to terrestrial ecosystems combined	2.4%	17 times higher	40%

Note that the share of freshwater ecosystem in the value of all terrestrial ecosystems combined is very high, but their overall global areas comprise a very small proportion of all global terrestrial ecosystems. Data adapted from Ref. 15.

and Balmford et al. (16) estimated the current annual expenditure on maintaining and managing the existing global network of protected areas to be USD 6.5 billion (2000 dollar values). When adding the funding shortfalls in managing the existing network, the cost of expanding it for providing optimal protection to all global ecosystem types and managing the added areas, and the opportunity costs incurred by people living in or near these areas, the overall annual cost amounted to USD 45 billion (16). When adding the cost of applying protection measures to ecosystems in areas other than the protected ones, the overall annual cost of conservation went up to USD 317 billion (18). Yet this is only a tiny fraction of the benefit derived from global ecosystems whose mean value, estimated by Costanza et al. (15), amounts to USD 38 trillion (updated to USD 2000 value) (16). Thus, the overall benefit–cost ratio of an effective global program for securing ecosystem services of noncultivated and nonurban ecosystems is around 100:1. Nevertheless, though biodiversity conservation costs are extremely low relative to the benefits of ecosystem services (in whose provision biodiversity is instrumental), current conservation is insufficient and threats to ecosystem services is mounting, including those generated by water resource development.

33.2.4 The Impact of Water Resource Development on Ecosystem Services

33.2.4.1 Water Resource Development: Ecosystem Changes and Transformations   Just as soil comprises the infrastructure of terrestrial ecosystems, water comprises the infrastructure of freshwater ones. However, unlike soil that is used, and its renewability is extremely low, freshwater is a consumed but renewable ecosystem good, provisioned by freshwater ecosystems. It is conceivable that prior to the adoption of agriculture and urbanization, humans' use of the water provision service of freshwater ecosystems had no impact on the functionality of these ecosystems. Later however, the increased demand drove people to practice. Water resource development," meaning management of freshwater ecosystems for intensifying their water provision service, through increasing the rate of water extraction from these ecosystems and conveying the extracted water to other ecosystems. These include installing pumping devices or enlarging outlets intensifying water withdrawal, and construction of water conveyance infrastructures transporting the water to other ecosystems or to artificial water storages that function as constructed freshwater ecosystems.

All these actions bring about just an ecosystem change and/or an ecosystem transformation, either to the freshwater ecosystem or to the ecosystems to which the water is conveyed, or both. Pumping from a lake at a rate that is faster than the natural inflow brings about an ecosystem change and an ecosystem transformation. The change is that of reducing the depth and the surface area of the freshwater ecosystem. The latter entails a transformation, whereby part of the freshwater ecosystem, exposed due to the shrinkage, is transformed to a terrestrial ecosystem. Damming a perennial river, a common water resource development practice, constitutes transformation of the freshwater ecosystem behind the dam, whereby a river ecosystem is transformed to a lake ecosystem. It also constitutes ecosystem changes, through reducing the flow velocity of the downstream section of the river, a change that affects both the stream and the riparian ecosystems. The magnitude of ecosystem changes and transformations driven by dams is astounding, given that during the second half of the twentieth century the number of large dams increased from 5000 to 45,000, and the number of small dams (below 15 m in height) was much higher (their collective volume has been greater than that of all large dams combined) (19).

33.2.4.2 The Extent of Ecosystem Changes Driven by Water Resource Development   All measures of water resource development combined generate annually (based on year 2000) 70–80% of the world's surface freshwater resources (all large dams combined provide water for 30–40% of irrigated cultivated land), and produce 40% of the world food crops (20). More important, water resource development drives the expansion of irrigated agriculture. Though water resource development often changes rainfed, cultivated ecosystems to irrigated ones, the major impact is that of transforming natural ecosystems, such as woodland, grassland, savanna, and even desert ecosystems, into irrigated agriculture ecosystems. Thus, water resource development has driven an increase in the extent of global irrigated agricultural land (from ca. 138 to 271 million ha in 1961 and 2000, respectively) (21, 22).

While water resource development constitutes physical interventions that directly drive ecosystems' change and transformation, there are demographic, social, and economic processes that indirectly drive it. People turning to water resource development are motivated by aspirations of improving their well-being, expecting that the intensified water provision service of freshwater ecosystems would promote the food production service of the ecosystems transformed to cultivated ones. However, ecosystem transformation always involves service trade-offs, in which the promotion of the water and food provision services are often at the expense of an unintended degradation of other services, which compromises rather than improves human well-being. People may then respond to their reduced well-being in taking actions that indirectly drive further ecosystem change (Fig. 33.3).

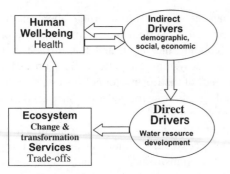

FIGURE 33.3   The linkages between ecosystem services, human well-being, and their direct and indirect drivers, in the context of water and health. Rectangles are states and trends, circles are drivers of states and trends. (After Ref. 1.)

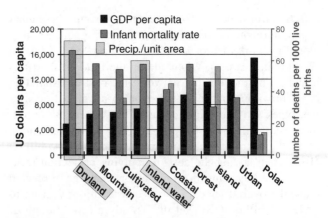

FIGURE 33.5   Measures of human well-being in the drylands in relation to mean annual precipitation, compared to those of other ecosystems. (Adapted from Ref. 1.)

Water resource development promotes not only agricultural but also urban expansion, whereby natural ecosystems are transformed to urban ones and water is mostly used for domestic and industrial purposes. However, the water and land appropriation involved are far smaller in urban than in cultivated ecosystems.

33.2.4.3 Water Resource Development—Dryland Versus Nondryland Ecosystems

At the global scale, the indirect drivers of water resource development are human population growth, associated with the increase in per capita consumption of earth's finite resources—land and water. However, there are differences in the rates of these two drivers of ecosystem change between dryland and nondryland areas. Drylands, which comprise 41% of global land, are subjected to climatic conditions whereby mean

annual potential evapotranspiration is about 1.5 times greater than the mean annual rainfall. This potential water deficit makes water the factor that limits biological productivity in the drylands (23). Hence, biological productivity of drylands is the lowest, compared to that of all other nondryland ecosystems, save the polar ones where productivity is limited by light and temperature (Fig. 33.4). As expected, human population size in the drylands is relatively smaller than in the nondrylands −31.7% of global population resided (in year 2000) in the 41.3% of global land which is dryland (24). However, population growth rate is unexpectedly higher in drylands than in other ecosystems (Fig. 33.4). Given the low productivity and the high population growth rate, it is not surprising that human well-being in the drylands is lower than in nondrylands (Fig. 33.5). It follows that a relatively low population growth but relatively high per capita consumption in the nondrylands, and relatively high population growth but relatively low per capita consumption in drylands, jointly contribute to the increasing global food demand (Fig. 33.6).

This increasing global food demand drives water resource development differently in the drylands than in the nondrylands. Since in the drylands productivity is constrained by water, effectively increasing food production in the drylands requires irrigation. Yet in the drylands freshwater ecosystems are scarce, and irrigation water is readily lost through evaporation. In the nondrylands though, productivity is often constrained by nutrients or temperature, and where irrigation can boost productivity, freshwater ecosystems abound, and evaporation losses are smaller. Indeed, in the nondrylands water resource development mostly targets urban domestic and industrial needs, whereas in the drylands it mostly targets rural irrigation needs (Fig. 33.6). To conclude,

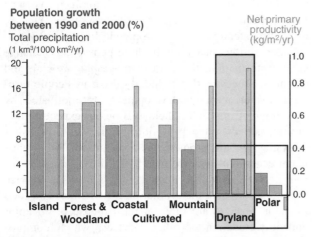

FIGURE 33.4   Comparing drylands to nondrylands: precipitation, biological productivity and human population growth. (Adapted from Ref. 1.)

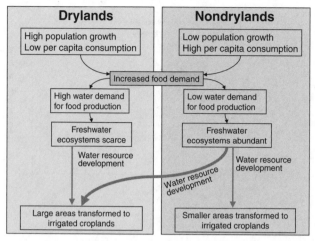

FIGURE 33.6 Drivers and impacts of water resource development in drylands and nondrylands.

global food demand drives water resource development mainly for irrigating the drylands, but since the drylands' freshwater ecosystems can rarely respond to demand, nondryland freshwater ecosystems may be tapped for irrigated food production in the drylands. Nevertheless, in both the drylands and the nondrylands, water resource development drives ecosystem change and transformation (Fig. 33.7).

33.2.4.4 Causes and Implications of Biodiversity Changes
To the farmer, the loss of services provided

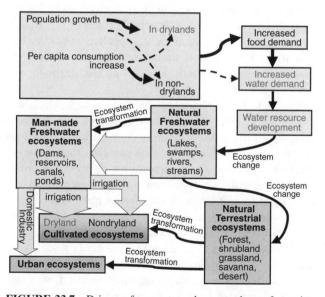

FIGURE 33.7 Drivers of ecosystem change and transformation through water resource development. Black arrows show movement of water.

by the ecosystem prior to its transformation to his farmland is not an adversity, since his expectation from this transformation only pertains to the biological productivity service, with which he is likely to be satisfied. However, the transformation may affect biodiversity, and since biodiversity is instrumental in service provision, water resource development aimed at intensifying certain services may reduce the flow of others. When a natural ecosystem is transformed to a cultivated one, many of the original biodiversity components are lost. Moreover, risks to biodiversity do not stop at the edge of the agricultural field. Rather, it permeates to other ecosystems, including the remaining nontransformed ones. This is because the survival of a species depends on its population size and the smaller the population the risks of its local extinction increase. Since population size is correlated with the size of the species' ecosystem area, the transformation to an agricultural (or urban) ecosystem reduces the area required for maintaining a functional population of the species, leading to a population decline, which is detrimental to the provision of services in which this species is involved. Further decline would lead to local extinction, which is in many cases irreversible (25). Similarly, as the transformation-driven area contraction reaches an ecosystem-specific threshold, the number of species declines (26); hence, service degradation affects not only the transformed ecosystems but also others that have been intact.

An added adversity to the functionality of biodiversity is the fragmentation of the "natural" ecosystem inflicted by transformation. Even when it reduces the overall habitat size of a species by only a small fraction, this transformation often fragments the formerly noncontiguous nontransformed, "natural" ecosystem into several noncontiguous patches of nontransformed ecosystem. The populations of each of the species confined to each fragment are then small and, hence, at risk of further reductions and eventual local extinction. Furthermore, once the transformation to cultivated ecosystems has taken place, pesticides and fertilizers (often associated with the irrigation package) find their way into adjacent and distant nontransformed ecosystems (9). Insecticides and herbicides are often concentrated at top levels of the food chain, sometimes reaching lethal concentrations in top predators, whose decline or extinction may dramatically change the nature of the local biodiversity and, hence, impair a whole suite of services. In addition, pesticides and fertilizers are runoff-, drainage-, and wind-transported from cultivated ecosystems to freshwater ones; while pesticides can be concentrated through links of the food web and damage top-level species, fertilizers cause eutrophication and its associated biodiversity loss and service degradation. Thus, part of the water provided by freshwater ecosystems to irrigate agriculture is redirected to the freshwater ecosystem with a contamination load that damages biodiversity, which impairs their service provision, including that of water provision for agricultural use.

Water provision to agriculture depends on the water renewability in the tapped freshwater ecosystems. Terrestrial natural ecosystems regulate the water supply to freshwater ecosystems. Since vegetation cover of the cultivated ecosystems, often provided by a seasonal single-species crop, is structurally simple, the water provision service of these ecosystems is inferior to that of the perennial and structurally diverse vegetation cover of the natural ecosystems they replaced. The resulting reduced renewability of the freshwater ecosystems thus reduces irrigation water that water resource development had set out to increase. Furthermore, the expansive construction-driven vegetation removal around freshwater ecosystems may entail changes in albedo, evaporation, cloud formation, and rainfall distribution, which individually or jointly affect water renewability.

Finally, while driven by water resource development, transformed ecosystems gain water, other nontransformed terrestrial ones lose this water. Thus, dams not only change and transform the dammed freshwater ecosystem and drive transformation of other often distant terrestrial ecosystems. Rather, dams also change nontargeted terrestrial ecosystems. For example, dams deny from downstream terrestrial ecosystems' floodwater that contributes to biodiversity and hence to the productivity of terrestrial ecosystems, through vertical infiltration, concomitant with lateral redistribution. Arresting moderate floods, dams reduce leaching of salts and the deposition of nutrient-rich soil in natural dryland ecosystems, thus damaging their biodiversity. In desert watersheds, for example, the channel is the only landscape component with perennial vegetation. Desert dams reduce the subsurface runoff in the channel, which lasts longer than the surface runoff and is critical for the persistence of the channel vegetation and its animal biodiversity.

33.2.4.5 Water Resource Development, Desertification, and Climate Change

By year 2000, 38% of the global land area has gone through ecosystem transformation, more than half of which to cultivated ecosystems (24%), and much less to urban ecosystems (3%) (1). The highest percentage of transformed ecosystem is that of the temperate forest, 67% of which has been probably transformed to cultivated land, mostly likely irrigated (which is not necessarily the case for the 34% of transformed tropical and subtropical forests, Fig. 33.8). By area, however, the transformed temperate forest only amounted to 4.2 million km^2. On the other hand, the

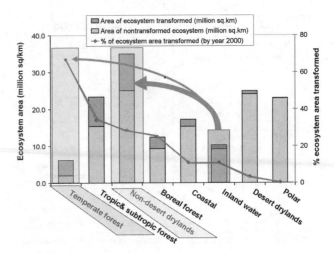

FIGURE 33.8 Ecosystem transformation at the global scale. Columns: areas in million sq. km. Curve: proportion transformed of each ecosystem type. Arrows indicate the flow of water from the freshwater ecosystems (in the "inland water" category) to other ecosystems. (Adapted from Ref. 1.)

ecosystem of the largest transformed area is that of the nondesert drylands, where 10 million km^2 of the nondesert drylands have been transformed. Yet these transformed nondesert dryland areas comprise only 29% of the area of this ecosystem. However, it is highly likely that much of this transformed area is that of irrigated cultivation (Fig. 33.8). It is highly likely that this water resource development-driven transformation contributes to the process of desertification. All drylands, and especially many of the nondesert ones, are at risk of desertification, defined as persistent reduction in biological productivity relative to a dryland potential (27). Rather paradoxically, cultivation, and especially an intensive one driven by irrigation, dramatically increases this risk, mostly through soil salinization, but also through soil erosion (23, 28).

Desertification, restricted (by definition) to drylands, exacerbates global climate change through two different processes. First, a direct driver of desertification is the loss of the soil vegetation cover, thus reducing the photosynthesis-driven removal of carbon dioxide from the atmosphere. Second, desertification does not only mean the permanent loss of this sink function of plant biodiversity, but the eroded soil loses its carbon reserve stored in its organic matter, which is emitted to the atmosphere when oxidized while being carried away by water or wind. Thus, since much of dryland cultivation is driven by water resource development targeting intensification of water and food provision, it is likely to contribute to both desertification and global warming, as well as to biodiversity losses, processes that are mutually reinforcing (Fig. 33.9).

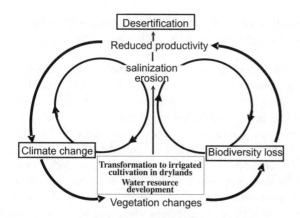

FIGURE 33.9 Interlinkages and mutually reinforcing processes of desertification, biodiversity loss and climate change, driven by water resource development. (Adapted from Ref. 24.)

33.3 THE SERVICE OF DISEASE REGULATION

The previous section highlighted the effect of water resource development on ecosystem services, especially those of water provision and regulation, and food provision. Disease regulation is another ecosystem regulating service (Fig. 33.1), all of which jointly contribute to human well-being, but this specific one is directly linked to one component of human well-being, that of human health (Fig. 33.3). Human health can be compromised by noninfectious as well as by infectious (i.e., parasitic) diseases. Human susceptibility to noninfectious diseases is often affected by ecosystem services such as food provision, water purification, and climate regulation, but only the infectious diseases (accounting for 29 of the 96 major causes of human morbidity and representing 24% of the global burden of disease) (29) are regulated by ecosystems. This specific ecosystem service of disease regulation is elaborated in this section.

33.3.1 The Agents of Infectious Diseases

It is parasitic organisms, the pathogens, which at least during part of their life cycle are external to their human host, that cause infectious diseases. The pathogen organisms interact within ecosystems with other species serving as intermediate hosts, who function as indirect agents of infectious disease. All these disease agent organisms combined comprise a component of nature's biodiversity; hence, they are an integral part of ecosystems. The route of the pathogenic parasite to its human host can be direct, but it is more often indirect through the intermediate (or secondary) host, which is an organism belonging to another species that may function as a vector transmitting the pathogenic parasite to the human (primary) host. This vector, too, is often a parasite transmitting the pathogen

during the act of parasitizing, for example, feeding on blood of the primary host species. In some cases the vector is not a parasite, whereby the transmission of the pathogen is through nonparasitic contact, for example, with infected excretion with the secondary host. In addition, such secondary hosts, or even a third species can function as a reservoir for the pathogenic species, in whose population the pathogen organisms are maintained until they are drawn for infecting the primary host, that is, humans. Finally, pathogens of infectious diseases can infect humans directly, with no other agent species involved (e.g., the *Vibrio* bacteria causing cholera) (30).

Whereas many pathogenic species are host-specific parasites, most vectors are nonpathogenic parasites that feed on a wide range of species. Thus, while human infectious diseases are mostly caused by human-specific, relatively small invertebrates and often microscopic species (of which 1415 are known) (31), the vectors that transmit these parasites to primary hosts often feed on several much larger species, mainly vertebrates, with whom they share the same ecosystems. Population sizes of the parasitic species, the pathogen, and its vector are regulated through parasite–host interactions, which are a subset of the generic prey–predator dynamics, one of the interactions that mold the relative abundance of species in all ecosystems, including those species serving just as reservoirs (Fig. 33.10). For example, the malaria *plasmodium* pathogen, jointly with its mosquito vector, can reduce the populations of their hosts, which in turn may

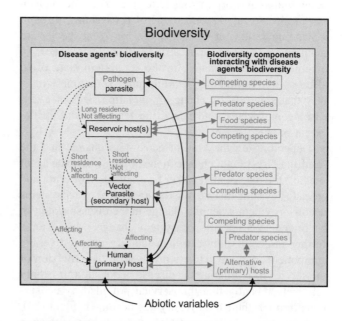

FIGURE 33.10 Ecosystem and biodiversity components involved in the provision of disease regulation service. Dashed arrows show routes of pathogen organism to human host. Solid arrows show interactions, gray text refers to the pathogen.

bring down the number of their own populations, an interaction that can lead to a dynamic equilibrium that maintains the parasite, the vector, and the host in specific population sizes, which may be stable or undergoing bounded fluctuations. At the same time, this equilibrium is affected by another suite of prey–predator interaction, such as insectivorous bird species preying on mosquitoes, and plant species that serve as food of herbivore reservoir species. On top of these complex prey–predator interactions, population sizes of all the species involved in human parasitic diseases (i.e., the disease agents) are also modulated by competitive interactions that regulate the use of their common "ecological niche," that is, a host that can be shared by several parasite species and hence is competed for by them (32).

33.3.2 The Community Equilibrium Mechanism of Disease Regulation

The highly complex biodiversity interactions (Fig. 33.10) are modulated by the other component of ecosystems, namely, the abiotic, physical, and chemical variables, mostly driven by the climate. Thus, both the biotic and abiotic attributes of an ecosystem jointly mold the dynamic equilibrium between all species comprising the ecological community of the ecosystem. This community equilibrium is maintained by all biodiversity components within the ecosystem, including the component of disease agents—those species that directly and indirectly are involved in generating human infectious diseases. This community equilibrium constitutes the major mechanism of the disease regulation service of ecosystems, through which populations of the human disease agent species are maintained in ecosystems at a certain typical size. An externally induced change in the ecosystem state often affects the community equilibrium, sometimes in a way that significantly changes population sizes of a disease-agent species, and hence the temporal and spatial extent of the disease.

Depending on the kind of ecosystem change and the resulting direction of change in the population of the agent species, the disease may either be eradicated, or erupt and spread. Thus, the disease-regulating service of ecosystems becomes evident when the change in the disease state can be linked to a specific ecosystem change. Moreover, a further in-depth study can then identify the biodiversity components that are actively involved in the provision of the service of regulating the disease in question. It is only relatively recently that this service provided by ecosystems through their biodiversity has become evident (33), through the elucidation of linkages between a disease spread or emergence and human-induced ecosystem changes. These changes, the mechanisms in which they disrupt the ecosystem and species' community dynamic

equilibriums, and their effects on human infectious diseases are presented in the following sections.

33.3.2.1 Ecosystem Transformation, Contraction, and Fragmentation

As depicted by Fig. 33.10, the disease regulation service depends on a very rich biodiversity, maintained in a dynamic equilibrium of species' population sizes. However, population size depends on habitat size, and when a species' habitat shrinks, its population may decline to the point of risking a local random extinction (see Section 33.2.4.4). Such local extinctions reduce the ecosystem's biodiversity, hence, destabilizes the community equilibrium that, among other things, regulated human infectious diseases. Ecosystems of wide expanses support diverse biodiversity by including many different habitat types and large areas of each. Yet, most human development involves transforming large areas of natural intact ecosystems to cultivated lands or urban areas, or fragmenting large intact areas by infrastructures and human settlements, thus making "island" habitats supporting each only small populations, hence, at high risk of extinction. This ecosystem shrinkage and fragmentation reduces biodiversity, leading to impaired provision of the disease regulation service of both freshwater and terrestrial ecosystems. This results in disease outbreaks in the cultivated and urban ecosystems, the ones who brought about the shrinkage and fragmentation of the natural intact ones.

Water resource development reduces the size of freshwater ecosystems through water extraction and at the same time, this water is used to construct small and fragmented freshwater ecosystems in the form of reservoirs maintained by dams, canals, and water-covered rice fields. In these three freshwater ecosystem types, biodiversity is much lower than that of the freshwater ecosystems from which the water for creating them has been withdrawn. These man-made freshwater ecosystems of relatively low biodiversity are often rich breeding grounds for vector species that drive eruptions of large-scale infectious diseases. It can be, therefore, inferred that the disease agents in these small-sized ecosystems have been released from the regulation affected by richer biodiversity of their larger-sized ecosystems of origin in which their populations are under control and disease, if occurs, is mostly endemic. This assertion is corroborated by the observation that transmission of diseases, such as schistosomiasis, onchocerciasis, lymphatic filariasis, dracunculiasis, and malaria, are associated in many African countries with small dams more frequently than in larger ones (34). In addition, the incidence of other vector-transmitted diseases increases when dams and canals are constructed (e.g., the Rift Valley Fever of Africa, caused by Phlebovirus transmitted by *Aedes* and *Culex* insect vectors) (35). Similarly, there are many examples linking increased prevalence of infectious disease in irrigated, water-flooded rice fields, whereby such diseases are absent or rare around large freshwater ecosystems.

A case of an infectious disease regulated by forest ecosystem community equilibrium but emerging when regulation is impaired by forest fragmentation is that of Lyme disease of eastern U.S. oak forest ecosystems (36). The pathogen is spirochete bacteria, a tick is the vector, and a mouse and a deer are reservoir hosts. The incidence of disease is linked to the availability of preferred deer food, oak acorns, which fluctuate between years, depending on both climate and the dynamics of an introduced exotic defoliating insect—the gypsy moth whose pupae are preyed upon by the mouse. Thus, the disease is regulated by the complex community interactions that affect the population dynamics of the pathogen. However, fragmentation-driven forests' transformation to urban and agricultural ecosystems cause the mice hosts in the remaining isolated forest fragment to lose many of their predatory and competing species, thus increasing disease risk through the increase in both the absolute and relative density of the mouse, the main reservoir species (37).

Whereas most cases of disease regulation provided by freshwater ecosystems involve regulation of vectors, disease regulation of terrestrial ecosystems often involves the regulation of reservoir hosts of the pathogen. These hosts are usually mammals that transmit the pathogen to humans not by parasitizing them but rather incidentally, for example, when people make contact with the pathogen-infected excretions or secretions. About 60% of all pathogen species of infectious diseases cause the resulting diseases (called "zoonoses"), and 75% of the species considered to be newly emerging pathogens are zoonotic (31). Many of the currently important human emerging infectious diseases (EIDs) are driven by human-induced changes of natural or mildly managed ecosystems. In many cases, but not always, these induced changes to ecosystems are driven by water resource development. These changes not only affect the mammalian hosts so that they escape regulation and increase in numbers, thus increasing the incidence of human contact with the hosted pathogen they shed, but often also affect the pathogen. This occurs when mutants of the pathogen population that happen to be resistant or adaptable to the human-induced new ecosystem state are selected for and expand their niche to include new host species, human included (38).

A case of a zoonotic disease whose emergence can be attributed to a contraction of a Malaysian rainforest ecosystem (though not necessarily driven by water resource development) is that of the Nipah Encephalitis, whose Nipah virus pathogen is hosted by *Pteropus vampyrus* fruit bats ("flying fox"), the reservoir host (39). A combination of reduction in forest fruits due to ENSO-caused drought, and a forest contraction through transformation

of forest ecosystem to farmland ecosystem, which brought cultivated fruits into within the foraging range of the bats in the remaining forest, enabled the farms' domestic pigs to consume fruit remains of the bats' meal, infected with the bat virus-laden saliva. The pigs have thus become an "amplifier" reservoir host, shedding the Nipah virus in respiratory secretions and saliva, thus infecting humans (40). Thus, the Nipha pathogenic virus, previously unknown as a human pathogen, emerged from its natural reservoir hosts (fruit bats) via the domestic animal (pig) amplifier hosts. The mechanism of disease regulation has been that of ecosystem community equilibrium, which regulated the bat population by fruit availability, matching the bats' foraging range, and the populations of other mammals at low level, which prevented them to become infected and function as reservoir hosts. The regulation collapsed when the forest ecosystem contracted, yet the transformed areas, still within the bats' foraging range provided a greater availability of bat food and larger populations of mammalian species, compared with what the remaining forest ecosystem could provide.

Rabies is another zoonotic disease, whereby the pathogen is a virus infecting many wild mammalian species, all of which function as both hosts and vectors. The emergence and reemergence of this disease in humans is similar to that of the Nipah encephalitis, in that domestic animals are the hosts, to whom the pathogen is transmitted by contact with wild mammalian hosts. Such contacts are always a result of ecosystem transformation or contraction. These enable mammalian species to escape regulation when they dart into the human-transformed areas adjacent to the natural ecosystems, thus expanding their niche (feeding on garbage, livestock, and pets), resulting in increasing population size, hence, the probability of encountering pets and humans to be infected. The same ecosystem change also drives domestic dogs to increase their foraging range into the natural ecosystem, where they become infected.

Among the mammalian biodiversity component that functions as a rabies pathogen host, most notable are the vampire bat species of South America, where in some areas these bats are second only to dogs in rabies transmission to humans (41). Unlike the case of the Nipah virus and the fruit bats, the vampire bats transmit the virus directly through feeding on blood of other mammal species (thus, acting like parasites). The bat-transmitted disease outbreaks occur when the bats expand their foraging range, from their ecosystems to the adjacent, human-transformed ones, in search of food—domestic fruits (fruit bats) and domestic mammals and humans (vampire bats). However, unlike rabies, whereby wild mammalian host species increase their population size due to human-induced ecosystem changes often driven by water resource development, the vampire bat is drawn to humans when the availability of its large mammalian hosts, a component of its ecosystem's biodiversity, declines. The decline in population size of large forest mammalian species is attributed to human encroachment, including hunting for food. Deprived of their abundant and readily obtained mammalian food sources, vampire bats seek alternative hosts (humans) to feed on, thus making human the primary host (42). The specific disease regulation here is that of regulating the mammalian component of biodiversity (i.e., the species that functions as a bat host) by either their food species or their predators in such numbers that sustainably supports the vampire bat population.

33.3.2.2 Watershed Management-Induced Ecosystem Change

The ecosystem service of disease regulation is also instrumental at the landscape scale, when a whole watershed in which a freshwater ecosystem such as a river interacts with a terrestrial ecosystem through which it flows. An example of loss of disease regulation in a watershed due to changes to both the freshwater (river) and the terrestrial (forest) ecosystems is that of freshwater resource development executed in the Mahaweli River watershed in Sri Lanka. A series of dams constructed along the river not only created small and fragmented freshwater ecosystems but also reduced the river flow in the interdam sections. This freshwater ecosystem change drove the transformation of 1,650 km^2 of relatively dry woodland ecosystem to irrigated rice ecosystems. The endemic, that is, ecosystem-regulated, malaria of this watershed not only increased two- to five-fold in around the valley's rice fields but it also expanded upstream into villages along the river banks, as the reduced river flow and the abundance of the small dams provided new breeding grounds for mosquitoes. These ecosystem changes not only increased the population size of the *Anopheles* species known from this watershed but another *Plasmodium*-vector *Anopheles*, whose population probably was controlled at extremely low levels, erupted, and contributed to the prevalence of malaria. Furthermore, two *Culex* mosquito species that apparently escaped regulation invaded the rice paddies, where the fertilizers applied to the paddies for boosting rice production triggered their population growth. This eruption coincided with the introduction to the rice-based farms' domestic pigs hosting the Japanese encephalitis virus. This reservoir host provided for transmission of the virus from pigs to humans, by the two *Culex* vector species.

The increased incidence of malaria and the emergence of Japanese encephalitis were not only of a spatial but also of a temporal scale—the annual wet/dry seasonality of the lowland forest ecosystem was obliterated by the year-round irrigation, promoting year-round disease incidence. Thus, the transformation of the forest ecosystem to cultivated wetlands (rice paddies), and the river ecosystem to small artificial water bodies (dammed lakes, pools, canals)

not only removed much of the indigenous rich biodiversity but also replaced it with an exotic poor one, based on introduced domestic species—rice and pigs. Furthermore, the reduction in the overall biodiversity was associated with an increase of disease-agent biodiversity. Apparently, the mosquito vectors were regulated as long as the watershed ecosystems remained intact, regulation that has been lost once water resource development dramatically changed biodiversity (32, 43).

33.3.2.3 Irrigation-Driven Ecosystem Change in Drylands

Arid and semiarid drylands are frequently used as rangelands and as long as their pastoral livelihood is maintained, small dams and other runoff harvesting structures are employed to store water during the rainy season for use in the dry season, mainly for watering livestock. These small ephemeral freshwater ecosystems support aquatic and riparian biodiversity that include disease-agent species. In the Thar Desert of Rajasthan, for example, these ecosystems include *Anopheles* species, yet malaria used to be rare and endemic (44). In the savanna region of Ghana too, mosquito vectors of a freshwater nematode, pathogen of lymphatic filariasis ("Elephantiasis") inhabit such freshwater ecosystems, but the rate of infection has been low (45). In both countries, ecosystem transformation to irrigated agriculture not only increases the number of small freshwater ecosystems through damming or the emergence of small ponds when irrigation elevates the water table but it also obliterates the wet/dry seasonality. These dryland irrigation-driven changes in the suite of small freshwater ecosystems often disrupts the service of disease regulation, apparently operational through the constraints imposed on ecosystem functions and biodiversity interactions by the brevity of the wet season. Thus, in the Rajasthan irrigated desert areas a fourfold increase in malaria since the 1960s of the twentieth century occurred (44), and in Ghana the transmission rate of the lymphatic filariasis infection was much larger in the irrigated savanna than in the nontransformed and hence, not irrigated ecosystems (45).

Irrigation-driven large-scale agricultural development takes place also in grassland and shrubland ecosystems transformed to cultivated ones, in the Argentinian pampas and other South American countries (46). The pathogens are rodent-borne hemorrhagic viruses—three different *arenaviruses*, transmitted by several members of three genera of new world rodents, hosts and reservoirs of these viruses, which infect humans through contact with the rodents' urine and feces. Thus, the small mammal hosts are not parasites, yet they function as vectors since their density in agricultural areas is likely to be much higher than it had been in the grasslands, prior to their transformation to cultivated ecosystems. Due to the rodents' high densities the frequency of contacts between farmers and the rodents' excretions are higher than would have been in the natural ecosystems prior to their change. Biodiversity of the intact ecosystems has been much greater and different from the monocultural agrodiversity, such that population sizes of the rodents, and hence the pathogen were regulated through the mechanism of community equilibrium. Furthermore, cultivation increased the availability of rodents' food, making rodents an integral component of the cultivated ecosystems. Thus, the hemorrhagic fever disease was apparently regulated by the grassland and shrubland ecosystems' community equilibrium mechanism through controlling population sizes of the *arenavirus* rodent hosts. The transformation of this ecosystem to farmland made these hosts effective vectors, since once escaping from the natural regulation of their host, their contact with humans has been dramatically facilitated.

33.3.2.4 Urbanization-Driven Transformation of Forest Ecosystem
While the rainforests of the Amazon basin support many mosquito species, none of them are very common, since permanent forest freshwater bodies abound with larvae predators, and malaria is rare. For example, the malaria vector *Anopheles darlingi* was absent in forest sites sampling, while it occurred in most forest-transformed sites (47). However, as the city of Belem in Brazil emerged from the rainforest, malaria appeared within the city limits. As urbanization progressed at the expense of the forest, and the city population increased (from around 200,000 to 1,400,000 during the last half century), the number of malaria vector species also increased, from 2 to 10 species and the number of malaria cases increased from hundreds to thousands in the 1990s (48). It is conceivable that with the city expansion, more forest habitat types, each of them inhabited by a different mosquito species have been encountered and their species were incorporated into the city's biodiversity. Alternatively, the increased human population density, serving as host to the vectors, could have brought about a population increase of the naturally rare species, otherwise controlled by their forest mammalian hosts. Either way, the natural rainforest ecosystem, inhabited by a sparse human indigenous population, regulated malaria at a very low incidence by keeping all vector species at very low population sizes. Yet urbanization, which brought a huge human population into contact with the forest ecosystem, released the vectors from their forest ecosystem controlling biodiversity components. Similarly, in the Peruvian Amazon forest transformation to farmland, biting rates of humans by malaria vector mosquitoes have been more abundant in deforested than in forested areas, even after adjusting for the difference between the sites in human population size (49).

33.3.3 Disease Regulation Through Agent's Competitive Interaction

This is a subset of the generic community equilibrium mechanism of disease regulation, whereby a specific community interaction, that of interspecific competition, regulates the disease. This regulation is affected when a potential disease agent species is kept in low numbers by a closely related but nonagent or less virulent species having a competitive advantage. This mechanism becomes evident when human-induced ecosystem change reduces the competitive advantage of the nonagent species, thus enabling the potential agent to "invade the ecological niche" and turn into a virulent disease agent. The changes affecting these competitive interactions and their effects on human infectious diseases are presented in the following sections.

33.3.3.1 Irrigation-Driven Desert Ecosystem Change
The eruption of malaria in the Thar Desert following an irrigation-driven ecosystem change is that of destabilizing competitive interactions between two *Anopheles* species that "shared a niche" in the small freshwater bodies in that desert. One species is a common but poor vector; the other is a very rare but an effective vector. The ecosystem change turned the formerly rare and aggressive vector into the commonest one, resulting in *Plasmodium* infections rising in the irrigated areas from 12% in 1986 to 63% in 1994 (44). Apparently, the more aggressive vector was outcompeted by the less aggressive one in the nontransformed ecosystems in which a short wet season is succeeded by a long dry one. The irrigation-driven transformation of ephemeral to perennial freshwater ecosystems provided the more efficient vector with a competitive edge. Thus, competitive interactions between vector species, maintained by the ecosystem's typical dryland seasonality, constitute the disease regulation service.

33.3.3.2 Deforestation-Driven Ecosystem Change
Several closely related species of *Simulium* (blackflies) are of varying capacities to act as vectors of the nematode filarial worm (*Onchocerca volvulus*), the pathogen of Onchocerciasis ("river blindness") disease. These blackflies inhabit the freshwater ecosystems of fast-flowing rivers of the West African Shaelian savanna where the river blindness affects rural communities living in the riparian ecosystems. One of these *Simulium* species, apparently not an effective vector, inhabits the forests located south of the savanna. These forests, subjected to extensive deforestation followed by cultivation, have been invaded by the closely related but effective vector blackfly species, thus increasing the risk of Onchocerciasis invasion (50). Since the forest and the savanna are entirely different ecosystems, yet the savanna blackfly species managed to invade the forest, it suggests that it has been competitively excluded from the forest prior to its human-induced deforestation, by the forest blackfly species. Thus, deforestation and cultivation vacated the niche of the local *Simulium*, to be appropriated by the invading effective vector.

The collapse of this disease regulation mechanism in the forest ecosystem not only put at risk the people there but even more so the people of the savanna. This is because the invading vector species population in the deforested areas served as a reservoir for reinvading savanna areas in which disease control projects had already eradicated that species (51). Thus, the river blindness case demonstrates how an ecosystem change that impairs the ecosystem service of disease regulation in one ecosystem type and region also impairs the eradication of the same disease in another ecosystem type at another region.

33.3.3.3 Disease Control-Driven Ecosystem Change
The threat of erupting river blindness disease in West African forests triggered the widespread distribution of a new microfilaricidal drug (Ivermectin) in the forest regions. On top of the ecosystem change that brought the invasion of the simulid vector, the application of a drug targeting the pathogen constitutes an additional ecosystem change, since it resulted in the emergence "from obscurity" of another lethal filarial worm, *Loa loa* (52). This "new" pathogen has been outcompeted by the *O. volvulus* pathogen, but the drug brought out a change in these competitive interactions. Thus, both the vector and the pathogen it transmits are regulated in the forest by competitive interactions among vectors and among pathogens.

33.3.4 Infectious Diseases Whose Ecosystem Regulation Is not Known

Though all pathogen species have their roots in natural ecosystems, hence, are directly and indirectly affected and regulated by ecosystem functions in which biodiversity is involved, for a large number of infectious diseases the ecosystem regulation remains obscure. For example, the cholera outbreaks are localized in time and space, yet their *Vibrio* bacterial pathogens are an integral biodiversity component in marine, coastal, and estuarine ecosystems, covering a large geographical extent, and at all times. Chitinaceous marine zooplankton species and estuarine and river phytoplankton and aquatic plants species function as "hosts" or a reservoir (though the pathogens do not penetrate but are attached externally), thus maintaining them in natural ecosystems (30). Yet, the ecosystem changes that trigger the cholera eruptions are not known, and hence, the cholera regulation service of these ecosystems, as well as of other infectious diseases, remains obscure.

33.4 WATER RESOURCES—MANAGEMENT RATHER THAN DEVELOPMENT

33.4.1 The Nonsustainability of Water Resource Development

The global environment is composed of ecosystems, intact or used and managed by people, who derive benefits from

their functions—the "ecosystem services." Many of these services, such as food provision and disease regulation, are not just "benefits" but are of survival value. Biodiversity, the living component of all ecosystems, is involved in the provision of all services; hence, any human development activity that results in ecosystem change also causes biodiversity changes. These are succeeded by changes in ecosystem service provision, often associated with a trade-off, in which the targeted intensification of some services often results in inadvertent degradation of other services (Fig. 33.3). Since the services to be intentionally promoted and the ones that may be unintentionally degraded are beneficial to people and contribute to their well-being, development policies need to strike a balance between the aspired promotion and the resulting degradation of different ecosystem services, in order to optimize the use of ecosystems so that human well-being is sustainably maximized.

These considerations apply to all development actions, ecosystems, and services, yet water resource development, freshwater ecosystems, and the service of disease regulation merit special attention. This is because water resource development capitalizes on the water provision service of freshwater ecosystems, which mostly drive the transformation of natural and mildly used ecosystems (such as rangelands) to cultivated ecosystems. The trade-off services are food provision, which is promoted in cultivated ecosystems, at the cost of degrading other services, such as, for example, the soil conservation service of these ecosystems and the cultural services of the utilized freshwater ecosystems. Inasmuch as the degradation of such services is worrying, the degradation of the disease regulation service that often (but not always) is driven by water resource development needs to cause utmost concern, since it often leads to the eruption, reemergence and emergence of human infectious diseases (Fig. 33.11).

When considering the degradation of the disease regulation service jointly with the degradation of many other services, it emerges that water resource development has the potential, too often realized, to undermine its own sustainability. The Thar Desert case (Sections 33.3.2.3 and 33.3.3.1) substantiates this sober conclusion. Irrigation intensified the food provision service of this desert ecosystem, which promoted the well-being of its land users. However, the eruption of malaria associated with the ecosystem change driven by this water resource development risked the most significant component of human well-being, namely health, of the same land users. This and other examples discussed in this chapter call for a paradigm shift: rather than *developing* water resource that leads to nonsustainability, *managing* them for attaining sustainability of human well-being needs to be considered.

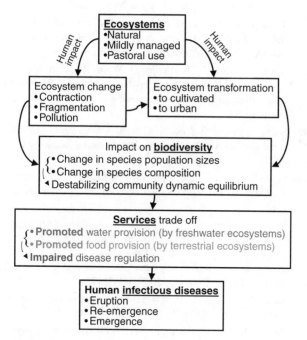

FIGURE 33.11 Ecosystem change and human infectious diseases.

33.4.2   The Alternative of Water Resource Management

The most common practice of water resource development is dam construction. Since this is a costly operation, it is usually implemented only when the construction cost is much smaller than the calculated aspired benefits. "Externalities," expressed in the service trade-offs, especially including the degradation of the disease regulation service, are not included in these cost–benefit calculations, neither with respect to large dams (19), nor when the alternative of a small dam is advocated (53).

For water resource development to become sustainable, it is necessary to include all "externalities" in the project's costs—especially the expected trade-offs. This precautionary practice is likely to encourage considering management rather than "development" of water resources. Namely, instead of engaging in projects that constitute large-scale interventions in the structure and function of freshwater ecosystem, leading to more intensive and extensive intervention in terrestrial ecosystems, measures to better manage water resources already appropriated for agricultural and urban uses are to be taken. This includes reducing conveyance and evaporation losses, increasing water-use efficiency, treating domestic wastewater for agricultural reuse, storing water in aquifers rather than in open dams, and developing agrotechnologies and crop preferences that reduce water use. Concurrently with this supply management, the demand for water needs to be managed too. This pertains to social and cultural changes that address population growth and consumption rates, which drive water resource development. These need to be managed too, otherwise no management of water resources would be sustainable.

33.4.3   Management for Securing the Disease Regulation Service of Ecosystems

Only in few cases, the specific mechanism of disease regulation has been identified. For most infectious diseases, it is the overall community equilibrium maintained by the ecosystem's biodiversity that generically provides for the infectious disease regulation, including those diseases regulated by a known specific mechanism. Therefore, for enabling ecosystems to provide the best disease regulation service, their indigenous biodiversity needs to be protected from human impacts. However, since hardly any ecosystem is now free from human impact, ecosystems, just like water resources, need to be managed for optimizing service provision, disease regulation included. The objectives of such management, of either freshwater or terrestrial ecosystems, are to maintain as much as possible the full suite of species indigenous to that ecosystem, in numbers that are close to their inherent population sizes, and to provide for them large contiguous areas, preferably connected by corridors to other ecosystems of the same type. To achieve these demanding objectives, an ecosystem and biodiversity-monitoring program needs to accompany any development project. This enables detecting deviations from the indigenous state, which would prompt studying the drivers and implications of these changes for service provision, thus enabling revising and updating the management practice.

REFERENCES

1. Millennium Ecosystem Assessment (MA). *Ecosystems and Human Well-Being: Synthesis.* Washington, DC: Island Press, 2005.

2. Lambeck RJ. Biodiversity—the variety of life. Retaining conservation in agricultural regions: a case study from the Wheatbelt of Western Australia. Biodiversity Technical Paper, No. 2. CSIRO Division of Wildlife and Ecology Commonwealth of Australia, 1999. Available at http://www.environment.gov.au/biodiversity/publications/technical/landscape/chapter2a.html.

3. Loreau M, Naeem S, Inchausti P, Bengtsson J, Grime JP, Hector A, Hooper DU, Huston MA, Raffaelli D, Schmid B, Tilman D, Wardle DA. Biodiversity and ecosystem functioning: current knowledge and future challenges. *Science* 2001;294:804–808.

4. Heywood VH, Bastge I. Introduction. In: Heywood VH, editor. *Global Biodiversity Assessment.* Cambridge: Cambridge University Press, 1995.

5. Schlesinger WH, Reynolds JF, Cunningham GL, Huennuke LF, Jarell WM, Virginia RA, Whitford WG. Biological feedbacks in global desertification. *Science* 1990;247:1043–1048.

6. Boeken B, Shachak M. Desert plant communities in human-made patches, implications for management. *Ecol Appl* 1994;4:702–716.

7. Stanhill G. Effects of land use on the water balance of Israel. In: Graber M, Cohen A, Magaritz M,editors. *Regional Implications of Future Climate Change.* Jerusalem: Israel Academy of Sciences and Humanities, 1993, pp. 200–216.

8. Sandstrom K. Can forests "provide" water—widespread myth or scientific reality? *Ambio* 1998;27:132–138.

9. Zhang W, Ricketts TH, Kremen C, Carney K, Swinton SM. Ecosystem services and disservices to agriculture. *Ecol Econ* 2007;64:253–260.

10. Mooney HA, Lubchenco J, Dirzo R, Sala OE. Biodiversity and ecosystem functioning: ecosystem analyses. In: Heywood VH, editor. *Global Biodiversity Assessment.* Cambridge: Cambridge University Press, 1995.

11. National Research Council (NRC). *Restoration of Aquatic Ecosystems.* Washington, DC: National Academy Press, 1992.

12. Mitsch WJ, Gosselink JG. The value of wetlands: importance of scale and landscape setting. *Ecol Econ* 2000;35:25–33.

13. United States Environmental Protection, Agency (U.S. EPA). *Constructed Wetlands for Wastewater Treatment and Wildlife Habitat* (EPA832-R-93-005) Washington, DC: U.S. Environmental Protection Agency, 1993.

14. Safriel U. Deserts and the planet—linkages between deserts and non-deserts. In: Ezcurra E,editor. *Global Deserts Outlook.* Nairobi: UNEP, 2006.

15. Costanza R, d'Arge R, de Groot R, Farber S, Grasso M, Hannon B, Limburg K, Naeem S, O'Neill RV, Paruelo J, Raskin RG, Sutton P, van den Belt M. The value of the world's ecosystem services and natural capital. *Nature* 1997;387:253–260.

16. Balmford A, Bruner A, Cooper P, Costanza R, Farber S, Green RE, Jenkins M, Jefferiss P, Jessamy V, Madden J, Munro K, Myers N, Naeem S, Paavola J, Rayment M, Rosendo S, Roughgarden J, Trumper K, Turner RK. Economic reasons for conserving wild nature. *Science* 2002;297:950–953.

17. Perrings C. Ecological economics after the Millennium Assessment. *Int J Ecol Econ Stat* 2006;6:8–22.

18. James A, Gaston KJ, Balmford A. Can we afford to conserve biodiversity? *Bioscience* 2001;51:43–52.

19. World Commission on Dams (WCD). *Dams and Development: A New Framework for Decision-Making.* UK: Earthscan, 2000.

20. Levy M, Babu S, Hamilton K. Ecosystem conditions and human well-being, In: Hassan R, Scholes R, Ash N, editors. *Ecosystems and Human Well-Being: Current State and Trends,* Vol. 1. Washington: Island Press, 2005, pp. 123–164.

21. Brown LR, Renner M, Flavin C. *Vital Signs.* New York: W.W. Norton & Company, 1997.

22. United Nations Environment Program (UNEP). *Global Environment Outlook—3.* London: Earthscan Publications, 2002.

23. Safriel U, Adeel Z. Dryland systems. In: Hassan R, Scholes R, Ash N,editors. *Ecosystems and Human Well-Being: Current State and Trends.* Washington: Island Press, 2005.

24. Adeel Z, Safriel U, Niemeijer D, White R. *Ecosystems and Human Well-Being: Desertification Synthesis.* Washington DC: Millennium Ecosystem Assessment, World Resource Institute, 2005.

25. National Research, Council (NRC). *Science and the Endangered Species Act.* Washington, DC: National Academy Press, 1995.

26. Soule ME, editor. 1986. *Conservation Biology: The Science of Scarcity and Diversity.* Sutherland, MA: Sinauer Associates, 1986.

27. Safriel U. Status of desertification in the Mediterranean Region. In: Rubio JL, Safriel U, Daussa R, Blum WEH, Pedrazzini F, editors. *Water Scarcity, Land Degradation and Desertification in the Mediterranean Region: Environmental and Security Aspects.* Berlin: Springer, 2009, pp. 33–73.

28. Safriel UN, Adeel Z. 2008. Development paths of drylands—is sustainability achievable? *Sustain Sci J* 2008;3:117–123.

29. World Health Organization (WHO). *World Health Report, 2004: Changing History.* Geneva: WHO, 2004.

30. Colwell RR. Global climate and infectious disease: the cholera paradigm. *Science* 1996;274:2025–2031.

31. Taylor LH, Latham SM, Woolhouse MEJ. Risk factors for human disease emergence. *Phil Trans Soc Lond B* 2001; 356:983–989.

32. Patz JA, Confalonieri UEC. Human health: ecosystem regulation of infectious diseases. In: Hassan R, Scholes R, Ash N, editors. *Ecosystems and Human Well-Being: Current State and Trends,* Vol. 1. Washington: Island Press, 2005, pp. 390–415.

33. Chivian E. Species loss and ecosystem disruption: the implications for human health. *Can Med Assoc J* 2001;164:66–69.

34. Hunter JM, Rey LK, Chu Y, Adekolu-John EO, Mott KE. *Parasitic Diseases in Water Resources Development: The Need for Intersectoral Negotiation.* Geneva: World Health Organization, 1993, 152 pp.

35. Lefevre PC. Current status of rift valley fever. What lessons to deduce from the epidemics of 1977 and 1987? *Med Trop* 1997;57(Suppl 3):61–64.

36. Jones CG, Ostfeld RS, Richard MP, Schauber EM, Wolff JO. Chain reactions linking acorns to gypsy moth outbreaks and Lyme disease risk. *Science* 1998;279:1023–1026.

37. Ostfeld SR, Keesing F. The function of biodiversity in the ecology of vector-borne zoonotic diseases. *Can J Zool* 2000;78:2061–2078.

38. Cunningham AA, Daszak P, Rodríguez JP. 2003. Pathogen pollution: defining a parasitological threat to biodiversity conservation. *J Parasitol* 2003;89:s78–s83.

39. Anonymous. *Nipah Virus Infection.* The Center of Food Security and Public Health, Iowa State University, 2007. Available at http://www.cfsph.iastate.edu/Factsheets/pdfs/nipah.pdf.

40. Chua KB, Chua BH, Wang CW. Anthropogenic deforestation, El Nino and the emergence of Nipah virus in Malaysia. *Malays J Pathol* 2002;24:15–21.

41. Schneider MC, Burgoa CS. Algunas consideraciones sobre la rabiahumana transmitida por murciélago. *México Rev Salud Publ* 1995;37:354–362.

42. Costa MB, Bonito RF, Nishioka AS. An outbreak of vampire bat bite in a Brazilian Village. *Trop Med Parasitol* 1993; 44:1219–1220.

43. Wijesundera Mde S. Malaria outbreaks in new foci in Sri Lanka. *Parasitol Today* 1988;4:147–150.

44. Tyagi BK. *Malaria in the Thar Desert: Facts, Figures and Future.* India: Agrobis, 2002.

45. Appawu MA, Dadzie SK, Baffoe-Wilmot A, Wilson MD. Lymphatic filariasis in Ghana: entomological investigation of transmission dynamics and intensity in communities served by irrigation systems in the Upper East Region of Ghana. *Trop Med Int Health* 2001;6:511–516.

46. de Manzione N, Salas RA, Paredes H, Godoy O, Rojas L, et al. Venezuelan hemorrhagic fever: clinical and epidemiological studies of 165 cases. *Clin Infect Dis* 1998;26:308.

47. Tadei WP, Thatcher BD, Santos JMH, Scarpassa VM, Rodrigues IB, Rafael MS. Ecologic observations on anopheline vectors of malaria in the Brazilian Amazon. *Am J Trop Med Hyg* 1998;59:325–335.

48. Povoa MM, Conn JE, Schlichting CD, Amaral JC, Segura MN, et al. Malaria vectors and the re-emergence of *Anopheles darlingi* in Belém, Pará, Brazil. *J Med Entomol* 2003;40:379–386.

49. Vittor AY, Gilman RH, Tielcsh J, Glass G, Shields T, Wagner SL, Piendo-Cancino V, Patz JA. The effect of deforestation on the human-biting rate of *Anopheles darlingi,* the primary vector of *Falciparum* malaria in the Peruvian Amazon. *Am J Trop Med Hyg* 2006;74:3–11.

50. Wilson MD, et al. Deforestation and the spatio-temporal distribution of savannah and forest members of the *Simulium damnosum* complex in southern Ghana and south-western Togo. *Trans R Soc Trop Med Hyg* 2002;96:632–639.

51. Baker RP, Guillet A, Seketeli P, Poudiougo D, Boakye Wilson MD, Bissan Y. Progress in controlling the reinvasion of windborne vectors into the western area of the onchocerciasis control program in West Africa. *Phil Trans R Soc Lond Ser B Biol Sci* 1990;328:731–750.

52. Thomson MC, Connor SJ. Environmental information systems for the control of arthropod vectors of disease. *Med Vet Entomol* 2000;14:227–244.

53. Jewsbury JM, Imevbore AMA. Small dam health statistics. *Parasitol Today* 1988;4:57–58.

34

OCEAN POLLUTION: HEALTH AND ENVIRONMENTAL IMPACTS OF BROMINATED FLAME RETARDANTS

SUSAN D. SHAW AND KURUNTHACHALAM KANNAN

34.1 INTRODUCTION

The oceans are global sinks for persistent organic pollutants (POPs), including legacy compounds such as polychlorinated biphenyls (PCBs) and chlorinated pesticides (DDTs, chlordanes), which were banned 30 years ago. The concentrations of several legacy pollutants are slowly declining in marine food webs across the world. Other halogenated chemicals including the brominated flame retardants (BFRs) polybrominated diphenyl ethers (PBDEs) are contaminants of emerging concern, as levels have been increasing in marine ecosystems and people for over 30 years. Introduced in the 1970s, PBDE commercial mixtures were added to a variety of polymers such as polystyrene and polyurethane foams, high-impact polystyrene, and epoxy resins that were applied to consumer products including electronic equipment (circuit boards, computers, monitors, televisions), textiles, commercial and residential construction materials, insulation, mattresses and foam cushions in furniture, baby products, and automobile and aircraft interiors to increase their flame resistance (1, 2). As additive flame retardants, PBDEs are not chemically bound (e.g., covalently bound) but are physically blended with polymers, thus over time, the chemicals have leached out of products, accumulated in indoor air and dust, and eventually entered the natural environment (3).

Although many PBDEs were known to have toxic properties, the potential adverse health or environmental impacts of the chemicals were not considered when flammability standards requiring their use were implemented. Recent studies show that PBDEs elicit a wide range of adverse health effects in animals and humans including endocrine disruption, immunotoxicity, reproductive toxicity, and effects on fetal/child development and neurologic function (1, 4–8).

By the 1990s, BFRs were recognized as a global contamination problem, as large amounts have migrated from consumer and household products to the environment and have been detected in marine and riverine systems throughout the world and in all matrices examined—air, water, soil, sediment, sludge, dust, bivalves, crustaceans, fish, birds, terrestrial and marine mammals, and human tissues (5, 9–16). PBDE exposure is particularly significant for people and biota near densely populated coastal areas. These hydrophobic compounds enter coastal waters directly through municipal and industrial wastewater outfalls, landfill leachate, and indirectly via atmospheric deposition from multiple sources. Lipophilic PBDEs readily accumulate in aquatic food webs and are found at higher levels in marine organisms than in terrestrial biota (11).

The detection of PBDEs in diverse deep-sea organisms including sperm whales (17), cephalopods (18), and deep-sea fishes (19–22) confirms that these contaminants have reached the deep-ocean environment. The finding of higher levels of PBDEs in organisms from deep seas compared to those from shallow waters suggests a vertical distribution of the compounds throughout the water column due to their association with organic particles that finally reach the sea bottom. Thus, sediment and biota from deeper waters act as a depot for PBDEs, as with other persistent halogenated contaminants that cycle through marine food webs and biomagnify with increasing trophic level. As top predators, marine mammals

Water and Sanitation-Related Diseases and the Environment: Challenges, Interventions, and Preventive Measures, First Edition.
Edited by Janine M. H. Selendy.

FIGURE 34.1 Seals and other marine top predators carry high loads of persistent organic chemicals that migrate from land to the sea. Photograph courtesy of Bowman Gray.

accumulate extremely high levels of PBDEs and other POPs over a long lifetime, are vulnerable to numerous contaminant-related health effects, and are important "sentinels" for exposure and effects of POPs in humans (23–26). Like marine mammals, people are exposed to PBDEs via consumption of contaminated seafood and exhibit similar toxic responses (Fig. 34.1).

The extent of PBDE contamination in marine ecosystems of North America is of special concern. For three decades, North America has consumed more than half the global production of PBDEs and 95% or more of the penta-BDE product, the constituents of which are more persistent and bioaccumulative than those of the octa- or deca-BDE mixtures (27). Accordingly, PBDE levels in the general population and biota from North America are higher than the rest of the world, and North American concentrations are increasing (5, 10).

Burning municipal and industrial waste dumps in developing countries and electronic waste (e-waste) recycling areas in China and southeast Asia are severely contaminated by PBDEs and their combustion products, polybrominated dibenzo-*p*-dioxins (PBDDs) and dibenzo-furans (PBDFs) (28). People living and/or working in these areas typically work without the use of protective equipment and are high-risk human populations for PBDE exposure and related health problems.

This chapter examines consequences to health and the environment of the global production and use of PBDEs. We outline potential future impacts of large existing stores of banned PBDEs in consumer products, the vast reservoirs of deca-BDE in marine sediments, the mounting evidence of harmful effects resulting from exposure to PBDEs and their toxic combustion products, especially among infants and children, and the potential threat of halogenated chemicals currently marketed as replacements for banned PBDE mixtures.

34.2 CHEMICAL AND PHYSICAL PROPERTIES

PBDEs are brominated aromatic compounds consisting of two phenyl rings linked by an ether bond and having variable hydrogen to bromine substitutions. PBDEs are structurally similar to other halogenated compounds including PCBs, dioxins and furans, and polybrominated biphenyls (PBBs) (Fig. 34.2). These chemicals are also similar in structure to thyroid hormones and thus have the potential to interact with thyroid hormones and their receptors. Analogous to PCBs, 209 distinct PBDE isomers are theoretically possible; however, each commercial mixture contains only a limited number of congeners from each homologue group. PBDEs have been produced since the 1960s as three commercial mixtures (penta-, octa-, and deca-BDE) that vary in degree of bromination. The major constituents of penta-BDE, BDE-47, -99, and -100, with minor contributions from BDE-153, -154, and -85 (29), are highly lipophilic, persistent and bioaccumulative, and subject to long-range transport (2, 27). The major congener found in octa-BDE is hepta-BDE-183; other constituents are nona-BDE-203 and several octa- and nona-BDEs, whereas deca-BDE contains primarily (97%) BDE-209 and low levels of nona-BDEs (2, 29).

Despite mounting toxicological evidence and the ubiquitous presence of PBDEs in the environment, at present only a limited amount of experimental data related to biological activities and physico-chemical properties are available (30). PBDEs have low vapor pressures, very low water solubility, and high log octanol/water partition coefficient (K_{ow}) values. The vapor pressures of penta-BDE, octa-BDE, and deca-BDE are 4.69×10^{-5}, 6.59×10^{-6}, and 2.59×10^{-9} Pa at 21°C (http://www.ec.gc.ca/CEPARegistry). Water solubilities of PBDE congeners are in the low µg/L levels with deca-BDE having the lowest water solubility (<0.1 µg/L). It is expected that PBDEs entering the environment will tend to bind to the organic fraction of particulate matter. Level III fugacity modeling (*EPI v. 3.10, Syracuse Research Corporation*) indicates that much of the substance would be expected to partition to sediment (approximately 59%), followed by soils (approximately 40%), water (1.2%), and air (0.2%). (http://www.ec.gc.ca/CEPARegistry/documents/subs_list/PBDE_draft/PBDE_P3.cfm). Less highly brominated congeners are relatively more water soluble than more highly brominated congeners. Similarly, lower brominated congeners can be transported to remote areas by long range atmospheric transport. Measured data indicate that tetra-, penta-, and hexa-BDE congeners are highly bioaccumulative, with bioconcentration factors (BCFs) exceeding 5,000 for aquatic species. A bioaccumulation factor (BAF) of 1.4×10^{6} was reported for penta-BDE in blue mussels (*Mytilus edulis*) exposed for 44 days (31). The same study reported BCFs of 1.3×10^{6} for tetra-BDE and 2.2×10^{5} for hexa-BDE in blue mussels. High rates of accumulation in marine biota are supported by high log K_{ow} values for PBDEs

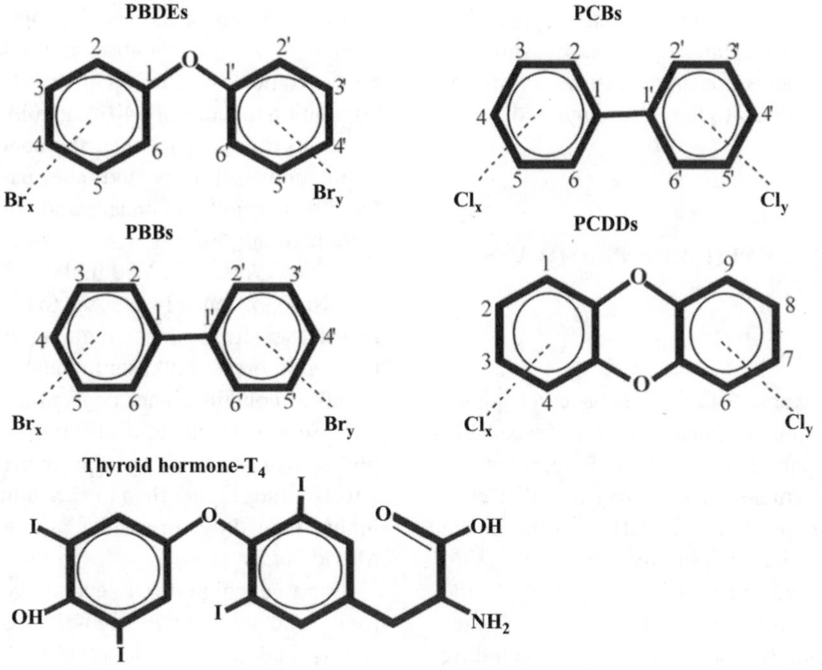

FIGURE 34.2 Chemical structures of PBDEs, PCBs, PBBs, PCDDs, and thyroid hormones (thyroxine, T₄).

and reports of biomagnification of tetra-BDE and penta-BDE in aquatic food chains (11, 32).

34.3 PBDE USES AND REGULATORY RESTRICTIONS

The major uses of halogenated flame-retardant chemicals in North America are in (i) electronics, (ii) building insulation, (iii) transportation, and (iv) home furnishings (2). The chemicals are commonly used at levels up 5% of the weight of polyurethane foam and 15% of the weight of the plastic of electronic housings (33). Penta-BDE mixtures were mainly used in polyester and flexible polyurethane foam formulations. The main use of octa-BDE was in a variety of thermoplastic resins, in particular ABS (acrylonitrile-butadiene-styrene) plastic, which can contain up to 12% by weight octa-BDE. Deca-BDE, the most widely used PBDE globally, was added to various plastic polymers such as polyvinyl chloride, polycarbonates, and high-impact polystyrene, as well as back coating for textiles (commercial furniture, automobile fabrics, and carpets).

Because of environmental and public health concerns, the penta- and octa-BDE commercial formulations were banned in California in 2003, in Europe in 2004, and withdrawn from commerce in the United States in 2004 (34). In May 2009, the Stockholm Convention included penta- and octa-BDEs in the list of persistent organic pollutants (POPs) that are environmentally persistent, bioaccumulative, and toxic to humans

and the environment (35). Despite restrictions, large amounts of foam furniture, juvenile products, and plastics containing PBDEs are still in use and will be disposed of after their lifetimes, creating second-tier outdoor reservoirs (e.g., landfills, wastewater treatment plants, electronic waste recycling facilities or stockpiles of hazardous wastes) for the future dispersal of PBDEs to the ocean environment. Moreover, banned PBDE mixtures have been replaced by halogenated counterparts that are similar in structure and likely to pose similar risks to health and the environment (5, 36).

Deca-BDE, the most widely used PBDE in all markets, accounting for 80% of worldwide production (37), was extensively used in various plastic polymers such as polyvinyl chloride, polycarbonates, and high-impact polystyrene, as well as back coating for textiles (commercial furniture, automobile fabrics, and carpets) (2). Massive releases of deca-BDE from industrial sources directly to the environment have been documented (3), and further concerns have been raised about the abiotic and biotic debromination of BDE-209, the major constituent of deca-BDE, to more bioaccumulative and potentially toxic congeners (27, 38–41). Amidst mounting evidence of the developmental neurotoxicity of BDE-209, deca-BDE was banned in Sweden in 2007, followed by partial bans in U.S. states (Washington, Maine, Oregon, and Vermont), the European Union (EU) (2008), and Canada (2009). In December 2009, the U.S. Environmental Protection Agency (EPA) announced a negotiated 3-year phase-out of this flame retardant in consumer products by the three major global producers (42). However, deca-BDE is still in high-volume use in some applications (e.g., as an additive flame retardant in plastic shipping pallets) and marine sediments constitute vast reservoirs for continuing inputs of deca-BDE to the environment for years to come.

34.4 PBDE SOURCES, EXPOSURE PATHWAYS, LEVELS, AND TEMPORAL TRENDS

34.4.1 Human Exposure

The sources of human exposure to PBDEs have been associated with the indoor environment and diet, especially fish (33, 43–51). In contrast to the pattern reported for PCBs (52, 53), several studies have shown that diet contributes only a small proportion to PBDE exposure in the general population (45, 54, 55). PBDE concentrations in foodstuffs from the United States were comparable to the concentrations reported for several European countries and Japan (56–61). Sources other than the diet, including house dust, were found to be important contributors to the elevated levels of PBDEs in the United States and Canada (33, 38, 44, 47–49, 62). Elevated concentrations (on the order of several tens to hundreds of µg/g) of PBDEs in house dust samples from the United States and Canada coupled with a daily dust ingestion rate of 20–200 mg by the U.S. general population (43) suggest the significance of dust as a source of human exposure (38, 63). Recent studies have documented the occurrence of PBDEs at several tens of µg/g concentrations in dust from U.S. automobiles (64) and garages (65), hundreds to thousands of pg/m^3 concentrations of PBDEs in indoor air (16), and tens to hundreds of pg/m^3 concentrations in automobile air (66). A significant positive correlation between PBDE concentrations in house dust and breast milk has been shown (63). However, a recent study showed that dietary sources contributed significantly to human exposure to PBDEs in the United States (67). Levels of PBDEs in serum of vegetarians were 23–27% lower than levels found in non-vegetarians (67), suggesting the significance of diet as a source of PBDEs in American people.

Human exposure to PBDEs, with particular focus on external exposure routes (e.g., dust, diet, and air) and the resulting internal exposure (e.g., breast milk and blood) has been reviewed recently (16, 68). Overall, populations in the United States and Canada are exposed to higher levels of PBDEs by various pathways than the general populations in the rest of the world. Lorber (48) estimated that 80–90% of the exposure of Americans to PBDEs results from dust ingestion and inhalation. Infants in the United States are exposed at higher levels than adults through the ingestion of breast milk and ingestion of dust due to frequent hand-to-mouth contact (16, 62).

PBDE levels in tissues of people from the general population of various countries are shown in Table 34.1. The concentrations of PBDEs in the North American general population (mean ~30–40 ng/g lipid weight, lw in serum) are 10 to 40 times higher than the concentrations reported for populations in Europe and other parts of the world (10, 48, 69–72). The highest concentration of PBDEs (9,630 ng/g lw) in human adipose fat was reported for a sample collected from New York (69). Within the United States, Californians are disproportionately exposed to PBDEs, presumably due to the state's stringent fire regulations and high BFR usage. California breast milk, serum, and house dust samples were found to contain higher concentrations of PBDEs (49, 73, 74). Serum levels in California children are five times higher than in similar-aged children across the United States, 2 to 10 times higher than in U.S. adults, and 10 to 100 times higher than those in similar aged children in Europe and Mexico (75).

Extremely high PBDE levels have been detected in people living and working in high-exposure regions where local waters and aquatic biota are severely contaminated by PBDEs (Table 34.2). PBDE concentrations in serum of children living and working at a smoldering waste disposal site (WDS) in Managua, Nicaragua, were among the highest ever reported (mean 438 ng/g lw) (96). High PBDE serum levels (mean 580 ng/g lw) have also been reported in people from Guiyu City in southeast China who are involved in the dismantling of electronic wastes (97) and in residents of Laizhou Bay, China, a major production area for PBDEs and other brominated flame retardants (BFRs) (mean 842 and 384 ng/g lw in females and males, respectively) (98). These observations highlight the need for worldwide exposure assessment of PBDEs, not just in the developed countries (Fig. 34.3).

34.4.2 Exposure of Marine Mammals

Marine mammals such as seals and dolphins are long-lived, apex predators that accumulate high concentrations of POPs via consumption of contaminated fish, including many species commonly consumed by people. For adult animals, the main route of exposure is through diet, reflecting food web contamination via various source inputs (local point and nonpoint sources and/or long-range atmospheric transport), whereas young animals are exposed to lipophilic PBDEs *in utero* and in breast milk. High PBDE levels (>1,000 ng/g lw) have been reported in marine mammals from urbanized coastal areas of North America, and, akin to the pattern in their fish prey, congener patterns exhibit a penta-BDE signature (5, 10). Highly brominated BDEs have also been detected in marine mammals, suggesting exposure to the octa- and deca-BDE formulations.

Similar to the distribution pattern observed in people, marine mammals from the California coast contain the

TABLE 34.1 PBDE Concentrations (Mean and Range, ng/g Lipid Weight) in Human Tissues (General Population)

Country	City/State	Tissue	Year	Mean Age	Sex	N	No. of Cong.	ΣPBDEs Mean	ΣPBDEs Min	ΣPBDEs Max	References
United States	California	Breast adipose	1996–1998	47	F	32	1	29	5.2	200	74
	California	Breast adipose	1996–1999	47	F	23	5	41	17	460	73
	California[a]	Plasma	1999–2001	26	F	24	7	21	5.3	320	76
	California	Serum	1997–1999	31	F	50	1	51	<10	510	74
	California	Serum	2003–2004	12–60+	F/M	276	6	62			49
	California	Serum	2003–2005	2–5	F/M	94	11	210			75
	Indiana	Cord serum	2001			12	6	39	14	460	77
	Indiana	Serum	2001	26	F	12	6	37	15	580	77
	Maryland	Cord serum	2004–2005			297	8	27			78
	Massachusetts	Milk	2004–2005	19–41	F	46	12	30	4.3	260	63
	Massachusetts	Milk	2004	35	F	38	17	20	0.06	1910	79
	Mississippi	Whole blood	2003	49	F/M	29	13	31	4.7	360	80
	New York	Adipose	2003–2004	32	F/M	52	11	400	17	9630	69
	New York	Whole blood	2003	42	F/M	10	13	25	4.6	140	80
	North Carolina	Milk	2004–2006		F	301	8	89	1	2010	81
	Pacific Northwest[b]	Milk	2003	30.5	F	40	12	50	6.3	320	82
	Tennessee	Serum	1985–1989			9	6	9.6	4.6	74	83
	Tennessee	Serum	1990–1994			14	6	48	7.5	86	83
	Tennessee/ Washington	Serum	1995–1999			10	6	71	42	120	83
	Tennessee/ Washington	Serum	2000–2002			7	6	61	47	160	83
	Texas	Serum	2007	32	F	29	9	58	6.7	500	61
	Texas	Milk	2001–2004	29	F	47	13	34	6.2	420	84
	Texas	Milk	2007	32	F	29	9	76	13	580	61
	Texas	Liver	2004	Fetus		11	13	23	4	99	85
Canada	Ontario/Quebec	Milk	1992		F	10	6	3.1	0.79	29	86
	Quebec City[c]	Plasma	2003–2004	48–76	F	110	4	64	0.81	3.1	87
China	Tianjin	Serum	2006	31	F/M	115	21	46	0.48	1980	88
Europe	Antwerp, Belgium	Adipose	2003–2005	37	F/M	25	7	5.3	1.4	13	20
	Antwerp, Belgium	Liver	2003–2005	37	F/M	25	7	3.6	1.0	10	20
	Czech Republic	Adipose		36	F/M	98	10	4.4			89
	London, England	Milk	2001–2003	24–34	F	54	15	6.6	0.3	69.0	90
	Berlin, Germany	Milk	2001–2004	32	F	89	9	2.5		18	91
	Rome, Italy	Milk	2000–2001	34	F	10	11	4.1			92
	Menorca Island, Spain	Cord serum	1997–1998		F/M	92	13	6.2			93
	Menorca Island, Spain	Serum	1997–1998	4	F/M	244	13	4.3			93
	Grenada, Spain	Adipose	2003	60	F	20	11	3.9	1.4	11	94
	Örebro, Sweden	Adipose	2007	46	F/M	10	20	4.5	1.2	7.5	95
Brazil	Porto Alegre	Breast adipose	2004–2005	56	F	25	11	6.6	0.19	130	70
	Chihuahua	Serum	2006	6–13	F/M	27		2.7			72
	El Refugio	Serum	2006	6–13	F/M	15		16			72
	Juarez	Serum	2006	6–13	F/M	43		4.8			72
Mexico	Milpillas	Serum	2006	6–13	F/M	52		8.6			72
	San Juan Tilapa	Serum	2006	6–13	F/M	20		3.7			72
	San Luis Potosi	Serum	2006	6–13	F/M	16		7.3			72

[a] Salinas Valley mothers of Mexican descent.
[b] 10 samples each from Portland, OR; Seattle, WA; Missoula, MT; Vancouver, BC, Canada.
[c] Postmenopausal women.

TABLE 34.2 PBDE Concentrations (Mean and Range, ng/g Lipid Weight) in People from High Exposure Regions

Country	City	Tissue	Year	Mean Age	Sex	N	No. of Cong.	ΣPBDEs Mean	ΣPBDEs Min	ΣPBDEs Max	References
Nicaragua	Managua WDS[a]	Serum	2002	14	F/M	11	10	440			96
	Managua WDS[b]	Serum	2002	14	F/M	23	10	69			96
	Urban Managua	Serum	2002	14	F/M	37	10	34			96
	Rural Managua	Serum	2002	15–20	F	8	10	22			96
	Urban Managua	Serum	2002	18–25	F	3	10	75			96
	Urban Managua	Serum	2002	42–44	F	4	10	65			96
China	Guiyu[c]	Cord blood	2007			102	8	14	1.1	510	99
	Chaonan[d]	Cord blood	2007			51	8	5.2	0.29	360	99
	Guiyu[c]	Serum	2005	18–81	F/M	26	16	580	140	8500	97
	Haojiang[d]	Serum	2005	18–81	F/M	21	16	190	16	490	97
	Southeast[c]	Serum		35	F/M	23		380	77	8450	100
	Southeast[d]	Serum		37	F/M	26		160	18	440	100
	Laizhou Bay[e]	Serum	2007	36.5	M	71	8	380			98
	Laizhou Bay[e]	Serum	2007	40.4	F	85	8	840			98
	Laizhou Bay[e]	Milk	2007	29.6	F	15	8	86			98

[a] Children living and working at a waste disposal site (WDS).
[b] Children working at the WDS and living in surrounding city.
[c] E-waste recycling region.
[d] Reference site.
[e] PBDE production area.

highest reported PBDE levels in the world (5). The highest concentrations on record were detected in adult male sea lions (mean 55,300 ng/g lw; range 3430–194,000 ng/g lw) and sea lion pups (51,100 ng/g lw; range 2900–236,000) from southern California (101). California sea lions are near-shore feeders inhabiting contaminated coastal embayments (e.g., the Los Angeles/Long Beach Harbor complex). High PBDE concentrations were also reported in transient killer whales (*Orcinus orca*) off the California coast (up to 12,600 ng/g lw) and in resident killer whales from Puget Sound–Strait of Georgia Basin (mean 7500 and 6800 ng/g lw in females and males, respectively) (102, 103). In addition to proximity to source, PBDE concentrations in transient killer whales are influenced by their high trophic level diet of seals and porpoises.

Along the U.S. Atlantic coast, urban development has produced some of the densest concentrations of human populations in North America, and POP contamination has been a concern since the 1950s. Relatively high concentrations of PBDEs (mean 3000–4000 ng/g lw, range 80–25,700 ng/g lw) were reported in harbor seal pups (*Phoca vitulina concolor*) along the northwest Atlantic (from the Gulf of Maine to New York) (104) and in juvenile bottlenose dolphins from the Charleston Harbor estuary, North Carolina (mean 6000–8000 ng/g lw, range 1756–22,800 ng/g lw) (105) (Fig. 34.4). Charleston Harbor is a highly urbanized estuary with multiple inputs of PBDEs from sources such as wastewater treatment facilities. Given their high body burdens, the PBDE

concentrations in North American marine mammals are of concern (Fig. 34.5).

34.4.3 Temporal Trends

Temporal trend studies have indicated that PBDE levels in marine biota and people in North America have been increasing exponentially over the past 30 years, in some cases doubling as rapidly as every 2–5 years (Fig. 34.6). Following restrictions on penta-BDE production in the 1990s, PBDE levels began to decline in Europeans (106) and may be leveling off in the United States (71). However, current PBDE concentrations in tissues of North Americans are ~20 to 100 times higher than levels in people anywhere else in the world, reflecting the use of 95% of the global production of the penta-BDE product and 44% of deca-BDE over decades (37).

34.5 HEALTH EFFECTS OF PBDEs IN ANIMALS AND HUMANS

34.5.1 Effects in Laboratory Animals

PBDEs have the potential to disrupt the endocrine system at multiple target sites in amphibians, birds, fish, mice, and rats resulting in effects on thyroid, ovarian, and androgen function (1, 5, 113–116). Like other BFRs, the primary toxic effect of PBDEs is thought to be disruption of thyroid hormone homeostasis and several mechanisms have been

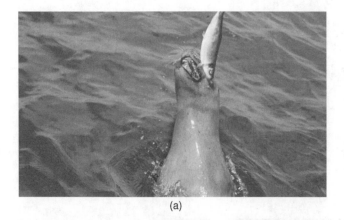

(a)

(b)

FIGURE 34.3   (a) Seal eating fish. Photograph by John Fuller. (b) School of fish. Photograph courtesy of MERI.

FIGURE 34.4   Harbor seal mother and pup. Photograph by Cynthia Stroud.

FIGURE 34.5   Harbor seals on ledge. Photograph by Cynthia Shroud.

proposed: interference in the transport of T_4 via competitive binding to thyroid transport proteins (TTR) and thyroid hormone receptors; induction of thyroid hormone metabolic activity; and interference with the hypothalamus–pituitary–thyroid axis (117). Disruption of thyroid homeostasis during development is of special concern because small changes in maternal and fetal thyroid homeostasis cause neurologic impairments, including small decreases in IQ of offspring (1, 4).

Many of the adverse health effects of PBDEs result from prenatal or neonatal exposure including effects on liver enzymes (118), endocrine disruption (altered thyroid hormone levels) (119), reproductive damage (120–122), immunotoxicity (113, 123), and neurotoxic effects (114, 124, 125). Mice developmentally exposed either to penta- or higher BDEs (124–129) and rats exposed to BDE-209 (130, 131) during the period of brain growth spurt have shown neurotoxic effects including impairment of spontaneous behavior, cholinergic transmitter susceptibility, and habituation capability. Deficits in learning and memory were observed to last into adulthood and worsen with age. Rice et al. (132) found that deca-BDE exposure in neonatal mice resulted in developmental delays, changes in spontaneous locomotor activity, and a dose-related reduction in serum T_4 concentrations. These results suggest that the neurodevelopmental effects of PBDEs are related to perturbations in thyroid hormone homeostasis in the neonate.

As endocrine disruptors, PBDEs may also cause disturbances in adipocyte metabolism that predispose animals to obesity and diabetes (133). Four weeks of *in vivo* exposure of young male rats to a penta-BDE mixture resulted in a significant increase in lipolysis and a significant decrease in insulin-stimulated metabolism in rat adipocytes, both hallmark features of metabolic obesity and type 2 diabetes. Plasma T_4 levels were also significantly decreased in the penta-BDE-treated rats. Further research on the

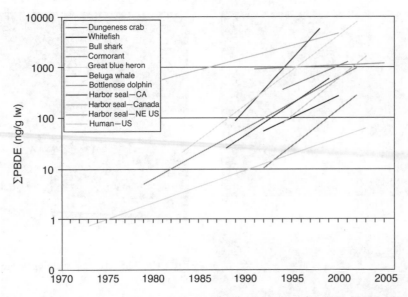

FIGURE 34.6   PBDE concentrations have been increasing in North American wildlife and human tissues over the past 30 years. Trend lines show average PBDE concentrations reported for the first and last year of the studies. References: Dungeness crab (Ref. 107); whitefish (Ref. 108); bull shark (Ref. 109); cormorant (Ref. 110); great blue heron (Ref. 110); beluga whale (Ref. 111); bottlenose dolphin (Ref. 109); harbor seal—San Francisco Bay, CA (Ref. 73); harbor seal—British Columbia, Canada (Ref. 112); harbor seal—NW Atlantic United States (Ref. 104); humans—United States (Ref. 80). (*See insert for color representation of this figure.*)

effects of penta-BDE on human adipocyte function is needed (Fig. 34.7).

34.5.2   Effects in Humans

PBDEs are associated with a wide range of adverse effects in humans including endocrine disruption, reproductive effects, diabetes, and effects on fetal/child development (1, 5). A major concern regarding potential adverse health outcomes of PBDEs relates to their neurodevelopmental effects. A recent study (8) showed that exposure to PBDEs in umbilical cord blood is associated with neurodevelopmental effects in children including substantial IQ deficits. Children with

higher prenatal exposure to PBDE congeners -47, -99, and -100 scored significantly lower on tests of mental and physical development at 1–4 and 6 years. Children in the highest 20% of the exposure distribution showed significantly lower IQ performance scores (ranging from 5 to 8 points lower) at all age points.

Several recent studies have reported adverse human reproductive/developmental outcomes related to PBDE exposure. Harley et al. (6) reported an association between PBDE exposure and reduced fertility in women from a predominantly Mexican-immigrant community in California. Increasing serum levels of BDE-47, -99, -100, and -153 and the sum of the four congeners were all significantly associated

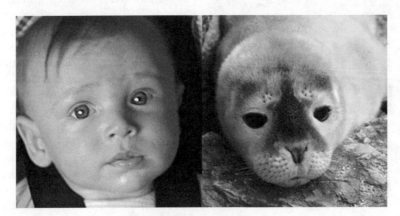

FIGURE 34.7   Human infants and seal pups are exposed to large quantities of toxic organic chemicals through mother's milk. Photo credit: MERI.

with longer time to pregnancy. Prenatal PBDE exposure of the infants of these women was associated with low birth weight, altered cognitive behavior, and significantly reduced plasma levels of thyroid stimulating hormone (TSH) (134). A study by Wu et al. (99) reported that elevated PBDE concentrations in umbilical cord blood were associated with adverse birth outcomes such as premature delivery, low birth weight, and stillbirth among the infants of pregnant women involved in electronic waste (e-waste) recycling in Guiyu, China. Whereas most human tissues show a penta-BDE "signature" dominated by tetra-BDE-47, deca-BDE (BDE-209) dominated the PBDE congener profiles in the Guiyu women. An earlier study reported that elevated levels of PBDEs in breast milk of pregnant Taiwanese women were significantly associated with adverse birth outcomes including weight, length, and chest circumference of their infants (135). In both studies, effects were observed at lower levels than average PBDE levels in the adult U.S. population.

In a study of mother–son pairs from Denmark and Finland, elevated PBDE levels in breast milk were correlated with cryptorchidism (undescended testicles) in the children (136). The PBDE levels associated with cryptorchidism were also positively correlated with serum lutenizing hormone (LH) concentrations in the infants, which suggested a possible compensatory mechanism to achieve normal testosterone levels and is consistent with the anti-androgenic effects of PBDEs observed in experimental animals. A pilot study conducted by Japanese researchers reported that elevated blood levels of BDE-153 were correlated with decreased sperm count and decreased testes size (7). The importance of house dust as a major exposure route for PBDEs in humans has been highlighted (16, 48, 71), and a recent study reported a relationship between altered hormone levels in American men and PBDE levels in house dust (137). Findings included significant inverse associations between PBDEs in house dust and serum concentrations of the free androgen index, lutenizing hormone (LH), and follicle-stimulating hormone (FSH) and positive associations between PBDEs and sex hormone binding globulin (SHBG) and free T_4.

Significant relationships between PBDEs and elevated thyroid hormone levels have been reported in several human studies. Turyk et al. (138) reported an association between PBDEs and elevated T_4 levels and thyroglobulin antibodies in the blood of adult male consumers of Great Lakes sport fish. Effects were observed at PBDE levels comparable to those found in the general U.S. population, and were independent of PCB exposure and sport fish consumption. A recent study of Inuit adults (139) reported that plasma concentrations of BDE-47 were related to increasing total triiodothyronine (T_3) levels. Yuan et al. (100) found an association between serum levels of PBDEs and elevated TSH in people living near or working at an e-waste dismantling site in the highly contaminated Pearl River Delta area of southeast China. Elevated TSH levels may be a compensation for the reduction of circulating thyroid hormones and are indicative of stress on the thyroid system.

Some PBDEs are reported to cause disturbances in glucose and lipid metabolism in adipose tissue, which are characteristic of metabolic obesity and type 2 diabetes (133). Turyk et al. (138) reported a non-significant association between PBDE exposure and diabetes in Great Lakes sport fish consumers with hypothyroid disease. A recent study examined associations between diabetes and PBDEs in U.S. adults (140). Serum concentrations of the hexa-BDE 153 were significantly related to metabolic (obesity) syndrome and diabetes prevalence at background concentrations, suggesting that PBDEs may contribute to diabetes in the general population.

The carcinogenic potential of PBDEs has not been adequately addressed in animals or humans. In a long-term feeding study, a significant increase in the incidence of follicular-cell hyperplasia of the thyroid gland was observed in mice (male and female) exposed to high doses of BDE-209 (141). A study by Hardell et al. (142) reported an association between BDE-47 concentrations and an increased risk for non-Hodgkin's lymphoma (NHL). In the highest risk, highest exposure group, BDE-47 was also significantly correlated with elevated titers to Epstein–Barr IgG, a human herpes virus that has been associated with some subgroups of NHL. The incidence of thyroid cancer has been increasing in the United States during the past several decades, especially among women, and it is hypothesized that part of the observed increase in thyroid cancer rates may be related to the increasing population exposure to PBDEs and other thyroid hormone disrupting compounds (117) (Fig. 34.8).

34.5.3 Marine Mammals as Wildlife Sentinels

PBDEs have been associated with adverse effects including altered thyroid hormone homeostasis, reproductive failure, developmental effects, and immunotoxicity in numerous aquatic and marine wildlife species such as fish, piscivorous birds, mink, otters, seals, and porpoises (5). In sensitive species such as mink (*Mustela vison*), PBDEs have been associated with complete reproductive failure, thyroid alterations, and developmental effects at doses that had no effect in rodents (123, 143). DDT-like effects (eggshell thinning, infertility) and thyroid alterations have been associated with exposure to environmentally relevant levels of PBDEs in kestrels (*Falco sparverius*), peregrine falcons (*Falco peregrinus*), and osprey (*Pandion haliaetus*) (144–146). PBDEs have also been shown to affect the immune system in fish and mink (113, 123). A recent study showed that dietary exposure of Chinook salmon (*Oncorhynchus tshawytscha*) to environmentally relevant concentrations (several ten to hundreds of ng/g) of PBDEs increased susceptibility to pathogenic microorganisms (113) (Fig. 34.9).

FIGURE 34.8   Marine mammal die-offs are increasingly common—and most are linked with immunotoxic pollutants that diminish immune resilience to disease. Danish researchers discovered dead harbor seals on Anholt Island beach during an epidemic that decimated European seals in 2007. Photograph by Theo Harkonen.

For marine mammals inhabiting coastal waters or semi-enclosed seas near industrialized areas, tissue burdens of complex mixtures of pollutants can reach staggering levels and many populations are currently at risk for adverse effects of PBDEs and other POPs (5, 147). Studies have shown that co-exposure to PBDEs and PCBs is associated with thyroid hormone alterations in gray seals (*Halichoerus grypus*) (148) and harbor seals (*Phoca vitulina*) (149) and with thymic atrophy and splenic depletion in harbor porpoises (*Phocoena phocoena*) from the North and Baltic Seas (150). A study of infectious diseases in California sea otters (*Enhydra lutris*) co-exposed to PCBs and PBDEs also suggested possible synergistic interactions between these contaminant groups (151). However, a recent study reported that levels of PBDEs alone significantly reduced the probability of first year survival in gray seals (152).

Since the 1980s, infectious disease outbreaks and mass mortalities associated with exposure to immunotoxic chemicals have been sharply increasing among marine mammals (147, 153). Tens of thousands of seals, dolphins, porpoises, and small whales inhabiting the Baltic Sea, Caspian Sea, Mediterranean, the northwest Atlantic, and other polluted waters have succumbed to sweeping epidemics (154–156). Some of these mortalities have been attributed to algal toxins, low-frequency sonar, or natural causes—but most are linked with immunotoxic pollutants that compromise the animals' normal resistance to disease (153). These events suggest that many populations are living at the edge of their physiological tolerance range. For contaminant-stressed marine mammals, the added stress of climate change may pose a threat to survival, as climate change alterations in food webs, lipid dynamics, ice and snow melt, organic carbon cycling, and severe storm events are expected to increase both the distribution and toxicity of persistent organic pollutants in coastal and oceanic environments (157, 158). As large reservoirs of PBDEs migrate from land-based sources to the ocean environment, marine mammals are expected to become wildlife sentinels at the highest risk for PBDE exposure and related adverse health outcomes including large-scale disease outbreaks and population declines. Human populations such as indigenous people in the Arctic who subsist on a traditional diet of seafood and, in some cases, marine mammals may also face increasing health risks over time (139).

FIGURE 34.9   Harbor seals on Maine ledge. Photograph by Bowman Gray.

34.6 THE LEGACY OF PBDEs: REPLACEMENT FLAME RETARDANTS AND PBDE COMBUSTION PRODUCTS

34.6.1 Halogenated Replacement Flame Retardants

The major replacement chemicals for penta-BDE commercial mixtures in furniture and children's foam products (nursing pillows, baby carriers, high chairs) are Firemaster 550® and TDCP or chlorinated Tris [tris (1,3-dichloro-2-propyl) phosphate] (159), both of which share some of the undesirable properties of penta-BDE. The most commonly used substitute, Firemaster 550®, contains the brominated components: (i) triphenyl phosphate (TPP); (ii) triaryl phosphate isopropylated; (iii) 2-ethylhexyl-2,3,4,5-tetra-bromobenzoate (TBB); and (4) bis (2-ethylhexyl) tetrabromophthalate (TBPH) (160). Firemaster 550 components have been found in dust and sewage sludge and marine sediments in California (161) as well as in tissues of dolphins and porpoises in the South China Sea near flame retardant production facilities (162). Although removed from use in children's sleepwear in 1978, chlorinated Tris (TDCP) is currently the second most used flame retardant in furniture and juvenile product foams in amounts up to 5% by weight. Recent studies showed penta-BDE, TDCP, and Firemaster 550 components can migrate from foam products into indoor house dust, and were found at levels comparable to or greater than levels of PBDEs in dust (159). Inhalation and ingestion of contaminated dust has been shown to be a major route of human exposure, especially for children (16, 48, 63, 159).

Although Firemaster 550 and chlorinated Tris are in high-volume use, little information is available on the health effects of these chemicals in animals or humans. Firemaster 550 components TBB and TBPH are genotoxic in fish, causing significant DNA damage (increased DNA strand breaks from liver cells) following dietary exposure (163). Triphenyl phosphate is toxic to aquatic organisms including Daphnia (italicized) (164), rainbow trout, and fathead minnows (165). Triaryl phosphate isopropylated is a reproductive/developmental toxin at mid- to high doses in rats (165). Histopathologic changes were observed in female reproductive organs and adrenals at all doses. Studies have reported that TDCP is mutagenic (166, 167) and carcinogenic in rats (166, 168). TDCP is also absorbed by humans (169) and was removed from use in sleepwear in 1978. The U.S. Consumer Product Safety Commission (CPSC) considers TDCP a probable human carcinogen and estimates the lifetime cancer risk from TDCP-treated furniture foam is up to 300 cancer cases/million (166). The EPA considers TDCP a moderate hazard for cancer and reproductive/developmental effects (170).

A recent study showed that men living in homes with higher amounts of organophosphate (OP) flame retardants, TDCP and TPP, in household dust had reduced sperm counts and altered levels of hormones related to fertility and thyroid function (171). Higher levels of TPP in dust were associated with a substantial (19%) reduction of sperm concentrations and a 10% increase in prolactin levels. Increased prolactin is considered a marker of decreased neuroendocrine/dopamine activity and also may be associated with erectile dysfunction (172). Higher levels of TDCP in dust were associated with a 17% increase in prolactin and a 3% decline in free thyroid hormone levels. Given that human exposure to these replacement flame retardants in house dust may now be comparable to or exceed exposure to PBDEs (159, 160), these data are cause for concern.

A major replacement for deca-BDE, decabromodiphenyl ethane (DBDPE), has very similar physical and chemical properties and may be more persistent and bioaccumulative than BDE-209 (173). An international survey found DBDPE in sewage sludge from 12 countries on 3 continents (174), suggesting that this fire retardant is a global concern. DBDPE has been detected up to hundreds of parts per billion levels in wildlife species including Great Lakes herring gulls (175), Chinese waterbirds (176), and giant and red pandas (177). The compound was also found at relatively high concentrations in U.S. house dust samples (160) and in children's toys purchased from south China (178), suggesting that people including young children, may be exposed to it. Given the neurodevelopmental properties of deca-BDE, the exposure of infants and toddlers to DBDPE is of concern.

Another alternative to deca-BDE in plastic enclosures is hexabromocyclododecane (HBCD), presently banned in Norway, listed by the European Union and under the Stockholm Convention as one of several chemicals for priority action (179), and identified for risk assessment in Canada, Australia, and Japan,[1] (14). HBCDs are also added to polystyrene foams used as thermal insulation building materials, and to a minor extent in upholstery textile coatings, cable, latex binders, and electrical equipment (2). In recent years, the global demand for HBCDs has increased exponentially in Europe and Asia, and temporal studies show that levels of the most persistent stereoisomer, α-HBCD, are increasing in marine biota and humans (13, 14, 16, 180).

34.6.2 PBDE Combustion Byproducts: Brominated Dioxins and Furans

When PBDEs and other bromine-containing organic chemicals are combusted, as in most e-waste recycling operations, toxic polybrominated dibenzo-p-dioxins (PBDDs), and dibenzofurans (PBDFs) are released (181). PBDD/Fs can be unintentionally formed as by-products in the production of

---

[1]The EU has banned HBCD - announced Feb 17, 2011 -under the REACH program (Registration, Evaluation, Authorisation and Restriction of Chemical substances). Available at http://ec.europa.eu/environment/chemicals/reach/reach_intro.htm

PBDE mixtures (182), from metallurgic processes (183), and from photolytic (184, 185) and thermal processes (186), resulting in subsequent release to the atmosphere, deposition, and accumulation in food webs and people. Under certain combustion conditions such as accidental fires and uncontrolled burning of residential (187) and electronic wastes (28, 188, 189) as well as gasification/pyrolysis, considerable amounts of PBDD/Fs are formed from precursor molecules, particularly PBDEs (186). Firefighters are potentially exposed to high levels of these and other pyrolysis and combustion products at the fire scene and during clean-up after fires (190, 191). Studies of e-waste recycling areas in China have shown that brominated and mixed bromo–chloro dioxin/furans (PXDD/Fs) have severely contaminated the environment (28, 188, 189, 192), including air, soil, and drinking water. In these studies, the levels of PBDF/PXDD/Fs exceeded all soil standards for dioxin worldwide. PBDEs were shown to be the main precursors and responsible compounds for the largest part of dioxin/furan contamination. The primitive recycling and end-of-life treatment ("open burning") of the thousands of tons of PBDEs contained in e-waste is predicted to result in the formation and release of PXDD/PXDFs on the scale of tons (189). Similar primitive techniques are commonly used in developing countries that currently receive most of the world's e-waste including a large share of total PBDEs.

PBDD/Fs and PXDD/Fs have been shown to have similar toxicities as their chlorinated counterparts (polychlorinated dioxins and furans, PCDD/Fs) in cell lines of humans and mammalian species (181, 193–195). Like the PCDD/Fs, PBDD/Fs, and PXDD/Fs bind to the cytosolic aryl hydrocarbon receptor (AhR) and induce hepatic microsomal enzymes, but some mixed PXDD/F congeners are more potent AhR agonists than PCDD/Fs (181, 196). In animal models, all the classic effects demonstrated for PCDD/Fs—thymic atrophy, wasting of body mass, lethality, teratogenesis, reproductive effects, chloracne, immunotoxicity, enzyme induction, decreases in T_4 and vitamin A, and increased hepatic porphyrins—have been observed in studies of PBDD/Fs (181, 195, 197). *In vitro* responses are also similar including enzyme induction, anti-estrogen activity in human breast cancer cells, and transformation of mouse macrophages into tumor cells (197). Recent reports confirm that PBDD/Fs are present at significant levels in house dust (198, 199), fish (200), and human tissues (94, 95, 201, 202). A recent study reported levels of PBDD/Fs in adipose tissue samples from individuals from the general Swedish population contributed significantly, up to 14%, to the total amount of dioxin toxic equivalency (TEQ) of POPs (95).

Firefighters are exposed to complex mixtures of contaminants including pyrolysis and combustion products at the fire scene and during clean-up after fires (190, 191), and a large number of studies show an association between firefighting and significantly elevated rates of specific types of cancer, including multiple myeloma, non-Hodgkin's lymphoma, and prostate and testicular cancer (reviewed in Refs (190, 203). These four types of cancer are potentially related to exposure to dioxins and/or furans (PXDD/Fs) formed via combustion processes during and after fires (190).

E-waste recycling workers including children in China and southeast Asia are highly exposed to brominated and brominated–chlorinated dioxin/furans (PXDD/Fs) and other toxic organic and inorganic pollutants (28, 188, 189, 192). Studies are needed on the health effects of PBDEs and their toxic combustion products in highly exposed people in developing countries.

34.7 CONCLUSIONS AND RECOMMENDATIONS

Since their introduction in the 1970s, PBDEs have been increasing in the environment, biota, and humans and are beginning to rival PCBs as the predominant contaminants in wildlife. They were produced to retard fires and theoretically could have a direct benefit, yet mounting evidence of their persistence, bioaccumulation, and toxicity in animals and in humans is of concern.

Despite recent controls on the production and use of PBDE commercial mixtures, penta-BDE containing products will remain a reservoir for PBDE releases to the oceans for years to come. For example, the average lifetime for foam-containing household furniture and automobile padding has been estimated at 10 years (204). PBDEs are semivolatile and can form thin films on walls and windows (205). It is the entry of these chemicals into domestic dust through the aging/disposal of consumer products that supports the ultimate transport of PBDEs to surface and coastal waters by municipal wastewater systems (5). It is estimated that 80% of human exposure to PBDEs comes from inhalation/ingestion of PBDE-laden indoor dust, implying that only a fraction of the total PBDEs produced have reached the outdoor environment (48).

The indoor reservoir of PBDEs has been termed an environmental "time bomb" (205) because these chemicals are slowly leaching into coastal and marine waters and contaminating ocean food webs. It is predicted that the main exposure route for humans will eventually shift from the indoor environment to the food web (205), signifying the delivery of significant quantities of PBDEs to the oceans. The magnitude and timing of this shift cannot be predicted since they depend upon unknown factors such as the amount of PBDEs yet to be released (from products containing the compounds) and the uptake, incorporation, and magnification of PBDEs into marine and terrestrial food webs.

Concerted action is needed not only to globally enforce current bans on production and use of PBDEs but to find ways of reducing existing indoor reservoirs and managing the end-of-life of PBDE-containing products. However, even if inputs of PBDEs can be controlled/reduced in the terrestrial environment, once these compounds have reached the oceans, it is unlikely that levels will decline any time soon. PBDE emissions are repeating the pattern of PCBs, such that we are now at the same point reached for PCBs in the late 1960s (206, 207). PBDE discharges continue to increase, and these compounds are loading into all compartments of the environment. The oceans are the ultimate sink for PBDEs and the evidence suggests that if all PBDE releases to the environment were controlled today, it will take decades after the end of discharge for marine sediments to bury them (207).

A major emerging concern is the introduction of alternative flame retardant chemicals that are being marketed as replacements for banned PBDE mixtures. Halogenated replacement chemicals such as Firemaster 550 components, chlorinated Tris, and DBDPE are becoming widespread contaminants in the environment and in wildlife. The detection of Firemaster 550 components in cetaceans from the South China Sea confirms that these chemicals have permeated the ocean food web. These potentially toxic brominated and chlorinated flame retardant chemicals are used in consumer products in close contact with humans without adequate consideration of their health and environmental impacts. The continued use of halogenated replacement chemicals for PBDEs in consumer products requires rigorous scrutiny as current research suggests that they have the potential to contribute to serious environmental degradation and long-term health problems in wildlife and people.

ACKNOWLEDGMENTS

The authors thank Michelle Berger for her invaluable assistance in the preparation of this manuscript.

REFERENCES

1. Birnbaum LS, Staskal DF. Brominated flame retardants: cause for concern? *Environ Health Perspect* 2004;112:9–17.

2. Alaee M, Arias P, Sjodin A, Bergman Å. An overview of commercially used brominated flame retardants, their applications, their use patterns in different countries/ regions and possible modes of release. *Environ Int* 2003;29:683–689.

3. Hale RC, La Guardia MJ, Harvey E, Gaylor MO, Mainor TM. Brominated flame retardant concentrations and trends in abiotic media. *Chemosphere* 2006;64:181–186.

4. Costa LG, Giordano G. Developmental neurotoxicity of polybrominated diphenyl ether (PBDE) flame retardants. *NeuroToxicology* 2007;28:1047–1067.

5. Shaw SD, Kannan K. Polybrominated diphenyl ethers in marine ecosystems of the American continents: foresight from current knowledge. *RevEnviron Health* 2009;24:157–229.

6. Harley KG, Marks AR, Chevrier J, Bradman A, Sjödin A, Eskenazi B. PBDE concentrations in women's serum and fecundability. *Environ Health Perspect* 2010;118:699–704.

7. Akutsu K, Takatori S, Nozawa S, Yoshiike M, Nakazawa H, Hayakawa K, Makino T, Iwamoto T. Polybrominated diphenyl ethers in human serum and sperm quality. *Bull Environ Contam Toxicol* 2008;80:345–350.

8. Herbstman JB, Sjodin A, Kurzon M, Lederman SA, Jones RS, Rauh V, Needham LL, Tang D, Niedzwiecki M, Wang RY, Perera F. Prenatal exposure to PBDEs and neurodevelopment. *Environ Health Perspect* 2010;118:712–719.

9. Loganathan BG, Kannan K, Watanabe I, Kawano M, Irvine K, Kumar S, Sikka HC. Isomer-specific determination and toxic evaluation of polychlorinated biphenyls (PCBs), polychlorinated/brominated dibenzo-*p*-dioxins (PCDDs/PBDDs), dibenzofurans (PCDFs/PBDFs), polybrominated biphenyl ethers (PBBEs) and extractable organic halogen (EOX) in carp from the Buffalo River, New York. *Environ Sci Technol* 1995;29:1832–1838.

10. Hites RA. Polybrominated diphenyl ethers in the environment and in people: a meta-analysis of concentrations. *Environ Sci Technol* 2004;38:945–956.

11. de Wit CA. An overview of brominated flame retardants in the environment. *Chemosphere* 2002;46:583–624.

12. Law RJ, Allchin CR, de Boer J, Covaci A, Herzke D, Lepom P, Morris S, Tronczynski J, de Wit CA. Levels and trends of brominated flame retardants in the European environment. *Chemosphere* 2006;64:187–208.

13. Tanabe S, Ramu K, Isobe T, Takahashi S. Brominated flame retardants in the environment of Asia-Pacific: an overview of spatial and temporal trends. *J Environ Monit* 2008;10:188–197.

14. Covaci A, Gerecke AC, Law RJ, Voorspoels S, Kohler M, Heeb NV, Leslie H, Allchin CR, De Boer J. Hexabromocyclododecanes (HBCDs) in the environment and humans: a review. *Environ Sci Technol* 2006;40:3679–3688.

15. Kimbrough KL, Johnson WE, Lauenstein GG, Christensen JD, Apeti DA. 2009. *An Assessment of Polybrominated Diphenyl Ethers (PBDEs) in Sediments and Bivalves of the U.S. Coastal Zone.* NOS NCCOS 94.

16. Johnson-Restrepo B, Kannan K. An assessment of sources and pathways of human exposure to polybrominated diphenyl ethers in the United States. *Chemosphere* 2009;76:542–548.

17. de Boer J, Wester PG, Klamer HJ, Lewis WE, Boon JP. Do flame retardants threaten ocean life? *Nature* 1998;394:28–29.

18. Unger MA, Harvey E, Vadas GG, Vecchione M. Persistent pollutants in nine species of deep-sea cephalopods. *Mar Pollut Bull* 2008;56:1498–1500.

19. Ramu K, Kajiwara N, Mochizuki H, Miyasaka H, Asante KA, Takahashi S, Ota S, Yeh H-S, Nishida S, Tanabe S. Occurrence of organochlorine pesticides, polychlorinated biphenyls and polybrominated diphenyl ethers in deep-sea fishes from the Sulu Sea. *Mar Pollut Bull* 2006;52:1827–1832.

20. Covaci A, Voorspoels S, Roosens L, Jacobs W, Blust R, Neels H. Polybrominated diphenyl ethers (PBDEs) and polychlorinated biphenyls (PCBs) in human liver and

adipose tissue samples from Belgium. *Chemosphere* 2008;73:170–175.

21. Oshihoi T, Isobe T, Takahashi S, Kubodera T, Tanabe S. Contamination status of organohalogen compounds in deep-sea fishes in northwest Pacific Ocean, off-Tohoku, Japan. In: Obayashi Y, Isobe T, Subramanian A, Suzuki S, Tanabe S, editors. *Interdisciplinary Studies on Environmental Chemistry—Environmental Research in Asia.* Terrapub, Tokyo, Japan, 2009, pp. 67–72.

22. Toyoshima S, Isobe T, Ramu K, Miyasaka H, Omori K, Takahashi S, Nishida S, Tanabe S. Organochlorines and brominated flame retardants in deep-sea ecosystem of Sagami Bay. In: Obayashi Y, Isobe T, Subramanian A, Suzuki S, Tanabe S,editors. *Interdisciplinary Studies on Environmental Chemistry—Environmental Research in Asia.* Terrapub, Tokyo, Japan, 2009, pp. 83–90.

23. De Swart RL, Ross PS, Vedder LJ, Timmerman HH, Hoisterkamp S, Van Loveren H, Vos JG, Reijnders PJH, Osterhaus ADME. Impairment of immune function in harbor seals (*Phoca vitulina*) feeding on fish from polluted waters. *Ambio* 1994;23:155–159.

24. Ross PS, De Swart RL, Addison R, Van Loveren H, Vos JG, Osterhaus ADME. Contaminant-induced immunotoxicity in harbour seals: wildlife at risk? *Toxicology* 1996;112:157–169.

25. Kannan K, Blankenship AL, Jones PD, Giesy JP. Toxicity reference values for the toxic effects of polychlorinated biphenyls to aquatic mammals. *Hum Ecol Risk Assess* 2000;6:181–201.

26. Shaw SD, Brenner D, Bourakovsky A, Mahaffey CA, Perkins CR. Polychlorinated biphenyls and chlorinated pesticides in harbor seals (*Phoca vitulina concolor*) from the northwestern Atlantic coast. *Mar Pollut Bull* 2005;50:1069–1084.

27. Hale RC, Alaee M, Manchester-Neesvig JB, Stapleton HM, Ikonomou MG. Polybrominated diphenyl ether flame retardants in the North American environment. *Environ Int* 2003;29:771–779.

28. Ma J, Addink R, Yun SH, Cheng J, Wang W, Kannan K. Polybrominated dibenzo-*p*-dioxins/dibenzofurans and polybrominated diphenyl ethers in soil, vegetation, workshop-floor dust, and electronic shredder residue from an electronic waste recycling facility and in soils from a chemical industrial complex in eastern China. *Environ Sci Technol* 2009;43:7350–7356.

29. La Guardia MJ, Hale RC, Harvey E. Detailed polybrominated diphenyl ether (PBDE) congener composition of the widely used penta- octa- and deca-PBDE technical flame-retardant mixtures. *Environ Sci Technol* 2006;40: 6247–6254.

30. Papa E, Kovarich S, Gramatica P. Development, validation and inspection of the applicability domain of QSPR models for physicochemical properties of polybrominated diphenyl ethers. *QSAR Comb Sci* 2009;28:790–796.

31. Gustafsson K, Björk M, Burreau S, Gilek M. Bioaccumulation kinetics of brominated flame retardants (polybrominated di-

phenyl ethers) in blue mussels (*Mytilus edulis*). *Environ Toxicol Chem* 1999;18:1218–1224.

32. Wenning RJ. Uncertainties and data needs in risk assessment of three commercial polybrominated diphenyl ethers: probabilistic exposure analysis and comparison with European Commission results. *Chemosphere* 2002;46:779–796.

33. Allen JG, McClean MD, Stapleton HM, Webster TF. Critical factors in assessing exposure to PBDEs via house dust. *Environ Int* 2008;34:1085–1091.

34. Betts K. New thinking on flame retardants. *Environ Health Perspect* 2008;116:A210–A213.

35. *Stockholm Convention on Persistent Organic Pollutants.* 2009. Available at http://chm.pops.int/

36. Blum A. The fire retardant dilemma. *Science* 2007;318:194–195.

37. Bromine Science and Environmental Forum (BSEF). 2003. Major brominated flame retardants volume estimates: total market demand by region in 2001. Available at www.bsef.com.

38. Stapleton HM, Dodder NG. Photodegredation of decabromodiphenyl ether in natural sunlight. *Environ Toxicol Chem* 2008;27:306–312.

39. Stapleton HM, Alaee M, Letcher RJ, Baker JE. Debromination of the flame retardant decabromodiphenyl ether by juvenile carp (*Cyprinus carpio*) following dietary exposure. *Environ Sci Technol* 2004;38:112–119.

40. La Guardia MJ, Hale RC, Harvey E. Evidence of debromination of decabromodiphenyl ether (BDE-209) in biota from a wastewater receiving stream. *Environ Sci Technol* 2007;41:6663–6670.

41. Shaw SD, Berger ML, Brenner D, Kannan K, Lohmann N, Päpke O. Bioaccumulation of polybrominated diphenyl ethers and hexabromocyclododecane in the northwest Atlantic marine food web. *Sci Total Environ* 2009;407:3323–3329.

42. US Environmental Protection Agency (US EPA). Announcements, 2009. Available at www.epa.gov/oppt/pbde/

43. Jones-Otazo HA, Clarke JP, Diamond ML, Archbold JA, Ferguson G, Harner T, Richardson GM, Ryan JJ, Wilford B. Is house dust the missing exposure pathway for PBDEs? An analysis of the urban fate and human exposure to PBDEs. *Environ Sci Technol* 2005;39:5121–5130.

44. Wilford BH, Shoeib M, Harner T, Zhu J, Jones KC. Polybrominated diphenyl ethers in indoor dust in Ottawa, Canada: implications for sources and exposure. *Environ Sci Technol* 2005;39:7027–7036.

45. Schecter A, Päpke O, Harris TR, Tung KC, Musumba A, Olson J, Birnbaum LS. Polybrominated diphenyl ether (PBDE) levels in an expanded market basket survey of U.S. food and estimated PBDE dietary intake by age and sex. *Environ Health Perspect* 2006;114:1515–1520.

46. Schecter A, Harris TR, Shah N, Musumba A, Päpke O. Brominated flame retardants in US food. *Mol Nutr Food Res* 2008;52:266–272.

47. Harrad SJ, Ibarra C, Diamond ML, Melymuk L, Robson M, Douwes J, Roosens L, Dirtu AC, Covaci A. Polybrominated

diphenyl ethers in domestic indoor dust from Canada, New Zealand, United Kingdom and United States. *Environ Int* 2008;34:232–238.

48. Lorber M. Exposure of Americans to polybrominated diphenyl ethers. *J Expo Sci Environ Epidemiol* 2008;18:2–19.

49. Zota AR, Rudel RA, Morello-Frosch RA, Brody JG. Elevated house dust and serum concentrations of PBDEs in California: unintended consequences of furniture flammability standards? *Environ Sci Technol* 2008;42:8158–8164.

50. Stapleton HM, Kelly SM, Allen JG, McClean MD, Webster TF. Measurement of polybrominated diphenyl ethers on hand wipes: estimating exposure from hand-to-mouth contact. *Environ Sci Technol* 2008a;42:3329–3334.

51. Toms L-ML, Harden FA, Paepke O, Hobson P, Ryan JJ, Mueller JF. Higher accumulation of polybrominated diphenyl ethers in infants than adults. *Environ Sci Technol* 2008;42:7510–7515.

52. Kannan K, Tanabe S, Ramesh A, Subramanian A, Tatsukawa R. Persistent organochlorine residues in foodstuffs from India and their implications on human dietary exposure. *J Agric Food Chem* 1992;40:518–524.

53. Kannan K, Tanabe S, Giesy JP, Tatsukawa R. Organochlorine pesticides and polychlorinated biphenyls in foodstuffs from Asian and oceanic countries. *Rev Environ Contam Toxicol* 1997;152:1–55.

54. Schecter A, Päpke O, Tung K-C, Staskal DF, Birnbaum LS. Polybrominated diphenyl ethers contamination of United States food. *Environ Sci Technol* 2004;38:5306–5311.

55. Huwe JK, Larsen GI. Polychlorinated dioxins, furans and biphenyls and polybrominated diphenyl ethers in a U. S. meat market basket and estimates of dietary intake. *Environ Sci Technol* 2005;39:5606–5611.

56. Bakker MI, de Winter-Sorkina R, de Mul A, Boon PE, van Donkersgoed G, van Klaveren JD, Baumann BA, Hijman WC, van Leeuwen SPJ, de Boer J, Zeilmaker MJ. Dietary intake and risk evaluation of polybrominated diphenyl ethers in The Netherlands. *Mol Nutr Food Res* 2008;52:204–216.

57. Bocio A, Llobet J, Domingo J, Corbella J, Teixido A, Casas C. Polybrominated diphenyl ethers (PBDEs) in foodstuffs: human exposure through the diet. *J Agric Food Chem* 2003;51:3191–3195.

58. Darnerud PO, Atuma S, Aune M, Bjerselius R, Glynn A, Petersson Grawe K, Becker W. Dietary intake estimations of organohalogen contaminants (dioxins, PCB, PBDE and chlorinated pesticides, e.g. DDT) based on Swedish market basket data. *Food Chem Toxicol* 2006;44:1597–1606.

59. Domingo JL, Marti-Cid R, Castell V, Llobet JM. Human exposure to PBDEs through the diet in Catalonia, Spain: temporal trend: a review of recent literature on dietary PBDE intake. *Toxicology* 2008;248:25–32.

60. Ohta S, Ishizuka D, Nishimura H, Nakao T, Aozasa O, Shimidzu Y, Ochiai F, Kida T, Nishi M, Miyata H. Comparison of polybrominated diphenyl ethers in fish, vegetables, and meats and levels in human milk of nursing women in Japan. *Chemosphere* 2002;46:689–696.

61. Schecter A, Colacino J, Haffner D, Patel K, Opel M, Päpke O, Birnbaum L. Perfluorinated compounds, polychlorinated biphenyl, and organochlorine pesticide contamination in composite food samples from Dallas, Texas. *Environ Health Perspect* 2010;118:796–802.

62. Fischer D, Hooper K, Athanasiadou M, Athanasiadis M, Bergman Å. Children show highest levels of polybrominated diphenyl ethers in a California family of four: a case study. *Environ Health Perspect* 2006;114:1581–1584.

63. Wu N, Herrmann T, Paepke O, Tickner J, Hale R, Harvey E, La Guardia M, McClean MD, Webster TF. Human exposure to PBDEs: associations of PBDE body burdens with food consumption and house dust concentrations. *Environ Sci Technol* 2007;41:1584–1589.

64. Lagalante AF, Oswald TD, Calvosa FC. Polybrominated diphenyl ether (PBDE) levels in dust from previously owned automobiles at United States dealerships. *Environ Int* 2009;35:539–544.

65. Batterman SA, Chernyak S, Jia C, Godwin C, Charles S. Concentrations and emissions of polybrominated diphenyl ethers from U. S. houses and garages. *Environ Sci Technol* 2009;43:2693–2700.

66. Mandalakis M, Atsarou V, Stephanou EG. Airborne PBDEs in specialized occupational settings, houses and outdoor urban areas in Greece. *Environ Pollut* 2008;155:375–382.

67. Fraser AJ, Webster TF, McClean MD. Diet contributes significantly to the body burden of PBDEs in the general U. S. population. *Environ Health Perspect* 2009;117:1520–1525.

68. Frederiksen M, Vorkamp K, Thomsen M, Knudsen LE. Human internal and external exposure to PBDEs—a review of levels and sources. *Int J Hyg Environ Health* 2008;212:109–134.

69. Johnson-Restrepo B, Kannan K, Rapaport D, Rodan B. Polybrominated diphenyl ethers and polychlorinated biphenyls in human adipose tissue from New York. *Environ Sci Technol* 2005;39:8243–8250.

70. Kalantzi OI, Brown FR, Caleffi M, Goth-Goldstein R, Petreas M. Polybrominated diphenyl ethers and polychlorinated biphenyls in human breast adipose samples from Brazil. *Environ Int* 2009;35:113–117.

71. Sjödin A, Wong L-Y, Jones RS, Park A, Zhang Y, Hodge C, Dipietro E, McClure C, Turner W, Needham LL, PattersonJr. DG. Serum concentrations of polybrominated diphenyl ethers (PBDEs) and polybrominated biphenyl (PBB) in the United States population: 2003–2004. *Environ Sci Technol* 2008;42:1377–1384.

72. Pérez-Maldonado IN, Ramírez-Jiménez MDR, Martínez-Arévalo LP, López-Guzmán OD, Athanasiadou M, Bergman Å, Yarto-Ramírez M, Gavilán-García A, Yáñez L, Díaz-Barriga F. Exposure assessment of polybrominated diphenyl ethers (PBDEs) in Mexican children. *Chemosphere* 2009;75:1215–1220.

73. She J, Petreas M, Winker J, Visita P, McKinney M, Kopec D. PBDEs in the San Francisco Bay area: measurements in harbor seal blubber and human breast adipose tissue. *Chemosphere* 2002;46:697–707.

74. Petreas M, She J, Brown FR, Winkler J, Windham G, Rogers E, Zhao G, Bhatia R, Charles MJ. High body burdens of 2,2′,4, 4′-tetrabromodiphenyl ether (BDE-47) in California women. *Environ Health Perspect* 2003;111:1175–1180.

75. Rose M, Bennett DH, Bergman Å, Fangstrom BF, Pessah IN, Hertz-Picciotto I. PBDEs in 2-5-year-old children from California and associations with diet and indoor environment. *Environ Sci Technol* 2010;44:2648–2653.

76. Bradman A, Fenster L, Sjodin A, Jones RS, PattersonJr.DG, Eskenazi B. Polybrominated diphenyl ether levels in the blood of pregnant woman living in an agricultural community in California. *Environ Health Perspect* 2007;115:71–74.

77. Mazdai A, Dodder NG, Abernathy MP, Hites RA, Bigsby RM. Polybrominated diphenyl ethers in maternal and fetal blood samples. *Environ Health Perspect* 2003;111:1249–1252.

78. Herbstman JB, Sjodin A, Apelberg BJ, Witter FR, Patterson Jr.DG, Halden RU, Jones RS, Park A, Zhang Y, Heidler J, Needham LL, Goldman LR. Determinants of prenatal exposure to polychlorinated biphenyls (PCBs) and polybrominated diphenyl ethers (PBDEs) in an urban population. *Environ Health Perspect* 2007;115:1794–1800.

79. Johnson-Restrepo B, Addink R, Wong C, Arcaro K, Kannan K. Polybrominated diphenyl ethers and organochlorine pesticides in breast milk from Massachusetts, USA. *J Environ Monit* 2007;9:1205–1212.

80. Schecter A, Päpke O, Tung KC, Joseph J, Harris TR, Dahlgren J. Polybrominated diphenyl ether flame retardants in the U.S. population: current levels, temporal trends, and comparison with dioxins, dibenzofurans, and polychlorinated biphenyls. *J Occup Environ Med* 2005;47:199–211.

81. Daniels JL, Pan I-J, Jones R, Anderson S, PattersonJr.DG, Needham LL, Sjödin A. Individual characteristics associated with PBDE levels in U. S. human milk samples. *Environ Health Perspect* 2010;118:155–160.

82. She J, Holden A, Sharp M, Tanner M, Williams-Derry C, Hooper K. Polybrominated diphenyl ethers (PBDEs) and polychlorinated biphenyls (PCBs) in breast milk from the Pacific Northwest. *Chemosphere* 2007;67:S307–S317.

83. Sjödin A, Jones RS, Focant J-F, Lapeza C, Wang RY, McGahee EE, Zhang Y, Turner WE, Slazyk B, Needham LL, Patterson DGJ. Retrospective time-trend study of polybrominated diphenyl ether and polybrominated and polychlorinated biphenyl levels in human serum from the United States. *Environ Health Perspect* 2004;112:654–658.

84. Schecter A, Pavuk M, Päpke O, Ryan JJ, Birnbaum LS, Rosen R. Polybrominated diphenyl ethers (PBDEs) in US mother's milk. *Environ Health Perspect* 2003;111:1723–1729.

85. Schecter A, Johnson-Welch S, Tung KC, Harris TR, Päpke O, Rosen R. Polybrominated diphenyl ether (PBDE) levels in livers of U. S. human fetuses and newborns. *J Toxicol Environ Health Part A* 2007;70:1–6.

86. Ryan JJ, Patry B. Determination of brominated diphenyl ethers (BDEs) and levels in Canadian human milks. *Organohalogen Compounds* 2000;47:57–60.

87. Sandanger TM, Sinotte M, Dumas P, Marchand M, Sandau CD, Pereg D, Bérubé S, Brisson J, Ayotte P. Plasma concentrations of selected organobromine compounds and polychlorinated biphenyls in postmenopausal women of Québec. *Environ Health Perspect* 2007;115:1429–1434.

88. Zhu L, Ma B, Hites RA. Brominated flame retardants in serum from the general population in Northern China. *Environ Sci Technol* 2009;43:6963–6968.

89. Pulkrabová J, Hrádková P, Hajšlová J, Poustka J, Nápravníková M, Poláček V. Brominated flame retardants and other organochlorine pollutants in human adipose tissue samples from the Czech Republic. *Environ Int* 2009;35:63–68.

90. Kalantzi OI, Martin FL, Thomas GO, Alcock RE, Tang HR, Drury SC, Carmichael PL, Nicholson JK, Jones KC. Different levels of polybrominated diphenyl ethers (PBDEs) and chlorinated compounds in breast milk from two U.K. regions. *Environ Health Perspect* 2004;112:1085–1091.

91. Vieth B, Rüdiger T, Ostermann B, Mielke H. Residues of flame retardants in breast milk from Germany with specific regard to polybrominated diphenyl ethers (PBDEs). UFOPLAN Ref. No. 202 61 218/03. Federal Institute for Risk Assessment, 2005.

92. Ingelido AM, Ballard T, Dellatte E, di Domenico A, Ferri F, Fulgenzi AR, Herrmann T, Iacovella N, Miniero R, Papke O, Porpora MG, Felip ED. Polychlorinated biphenyls (PCBs) and polybrominated diphenyl ethers (PBDEs) in milk from Italian women living in Rome and Venice. *Chemosphere* 2007;67: S301–S306.

93. Carrizo D, Grimalt JO, Ribas-Fitó N, Sunyer J, Torrent M. Influence of breastfeeding in the accumulation of polybromodiphenyl ethers during the first years of child growth. *Environ Sci Technol* 2007;41:4907–4912.

94. Fernandez MF, Araque P, Kiviranta H, Molina-Molina JM, Rantakokko P, Laine O, Vartiainen T, Olea N. PBDEs and PBBs in the adipose tissue of women from Spain. *Chemosphere* 2007;66:377–383.

95. Jogsten IE, Hagberg J, Lindström G, van Bavel B. Analysis of POPs in human samples reveal a contribution of brominated dioxin of up to 15% of the total dioxin TEQ. *Chemosphere* 2010;78:113–120.

96. Athanasiadou M, Cuadra SN, Marsh G, Bergman Å, Jakobsson K. Polybrominated diphenyl ethers (PBDEs) and bioaccumulative hydroxy PBDE metabolites in young humans from Managua, Nicaragua. *Environ Health Perspect* 2008;116:400–408.

97. Bi X, Thomas G, Jones KC, Qu W, Sheng G, Martin FL, Fu J. Exposure of electronics dismantling workers to polybrominated diphenyl ethers, polychlorinated biphenyls, and organochlorine pesticides in South China. *Environ Sci Technol* 2007;41:5647–5653.

98. Jin J, Wang Y, Yang C, Hu J, Liu W, Cui J, Tang X. Polybrominated diphenyl ethers in the serum and breast milk of the resident population from production area, China. *Environ Int* 2009;35:1048–1052.

99. Wu K, Xu X, Liu J, Guo Y, Li Y, Huo X. Polybrominated diphenyl ethers in umbilical cord blood and relevant factors in neonates from Guiyu, China. *Environ Sci Technol* 2010;44:813–819.

100. Yuan J, Chen L, Chen D, Guo H, Bi X, Ju Y, Jiang P, Shi J, Yu Z, Yang J, Li L, Jiang Q, Sheng G, Fu J, Wu T, Chen X. Elevated serum polybrominated diphenyl ethers and thyroid-stimulating hormone associated with lymphocytic micronuclei in Chinese workers from an e-waste dismantling site. *Environ Sci Technol* 2008;42:2195–2200.

101. Meng X-Z, Blasius ME, Gossett RW, Maruya KA. Polybrominated diphenyl ethers in pinnipeds stranded along the southern California coast. *Environ Pollut* 2009;157:2731–2736.

102. Rayne S, Ikonomou MG, Ross PS, Ellis GM, Barrett-Lennard LG. PBDEs, PBB, and PCNs in three communities of free-ranging killer whales (*Orcinus orca*) from the northeastern Pacific ocean. *Environ Sci Technol* 2004;39:4293–4299.

103. Krahn MM, Hanson MB, Baird RW, Boyer RH, Burrows DG, Emmons CK, Ford JKB, Jones LL, Noren DP, Ross PS, Schorr GS, Collier TK. Persistent organic pollutants and stable isotopes in biopsy samples (2004/2006) from southern resident killer whales. *Mar Pollut Bull* 2007;54:1903–1911.

104. Shaw SD, Brenner D, Berger ML, Fang F, Hong C-S, Addink R, Hilker D. Bioaccumulation of polybrominated diphenyl ethers in harbor seals from the northwest Atlantic. *Chemosphere* 2008;73:1773–1780.

105. Fair PA, Mitchum GB, Hulsey TC, Adams J, Zolman ES, McFee W, Wirth E, Bossart GD. Polybrominated diphenyl ethers (PBDEs) in blubber of free-ranging bottlenose dolphins (*Tursiops truncatus*) from two southeast Atlantic estuarine areas. *Arch Environ Contam Toxicol* 2007;53:483–494.

106. Guvenius DM, Noren K. Polybrominated diphenyl ethers in Swedish human milk. The follow-up study. Second International Workshop on Brominated Flame Retardants. University of Stockholm, Sweden, May 14–16, 2001.

107. Ikonomou MG, Fernandez MP, Hickman ZL. Spatio-temporal and species-specific variation in PBDE levels/patterns in British Columbia's coastal waters. *Environ Pollut* 2006;140:355–363.

108. Rayne S, Ikonomou MG, Antcliffe B. Rapidly increasing polybrominated diphenyl ether concentrations in the Columbia River system from 1992 to 2000. *Environ Sci Technol* 2003;37:2847–2854.

109. Johnson-Restrepo B, Kannan K, Addink R, Adams DH. Polybrominated diphenyl ethers and polychlorinated biphenyls in a marine food web of coastal Florida. *Environ Sci Technol* 2005;39:8243–8250.

110. Elliott JE, Wilson LK, Wakeford B. Polybrominated diphenyl ether trends in eggs of marine and freshwater birds from British Columbia, Canada, 1979–2002. *Environ Sci Technol* 2005;39:5584–5591.

111. Lebeuf M, Gouteux B, Measures L, Trottier S. Levels and temporal trends (1988–1999) of polybrominated diphenyl ethers in beluga whales (*Delphinapterus leucas*) from the St. Lawrence Estuary, Canada. *Environ Sci Technol* 2004;38:2971–2977.

112. Ikonomou MG, Rayne S, Fischer M, Fernandez MP, Cretney W. Occurrence and congener profiles of polybrominated diphenyl ethers (PBDEs) in environmental samples from coastal British Columbia, Canada. *Chemosphere* 2002;46:649–663.

113. Arkoosh MR, Boylen D, Dietrich J, Anulacion BF, Ylitalo GM, Bravo CF, Johnson LL, Loge FJ, Collier TK. Disease susceptibility of salmon exposed to polybrominated diphenyl ethers (PBDEs). *Aquat Toxicol* 2010; in press. doi: 10.1016/j.aquatox.2010.01.013.

114. Chou C-T, Hsiao Y-C, Ko F-C, Cheng J-O, Cheng Y-M, Chen T-H. Chronic exposure of 2,2′,4,4′-tetrabromodiphenyl ether (PBDE-47) alters locomotion behavior in juvenile zebrafish (*Danio rerio*). *Aquat Toxicol* 2010;98:51–59.

115. Darnerud PO. Brominated flame retardants as possible endocrine disrupters. *Int J Androl* 2008;31:152–160.

116. Legler J. New insights into the endocrine disrupting effects of brominated flame retardants. *Chemosphere* 2008;73:216–222.

117. Zhang Y, Guo GL, Han X, Zhu C, Kilfoy BA, Zhu Y, Boyle P, Zheng T. Do polybrominated diphenyl ethers (PBDE) increase the risk of thyroid cancer? *Biosci Hypo* 2008;1:195–199.

118. Lundgren M, Darnerud PO, Molin Y, Lilienthal H, Blomberg J, Ilbäck N-G. Viral infection and PBDE exposure interact on CYP gene expression and enzyme activities in the mouse liver. *Toxicology* 2007;242:100–108.

119. Kuriyama SN, Wanner A, Fidalgo-Neto AA, Talsness CE, Koerner W, Chahoud I. Developmental exposure to low-dose PBDE-99: tissue distribution and thyroid hormone levels. *Toxicology* 2007;242:80–90.

120. Lilienthal H, Hack A, Roth-Härer A, Grande SW, Talsness CE. Effects of developmental exposure to 2,2′,4,4′,5-pentabromodiphenyl ether (PBDE-99) on sex steroids, sexual development, and sexually dimorphic behavior in rats. *Environ Health Perspect* 2006;114:194–201.

121. McDonald T. Polybrominated diphenyl ether levels among United States residents: daily intake and risk of harm to the developing brain and reproductive organs. *Integr Environ Assess Manage* 2005;1:343–354.

122. Talsness CE. Overview of toxicological aspects of polybrominated diphenyl ethers: a flame-retardant additive in several consumer products. *Environ Res* 2008;108:158–167.

123. Martin PA, Mayne GJ, Bursian SJ, Tomy GT, Palace VP, Pekarik C, Smits J. Immunotoxicity of the commercial polybrominated diphenyl ether mixture DE-71 in ranch mink (*Mustela vision*). *Environ Toxicol Chem* 2007;26:988–997.

124. Eriksson P, Jakobsson E, Fredriksson A. Brominated flame retardants: a novel class of developmental neurotoxicants in our environment? *Environ Health Perspect* 2001;109:903–908.

125. Eriksson P, Viberg H, Jakobsson E, Orn U, Fredriksson A. A brominated flame retardant, 2, 2′,4,4′,5-pentabromodiphenyl ether: uptake, retention, and induction of neurobehavioral

alterations in mice during a critical phase of neonatal brain development. *Toxicol Sci* 2002;67:98–103.

126. Viberg H, Fredriksson A, Eriksson P. Neonatal exposure to polybrominated diphenyl ether (PBDE 153) disrupts spontaneous behaviour, impairs learning and memory, and decreases hippocampal cholinergic receptors in adult mice. *Toxicol Appl Pharmacol* 2003;192:95–106.

127. Viberg H, Fredriksson A, Jakobsson E, Orn U, Eriksson P. Neurobehavioral derangements in adult mice receiving decabrominated diphenyl ether (PBDE 209) during a defined period of neonatal brain development. *Toxicol Sci* 2003;76:112–120.

128. Viberg H, Fredriksson A, Eriksson P. Neonatal exposure to the brominated flame-retardant, 2,2′,4,4′,5-pentabromodiphenyl ether, decreases cholinergic nicotinic receptors in hippocampus and affects spontaneous behaviour in the adult mouse. *Environ Toxicol Pharmacol* 2004;17:61–65.

129. Viberg H, Johansson N, Fredriksson A, Eriksson J, Marsh G, Eriksson P. Neonatal exposure to higher brominated diphenyl ethers, hepta-, octa-, or nonabromodiphenyl ether, impairs spontaneous behavior and learning and memory functions of adult mice. *Toxicol Sci* 2006;92:211–218.

130. Viberg H, Fredriksson A, Eriksson P. Changes in spontaneous behaviour and altered response to nicotine in the adult rat, after neonatal exposure to the brominated flame retardant, decabrominated diphenyl ether (PBDE 209). *NeuroToxicology* 2007;28:136–142.

131. Johansson N, Viberg H, Fredriksson A, Eriksson P. Neonatal exposure to deca-brominated diphenyl ether (PBDE 209) causes dose–response changes in spontaneous behaviour and cholinergic susceptibility in adult mice. *NeuroToxicology* 2008;29:911–919.

132. Rice DC, Reeve EA, Herlihy A, Zoeller RT, Thompson WD, Markowski VP. Developmental delays and locomotor activity in the C57BL6/J mouse following neonatal exposure to the fully-brominated PBDE, decabromodiphenyl ether. *Neurotoxicol Teratol* 2007;29:511–520.

133. Hoppe AA, Carey GB. Polybrominated diphenyl ethers as endocrine disruptors of adipocyte metabolism. *Obesity* 2007;15:2942–2950.

134. Harley KG, Chevrier J, Bradman A, Sjodin A, Eskenazi B. Associations between maternal PBDE serum concentrations and birth weight and duration of gestation. *Organohalogen Compounds* 2009;71:2251.

135. Chao H-R, Wang S-L, Lee W-J, Wang Y-F, Päpke O. Levels of polybrominated diphenyl ethers (PBDEs) in breast milk from central Taiwan and their relation to infant birth outcome and maternal menstruation effects. *Environ Intl* 2007;33:239–245.

136. Main KM, Kiviranta H, Virtanen HE, Sundqvist E, Tuomisto JT, Tuomisto J, Vartiainen T, Skakebaek NE, Toppari J. Flame retardants in placenta and breast milk and cryptorchidism in newborn boys. *Environ Health Perspect* 2007;115:1519–1526.

137. Meeker JD, Johnson PI, Camann D, Hauser R. Polybrominated diphenyl ether (PBDE) concentrations in house dust are related to hormone levels in men. *Sci Total Environ* 2009;407:3425–3429.

138. Turyk ME, Persky VW, Imm P, Knobeloch L, Chatterton RJ, Anderson HA. Hormone disruption by PBDEs in adult male sport fish consumers. *Environ Health Perspect* 2008;116:1635–1641.

139. Dallaire R, Dewailly E, Pereg D, Dery S, Ayotte P. Thyroid function and plasma concentrations of polyhalogenated compounds in Inuit adults. *Environ Health Perspect* 2009;117:1380–1386.

140. Lim J-S, Lee D-H, Jacobs DRJ. Association of brominated flame retardants with diabetes and metabolic syndrome in the U.S. population, 2003–2004. *Diabetes Care* 2008;31:1802–1807.

141. National Toxicology Program (NTP). Toxicology and carcinogenesis studies of decabromodiphenyl oxide (CAS No. 1163-19-5) in F344/N rats and B6C3F1 mice (feed studies). TR-309. Research Triangle Park, NC, 1986.

142. Hardell L, Eriksson M, Lindstrom G, Van Bavel B, Linde A, Carlberg M, Liljegren G. Case-control study on concentrations of organohalogen compounds and titers of antibodies to Epstein–Barr virus antigens in the etiology of non-Hodgkin lymphoma. *Leuk Lymphoma* 2001;42:619–629.

143. Zhang S, Bursian S, Martin PA, Chan HM, Martin JW. Dietary accumulation, disposition and metabolism of technical pentabrominated diphenyl ether (DE-71) in pregnant mink (*Mustela vision*) and their offspring. *Environ Toxicol Chem* 2008b;27:1184–1183.

144. Fernie KJ, Shutt JL, Letcher RJ, Ritchie IJ, Bird DM. Environmentally relevant concentrations of DE-71 and HBCD alter eggshell thickness and reproductive success of American kestrels. *Environ Sci Technol* 2009;43:2124–2130.

145. Johansson AK, Sellstrom U, Lindberg P, Bignert A, de Wit CA. Polybrominated diphenyl ether congener patterns, hexabromocyclododecane, and brominated biphenyl 153 in eggs of peregrine falcons (*Falco peregrinus*) breeding in Sweden. *Environ Toxicol Chem* 2009;28:9–17.

146. McKernan MA, Rattner BA, Hale RC, Ottinger MA. Toxicity of polybrominated diphenyl ethers (DE-71) in chicken (*Gallus gallus*), mallard (*Anas platyrhynchos*), and American kestral (*Falco sparverius*) embryos and hatchlings. *Environ Toxicol Chem* 2009;28:1007–1012.

147. Ross PS. The role of immunotoxic environmental contaminants in facilitating the emergence of infectious diseases in marine mammals. *Hum Ecol Risk Assess* 2002;8:277–292.

148. Hall AJ, Kalantzi OI, Thomas GO. Polybrominated diphenyl ethers (PBDEs) in grey seals during their first year of life—are they thyroid hormone endocrine disruptors? *Environ Pollut* 2003;126:29–37.

149. Hall AJ, Thomas GO. Polychlorinated biphenyls, DDT, polybrominated diphenyl ethers and organic pesticides in United Kingdom harbor seals (*Phoca vitulina*)-mixed exposures and thyroid homeostasis. *Environ Toxicol Chem* 2007;26:851–861.

150. Beineke A, Siebert U, McLachlan M, Bruhn R, Thron K, Failing K, Müller G, Baumgärtner W. Investigations of the

potential influence of environmental contaminants on the thymus and spleen of harbor porpoises (*Phocoena phocoena*). *Environ Sci Technol* 2005;39:3933–3938.

151. Kannan K, Perrotta E, Thomas NJ, Aldous KM. A comparative analysis of polybrominated diphenyl ethers and polychlorinated biphenyls in southern sea otters that died of infectious diseases and noninfectious causes. *Arch Environ Contam Toxicol* 2007;53:293–302.

152. Hall AJ, Thomas GO, McConnell BJ. Exposure to persistent organic pollutants and first-year survival probability in gray seal pups. *Environ Sci Technol* 2009;43:6364–6369.

153. Van Bressem M-F, Raga JA, De Guardo G, Jepson PD, Duignan PJ, Siebert U, Barrett T, de Oliveira Santos MC, Moreno IB, Siciliano S, Aguilar A, Van Waerebeek K. Emerging infectious diseases in cetaceans worldwide and the possible role of environmental stressors. *Dis Aquat Organ* 2009;86:143–157.

154. Jensen T, van de Bildt M, Dietz HH, Andersen TH, Hammer AS, Kuiken T, Osterhaus A. Another phocine distemper outbreak in Europe. *Science* 2002;297:209.

155. Kajiwara N, Watanabe M, Wilson S, Eybatov T, Mitrofanov IV, Aubrey DG, Khuraskin LS, Miyazaki N, Tanabe S. Persistent organic pollutants (POPs) in Caspian seals of unusual mortality event during 2000 and 2001. *Environ Pollut* 2008;152:431–442.

156. Kannan K, Tanabe S, Borrell A, Aguilar A, Focardi S, Tatsukawa R. Isomer-specific analysis and toxic evaluation of polychlorinated biphenyls in striped dolphins affected by an epizootic in the Western Mediterranean Sea. *Arch Environ Contam Toxicol* 1993;25:227–233.

157. Noyes PD, McElwee MK, Miller HD, Clark BW, Van Tiem LA, Walcott KC, Erwin KN, Levin ED. The toxicology of climate change: environmental contaminants in a warming world. *Environ Int* 2009;35:971–986.

158. Schipper J, Chanson JS, Chiozza F, Cox NA, Hoffmann M, Katariya V, Lamoreux J, Rodrigues ASL, Stuart SN, Temple HJ, Baillie J, Boitani L, Jr. TEL, Mittermeier RA, Smith AT, et al. The status of the world's land and marine mammals: diversity, threat, and knowledge. *Science* 2008;322:225–230.

159. Stapleton HM, Klosterhaus S, Eagle S, Fuh J, Meeker JD, Blum A, Webster TF. Detection of organophosphate flame retardants in furniture foam and U. S. house dust. *Environ Sci Technol* 2009;43:7490–7495.

160. Stapleton HM, Allen JG, Kelly SM, Konstantinov A, Klosterhaus S, Watkins D, McClean MD, Webster TF. Alternate and new brominated flame retardants detected in U. S. house dust. *Environ Sci Technol* 2008;42:6910–6916.

161. Klosterhaus S, Konstantinov A, Stapleton H. Characterization of the brominated chemicals in a pentaBDE replacement mixture and their detection in biosolids collected from two San Francisco Bay area wastewater treatment plants. 2008. Available at http://www.sfei.org/rmp/posters/08BFR_Poster_klosterhaus_shrunk.pdf.

162. Lam JCW, Lau RKF, Murphy MB, Lam PKS. Temporal trends of hexabromocyclododecanes (HBCDs) and polybrominated diphenyl ethers (PBDEs) and detection of two novel flame retardants in marine mammals from Hong Kong, South China. *Environ Sci Technol* 2009;43:6944–6949.

163. Bearr JS, Stapleton HM, Mitchelmore CL. Accumulation and DNA damage in fathead minnows (*Pimephales promelas*) exposed to 2 brominated flame-retardant mixtures, Firemaster 550 and Firemaster BZ-54. *Environ Toxicol Chem* 2010;29:1–8.

164. Lin K. Joint acute toxicity of tributyl phosphate and triphenyl phosphate to *Daphnia magna*. *Environ Chem Lett* 2009;7:309–312.

165. Chemtura. Material Safety Data Sheet #694, 2006. Product: Durad 150. Available at www.chemtura.com.

166. Babich MA, Thomas TA, Hatlelid KM. CPSC staff preliminary risk assessment of flame retardant (FR) chemicals in upholstered furniture foam. January 30, 2006. Consumer Product Safety Commission.

167. Gold MD, Blum A, Ames BN. Another flame retardant, tris-(1,3-dichloro-2-propyl)-phosphate, and its expected metabolites are mutagens. *Science* 1978;200:785–787.

168. National Toxicology Program (NTP). Tris(2 3-dibromopropyl) phosphate CAS No. 126-72-7. Report on Carcinogens, 11th edition. U.S. Department of Health and Human Services, Research Triangle Park, NC, 2005.

169. Hudec T, Thean J, Kuehl D, Dougherty RC. Tris(dichloropropyl)phosphate, a mutagenic flame retardant: frequent occurrence in human seminal plasma. *Science* 1981;211:951–952.

170. U.S. Environmental Protection Agency, (U.S., EPA). 2005. Furniture flame retardancy partnership: environmental profiles of chemical flame-retardant alternatives for low-density polyurethane foam. Available at http://www.epa.gov/dfe/pubs/index.htm.

171. Meeker JD, Stapleton HM. House dust concentrations of organophosphate flame retardants in relation to hormone levels and semen quality parameters. *Environ Health Perspect* 2010;118:318–323.

172. Betts K. Endocrine damper? Flame retardants linked to male hormone, sperm count changes. *Environ Health Perspect* 2010;118:A130.

173. Betts K. Glut of data on "new" flame retardant documents its presence all over the world. *Environ Sci Technol* 2009;43:236–237.

174. Ricklund N, Kierkegaard A, McLachlan MS. An international survey of decabromodiphenyl ethane (deBDethane) and decabromodiphenyl ether (decaBDE) in sewage sludge samples. *Chemosphere* 2008;73:1799–1804.

175. Gauthier LT, Hebert CE, Weseloh DVC, Letcher RJ. Dramatic changes in the temporal trends of polybrominated diphenyl ethers (PBDEs) in herring gull eggs from the Laurentian Great Lakes: 1982–2006. *Environ Sci Technol* 2008;42:1524–1530.

176. Gao Z, Xu J, Xian Q, Feng J, Chen X, Yu H. Polybrominated diphenyl ethers (PBDEs) in aquatic biota from the lower reach of the Yangtze River, East China. *Chemosphere* 2009;75:1273–1279.

177. Hu G-C, Luo X-J, Dai J-Y, Zhang X-L, Wu H, Zhang C-L, Guo W, Xu M-Q, Mai B-X, Wei F-W. Brominated flame retardants, polychlorinated biphenyls, and organochlorine pesticides in captive Giant Panda (*Ailuropoda melanoleuca*) and Red Panda (*Ailurus fulgens*) from China. *Environ Sci Technol* 2008;42:4704–4709.

178. Chen S-J, Ma Y-J, Wang J, Chen D, Luo X-J, Mai B-X. Brominated flame retardants in children's toys: concentration, composition, and children's exposure and risk assessment. *Environ Sci Technol* 2009;43:4200–4206.

179. National Chemicals Inspectorate (KEMI). Draft of the EU Risk Assessment Report on Hexabromocyclododecane. Sundbyberg, Sweden, 2005.

180. Stapleton HM, Dodder NG, Kucklick JR, Reddy CM, Schantz MM, Becker PR, Gulland F, Porter BJ, Wise SA. Determination of HBCD, PBDEs and MeO-BDEs in California sea lions (*Zalophus californianus*) stranded between 1993 and 2003. *Mar Pollut Bull* 2006;52:522–531.

181. Birnbaum LS, Staskal DF, Diliberto JJ. Health effects of polybrominated dibenzo-*p*-dioxins (PBDDs) and dibenzofurans (PBDFs). *Environ Int* 2003;29:855–860.

182. Sakai S, Watanabe J, Honda Y, Takatsuki H, Aoki I, Futamatsu M, Shiozaki K. Combustion of brominated flame retardants and behavior of its by-products. *Chemosphere* 2001;42:519–531.

183. Wang L-C, Wang Y-F, Hsi H-C, Chang-Chien G-P. Characterizing the emissions of polybrominated diphenyl ethers (PBDEs) and polybrominated dibenzo-*p*-dioxins and dibenzofurans (PBDD/Fs) from metallurgical processes. *Environ Sci Technol* 2010;44:1240–1246.

184. Steen PO, Grandbois M, McNeill K, Arnold WA. Photochemical formation of halogenated dioxins from hydroxylated polybrominated diphenyl ethers (OH-PBDEs) and chlorinated derivatives (OH-PBCDEs). *Environ Sci Technol* 2009;43:4405–4411.

185. Hagberg J, Olsman H, van Bavel B, Engwall M, Lindström G. Chemical and toxicological characterisation of PBDFs from photolytic decomposition of decaBDE in toluene. *Environ Int* 2006;32:851–857.

186. Weber R, Kuch B. Relevance of BFRs and thermal conditions on the formation pathways of brominated and brominated-- chlorinated dibenzodioxins and dibenzofurans. *Environ Int* 2003;29:699–710.

187. Gullett BK, Wyrzykowska B, Grandesso E, Touati A, Tabor DG, Ochoa GS. PCDD/F, PBDD/F, and PBDE emissions from open burning of a residential waste dump. *Environ Sci Technol* 2010;44:394–399.

188. Yu X, Zennegg M, Engwall M, Rotander A, Larsson M, Wong MH, Weber R. E-waste recycling heavily contaminates a Chinese city with chlorinated, brominated and mixed halogenated dioxins. *Organohalogen Compounds* 2008;70:813–816.

189. Zennegg M, Xiezhi Y, Wong MH, Weber RR. Fingerprints of chlorinated, brominated and mixed halogenated dioxins at two e-waste recycling sites in Guiyu, China. *Organohalogen Compounds* 2009;71:2263–2267.

190. LeMasters GK, Genaidy AM, Succop P, Deddens JA, Sobeih T, Berriera-Viruet H, Dunning K, Lockey J. Cancer risk among firefighters: a review and meta-analysis of 32 studies. *J Occup Environ Med* 2006;48:1189–1202.

191. Brandt-Rauf PW, Fallon LFJ, Tarantini T, Idema C, Andrews L. Health hazards of fire fighters: exposure assessment. *Br J Ind Med* 1988;45:606–612.

192. Li H, Yu L, Sheng G, Fu J, Peng P. Severe PCDD/F and PBDD/F pollution in air around an electronic waste dismantling area in China. *Environ Sci Technol* 2007;41:5641–5646.

193. Olsman H, Engwall M, Kammann U, Klempt M, Otte J, van Bavel B, Hollert H. Relative differences in aryl hydrocarbon receptor-mediated response for 18 polybrominated and mixed halogenated dibenzo-*p*-dioxins and -furans in cell lines from four different species. *Environ Toxicol Chem* 2007;26:2448–2454.

194. Samara F, Gullett BK, Harrison RO, Chu A, Clark GC. Determination of relative assay response factors for toxic chlorinated and brominated dioxins/furans using an enzyme immunoassay (EIA) and a chemically-activated luciferase gene expression cell bioassay (CALUX). *Environ Int* 2009;35:588–593.

195. Weber LWD, Greim H. The toxicity of brominated and mixed-halogenated dibenzo-*p*-dioxins and dibenzofurans: an overview. *J Toxicol Environ Health* 1997;50:195–215.

196. Behnisch PA, Hosoe K, Sakai S. Brominated dioxin-like compounds: *in vitro* assessment in comparison to classical dioxin-like compounds and other polyaromatic compounds. *Environ Int* 2003;29:861–877.

197. World Health Organization (WHO). Polybrominated dibenzo-*p*-dioxins and dibenzofurans. Environmental Health Criteria 205. 1998. Geneva. Available at http://www.inchem.org/documents/ehc/ehc/ehc205.htm.

198. Franzblau A, Zwica L, Knutson K, Chen Q, Lee S-Y, Hong B, Adriaems P, Demond A, Garabrant D, Gillespie B, Lepkowski J, Luksemburg WJ, Maier M, Towey T. An investigation of homes with high concentrations of PCDDs, PCDFs, and/or dioxin-Like PCBs in house dust. *J Occup Environ Hyg* 2009;6:188–199.

199. Takigami H, Sozuki G, Hirai Y, Sakai S-I. Transfer of brominated flame retardants from components into dust inside television cabinets. *Chemosphere* 2008;73:161–169.

200. Ashizuka Y, Nakagawa R, Hori T, Yasutake D, Tobiishi K, Sasaki K. Determination of brominated flame retardants and brominated dioxins in fish collected from three regions of Japan. *Mol Nutr Food Res* 2008;52:273–283.

201. Choi J-W, Fujimaki S, Kitamura K, Hashimoto S, Ito H, Suzuki N, Sakai S-I, Morita M. Polybrominated dibenzo-*p*-dioxins, dibenzofurans, and diphenyl ethers in Japanese human adipose tissue. *Environ Sci Technol* 2003;37:817–821.

202. Kotz A, Malisch R, Kypke K, Oehme M. PBDE, PBDD/F and mixed chlorinated-brominated PXDD/F in pooled human milk samples from different countries. *Organohalogen Compounds* 2009;71:1540–1544.

203. Kang D, Davis LK, Hunt P, Kriebel D. Cancer incidence among male Massachusetts firefighters, 1987–2003. *Am J Ind Med* 2008;51:329–335.

204. Prevedourous K, Jones KC, Sweetman AJ. Estimation of the production, consumption, and atmospheric emissions of pentabrominated diphenyl ether in Europe between 1970 and 2000. *Environ Sci Technol* 2004;38:3224–3231.

205. Weschler CJ, Nazaroff WW. Semivolatile organic compounds in indoor environments. *Atmos Environ* 2008;42:9018–9040.

206. Harrad SJ, Diamond ML. New directions: exposure to polybrominated diphenyl ethers (PBDEs) and polychlorinated biphenyls (PCBs): current and future scenarios. *Atmos Environ* 2006;40:1187–1188.

207. Johannessen SC, Macdonald RW, Wright CA, Burd B, Shaw DP, van Roodselaar A. Joined by geochemistry, divided by history: PCBs and PBDEs in Strait of Georgia sediments. *Mar Environ Res* 2008;66:S112–S120.

SECTION VIII

CONCLUSION

What has been and what can be achieved in the control of water-related diseases if one adheres to the multifactorial approach has been covered throughout this book. This final section highlights examples of preventive measures and of successful interventions drawn from a diversity of locations in both developed and developing countries.

35

ASSESSMENT OF PROGRESS IN MEETING THE UN MILLENNIUM DEVELOPMENT GOALS

Gretchen Loeffler Peltier, Unni Krishnan Karunakara, and Anson Elisabeth Wright

35.1 INTRODUCTION

In 2000, world leaders adopted the United Nations Millennium Declaration. The declaration highlighted six values that are essential to international relations: freedom, equality, solidarity, tolerance, respect for nature, and shared responsibility. These values are highlighted in the Secretary-General of the United Nations Ban Ki-moon's challenge in the 2009 Millennium Development Goals Report.

> The global community cannot turn its back on the poor and the vulnerable. We must strengthen global cooperation and solidarity, and redouble our efforts to reach the MDGs and advance the broader development agenda. Nothing less than the viability of our planet and the future of humanity are at stake. (1)

These values were translated into eight specific actions and objectives (Table 35.1) called the Millennium Development Goals (MDGs). In total, 189 UN Member States adopted the Millennium Declaration with an expected target for achieving these goals by 2015. The eight MDGs are targets to ensure that the commitments to the Millennium Declaration are met. Achievement of the targets outlined in all eight of the MDGs depends on improvements in health and in water quality and quantity, sanitation, and hygiene. However, two of the Millennium Development Goals (Goal 4, reducing child mortality and Goal 7, ensuring environmental sustainability) are directly related to water quality, sanitation, and hygiene. These two goals are explained in greater detail in Table 35.2.

35.2 UN MILLENNIUM DECLARATION GOAL 4: REDUCING CHILD MORTALITY

The MDG target for child mortality seeks to reduce child mortality by two-thirds between 1990 and 2015. Globally, child mortality has decreased from 93 deaths per 1000 live births in 1990 to 67 deaths per 1000 live births in 2007. Pneumonia, diarrhea, and malaria are the leading causes of death in children under 5, and over 50% of all under-5 deaths are linked to malnutrition (2). The effects of malnutrition directly result in higher mortality, underweight status, and stunting in children. Malnutrition also leads to important indirect effects such as increased vulnerability to diseases and repeated infections. Diarrheal disease episodes occur with increased frequency, longer duration, and more severity in children that are malnourished (3).

Water quality, water quantity, sanitation, and hygiene practices in a community directly influence the burden of child mortality and morbidity through malnutrition, enteric and respiratory infections, and other related infections (Fig. 35.1). Current research suggests that handwashing has the potential to reduce diarrheal disease (4) as well as the incidence of upper respiratory infections (5) and lower respiratory infections (6). Tropical enteropathy, a disorder of the small intestine caused by ingestion of large quantities of fecal bacteria, is linked with diarrheal disease, malnutrition, underweight, and stunting (7). Neglected tropical diseases, particularly soil-transmitted helminths (STHs), are also linked to inadequate sanitation,

Water and Sanitation-Related Diseases and the Environment: Challenges, Interventions, and Preventive Measures, First Edition.
Edited by Janine M. H. Selendy.
© 2011 Wiley-Blackwell. Published 2011 by John Wiley & Sons, Inc.

TABLE 35.1   Millennium Development Goals (1)

Goal	Description
1	Eradicate extreme hunger and poverty
2	Achieve universal primary education
3	Promote gender equality and empower women
4	Reduce child mortality
5	Improve maternal health
6	Combat HIV/AIDS, malaria, and other diseases
7	Ensure environmental sustainability
8	Develop a global partnership for development

anemia, and malnutrition (8). Clearly, a comprehensive approach for addressing the causes of these diseases leading to child morbidity and mortality is warranted.

35.3   UN MILLENNIUM DECLARATION GOAL 7: ENSURE ENVIRONMENTAL SUSTAINABILITY

The MDG target for water and sanitation seeks to halve by 2015 the proportion of people without sustainable access to safe drinking water and basic sanitation. An improved drinking-water source is defined as one that is protected from contamination from human and animal waste. Piped water into dwelling, public tap, tubewell/borehole, protected dug

well, protected spring, and rainwater are all examples of improved drinking-water sources. An improved sanitation facility is defined as one that hygienically separates human waste from human contact. Examples include ventilated improved pit (VIP) latrine, pit latrine with slab, composting toilet, or flush/pour-flush to piped sewer system, septic tank, or pit latrine.

Strategies to improve sanitation must consider environmental impacts of waste disposal to minimize distribution of human waste into ground and surface water supplies. Lack of adequate improvements and advances in drinking-water quality, sanitation, and hygiene are primary contributors to diarrhea morbidity, tropical enteropathy, and these persistent conditions are what put children most at risk for impaired growth, impaired cognition, and decreased fitness (7, 9). The challenge to develop and implement plans for improving water, sanitation, and hygiene becomes even greater as communities contend with increasing population size and distribution, changes in the timing and quantity of rainfall, and ecosystem degradation.

35.3.1   Behavior Change

Access to improved water sources and sanitation are the indicators being tracked for progress toward the MDGs and these goals may or may not be met by 2015. Ultimately we are

TABLE 35.2   Targets and Indicators for MDGs 4 and 7

Goal	Target	Indicator
MDG 4: Reduce child mortality	Target 4.A: Reduce by two-thirds, between 1990 and 2015, the under-5 mortality rate	4.1 Under-5 mortality rate 4.2 Infant mortality rate 4.3 Proportion of 1-year-old children immunized against measles
MDG 7: Ensure environmental sustainability	Target 7.A: Integrate the principles of sustainable development into country policies and programs and reverse the loss of environmental resources Target 7.B: Reduce biodiversity loss, achieving, by 2010, a significant reduction in the rate of loss Target 7.C: Halve, by 2015, the proportion of people without sustainable access to safe drinking water and basic sanitation Target 7.D: By 2020, to have achieved a significant improvement in the lives of at least 100 million slum dwellers	7.1 Proportion of land area covered by forest 7.2 CO_2 emissions, total, per capita and per USD 1 GDP (PPP) 7.3 Consumption of ozone-depleting substances 7.4 Proportion of fish stocks within safe biological limits 7.5 Proportion of total water resources used 7.6 Proportion of terrestrial and marine areas protected 7.7 Proportion of species threatened with extinction 7.8 Proportion of population using an improved drinking-water source 7.9 Proportion of population using an improved sanitation facility 7.10 Proportion of urban population living in slums[a]

Indicators are used to measure progress to meeting targets and goals (32).

[a] The actual proportion of people living in slums is measured by a proxy, represented by the urban population living in households with at least one of the four characteristics: (a) lack of access to improved water supply; (b) lack of access to improved sanitation; (c) overcrowding (three or more persons per room); and (d) dwellings made of nondurable material.

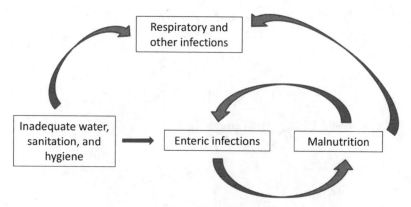

FIGURE 35.1 Pathways linking water, sanitation, and hygiene with human health. (Adapted from Ref. 31.)

striving to reduce global morbidity and mortality, improve living conditions and livelihoods, and protect the natural environment. In order for this to happen, technologies and resources must be available, but behavior must change as well. If hardware (e.g., latrines and water delivery systems) are built, but the community does not support these improvements either by not performing routine maintenance or underutilizing the new systems, it is unlikely that a reduction in morbidity will occur. The software (e.g., education and behavior change) piece presents a significant challenge and many approaches are trying to get address this through techniques, including social marketing and community mobilization. Later in this chapter, we highlight some examples of initiatives that have successfully incorporated social marketing and community mobilization into water and sanitation interventions. Education and access are not enough to elicit behavior change and without investment in understanding behavior and culture, any intervention will have limited, if any, long-term success.

35.4 PROGRESS

35.4.1 Reducing Child Mortality

There have been some successes in reducing global child mortality. In 1990, a total of 12.6 million children died across the globe, but by 2007 this figure declined to 9 million despite a global population increase (1). Compared to the baseline of 1990, 10,000 fewer children died each day in 2008 and the rate of decline in under-5 mortality increased between 2000 and 2008 (2.3%) compared to the period from 1990 to 2000 (1.4%). Significant progress in health interventions such as coverage for insecticide-treated bed nets for malaria, Haemophilus influenzae type B vaccine, and vitamin A supplementation are likely accelerating the reduction of child mortality and the effects will be seen in the next few years (10).

Despite these apparent improvements, progress toward the goal of reducing under-5 mortality by two-thirds has been very limited in some regions of the world (Fig. 35.2). Developed regions saw reductions in child mortality from 11 to 6 deaths per 1000 live births compared to 103 to 74 deaths per 1000 live births in developing regions. Very little progress has been made in countries within sub-Saharan Africa and southern Asia (1), and the rate of decline is not sufficient reach the MDG by 2015. Among the 67 countries with mortality rates of 40 or more per 1000, only 10 are on track to meet MDG 4 (10). As progress is clearly being made with malaria, the global health community must now focus on pneumonia and diarrhea as two of the three most important causes of under-5 mortality.

35.4.2 Ensure Environmental Sustainability

Between 1990 and 2006, the percentage of the population with access to improved drinking water increased from 77% to 87% (11). During this same period, 1.1 billion people gained access to improved sanitation facilities. However, 2.6 billion people lack basic sanitation worldwide and close to 1 billion utilize unimproved drinking-water sources. As of 2007, 18% of the world's population, or 1.2 billion people, practiced open defecation (1). In part because of this, the MDG of halving the proportion of people without sustainable access to safe drinking water and basic sanitation is especially challenging, particularly in rural areas.

The drinking-water target will likely be achieved by 2015 and current trends suggest that more than 90% of the global population will be using improved drinking-water sources by 2015. Countries in sub-Saharan Africa face the greatest challenges in drinking water (Figs. 35.3 and 35.4). Improved water includes public taps or standpipes, tubewells or boreholes, protected dug wells, protected springs, collected rainwater, and piped water to the household.

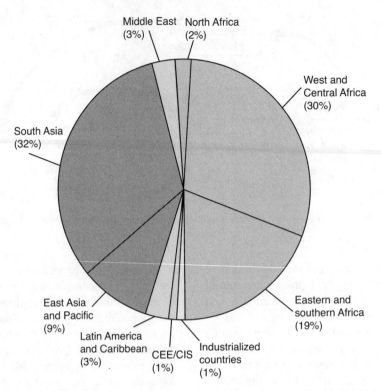

FIGURE 35.2   Regional distribution of children who died before they reached their fifth birthday in 2008. CEE/CIS: Central and Eastern Europe/Commonwealth of Independent States. (Adapted from Ref. 10.)

However, the quality or quantity of water available through improved sources is not guaranteed. In fact, drinking-water quality at many improved water sources has not met the microbiological standards set by the World Health Organization (WHO) (1). WHO drinking-water standards can be found online: http://www.who.int/water_sanitation_health/dwq/guidelines/en/index.html.

Based on current trends, the sanitation target will not be met by 2015. The lowest improved sanitation coverage can be found in sub-Saharan Africa and southern Asia

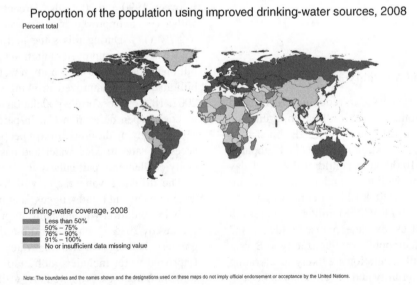

FIGURE 35.3   Global map of drinking-water coverage in 2008. *Source*: WHO/UNICEF JMP, 2010. (*See insert for color representation of this figure.*)

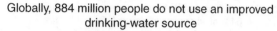

Globally, 884 million people do not use an improved
drinking-water source

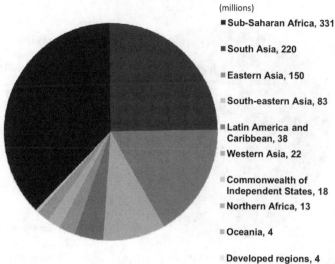

(millions)

- Sub-Saharan Africa, 331
- South Asia, 220
- Eastern Asia, 150
- South-eastern Asia, 83
- Latin America and Caribbean, 38
- Western Asia, 22
- Commonwealth of Independent States, 18
- Northern Africa, 13
- Oceania, 4
- Developed regions, 4

FIGURE 35.4 Proportion of global population lacking improved water in 2008. *Source*: WHO/UNICEF JMP, 2010.

(Figs. 35.5 and 35.6). For the outlook to change, a minimum of 173 million people per year will need to begin using improved sanitation facilities. In the 2008 WHO/Joint Monitoring Program Report, sanitation coverage is categorized as four distinct groups or steps in the sanitation ladder: open defecation, unimproved sanitation, shared facilities, and improved sanitation (11). The sanitation ladder approach acknowledges improvements from one step of the ladder to another, even if improved sanitation is not immediately achieved. The MDG indicator of improved sanitation does not consider moving up

the sanitation ladder to be progress. However, reduction of open defecation is increasingly recognized as an important goal, even if it is not immediately replaced with improved sanitation facilities such as VIP latrines. Community-led total sanitation (CLTS) is based on this idea and therefore is an important distinction. CLTS is a locally acceptable, sustainable approach that empowers the local community to transform sanitation conditions at the grassroots level (see Case Study 1).

Assessment of the progress toward meeting the sanitation and water targets is challenging. Solely quantifying whether

Proportion of the population using improved sanitation facilities, 2008

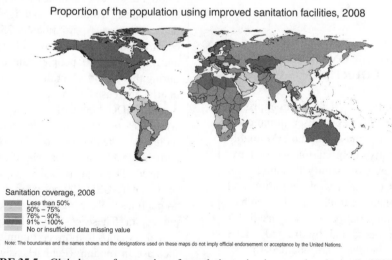

Sanitation coverage, 2008
- Less than 50%
- 50% − 75%
- 76% − 90%
- 91% − 100%
- No or insufficient data missing value

Note: The boundaries and the names shown and the designations used on these maps do not imply official endorsement or acceptance by the United Nations.

FIGURE 35.5 Global map of proportion of population using improved sanitation in 2008. *Source*: WHO/UNICEF JMP, 2010.

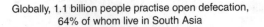

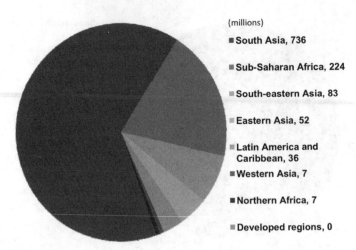

FIGURE 35.6 Distribution of global population practicing open defecation in 2008. *Source*: WHO/ UNICEF JMP, 2010.

or not the targets are met presents a distinctively different picture than when the initial regional conditions and improvements are considered. For example, many countries, particularly in sub-Saharan Africa and southern Asia may not halve the proportion of the population without sustainable access to basic sanitation by 2015, as defined by the MDG expectations. However, they may have reduced the proportion of their population practicing open defecation, the lowest rung on the sanitation ladder. Likewise the target of halving the proportion of the population without access to improved drinking water does not include a discussion of water quality. Therefore, the MDG target could be achieved, but the quality of the improved source may not meet WHO standards.

BOX 35.1 CASE STUDY 1: COMMUNITY-LED TOTAL SANITATION (COURTESY OF UNICEF)

Community-led total sanitation (CLTS) is an approach that empowers local communities to discontinue open defecation and focuses on behavioral and sustainable change through community mobilization without external hardware subsidies. Community dynamics and perceptions are seen as the drivers of sanitation and behavior change rather than the traditional approach of emphasizing construction quality and the need for subsidies. Through a process led by trained CLTS facilitators, community members are enabled to see, and feel, the negative aspects of open defecation.

UNICEF, in conjunction with the government of Zambia, piloted the CLTS approach in the southern province of Zambia, where sanitation coverage was 40%, to determine whether CLTS could be an effective strategy for rural sanitation implementation in the country. Twelve communities with low sanitation coverage and no history of subsidized sanitation projects in the past were selected for the pilot project. Sanitation coverage for a rural community of approximately 4,500 increased from 23% to 88% 2 months after the CLTS program began (12).

After the success of the pilot project in 12 communities, the program was scaled up to the entire district. Between November 2007 and July 2009, 635 villages in Choma district of Zambia were targeted using the CLTS approach and a total of 551 villages were verified as open defecation free. Approximately 25,000 toilets have been constructed by households with zero hardware subsidy, and over 150,000 people gained access to sanitation during this period. Overall sanitation coverage across the district increased from 27% to 67%, meaning that the district MDG target of 64% has been surpassed, even while 20% of the communities in the district are yet to be included in the intervention. Even though CLTS involves zero hardware subsidies, significant investment is still required. The cost of CLTS in Choma district was approximately USD 400 per open defecation village, USD 14 per household using improved sanitation, and USD 2.3 per capita. These costs will likely decrease as the approach expands to more communities without significant increase in funding (Peter Harvey, UNICEF). For more information, see http://www.unicef.org/wes.

BOX 35.2 CASE STUDY 2: HANDWASHING WITH SOAP (COURTESY OF THE WORLD BANK)

The Handwashing with Soap (HWWS) project was launched by the World Bank in 2007 in Peru, Senegal, Tanzania, and Vietnam with the aim of better understanding and quantifying the health impacts of improving handwashing practices among mothers of children under 5 and school children. The interventions are based on formative research conducted among the target group, which helps define the barriers and facilitators to handwashing practices (e.g., most people know they should wash their hands, but very few do. Why is that, and what else is needed to turn knowledge into action?). The research is evaluated through a behavior-change framework called FOAM (focus, opportunity, ability, and motivation) that highlights the critical determinants of behavior change, which in turn helps define the intervention design.

In Peru, HWWS builds on earlier work conducted under the Public-Private Partnership for Handwashing (PPPHW—launched in 2004), which has evolved into the National Handwashing Initiative (NHI). The Water and Sanitation Program (WSP) supports the NHI with national and regional authorities, private sector, and civil society organizations to promote HWWS countrywide: 23 regions out of 24 in the country, and 800 districts. (This number will increase by 43 districts in the coming months as new partners engage at operational level.) The NHI aims to improve the handwashing behaviors among almost 1.3 million school children and mothers of children under 5. The interventions are delivered via a mix of mass media, interpersonal communication, and direct consumer contact. Additionally, a robust impact evaluation is being conducted to measure the ultimate impacts of this work.

Thus far, through interpersonal communication (250,000), direct consumer contact (170,000), and mass media, the National Handwashing Initiative has reached over 4 million (this is only for mass media) women and children about the importance of washing their hands with soap at critical times. Also, 13,000 front-line workers have been trained on promoting handwashing with soap from December 2008 to December 2009. Currently, the program has been provided in 1900 primary schools in rural and periurban areas of the country. It is difficult at this time to specify how many people are practicing improved handwashing, but with the impact evaluation endline, scheduled for 2010, the NHI will be able to quantify the diarrheal and respiratory disease reduction, effects on child growth and cognitive development, as well as poverty indicators.

For more information, please visit www.wsp.org.

BOX 35.3 CASE STUDY 3: HOUSEHOLD TREATMENT AND STORAGE OF DRINKING WATER (COURTESY OF CDC)

The Safe Water System (SWS) is a water quality intervention developed by the Centers for Disease Control and Prevention (CDC) and the Pan American Health Organization. The objective is to reduce the morbidity and mortality associated with diarrheal disease. The SWS includes three elements: (i) water treatment with dilute hypochlorite solution at the point-of-use; (ii) storage of water in a safe container; and (iii) education to improve hygiene and water and food handling practices. Use of the SWS has been proven to improve the microbiological quality of stored water and reduce diarrhea, and SWS projects have been implemented in over 30 countries worldwide.

Population Services International (PSI) is currently the largest SWS-implementing organization and has sold over 50 million bottles of hypochlorite solution in 19 countries over the past 10 years, potentially treating 50 billion liters of water. PSI is a social marketing nongovernmental organization that sells affordable health products in developing countries through wholesale and retail commercial networks. Users are instructed to add one full bottle cap of the solution to clear water (or two caps to turbid water) in a standard-sized (20 L) storage container, agitate, and wait 30 min before drinking. This is enough to last one family of 5–6 persons approximately 50 days. The largest national social marketing programs—in Kenya, Madagascar, and Zambia—sell SWS products that are used by ~10% of the population as verified by nationwide survey.

To increase usage from 10% of the nationwide population, local promotion—via schools, clinics, religious institutions, or women's groups—is used. The Jolivert Safe Water for Families project in northern Haiti began in September 2002 based in a local clinic with technicians who manage the program and complete behavior change trainings and household visits. Four months after pilot program initiation, a reduction in diarrheal disease incidence of 55% among users was noted. Based on these successful results, the project expanded throughout neighboring communities, and now has over 4,000 families enrolled. Project income from chlorine sales fully supports technical salaries, although a subsidy is needed for the bucket and tap necessary to enroll families in the programs. An evaluation in June 2007 revealed that 76% of tests conducted by technicians during household visits showed positive chlorine residual, and there was no significant decrease in correct use after more than 3 years of families' entrance into the program.

The main challenges to the long-term success and sustainability of SWS programs are maintaining the community programming aspects while increasing nationwide distribution, and developing behavior change communications to encourage correct and consistent use of the household water treatment products.

For more information on SWS projects and research, visit www.cdc.gov/safewater or www.ehproject.org.

35.5   MOVING FORWARD

35.5.1   Impact on All MDGs

Progress toward improving and sustaining improved water, sanitation, and hygiene on a local, regional, national, and global level directly and indirectly impacts each of the eight Millennium Development Goals (Table 35.3). Water and sanitation interventions lack funding, political, and programming support at all levels. However, increasing the proportion of the world's population with sustainable access to water and sanitation significantly contributes to efforts to reduce poverty, hunger, child mortality, infectious diseases, maternal health, and improving equality (13).

35.5.2   Long-Term Investment

The MDGs provide a target for countries to improve the health, environment, and welfare of their citizens. However, even when these targets are achieved, more is needed. Resources must continue to be directed to the ongoing effort to maintain the newly established systems and infrastructure. Latrines must be maintained and repaired. Pumps, pipes, and filters must be repaired and replaced. Protected springs must be renovated and activities in spring watersheds must continually be monitored. In addition, underserved communities in countries that have already achieved the MDG target that are still lacking improved water and sanitation should not be ignored. If current trends continue, more than 90% of the global population will have access to an improved water source (11). However, the remaining 10% of the population—approximately 680 million people—cannot be forgotten.

35.5.3   Overcoming Barriers

Limitations of the original goals, lack of funding at all levels, and lack of programming support are three significant barriers to achieving the MDGs, but they can be overcome with concerted effort. As discussed earlier, achievement of the MDGs for water and sanitation does not ensure the quality or quantity of water available, nor does it ensure the functionality of the improved sanitation system. Likewise it does not ensure clear reductions in child mortality and morbidity. However, by using the MDGs as starting points, communities

can work toward improving long-term access and sustainability of high-quality water sources and sanitation facilities.

Greater local, regional, and international financial support must be channeled into installation, implementation, and maintenance of water and sanitation improvements. The total estimated spending required for meeting the individual water and sanitation goals in developing countries is approximately USD 36 billion per year between 2005 and 2014. The combined annual spending of approximately USD 72 billion needed per year includes maintenance of existing systems, replacement of aging facilities or structures, and accounts for future population increases. The magnitude of the spending for new coverage highlights a disparity between the two sectors (USD 4.2 billion/year for water versus USD 14.2 billion/year for sanitation) illustrated in Table 35.4 (13, 14). This difference is largely explained by the larger number of persons or households currently without access to adequate sanitation as compared to those without access to improved water supplies. Funding for increased coverage and investment in water and sanitation is particularly important in rural areas in Asia and Africa compared to spending to support existing coverage in urban areas across all regions (13).

As comparison, total annual development assistance for health (DAH) spending in 2007 was USD 21.8 billion. In that same year, spending for HIV/AIDS, tuberculosis, and malaria was USD 4.9, USD 0.6, and USD 0.7 billion, respectively. Development assistance for health was defined as the total of all assistance for health channeled through public and private institutions whose primary purpose is to advance development in developing countries (15). Clearly the costs associated with improving water and sanitation require significant attention and investment from local, national, and international groups.

Frequently there is a lack of programming support specifying how water, sanitation, and hygiene needs will be realistically addressed. Setting water, sanitation, and hygiene priorities at the local, regional, and national level often involves the cooperation and agreement of multiple sectors and institutions (14). Lack of accountability to only one sector or institution contributes to inadequate attention and financial investment in sanitation, water, and hygiene policy development and implementation. One argument is that water and sanitation should be a stand-alone sector with its own funding stream and policy initiatives. Others argue that water and sanitation should be a priority across all sectors and each should determine the best way to address based on their needs. Both are valid arguments and both may be correct. Regardless of what approach(es) are taken, the outcome is clear—water, sanitation, and hygiene conditions must improve across the globe or equality, health, livelihoods, and the natural environment will continue to suffer. Water and sanitation have to be seen as a priority at the national and international level, otherwise they won't be allocated the necessary funding. Initiatives such as the International Year

TABLE 35.3   Influence of Water and Sanitation on all Eight Millennium Development Goals

Millennium Development Goal	Why Governments Should Act	How Governments Should Act
1. Eradicate extreme poverty and hunger	The lack of clean water and adequate sanitation is a significant cause of poverty and malnutrition.	Bring water and sanitation into the mainstream of national strategies for achieving the MDGs.
2. Achieve universal primary education	Collecting water and carrying it over long distances keep girls out of school, promoting a future of illiteracy and restricted choice. The absence of adequate sanitation and water in schools is a major reason that girls withdraw from school.	Ensure that every school has adequate water and sanitation facilities, including separate facilities for boys and girls.
3. Promote gender equality and empower women	Women bear the brunt of collecting water, which amounts to as much as 4 h a day. Deprivation in water and sanitation perpetuates gender inequality.	Demand gender equity in water and sanitation be at the center of national poverty reduction strategies and include representation from both men and women.
4. Reduce child mortality	Contaminated water and poor sanitation account for the vast majority of the 1.8 million child deaths each year from diarrhea—almost 5,000 every day—making it the second largest cause of child mortality.	Treating child deaths from water and sanitation as a national crisis, one that violates basic human rights. Establishing explicit linkages between national targets for lowering child mortality and targets for expanding access to water and sanitation.
5. Improve maternal health	The provision of water and sanitation reduces the incidence of diseases and health conditions that undermine maternal health.	Recognize water and sanitation as key components in strategies for health promotion and empower women to shape decisions on water and sanitation at the household, local, and national levels.
6. Combat HIV/AIDS, malaria, and other diseases	Inadequate access to water and sanitation puts those with HIV/AIDS at increased risk for infection. Poor sanitation and drainage contribute to malaria and other neglected diseases.	Integrate water and sanitation into national and global strategies for combating malaria and improving living conditions of people living with HIV/AIDS. Investing in drainage and sanitation facilities that reduce the presence of flies, mosquitoes, and contact with human waste.
7. Ensure environmental sustainability	Based on current trends, the goal of halving the proportion of people without access to basic sanitation will be missed by 2015. The exploitation of water resources represents a growing threat to human development, generating an ecological crisis for future generations.	Providing national and international political leadership to overcome the deficits in water and sanitation. Introducing integrated water resources management policies that constrain water use within the limits of environmental sustainability.
8. Develop a global partnership for development	There is no effective partnership for water and sanitation. National governments are failing to put in place the policies and financing needed to accelerate progress.	Initiate a global plan of action to galvanize political action, placing water and sanitation on the forefront, mobilizing resources, and supporting nationally owned planning processes. Empower local governments and local communities through decentralization, capacity development, and appropriate financing.

Adapted from Ref. 33.

of Sanitation have made some progress, but the topic of sanitation in particular is difficult for people to discuss and therefore suffers from lack of public and political attention.

35.5.4   Critical Messages

Clear steps can be taken to move toward reducing child morbidity and mortality. By focusing on these priorities we can have an impact on child morbidity and mortality even if there is not a complete eradication of disease. For example, water quality improvements are important for reduction of diarrheal disease, but many of the pathogens associated with diarrheal disease are transmitted through other pathways, including indirect or direct contact with feces and ingestion of contaminated food (16). In communities where open defecation is prevalent and hygiene is poor, water quality

TABLE 35.4   Distribution of Total Annual Spending to Meet MDG for Water and Sanitation (13)

	Distribution			
	Rural (%)[a]	Urban (%)[a]	Existing Coverage (%)[b]	New Coverage (%)[b]
Water	32	68	88	12
Sanitation	41	59	60	40

[a] Rural and urban indicates the percentage of total spending that should be distributed to urban and rural areas to meet the MDG.
[b] Existing and new coverage indicates the percentage of total spending that should be distributed to establish new coverage or maintain existing coverage to meet the MDG.

interventions in these "fecal-rich" environments may have limited success in reducing diarrheal disease without sanitation and hygiene improvements (17).

The F-diagram (Fig. 35.7), used frequently in public health, illustrates the different pathways that fecal material can move throughout the environment and highlights opportunities for breaking the transmission pathway (18, 19). Eisenberg and colleagues developed a dynamic version of the F-diagram and determined that when sanitation conditions are poor, water quality improvements may have minimal impact regardless of amount of water contamination. If each transmission pathway alone is sufficient to maintain diarrheal disease, single-pathway interventions, such as water or sanitation only, will have minimal benefit. Ultimately an intervention will be successful only if all potential transmission pathways are eliminated (20). We have specifically not argued for a hierarchy of priorities in part because each community will have unique conditions and transmission pathways. Instead we focus on a suite of priorities to eliminate the introduction and transmission of fecal material to the community.

These priorities include: (i) promotion of handwashing with soap (see Case Study 2), (ii) reduce open defecation and move up the sanitation ladder, (iii) point-of-use treatment and safe storage (see Case Study 3), and (iv) source water protection and monitoring. Handwashing with soap has shown to reduce diarrheal disease by approximately 44% (4), upper respiratory infections by 23% (5), and pneumonia by 50% (6). Point-of-use treatment (e.g., chlorination, filtration, flocculation/disinfection) and safe storage are critical to protecting drinking water from bacterial contamination, which can occur either during transport from the source to the household or while drinking water is being stored in the home (16, 21, 22). Discussion over the appropriateness of scaling up of household treatment is ongoing (22–24), but despite this both the sales of household treatment products as well as the evidence that it is effective continue to increase (25).

Source water protection has limited success in ensuring safe drinking water when water is collected, transported, and stored in the home rather than directly at the source. A review of current literature measuring bacterial counts in source and household water identified significant fecal contamination after collection from the water source (21). For this reason, each community should individually evaluate the effectiveness of source water protection for protection of drinking water. If water is always transported and stored in another location for consumption, point-of-use treatment will be a more effective method of protecting drinking water.

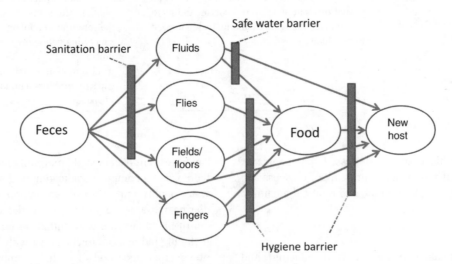

FIGURE 35.7   F-diagram.

However, if water is consumed from the source directly by a proportion of the community, source protection and point-of-use treatment are both critical.

35.5.5 Incorporation of Behavior Change

The idea of social marketing, applying marketing to achieve behavior change for a social good, is not new to the field of public health. Large-scale changes such as bed nets to prevent malaria, household disinfectant to treat drinking water, and condoms to prevent HIV are largely a result of social marketing campaigns (26). Social marketing, community mobilization, and motivational interviewing were effective in increasing product adoption of the Safe Water System, particular when these techniques were used in combination (27). Zane and Kremer (28) suggest that research efforts should be focused on understanding effective promotion strategies and how to support long-term behavior change rather than the effectiveness of handwashing and point-of-use technologies. Clearly, no universal solution exists for long-term change in all communities, which makes implementation even more challenging. However, a template or tool-kit approach would provide a guide for communities that could then be tailored to the specific needs and resources available.

Improvement of public health and reduction of disease are typically the drivers for water, sanitation, and hygiene interventions that are implemented in developing countries. However, health is often not a primary driver for sanitation and hygiene as illustrated by the top reasons (lack of smell and flies, gain prestige, privacy, cleaner surroundings) rural householders were satisfied about their new latrines in the Philippines and Benin (29). A study in Ghana highlighted motivations for hygiene behavior and grouped them into three categories: desires to nurture, desires to void disgust, and the desire to gain social status (30). Identifying and understanding the specific needs, perspectives, and motivations of a community is critical before embarking on any behavior change initiatives.

35.6 CONCLUSION

Improvements in water quality, sanitation, and hygiene both directly and indirectly influence our ability to achieve all eight of the MDGs. The ability to ensure improvements in water/sanitation/hygiene—a crucial component of the MDGs—requires a comprehensive understanding of the factors that influence water, health, and nutrition. These factors include long-term adoption of water source protection, point-of-use treatment, and improved sanitation and hygiene. Cross-sector collaboration is also critical toward achievement of the MDGs. In recent years, global health has been moving away from single vertical interventions toward more integrated systems approaches incorporating several cross-sectoral interventions.

A pragmatic approach to improved water, sanitation, and hygiene must consider affordability, long-term sustainability, acceptability, scalability, and effectiveness of the methods (16). This requires extensive political and financial support and commitment for the installation, implementation, and maintenance of water and sanitation improvements. Incorporation of the factors that positively influence adoption into future water, sanitation, and hygiene interventions will lead to the long-term success of these projects and the sustainability of intervention programs and achievement of the MDGs.

ACKNOWLEDGMENTS

We would like to thank several individuals from UNICEF, CDC, and the World Bank for providing information on case studies and current water and sanitation targets. They include Thérèse Dooley, Rolf Luyendjik, Peter van Maanen, Peter Harvey from UNICEF, Eric Mintz and Daniele Lantagne from CDC, and Nathaniel Paynter and Rocio Florez from the World Bank.

REFERENCES

1. United, Nations. *The Millennium Development Goals Report 2009*. New York: United Nations, 2009.

2. Bryce J, Boschi-Pinto C, Shibuya K, Black RE. WHO estimates of the causes of death in children. *Lancet* 2005;365:1147–1152.

3. Guerrant RL, Oria RB, Moore SR, Oria M, Lima A. Malnutrition as an enteric infectious disease with long-term effects on child development. *Nutr Rev* 2008;66(9):487–505.

4. Curtis V, Cairncross S. Effect of washing hands with soap on diarrhoea risk in the community: a systematic review. *Lancet Infect Dis* 2003;3:275–281.

5. Rabie T, Curtis V. Handwashing and risk of respiratory infections: a quantitative systematic review. *Trop Med Int Health* 2006;11:258–267.

6. Luby SP, Agboatwalla M, Feikin DR, Painter J, Billhimer W, Altaf A, Hoekstra, RM. Effect of handwashing on child health: a randomised controlled trial. *Lancet* 2005;366:225–233.

7. Humphrey, JH. Child undernutrition, tropical enteropathy, toilets, and handwashing. *Lancet* 2009;374:1032–1035.

8. Hotez P. Hookworm and poverty. In: Reducing the Impact of Poverty on Health and Human Development: Scientific Approaches. *Ann N Y Acad Sci* 2008;1136:38–44.

9. Guerrant RL, Kosek M, Moore S, Lorntz B, Brantley R, Lima AAM. Magnitude and impact of diarrheal diseases. *Arch Med Res* 2002;33:351–355.

10. You D, Wardlaw T, Salama P, Jones G. Levels and trends in under-5 mortality, 1990–2008. *Lancet* 2009. DOI: 10.1016/S0140-6736(09)61601-9.

11. World Health Organization and United Nations Children's Fund Joint Monitoring Programme for Water Supply and, Sanitation., *Progress on Drinking Water and Sanitation: Special Focus on Sanitation*. New York/Geneva: UNICEF/ WHO, 2008.

12. United Nations Children's, Fund., *Soap, Toilets, and Taps— A Foundation for Healthy Children*. New York, NY: United Nations Children's Fund, 2009.

13. Hutton G, Bartram J. Global costs of attaining the Millennium Development Goal for water supply and sanitation. *B WHO* 2008;86:13–19.

14. Prüss-Üstün A, Bos R, Gore F, Bartram J. *Safer Water, Better Health: Costs, Benefits and Sustainability of Interventions to Protect and Promote Health*. Geneva: World Health Organization, 2008.

15. Institute for Health Metrics and, Evaluation., *Financing Global Health 2009: Tracking Development Assistance for Health*. IHME, University of Washington, 2009.

16. Clasen TF, Roberts IG, Rabie T, Schmidt W-P, Cairncross S. Interventions to improve water quality for preventing diarrhoea. Cochrane Database of Systematic Reviews 2006, Issue 3. Art. No.: CD004794. DOI: 10.1002/14651858.CD004794.pub2.

17. Briscoe J. Intervention studies and the definition of dominant transmission routes. *Am J Epidemiol* 1984;120:449–455.

18. Wagner EG, Lanoix JN. Excreta disposal for rural areas and small communities. *Monogr Ser World Health Organ* 1958;39:1–182.

19. Kawata K. Water and other environmental interventions— minimum investment concept. *Am J Clin Nutr* 1978;31: 2114–2123.

20. Eisenberg JNS, Scott JC, Porco T. Integrating disease control strategies: balancing water sanitation and hygiene interventions to reduce diarrheal disease burden. *Am J Public Health* 2007;97:846–852.

21. Wright J, Gundry S, Conroy R. Household drinking water in developing countries: a systematic review of microbiological contamination between source and point-of-use. *Trop Med Int Health* 2004;9(1):106–117.

22. Clasen T, Bartram J, Colford J, Luby S, Quick R, Sobsey M. Comment on *Household Water Treatment in Poor Popula-*

tions: Is There Enough Evidence for Scaling up Now?. *Environ Sci Technol* 2009;43:5542–5544.

23. Hunter PR. Household water treatment in developing countries: comparing different intervention types using meta-regression. *Environ Sci Technol* 2009;43:8991–8997.

24. Schmidt WP, Cairncross S. Household water treatment in poor populations: is there enough evidence for scaling up now? *Environ Sci Technol* 2009;43:986–992.

25. Mintz E. Personal communication. Centers for Disease Control, Atlanta, GA, December 1, 2009.

26. Jenkins MW, Scott B. Behavioral indicators of household decision-making and demand for sanitation and potential gains from social marketing in Ghana. *Soc Sci Med* 2007;64:2427–2442.

27. Quick R. Changing community behaviour: experience from three African countries. *Int J Environ Health Res* 2003;13: S115–S121.

28. Zane A, Kremer M. What works in fighting diarrheal diseases in developing countries? *A critical review*. *World Bank Res Obser* 2007;22:1–24.

29. Cairncross S. *The Case for Marketing Sanitation*. Water & Sanitation Programme. Nairobi: World Bank, 2004.

30. Scott B, Curtis V, Rabie T, Garbrah-Aidoo N. Health in our hands, but not in our heads: understanding hygiene motivation in Ghana. *Health Policy Plann* 2007;22:225–233.

31. Fewtrell L, Prüss-Üstün A, Bos R, Gore F, Bartram J. Water, sanitation and hygiene: quantifying the health impact at national and local levels in countries with incomplete water supply and sanitation coverage. WHO Environmental Burden of Disease Series, No. 15. Geneva: World Health Organization, 2007.

32. UN, Statistics, Division., Official list of MDG indicators, 2008. Available at http://mdgs.un.org/unsd/mdg/Host.aspx? Content=Indicators/OfficialList.htm. Accessed on November 15, 2009.

33. UN, Human Development, Report., *Beyond Scarcity: Power, Poverty, and the Global Health Crisis*. New York: United Nations, 2006.

36

EXTENDING THE RIGHT TO HEALTH CARE AND IMPROVING CHILD SURVIVAL IN MEXICO

Julio Frenk and Octavio Gómez-Dantés

36.1 INTRODUCTION

Mexico is one of the few countries on track to achieve Millennium Development Goal (MDG) 4, which has set the target of reducing the 1990 child mortality rate by two-thirds no later than 2015 (1). Progress in Mexico has been the result of the combined implementation, over the past 25 years, of several efforts to strengthen the health system and various disease-oriented strategies. This chapter describes in detail this achievement and its determinants, with emphasis on the recent health system reform, which created the conditions to guarantee universal coverage of high-quality health services with financial protection for all.

36.2 CHILD SURVIVAL IN MEXICO: THE DIAGONAL APPROACH

In the last two decades of the past century, through the explicit adoption of the primary health care paradigm, Mexico designed several initiatives to extend access to essential health care by strengthening the public supply of clinical services (2). These efforts were complemented with the implementation of major community-based programs to address the challenges related to common infections, malnutrition, and maternal and child health.

Health system strengthening efforts started in 1979 with the creation of a program to expand the network of health care units in rural areas. This program, originally called *IMSS-Coplamar* and recently renamed *IMSS-Oportunidades*, built,

in its initial phase, more than 3000 ambulatory units and 64 rural hospitals (3).

In the early 1980s, two other programs were launched, one to expand the health coverage of the urban poor and the other one to care for the health needs of the rural population (4).

In the early 1990s, the Program to Strengthen Health Services for the Uninsured Population (PASSPA), cofinanced with federal resources and a loan from the World Bank, was implemented to improve and extend the health service infrastructure for the uninsured population in the four poorest states of the country (5).

Finally, in 1996, a program to further expand the coverage of essential care in the rural population nationwide was created. The Program for Extension of Coverage (PAC) was designed to provide on a regular basis 12 essential interventions: (i) basic household sanitary measures; (ii) family planning services; (iii) prenatal, perinatal, and postnatal care; (iv) nutrition and growth surveillance; (v) immunization; (vi) treatment of diarrhea; (vii) treatment of common parasitic diseases; (viii) treatment of acute respiratory infections; (ix) prevention and treatment of tuberculosis; (x) prevention and control of hypertension and diabetes; (xi) initial treatment of injuries; and (xii) community training for health promotion (6).

These efforts to strengthen the provision of personal health services were complemented with the implementation of several disease-oriented initiatives, mostly community-based interventions whose main objective was to reduce the burden of malnutrition and common infections in children, basically diarrheal diseases and respiratory infections.

Water and Sanitation-Related Diseases and the Environment: Challenges, Interventions, and Preventive Measures, First Edition.
Edited by Janine M. H. Selendy.

The first major initiative of this period was the introduction of the oral rehydration therapy (ORT) in public hospitals in 1984 as part of the National Program to Control Diarrheal Diseases (7). Eventually other measures to control these diseases were widely disseminated: the surveillance of the municipal chlorination of drinking water, the promotion of breast-feeding, and the training of mothers in the use of ORT in cases of diarrhea and in the early detection of alarm signs of dehydration.

In this period there were also important improvements in access to water and sanitation in Mexico (Table 36.1). In fact, target 3 of MDG 7 (to halve, by 2015, the proportion of the population without sustainable access to safe drinking water and basic sanitation) will certainly be met.

All these measures combined had an important impact on the incidence of diarrhea and the number of deaths due to diarrheal diseases. Between 1985 and 2000, diarrhea mortality rates in Mexico in children under 5 years declined from 11.5 per 1000 live births to only 1.6 (8).

The second major public health measure adopted in this period was the introduction, in 1991, of the Universal Immunization Program. Mexico had established a National Immunization Program in 1973, but it was until two decades later that this program, launched as a vertical initiative, was placed at the center of the national health agenda. It was implemented in response to a major measles epidemic, the results of a national vaccination survey showing that complete vaccination coverage in children under 5 years was below 50%, and the commitments made by Mexico in the 1990 World Summit for Children (9). The program initially offered six vaccines (BCG, TDP, OPV, and antimeasles), which were traditionally applied during the National Health Weeks, a strategy established in 1993. By 2000 the immunization scheme included 10 vaccines (BCG, TDP, OPV, MRM, hepatitis B, and Hib) and had reached complete coverage of more than 90% in children under 1 year and more than 95% in children under 5 years (10).

The effects of the Universal Immunization Program were dramatic and immediate. The last cases of polio and diphtheria were notified in 1990 and 1991, respectively, while the last case of autochthonous measles was reported in 1996 (11–13). Important as they are, the effects of this program were not limited to particular diseases; child survival as a whole was positively affected, as will be discussed further ahead in this chapter.

National Health Weeks, now held three times a year, were used not only to offer all the vaccines of the WHO Expanded Program on Immunization but also to provide a package of interventions particularly useful for children: the promotion of the use of oral rehydration salts and the free distribution of mega-doses of oral vitamin A and the antihelminthic drug albendazol.

Finally, in 1997, the federal government launched a program to enhance basic capabilities of families living in extreme poverty, which has had proven impacts in the health and nutritional status of the poor population (14–17). Initially called *PROGRESA* and later renamed *Oportunidades*, this program, which is still active and benefiting over 5 million households, offers incentives for families to invest in their children's human capital through conditional cash transfers. In order to receive these subsidies, families are required to send their children to school rather than work, attend a clinic to receive a specified package of health promotion and disease prevention interventions, and provide a specially formulated nutritional supplement to pregnant and lactating women, all children aged 6–23 months, and low-weight children aged 2–5 years.

This last program has contributed to the improvement of malnutrition in children in Mexico. Advances in average height-for-age and prevalence of stunting were already detected in the 1999 National Nutrition Survey, and they could hardly be attributed to this intervention. However, additional and more prominent improvements in these two indicators were also recorded for the period 1999–2006, especially among poor and indigenous children (18).

To improve decision making, several measures were implemented. Salient among them was the creation, in 1987, of the National Institute of Public Health. This teaching and research institution, in collaboration with the main health care delivery organizations, implemented regular surveys to monitor progress in health conditions and health care, developed broad research initiatives that generated inputs for public health programs, and built the public health manpower that the national health system was demanding.

The impact of the measures implemented between 1980 and 2000 in child survival were enormous. However,

TABLE 36.1 Changes in Access to Sanitation Facilities and Drinking Water Sources, Mexico 1990–2008

	Percentage of the Population			Number of People Who Gained Access 1990–2008 (thousand)
	1990	2000	2008	
Access to improved sanitation facilities	66	76	85	37,226
Access to improved drinking water sources	85	90	94	31,149

Source: World Health Organization, UNICEF. Progress on sanitation and drinking water, 2010 update. Geneva: WHO, 2010.

additional innovations had to be designed in order to guarantee the sustainability and expansion of these efforts and meet the health-related MDGs.

36.2.1   The Mexican Health Reform

A constitutional amendment establishing the right to the protection of health was passed in Mexico in 1983 (19). However, in practice not all individuals had been equally able to exercise this right. Those working in the formal sector of the economy and their families, around half of the population, had social health insurance. The other half was left without access to any form of social protection in health. This meant that almost 50 million people in the country were receiving health care on an assistential basis, including 2.5 million families from the poorest segments of the population who received only the set of essential interventions included in PAC.

This reality was well documented. Several national health accounts studies developed in the early 1990s had revealed that more than 50% of health expenditure in Mexico was out-of-pocket. This was due to the fact that half of the population lacked health insurance.

The high levels of out-of-pocket expenditure were exposing Mexican families to ruinous financial episodes. In fact, in 2000, nearly 3 million Mexican households suffered catastrophic health expenditures (20). Not surprisingly, Mexico performed poorly on the international comparative analysis of fair financing developed by the World Health Organization as part of the *World Health Report 2000* (21). These results prompted the development of additional analysis that showed that catastrophic and impoverishing health expenditures were endemic among the poor and uninsured households.

The products of this analysis generated the advocacy tools to promote a legislative reform that established the System for Social Protection in Health (SSPH), approved by a large majority of the Mexican Congress in the spring of 2003. This system is mobilizing public resources by a full percentage point of GDP over a period of 7 years to provide health insurance through a new scheme called *Seguro Popular* to all those ineligible for social security: the self-employed, those out of the labor market, and those working in the informal sector of the economy. This public insurance scheme guarantees regular access to two sets of entitlements: (i) a package of more than 250 essential interventions, which include all services offered in ambulatory units and general hospitals and (ii) a package of 18 costly interventions, which includes, among other interventions, treatment for cancer in children, cervical and breast cancer, and HIV/AIDS. By June of 2009, 30 million people were enrolled in the new insurance program (22).

Through the SSPH a protected fund for community health services targeting health promotion and disease prevention interventions was also created. This fund now finances on a regular basis all public health interventions, including those that were critical to reduce child mortality in the last two decades of the past century.

In sum, the formal recognition of health care as a social right was not enough to guarantee regular access to comprehensive health care. As Lynn Hunt has written, "human rights are still easier to endorse than to enforce" (23). What was lacking was the definition of the entitlements that ensued from such a claim, and the financial and organizational vehicles to translate them into personal and public health services accessible to all. And that is exactly what the SSPH did for half of the Mexican population: create the conditions for the effective exercise of the right to health care.

36.2.2   Meeting MDG 4

A structural reform addressing the broad challenge of financing universal coverage of high-quality services was also demanding a clear sense of priorities. This is an imperative not only in terms of resource allocation, but also to gather public support by relating abstract financial and managerial notions to concrete deliverables. Every reform must have a limited number of "flagship initiatives" to focus attention on its specific benefits. From the outset, it was decided that those interventions needed to meet the health-related MDGs would be included among those initiatives.

In order to reduce maternal and neonatal mortality and disability, a novel initiative was launched under the name "Fair Start in Life" (APV or *Arranque Parejo en la Vida*). The name of this initiative was meant to underscore the need for equality of opportunity, a core explicit value of the Mexican health reform (24).

The maternal component of this program included the mobilization of resources to strengthen health care networks, equipment, and the supply of drugs and other inputs, most notably, safe blood. The supply of skilled human resources was improved through the reintroduction of obstetric nurses and the training of traditional birth attendants. Efforts were made to extend coverage of antenatal care, with emphasis on early identification of high-risk pregnancies, institutional deliveries, and timely diagnosis and treatment of obstetric emergencies.

The neonatal component of APV included the administration of folic acid to pregnant women for the prevention of neural tube defects, preventive neonatal screening of congenital metabolic diseases, and the strict adherence to the vaccination scheme for this age group.

These measures produced an increase in effective coverage of antenatal care; an expansion of effective coverage of births attended by skilled health workers, which reached 93% in 2006, with small variations among states; improvements in immunization coverage in children under 1 year; and, most importantly, a significant acceleration in the rate of

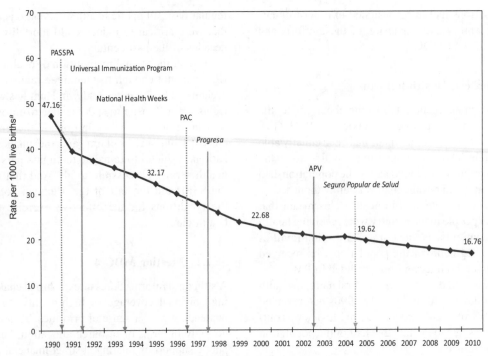

FIGURE 36.1   Evolution of under-5 mortality rate in Mexico, 1990–2010. Source: Ministry of Health Mexico. Estimates based on projections from the National Population Council (CONAPO) 1/ Probability of dying by age 5 per 1000 live births

decline of maternal and neonatal mortality (25, 26). Between 2001 and 2006, neonatal mortality dropped over 15%, the highest decline in the last five quinquennia (27).

Immunization efforts were also bolstered through the expansion of the number of vaccines in the basic scheme and the strengthening of immunization campaigns. By 2006 Mexican children had free access to one of the most comprehensive immunization schemes in the world, which includes vaccines against hepatitis B, diphtheria, tetanus, pertussis, *Haemophilus influenzae* type b, polio, measles, mumps, rubella, influenza, TB, and meningitis. That same year, complete immunization coverage reached 95% in children under 1 year and 98% in children under 5 years old, one of the highest immunization coverage levels in the Americas (28). Effective coverage for measles immunization in children aged 18–59 months reached 92.1% (25).

These continuous and increasing efforts explain the positive recent evolution of under-5 mortality rate in Mexico, which declined from 47.1 per 1000 live births in 1990 to 16.7 in 2010 (Fig. 36.1). In fact, projections developed by the Ministry of Health and the National Population Council indicate that Mexico will meet the MDG 4 target ahead of time. In order to meet this target Mexico would need to reach an under-5 mortality rate of 15.7 per 1000 live births by 2015, a level that will very probably be reached by 2013.

36.3 CONCLUSIONS

The consistent decline of child mortality over a period of 25 years, which places Mexico on route to achieve MDG 4, offers several lessons for other countries.

First of all, this achievement was the product of the combined implementation of health system strengthening efforts and disease-oriented interventions, which proves that the dilemma between horizontal and vertical approaches is a false one (29). In the past decades, Mexico implemented several strategies to upgrade general health services while also focusing on cost-effective interventions for specific priorities, such as child mortality, that reinforced the overall structure of the health system.

Second, improvements in child mortality were also strongly dependent on the development and continuous expansion of a base of evidence for policy design, policy-making, and policy evaluation. Most of the initiatives implemented in this period to confront the challenges related to common infections, malnutrition, and maternal and child health were supported by evidence generated by large-scale health surveys. The development of institutions such as the National Institute of Public Health was also crucial. This center of excellence has produced relevant research and policy analysis, trained researchers who occupy key policy-making positions, carried out independent and credible evaluations, and greatly enriched the quality of information.

Finally, stability of leadership in the Ministry of Health also contributed to this achievement (2, 8). The story of Mexico in this regard may be unique: since the 1940s, most ministers have served an entire term of office; in many cases, they were selected for their technical expertise rather than their political affiliation; and, especially in the past 25 years, they have preserved and even expanded policies from preceding administrations when proved effective.

REFERENCES

1. Global Health, Council. Millennium Development Goals: measuring progress. Available at www.globalhealth.org/child_health/measuring_progress/. Accessed on March 22, 2010.

2. Frenk J, Sepúlveda J, Gómez-Dantés O, Knaul F. Evidence-based health policy: three generations of reform in Mexico. *Lancet* 2003;362:1667–1671.

3. Soberón G. Desarrollo de las políticas públicas. In: Urbina-Fuentes M, Moguel-Ancheita A, Muñiz-Martelon ME, Solís-Urdaibay JA, editors. *La Experiencia Mexicana en Salud Pública. Oportunidad y Rumbo para el Tercer Milenio*. Mexico City: Fondo de Cultura Económica, 2006, pp. 543–566.

4. Rodríguez-Domínguez J. Salud para todos. Atención primaria de la salud. In: Urbina-Fuentes M, Moguel-Ancheita A, Muñiz-Martelon ME, Solís-Urdaibay JA, editors. *La Experiencia Mexicana en Salud Pública. Oportunidad y Rumbo para el Tercer Milenio*. Mexico City: Fondo de Cultura Económica, 2006, pp. 577–594.

5. Gómez-Dantés O, Garrido-Latorre F, López-Moreno S, Villa B, López-Cervantes M. Evaluación de un programa de salud para población no asegurada. *Saúde Publica* 1999;33:401–412.

6. Secretaría de, Salud., *Programa de Ampliación de Cobertura. Lineamientos de operación*. Mexico City: Secretaría de Salud, 1996.

7. Fajardo-Ortiz G. De 1982 and 2001: Tiempos de reformas y nuevos avances. In: Fajardo-Ortiz G, Carrillo AM, Neri-Vela R, editors. *Perspectiva histórica de atención a la salud en México 1902-2002*. Mexico City: PAHO, UNAM, Sociedad Mexicana de Historia y Filosofía de la Medicina, 2002, pp. 101–123.

8. Sepúlveda J, Bustreo F, Tapia R, et al. Improvement of child survival in Mexico: the diagonal approach. *Lancet* 2006;368:2017–2027.

9. Valdespino-Gómez JL, García-García ML. 30 aniversario del Programa Nacional de Vacunación contra sarampión en México. Los grandes beneficios y los riesgos potenciales. *Gac Med Mex* 2004;140:639–641.

10. Santos JI. Cambios en los esquemas de vacunación y la vacunación en adultos. In: Urbina-Fuentes M, Moguel-Ancheita A, Muñiz-Martelon ME, Solís-Urdaibay JA, editors. *La Experiencia Mexicana en Salud Pública. Oportunidad y Rumbo para el Tercer Milenio*. Mexico City: Fondo de Cultura Económica, 2006, pp. 191–223.

11. Organización Panamericana de la, Salud. Poliomileitis: cronología 1909–2007. Available at http://www.mex.ops-oms.org/contenido/polio/crono_polio.htm. Accessed on March 24, 2010.

12. Santos JI. Nuevo esquema de vacunación en México. *Salud Publica Mex* 1999;41:1–2.

13. Santos JI. El Programa Nacional de Vacunación: orgullo de México. *Rev Fac Med UNAM* 2002;45:142–153.

14. Rivera J, Sotres-Alvarez D, Habicht JP, Shamah T, Villalpando S. Impact of the Mexican Program for Education, Health, and Nutrition (*Progresa*) on rates of growth and anemia in infants and young children. A randomized effectiveness study. *JAMA* 2004;219:2563–2570.

15. Fernald LCH, Gertler P, Neufeld L. Role of cash in conditional cash transfer programmes for child health, growth, and development: an analysis of Mexicós *Oportunidades*. *Lancet* 2008;371:828–837.

16. Leroy JL, García Guerra A, García R. Domínguez C, Rivera J, Neufeld LM. *Oportunidades* Program increases the linear growth of children enrolled at young ages in urban Mexico. *J Nutr* 2008;138:793–798.

17. Secretaría de Desarrollo Social. Oportunidades. External evaluation site. Available at http://evaluacion.oportunidades.gob.mx:8010/index2.php?a=800. Accessed on March 23, 2010.

18. González de Cossio T, Rivera J, González-Castell D, Unar-Munguía M, Monterrubio E. Child malnutrition in Mexico in the last two decades: prevalence using the new WHO 2006 growth standards. *Salud Publica Mex* 2009;51 (Suppl 4): S494–S506.

19. Soberón G. El cambio estructural en la salud. *Salud Publica Mex* 1987;29:127–140.

20. Secretaría de Salud. Programa Nacional de Salud 2001–2006. La democratización de la salud en México. Hacia un sistema universal de salud. Mexico City: Secretaría de Salud, 2001, p. 57.

21. World Health, Organization., *World Health Report 2000. Health Systems: Improving Performance*. Geneva: WHO, 2000.

22. Secretaría de Salud. Sistema de Protección Social en Salud. Informe de resultados. Primer semestre de 2009. Mexico City: Secretaría de Salud, 2009.

23. Hunt L. *Inventing Human Rights. A History*. New York and London: WW Norton & Company, 2007, p. 208.

24. Frenk J, Gómez-Dantés O. Ideas and ideals: ethical basis of health reform in Mexico. *Lancet* 2009;373:1406–1408.

25. Lozano R, Soliz P, Gakidou E, et al. Benchmarking of performance of Mexican states with effective coverage. *Lancet* 2006;368:1729–1741.

26. Gakidou E, Lozano R, González-Pier E, et al. Assessing the effect of the 2001–2006 Mexican health reform: an interim report card. *Lancet* 2006;368:1920–1935.

27. Secretaría de Salud. Salud: México 2006. Información para la rendición de cuentas. Mexico City: Secretaría de Salud, 2007, p. 19.

28. Secretaría de Salud. Salud: México 2006. Información para la rendición de cuentas. Mexico City: Secretaría de Salud, 2007, p. 190.

29. Sepúlveda J. Foreword.In: Jamison DT, Breman JG, Measham AR, et al., editors. *Disease Control Priorities in Developing Countries*, 2nd edition. Washington, DC: Oxford University Press, 2006, pp. xiii–xv.

37

USING KINSHIP STRUCTURES IN HEALTH PROGRAMMING—AN EXAMPLE OF PREVENTIVE MEASURES AND SUCCESSFUL INTERVENTIONS

Moses N. Katabarwa

37.1 INTRODUCTION

Public health has been transforming less advantaged rural communities in sub-Saharan Africa with a focus on improving technologies, and providing skilled manpower to manage and maintain them. Meanwhile, development and health indicators have continued to slide or remain below optimal levels. This failure is blamed on factors such as civil conflict, corruption, and conservativeness of traditional communities, to mention a few. However, ignorance and lack of involving the traditional African kinship system is rarely cited as a barrier to success. Indeed, research suggests it is a resource, particularly in rural communities, that has not yet been harnessed for socioeconomic and cultural development (1, 2).

37.2 THE TRADITIONAL AFRICAN KINSHIP SYSTEM

The traditional African kinship system is a central social structure that defines how people interact with one another, perceive their relationships, understand their origin, and view expectations that guide their behavior. Kinship refers intuitively to "blood relationships," and the essential strands of kinship are successive relations between parents and their children (1, 3).

Although it refers to mainly an extended family of blood-related individuals, women are included in their respective kinships by marriage. In rural sub-Saharan Africa, this group of related persons often owns and occupies a specific geographical area, comprising a part, if not all, of the administrative unity commonly referred to as a community.

Kinships serve as a model for relationships even to nonrelatives. Dealing with traditional communities requires an understanding of the kinship system in order to make sense of almost anything else. A person without kinsmen almost has no hope for a normal life, a valuable marriage, the ability to meet subsistence needs, or to find care when sick, injured, or elderly. The kinship structure forms the basis of political, economic, and even religious organization (1, 3).

Kinship is the main structure that produces and shapes patterns of behavior, and maintains social legal systems. It is a stage where interactions determine which behaviors become the norms and values of communities (4). These are passed on and modified from generation to generation. Noncompliant individuals may find themselves "boxed in" by sanctions predetermined by the respective traditional kinship's social legal system. They are compelled to comply or risk being criticized and even excommunicated (3, 5, 6). The kinship structure under different names exists in all rural societies in sub-Saharan Africa, and many urban dwellers still pay allegiance to their respective kinships back in their rural villages. Kinship is not a unique characteristic to sub-Sahara Africa, it exists everywhere; but it is more apparent in North Africa, the Middle East, and parts of Asia. It is through the kinship system that people have made sense of and coped with disasters, diseases, and other adversities over many centuries. One would expect such a system to have a synergetic effect with the current advancement in science and technology, but this resource is largely untapped.

Water and Sanitation-Related Diseases and the Environment: Challenges, Interventions, and Preventive Measures, First Edition.
Edited by Janine M. H. Selendy.
© 2011 Wiley-Blackwell. Published 2011 by John Wiley & Sons, Inc.

37.2.1 Using the Traditional African Kinship System in Health Programming

Evidence suggests that it is vital to understand and utilize the kinship structure in order to convey the skills and information necessary to realize affordable, equitable, and quality health services to underprivileged communities. The kinship structure is one of the main pillars of successful roots-up health programming as it enhances involvement of individual community members, creates a demand for essential services, and entrenches desired ethics and standards for disease prevention and control (1, 6).

Identification of the kinship structure as critical for successful health programming was unintentional. It happened at the time when health workers and donor agencies were in favor of monetary incentives to community members who were then referred to as "volunteers." Monitoring data showed that the less monetary incentives provided, the better the performance of community-selected ivermectin distributors for onchocerciasis control (6–8). Since the mid-1990s, this observation has confounded the assumption that incentives to community participants would improve their performance. When monetary incentives were not provided, some ivermectin distributors continued to distribute ivermectin with stunning success, while others registered poor performance. Studies have consistently shown that ivermectin distributors who continue to provide quality service without demanding incentives are those who serve their kinsmen. However, outside their kinships, performance remains below the expected level even when incentives are provided (1, 7, 8).

In the kinship structure, serving ones' kin is a "joyful" obligation where rewards are not expected. The kinsmen are the distributor's insurance in case of illness, urgent need for labor support in his or her garden, construction of a house, and even support to have a respectable marriage. Expecting any form of incentives from ones' relatives in return for the service provided is sacrilegious, and swiftly triggers their wrath (1, 3, 5). It was for this reason that some Carter Center-assisted onchocerciasis control programs have adopted the traditional kinship structure for community-directed treatment with ivermectin (CDTI) for onchocerciasis control, and other mass drug administration programs.

The traditional kinship structure improves attendance of health education sessions and creates a more informed, inclusive, and effective decision-making process. It also makes community mobilization and education easier than in communities where it is not utilized. This has improved community skills and confidence to tackle public health challenges. High treatment coverage (≥90%) of eligible populations has been attained and sustained (1, 7, 8). Kinship-based health workers walk short distances and have lighter workloads that are completed in a short period, leaving more time for domestic chores. Kinship groups have also tended to select many interested persons for training as ivermectin distributors compared to other community structures (1, 9). Even when the ratio of at least one distributor per six households (about 42 persons) has been attained in many kinships, training of new distributors continues.

It has also been observed that kinship-based health workers are more likely to be involved in multidisease control compared to those selected outside this structure (1, 9). Since they experience fewer social barriers, most health topics are on the table for discussion and action. This includes diseases with stigma such as sexually transmitted diseases (STDs), HIV/AIDS, and tuberculosis. Kinsmen are able to share their own experiences, however embarrassing, so that health challenges can be overcome as survival and success are the main motivation. The knowledge acquired and the successes attained become engrained in every kinship member, including children. This is life changing for the entire kinship and the community. On this foundation values are shaped, and confidence to tackle more challenging situations becomes the norm. It is within the kinship structure that community safeguards with additional resources apart from those outside the community are mobilized to prevent and control diseases such as dracunculiasis, schistosomiasis, and other preventable water-related diseases that usually constitute at least 70% of their public health challenges.

One of the frustrations for public health programs in sub-Sahara Africa is minimal involvement of women. Where the traditional African kinship structure has been utilized, women's attendance of health education sessions and their involvement as kinship-selected community health workers has greatly improved. Women's involvement has been appreciated by their communities, and their performance rated as better than that of their male counterparts (10). Women have always been caregivers in their communities, and having them back in public health where historically they have excelled is not "rocket" science. The kinship structure and its social legal systems naturally allowed this to happen. It was ill-fated that biomedical culture resisted coexistence with the traditional kinship structure. Anything deemed traditional was ignored in health programming, and persistently undervalued. This tendency is significantly responsible for the discouragement of women's involvement in public health programs.

It has been gratifying to see women selected as community health workers using the traditional kinship structure in Abu Hamad in north Sudan, a traditional and conservative Islamic area. As long as they operate within their respective kinships, they educate and debate their kinsmen without hurting the male self-worth. I have seen women reject openly what they view as not beneficial to women, children, and the entire kinship. Such a behavior in most rural communities is tolerated within the kinship, and rarely outside this age-old structure. Their excitement and involvement during health education was never considered untraditional or unIslamic.

Instead, their contribution to health care and building of viable families has been valued, and their protection assured. This has been observed in similar traditional and religious settings in Carter Center assisted river blindness control programs in Cameroon, Nigeria, Sudan, and Uganda.

It is through the traditional kinship structure that issues like the marginalization of women in education, forced early marriages, female circumcision, family planning, and maternal and child health can be addressed. It is through this structure that threats against individuals and their families become visible, and a concern that can be translated into action. There is a saying that "blood is thicker than water." In some sub-Saharan cultures, it is even said that "a hyena from your home eats you better that a hyena from a distant place." A hyena has a ferocious bite no matter where it's coming from. However, this popular saying emphasizes the importance of the kinship structure.

One of the characteristics of the traditional kinship system is that it presents a mechanism for competition that critics consistently fail to recognize. The example they present is the principle of the "leveling mechanism" that exists in many traditional communities (2, 11). A conspicuous example is huts in a village that are built almost with the same size and design. A different size and design by a kinsman could be viewed as a threat to the entire kinship. In some cultures, if one kinsman is well off, the other kinsmen and other people come for favors, stripping the household of its wealth and keeping it at the same level with others in the same geographical area. The principle of leveling mechanism was to keep all members in check and under the control of their kinship social legal system. A well-off family could decide not to depend on other families and opt out of obligations to other kindred families. Such a behavior was frightening and could endanger the cohesion and existence of the kinship in question. It is due to the leveling mechanism that policy and decision makers tended to criticize the traditional kinship structure as a vehicle for stifling competition and progress.

However, the kinship structure is proving adaptable to current needs and continues to confound its critics. In some regions, it has improved education, farming, commerce, and construction of family houses. The Bameleke people of west region of Cameroon are a good example of what kinship structures can contribute to community and regional transformation. Through their kinship structure, the Bameleke people have organized an effective credit system outside the "modern" banking system. The members only expect the recipient to succeed and pay back what was given. When that happens, other kinsmen are assisted. Through this credit system, the Bameleke have become a dominant force in commerce, improving education and family houses in their respective villages compared to other people of Cameroon. The traditional kinship structure's supportive legal system ensures member compliance in paying back or face the wrath of fellow kinsmen. When drastic action is taken, the official government legal system may not be consulted, and the punishment imposed may not be necessarily illegal, but its impact in most cases is effective in dealing with noncompliant individuals.

The Bakiga of southwestern Uganda have a kinship structure referred to as "*Engozi*" that ensures family access to adequate labor during farming seasons, and domestic support when the provider of a member family is faced with unique challenges such as ill health within his or her household (12, 13). Through this system, the kinship has ensured adequate food reserves for the families concerned, and taken care of their weak or physically challenged members.

The importance of the kinship structure was also observed in Guinea in 1981 where an agricultural development program supported by USAID tripled its cost from USD 4 million to USD 15 million without realizing expected goals. Research revealed that the failure was due to lack of understanding and utilization of Malinke kinship structure. Land, labor, and economic cooperation among the Malinke society was controlled through the kinship structure. It was only after this strong connection between the kinship and the agricultural systems had been revealed that appropriate changes were made, and success realized in the multimillion-dollar USAID project (14). All these unique and positive features of the kinship structure are assets that policy and decision makers can exploit for successful health programming.

37.3 CONCLUSION

Though often limited to geographical areas, the kinship structure encourages interactions across kinships in order to ensure security, exchange of goods and services, and useful relationship through marriage. Community projects such as construction of schools, health units, convenient safe water points, and even places of worship are attained due to successful interkinship collaboration. The kinship structure is still strong and well respected; therefore, understanding and utilizing it is vital for successful health programming.

REFERENCES

1. Katabarwa NM, Habomugisha P, Agunyo S, Mckelvery CA, Ogweng N, Kwebiiha S, Byenume F, Male B, McFarland D. Traditional kinship system enhanced classic community-directed treatment with ivermectin (CDTI) for onchocerciasis control in Uganda. *Trans R Soc Trop Med Hyg* 2009;104:265–272.
2. Katabarwa NM, Habomugisha P, Agunyo S. Involvement of women in community-directed treatment with ivermectin for the control of onchocerciasis in Rukungiri district, Uganda: a knowledge, attitude and practice study. *Ann Trop Med Parasitol* 2001;95(5):485–494.

3. Keesing RM, Strathern AJ. *Cultural Anthropology. A Contemporary Perspective*, 3rd edition. New York: Harcourt Brace College Publishers, 1998.

4. Loustaunau OM, Sobo JE. *The Cultural Context of Health, Illness, and Medicine*. Westport, CT: Bergin & Garvey, 1997.

5. Howard MC. *Contemporary Cultural Anthropology*, 3rd edition. Harper Collins Publishers, 1998.

6. Katabarwa NM, Richards F.O. Jr., Ndyomugyenyi R. In rural Ugandan communities the traditional kinship/clan system is vital to the success and sustainment of the African Programme for Onchocerciasis Control. *Ann Trop Med Parasitol* 2000;94:485–495.

7. Katabarwa MN, Richards FO Jr. Community-directed health (CDH) workers enhance the performance and sustainability of CDH programmes: experience from ivermectin distribution in Uganda. *Ann Trop Med Parasitol* 2001;95(3): 275–286.

8. Katabarwa NM, Habomugisha P, Richards FO. Implementing community-directed treatment with ivermectin for the control of onchocerciasis in Uganda (1999–2002): an evaluation. *Ann Trop Med Parasitol* 2002;96(1):61–73.

9. Katabarwa M, Habomugisha P, Eyamba A, Agunyo S, Mentou A. Monitoring ivermectin distributors involved in integrated health care services through community-directed interventions—a comparison of Cameroon and Uganda experiences over a period of three years (2004–2006). *Trop Med Int Health* 2009;15(2):216–223.

10. Katabarwa NM, Habomugisha P, Agunyo S. Involvement and performance of women in community-directed treatment with ivermectin for onchocerciasis control in Rukungiri District, Uganda. *J Health Soc Care Community* 2002;10:382–393.

11. Salzaman PC. *The Anthropology of Real Life. Events in Human Experience*. Prospect Heights, IL: Waveland Press, 1991.

12. Katabarwa MN, Habomugisha P, Richards FO Jr. Implementing community-directed treatment with ivermectin for the control of onchocerciasis in Uganda (1997–2000): an evaluation. *Ann Trop Med Parasitol* 2002;96(1):61–73.

13. Katabarwa M. Modern health services versus traditional engozi system in Uganda. *Lancet* 1999;24,354(9175):343.

14. Ferraro G. *Cultural Anthropology, An Applied Perspective*, 3rd edition. Belmont, CA: An International Thomson Publishing Company, 1998.

INDEX

Water and Sanitation-Related Diseases and the Environment: Challenges, Interventions, and Preventive Measures, First Edition.
Edited by Janine M. H. Selendy.
© 2011 Wiley-Blackwell. Published 2011 by John Wiley & Sons, Inc.